# Intermediate Algebra, 10e

Custom Edition for Hudson Valley Community College

MATH 110

Kaufmann | Schwitters | Ewen

CENGAGE
Learning

Australia • Brazil • Japan • Korea • Mexico • Singapore • Spain • United Kingdom • United States

**Intermediate Algebra, 10e, Custom Edition for Hudson Valley Community College, MATH 110**

Senior Manager, Student Engagement:

Linda deStefano

Janey Moeller

Manager, Student Engagement:

Julie Dierig

Marketing Manager:

Rachael Kloos

Manager, Production Editorial:

Kim Fry

Manager, Intellectual Property Project Manager:

Brian Methe

Senior Manager, Production and Manufacturing:

Donna M. Brown

Manager, Production:

Terri Daley

Printed in the United States of America

Intermediate Algebra, 10th Edition
Jerome E. Kaufmann, Karen L. Schwitters



Elementary Technical Mathematics, 11th Edition
Dale Ewen, C. Robert Nelson



ISBN-13: 978-1-305-00640-9

ISBN-10: 1-305-00640-2

**WCN: 01-100-101**

**Cengage Learning**

5191 Natorp Boulevard
Mason, Ohio 45040
USA

Cengage Learning is a leading provider of customized learning solutions with office locations around the globe, including Singapore, the United Kingdom, Australia, Mexico, Brazil, and Japan. Locate your local office at: **international.cengage.com/region.**

Cengage Learning products are represented in Canada by Nelson Education, Ltd.
For your lifelong learning solutions, visit **www.cengage.com/custom.**
Visit our corporate website at **www.cengage.com.**

# Brief Contents

WebAssign.

# WebAssign Student Quick Start Guide

You can use WebAssign to access your homework, quizzes, and tests — whatever your instructor chooses — at any time of day or night, from any computer with a connection to the Internet and a Web browser. Your instructor creates your assignments, schedules them, and decides how many submissions you get. Your instructor also determines if you can have an extension, if you can save your work without submitting it at the time, and how much feedback you get after you submit an assignment.

The WebAssign support staff cannot change your username or password, give extensions, change your score, give you extra submissions, or help you with the content of your assignments.

## Logging In

You can log in to WebAssign using any Web browser connected to the Internet. There are two different ways to log in to WebAssign. Each requires information from your teacher. If you are unsure about how to log in, please check with your teacher or another student in your class.

Go to the login page at http://webassign.net/login.html or the web address provided by your teacher. The way you log in depends on how your instructor set up the class:

- If your teacher created a WebAssign account for you, they will provide you with a **Username**, an **Institution** code and a **Password**. Simply enter this information in the boxes provided and click the **Log In** button.

WebAssign.

- If your teacher wants you to **Self-Enroll** in the WebAssign course they will provide you with a **Class Key**. You will create your own username and password. It is important that you remember this information so you can log in for the remainder of the class. In this case, just click the **I have a Class Key** button. You don't need to enter any other information on this page.

  Then, enter the **Class Key** your instructor provided and click **Submit.** Verify you are enrolling in the correct class on the next page.

**Class Key**

Enter the Class Key that you received from your instructor. You will only need to complete this once. After you have created your account, you can log in on the main page.

**Class Key**

*Class Keys generally start with an institution code, followed by two sets of four digits.*

Submit

- Enter your preferred Login and Student information.
- Click the **Create My Account** button to complete the enrollment process.
- A review screen will display, showing your username, institution code, and password. **Retain a copy of this information.** You will need it to log into WebAssign.

**Log In Information**

Required fields are marked with an asterisk (*).

Preferred Username * Check Availability

*Your username may contain letters, numbers, and the following characters: underscore (_), hyphen (-), period (.)*

Institution Code **webassign**

Password *

Re-Enter Password *

*Passwords are case-sensitive.*

**Student Information**

Required fields are marked with an asterisk (*).

First Name *

Last Name *

Email Address *

Student ID Number

Create My Account

WebAssign.

## Access Codes

Logout

Home | My Assignments | Grades | Communication | Calendar

Guide | Help | My Options

Calculus, Fall 2008

Home

Chris Hall
Instructor: Chris Hall
WebAssign University

WebAssign Notices

Note: The following message is shown to your students. As WebAssign faculty you are not required to enter an access code. According to our records you have not yet entered an access code for this class. The grace period will end Friday, August 1, 2008 at 11:06 AM EDT. After that date you will no longer be able to see your WebAssign assignments or grades, until you enter an access code.

If you have an access code (purchased with your textbook or from your bookstore) choose the appropriate prefix from the menu below. If your access code is not listed please contact your instructor.

Choose your access code prefix | Go

-or-

You may purchase an access code online.

Purchase an access code online

Continue without entering an access code
(13 days remaining)

Once you log in, you may see a WebAssign Notice about entering an access code for your class. You can get an Access Code from any of the following places if you need to use one:

- A new textbook you purchased for the class.
- Your bookstore, which may sell Access Code cards.
- Online, where you can purchase an access code with a credit card.

You have a 14 day grace period to use WebAssign, starting with the WebAssign class start date. During this time you can work on and view your WebAssign assignments without registering a code.

After the grace period is over you will only see the code registration message until you submit or purchase a code.

There are two types of WebAssign access code cards. The small card requires you to scratch off the silver surface in order to reveal the complete access code.

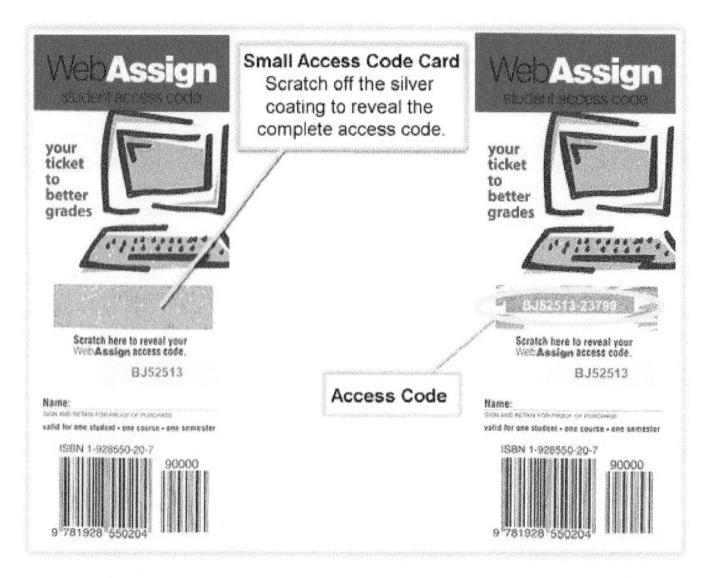

WebAssign.

The larger security envelope card requires you to open the card to reveal the access code number.

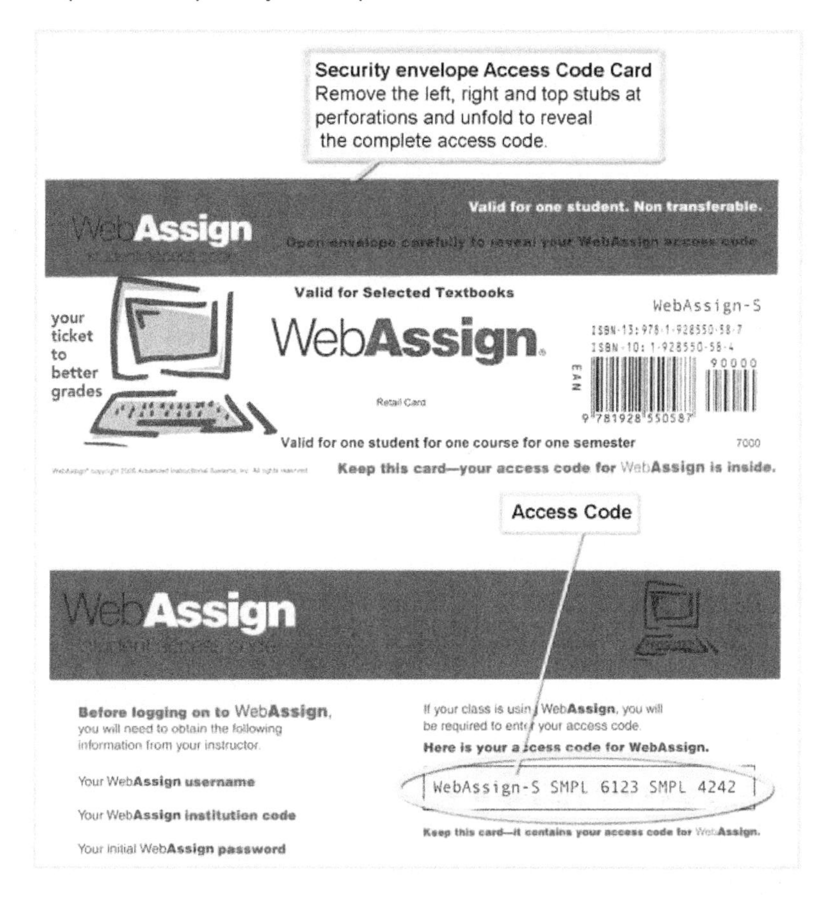


If you would like to purchase an access code directly from WebAssign online, you may do so with a credit card. Your code will be automatically registered to your WebAssign account as soon as the transaction is complete. You will receive an email confirmation. Please keep a copy for your records.

## Your WebAssign Home Page

Once you have successfully logged in you will see your WebAssign homepage. If you are taking more than one WebAssign class, you will need to select which class you wish to view first.

The upper right corner features links to a complete Student **Guide**, as well as a link to WebAssign Technical Support under **Help**. If you want to change your password or add or update your email address, simply click **My Options** in the upper right hand corner.

You will see your assignments and due dates listed, as well as any Communications, Grades, and Announcements posted by your teacher.

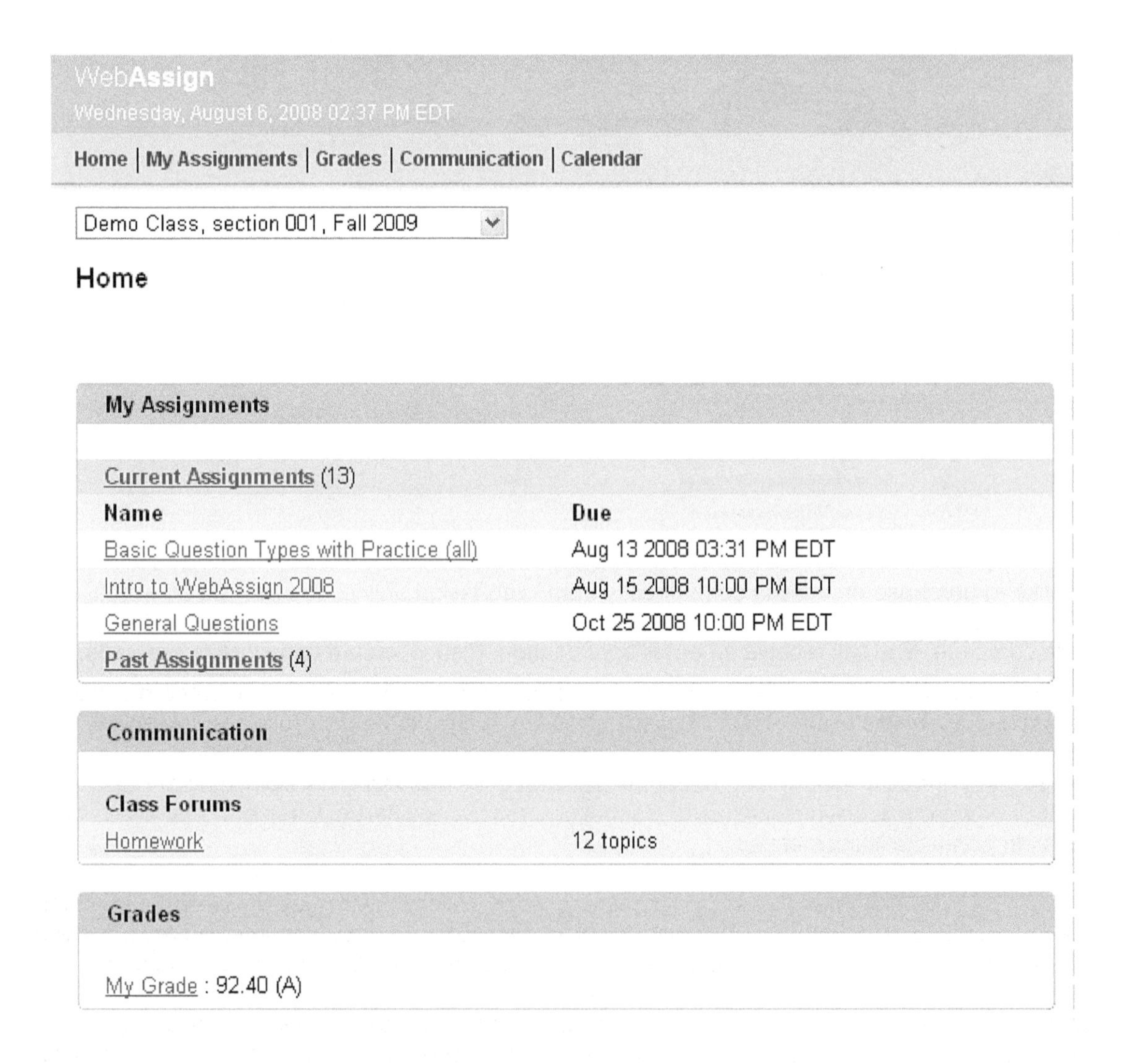

![WebAssign]

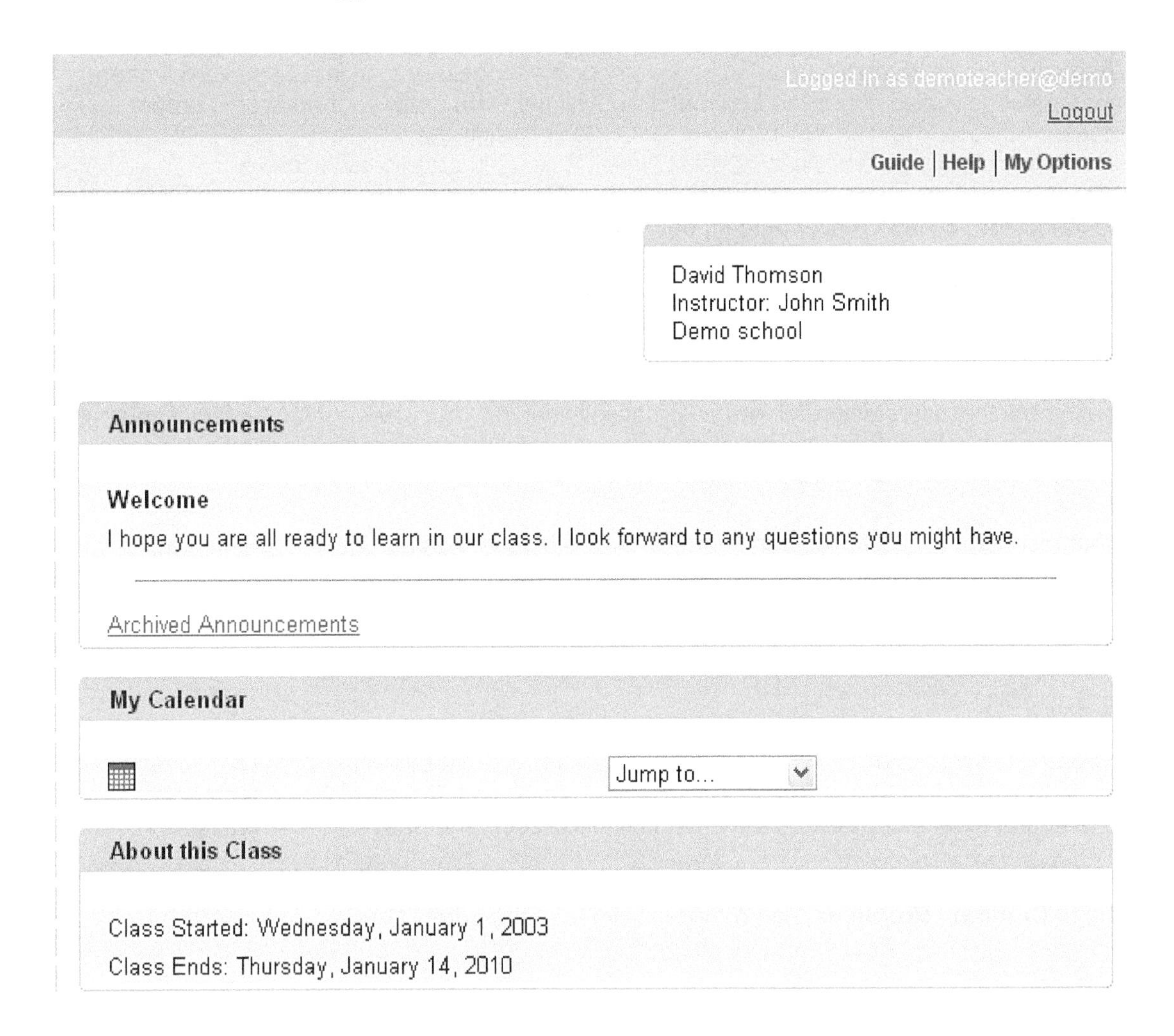


## Answering Questions

WebAssign has a variety of different question types, ranging from multiple choice to fill-in-the-blank to symbolic questions. Here are some things to keep in mind as you work through your assignments:

- Some questions may include numbers or words that appear in red. This signifies that the number or word has been randomized, so that you receive a different version of the same basic question from your classmates.

- Some WebAssign questions check the number of significant figures in your answer. If you enter the correct value with the wrong number of significant figures, you will not receive credit, but you will receive a hint that your number does not have the correct number of significant figures.

- Some questions require entering symbolic notation. Answer symbolic questions by using calculator notation. You must use the exact variables specified in the questions. The order is not important as long as it is mathematically correct. Clicking on the eye button previews the expression you enter in proper mathematical notation. Clicking on the symbolic formatting help button provides tips for using the correct keystrokes.

WebAssign®

- When you click on some WebAssign chemistry or math questions an input palette will open. These palettes, called chemPad and mathPad, will help you enter your answer in proper notation.

- Some questions may require the use of an Active Figure simulation. Active Figures require the free Macromedia Flash Player plug-in, downloadable from www.macromedia.com.

- If your instructor allows it, you can save your work without grading by selecting the Save Work button at the end of the question. After you save your work, it will be available to you the next time you click the assignment.

- Please note that WebAssign will **not** automatically submit your answers for scoring if you only **Save** your work. Your teacher will not be able to see your submissions. Please be sure to **Submit** prior to the due date and time.

- If your instructor allows it, you can submit answers by question part or for the entire assignment. To submit an individual question answer for grading, click the **Submit New Answers to Question** __ button at the bottom of each question. To submit the entire assignment for grading, click the **Submit All New Answers** button at the end of the assignment.

## Technical Support

If you are having difficulty logging in, please be sure to check with your teacher and verify whether an account has been created for you or whether you need to self-enroll. In either case your teacher needs to provide the appropriate information (username, institution code and password OR Class Key).

To email WebAssign Support go to http://www.webassign.net/info/support/report.html. This page also lists answers to **Common Problems**, and provides links to the **Student Guide**.

August 7, 2008

# 1 Basic Concepts and Properties

Paul Bradbury/OJO Images/Getty Images

## Study Skill Tip

*"Before beginning, prepare carefully"*
MARCUS TULLIUS CICERO, ANCIENT ROMAN SCHOLAR

There are many factors that affect success in a math course, such as the instructor, the textbook, your motivation, time of the day for class, etc. However, one of the most important factors for success is being placed in the right course for you. Can you imagine taking French II without having taken French I? What would be the likelihood of being successful? If at the beginning of this course, you think the material is way too difficult or way too easy, talk to your instructor regarding your placement in this course.

Two other factors that are extremely important for success in a math course are attending class regularly and doing the homework. If at all possible, don't ever miss class. However, take action right now and find a classmate whom you can contact in case you miss class. Get the names and college email addresses of several fellow students whom you could possibly contact to get the class notes in case you miss class.

Also, know the resources available if you need help with the homework. Become aware of your instructor's office hours and the location of any tutoring centers on campus. Also consider utilizing websites for additional help with your math course. Your instructor or fellow classmates can usually suggest appropriate websites for Intermediate Algebra.

***Are you prepared enough to feel confident about your success in this algebra class?***

## Chapter Preview

An Intermediate Algebra course assumes that you have basic arithmetic skills, including fractions and basic algebra skills. This chapter includes a review of fractions in Section 1.0. The section is written with diagnostic problems to help you determine whether you have mastery of the basic operations with fractions. I encourage you to try the four sets of mastery problems in Section 1.0 even if your instructor does not assign Section 1.0. The answers to the mastery sets are in the back of the book.

Algebra is often described as *generalized arithmetic*. That description does convey an important idea: A good understanding of arithmetic provides a sound basis for the study of algebra. In this chapter we use the concepts of *numerical expression* and *algebraic expression* to review some ideas from arithmetic and begin the transition to algebra. Be sure you thoroughly understand the basic concepts reviewed in this first chapter.

## 1.0 Review of Fractions

As with any math course, you need the prerequisite skills in order to be successful with the new material presented. Students enter into Intermediate Algebra with varying levels of math proficiency. Some students have a strong background and come into the course fully prepared. Other students may have a weak background or have not been enrolled in a math course for a while and have forgotten the prerequisite skills.

Throughout this section, problems will be presented for you to determine your mastery of some of the prerequisite arithmetic and algebra skills. There will be ten problems. Answers to these problems are in the back of the book. Use the following legend as a guide to direct your studying of the material immediately following the problems.

| Number of problems correct | Prescription |
|---|---|
| 10 or 9 | You probably have mastery of this skill and can go on to the next topic without reviewing. |
| 8 or 7 | You have some basic mastery of this skill but do need to review. Read the material and do the corresponding problems in the problem set to gain mastery of this skill. |
| 6 or less | You have not mastered this skill. Read the material and do the corresponding problems in the problem set to gain mastery of this skill. If after that you are still not proficient in this skill, ask your instructor for additional study materials. |

Try the following problems to help determine your proficiency with prime numbers.

### Mastery Set 1 Prime Numbers

For Problems 1–4, label the number as prime or composite.

**1.** 9 **2.** 30 **3.** 41 **4.** 57

For Problems 5–7, factor the composite number into a product of prime numbers.

**5.** 40 **6.** 84 **7.** 210

For Problems 8–10, find the least common multiple of the given numbers.

**8.** 4 and 18 **9.** 6, 18, and 21 **10.** 4, 10, and 15

## Prime Numbers

Because prime numbers and prime factorization play an important role in the operations with fractions, let's begin by considering two special kinds of whole numbers: prime numbers and composite numbers.

### Definition 1.1

A **prime number** is a whole number greater than 1 that has no factors (divisors) other than itself and 1. Whole numbers greater than 1 that are not prime numbers are called **composite numbers**.

The prime numbers less than 50 are 2, 3, 5, 7, 11, 13, 17, 19, 23, 29, 31, 37, 41, 43, and 47. Note that each of these has no factors other than itself and 1. We can express every composite number as the indicated product of prime numbers. Consider the following examples:

$$4 = 2 \cdot 2 \qquad 6 = 2 \cdot 3 \qquad 8 = 2 \cdot 2 \cdot 2 \qquad 10 = 2 \cdot 5 \qquad 12 = 2 \cdot 2 \cdot 3$$

In each case we express a composite number as the indicated product of prime numbers. This form is called the prime-factored form of the number. There are various procedures to find the prime factors of a given composite number. For our purposes, the simplest technique is to factor the given composite number into any two easily recognized factors and then continue to factor each of these until we obtain only prime factors. Consider these examples:

$$18 = 2 \cdot 9 = 2 \cdot 3 \cdot 3 \qquad 27 = 3 \cdot 9 = 3 \cdot 3 \cdot 3$$

$$24 = 4 \cdot 6 = 2 \cdot 2 \cdot 2 \cdot 3 \qquad 150 = 10 \cdot 15 = 2 \cdot 5 \cdot 3 \cdot 5$$

It does not matter which two factors we choose first. For example, we might start by expressing 18 as $3 \cdot 6$ and then factor 6 into $2 \cdot 3$, which produces a final result of $18 = 3 \cdot 2 \cdot 3$. Either way, 18 contains two prime factors of 3 and one prime factor of 2. The order in which we write the prime factors is not important.

## Least Common Multiple

It is sometimes necessary to determine the smallest common nonzero multiple of two or more whole numbers. We call this nonzero number the **least common multiple**. In our work with fractions, there will be problems for which it will be necessary to find the least common multiple of some numbers—usually the denominators of fractions. So let's review the concepts of multiples. The set of all whole numbers that are multiples of 5 consists of 0, 5, 10, 15, 20, 25, and so on. In other words, 5 times each successive whole number ($5 \cdot 0 = 0$, $5 \cdot 1 = 5$, $5 \cdot 2 = 10$, $5 \cdot 3 = 15$, and so on) produces the multiples of 5. In a like manner, the set of multiples of 4 consists of 0, 4, 8, 12, 16, and so on. We can find the least common multiple of 5 and 4 by using a simple listing of the multiples of 5 and the multiples of 4.

Multiples of 5 are 0, 5, 10, 15, 20, 25, 30, 35, 40, 45, . . .

Multiples of 4 are 0, 4, 8, 12, 16, 20, 24, 28, 32, 36, 40, 44, 48, . . .

The nonzero numbers in common on the lists are 20 and 40. The least of these, 20, is the least common multiple. Stated another way, 20 is the smallest nonzero whole number that is divisible by both 4 and 5.

From your knowledge of arithmetic, you will often be able to determine the least common multiple by inspection. For instance, the least common multiple of 6 and 8 is 24.

Therefore, 24 is the smallest nonzero whole number that is divisible by both 6 and 8. If we cannot determine the least common multiple by inspection, then using the prime-factored form of composite numbers is helpful. The procedure is as follows.

**Step 1** Express each number as a product of prime factors.

**Step 2** The least common multiple contains each different prime factor. For each different factor, determine the most times each different factor is used in any of the factorizations. Those factors will then be used that number of times in the least common multiple. (For example, if the factor 2 occurs at most three times in any factorization, then the least common multiple will have the factor 2 used three times.)

The following examples illustrate this technique for finding the least common multiple of two or more numbers.

**Classroom Example**
Find the least common multiple of 8 and 30.

**EXAMPLE 1** Find the least common multiple of 24 and 36.

### Solution

Let's first express each number as a product of prime factors.

$$24 = 2 \cdot 2 \cdot 2 \cdot 3$$
$$36 = 2 \cdot 2 \cdot 3 \cdot 3$$

There are two different factors, 2 and 3, in the prime-factored forms.

The prime factor 2 occurs the most times (three times) in the factorization of 24. Because the factorization of 24 contains three 2s, the least common multiple must have three 2s.

The prime factor 3 occurs the most times (two times) in the factorization of 36. Because the factorization of 36 contains two 3s, the least common multiple must have two 3s.

The least common multiple of 24 and 36 is therefore $2 \cdot 2 \cdot 2 \cdot 3 \cdot 3 = 72$.

**Classroom Example**
Find the least common multiple of 42 and 60.

**EXAMPLE 2** Find the least common multiple of 48 and 84.

### Solution

$$48 = 2 \cdot 2 \cdot 2 \cdot 2 \cdot 3$$
$$84 = 2 \cdot 2 \cdot 3 \cdot 7$$

There are three different factors, 2, 3, and 7, in the prime-factored forms.

The most number of times that 2 occurs is four times in the factored form of 48.

The factors 3 and 7 only occur once in each factored form, so we need one factor of each for the least common multiple.

The least common multiple of 48 and 84 is $2 \cdot 2 \cdot 2 \cdot 2 \cdot 3 \cdot 7 = 336$.

**Classroom Example**
Find the least common multiple of 10, 15, and 24.

**EXAMPLE 3** Find the least common multiple of 12, 18, and 28.

### Solution

$$28 = 2 \cdot 2 \cdot 7$$
$$18 = 2 \cdot 3 \cdot 3$$
$$12 = 2 \cdot 2 \cdot 3$$

There are three different factors, 2, 3, and 7, in the prime-factored forms.

The most number of times that 2 occurs in any factored form is twice, so we need two factors of 2 in the least common multiple.

The most number of times that 3 occurs in any factored form is twice, so we need two factors of 3 in the least common multiple.

The factor, 7, only occurs once in the factored forms, so we need one factor of 7 for the least common multiple.

The least common multiple is $2 \cdot 2 \cdot 3 \cdot 3 \cdot 7 = 252$.

**Classroom Example**
Find the least common multiple of 6 and 25.

**EXAMPLE 4** Find the least common multiple of 8 and 9.

### Solution

$9 = 3 \cdot 3$

$8 = 2 \cdot 2 \cdot 2$

There are two different factors, 2 and 3, in the prime-factored forms.

The most number of times that 2 occurs in any factored form is three times, so we need three factors of 2 in the least common multiple.

The most number of times that 3 occurs in any factored form is twice, so we need two factors of 3 in the least common multiple.

The least common multiple is $2 \cdot 2 \cdot 2 \cdot 3 \cdot 3 = 72$.

Try the following problems to help determine your proficiency with reducing, multiplying, and dividing fractions.

## Mastery Set 2 Reducing, Multiplying, and Dividing Fractions

For Problems 1–3, reduce the fraction to lowest terms.

**1.** $\frac{10}{25}$ **2.** $\frac{12}{34}$ **3.** $\frac{14}{42}$

For Problems 4–7, multiply the fractions and express the answer in lowest terms.

**4.** $\frac{1}{5} \cdot \frac{2}{7}$ **5.** $\frac{4}{3} \cdot \frac{9}{24}$ **6.** $\frac{12}{5} \cdot \frac{1}{9}$ **7.** $\frac{7}{2} \cdot \frac{12}{7}$

For Problems 8–10, divide the fractions and express the answer in lowest terms.

**8.** $\frac{6}{5} \div \frac{2}{3}$ **9.** $\frac{21}{5} \div \frac{14}{15}$ **10.** $4 \div \frac{1}{5}$

### Reducing Fractions

Before we proceed too far with operations on fractions, we need to learn about reducing fractions. The following property is applied throughout our work with fractions. We call this property the fundamental property of fractions.

**Fundamental Property of Fractions**

If $b$ and $k$ are nonzero integers, and $a$ is any integer, then $\frac{a \cdot k}{b \cdot k} = \frac{a}{b}$.

The fundamental property of fractions provides the basis for what is often called reducing fractions to lowest terms, or expressing fractions in simplest or reduced form. Let's apply the property to a few examples.

Classroom Example
Reduce $\frac{25}{35}$ to lowest terms.

**EXAMPLE 5** Reduce $\frac{12}{18}$ to lowest terms.

### Solution

$$\frac{12}{18} = \frac{2 \cdot \not{6}}{3 \cdot \not{6}} = \frac{2}{3}$$ A common factor of 6 has been divided out of both numerator and denominator

Classroom Example
Change $\frac{18}{50}$ to simplest form.

**EXAMPLE 6** Change $\frac{14}{35}$ to simplest form.

### Solution

$$\frac{14}{35} = \frac{2 \cdot \not{7}}{5 \cdot \not{7}} = \frac{2}{5}$$ A common factor of 7 has been divided out of both numerator and denominator

Classroom Example
Reduce $\frac{24}{28}$.

**EXAMPLE 7** Reduce $\frac{72}{90}$.

### Solution

$$\frac{72}{90} = \frac{\not{2} \cdot 2 \cdot 2 \cdot \not{3} \cdot \not{3}}{\not{2} \cdot \not{3} \cdot \not{3} \cdot 5} = \frac{4}{5}$$ The prime-factored forms of the numerator and denominator may be used to find common factors

## Multiplying Fractions

We are now ready to consider multiplication problems with the understanding that the final answer should be expressed in reduced form. Study the following examples carefully; we use different methods to simplify the problems.

We can define the multiplication of fractions in common fractional form as follows.

### Multiplying Fractions

If $a$, $b$, $c$, and $d$ are integers, with $b$ and $d$ not equal to zero, then $\frac{a}{b} \cdot \frac{c}{d} = \frac{a \cdot c}{b \cdot d}$.

To multiply fractions in common fractional form, we simply multiply numerators and multiply denominators. The following examples illustrate the multiplying of fractions.

$$\frac{1}{3} \cdot \frac{2}{5} = \frac{1 \cdot 2}{3 \cdot 5} = \frac{2}{15}$$

$$\frac{3}{4} \cdot \frac{5}{7} = \frac{3 \cdot 5}{4 \cdot 7} = \frac{15}{28}$$

$$\frac{3}{5} \cdot \frac{5}{3} = \frac{15}{15} = 1$$

The last of these examples is a very special case. If the product of two numbers is 1, then the numbers are said to be reciprocals of each other.

Classroom Example
Multiply $\left(\frac{18}{5}\right)\left(\frac{15}{14}\right)$.

**EXAMPLE 8** Multiply $\left(\frac{9}{4}\right)\left(\frac{14}{15}\right)$.

### Solution

$$\left(\frac{9}{4}\right)\left(\frac{14}{15}\right) = \frac{\cancel{3} \cdot 3 \cdot \cancel{2} \cdot 7}{\cancel{2} \cdot 2 \cdot \cancel{3} \cdot 5} = \frac{21}{10}$$

Factor each numerator and denominator and then reduce

Classroom Example
Find the product of $\frac{4}{7}$ and $\frac{21}{8}$.

**EXAMPLE 9** Find the product of $\frac{8}{9}$ and $\frac{18}{24}$.

### Solution

$$\frac{\overset{1}{\cancel{8}}}{\underset{1}{\cancel{9}}} \cdot \frac{\overset{2}{\cancel{18}}}{\underset{3}{\cancel{24}}} = \frac{2}{3}$$

A common factor of 8 has been divided out of 8 and 24, and a common factor of 9 has been divided out of 9 and 18

## Dividing Fractions

The next example motivates a definition for division of rational numbers in fractional form:

$$\frac{\frac{3}{4}}{\frac{2}{3}} = \left(\frac{\frac{3}{4}}{\frac{2}{3}}\right)\left(\frac{\frac{3}{2}}{\frac{3}{2}}\right) = \frac{\left(\frac{3}{4}\right)\left(\frac{3}{2}\right)}{1} = \left(\frac{3}{4}\right)\left(\frac{3}{2}\right) = \frac{9}{8}$$

Note that $\left(\frac{\frac{3}{2}}{\frac{3}{2}}\right)$ is a form of 1, and $\frac{3}{2}$ is the reciprocal of $\frac{2}{3}$. In other words, $\frac{3}{4}$ divided by $\frac{2}{3}$ is equivalent to $\frac{3}{4}$ times $\frac{3}{2}$. The following definition for division now should seem reasonable.

> **Division of Fractions**
>
> If $b$, $c$, and $d$ are nonzero integers, and $a$ is any integer, then $\frac{a}{b} \div \frac{c}{d} = \frac{a}{b} \cdot \frac{d}{c}$.

Note that to divide $\frac{a}{b}$ by $\frac{c}{d}$, we multiply $\frac{a}{b}$ times the reciprocal of $\frac{c}{d}$, which is $\frac{d}{c}$. The next examples demonstrate the important steps of a division problem.

$$\frac{2}{3} \div \frac{1}{2} = \frac{2}{3} \cdot \frac{2}{1} = \frac{4}{3}$$

$$\frac{5}{6} \div \frac{3}{4} = \frac{5}{6} \cdot \frac{4}{3} = \frac{5 \cdot 4}{6 \cdot 3} = \frac{5 \cdot \cancel{2} \cdot 2}{\cancel{2} \cdot 3 \cdot 3} = \frac{10}{9}$$

$$\frac{6}{7} \div 2 = \frac{6}{7} \div \frac{2}{1} = \frac{\overset{3}{\cancel{6}}}{7} \cdot \frac{1}{\underset{1}{\cancel{2}}} = \frac{3}{7}$$

Classroom Example

Divide $\frac{2}{5} \div \frac{2}{3}$.

**EXAMPLE 10** Divide $\frac{4}{9} \div \frac{3}{2}$.

### Solution

$\frac{4}{9} \div \frac{3}{2} = \frac{4}{9} \cdot \frac{2}{3} = \frac{8}{27}$ Multiply by the reciprocal of $\frac{3}{2}$

Classroom Example

Divide $\frac{21}{4} \div 3$.

**EXAMPLE 11** Divide $\frac{7}{2} \div 3$.

### Solution

$\frac{7}{2} \div 3 = \frac{7}{2} \div \frac{3}{1}$ Rewrite 3 as $\frac{3}{1}$

$= \frac{7}{2} \cdot \frac{1}{3} = \frac{7}{6}$ Multiply by $\frac{1}{3}$, the reciprocal of $\frac{3}{1}$

Try the following problems to help determine your proficiency with addition and subtraction of fractions.

## Mastery Set 3 Adding and Subtracting Fractions

For Problems 1–5, perform the addition. Express the answer in lowest terms.

**1.** $\frac{1}{8} + \frac{5}{8}$ **2.** $\frac{3}{4} + \frac{1}{8}$ **3.** $\frac{3}{5} + \frac{2}{3}$ **4.** $\frac{7}{12} + \frac{3}{8}$ **5.** $\frac{11}{60} + \frac{13}{24}$

For Problems 6–9, perform the subtraction. Express the answer in lowest terms.

**6.** $\frac{6}{7} - \frac{2}{7}$ **7.** $\frac{11}{12} - \frac{1}{4}$ **8.** $\frac{7}{8} - \frac{4}{5}$ **9.** $\frac{7}{18} - \frac{5}{24}$

**10.** If Jessica ate $\frac{3}{8}$ of a pepperoni pizza and $\frac{1}{4}$ of a cheese pizza, what was her total portion of pizza eaten?

## Adding and Subtracting Fractions

Suppose that it is one-fifth of a mile between your dorm and the union and two-fifths of a mile between the union and the library along a straight line, as indicated in Figure 1.1. The total distance between your dorm and the library is three-fifths of a mile, and we write $\frac{1}{5} + \frac{2}{5} = \frac{3}{5}$.

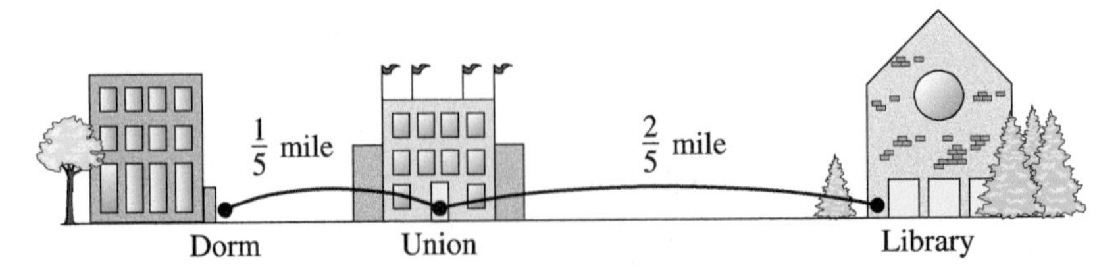


**Figure 1.1**

A pizza is cut into seven equal pieces and you eat two of the pieces (see Figure 1.2). How much of the pizza remains? We represent the whole pizza by $\frac{7}{7}$ and conclude that $\frac{7}{7} - \frac{2}{7} = \frac{5}{7}$ of the pizza remains.

**Figure 1.2**

These examples motivate the following definition for addition and subtraction of rational numbers in $\frac{a}{b}$ form.

### Addition and Subtraction of Fractions

If $a$, $b$, and $c$ are integers, and $b$ is not zero, then

$$\frac{a}{b} + \frac{c}{b} = \frac{a + c}{b} \quad \text{Addition}$$

$$\frac{a}{b} - \frac{c}{b} = \frac{a - c}{b} \quad \text{Subtraction}$$

We say that fractions with common denominators can be added or subtracted by adding or subtracting the numerators and placing the results over the common denominator. Consider the following examples:

$$\frac{3}{7} + \frac{2}{7} = \frac{3 + 2}{7} = \frac{5}{7}$$

$$\frac{7}{8} - \frac{2}{8} = \frac{7 - 2}{8} = \frac{5}{8}$$

$$\frac{5}{6} - \frac{1}{6} = \frac{5 - 1}{6} = \frac{4}{6} = \frac{2}{3} \quad \text{We agree to reduce the final answer}$$

How do we add or subtract if the fractions do not have a common denominator? We use the fundamental principle of fractions, $\frac{a \cdot k}{b \cdot k} = \frac{a}{b}$, to get equivalent fractions that have a common denominator. **Equivalent fractions** are fractions that name the same number. Consider the next example, which shows the details.

Classroom Example
Add $\frac{2}{3} + \frac{4}{7}$.

**EXAMPLE 12** Add $\frac{1}{4} + \frac{2}{5}$.

### Solution

$$\frac{1}{4} = \frac{1 \cdot 5}{4 \cdot 5} = \frac{5}{20} \quad \frac{1}{4} \text{ and } \frac{5}{20} \text{ are equivalent fractions}$$

$$\frac{2}{5} = \frac{2 \cdot 4}{5 \cdot 4} = \frac{8}{20} \quad \frac{2}{5} \text{ and } \frac{8}{20} \text{ are equivalent fractions}$$

$$\frac{5}{20} + \frac{8}{20} = \frac{13}{20}$$

Note that in Example 12 we chose 20 as the common denominator, and 20 is the least common multiple of the original denominators 4 and 5. (Recall that the least common multiple is the smallest nonzero whole number divisible by the given numbers.) In general, we use the least common multiple of the denominators of the fractions to be added or subtracted as a **least common denominator** (LCD).

Classroom Example
Subtract $\frac{17}{10} - \frac{2}{15}$.

**EXAMPLE 13** Subtract $\frac{5}{8} - \frac{7}{12}$.

### Solution

By inspection the LCD is 24.

$$\frac{5}{8} - \frac{7}{12} = \frac{5 \cdot 3}{8 \cdot 3} - \frac{7 \cdot 2}{12 \cdot 2}$$ Change each fraction to an equivalent fraction with 24 as the denominator

$$= \frac{15}{24} - \frac{14}{24} = \frac{1}{24}$$

If the LCD is not obvious by inspection, then we can use the technique of prime factorization to find the least common multiple.

Classroom Example
Add $\frac{17}{18} + \frac{5}{12}$.

**EXAMPLE 14** Add $\frac{5}{18} + \frac{7}{24}$.

### Solution

If we cannot find the LCD by inspection, then we can use the prime-factored forms.

$$\left.\begin{aligned} 18 &= 2 \cdot 3 \cdot 3 \\ 24 &= 2 \cdot 2 \cdot 2 \cdot 3 \end{aligned}\right\} \longrightarrow \text{LCD} = 2 \cdot 2 \cdot 2 \cdot 3 \cdot 3 = 72$$

$$\frac{5}{18} + \frac{7}{24} = \frac{5 \cdot 4}{18 \cdot 4} + \frac{7 \cdot 3}{24 \cdot 3}$$ Change each fraction to an equivalent fraction with 72 as the denominator

$$= \frac{20}{72} + \frac{21}{72} = \frac{41}{72}$$

Classroom Example
Chester, the cat, gets $\frac{5}{8}$ cup of Organics Cat Chow and $\frac{2}{3}$ cup of Mighty Fine cat food daily. What is the total amount of cat food Chester is receiving daily?

**EXAMPLE 15**

Marcey put $\frac{5}{8}$ pound of chemicals in the spa to adjust the water quality. Michael, not realizing Marcey had already put in chemicals, put $\frac{3}{14}$ pound of chemicals in the spa. The chemical manufacturer states that you should never add more than 1 pound of chemicals. Have Marcey and Michael together put in more than 1 pound of chemicals?

### Solution

Add $\frac{5}{8} + \frac{3}{14}$.

$$\left.\begin{aligned} 8 &= 2 \cdot 2 \cdot 2 \\ 14 &= 2 \cdot 7 \end{aligned}\right\} \longrightarrow \text{LCD} = 2 \cdot 2 \cdot 2 \cdot 7 = 56$$

$$\frac{5}{8} + \frac{3}{14} = \frac{5 \cdot 7}{8 \cdot 7} + \frac{3 \cdot 4}{14 \cdot 4} = \frac{35}{56} + \frac{12}{56} = \frac{47}{56}$$

No, Marcey and Michael have not added more than 1 pound of chemicals.

Try the following problems to help determine your proficiency with simplifying numerical expressions involving fractions.

## Mastery Set 4 Simplifying Numerical Expressions

For Problems 1–10, simplify the expression.

**1.** $\frac{3}{4} + \frac{5}{6} \div \frac{2}{3}$ **2.** $\frac{7}{8} \cdot \frac{6}{5} + \frac{3}{4}$ **3.** $\frac{6}{7} \div \frac{2}{3} \cdot \frac{1}{4}$

**4.** $\frac{7}{10} + \frac{4}{5} - \frac{11}{15} + \frac{2}{3}$ **5.** $\frac{15}{4} \div \frac{5}{2} + \frac{1}{8} \div \frac{3}{4}$

**6.** $\frac{7}{2} - \frac{1}{2} \cdot \frac{6}{5}$ **7.** $4\left(\frac{8}{3} - \frac{5}{12}\right)$ **8.** $\frac{1}{2}\left(\frac{5}{3} + \frac{1}{2}\right)$

**9.** $\frac{5}{8}\left(\frac{2}{3} - \frac{1}{4} + \frac{5}{6}\right)$ **10.** $24\left(\frac{3}{8} + \frac{3}{4} - \frac{5}{12}\right)$

## Simplifying Numerical Expressions

We now consider simplifying numerical expressions that contain fractions. Because of the mixed operations shown, you will have to use the order of operations agreement.

1. First perform any operations in grouping symbols.
2. Then perform any multiplications and divisions as they appear from left to right. Note that multiplication is done before division only if it occurs first when reading the problem from left to right.
3. Lastly perform any additions or subtractions as they appear from left to right.

Study the following examples paying attention to the order of operations.

**Classroom Example**
Simplify $\frac{7}{10} + \frac{5}{9} \div \frac{2}{3} - \frac{1}{3} \cdot \frac{6}{5}$.

### EXAMPLE 16

Simplify $\frac{3}{4} + \frac{2}{3} \cdot \frac{3}{5} - \frac{1}{2} \cdot \frac{1}{5}$.

**Solution**

$$\frac{3}{4} + \frac{2}{3} \cdot \frac{3}{5} - \frac{1}{2} \cdot \frac{1}{5} = \frac{3}{4} + \frac{2}{5} - \frac{1}{10}$$ Perform the two multiplications

$$= \frac{15}{20} + \frac{8}{20} - \frac{2}{20}$$ Change to equivalent fractions with 20 as the LCD

$$= \frac{15 + 8 - 2}{20} = \frac{21}{20}$$

**Classroom Example**
Simplify $\frac{3}{4}\left(\frac{7}{2} - \frac{13}{6}\right)$.

### EXAMPLE 17

Simplify $\frac{5}{8}\left(\frac{1}{2} + \frac{1}{3}\right)$.

**Solution**

$$\frac{5}{8}\left(\frac{1}{2} + \frac{1}{3}\right) = \frac{5}{8}\left(\frac{3}{6} + \frac{2}{6}\right)$$ Change to equivalent fractions with 6 as the LCD

$$= \frac{5}{8}\left(\frac{5}{6}\right) = \frac{25}{48}$$ Add the fractions in parentheses, then multiply

## Problem Set 1.0

For Problems 1–12, factor each composite number into a product of prime numbers; for example, $18 = 2 \cdot 3 \cdot 3$.

**1.** 26
**2.** 16
**3.** 36
**4.** 80
**5.** 49
**6.** 92
**7.** 56
**8.** 144
**9.** 120
**10.** 84
**11.** 135
**12.** 98

For Problems 13–24, find the least common multiple of the given numbers.

**13.** 6 and 8
**14.** 8 and 12
**15.** 12 and 16
**16.** 9 and 12
**17.** 28 and 35
**18.** 42 and 66
**19.** 49 and 56
**20.** 18 and 24
**21.** 8, 12, and 28
**22.** 6, 10, and 12
**23.** 9, 15, and 18
**24.** 8, 14, and 24

For Problems 25–30, reduce each fraction to lowest terms.

**25.** $\frac{8}{12}$
**26.** $\frac{12}{16}$
**27.** $\frac{16}{24}$
**28.** $\frac{18}{32}$
**29.** $\frac{15}{9}$
**30.** $\frac{48}{36}$

For Problems 31–36, multiply or divide as indicated, and express answers in reduced form.

**31.** $\frac{3}{4} \cdot \frac{5}{7}$
**32.** $\frac{4}{5} \cdot \frac{3}{11}$
**33.** $\frac{2}{7} \div \frac{3}{5}$
**34.** $\frac{5}{6} \div \frac{11}{13}$
**35.** $\frac{3}{8} \cdot \frac{12}{15}$
**36.** $\frac{4}{9} \cdot \frac{3}{2}$

**37.** A certain recipe calls for $\frac{3}{4}$ cup of milk. To make half of the recipe, how much milk is needed?

**38.** John is adding a diesel fuel additive to his fuel tank, which is half full. The directions say to add $\frac{1}{3}$ of the bottle to a full fuel tank. What portion of the bottle should he add to the fuel tank?

**39.** Mark shares a computer with his roommates. He has partitioned the hard drive in such a way that he gets $\frac{1}{3}$ of the disk space. His part of the hard drive is currently $\frac{2}{3}$ full. What portion of the computer's hard drive space is he currently taking up?

**40.** Angelina teaches $\frac{2}{3}$ of the deaf children in her local school. Her local school educates $\frac{1}{2}$ of the deaf children in the school district. What portion of the school district's deaf children is Angelina teaching?

For Problems 41–57, add or subtract as indicated, and express answers in lowest terms.

**41.** $\frac{2}{7} + \frac{3}{7}$
**42.** $\frac{3}{11} + \frac{5}{11}$
**43.** $\frac{7}{9} - \frac{2}{9}$
**44.** $\frac{11}{13} - \frac{6}{13}$
**45.** $\frac{3}{4} + \frac{9}{4}$
**46.** $\frac{5}{6} + \frac{7}{6}$
**47.** $\frac{11}{12} - \frac{3}{12}$
**48.** $\frac{13}{16} - \frac{7}{16}$
**49.** $\frac{5}{24} + \frac{11}{24}$
**50.** $\frac{7}{36} + \frac{13}{36}$
**51.** $\frac{1}{3} + \frac{1}{5}$
**52.** $\frac{1}{6} + \frac{1}{8}$
**53.** $\frac{15}{16} - \frac{3}{8}$
**54.** $\frac{13}{12} - \frac{1}{6}$
**55.** $\frac{7}{10} + \frac{8}{15}$
**56.** $\frac{7}{12} + \frac{5}{8}$
**57.** $\frac{11}{24} + \frac{5}{32}$

**58.** Alicia and her brother Jeff shared a pizza. Alicia ate $\frac{1}{8}$ of the pizza, while Jeff ate $\frac{2}{3}$ of the pizza. How much of the pizza has been eaten?

**59.** Rosa has $\frac{1}{3}$ pound of blueberries, $\frac{1}{4}$ pound of strawberries, and $\frac{1}{2}$ pound of raspberries. If she combines these for a fruit salad, how many pounds of these berries will be in the salad?

**60.** A chemist has $\frac{11}{16}$ of an ounce of dirt residue to perform crime lab tests. He needs $\frac{3}{8}$ of an ounce to perform a test for iron content. How much of the dirt residue will be left for the chemist to use in other testing?

For Problems 61–68, simplify each numerical expression, expressing answers in reduced form.

**61.** $\frac{1}{4} - \frac{3}{8} + \frac{5}{12} - \frac{1}{24}$

**62.** $\frac{3}{4} + \frac{2}{3} - \frac{1}{6} + \frac{5}{12}$

**63.** $\frac{5}{6} + \frac{2}{3} \cdot \frac{3}{4} - \frac{1}{4} \cdot \frac{2}{5}$

**64.** $\frac{2}{3} + \frac{1}{2} \cdot \frac{2}{5} - \frac{1}{3} \cdot \frac{1}{5}$

**65.** $\frac{3}{4} \cdot \frac{6}{9} - \frac{5}{6} \cdot \frac{8}{10} + \frac{2}{3} \cdot \frac{6}{8}$

**66.** $\frac{3}{5} \cdot \frac{5}{7} + \frac{2}{3} \cdot \frac{3}{5} - \frac{1}{7} \cdot \frac{2}{5}$

**67.** $\frac{7}{13}\left(\frac{2}{3} - \frac{1}{6}\right)$

**68.** $48\left(\frac{5}{12} - \frac{1}{6} + \frac{3}{8}\right)$

**69.** Blake Scott leaves $\frac{1}{4}$ of his estate to the Boy Scouts, $\frac{2}{5}$ to the local cancer fund, and the rest to his church. What fractional part of the estate does the church receive?

**70.** Franco has $\frac{7}{8}$ of an ounce of gold. He wants to give $\frac{3}{16}$ of an ounce to his friend Julie. He plans to divide the remaining amount of his gold in half to make two rings. How much gold will he have for each ring?

## 1.1 Sets, Real Numbers, and Numerical Expressions

OBJECTIVES

1. Identify certain sets of numbers
2. Apply the properties of equality
3. Simplify numerical expressions

In arithmetic, we use symbols such as 6, $\frac{2}{3}$, 0.27, and $\pi$ to represent numbers. The symbols $+$, $-$, $\cdot$, and $\div$ commonly indicate the basic operations of addition, subtraction, multiplication, and division, respectively. Thus we can form specific **numerical expressions**. For example, we can write the indicated sum of six and eight as $6 + 8$.

In algebra, the concept of a variable provides the basis for generalizing arithmetic ideas. For example, by using $x$ and $y$ to represent any numbers, we can use the expression $x + y$ to represent the indicated sum of any two numbers. The $x$ and $y$ in such an expression are called **variables**, and the phrase $x + y$ is called an **algebraic expression**.

We can extend to algebra many of the notational agreements we make in arithmetic, with a few modifications. The following chart summarizes the notational agreements that pertain to the four basic operations.

| Operation | Arithmetic | Algebra | Vocabulary |
|---|---|---|---|
| Addition | $4 + 6$ | $x + y$ | The *sum* of $x$ and $y$ |
| Subtraction | $14 - 10$ | $a - b$ | The *difference* of $a$ and $b$ |
| Multiplication | $7 \cdot 5$ or $7 \times 5$ | $a \cdot b$, $a(b)$, $(a)b$, $(a)(b)$, or $ab$ | The *product* of $a$ and $b$ |
| Division | $8 \div 4$, $\frac{8}{4}$, or $4\overline{)8}$ | $x \div y$, $\frac{x}{y}$, or $y\overline{)x}$ | The *quotient* of $x$ and $y$ |

Note the different ways to indicate a product, including the use of parentheses. The $ab$ form is the simplest and probably the most widely used form. Expressions such as $abc$, $6xy$, and $14xyz$ all indicate multiplication. We also call your attention to the various forms that indicate division; in algebra, we usually use the fractional form $\frac{x}{y}$, although the other forms do serve a purpose at times.

## Use of Sets

We can use some of the basic vocabulary and symbolism associated with the concept of sets in the study of algebra. A **set** is a collection of objects, and the objects are called **elements** or **members** of the set. In arithmetic and algebra the elements of a set are usually numbers.

The use of set braces, { }, to enclose the elements (or a description of the elements) and the use of capital letters to name sets provide a convenient way to communicate about sets. For example, we can represent a set $A$, which consists of the vowels of the alphabet, in any of the following ways:

| | |
|---|---|
| $A = \{\text{vowels of the alphabet}\}$ | Word description |
| $A = \{a, e, i, o, u\}$ | List or roster description |
| $A = \{x \mid x \text{ is a vowel}\}$ | Set builder notation |

We can modify the listing approach if the number of elements is quite large. For example, all of the letters of the alphabet can be listed as

$$\{a, b, c, \ldots, z\}$$

We simply begin by writing enough elements to establish a pattern; then the three dots indicate that the set continues in that pattern. The final entry indicates the last element of the pattern. If we write

$$\{1, 2, 3, \ldots\}$$

the set begins with the counting numbers 1, 2, and 3. The three dots indicate that it continues in a like manner forever; there is no last element. A set that consists of no elements is called the **null set** (written $\varnothing$).

**Set builder notation** combines the use of braces and the concept of a variable. For example, $\{x \mid x \text{ is a vowel}\}$ is read "the set of all $x$ such that $x$ is a vowel." Note that the vertical line is read "such that." We can use set builder notation to describe the set $\{1, 2, 3, \ldots\}$ as $\{x \mid x > 0 \text{ and } x \text{ is a whole number}\}$.

We use the symbol $\in$ to denote set membership. Thus if $A = \{a, e, i, o, u\}$, we can write $e \in A$, which we read as "$e$ is an element of $A$." The slash symbol, /, is commonly used in mathematics as a negation symbol. For example, $m \notin A$ is read as "$m$ is not an element of $A$."

Two sets are said to be *equal* if they contain exactly the same elements. For example,

$$\{1, 2, 3\} = \{2, 1, 3\}$$

because both sets contain the same elements; the order in which the elements are written doesn't matter. The slash mark through the equality symbol denotes "is not equal to." Thus if $A = \{1, 2, 3\}$ and $B = \{1, 2, 3, 4\}$, we can write $A \neq B$, which we read as "set $A$ is not equal to set $B$."

## Real Numbers

We refer to most of the algebra that we will study in this text as the **algebra of real numbers**. This simply means that the variables represent real numbers. Therefore, it is necessary for us to be familiar with the various terms that are used to classify different types of real numbers.

| | |
|---|---|
| $\{1, 2, 3, 4, \ldots\}$ | Natural numbers, counting numbers, positive integers |
| $\{0, 1, 2, 3, \ldots\}$ | Whole numbers, nonnegative integers |
| $\{\ldots -3, -2, -1\}$ | Negative integers |
| $\{\ldots -3, -2, -1, 0\}$ | Nonpositive integers |
| $\{\ldots -3, -2, -1, 0, 1, 2, 3, \ldots\}$ | Integers |

We define a **rational number** as follows:

**Definition 1.2 Rational Numbers**

A rational number is any number that can be written in the form $\frac{a}{b}$, where $a$ and $b$ are integers, and $b$ does not equal zero.

We can easily recognize that each of the following numbers fits the definition of a rational number.

$$\frac{-3}{4} \quad \frac{2}{3} \quad \frac{15}{4} \quad \text{and} \quad \frac{1}{-5}$$

However, numbers such as $-4, 0, 0.3$, and $6\frac{1}{2}$ are also rational numbers. All of these numbers could be written in the form $\frac{a}{b}$ as follows.

$-4$ can be written as $\frac{-4}{1}$ or $\frac{4}{-1}$

$0$ can be written as $\frac{0}{1} = \frac{0}{2} = \frac{0}{3} = \ldots$

$0.3$ can be written as $\frac{3}{10}$

$6\frac{1}{2}$ can be written as $\frac{13}{2}$

We can also define a rational number in terms of decimal representation. We classify decimals as terminating, repeating, or nonrepeating.

| Type | Definition | Examples | Rational numbers |
|---|---|---|---|
| Terminating | A terminating decimal ends. | 0.3, 0.46, 0.6234, 1.25 | Yes |
| Repeating | A repeating decimal has a block of digits that repeats indefinitely. | 0.66666 . . .<br>0.141414 . . .<br>0.694694694 . . .<br>0.23171717 . . . | Yes |
| Nonrepeating | A nonrepeating decimal does not have a block of digits that repeats indefinitely and does not terminate. | 3.1415926535 . . .<br>1.414213562 . . .<br>0.276314583 . . . | No |

A repeating decimal has a block of digits that can be any number of digits and may or may not begin immediately after the decimal point. A small horizontal bar (overbar) is commonly used to indicate the repeat block. Thus 0.6666 . . . is written as $0.\overline{6}$, and 0.2317171717 . . . is written as $0.23\overline{17}$.

In terms of decimals, we define a rational number as a number that has a terminating or a repeating decimal representation. The following examples illustrate some rational numbers written in $\frac{a}{b}$ form and in decimal form.

$$\frac{3}{4} = 0.75 \qquad \frac{3}{11} = 0.\overline{27} \qquad \frac{1}{8} = 0.125 \qquad \frac{1}{7} = 0.\overline{142857} \qquad \frac{1}{3} = 0.\overline{3}$$

We define an **irrational number** as a number that *cannot* be expressed in $\frac{a}{b}$ form, where $a$ and $b$ are integers, and $b$ is not zero. Furthermore, an irrational number has a nonrepeating and nonterminating decimal representation. Some examples of irrational numbers and a partial decimal representation for each follow.

$\sqrt{2} = 1.414213562373095\ldots$ $\qquad$ $\sqrt{3} = 1.73205080756887\ldots$

$\pi = 3.14159265358979\ldots$

The set of **real numbers** is composed of the rational numbers along with the irrational numbers. Every real number is either a rational number or an irrational number. The following tree diagram summarizes the various classifications of the real number system.

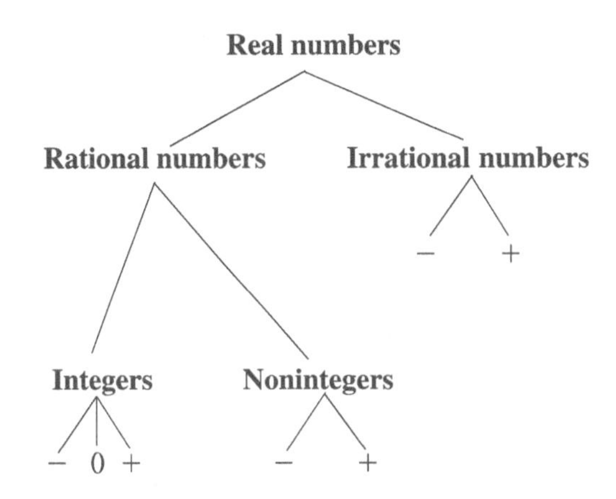


We can trace any real number down through the diagram as follows:

7 is real, rational, an integer, and positive

$-\frac{2}{3}$ is real, rational, noninteger, and negative

$\sqrt{7}$ is real, irrational, and positive

0.38 is real, rational, noninteger, and positive

**Remark:** We usually refer to the set of nonnegative integers, $\{0, 1, 2, 3, \ldots\}$, as the set of **whole numbers**, and we refer to the set of positive integers, $\{1, 2, 3, \ldots\}$, as the set of **natural numbers**. The set of whole numbers differs from the set of natural numbers by the inclusion of the number zero.

The concept of subset is convenient to discuss at this time. A set $A$ is a **subset** of a set $B$ if and only if every element of $A$ is also an element of $B$. This is written as $A \subseteq B$ and read as "$A$ is a subset of $B$." For example, if $A = \{1, 2, 3\}$ and $B = \{1, 2, 3, 5, 9\}$, then $A \subseteq B$ because every element of $A$ is also an element of $B$. The slash mark denotes negation, so if $A = \{1, 2, 5\}$ and $B = \{2, 4, 7\}$, we can say that $A$ is not a subset of $B$ by writing $A \nsubseteq B$. Figure 1.3 represents the subset relationships for the set of real numbers. Refer to Figure 1.3 as you study the following statements, which use subset vocabulary and subset symbolism.

**1.** The set of whole numbers is a subset of the set of integers.

$$\{0, 1, 2, 3, \ldots\} \subseteq \{\ldots, -2, -1, 0, 1, 2, \ldots\}$$

**2.** The set of integers is a subset of the set of rational numbers.

$$\{\ldots, -2, -1, 0, 1, 2, \ldots\} \subseteq \{x \mid x \text{ is a rational number}\}$$

**3.** The set of rational numbers is a subset of the set of real numbers.

$$\{x \mid x \text{ is a rational number}\} \subseteq \{y \mid y \text{ is a real number}\}$$

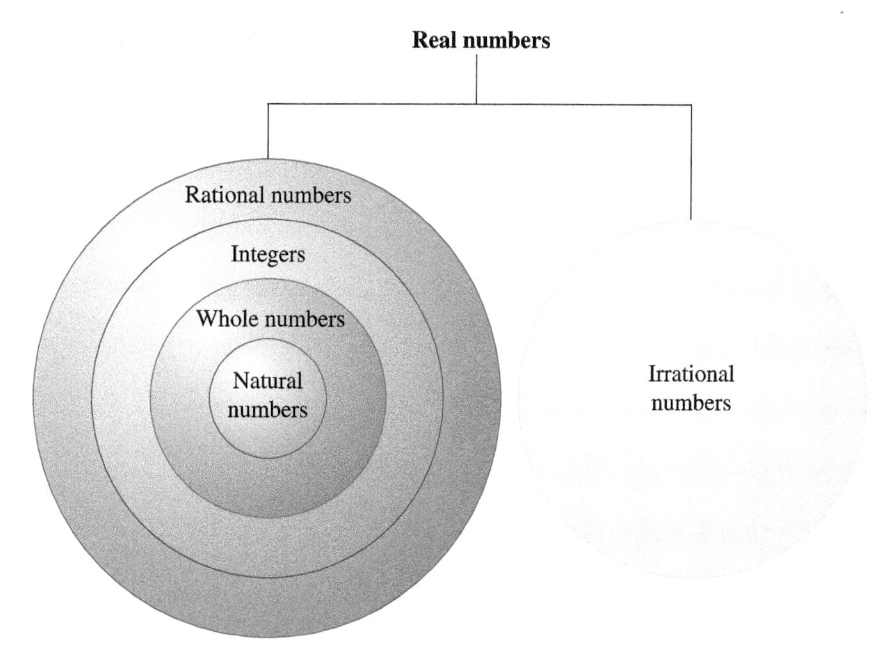


**Figure 1.3**

## Properties of Equality

The relation *equality* plays an important role in mathematics—especially when we are manipulating real numbers and algebraic expressions that represent real numbers. An **equality** is a statement in which two symbols, or groups of symbols, are names for the same number. The symbol $=$ is used to express an equality. Thus we can write

$$6 + 1 = 7 \qquad 18 - 2 = 16 \qquad 36 \div 4 = 9$$

(The symbol $\neq$ denotes *is not equal to.*) The following four basic properties of equality are self-evident, but we do need to keep them in mind. (We will expand this list in Chapter 2 when we work with solutions of equations.)

| Properties of equality | Definition: For real numbers $a$, $b$, and $c$ | Examples |
|---|---|---|
| Reflexive property | $a = a$ | $14 = 14$, $x = x$, $a + b = a + b$ |
| Symmetric property | If $a = b$, then $b = a$. | If $3 + 1 = 4$, then $4 = 3 + 1$. If $x = 10$, then $10 = x$. |
| Transitive property | If $a = b$ and $b = c$, then $a = c$. | If $x = 7$ and $7 = y$, then $x = y$. If $x + 5 = y$ and $y = 8$, then $x + 5 = 8$. |
| Substitution property | If $a = b$, then $a$ may be replaced by $b$, or $b$ may be replaced by $a$, without changing the meaning of the statement. | If $x + y = 4$ and $x = 2$, then we can replace $x$ in the first equation with the value 2, which will yield $2 + y = 4$. |

## Simplifying Numerical Expressions

Let's conclude this section by *simplifying some numerical expressions* that involve whole numbers. When simplifying numerical expressions, we perform the operations in the following order. Be sure that you agree with the result in each example.

1. Perform the operations inside the symbols of inclusion (parentheses, brackets, and braces) and above and below each fraction bar. Start with the innermost inclusion symbol.
2. Perform all multiplications and divisions in the order in which they appear from left to right.
3. Perform all additions and subtractions in the order in which they appear from left to right.

**Classroom Example**
Simplify $25 + 55 \div 11 \cdot 4$.

**EXAMPLE 1** Simplify $20 + 60 \div 10 \cdot 2$.

### Solution

First do the division.

$$20 + 60 \div 10 \cdot 2 = 20 + 6 \cdot 2$$

Next do the multiplication.

$$20 + 6 \cdot 2 = 20 + 12$$

Then do the addition.

$$20 + 12 = 32$$

Thus $20 + 60 \div 10 \cdot 2$ simplifies to 32.

**Classroom Example**
Simplify $4 \cdot 9 \div 3 \cdot 6 \div 8$.

**EXAMPLE 2** Simplify $7 \cdot 4 \div 2 \cdot 3 \cdot 2 \div 4$.

### Solution

The multiplications and divisions are to be done from left to right in the order in which they appear.

$$\begin{aligned} 7 \cdot 4 \div 2 \cdot 3 \cdot 2 \div 4 &= 7 \cdot 4 \div 2 \cdot 3 \cdot 2 \div 4 \\ &= 28 \div 2 \cdot 3 \cdot 2 \div 4 && \text{7 times 4 gives 28} \\ &= 14 \cdot 3 \cdot 2 \div 4 && \text{28 divided by 2 yields 14} \\ &= 42 \cdot 2 \div 4 && \text{14 times 3 equals 42} \\ &= 84 \div 4 && \text{42 times 2 is 84} \\ &= 21 && \text{84 divided by 4 gives 21} \end{aligned}$$

Thus $7 \cdot 4 \div 2 \cdot 3 \cdot 2 \div 4$ simplifies to 21.

**Classroom Example**
Simplify $3 \cdot 7 + 16 \div 4 - 3 \cdot 8 + 6 \div 2$.

**EXAMPLE 3** Simplify $5 \cdot 3 + 4 \div 2 - 2 \cdot 6 - 28 \div 7$.

### Solution

First we do the multiplications and divisions in the order in which they appear. Then we do the additions and subtractions in the order in which they appear.

$$\begin{aligned} 5 \cdot 3 + 4 \div 2 - 2 \cdot 6 - 28 \div 7 &= 5 \cdot 3 + 4 \div 2 - 2 \cdot 6 - 28 \div 7 \\ &= 15 + 2 - 12 - 4 \\ &= 17 - 12 - 4 \\ &= 5 - 4 \\ &= 1 \end{aligned}$$

Thus $5 \cdot 3 + 4 \div 2 - 2 \cdot 6 - 28 \div 7$ simplifies to 1.

**Classroom Example**
Simplify $(7 + 2)(3 + 8)$.

**EXAMPLE 4** Simplify $(4 + 6)(7 + 8)$.

### Solution

We use the parentheses to indicate the *product* of the quantities $4 + 6$ and $7 + 8$. We perform the additions inside the parentheses first and then multiply.

$$(4 + 6)(7 + 8) = (10)(15) = 150$$

**Classroom Example**
Simplify $(2 \cdot 5 + 3 \cdot 6) \cdot (7 \cdot 4 - 8 \cdot 3)$.

**EXAMPLE 5** Simplify $(3 \cdot 2 + 4 \cdot 5)(6 \cdot 8 - 5 \cdot 7)$.

### Solution

$$\begin{aligned}(3 \cdot 2 + 4 \cdot 5)(6 \cdot 8 - 5 \cdot 7) &= (3 \cdot 2 + 4 \cdot 5)(6 \cdot 8 - 5 \cdot 7) \\ &= (6 + 20)(48 - 35) && \text{Performed the multiplications inside the parentheses} \\ &= (26)(13) && \text{Performed the addition/subtraction inside the parentheses} \\ &= 338\end{aligned}$$

Thus $(3 \cdot 2 + 4 \cdot 5)(6 \cdot 8 - 5 \cdot 7)$ simplifies to 338.

**Classroom Example**
Simplify $3 + 9[2(5 + 4)]$.

**EXAMPLE 6** Simplify $6 + 7[3(4 + 6)]$.

### Solution

We use brackets for the same purposes as parentheses. In such a problem we need to simplify *from the inside out*; that is, we perform the operations in the innermost parentheses first. We thus obtain

$$\begin{aligned}6 + 7[3(4 + 6)] &= 6 + 7[3(10)] \\ &= 6 + 7[30] \\ &= 6 + 210 \\ &= 216\end{aligned}$$

**Classroom Example**
Simplify $\frac{7 \cdot 6 - 3 \cdot 3}{2 \cdot 6 \div 3 - 1}$.

**EXAMPLE 7** Simplify $\frac{6 \cdot 8 \div 4 - 2}{5 \cdot 4 - 9 \cdot 2}$.

### Solution

First we perform the operations above and below the fraction bar. Then we find the final quotient.

$$\frac{6 \cdot 8 \div 4 - 2}{5 \cdot 4 - 9 \cdot 2} = \frac{48 \div 4 - 2}{20 - 18} = \frac{12 - 2}{2} = \frac{10}{2} = 5$$

**Remark:** With parentheses we could write the problem in Example 7 as $(6 \cdot 8 \div 4 - 2) \div (5 \cdot 4 - 9 \cdot 2)$.

## Concept Quiz 1.1

For Problems 1–10, answer true or false.

1. The expression $ab$ indicates the sum of $a$ and $b$.
2. The set $\{1, 2, 3 \ldots.\}$ contains infinitely many elements.

3. The sets $A = \{1, 2, 4, 6\}$ and $B = \{6, 4, 1, 2\}$ are equal sets.
4. Every irrational number is also classified as a real number.
5. To evaluate $24 \div 6 \cdot 2$, the first operation to be performed is to multiply 6 times 2.
6. To evaluate $6 + 8 \cdot 3$, the first operation to be performed is to multiply 8 times 3.
7. The number 0.15 is real, irrational, and positive.
8. If $4 = x + 3$, then $x + 3 = 4$ is an example of the symmetric property of equality.
9. The numerical expression $6 \cdot 2 + 3 \cdot 5 - 6$ simplifies to 21.
10. The number represented by $0.\overline{12}$ is a rational number.

## Problem Set 1.1

For Problems 1–10, identify each statement as true or false. (Objective 1)

1. Every irrational number is a real number.
2. Every rational number is a real number.
3. If a number is real, then it is irrational.
4. Every real number is a rational number.
5. All integers are rational numbers.
6. Some irrational numbers are also rational numbers.
7. Zero is a positive integer.
8. Zero is a rational number.
9. All whole numbers are integers.
10. Zero is a negative integer.

For Problems 11–18, from the list 0, 14, $\frac{2}{3}$, $\pi$, $\sqrt{7}$, $-\frac{11}{14}$, 2.34, $-19$, $\frac{55}{8}$, $-\sqrt{17}$, $3.2\overline{1}$, and $-2.6$, identify each of the following. (Objective 1)

11. The whole numbers
12. The natural numbers
13. The rational numbers
14. The integers
15. The nonnegative integers
16. The irrational numbers
17. The real numbers
18. The nonpositive integers

For Problems 19–28, use the following set designations.

$N = \{x|x \text{ is a natural number}\}$

$Q = \{x|x \text{ is a rational number}\}$

$W = \{x|x \text{ is a whole number}\}$

$H = \{x|x \text{ is an irrational number}\}$

$I = \{x|x \text{ is an integer}\}$

$R = \{x|x \text{ is a real number}\}$

Place $\subseteq$ or $\nsubseteq$ in each blank to make a true statement. (Objective 1)

19. $R$ ____ $N$
20. $N$ ____ $R$
21. $I$ ____ $Q$
22. $N$ ____ $I$
23. $Q$ ____ $H$
24. $H$ ____ $Q$
25. $N$ ____ $W$
26. $W$ ____ $I$
27. $I$ ____ $N$
28. $I$ ____ $W$

For Problems 29–32, classify the real number by tracing through the diagram in the text (see page 16). (Objective 1)

29. $-8$
30. 0.9
31. $-\sqrt{2}$
32. $\frac{5}{6}$

For Problems 33–42, list the elements of each set. For example, the elements of $\{x|x \text{ is a natural number less than } 4\}$ can be listed as $\{1, 2, 3\}$. (Objective 1)

33. $\{x|x \text{ is a natural number less than } 3\}$
34. $\{x|x \text{ is a natural number greater than } 3\}$
35. $\{n|n \text{ is a whole number less than } 6\}$
36. $\{y|y \text{ is an integer greater than } -4\}$
37. $\{y|y \text{ is an integer less than } 3\}$
38. $\{n|n \text{ is a positive integer greater than } -7\}$
39. $\{x|x \text{ is a whole number less than } 0\}$
40. $\{x|x \text{ is a negative integer greater than } -3\}$
41. $\{n|n \text{ is a nonnegative integer less than } 5\}$
42. $\{n|n \text{ is a nonpositive integer greater than } 3\}$

For Problems 43–50, replace each question mark to make the given statement an application of the indicated property of equality. For example, $16 = ?$ becomes $16 = 16$ because of the reflexive property of equality. (Objective 2)

**43.** If $y = x$ and $x = -6$, then $y = ?$ (Transitive property of equality)

**44.** $5x + 7 = ?$ (Reflexive property of equality)

**45.** If $n = 2$ and $3n + 4 = 10$, then $3(?) + 4 = 10$ (Substitution property of equality)

**46.** If $y = x$ and $x = z + 2$, then $y = ?$ (Transitive property of equality)

**47.** If $4 = 3x + 1$, then $? = 4$ (Symmetric property of equality)

**48.** If $t = 4$ and $s + t = 9$, then $s + ? = 9$ (Substitution property of equality)

**49.** $5x = ?$ (Reflexive property of equality)

**50.** If $5 = n + 3$, then $n + 3 = ?$ (Symmetric property of equality)

For Problems 51–78, simplify each of the numerical expressions. (Objective 3)

**51.** $16 + 9 - 4 - 2 + 8 - 1$

**52.** $18 + 17 - 9 - 2 + 14 - 11$

**53.** $9 \div 3 \cdot 4 \div 2 \cdot 14$

**54.** $21 \div 7 \cdot 5 \cdot 2 \div 6$

**55.** $7 + 8 \cdot 2$

**56.** $21 - 4 \cdot 3 + 2$

**57.** $9 \cdot 7 - 4 \cdot 5 - 3 \cdot 2 + 4 \cdot 7$

**58.** $6 \cdot 3 + 5 \cdot 4 - 2 \cdot 8 + 3 \cdot 2$

**59.** $(17 - 12)(13 - 9)(7 - 4)$

**60.** $(14 - 12)(13 - 8)(9 - 6)$

**61.** $13 + (7 - 2)(5 - 1)$

**62.** $48 - (14 - 11)(10 - 6)$

**63.** $(16 - 8) \div 4(3 - 1)$

**64.** $(27 - 9) \div 6(5 - 2)$

**65.** $(5 \cdot 9 - 3 \cdot 4)(6 \cdot 9 - 2 \cdot 7)$

**66.** $(3 \cdot 4 + 2 \cdot 1)(5 \cdot 2 + 6 \cdot 7)$

**67.** $7[3(6 - 2)] - 64$

**68.** $12 + 5[3(7 - 4)]$

**69.** $[3 + 2(4 \cdot 1 - 2)][18 - (2 \cdot 4 - 7 \cdot 1)]$

**70.** $3[4(6 + 7)] + 2[3(4 - 2)]$

**71.** $\dfrac{3 + 4 \cdot 6}{8 - 5}$

**72.** $\dfrac{20 - 2}{20 - 2 \cdot 7}$

**73.** $14 + 4\left(\dfrac{8 - 2}{12 - 9}\right) - 2\left(\dfrac{9 - 1}{19 - 15}\right)$

**74.** $12 + 2\left(\dfrac{12 - 2}{7 - 2}\right) - 3\left(\dfrac{12 - 9}{17 - 14}\right)$

**75.** $[7 + 2 \cdot 3 \cdot 5 - 5] \div 8$

**76.** $[27 - (4 \cdot 2 + 5 \cdot 2)][(5 \cdot 6 - 4) - 20]$

**77.** $\dfrac{3 \cdot 8 - 4 \cdot 3}{5 \cdot 7 - 34} + 19$

**78.** $\dfrac{4 \cdot 9 - 3 \cdot 5 - 3}{18 - 12}$

**79.** You must of course be able to do calculations like those in Problems 51–78 both with and without a calculator. Furthermore, different types of calculators handle the priority-of-operations issue in different ways. Be sure you can do Problems 51–78 with *your* calculator.

## Thoughts Into Words

**80.** Explain in your own words the difference between the reflexive property of equality and the symmetric property of equality.

**81.** Your friend keeps getting an answer of 30 when simplifying $7 + 8(2)$. What mistake is he making and how would you help him?

**82.** Do you think $3\sqrt{2}$ is a rational or an irrational number? Defend your answer.

**83.** Explain why every integer is a rational number but not every rational number is an integer.

**84.** Explain the difference between $1.\overline{3}$ and $1.3$.

### Answers to the Concept Quiz

**1.** False **2.** True **3.** True **4.** True **5.** False **6.** True **7.** False **8.** True **9.** True **10.** True

## 1.2 Operations with Real Numbers

**OBJECTIVES**

1. Review the real number line
2. Find the absolute value of a number
3. Add real numbers
4. Subtract real numbers
5. Multiply real numbers
6. Divide real numbers
7. Simplify numerical expressions
8. Use real numbers to represent problems

Before we review the four basic operations with real numbers, let's briefly discuss some concepts and terminology we commonly use with this material. It is often helpful to have a geometric representation of the set of real numbers as indicated in Figure 1.4. Such a representation, called the **real number line**, indicates a one-to-one correspondence between the set of real numbers and the points on a line. In other words, to each real number there corresponds one and only one point on the line, and to each point on the line there corresponds one and only one real number. The number associated with each point on the line is called the **coordinate** of the point.

**Figure 1.4**

Many operations, relations, properties, and concepts pertaining to real numbers can be given a geometric interpretation on the real number line. For example, the addition problem $(-1) + (-2)$ can be depicted on the number line as in Figure 1.5.

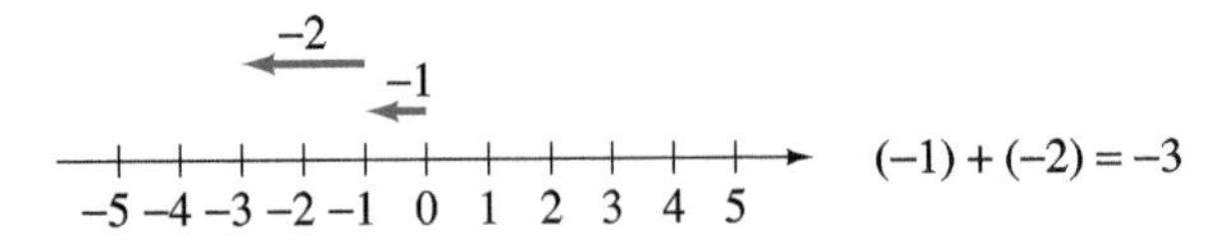


**Figure 1.5**

**Figure 1.6**

The inequality relations also have a geometric interpretation. The statement $a > b$ (which is read "$a$ is greater than $b$") means that $a$ is to the right of $b$, and the statement $c < d$ (which is read "$c$ is less than $d$") means that $c$ is to the left of $d$ as shown in Figure 1.6. The symbol $\leq$ means *is less than or equal to*, and the symbol $\geq$ means *is greater than or equal to*.

The property $-(-x) = x$ can be represented on the number line by following the sequence of steps shown in Figure 1.7.

1. Choose a point that has a coordinate of $x$.
2. Locate its opposite, written as $-x$, on the other side of zero.
3. Locate the opposite of $-x$, written as $-(-x)$, on the other side of zero.

Therefore, we conclude that *the opposite of the opposite of any real number is the number itself*, and we symbolically express this by $-(-x) = x$.

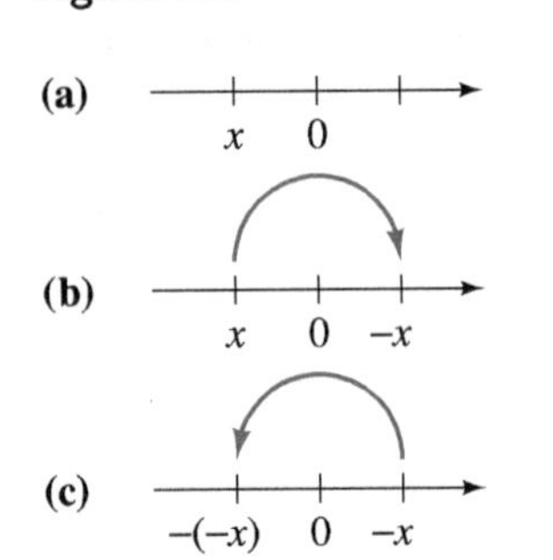


**Figure 1.7**

**Remark:** The symbol $-1$ can be read "negative one," "the negative of one," "the opposite of one," or "the additive inverse of one." The opposite-of and additive-inverse-of terminology is especially meaningful when working with variables. For example, the symbol $-x$, which is read "the opposite of $x$" or "the additive inverse of $x$," emphasizes an important issue. Because $x$ can be any real number, $-x$ (the opposite of $x$) can be zero, positive, or negative. If $x$ is positive, then $-x$ is negative. If $x$ is negative, then $-x$ is positive. If $x$ is zero, then $-x$ is zero.

## Absolute Value

We can use the concept of *absolute value* to describe precisely how to operate with positive and negative numbers. Geometrically, the **absolute value** of any number is the distance between the number and zero on the number line. For example, the absolute value of 2 is 2. The absolute value of $-3$ is 3. The absolute value of 0 is 0 (see Figure 1.8).

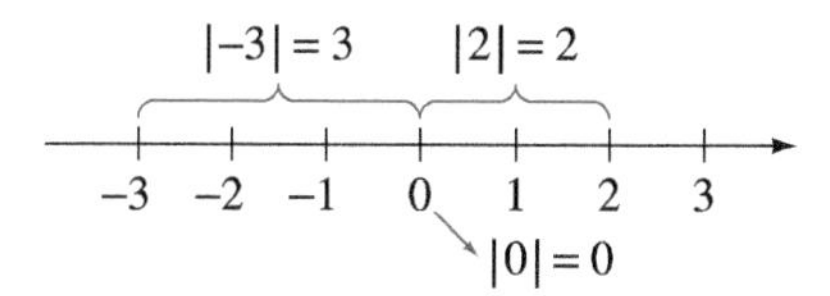


**Figure 1.8**

Symbolically, absolute value is denoted with vertical bars. Thus we write

$$|2| = 2 \qquad |-3| = 3 \qquad |0| = 0$$

More formally, we define the concept of absolute value as follows:

### Definition 1.3

For all real numbers $a$,

1. If $a \geq 0$, then $|a| = a$.
2. If $a < 0$, then $|a| = -a$.

According to Definition 1.3, we obtain

| | |
|---|---|
| $|6| = 6$ | By applying part 1 of Definition 1.3 |
| $|0| = 0$ | By applying part 1 of Definition 1.3 |
| $|-7| = -(-7) = 7$ | By applying part 2 of Definition 1.3 |

Note that the absolute value of a positive number is the number itself, but the absolute value of a negative number is its opposite. Thus the absolute value of any number except zero is positive, and the absolute value of zero is zero. Together these facts indicate that the absolute value of any real number is equal to the absolute value of its opposite. We summarize these ideas in the following properties.

### Properties of Absolute Value

The variables $a$ and $b$ represent any real number.

1. $|a| \geq 0$
2. $|a| = |-a|$
3. $|a - b| = |b - a|$    $a - b$ and $b - a$ are opposites of each other

## Adding Real Numbers

We can use various physical models to describe the addition of real numbers. For example, profits and losses pertaining to investments: A loss of \$25.75 (written as $-25.75$) on one investment, along with a profit of \$22.20 (written as 22.20) on a second investment, produces an overall loss of \$3.55. Thus $(-25.75) + 22.20 = -3.55$. Think in terms of profits and losses for each of the following examples.

$50 + 75 = 125$ $\quad$ $20 + (-30) = -10$

$-4.3 + (-6.2) = -10.5$ $\quad$ $-27 + 43 = 16$

$\frac{7}{8} + \left(-\frac{1}{4}\right) = \frac{5}{8}$ $\quad$ $-3\frac{1}{2} + \left(-3\frac{1}{2}\right) = -7$

Though all problems that involve addition of real numbers could be solved using the profit-loss interpretation, it is sometimes convenient to have a more precise description of the addition process. For this purpose we use the concept of absolute value.

### Addition of Real Numbers

**Two Positive Numbers** The sum of two positive real numbers is the sum of their absolute values.

**Two Negative Numbers** The sum of two negative real numbers is the opposite of the sum of their absolute values.

**One Positive and One Negative Number** The sum of a positive real number and a negative real number can be found by subtracting the smaller absolute value from the larger absolute value and giving the result the sign of the original number that has the larger absolute value. If the two numbers have the same absolute value, then their sum is 0.

**Zero and Another Number** The sum of 0 and any real number is the real number itself.

Now consider the following examples in terms of the previous description of addition. These examples include operations with rational numbers in common fraction form.

**Classroom Example**
Find the sum:
**(a)** $-4.5 + 6$
**(b)** $4\frac{2}{3} + \left(-1\frac{1}{4}\right)$
**(c)** $21 + (-57)$
**(d)** $-36.2 + 36.2$

**EXAMPLE 1** Find the sum of the two numbers:

**(a)** $(-6) + (-8)$ **(b)** $6\frac{3}{4} + \left(-2\frac{1}{2}\right)$ **(c)** $14 + (-21)$ **(d)** $-72.4 + 72.4$

### Solution

**(a)** $(-6) + (-8) = -(|-6| + |-8|) = -(6 + 8) = -14$

**(b)** $6\frac{3}{4} + \left(-2\frac{1}{2}\right) = \left(\left|6\frac{3}{4}\right| - \left|-2\frac{1}{2}\right|\right) = \left(6\frac{3}{4} - 2\frac{1}{2}\right) = \left(6\frac{3}{4} - 2\frac{2}{4}\right) = 4\frac{1}{4}$

**(c)** $14 + (-21) = -(|-21| - |14|) = -(21 - 14) = -7$

**(d)** $-72.4 + 72.4 = 0$

## Subtracting Real Numbers

We can describe the subtraction of real numbers in terms of addition.

### Subtraction of Real Numbers

If $a$ and $b$ are real numbers, then

$$a - b = a + (-b)$$

It may be helpful for you to read $a - b = a + (-b)$ as "$a$ minus $b$ is equal to $a$ plus the opposite of $b$." In other words, every subtraction problem can be changed to an equivalent addition problem. Consider the following example.

**Classroom Example**
Find the difference:
(a) $6 - 10$
(b) $-3 - (-15)$
(c) $11.3 - (-8.7)$
(d) $-\frac{5}{9} - \left(-\frac{2}{3}\right)$

**EXAMPLE 2** Find the difference between the two numbers:

**(a)** $7 - 9$ **(b)** $-5 - (-13)$ **(c)** $6.1 - (-14.2)$ **(d)** $-\frac{7}{8} - \left(-\frac{1}{4}\right)$

### Solution

**(a)** $7 - 9 = 7 + (-9) = -2$

**(b)** $-5 - (-13) = -5 + 13 = 8$

**(c)** $6.1 - (-14.2) = 6.1 + 14.2 = 20.3$

**(d)** $-\frac{7}{8} - \left(-\frac{1}{4}\right) = -\frac{7}{8} + \frac{1}{4} = -\frac{7}{8} + \frac{2}{8} = -\frac{5}{8}$

It should be apparent that addition is a key operation. To simplify numerical expressions that involve addition and subtraction, we can first change all subtractions to additions and then perform the additions.

**Classroom Example**
Simplify
$3 - 19 - 2 + 16 - 4 + 5.$

**EXAMPLE 3** Simplify $7 - 9 - 14 + 12 - 6 + 4$.

### Solution

Begin by changing all subtractions to adding the opposite.

$$\begin{aligned} 7 - 9 - 14 + 12 - 6 + 4 &= 7 + (-9) + (-14) + 12 + (-6) + 4 \\ &= -6 \end{aligned}$$

**Classroom Example**
Simplify
$-3\frac{2}{3} + \frac{7}{12} - \left(-\frac{1}{4}\right) + \frac{1}{12}.$

**EXAMPLE 4** Simplify $-2\frac{1}{8} + \frac{3}{4} - \left(-\frac{3}{8}\right) - \frac{1}{2}$.

### Solution

$$\begin{aligned} -2\frac{1}{8} + \frac{3}{4} - \left(-\frac{3}{8}\right) - \frac{1}{2} &= -2\frac{1}{8} + \frac{3}{4} + \frac{3}{8} + \left(-\frac{1}{2}\right) && \text{Change subtractions to adding the opposite} \\ &= -\frac{17}{8} + \frac{6}{8} + \frac{3}{8} + \left(-\frac{4}{8}\right) && \text{Change to equivalent fractions with a common denominator of 8} \\ &= -\frac{12}{8} = -\frac{3}{2} \end{aligned}$$

**Classroom Example**
Simplify $\left(\frac{3}{8} - \frac{1}{3}\right) - \left(\frac{3}{4} - \frac{11}{12}\right).$

**EXAMPLE 5** Simplify $\left(\frac{2}{3} - \frac{1}{5}\right) - \left(\frac{1}{2} - \frac{7}{10}\right)$.

### Solution

$$\begin{aligned} \left(\frac{2}{3} - \frac{1}{5}\right) - \left(\frac{1}{2} - \frac{7}{10}\right) &= \left[\frac{2}{3} + \left(-\frac{1}{5}\right)\right] - \left[\frac{1}{2} + \left(-\frac{7}{10}\right)\right] \\ &= \left[\frac{10}{15} + \left(-\frac{3}{15}\right)\right] - \left[\frac{5}{10} + \left(-\frac{7}{10}\right)\right] && \text{Within the brackets, change to equivalent fractions with a common denominator} \\ &= \left(\frac{7}{15}\right) - \left(-\frac{2}{10}\right) \\ &= \left(\frac{7}{15}\right) + \left(+\frac{2}{10}\right) && \text{Change subtraction to adding the opposite} \end{aligned}$$

$$= \frac{14}{30} + \left(+\frac{6}{30}\right)$$ Change to equivalent fractions with a common denominator

$$= \frac{20}{30} = \frac{2}{3}$$

## Multiplying Real Numbers

To determine the product of a positive number and a negative number, we can consider the multiplication of whole numbers as repeated addition. For example, $4 \cdot 2$ means four 2s; thus $4 \cdot 2 = 2 + 2 + 2 + 2 = 8$. Applying this concept to the product of 4 and $-2$ we get the following,

$$4(-2) = -2 + (-2) + (-2) + (-2) = -8$$

Because the order in which we multiply two numbers does not change the product, we know the following

$$4(-2) = -2(4) = -8$$

Therefore, the product of a positive real number and a negative real number is a negative number.

Finally, let's consider the product of two negative integers. The following pattern using integers helps with the reasoning.

$4(-2) = -8$ $\quad 3(-2) = -6$ $\quad 2(-2) = -4$

$1(-2) = -2$ $\quad 0(-2) = 0$ $\quad (-1)(-2) = ?$

To continue this pattern, the product of $-1$ and $-2$ has to be 2. In general, this type of reasoning helps us realize that the product of any two negative real numbers is a positive real number. Using the concept of absolute value, we can describe the *multiplication of real numbers* as follows:

### Multiplication of Real Numbers

1. The product of two positive or two negative real numbers is the product of their absolute values.
2. The product of a positive real number and a negative real number (either order) is the opposite of the product of their absolute values.
3. The product of zero and any real number is zero.

The following example illustrates this description of multiplication.

**Classroom Example**
Find the product for each of the following:
**(a)** $(-3)(-8)$
**(b)** $(7)(-11)$
**(c)** $\left(-\frac{5}{6}\right)\left(\frac{2}{5}\right)$

**EXAMPLE 6** Find the product for each of the following:

**(a)** $(-6)(-7)$ **(b)** $(8)(-9)$ **(c)** $\left(-\frac{3}{4}\right)\left(\frac{1}{3}\right)$

### Solution

**(a)** $(-6)(-7) = |-6| \cdot |-7| = 6 \cdot 7 = 42$

**(b)** $(8)(-9) = -(|8| \cdot |-9|) = -(8 \cdot 9) = -72$

**(c)** $\left(-\frac{3}{4}\right)\left(\frac{1}{3}\right) = -\left(\left|-\frac{3}{4}\right| \cdot \left|\frac{1}{3}\right|\right) = -\left(\frac{3}{4} \cdot \frac{1}{3}\right) = -\frac{1}{4}$

Example 6 illustrates a step-by-step process for multiplying real numbers. In practice, however, the key is to remember that the product of two positive or two negative numbers is positive, and the product of a positive number and a negative number (either order) is negative.

## Dividing Real Numbers

The relationship between multiplication and division provides the basis for dividing real numbers. For example, we know that $8 \div 2 = 4$ because $2 \cdot 4 = 8$. In other words, the quotient of two numbers can be found by looking at a related multiplication problem. In the following examples, we used this same reasoning to determine some quotients that involve integers.

$$\frac{6}{-2} = -3 \quad \text{because } (-2)(-3) = 6$$

$$\frac{-12}{3} = -4 \quad \text{because } (3)(-4) = -12$$

$$\frac{-18}{-2} = 9 \quad \text{because } (-2)(9) = -18$$

$$\frac{0}{-5} = 0 \quad \text{because } (-5)(0) = 0$$

$$\frac{-8}{0} \text{ is undefined} \qquad \text{Remember that division by zero is undefined!}$$

A precise description for *division of real numbers* follows.

### Division of Real Numbers

1. The quotient of two positive or two negative real numbers is the quotient of their absolute values.
2. The quotient of a positive real number and a negative real number or of a negative real number and a positive real number is the opposite of the quotient of their absolute values.
3. The quotient of zero and any nonzero real number is zero.
4. The quotient of any nonzero real number and zero is undefined.

The following example illustrates this description of division. Again, for practical purposes, the key is to remember whether the quotient is positive or negative.

Classroom Example
Find the quotient for each of the following:

(a) $\frac{-18}{-9}$

(b) $\frac{36}{-4}$

(c) $\frac{-5.2}{4}$

(d) $\frac{0}{\frac{5}{9}}$

**EXAMPLE 7** Find the quotient for each of the following:

**(a)** $\frac{-16}{-4}$ **(b)** $\frac{28}{-7}$ **(c)** $\frac{-3.6}{4}$ **(d)** $\frac{0}{\frac{7}{8}}$

### Solution

**(a)** $\frac{-16}{-4} = \frac{|-16|}{|-4|} = \frac{16}{4} = 4$

**(b)** $\frac{28}{-7} = -\left(\frac{|28|}{|-7|}\right) = -\left(\frac{28}{7}\right) = -4$

**(c)** $\frac{-3.6}{4} = -\left(\frac{|-3.6|}{|4|}\right) = -\left(\frac{3.6}{4}\right) = -0.9$

**(d)** $\frac{0}{\frac{7}{8}} = 0$

Now let's simplify some numerical expressions that involve the four basic operations with real numbers. Remember that multiplications and divisions are done first, from left to right, before additions and subtractions are performed.

**Classroom Example**
Simplify:
$-4\frac{1}{2} - 3\left(\frac{1}{6}\right) - (-2)\left(-\frac{3}{4}\right)$

**EXAMPLE 8** Simplify $-2\frac{1}{3} + 4\left(-\frac{2}{3}\right) - (-5)\left(-\frac{1}{3}\right)$.

### Solution

$$-2\frac{1}{3} + 4\left(-\frac{2}{3}\right) - (-5)\left(-\frac{1}{3}\right) = -2\frac{1}{3} + \left(-\frac{8}{3}\right) - \left(\frac{5}{3}\right)$$ Perform the multiplications

$$= -2\frac{1}{3} + \left(-\frac{8}{3}\right) + \left(-\frac{5}{3}\right)$$ Change subtraction to adding the opposite

$$= -\frac{7}{3} + \left(-\frac{8}{3}\right) + \left(-\frac{5}{3}\right)$$ Change to an improper fraction

$$= -\frac{20}{3}$$

**Classroom Example**
Simplify $21 \div (-3) + 7(-2)$.

**EXAMPLE 9** Simplify $-24 \div 4 + 8(-5) - (-5)(3)$.

### Solution

To begin perform the multiplications and divisions as they occur left to right.

$$-24 \div 4 + 8(-5) - (-5)(3) = -6 + (-40) - (-15)$$
$$= -6 + (-40) + 15$$ Change subtraction to adding the opposite
$$= -31$$

**Classroom Example**
Simplify
$-3.8 - 4[-2.7(1 - (-4))]$.

**EXAMPLE 10** Simplify $-7.3 - 2[-4.6(6 - 7)]$.

### Solution

$$-7.3 - 2[-4.6(6 - 7)] = -7.3 - 2[-4.6(-1)]$$
$$= -7.3 - 2[4.6]$$
$$= -7.3 - 9.2$$
$$= -7.3 + (-9.2)$$ Change subtraction to adding the opposite
$$= -16.5$$

**Classroom Example**
Simplify:
$[5(-2) - 6(4)][4(-2) + 7(1)]$

**EXAMPLE 11** Simplify $[3(-7) - 2(9)][5(-7) + 3(9)]$.

### Solution

$$[3(-7) - 2(9)][5(-7) + 3(9)] = [-21 - 18][-35 + 27]$$
$$= [-21 + (-18)][-35 + 27]$$
$$= [-39][-8]$$
$$= 312$$

Eray Haciosmanoglu/Shutterstock.com

**EXAMPLE 12** **Apply Your Skill**

On a flight from Orlando to Washington, D.C., the airline sold 52 economy seats, 25 business-class seats, and 12 first-class seats, and there were 20 empty seats. The airline has determined that it makes a profit of \$550 per first-class seat and \$100 profit per business-class seat. However, the airline incurs a loss of \$20 per economy seat and a loss of \$75 per empty seat. Determine the profit (or loss) for the flight.

**Classroom Example**
On a flight from Chicago to San Francisco, an airline sold 65 economy seats, 32 business-class seats, and 15 first-class seats, and there were 8 empty seats. The airline has determined that it makes a profit of \$475 per first-class seat and \$120 profit per business-class seat. However, the airline incurs a loss of \$25 per economy seat and a loss of \$80 per empty seat. Determine the profit (or loss) for the flight.

### Solution

Let the profit be represented by positive numbers and the loss be represented by negative numbers. Then the following expression would represent the profit or loss for this flight.

$$52(-20) + 25(100) + 12(550) + 20(-75)$$

Simplify this expression as follows:

$$52(-20) + 25(100) + 12(550) + 20(-75)$$
$$= -1040 + 2500 + 6600 - 1500 = 6560$$

Therefore, the flight had a profit of \$6560.

## Concept Quiz 1.2

For Problems 1–10, answer true or false.

1. The product of two negative real numbers is a positive real number.
2. The quotient of two negative integers is a negative integer.
3. The quotient of any nonzero real number and zero is zero.
4. If $x$ represents any real number, then $-x$ represents a negative real number.
5. The product of three negative real numbers is a negative real number.
6. The statement $|6 - 4| = |4 - 6|$ is a true statement.
7. The absolute value of every real number is a positive real number.
8. The absolute value of zero does not exist.
9. The sum of a positive number plus a negative number is always a negative number.
10. Every subtraction problem can be changed to an equivalent addition problem.

## Problem Set 1.2

1. Graph the following points and their opposites on the real number line: 1, $-2$, and 4.
2. Graph the following points and their opposites on the real number line: $-3$, $-1$, and 5.
3. Find the following absolute values: **(a)** $|-7|$ **(b)** $|0|$ **(c)** $|15|$
4. Find the following absolute values: **(a)** $|2|$ **(b)** $|-1|$ **(c)** $|-10|$

For Problems 5–54, perform the following operations with real numbers. (Objectives 3–6)

5. $8 + (-15)$
6. $9 + (-18)$
7. $(-12)+(-7)$
8. $(-7)+(-14)$
9. $-8 - 14$
10. $-17 - 9$
11. $9 - 16$
12. $8 - 22$
13. $(-9)(-12)$
14. $(-6)(-13)$
15. $(5)(-14)$
16. $(-17)(4)$
17. $(-56) \div (-4)$
18. $(-81) \div (-3)$
19. $\frac{-112}{16}$
20. $\frac{-75}{5}$
21. $-2\frac{3}{8} + 5\frac{7}{8}$
22. $-1\frac{1}{5} + 3\frac{4}{5}$
23. $4\frac{1}{3} - \left(-1\frac{1}{6}\right)$
24. $1\frac{1}{12} - \left(-5\frac{3}{4}\right)$
25. $\left(-\frac{1}{3}\right)\left(\frac{2}{5}\right)$
26. $(-8)\left(\frac{1}{3}\right)$
27. $\frac{1}{2} \div \left(-\frac{1}{8}\right)$
28. $\frac{2}{3} \div \left(-\frac{1}{6}\right)$
29. $0 \div (-14)$
30. $(-19) \div 0$
31. $(-21) \div 0$
32. $0 \div (-11)$
33. $-21 - 39$
34. $-23 - 38$
35. $-17.3 + 12.5$
36. $-16.3 + 19.6$
37. $21.42 - 7.29$
38. $2.73 - 8.14$
39. $-21.4 - (-14.9)$
40. $-32.6 - (-9.8)$
41. $(5.4)(-7.2)$
42. $(-8.5)(-3.3)$

43. $\frac{-1.2}{-6}$

44. $\frac{-6.3}{0.7}$

45. $\left(-\frac{1}{3}\right) + \left(-\frac{3}{4}\right)$

46. $-\frac{5}{6} + \frac{3}{8}$

47. $-\frac{3}{2} - \left(-\frac{3}{4}\right)$

48. $\frac{5}{8} - \frac{11}{12}$

49. $-\frac{2}{3} - \frac{7}{9}$

50. $\frac{5}{6} - \left(-\frac{2}{9}\right)$

51. $\left(-\frac{3}{4}\right)\left(\frac{4}{5}\right)$

52. $\left(\frac{1}{2}\right)\left(-\frac{4}{5}\right)$

53. $\frac{3}{4} \div \left(-\frac{1}{2}\right)$

54. $\left(-\frac{5}{6}\right) \div \left(-\frac{7}{8}\right)$

For Problems 55–94, simplify each numerical expression. (Objective 7)

55. $9 - 12 - 8 + 5 - 6$

56. $6 - 9 + 11 - 8 - 7 + 14$

57. $-21 + (-17) - 11 + 15 - (-10)$

58. $-16 - (-14) + 16 + 17 - 19$

59. $7\frac{1}{8} - \left(2\frac{1}{4} - 3\frac{7}{8}\right)$

60. $-4\frac{3}{5} - \left(1\frac{1}{5} - 2\frac{3}{10}\right)$

61. $16 - 18 + 19 - [14 - 22 - (31 - 41)]$

62. $-19 - [15 - 13 - (-12 + 8)]$

63. $[14 - (16 - 18)] - [32 - (8 - 9)]$

64. $[-17 - (14 - 18)] - [21 - (-6 - 5)]$

65. $4\frac{1}{12} - \frac{1}{2}\left(\frac{1}{3}\right)$

66. $-\frac{4}{5} - \frac{1}{2}\left(-\frac{3}{5}\right)$

67. $-5 + (-2)(7) - (-3)(8)$

68. $-9 - 4(-2) + (-7)(6)$

69. $\frac{2}{5}\left(-\frac{3}{4}\right) - \left(-\frac{1}{2}\right)\left(\frac{3}{5}\right)$

70. $-\frac{2}{3}\left(\frac{1}{4}\right) + \left(-\frac{1}{3}\right)\left(\frac{5}{4}\right)$

71. $(-6)(-9) + (-7)(4)$

72. $(-7)(-7) - (-6)(4)$

73. $3(5 - 9) - 3(-6)$

74. $7(8 - 9) + (-6)(4)$

75. $(6 - 11)(4 - 9)$

76. $(7 - 12)(-3 - 2)$

77. $-6(-3 - 9 - 1)$

78. $-8(-3 - 4 - 6)$

79. $56 \div (-8) - (-6) \div (-2)$

80. $-65 \div 5 - (-13)(-2) + (-36) \div 12$

81. $-3[5 - (-2)] - 2(-4 - 9)$

82. $-2(-7 + 13) + 6(-3 - 2)$

83. $\frac{-6 + 24}{-3} + \frac{-7}{-6 - 1}$

84. $\frac{-12 + 20}{-4} + \frac{-7 - 11}{-9}$

85. $14.1 - (17.2 - 13.6)$

86. $-9.3 - (10.4 + 12.8)$

87. $3(2.1) - 4(3.2) - 2(-1.6)$

88. $5(-1.6) - 3(2.7) + 5(6.6)$

89. $7(6.2 - 7.1) - 6(-1.4 - 2.9)$

90. $-3(2.2 - 4.5) - 2(1.9 + 4.5)$

91. $\frac{2}{3} - \left(\frac{3}{4} - \frac{5}{6}\right)$

92. $-\frac{1}{2} - \left(\frac{3}{8} + \frac{1}{4}\right)$

93. $3\left(\frac{1}{2}\right) + 4\left(\frac{2}{3}\right) - 2\left(\frac{5}{6}\right)$

94. $2\left(\frac{3}{8}\right) - 5\left(\frac{1}{2}\right) + 6\left(\frac{3}{4}\right)$

95. Use a calculator to check your answers for Problems 55–94.

For Problems 96–104, write a numerical statement to represent the problem. Then simplify the numerical expression to answer the question. (Objective 8)

96. A scuba diver was 32 feet below sea level when he noticed that his partner had his extra knife. He ascended 13 feet to meet his partner, get the knife, and then dove down 50 feet. How far below sea level is the diver?

97. Jeff played 18 holes of golf on Saturday. On each of 6 holes he was 1 under par, on each of 4 holes he was 2 over par, on 1 hole he was 3 over par, on each of 2 holes he shot par, and on each of 5 holes he was 1 over par. How did he finish relative to par?

98. After dieting for 30 days, Ignacio has lost 18 pounds. What number describes his average weight change per day?

99. Michael bet \$5 on each of the 9 races at the racetrack. His only winnings were \$28.50 on one race. How much did he win (or lose) for the day?

**100.** Max bought a piece of trim molding that measured $11\frac{3}{8}$ feet in length. Because of defects in the wood, he had to trim $1\frac{5}{8}$ feet off one end, and he also had to remove $\frac{3}{4}$ of a foot off the other end. How long was the piece of molding after he trimmed the ends?

**101.** Natasha recorded the daily gains or losses for her company stock for a week. On Monday it gained 1.25 dollars; on Tuesday it gained 0.88 dollar; on Wednesday it lost 0.50 dollar; on Thursday it lost 1.13 dollars; on Friday it gained 0.38 dollar. What was the net gain (or loss) for the week?

**102.** On a summer day in Florida, the afternoon temperature was 96°F. After a thunderstorm, the temperature dropped 8°F. What would be the temperature if the sun came back out and the temperature rose 5°F?

**103.** In an attempt to lighten a dragster, the racing team exchanged two rear wheels for wheels that each weighed 15.6 pounds less. They also exchanged the crankshaft for one that weighed 4.8 pounds less. They changed the rear axle for one that weighed 23.7 pounds less but had to add an additional roll bar that weighed 10.6 pounds. If they wanted to lighten the dragster by 50 pounds, did they meet their goal?

**104.** A large corporation has five divisions. Two of the divisions had earnings of $2,300,000 each. The other three divisions had a loss of $1,450,000, a loss of $640,000, and a gain of $1,850,000, respectively. What was the net gain (or loss) of the corporation for the year?

## Thoughts Into Words

**105.** Explain why $\frac{0}{8} = 0$, but $\frac{8}{0}$ is undefined.

**106.** The following simplification problem is incorrect. The answer should be $-11$. Find and correct the error.

$$\begin{aligned} 8 \div (-4)(2) - 3(4) \div 2 + (-1) &= (-2)(2) - 12 \div 1 \\ &= -4 - 12 \\ &= -16 \end{aligned}$$

### Answers to the Concept Quiz

**1.** True **2.** False **3.** False **4.** False **5.** True **6.** True **7.** False **8.** False **9.** False **10.** True

# 1.3 Properties of Real Numbers and the Use of Exponents

OBJECTIVES

1. Review the properties of the real numbers
2. Apply properties to simplify expressions
3. Evaluate exponential expressions

At the beginning of this section we will list and briefly discuss some of the basic properties of real numbers. Be sure that you understand these properties because they not only facilitate manipulations with real numbers but also serve as the basis for many algebraic computations.

### Closure Property for Addition

If $a$ and $b$ are real numbers, then $a + b$ is a unique real number.

### Closure Property for Multiplication

If $a$ and $b$ are real numbers, then $ab$ is a unique real number.

We say that the set of real numbers is *closed* with respect to addition and also with respect to multiplication. That is, the sum of two real numbers is a unique real number, and the product of two real numbers is a unique real number. We use the word "unique" to indicate "exactly one."

### Commutative Property of Addition

If $a$ and $b$ are real numbers, then

$$a + b = b + a$$

### Commutative Property of Multiplication

If $a$ and $b$ are real numbers, then

$$ab = ba$$

We say that addition and multiplication are commutative operations. This means that the order in which we add or multiply two numbers does not affect the result. For example, $6 + (-8) = (-8) + 6$ and $(-4)(-3) = (-3)(-4)$. It is important to realize that subtraction and division are *not* commutative operations; order *does* make a difference. For example, $3 - 4 = -1$ but $4 - 3 = 1$. Likewise, $2 \div 1 = 2$ but $1 \div 2 = \frac{1}{2}$.

### Associative Property of Addition

If $a$, $b$, and $c$ are real numbers, then

$$(a + b) + c = a + (b + c)$$

### Associative Property of Multiplication

If $a$, $b$, and $c$ are real numbers, then

$$(ab)c = a(bc)$$

Addition and multiplication are **binary operations**. That is, we add (or multiply) two numbers at a time. The associative properties apply if more than two numbers are to be added or multiplied; they are grouping properties. For example, $(-8 + 9) + 6 = -8 + (9 + 6)$; changing the grouping of the numbers does not affect the final sum. This is also true for multiplication, which is illustrated by $[(-4)(-3)](2) = (-4)[(-3)(2)]$. Subtraction and division are *not* associative operations. For example, $(8 - 6) - 10 = -8$, but $8 - (6 - 10) = 12$. An example showing that division is not associative is $(8 \div 4) \div 2 = 1$, but $8 \div (4 \div 2) = 4$.

### Identity Property of Addition

If $a$ is any real number, then

$$a + 0 = 0 + a = a$$

Zero is called the identity element for addition. This means that the sum of any real number and zero is the same real number. For example, $-87 + 0 = 0 + (-87) = -87$.

### Identity Property of Multiplication

If $a$ is any real number, then

$$a(1) = 1(a) = a$$

We call 1 the identity element for multiplication. The product of any real number and 1 is the same real number. For example, $(-119)(1) = (1)(-119) = -119$.

### Additive Inverse Property

For every real number $a$, there exists a unique real number $-a$ such that

$$a + (-a) = -a + a = 0$$

The real number $-a$ is called the *additive inverse of a* or the *opposite of a*. For example, 16 and $-16$ are additive inverses, and their sum is 0. The additive inverse of 0 is 0.

### Multiplication Property of Zero

If $a$ is any real number, then

$$(a)(0) = (0)(a) = 0$$

The product of any real number and zero is zero. For example, $(-17)(0) = 0(-17) = 0$.

### Multiplication Property of Negative One

If $a$ is any real number, then

$$(a)(-1) = (-1)(a) = -a$$

The product of any real number and $-1$ is the opposite of the real number. For example, $(-1)(52) = (52)(-1) = -52$.

### Multiplicative Inverse Property

For every nonzero real number $a$, there exists a unique real number $\frac{1}{a}$ such that

$$a\left(\frac{1}{a}\right) = \frac{1}{a}(a) = 1$$

The number $\frac{1}{a}$ is called the *multiplicative inverse of a* or the *reciprocal of a*. For example, the reciprocal of 2 is $\frac{1}{2}$ and $2\left(\frac{1}{2}\right) = \frac{1}{2}(2) = 1$. Likewise, the reciprocal of $\frac{1}{2}$ is $\frac{1}{\frac{1}{2}} = 2$. Therefore, 2 and $\frac{1}{2}$ are said to be reciprocals (or multiplicative inverses) of each other. Because division by zero is undefined, zero does not have a reciprocal.

**Distributive Property**

If $a$, $b$, and $c$ are real numbers, then

$$a(b + c) = ab + ac$$

The distributive property ties together the operations of addition and multiplication. We say that *multiplication distributes over addition*. For example, $7(3 + 8) = 7(3) + 7(8)$. Because $b - c = b + (-c)$, it follows that *multiplication also distributes over subtraction*. This can be expressed symbolically as $a(b - c) = ab - ac$. For example, $6(8 - 10) = 6(8) - 6(10)$.

The following examples illustrate the use of the properties of real numbers to facilitate certain types of manipulations.

**Classroom Example**
Simplify $[57 + (-14)] + 14$.

**EXAMPLE 1** Simplify $[74 + (-36)] + 36$.

### Solution

In such a problem, it is much more advantageous to group $-36$ and 36.

$$\begin{aligned}[74 + (-36)] + 36 &= 74 + [(-36) + 36] \\ &= 74 + 0 = 74\end{aligned}$$

Change the grouping using the associative property of addition

**Classroom Example**
Simplify $5[(-20)(18)]$.

**EXAMPLE 2** Simplify $[(-19)(25)](-4)$.

### Solution

It is much easier to group 25 and $-4$. Thus

$$\begin{aligned}[(-19)(25)](-4) &= (-19)[(25)(-4)] \\ &= (-19)(-100) \\ &= 1900\end{aligned}$$

Change the grouping using the associative property of multiplication

**Classroom Example**
Simplify
$(-21) + 13 + 26 + (-14) + 30 + (-42) + (-8)$.

**EXAMPLE 3** Simplify $17 + (-14) + (-18) + 13 + (-21) + 15 + (-33)$.

### Solution

We could add in the order in which the numbers appear. However, because addition is commutative and associative, we could change the order and group in any convenient way. For example, we could add all of the positive integers and add all of the negative integers, and then find the sum of these two results. It might be convenient to use the vertical format as follows:

$$\begin{array}{rrr} & -14 & \\ 17 & -18 & \\ 13 & -21 & -86 \\ \underline{15} & \underline{-33} & \underline{45} \\ 45 & -86 & -41 \end{array}$$

Classroom Example
Simplify $-12(-3 + 20)$.

**EXAMPLE 4** Simplify $-25(-2 + 100)$.

### Solution

For this problem, it might be easiest to apply the distributive property first and then simplify.

$$\begin{aligned} -25(-2 + 100) &= (-25)(-2) + (-25)(100) && \text{Apply the distributive property} \\ &= 50 + (-2500) \\ &= -2450 \end{aligned}$$

Classroom Example
Simplify $(-21)(-32 + 28)$.

**EXAMPLE 5** Simplify $(-87)(-26 + 25)$.

### Solution

For this problem, it would be better not to apply the distributive property but instead to add the numbers inside the parentheses first and then find the indicated product.

$$\begin{aligned} (-87)(-26 + 25) &= (-87)(-1) \\ &= 87 \end{aligned}$$

Classroom Example
Simplify $4.9(20) + 4.9(-30)$.

**EXAMPLE 6** Simplify $3.7(104) + 3.7(-4)$.

### Solution

Remember that the distributive property allows us to change from the form $a(b + c)$ to $ab + ac$ or from the form $ab + ac$ to $a(b + c)$. In this problem, we want to use the latter conversion. Thus

$$\begin{aligned} 3.7(104) + 3.7(-4) &= 3.7[104 + (-4)] \\ &= 3.7(100) \\ &= 370 \end{aligned}$$

Examples 4, 5, and 6 illustrate an important issue. Sometimes the form $a(b + c)$ is more convenient, but at other times the form $ab + ac$ is better. In these cases, as well as in the cases of other properties, you should *think first* and decide whether or not the properties can be used to make the manipulations easier.

## Exponents

Exponents are used to indicate repeated multiplication. For example, we can write $4 \cdot 4 \cdot 4$ as $4^3$, where the "raised 3" indicates that 4 is to be used as a factor 3 times. The following general definition is helpful.

**Definition 1.4**

If $n$ is a positive integer and $b$ is any real number, then

$$b^n = \underbrace{bbb \cdots b}_{n \text{ factors of } b}$$

We refer to the $b$ as the **base** and to $n$ as the **exponent**. The expression $b^n$ can be read "$b$ to the $n$th power." We commonly associate the terms *squared* and *cubed* with exponents of 2 and 3, respectively. For example, $b^2$ is read "$b$ squared" and $b^3$ as "$b$ cubed." An exponent of 1 is usually not written, so $b^1$ is written as $b$. The following examples illustrate Definition 1.4.

$$2^3 = 2 \cdot 2 \cdot 2 = 8 \qquad \left(\frac{1}{2}\right)^5 = \frac{1}{2} \cdot \frac{1}{2} \cdot \frac{1}{2} \cdot \frac{1}{2} \cdot \frac{1}{2} = \frac{1}{32}$$

$$3^4 = 3 \cdot 3 \cdot 3 \cdot 3 = 81 \qquad (0.7)^2 = (0.7)(0.7) = 0.49$$

$$-5^2 = -(5 \cdot 5) = -25 \qquad (-5)^2 = (-5)(-5) = 25$$

Please take special note of the last two examples. Note that $(-5)^2$ means that $-5$ is the base and is to be used as a factor twice. However, $-5^2$ means that 5 is the base and that after it is squared, we take the opposite of that result.

Simplifying numerical expressions that contain exponents creates no trouble if we keep in mind that exponents are used to indicate repeated multiplication. Let's consider some examples.

**Classroom Example**
Simplify $7(-1)^2 + 3(-4)^2$.

**EXAMPLE 7** Simplify $3(-4)^2 + 5(-3)^2$.

**Solution**

$$\begin{aligned} 3(-4)^2 + 5(-3)^2 &= 3(16) + 5(9) && \text{Find the powers} \\ &= 48 + 45 \\ &= 93 \end{aligned}$$

**Classroom Example**
Simplify $(4 - 11)^2$.

**EXAMPLE 8** Simplify $(2 + 3)^2$.

**Solution**

$$\begin{aligned} (2 + 3)^2 &= (5)^2 && \text{Add inside the parentheses before applying the exponent} \\ &= 25 && \text{Square the 5} \end{aligned}$$

**Classroom Example**
Simplify $[6(-2) - 5(-3)]^3$.

**EXAMPLE 9** Simplify $[3(-1) - 2(1)]^3$.

**Solution**

Begin by performing the multiplications inside the brackets.

$$\begin{aligned} [3(-1) - 2(1)]^3 &= [-3 - 2]^3 \\ &= [-5]^3 \\ &= -125 \end{aligned}$$

**Classroom Example**
Simplify:
$6\left(\frac{1}{3}\right)^3 - 12\left(\frac{1}{3}\right)^2 + 21\left(\frac{1}{3}\right) - 4$

**EXAMPLE 10** Simplify $4\left(\frac{1}{2}\right)^3 - 3\left(\frac{1}{2}\right)^2 + 6\left(\frac{1}{2}\right) + 2$.

**Solution**

$$\begin{aligned} 4\left(\frac{1}{2}\right)^3 - 3\left(\frac{1}{2}\right)^2 + 6\left(\frac{1}{2}\right) + 2 &= 4\left(\frac{1}{8}\right) - 3\left(\frac{1}{4}\right) + 6\left(\frac{1}{2}\right) + 2 && \text{Apply the exponents} \\ &= \frac{4}{8} - \frac{3}{4} + \frac{6}{2} + 2 \\ &= \frac{1}{2} - \frac{3}{4} + 3 + 2 && \text{Reduce fractions} \\ &= \frac{2}{4} - \frac{3}{4} + \frac{12}{4} + \frac{8}{4} && \text{Change to equivalent fractions with 4 as the LCD} \\ &= \frac{19}{4} \end{aligned}$$

## Concept Quiz 1.3

For Problems 1–10, answer true or false.

1. Addition is a commutative operation.
2. Subtraction is a commutative operation.
3. Zero is the identity element for addition.
4. The multiplicative inverse of 0 is 0.
5. The numerical expression $(-25)(-16)(-4)$ simplifies to $-1600$.
6. The numerical expression $82(8) + 82(2)$ simplifies to 820.
7. Exponents are used to indicate repeated additions.
8. The numerical expression $65(7^2) + 35(7^2)$ simplifies to 4900.
9. In the expression $(-4)^3$, the base is 4.
10. In the expression $-4^3$, the base is 4.

## Problem Set 1.3

For Problems 1–14, state the property that justifies each of the statements. For example, $3 + (-4) = (-4) + 3$ because of the commutative property of addition. (Objective 1)

1. $[6 + (-2)] + 4 = 6 + [(-2) + 4]$
2. $x(3) = 3(x)$
3. $42 + (-17) = -17 + 42$
4. $1(x) = x$
5. $-114 + 114 = 0$
6. $(-1)(48) = -48$
7. $-1(x + y) = -(x + y)$
8. $-3(2 + 4) = -3(2) + (-3)(4)$
9. $12yx = 12xy$
10. $[(-7)(4)](-25) = (-7)[4(-25)]$
11. $7(4) + 9(4) = (7 + 9)4$
12. $(x + 3) + (-3) = x + [3 + (-3)]$
13. $[(-14)(8)](25) = (-14)[8(25)]$
14. $\left(\frac{3}{4}\right)\left(\frac{4}{3}\right) = 1$

For Problems 15–26, simplify each numerical expression. Be sure to take advantage of the properties whenever they can be used to make the computations easier. (Objective 2)

15. $36 + (-14) + (-12) + 21 + (-9) - 4$
16. $-37 + 42 + 18 + 37 + (-42) - 6$
17. $[83 + (-99)] + 18$
18. $[63 + (-87)] + (-64)$
19. $(25)(-13)(4)$
20. $(14)(25)(-13)(4)$
21. $17(97) + 17(3)$
22. $-86[49 + (-48)]$
23. $14 - 12 - 21 - 14 + 17 - 18 + 19 - 32$
24. $16 - 14 - 13 - 18 + 19 + 14 - 17 + 21$
25. $(-50)(15)(-2) - (-4)(17)(25)$
26. $(2)(17)(-5) - (4)(13)(-25)$

For Problems 27–54, simplify each of the numerical expressions. (Objectives 2 and 3)

27. $2^3 - 3^3$
28. $3^2 - 2^4$
29. $-5^2 - 4^2$
30. $-7^2 + 5^2$
31. $(-2)^3 - 3^2$
32. $(-3)^3 + 3^2$
33. $3(-1)^3 - 4(3)^2$
34. $4(-2)^3 - 3(-1)^4$
35. $7(2)^3 + 4(-2)^3$
36. $-4(-1)^2 - 3(2)^3$
37. $-3(-2)^3 + 4(-1)^5$
38. $5(-1)^3 - (-3)^3$
39. $(-3)^2 - 3(-2)(5) + 4^2$
40. $(-2)^2 - 3(-2)(6) - (-5)^2$
41. $2^3 + 3(-1)^3(-2)^2 - 5(-1)(2)^2$
42. $-2(3)^2 - 2(-2)^3 - 6(-1)^5$
43. $(3 + 4)^2$
44. $(4 - 9)^2$
45. $[3(-2)^2 - 2(-3)^2]^3$
46. $[-3(-1)^3 - 4(-2)^2]^2$

**47.** $2(-1)^3 - 3(-1)^2 + 4(-1) - 5$

**48.** $(-2)^3 + 2(-2)^2 - 3(-2) - 1$

**49.** $2^4 - 2(2)^3 - 3(2)^2 + 7(2) - 10$

**50.** $3(-3)^3 + 4(-3)^2 - 5(-3) + 7$

**51.** $3\left(\frac{1}{2}\right)^4 - 2\left(\frac{1}{2}\right)^3 + 5\left(\frac{1}{2}\right)^2 - 4\left(\frac{1}{2}\right) + 1$

**52.** $4(0.1)^2 - 6(0.1) + 0.7$

**53.** $-\left(\frac{2}{3}\right)^2 + 5\left(\frac{2}{3}\right) - 4$

**54.** $4\left(\frac{1}{3}\right)^3 + 3\left(\frac{1}{3}\right)^2 + 2\left(\frac{1}{3}\right) + 6$

**55.** Use your calculator to check your answers for Problems 27–54.

For Problems 56–64, use your calculator to evaluate each numerical expression. (Objective 3)

**56.** $2^{10}$

**57.** $3^7$

**58.** $(-2)^8$

**59.** $(-2)^{11}$

**60.** $-4^9$

**61.** $-5^6$

**62.** $(3.14)^3$

**63.** $(1.41)^4$

**64.** $(1.73)^5$

## Thoughts Into Words

**65.** State, in your own words, the multiplication property of negative one.

**66.** Explain how the associative and commutative properties can help simplify $[(25)(97)](-4)$.

**67.** Your friend keeps getting an answer of 64 when simplifying $-2^6$. What mistake is he making, and how would you help him?

**68.** Write a sentence explaining, in your own words, how to evaluate the expression $(-8)^2$. Also write a sentence explaining how to evaluate $-8^2$.

**69.** For what natural numbers $n$ does $(-1)^n = -1$? For what natural numbers $n$ does $(-1)^n = 1$? Explain your answers.

**70.** Is the set $\{0, 1\}$ closed with respect to addition? Is the set $\{0, 1\}$ closed with respect to multiplication? Explain your answers.

### Answers to the Concept Quiz

**1.** True **2.** False **3.** True **4.** False **5.** True **6.** True **7.** False **8.** True **9.** False **10.** True

# 1.4 Algebraic Expressions

OBJECTIVES

1. Simplify algebraic expressions
2. Evaluate algebraic expressions
3. Translate from English to algebra

Algebraic expressions such as

$$2x, \quad 8xy, \quad 3xy^2, \quad -4a^2b^3c, \quad \text{and} \quad z$$

are called *terms*. A **term** is an indicated product that may have any number of factors. The variables involved in a term are called **literal factors**, and the numerical factor is called the **numerical coefficient**. Thus in $8xy$, the $x$ and $y$ are literal factors, and 8 is the numerical coefficient. The numerical coefficient of the term $-4a^2bc$ is $-4$. Because $1(z) = z$, the numerical coefficient of the term $z$ is understood to be 1. Terms that have the same literal factors are called **similar terms** or **like terms**. Some examples of similar terms are

$3x$ and $14x$ $\quad$ $5x^2$ and $18x^2$

$7xy$ and $-9xy$ $\quad$ $9x^2y$ and $-14x^2y$

$2x^3y^2$, $3x^3y^2$, and $-7x^3y^2$

By the symmetric property of equality, we can write the distributive property as

$$ab + ac = a(b + c)$$

Then the commutative property of multiplication can be applied to change the form to

$$ba + ca = (b + c)a$$

This latter form provides the basis for simplifying algebraic expressions by *combining similar terms*. Consider the following examples.

$$3x + 5x = (3 + 5)x = 8x \qquad -6xy + 4xy = (-6 + 4)xy = -2xy$$

$$5x^2 + 7x^2 + 9x^2 = (5 + 7 + 9)x^2 = 21x^2 \qquad 4x - x = 4x - 1x = (4 - 1)x = 3x$$

More complicated expressions might require that we first rearrange the terms by applying the commutative property for addition.

$$\begin{aligned} 7x + 2y + 9x + 6y &= 7x + 9x + 2y + 6y \\ &= (7 + 9)x + (2 + 6)y && \text{Distributive property} \\ &= 16x + 8y \\ 6a - 5 - 11a + 9 &= 6a + (-5) + (-11a) + 9 \\ &= 6a + (-11a) + (-5) + 9 && \text{Commutative property} \\ &= [6 + (-11)]a + 4 && \text{Distributive property} \\ &= -5a + 4 \end{aligned}$$

As soon as you thoroughly understand the various simplifying steps, you may want to do the steps mentally. Then you could go directly from the given expression to the simplified form, as follows:

$$14x + 13y - 9x + 2y = 5x + 15y$$

$$3x^2y - 2y + 5x^2y + 8y = 8x^2y + 6y$$

$$-4x^2 + 5y^2 - x^2 - 7y^2 = -5x^2 - 2y^2$$

Applying the distributive property to remove parentheses, and then to combine similar terms, sometimes simplifies an algebraic expression (as Example 1 illustrates).

**Classroom Example**
Simplify the following:
(a) $2(m - 3) + 5(m + 1)$
(b) $-4(n + 3) - 7(n - 4)$
(c) $-3(m - 2n) + (m + 2n)$

**EXAMPLE 1** Simplify the following:

**(a)** $4(x + 2) + 3(x + 6)$ **(b)** $-5(y + 3) - 2(y - 8)$ **(c)** $5(x - y) - (x + y)$

## Solution

**(a)**
$$\begin{aligned} 4(x + 2) + 3(x + 6) &= 4(x) + 4(2) + 3(x) + 3(6) && \text{Apply distributive property} \\ &= 4x + 8 + 3x + 18 \\ &= 4x + 3x + 8 + 18 \\ &= (4 + 3)x + 26 \\ &= 7x + 26 \end{aligned}$$

**(b)**
$$\begin{aligned} -5(y + 3) - 2(y - 8) &= -5(y) - 5(3) - 2(y) - 2(-8) && \text{Apply distributive property} \\ &= -5y - 15 - 2y + 16 \\ &= -5y - 2y - 15 + 16 \\ &= -7y + 1 \end{aligned}$$

**(c)** $5(x - y) - (x + y) = 5(x - y) - 1(x + y)$ Remember, $-a = -1(a)$

$= 5(x) - 5(y) - 1(x) - 1(y)$

$= 5x - 5y - 1x - 1y$

$= 4x - 6y$

When we are multiplying two terms such as 3 and $2x$, the associative property for multiplication provides the basis for simplifying the product.

$3(2x) = (3 \cdot 2)x = 6x$

This idea is put to use in Example 2.

**Classroom Example**
Simplify $2(6m - 7n) - 5(3m - 4n)$.

**EXAMPLE 2** Simplify $3(2x + 5y) + 4(3x + 2y)$.

### Solution

$$\begin{aligned}3(2x + 5y) + 4(3x + 2y) &= 3(2x) + 3(5y) + 4(3x) + 4(2y)\\ &= (3 \cdot 2)x + (3 \cdot 5)y + (4 \cdot 3)x + (4 \cdot 2)y\\ &= 6x + 15y + 12x + 8y\\ &= 6x + 12x + 15y + 8y\\ &= 18x + 23y\end{aligned}$$

After you are sure of each step, a more simplified format may be used, as the following examples illustrate.

$$\begin{aligned}5(a + 4) - 7(a + 3) &= 5a + 20 - 7a - 21\\ &= -2a - 1\end{aligned}$$

Be careful with this sign. Consider this as distributing $-7$ to $a$ and 3.

$$\begin{aligned}3(x^2 + 2) + 4(x^2 - 6) &= 3x^2 + 6 + 4x^2 - 24\\ &= 7x^2 - 18\end{aligned}$$

$$\begin{aligned}2(3x - 4y) - 5(2x - 6y) &= 6x - 8y - 10x + 30y\\ &= -4x + 22y\end{aligned}$$

## Evaluating Algebraic Expressions

An algebraic expression takes on a numerical value whenever each variable in the expression is replaced by a real number. For example, if $x$ is replaced by 5 and $y$ by 9, the algebraic expression $x + y$ becomes the numerical expression $5 + 9$, which simplifies to 14. We say that $x + y$ has a value of 14 when $x$ equals 5 and $y$ equals 9. If $x = -3$ and $y = 7$, then $x + y$ has a value of $-3 + 7 = 4$. The following examples illustrate the process of finding a value of an algebraic expression; we commonly refer to the process as **evaluating algebraic expressions**.

**Classroom Example**
Find the value of $5a - 9b$ when $a = 4$ and $b = -2$.

**EXAMPLE 3** Find the value of $3x - 4y$ when $x = 2$ and $y = -3$.

### Solution

It is always a good practice to use parentheses when substituting the number for the variable.

$$\begin{aligned}3x - 4y &= 3(2) - 4(-3) \quad \text{when } x = 2 \text{ and } y = -3\\ &= 6 + 12\\ &= 18\end{aligned}$$

**Classroom Example**
Evaluate $s^2 - 4st + t^2$ for $s = -6$ and $t = 2$.

**EXAMPLE 4** Evaluate $x^2 - 2xy + y^2$ for $x = -2$ and $y = -5$.

### Solution

Notice the use of parentheses when substituting.

$$\begin{aligned} x^2 - 2xy + y^2 &= (-2)^2 - 2(-2)(-5) + (-5)^2 \quad \text{when } x = -2 \text{ and } y = -5 \\ &= 4 + 4(-5) + 25 \\ &= 4 - 20 + 25 \\ &= 9 \end{aligned}$$

**Classroom Example**
Evaluate $(x - y)^3$ for $x = -5$ and $y = -7$.

**EXAMPLE 5** Evaluate $(a + b)^2$ for $a = 6$ and $b = -2$.

### Solution

Note the change to brackets on the outside because the inner grouping symbols are parentheses.

$$\begin{aligned} (a + b)^2 &= [6 + (-2)]^2 \quad \text{when } a = 6 \text{ and } b = -2 \\ &= (4)^2 \\ &= 16 \end{aligned}$$

**Classroom Example**
Evaluate $(5m + 3n)(2m - 7n)$ for $m = -2$ and $n = 3$.

**EXAMPLE 6** Evaluate $(3x + 2y)(2x - y)$ for $x = 4$ and $y = -1$.

### Solution

$$\begin{aligned} (3x + 2y)(2x - y) &= [3(4) + 2(-1)][2(4) - (-1)] \quad \text{when } x = 4 \text{ and } y = -1 \\ &= (12 - 2)(8 + 1) \\ &= (10)(9) \\ &= 90 \end{aligned}$$

**Classroom Example**
Evaluate $-3x + 4y + 8x - 7y$ for $x = \frac{3}{5}$ and $y = -\frac{1}{9}$.

**EXAMPLE 7** Evaluate $7x - 2y + 4x - 3y$ for $x = -\frac{1}{2}$ and $y = \frac{2}{3}$.

### Solution

Let's first simplify the given expression.

$$7x - 2y + 4x - 3y = 11x - 5y$$

Now we can substitute $-\frac{1}{2}$ for $x$ and $\frac{2}{3}$ for $y$.

$$\begin{aligned} 11x - 5y &= 11\left(-\frac{1}{2}\right) - 5\left(\frac{2}{3}\right) \\ &= -\frac{11}{2} - \frac{10}{3} \\ &= -\frac{33}{6} - \frac{20}{6} \quad \text{Change to equivalent fractions with a common denominator of 6} \\ &= -\frac{53}{6} \end{aligned}$$

**Classroom Example**
Evaluate $-2(8x - 7) + 6(3x + 5)$ for $x = -9.1$.

**EXAMPLE 8** Evaluate $2(3x + 1) - 3(4x - 3)$ for $x = -6.2$.

### Solution

Let's first simplify the given expression.

$$\begin{aligned} 2(3x + 1) - 3(4x - 3) &= 6x + 2 - 12x + 9 && \text{Apply distributive property} \\ &= -6x + 11 \end{aligned}$$

Now we can substitute $-6.2$ for $x$.

$$\begin{aligned} -6x + 11 &= -6(-6.2) + 11 \\ &= 37.2 + 11 \\ &= 48.2 \end{aligned}$$

**Classroom Example**
Evaluate $5(x^3 - 1) - 8(x^3 + 2) - (x^3 + 3)$ for $x = -3$.

**EXAMPLE 9** Evaluate $2(a^2 + 1) - 3(a^2 + 5) + 4(a^2 - 1)$ for $a = 10$.

### Solution

Let's first simplify the given expression.

$$\begin{aligned} 2(a^2 + 1) - 3(a^2 + 5) + 4(a^2 - 1) &= 2a^2 + 2 - 3a^2 - 15 + 4a^2 - 4 \\ &= 3a^2 - 17 \end{aligned}$$

Substituting $a = 10$, we obtain

$$\begin{aligned} 3a^2 - 17 &= 3(10)^2 - 17 \\ &= 3(100) - 17 \\ &= 300 - 17 \\ &= 283 \end{aligned}$$

## Translating from English to Algebra

To use the tools of algebra to solve problems, we must be able to translate from English to algebra. This translation process requires that we recognize key phrases in the English language that translate into algebraic expressions (which involve the operations of addition, subtraction, multiplication, and division). Some of these key phrases and their algebraic counterparts are listed in the following table. The variable $n$ represents the number being referred to in each phrase. When translating, remember that the commutative property holds only for the operations of addition and multiplication. Therefore, order will be crucial to algebraic expressions that involve subtraction and division.

| English phrase | Algebraic expression |
|---|---|
| **Addition** | |
| The sum of a number and 4 | $n + 4$ |
| 7 more than a number | $n + 7$ |
| A number plus 10 | $n + 10$ |
| A number increased by 6 | $n + 6$ |
| 8 added to a number | $n + 8$ |

*(continued)*

| English phrase | Algebraic expression |
|---|---|
| **Subtraction** | |
| 14 minus a number | $14 - n$ |
| 12 less than a number | $n - 12$ |
| A number decreased by 10 | $n - 10$ |
| The difference between a number and 2 | $n - 2$ |
| 5 subtracted from a number | $n - 5$ |
| **Multiplication** | |
| 14 times a number | $14n$ |
| The product of 4 and a number | $4n$ |
| $\frac{3}{4}$ of a number | $\frac{3}{4}n$ |
| Twice a number | $2n$ |
| Multiply a number by 12 | $12n$ |
| **Division** | |
| The quotient of 6 and a number | $\frac{6}{n}$ |
| The quotient of a number and 6 | $\frac{n}{6}$ |
| A number divided by 9 | $\frac{n}{9}$ |
| The ratio of a number and 4 | $\frac{n}{4}$ |
| **Combination of operations** | |
| 4 more than three times a number | $3n + 4$ |
| 5 less than twice a number | $2n - 5$ |
| 3 times the sum of a number and 2 | $3(n + 2)$ |
| 2 more than the quotient of a number and 12 | $\frac{n}{12} + 2$ |
| 7 times the difference of 6 and a number | $7(6 - n)$ |

An English statement may not always contain a key word such as *sum, difference, product*, or *quotient*. Instead, the statement may describe a physical situation, and from this description we must deduce the operations involved. Some suggestions for handling such situations are given in the following examples.

**Classroom Example**
Caitlin can read 550 words per minute. How many words will she read in $n$ minutes?

**EXAMPLE 10**

Sonya can keyboard 65 words per minute. How many words will she keyboard in $m$ minutes?

**Solution**

The total number of words keyboarded equals the product of the rate per minute and the number of minutes. Therefore, Sonya should be able to keyboard $65m$ words in $m$ minutes.

**Classroom Example**
Greg has $n$ nickels and $q$ quarters. Express this amount of money in cents.

## EXAMPLE 11

Russ has $n$ nickels and $d$ dimes. Express this amount of money in cents.

### Solution

Each nickel is worth 5 cents and each dime is worth 10 cents. We represent the amount in cents by $5n + 10d$.

**Classroom Example**
The cost of a 20-pound bag of unpopped popcorn is $d$ dollars. What is the cost per pound for the popcorn?

## EXAMPLE 12

The cost of a 50-pound sack of fertilizer is $d$ dollars. What is the cost per pound for the fertilizer?

### Solution

We calculate the cost per pound by dividing the total cost by the number of pounds.

We represent the cost per pound by $\frac{d}{50}$.

The English statement we want to translate into algebra may contain some geometric ideas. Tables 1.1 and 1.2 contain some of the basic relationships that pertain to linear measurement in the English and metric systems, respectively.

**Table 1.1** English System

| | | |
|---|---|---|
| 12 inches | = | 1 foot |
| 3 feet | = | 1 yard |
| 1760 yards | = | 1 mile |
| 5280 feet | = | 1 mile |

**Table 1.2** Metric System

| | | |
|---|---|---|
| 1 kilometer | = | 1000 meters |
| 1 hectometer | = | 100 meters |
| 1 decameter | = | 10 meters |
| 1 decimeter | = | 0.1 meter |
| 1 centimeter | = | 0.01 meter |
| 1 millimeter | = | 0.001 meter |

**Classroom Example**
The distance between two buildings is $f$ feet. Express this distance in yards.

## EXAMPLE 13

The distance between two cities is $k$ kilometers. Express this distance in meters.

### Solution

Because 1 kilometer equals 1000 meters, the distance in meters is represented by $1000k$.

**Classroom Example**
The length of the outdoor mall is $k$ kilometers and $h$ hectometers. Express this length in meters.

## EXAMPLE 14

The length of a rope is $y$ yards and $f$ feet. Express this length in inches.

### Solution

Because 1 foot equals 12 inches, and 1 yard equals 36 inches, the length of the rope in inches can be represented by $36y + 12f$.

**Classroom Example**
The length of a rectangle is $l$ yards, and the width is $w$ yards. Express the perimeter in feet.

## EXAMPLE 15

The length of a rectangle is $l$ centimeters, and the width is $w$ centimeters. Express the perimeter of the rectangle in meters.

### Solution

A sketch of the rectangle may be helpful (Figure 1.9).

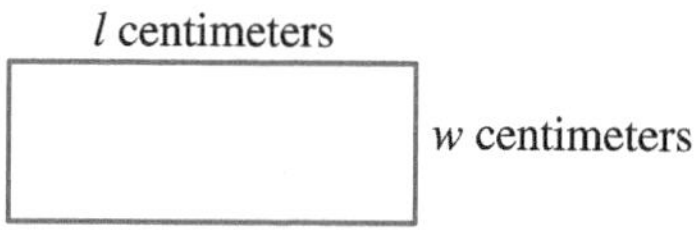

**Figure 1.9**

The perimeter of a rectangle is the sum of the lengths of the four sides. Thus the perimeter in centimeters is $l + w + l + w$, which simplifies to $2l + 2w$. Now because 1 centimeter equals 0.01 meter, the perimeter, in meters, is $0.01(2l + 2w)$. This could also be written as $\frac{2l + 2w}{100} = \frac{2(l + w)}{100} = \frac{l + w}{50}$.

## Concept Quiz 1.4

For Problems 1–10, answer true or false.

1. The numerical coefficient of the term $xy$ is 1.
2. The terms $5x^2y$ and $6xy^2$ are similar terms.
3. The algebraic expression $(3x - 4y) - (3x - 4y)$ simplifies to 0.
4. The algebraic expression $(x - y) - (x - y)$ simplifies to $2x - 2y$.
5. The value of $x^2 - y^2$ is 29 when $x = 5$ and $y = -2$.
6. The English phrase "4 less than twice the number $n$" translates into the algebraic expression $2n - 4$.
7. The algebraic expression for the English phrase "2 less than $y$" can be written as $y - 2$ or $2 - y$.
8. In the metric system, 1 centimeter = 10 millimeters.
9. If the length of a rectangle is $l$ inches and its width is $w$ inches, then the perimeter, in feet, can be represented by $24(l + w)$.
10. The value, in dollars, of $x$ five-dollar bills and $y$ ten-dollar bills can be represented by $5x + 10y$.

## Problem Set 1.4

Simplify the algebraic expressions in Problems 1–14 by combining similar terms. (Objective 1)

**1.** $-7x + 11x$ **2.** $5x - 8x + x$

**3.** $5a^2 - 6a^2$ **4.** $12b^3 - 17b^3$

**5.** $4n - 9n - n$ **6.** $6n + 13n - 15n$

**7.** $4x - 9x + 2y$ **8.** $7x - 9y - 10x - 13y$

**9.** $-3a^2 + 7b^2 + 9a^2 - 2b^2$

**10.** $-xy + z - 8xy - 7z$

**11.** $15x - 4 + 6x - 9$

**12.** $5x - 2 - 7x + 4 - x - 1$

**13.** $5a^2b - ab^2 - 7a^2b$

**14.** $8xy^2 - 5x^2y + 2xy^2 + 7x^2y$

Simplify the algebraic expressions in Problems 15–34 by removing parentheses and combining similar terms. (Objective 1)

**15.** $3(x + 2) + 5(x + 3)$ **16.** $5(x - 1) + 7(x + 4)$

**17.** $-2(a - 4) - 3(a + 2)$ **18.** $-7(a + 1) - 9(a + 4)$

**19.** $3(n^2 + 1) - 8(n^2 - 1)$ **20.** $4(n^2 + 3) + (n^2 - 7)$

**21.** $-6(x^2 - 5) - (x^2 - 2)$ **22.** $3(x + y) - 2(x - y)$

**23.** $5(2x + 1) + 4(3x - 2)$ **24.** $5(3x - 1) + 6(2x + 3)$

25. $3(2x - 5) - 4(5x - 2)$

26. $3(2x - 3) - 7(3x - 1)$

27. $-2(n^2 - 4) - 4(2n^2 + 1)$

28. $-4(n^2 + 3) - (2n^2 - 7)$

29. $3(2x - 4y) - 2(x + 9y)$

30. $-7(2x - 3y) + 9(3x + y)$

31. $3(2x - 1) - 4(x + 2) - 5(3x + 4)$

32. $-2(x - 1) - 5(2x + 1) + 4(2x - 7)$

33. $-(3x - 1) - 2(5x - 1) + 4(-2x - 3)$

34. $4(-x - 1) + 3(-2x - 5) - 2(x + 1)$

Evaluate the algebraic expressions in Problems 35–57 for the given values of the variables. (Objective 2)

35. $3x + 7y$, $x = -1$ and $y = -2$

36. $5x - 9y$, $x = -2$ and $y = 5$

37. $4x^2 - y^2$, $x = 2$ and $y = -2$

38. $3a^2 + 2b^2$, $a = 2$ and $b = 5$

39. $2a^2 - ab + b^2$, $a = -1$ and $b = -2$

40. $-x^2 + 2xy + 3y^2$, $x = -3$ and $y = 3$

41. $2x^2 - 4xy - 3y^2$, $x = 1$ and $y = -1$

42. $4x^2 + xy - y^2$, $x = 3$ and $y = -2$

43. $3xy - x^2y^2 + 2y^2$, $x = 5$ and $y = -1$

44. $x^2y^3 - 2xy + x^2y^2$, $x = -1$ and $y = -3$

45. $7a - 2b - 9a + 3b$, $a = 4$ and $b = -6$

46. $-4x + 9y - 3x - y$, $x = -4$ and $y = 7$

47. $(x - y)^2$, $x = 5$ and $y = -3$

48. $2(a + b)^2$, $a = 6$ and $b = -1$

49. $-2a - 3a + 7b - b$, $a = -10$ and $b = 9$

50. $3(x - 2) - 4(x + 3)$, $x = -2$

51. $-2(x + 4) - (2x - 1)$, $x = -3$

52. $-4(2x - 1) + 7(3x + 4)$, $x = 4$

53. $2(x - 1) - (x + 2) - 3(2x - 1)$, $x = -1$

54. $-3(x + 1) + 4(-x - 2) - 3(-x + 4)$, $x = -\frac{1}{2}$

55. $3(x^2 - 1) - 4(x^2 + 1) - (2x^2 - 1)$, $x = \frac{2}{3}$

56. $2(n^2 + 1) - 3(n^2 - 3) + 3(5n^2 - 2)$, $n = \frac{1}{4}$

57. $5(x - 2y) - 3(2x + y) - 2(x - y)$, $x = \frac{1}{3}$ and $y = -\frac{3}{4}$

For Problems 58–63, use your calculator and evaluate each of the algebraic expressions for the indicated values. Express the final answers to the nearest tenth. (Objective 2)

58. $\pi r^2$, $\pi = 3.14$ and $r = 2.1$

59. $\pi r^2$, $\pi = 3.14$ and $r = 8.4$

60. $\pi r^2 h$, $\pi = 3.14$, $r = 1.6$, and $h = 11.2$

61. $\pi r^2 h$, $\pi = 3.14$, $r = 4.8$, and $h = 15.1$

62. $2\pi r^2 + 2\pi rh$, $\pi = 3.14$, $r = 3.9$, and $h = 17.6$

63. $2\pi r^2 + 2\pi rh$, $\pi = 3.14$, $r = 7.8$, and $h = 21.2$

For Problems 64–78, translate each English phrase into an algebraic expression and use $n$ to represent the unknown number. (Objective 3)

64. The sum of a number and 4

65. A number increased by 12

66. A number decreased by 7

67. Five less than a number

68. A number subtracted from 75

69. The product of a number and 50

70. One-third of a number

71. Four less than one-half of a number

72. Seven more than three times a number

73. The quotient of a number and 8

74. The quotient of 50 and a number

75. Nine less than twice a number

76. Six more than one-third of a number

77. Ten times the difference of a number and 6

78. Twelve times the sum of a number and 7

For Problems 79–99, answer the question with an algebraic expression. (Objective 3)

79. Brian is $n$ years old. How old will he be in 20 years?

80. Crystal is $n$ years old. How old was she 5 years ago?

81. Pam is $t$ years old, and her mother is 3 less than twice as old as Pam. What is the age of Pam's mother?

82. The sum of two numbers is 65, and one of the numbers is $x$. What is the other number?

83. The difference of two numbers is 47, and the smaller number is $n$. What is the other number?

84. The product of two numbers is 98, and one of the numbers is $n$. What is the other number?

**85.** The quotient of two numbers is 8, and the smaller number is $y$. What is the other number?

**86.** The perimeter of a square is $c$ centimeters. How long is each side of the square?

**87.** The perimeter of a square is $m$ meters. How long, in centimeters, is each side of the square?

**88.** Jesse has $n$ nickels, $d$ dimes, and $q$ quarters in his bank. How much money, in cents, does he have in his bank?

**89.** Tina has $c$ cents, which is all in quarters. How many quarters does she have?

**90.** If $n$ represents a whole number, what represents the next larger whole number?

**91.** If $n$ represents an odd integer, what represents the next larger odd integer?

**92.** If $n$ represents an even integer, what represents the next larger even integer?

**93.** The cost of a 5-pound box of candy is $c$ cents. What is the price per pound?

**94.** Larry's annual salary is $d$ dollars. What is his monthly salary?

**95.** Mila's monthly salary is $d$ dollars. What is her annual salary?

**96.** The perimeter of a square is $i$ inches. What is the perimeter expressed in feet?

**97.** The perimeter of a rectangle is $y$ yards and $f$ feet. What is the perimeter expressed in feet?

**98.** The length of a line segment is $d$ decimeters. How long is the line segment expressed in meters?

**99.** The distance between two cities is $m$ miles. How far is this, expressed in feet?

**100.** Use your calculator to check your answers for Problems 35–57.

## Thoughts Into Words

**101.** Explain the difference between simplifying a numerical expression and evaluating an algebraic expression.

**102.** How would you help someone who is having difficulty expressing $n$ nickels and $d$ dimes in terms of cents?

**103.** When asked to write an algebraic expression for "8 more than a number," you wrote $x + 8$, and another student wrote $8 + x$. Are both expressions correct? Explain your answer.

**104.** When asked to write an algebraic expression for "6 less than a number," you wrote $x - 6$, and another student wrote $6 - x$. Are both expressions correct? Explain your answer.

### Answers to the Concept Quiz

**1.** True **2.** False **3.** True **4.** False **5.** False **6.** True **7.** False **8.** True **9.** False **10.** True

# Chapter 1 Summary

| OBJECTIVE | SUMMARY | EXAMPLE |
|---|---|---|
| Identify certain sets of numbers.<br>(Section 1.1/Objective 1) | A set is a collection of objects. The objects are called elements or members of the set. The sets of natural numbers, whole numbers, integers, rational numbers, and irrational numbers are all subsets of the set of real numbers. | From the list $-4, \frac{7}{5}, 0.35, \sqrt{2}$, and 0, identify the integers.<br>**Solution**<br>The integers are $-4$ and 0.<br>**Sample Problem 1**<br>From the list $-0.03, 8, 0$, and $\sqrt{5}$, identify the rational numbers. |
| Apply the properties of equality and the properties of real numbers.<br>(Section 1.1/Objective 2) | The properties of real numbers help with numerical manipulations and serve as a basis for algebraic computation. The properties of equality are listed on page 17, and the properties of real numbers are listed on pages 31–34. | State the property that justifies the statement, "If $x = y$ and $y = 7$, then $x = 7$."<br>**Solution**<br>The statement, "If $x = y$ and $y = 7$, *then* $x = 7$," is justified by the transitive property of equality.<br>**Sample Problem 2**<br>State the property that justifies the statement, "If $x = 7$, then $7 = x$." |
| Find the absolute value of a number.<br>(Section 1.2/Objective 2) | Geometrically, the absolute value of any number is the distance between the number and zero on the number line. More formally, the absolute value of a real number $a$ is defined as follows:<br>**1.** If $a \geq 0$, then $\lvert a \rvert = a$.<br>**2.** If $a < 0$, then $\lvert a \rvert = -a$. | Find the absolute value of the following:<br>**(a)** $\lvert -2 \rvert$ **(b)** $\left\lvert \frac{15}{4} \right\rvert$ **(c)** $\lvert -\sqrt{3} \rvert$<br>**Solution**<br>**(a)** $\lvert -2 \rvert = -(-2) = 2$<br>**(b)** $\left\lvert \frac{15}{4} \right\rvert = \frac{15}{4}$<br>**(c)** $\lvert -\sqrt{3} \rvert = -(-\sqrt{3}) = \sqrt{3}$<br>**Sample Problem 3**<br>Find the absolute value of the following:<br>**(a)** $\left\lvert -\frac{1}{2} \right\rvert$ **(b)** $\lvert 0 \rvert$ **(c)** $\lvert 4 \rvert$ |
| Addition of real numbers<br>(Section 1.2/Objective 3)<br>Subtraction of real numbers<br>(Section 1.2/Objective 4)<br>Multiplication and division of real numbers<br>(Section 1.2/Objectives 5 and 6) | The rules for addition of real numbers are on page 24.<br>Applying the principle $a - b = a + (-b)$ changes every subtraction problem to an equivalent addition problem.<br>**1.** The product (or quotient) of two positive numbers or two negative numbers is the product (or quotient) of their absolute values.<br>**2.** The product (or quotient) of one positive and one negative number is the opposite of the product (or quotient) of their absolute values. | Simplify:<br>**(a)** $-20 + 15 + (-4)$<br>**(b)** $40 - (-8)$<br>**(c)** $-3(-4)(-5)$<br>**Solution**<br>**(a)** $-20 + 15 + (-4) = -5 + (-4)$<br>$= -9$<br>**(b)** $40 - (-8) = 40 + (+8) = 48$<br>**(c)** $-3(-4)(-5) = 12(-5)$<br>$= -60$<br>**Sample Problem 4**<br>Simplify:<br>**(a)** $-11 + (-4) + 20$<br>**(b)** $-16 - (-4)$<br>**(c)** $-2(-1)(-3)$ |

Answers to **Sample Problems** are located in the back of the book.

| OBJECTIVE | SUMMARY | EXAMPLE |
|---|---|---|
| Evaluate exponential expressions.<br>(Section 1.3/Objective 3) | Exponents are used to indicate repeated multiplications. The expression $b^n$ can be read "$b$ to the $n$th power." We refer to $b$ as the base and $n$ as the exponent. | Simplify $2(-5)^3 + 3(-2)^2$.<br>**Solution**<br>$2(-5)^3 + 3(-2)^2$<br>$= 2(-125) + 3(4)$<br>$= -250 + 12$<br>$= -238$<br>**Sample Problem 5**<br>Simplify $-4(-3)^2 - 5(2)^2$. |
| Simplify numerical expressions.<br>(Section 1.1/Objective 3; Section 1.2/Objective 7) | We can evaluate *numerical expressions* by performing the operations in the following order.<br>**1.** Perform the operations inside the parentheses and above and below the fraction bars.<br>**2.** Evaluate all numbers raised to an exponent.<br>**3.** Perform all multiplications and divisions in the order they appear from left to right.<br>**4.** Perform all additions and subtractions in the order they appear from left to right. | Simplify $60 \div 2 \cdot 3 - (1 - 5)^2$.<br>**Solution**<br>$60 \div 2 \cdot 3 - (1 - 5)^2$<br>$= 60 \div 2 \cdot 3 - (-4)^2$<br>$= 60 \div 2 \cdot 3 - 16$<br>$= 30 \cdot 3 - 16$<br>$= 90 - 16$<br>$= 74$<br>**Sample Problem 6**<br>Simplify $6 + 4(1 - 3)^2 \div 2 \cdot 4$. |
| Simplify algebraic expressions.<br>(Section 1.3/Objective 2; Section 1.4/Objective 1) | Algebraic expressions such as $2x$, $3xy^2$, and $-4a^2b^3c$ are called *terms*. We call the variables in a term the literal factors, and we call the numerical factor the numerical coefficient. Terms that have the same literal factors are called similar or like terms. The distributive property in the form $ba + ca = (b + c)a$ serves as a basis for combining like terms. | Simplify $5x^2 + 3x - 2x^2 - 7x$.<br>**Solution**<br>$5x^2 + 3x - 2x^2 - 7x$<br>$= 5x^2 - 2x^2 + 3x - 7x$<br>$= (5 - 2)x^2 + (3 - 7)x$<br>$= 3x^2 + (-4)x$<br>$= 3x^2 - 4x$<br>**Sample Problem 7**<br>Simplify $4x^2 - 5x + 7 - 3x^2 + 2$. |
| Evaluate algebraic expressions.<br>(Section 1.3/Objective 3; Section 1.4/Objective 2) | An algebraic expression takes on a numerical value whenever each variable in the expression is replaced by a real number. The process of finding a value of an algebraic expression is referred to as *evaluating algebraic expressions*. | Evaluate $x^2 - 2xy + y^2$ when $x = 3$ and $y = -4$.<br>**Solution**<br>$x^2 - 2xy + y^2 =$<br>$(3)^2 - 2(3)(-4) + (-4)^2$<br>when $x = 3$ and $y = -4$.<br>$(3)^2 - 2(3)(-4) + (-4)^2 =$<br>$9 + 24 + 16 = 49$<br>**Sample Problem 8**<br>Evaluate $a^2 - b^2$ when $a = 5$ and $b = -2$. |

*(continued)*

| OBJECTIVE | SUMMARY | EXAMPLE |
|---|---|---|
| Translate from English to algebra.<br>(Section 1.4/Objective 3) | To translate English phrases into algebraic expressions, you must be familiar with key phrases that signal whether we are to find a sum, difference, product, or quotient. | Translate the English phrase *six less than twice a number* into an algebraic expression.<br>**Solution**<br>Let $n$ represent the number. "Six less than" means that 6 will be subtracted from twice the number. "Twice the number" means that the number will be multiplied by 2. The phrase *six less than twice a number* translates into $2n - 6$.<br>**Sample Problem 9**<br>Translate the English phrase *four subtracted from three times a number* into an algebraic expression. |
| Use real numbers to represent problems.<br>(Section 1.2/Objective 8) | Real numbers can be used to represent many situations in the real world. | A patient in the hospital had a body temperature of 106.7°. Over the next 3 hours his temperature fell 1.2° per hour. What was his temperature after the 3 hours?<br>**Solution**<br>$106.7 - 3(1.2)$<br>$= 106.7 - 3.6 = 103.1$<br>His temperature was 103.1°.<br>**Sample Problem 10**<br>Pear Inc. stock was selling at $109 a share before the price dropped $15 a day for 3 days. What was the price of the stock after the 3 days? |

# Chapter 1 Review Problem Set

**1.** From the list 0, $\sqrt{2}$, $\frac{3}{4}$, $-\frac{5}{6}$, $\frac{25}{3}$, $-\sqrt{3}$, $-8$, 0.34, $0.2\overline{3}$, 67, and $\frac{9}{7}$, identify each of the following.

**(a)** The natural numbers

**(b)** The integers

**(c)** The nonnegative integers

**(d)** The rational numbers

**(e)** The irrational numbers

For Problems 2–10, state the property of equality or the property of real numbers that justifies each of the statements. For example, $6(-7) = -7(6)$ because of the commutative property of multiplication; and if $2 = x + 3$, then $x + 3 = 2$ is true because of the symmetric property of equality.

**2.** $7 + [3 + (-8)] = (7 + 3) + (-8)$

**3.** If $x = 2$ and $x + y = 9$, then $2 + y = 9$.

**4.** $-1(x + 2) = -(x + 2)$

**5.** $3(x + 4) = 3(x) + 3(4)$

**6.** $[(17)(4)](25) = (17)[(4)(25)]$

**7.** $x + 3 = 3 + x$

**8.** $3(98) + 3(2) = 3(98 + 2)$

**9.** $\left(\frac{3}{4}\right)\left(\frac{4}{3}\right) = 1$

**10.** If $4 = 3x - 1$, then $3x - 1 = 4$.

For Problems 11–14, find the absolute value.

**11.** $|-6.2|$

**12.** $\left|\frac{7}{3}\right|$

**13.** $|-\sqrt{15}|$

**14.** $|-8|$

For Problems 15–26, simplify each of the numerical expressions.

**15.** $-8\frac{1}{4} + \left(-4\frac{5}{8}\right) - \left(-6\frac{3}{8}\right)$

**16.** $9\frac{1}{3} - 12\frac{1}{2} + \left(-4\frac{1}{6}\right) - \left(-1\frac{1}{6}\right)$

**17.** $-8(2) - 16 \div (-4) + (-2)(-2)$

**18.** $4(-3) - 12 \div (-4) + (-2)(-1) - 8$

**19.** $-3(2 - 4) - 4(7 - 9) + 6$

**20.** $[48 + (-73)] + 74$

**21.** $[5(-2) - 3(-1)][-2(-1) + 3(2)]$

**22.** $3 - [-2(3 - 4)] + 7$

**23.** $-4^2 - 2^3$

**24.** $(-2)^4 + (-1)^3 - 3^2$

**25.** $2(-1)^2 - 3(-1)(2) - 2^2$

**26.** $[4(-1) - 2(3)]^2$

For Problems 27–36, simplify each of the algebraic expressions by combining similar terms.

**27.** $3a^2 - 2b^2 - 7a^2 - 3b^2$

**28.** $4x - 6 - 2x - 8 + x + 12$

**29.** $\frac{1}{5}ab^2 - \frac{3}{10}ab^2 + \frac{2}{5}ab^2 + \frac{7}{10}ab^2$

**30.** $-\frac{2}{3}x^2y - \left(-\frac{3}{4}x^2y\right) - \frac{5}{12}x^2y - 2x^2y$

**31.** $3(2n^2 + 1) + 4(n^2 - 5)$

**32.** $-2(3a - 1) + 4(2a + 3) - 5(3a + 2)$

**33.** $-(n - 1) - (n + 2) + 3$

**34.** $3(2x - 3y) - 4(3x + 5y) - x$

**35.** $4(a - 6) - (3a - 1) - 2(4a - 7)$

**36.** $-5(x^2 - 4) - 2(3x^2 + 6) + (2x^2 - 1)$

For Problems 37–46, evaluate each of the algebraic expressions for the given values of the variables.

**37.** $-5x + 4y$ for $x = \frac{1}{2}$ and $y = -1$

**38.** $3x^2 - 2y^2$ for $x = \frac{1}{4}$ and $y = -\frac{1}{2}$

**39.** $-5(2x - 3y)$ for $x = 1$ and $y = -3$

**40.** $(3a - 2b)^2$ for $a = -2$ and $b = 3$

**41.** $a^2 + 3ab - 2b^2$ for $a = 2$ and $b = -2$

**42.** $3n^2 - 4 - 4n^2 + 9$ for $n = 7$

**43.** $3(2x - 1) + 2(3x + 4)$ for $x = 1.2$

**44.** $-4(3x - 1) - 5(2x - 1)$ for $x = -2.3$

**45.** $2(n^2 + 3) - 3(n^2 + 1) + 4(n^2 - 6)$ for $n = -\frac{2}{3}$

**46.** $5(3n - 1) - 7(-2n + 1) + 4(3n - 1)$ for $n = \frac{1}{2}$

For Problems 47–54, translate each English phrase into an algebraic expression, and use $n$ to represent the unknown number.

**47.** Four increased by twice a number

**48.** Fifty subtracted from three times a number

**49.** Six less than two-thirds of a number

**50.** Ten times the difference of a number and 14

**51.** Eight subtracted from five times a number

**52.** The quotient of a number and three less than the number

**53.** Three less than five times the sum of a number and 2

**54.** Three-fourths of the sum of a number and 12

For Problems 55–64, answer the question with an algebraic expression.

**55.** The sum of two numbers is 37, and one of the numbers is $n$. What is the other number?

**56.** Yuriko can type $w$ words in an hour. What is her typing rate per minute?

**57.** Harry is $y$ years old. His brother is 7 years less than twice as old as Harry. How old is Harry's brother?

**58.** If $n$ represents a multiple of 3, what represents the next largest multiple of 3?

**59.** Celia has $p$ pennies, $n$ nickels, and $q$ quarters. How much, in cents, does Celia have?

**60.** The perimeter of a square is $i$ inches. How long, in feet, is each side of the square?

**61.** The length of a rectangle is $y$ yards, and the width is $f$ feet. What is the perimeter of the rectangle expressed in inches?

**62.** The length of a piece of wire is $d$ decimeters. What is the length expressed in centimeters?

**63.** Joan is $f$ feet and $i$ inches tall. How tall is she in inches?

**64.** The perimeter of a rectangle is 50 centimeters. If the rectangle is $c$ centimeters long, how wide is it?

**65.** Kalya has the capacity to record 4 minutes of video on her cellular phone. She currently has $3\frac{1}{2}$ minutes of video clips. How much recording capacity will she have left if she deletes $2\frac{1}{4}$ minutes of clips and adds $1\frac{3}{4}$ minutes of recording?

**66.** During the week, the price of a stock recorded the following gains and losses: Monday lost \$1.25, Tuesday lost \$0.45, Wednesday gained \$0.67, Thursday gained \$1.10, and Friday lost \$0.22. What is the average daily gain or loss for the week?

**67.** A crime-scene investigator has 3.4 ounces of a sample. He needs to conduct four tests that each require 0.6 ounces of the sample, and one test that requires 0.8 ounces of the sample. How much of the sample remains after he uses it for the five tests?

**68.** For week 1 of a weight-loss competition, Team A had three members lose 8 pounds each, two members lose 5 pounds each, one member loses 4 pounds, and two members gain 3 pounds. What was the total weight loss for Team A in the first week of the competition?

# Chapter 1 Test

1. State the property of equality that justifies writing $x + 4 = 6$ for $6 = x + 4$.

2. State the property of real numbers that justifies writing $5(10 + 2)$ as $5(10) + 5(2)$.

For Problems 3–11, simplify each numerical expression.

3. $-4 - (-3) + (-5) - 7 + 10$

4. $7 - 8 - 3 + 4 - 9 - 4 + 2 - 12$

5. $5\left(-\frac{1}{3}\right) - 3\left(-\frac{1}{2}\right) + 7\left(-\frac{2}{3}\right) + 1$

6. $(-6) \cdot 3 \div (-2) - 8 \div (-4)$

7. $-\frac{1}{2}(3 - 7) - \frac{2}{5}(2 - 17)$

8. $[48 + (-93)] + (-49)$

9. $3(-2)^3 + 4(-2)^2 - 9(-2) - 14$

10. $[2(-6) + 5(-4)][-3(-4) - 7(6)]$

11. $[-2(-3) - 4(2)]^5$

12. Simplify $6x^2 - 3x - 7x^2 - 5x - 2$ by combining similar terms.

13. Simplify $3(3n - 1) - 4(2n + 3) + 5(-4n - 1)$ by removing parentheses and combining similar terms.

For Problems 14–20, evaluate each algebraic expression for the given values of the variables.

14. $-7x - 3y$ for $x = -6$ and $y = 5$

15. $3a^2 - 4b^2$ for $a = -\frac{3}{4}$ and $b = \frac{1}{2}$

16. $6x - 9y - 8x + 4y$ for $x = \frac{1}{2}$ and $y = -\frac{1}{3}$

17. $-5n^2 - 6n + 7n^2 + 5n - 1$ for $n = -6$

18. $-7(x - 2) + 6(x - 1) - 4(x + 3)$ for $x = 3.7$

19. $-2xy - x + 4y$ for $x = -3$ and $y = 9$

20. $4(n^2 + 1) - (2n^2 + 3) - 2(n^2 + 3)$ for $n = -4$

For Problems 21 and 22, translate the English phrase into an algebraic expression using $n$ to represent the unknown number.

21. Thirty subtracted from six times a number

22. Four more than three times the sum of a number and 8

For Problems 23–25, answer each question with an algebraic expression.

23. The product of two numbers is 72, and one of the numbers is $n$. What is the other number?

24. Tao has $n$ nickels, $d$ dimes, and $q$ quarters. How much money, in cents, does she have?

25. The length of a rectangle is $x$ yards and the width is $y$ feet. What is the perimeter of the rectangle expressed in feet?

# 2 Equations, Inequalities, and Problem Solving

Phil Boorman/Taxi/Getty Images

*"The man who thinks he can and the man who thinks he can't are both right."*
HENRY FORD

## Study Skill Tip

Class time is an intense study time. Start by being prepared physically and mentally for class. For the physical part, consider sitting in the area called the "golden triangle of success." That area is a triangle formed by the front row of the classroom to the middle seat in the back row. This is where the instructor focuses his/her attention. When sitting in the golden triangle of success, you will be apt to pay more attention and be less distracted.

To be mentally prepared for class and note taking, you should practice warming up before class begins. Warming up could involve reviewing the notes from the previous class session, reviewing your homework, preparing questions to ask, trying a few of the unassigned problems, or previewing the section for the upcoming class session. These activities will get you ready to learn during the class session.

Students often wonder if they should be taking notes or just listening. The answer is somewhat different for each student, but every student's notes should contain examples of problems, explanations to accompany those examples, and key rules and vocabulary for the example. The instructor will give clues as to when to write down given information. Definitely take notes when the instructor gives lists such as 1, 2, 3 or A, B, C, says this step is important, or says this problem will be on the test. Through careful listening, you will learn to recognize these clues.

*Do you think you can solve word problems?*

## Chapter Preview

This chapter focuses on solving equations, inequalities, and word problems. Students usually come into Intermediate Algebra being able to solve basic equations. The goal in this chapter is to become proficient at solving equations involving fractions or decimals. New to most students will be the solving of absolute value equations and inequalities presented in Section 2.7.

Word problems seem to be challenging for all students. Too often students decide that they cannot be successful at word problems, so they don't even try. I encourage you to adopt an attitude of trying the word problems. As with any new skill, word problems can be learned, but first you have to make an honest attempt to learn them.

# 2.1 Solving First-Degree Equations

OBJECTIVES

1 Solve first-degree equations

2 Use equations to solve word problems

In Section 1.1, we stated that an equality (equation) is a statement in which two symbols, or groups of symbols, are names for the same number. It should be further stated that an equation may be true or false. For example, the equation $3 + (-8) = -5$ is true, but the equation $-7 + 4 = 2$ is false.

**Algebraic equations** contain one or more variables. The following are examples of algebraic equations.

$$3x + 5 = 8 \qquad 4y - 6 = -7y + 9 \qquad x^2 - 5x - 8 = 0$$
$$3x + 5y = 4 \qquad x^3 + 6x^2 - 7x - 2 = 0$$

An algebraic equation such as $3x + 5 = 8$ is neither true nor false as it stands, and we often refer to it as an "open sentence." Each time that a number is substituted for $x$, the algebraic equation $3x + 5 = 8$ becomes a numerical statement that is true or false. For example, if $x = 0$, then $3x + 5 = 8$ becomes $3(0) + 5 = 8$, which is a false statement. If $x = 1$, then $3x + 5 = 8$ becomes $3(1) + 5 = 8$, which is a true statement. **Solving an equation** refers to the process of finding the number (or numbers) that make(s) an algebraic equation a true numerical statement. We call such numbers the **solutions** or **roots** of the equation, and we say that they *satisfy* the equation. We call the set of all solutions of an equation its **solution set**. Thus $\{1\}$ is the solution set of $3x + 5 = 8$.

In this chapter, we will consider techniques for solving **first-degree equations in one variable**. This means that the equations contain only one variable and that this variable has an exponent of 1. The following are examples of first-degree equations in one variable.

$$3x + 5 = 8 \qquad \frac{2}{3}y + 7 = 9 \qquad 7a - 6 = 3a + 4 \qquad \frac{x-2}{4} = \frac{x-3}{5}$$

**Equivalent equations** are equations that have the same solution set. For example,

**1.** $3x + 5 = 8$

**2.** $3x = 3$

**3.** $x = 1$

are all equivalent equations because $\{1\}$ is the solution set of each.

The general procedure for solving an equation is to continue replacing the given equation with equivalent but simpler equations until we obtain an equation of the form *variable* =

*constant* or *constant* = *variable*. Thus in the previous example, $3x + 5 = 8$ was simplified to $3x = 3$, which was further simplified to $x = 1$, from which the solution set $\{1\}$ is obvious.

To solve equations we need to use the various properties of equality. In addition to the reflexive, symmetric, transitive, and substitution properties we listed in Section 1.1, the following properties of equality are important for problem solving.

### Addition Property of Equality

For all real numbers $a$, $b$, and $c$,

$a = b$ if and only if $a + c = b + c$

### Multiplication Property of Equality

For all real numbers $a$, $b$, and $c$, where $c \neq 0$,

$a = b$ if and only if $ac = bc$

The addition property of equality states that when the same number is added to both sides of an equation, an equivalent equation is produced. The multiplication property of equality states that we obtain an equivalent equation whenever we multiply both sides of an equation by the same *nonzero* real number. The following examples demonstrate the use of these properties to solve equations.

Let's clarify another point. We stated the properties of equality in terms of only two operations, addition and multiplication. We could also include the operations of subtraction and division in the statements of the properties. That is, we could think in terms of subtracting the same number from both sides of an equation and also in terms of dividing both sides of an equation by the same nonzero number.

**Classroom Example**
Solve $3x - 5 = 16$.

**EXAMPLE 1** Solve $2x - 1 = 13$.

### Solution

$$2x - 1 = 13$$

$$2x - 1 + 1 = 13 + 1 \qquad \text{Add 1 to both sides}$$

$$2x = 14$$

$$\frac{1}{2}(2x) = \frac{1}{2}(14) \qquad \text{Multiply both sides by } \frac{1}{2}$$

$$x = 7$$

The solution set is $\{7\}$.

To check an apparent solution, we can substitute it into the original equation and see if we obtain a true numerical statement.

✔ **Check**

$$2x - 1 = 13$$

$$2(7) - 1 \stackrel{?}{=} 13$$

$$14 - 1 \stackrel{?}{=} 13$$

$$13 = 13$$

Now we know that {7} is the solution set of $2x - 1 = 13$. We will not show our checks for every example in this text, but do remember that checking is a way to detect arithmetic errors.

**Classroom Example**
Solve $-5 = -4a + 8$.

**EXAMPLE 2** Solve $-7 = -5a + 9$.

### Solution

$$-7 = -5a + 9$$
$$-7 - 9 = 5a + 9 - 9 \qquad \text{Subtract 9 from both sides}$$
$$-16 = -5a$$
$$\frac{-16}{-5} = \frac{-5a}{-5} \qquad \text{Divide both sides by } -5$$
$$\frac{16}{5} = a$$

The solution set is $\left\{\frac{16}{5}\right\}$.

Note that in Example 2 the final equation is $\frac{16}{5} = a$ instead of $a = \frac{16}{5}$. Technically, the symmetric property of equality (if $a = b$, then $b = a$) would permit us to change from $\frac{16}{5} = a$ to $a = \frac{16}{5}$, but such a change is not necessary to determine that the solution is $\frac{16}{5}$. Note that we could use the symmetric property at the very beginning to change $-7 = -5a + 9$ to $-5a + 9 = -7$; some people prefer having the variable on the left side of the equation.

**Classroom Example**
Solve $2x - 5 = -5$.

**EXAMPLE 3** Solve $6y + 17 = 17$.

### Solution

$$6y + 17 = 17$$
$$6y + 17 - 17 = 17 - 17 \qquad \text{Subtract 17 from both sides}$$
$$6y = 0$$
$$\frac{6y}{6} = \frac{0}{6} \qquad \text{Divide both sides by 6}$$
$$y = 0$$

The solution set is {0}.

The solution to the equation in Example 3 is zero. The next two examples show equations called contradictions and identities. Respectively, their solution sets are the empty set and the set of all real numbers. Study examples 3, 4, and 5 to know the differences between a solution of zero, no solution, and a solution of all real numbers.

**Classroom Example**
Solve $7m + 5 = 7m - 3$.

**EXAMPLE 4** Solve $3x + 10 = 3x - 4$.

### Solution

$$3x + 10 = 3x - 4$$
$$3x + 10 - 3x = 3x - 4 - 3x \qquad \text{Subtract } 3x \text{ from both sides}$$
$$10 = -4 \qquad \text{False statement}$$

Because we obtained an equivalent equation that is a false statement, there is no value of $x$ that will make the equation a true statement. When the equation is not true under any condition, then the equation is called a **contradiction**. The solution set for an equation that is a contradiction is the empty or null set, and it is symbolized by $\varnothing$.

Classroom Example
Solve $6n + 4 = 9n + 4 - 3n$.

**EXAMPLE 5** Solve $4x + 8 - x = 3x + 8$.

### Solution

$$4x + 8 - x = 3x + 8$$
$$3x + 8 = 3x + 8 \quad \text{Combine similar terms on the left side}$$
$$3x + 8 - 3x = 3x + 8 - 3x \quad \text{Subtract } 3x \text{ from both sides}$$
$$8 = 8 \quad \text{True statement}$$

Because we obtained an equivalent equation that is a true statement, any value of $x$ will make the equation a true statement. When an equation is true for any value of the variable, the equation is called an **identity**. The solution set for an equation that is an identity is the set of all real numbers. We will denote the set of all real numbers as {all reals}.

Classroom Example
Solve $8m - 7 = 5m + 8$.

**EXAMPLE 6** Solve $7x - 3 = 5x + 9$.

### Solution

$$7x - 3 = 5x + 9$$
$$7x - 3 + (-5x) = 5x + 9 + (-5x) \quad \text{Add } -5x \text{ to both sides}$$
$$2x - 3 = 9$$
$$2x - 3 + 3 = 9 + 3 \quad \text{Add 3 to both sides}$$
$$2x = 12$$
$$\frac{1}{2}(2x) = \frac{1}{2}(12) \quad \text{Multiply both sides by } \frac{1}{2}$$
$$x = 6$$

The solution set is $\{6\}$.

Classroom Example
Solve $2(x + 3) + 6(x - 4) = 5(x - 9)$.

**EXAMPLE 7** Solve $4(y - 1) + 5(y + 2) = 3(y - 8)$.

### Solution

$$4(y - 1) + 5(y + 2) = 3(y - 8)$$
$$4y - 4 + 5y + 10 = 3y - 24 \quad \text{Remove parentheses by applying the distributive property}$$
$$9y + 6 = 3y - 24 \quad \text{Simplify the left side by combining similar terms}$$
$$9y + 6 + (-3y) = 3y - 24 + (-3y) \quad \text{Add } -3y \text{ to both sides}$$
$$6y + 6 = -24$$
$$6y + 6 - 6 = -24 - 6 \quad \text{Subtract 6 from both sides}$$
$$6y = -30$$
$$\frac{6y}{6} = \frac{-30}{6} \quad \text{Divide both sides by 6}$$
$$y = -5$$

The solution set is $\{-5\}$.

We can summarize the process of solving first-degree equations in one variable as follows:

**Step 1** Simplify both sides of the equation as much as possible.

**Step 2** Use the addition property of equality to isolate a term that contains the variable on one side of the equation and a constant on the other side.

**Step 3** Use the multiplication property of equality to make the coefficient of the variable 1; that is, multiply both sides of the equation by the reciprocal of the numerical coefficient of the variable. The solution set should now be obvious.

**Step 4** Check each solution by substituting it in the original equation and verifying that the resulting numerical statement is true.

## Using Equations to Solve Problems

To use the tools of algebra to solve problems, we must be able to translate back and forth between the English language and the language of algebra. More specifically, we need to translate English sentences into algebraic equations. Such translations allow us to use our knowledge of equation solving to solve word problems. Let's consider an example.

**Classroom Example**
If we subtract 19 from two times a certain number, the result is 3. Find the number.

### EXAMPLE 8 Apply Your Skill

If we subtract 27 from three times a certain number, the result is 18. Find the number.

### Solution

Let $n$ represent the number to be found. The sentence "If we subtract 27 from three times a certain number, the result is 18" translates into the equation $3n - 27 = 18$. Solving this equation, we obtain

$$3n - 27 = 18$$
$$3n = 45 \quad \text{Add 27 to both sides}$$
$$n = 15 \quad \text{Divide both sides by 3}$$

The number to be found is 15.

We often refer to the statement "Let $n$ represent the number to be found" as **declaring the variable**. We need to choose a letter to use as a variable and indicate what it represents for a specific problem. This may seem like an insignificant exercise, but as the problems become more complex, the process of declaring the variable becomes even more important. Furthermore, it is true that you could probably solve a problem such as Example 8 without setting up an algebraic equation. However, as problems increase in difficulty, the translation from English to algebra becomes a key issue. Therefore, even with these relatively easy problems, we suggest that you concentrate on the translation process.

The next example involves the use of integers. Remember that the set of integers consists of $\{\ldots -2, -1, 0, 1, 2, \ldots\}$. Furthermore, the integers can be classified as even, $\{\ldots -4, -2, 0, 2, 4, \ldots\}$, or odd, $\{\ldots -3, -1, 1, 3, \ldots\}$.

**Classroom Example**
The sum of three consecutive even integers is six less than two times the largest of the three odd integers. Find the integers.

### EXAMPLE 9 Apply Your Skill

The sum of three consecutive integers is 13 greater than twice the smallest of the three integers. Find the integers.

### Solution

Because consecutive integers differ by 1, we will represent them as follows: Let $n$ represent the smallest of the three consecutive integers; then $n + 1$ represents the second largest, and $n + 2$ represents the largest.

The sum of the three consecutive integers — 13 greater than twice the smallest

$$\underbrace{n + (n + 1) + (n + 2)}_{\text{The sum of the three consecutive integers}} = \underbrace{2n + 13}_{\text{13 greater than twice the smallest}}$$
$$3n + 3 = 2n + 13$$
$$n = 10$$

The three consecutive integers are 10, 11, and 12.

To check our answers for Example 9, we must determine whether or not they satisfy the conditions stated in the original problem. Because 10, 11, and 12 are consecutive integers whose sum is 33, and because twice the smallest, 2(10), plus 13 is also 33, we know that our answers are correct. (Remember, in checking a result for a word problem, it is *not* sufficient to check the result in the equation set up to solve the problem; the equation itself may be in error!)

In the two previous examples, the equation formed was almost a direct translation of a sentence in the statement of the problem. Now let's consider a situation where we need to think in terms of a guideline not explicitly stated in the problem.

**Classroom Example**
Erik received a car repair bill for \$389. This included \$159 for parts, \$43 per hour for each hour of labor, and \$15 for taxes. Find the number of hours of labor.

## EXAMPLE 10 Apply Your Skill

Khoa received a car repair bill for \$412. This included \$175 for parts, \$60 per hour for each hour of labor, and \$27 for taxes. Find the number of hours of labor.

### Solution

See Figure 2.1. Let $h$ represent the number of hours of labor. Then $60h$ represents the total charge for labor.

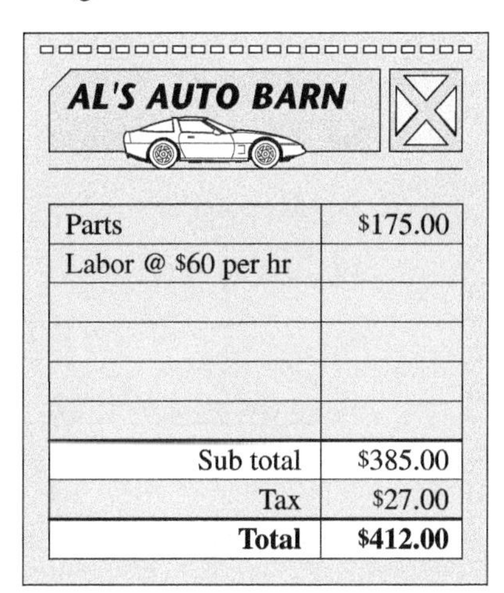


| Parts | \$175.00 |
|---|---|
| Labor @ \$60 per hr | |
| | |
| | |
| | |
| Sub total | \$385.00 |
| Tax | \$27.00 |
| **Total** | **\$412.00** |

**Figure 2.1**

We can use this guideline: *charge for parts plus charge for labor plus tax equals the total bill* to set up the following equation.

$$\underset{\text{Parts}}{175} + \underset{\text{Labor}}{60h} + \underset{\text{Tax}}{27} = \underset{\text{Total bill}}{412}$$

Solving this equation, we obtain

$$60h + 202 = 412$$
$$60h = 210 \qquad \text{Subtract 202 from both sides}$$
$$h = 3\frac{1}{2} \qquad \text{Divide both sides by 60}$$

Khoa was charged for $3\frac{1}{2}$ hours of labor.

## Concept Quiz 2.1

For Problems 1–10, answer true or false.

1. Equivalent equations have the same solution set.
2. $x^2 = 9$ is a first-degree equation.
3. The set of all solutions is called a solution set.
4. If the solution set is the null set, then the equation has at least one solution.
5. Solving an equation refers to obtaining any other equivalent equation.
6. If 5 is a solution, then a true numerical statement is formed when 5 is substituted for the variable in the equation.
7. Any number can be subtracted from both sides of an equation, and the result is an equivalent equation.
8. Any number can divide both sides of an equation to obtain an equivalent equation.
9. The equation $2x + 7 = 3y$ is a first-degree equation in one variable.
10. The multiplication property of equality states that an equivalent equation is obtained whenever both sides of an equation are multiplied by a nonzero number.

## Problem Set 2.1

For Problems 1–50, solve each equation. (Objective 1)

1. $3x + 4 = 16$
2. $4x + 2 = 22$
3. $5x + 1 = -14$
4. $7x + 4 = -31$
5. $-x - 6 = 8$
6. $8 - x = -2$
7. $4y - 3 = 21$
8. $6y - 7 = 41$
9. $3x - 4 = -4$
10. $5x + 1 = 12$
11. $-4 = 2x - 6$
12. $-2 = 3a - 2$
13. $-6y - 4 = 16$
14. $-8y - 2 = 18$
15. $4x - 1 = 4x + 7$
16. $9x - 3 = 6x + 18$
17. $5y + 2 = 2y - 11$
18. $9y + 3 = 9y - 10$
19. $3x + 4 = 5x - 2$
20. $2x - 1 = 6x + 15$
21. $-7a + 6 = -8a + 14$
22. $-6a - 4 = -7a + 11$
23. $5x + 3 - 2x = x - 15$
24. $4x - 2 - x = 5x + 10$
25. $6y + 18 + y = 2y + 3$
26. $5y + 14 + y = 3y - 7$
27. $4x - 3 + 2x = 8x - 3 - x$
28. $x - 4 - 4x = 6x + 9 - 8x$
29. $6n - 4 - 3n = 3n - 4$
30. $2n - 1 - 3n = 5n - 7 - 3n$
31. $4(x - 3) = -20$
32. $3(x + 2) = 3x + 6$
33. $-3(x - 2) = 11$
34. $-5(x - 1) = 5$
35. $5(2x + 1) = 4(3x - 7)$
36. $3(2x - 1) = 2(4x + 7)$
37. $5x - 4(x - 6) = -11$
38. $3x - 5(2x + 1) = 13$
39. $-2(3x - 1) - 3 = -4$
40. $-6(x - 4) - 10 = -12$
41. $-2(3x + 5) = -3(4x + 3)$
42. $-(2x - 1) = -5(2x + 9)$
43. $3(x - 4) - 7(x + 2) = -2(x + 13)$
44. $4(x - 2) - 3(x - 1) = 2(x + 6)$
45. $-2(3n - 1) + 3(n + 5) = -4(n - 4)$
46. $-3(4n + 2) + 2(n - 6) = -2(n + 1)$
47. $3(2a - 1) - 2(5a + 1) = 4(3a + 4)$
48. $4(2a + 3) - 3(4a - 2) = 5(4a - 7)$
49. $-2(n - 4) - (3n - 1) = -2 + (2n - 1)$
50. $-(2n - 1) + 6(n + 3) = -4 - (7n - 11)$

For Problems 51–66, use an algebraic approach to solve each problem. (Objective 2)

51. If 15 is subtracted from three times a certain number, the result is 27. Find the number.

**52.** If one is subtracted from seven times a certain number, the result is the same as if 31 is added to three times the number. Find the number.

**53.** Find three consecutive integers whose sum is 42.

**54.** Find four consecutive integers whose sum is $-118$.

**55.** Find three consecutive odd integers such that three times the second minus the third is 11 more than the first.

**56.** Find three consecutive even integers such that four times the first minus the third is six more than twice the second.

**57.** The difference of two numbers is 67. The larger number is three less than six times the smaller number. Find the numbers.

**58.** The sum of two numbers is 103. The larger number is one more than five times the smaller number. Find the numbers.

**59.** Angelo is paid double time for each hour he works over 40 hours in a week. Last week he worked 46 hours and earned $572. What is his normal hourly rate?

**60.** Suppose that a plumbing repair bill, not including tax, was $130. This included $25 for parts and an amount for 2 hours of labor. Find the hourly rate that was charged for labor.

**61.** Suppose that Maria has 150 coins consisting of pennies, nickels, and dimes. The number of nickels she has is 10 less than twice the number of pennies; the number of dimes she has is 20 less than three times the number of pennies. How many coins of each kind does she have?

**62.** Hector has a collection of nickels, dimes, and quarters totaling 122 coins. The number of dimes he has is 3 more than four times the number of nickels, and the number of quarters he has is 19 less than the number of dimes. How many coins of each kind does he have?

**63.** The selling price of a ring is $750. This represents $150 less than three times the cost of the ring. Find the cost of the ring.

**64.** In a class of 62 students, the number of females is one less than twice the number of males. How many females and how many males are there in the class?

**65.** An apartment complex contains 230 apartments, each having one, two, or three bedrooms. The number of two-bedroom apartments is 10 more than three times the number of three-bedroom apartments. The number of one-bedroom apartments is twice the number of two-bedroom apartments. How many apartments of each kind are in the complex?

**66.** Barry sells bicycles on a salary-plus-commission basis. He receives a weekly salary of $300 and a commission of $15 for each bicycle that he sells. How many bicycles must he sell in a week to have a total weekly income of $750?

## Thoughts Into Words

**67.** Explain why the solution set of the equation $x + 3 = x + 4$ is the null set.

**68.** Explain why the solution set of the equation $3(x + 5) = 3x + 15$ is the entire set of real numbers.

**69.** Why must potential answers to word problems be checked back into the original statement of the problem?

**70.** Suppose your friend solved the problem, *find two consecutive odd integers whose sum is 28* like this:

$$x + x + 1 = 28$$

$$2x = 27$$

$$x = \frac{27}{2} = 13\frac{1}{2}$$

She claims that $13\frac{1}{2}$ will check in the equation. Where has she gone wrong and how would you help her?

**71.** Make up an equation whose solution set is the null set and explain why this is the solution set.

**72.** Make up an equation whose solution set is the set of all real numbers and explain why this is the solution set.

## Further Investigations

**73.** Solve each of the following equations.

**(a)** $5x + 7 = 5x - 4$
**(b)** $4(x - 1) = 4x - 4$
**(c)** $3(x - 4) = 2(x - 6)$
**(d)** $7x - 2 = -7x + 4$
**(e)** $2(x - 1) + 3(x + 2) = 5(x - 7)$
**(f)** $-4(x - 7) = -2(2x + 1)$

**74.** Verify that for any three consecutive integers, the sum of the smallest and largest is equal to twice the middle integer. [*Hint*: Use $n$, $n + 1$, and $n + 2$ to represent the three consecutive integers.]

**Answers to the Concept Quiz**

**1.** True **2.** False **3.** True **4.** False **5.** False **6.** True **7.** True **8.** False **9.** False **10.** True

## 2.2 Equations Involving Fractional Forms

OBJECTIVES

1. Solve equations involving fractions
2. Distinguish between an expression and an equation
3. Solve word problems

To solve equations that involve fractions, it is usually easiest to begin by *clearing the equation of all fractions*. This can be accomplished by multiplying both sides of the equation by the least common multiple (LCM) of all the denominators in the equation. Remember that the LCM of a set of whole numbers is the smallest nonzero whole number that is divisible by each of the numbers. For example, the LCM of 2, 3, and 6 is 12. When working with fractions, we refer to the LCM of a set of denominators as the **least common denominator** (LCD). Let's consider some equations involving fractions.

**Classroom Example**
Solve $\frac{3}{8}x + \frac{1}{3} = \frac{7}{12}$.

**EXAMPLE 1** Solve $\frac{1}{2}x + \frac{2}{3} = \frac{3}{4}$.

**Solution**

$$\frac{1}{2}x + \frac{2}{3} = \frac{3}{4}$$

$$12\left(\frac{1}{2}x + \frac{2}{3}\right) = 12\left(\frac{3}{4}\right) \quad \text{Multiply both sides by 12, which is the LCM of 2, 3, and 4}$$

$$12\left(\frac{1}{2}x\right) + 12\left(\frac{2}{3}\right) = 12\left(\frac{3}{4}\right) \quad \text{Apply the distributive property to the left side}$$

$$6x + 8 = 9$$

$$6x = 1$$

$$x = \frac{1}{6}$$

The solution set is $\left\{\frac{1}{6}\right\}$.

✔ **Check**

$$\frac{1}{2}x + \frac{2}{3} = \frac{3}{4}$$

$$\frac{1}{2}\left(\frac{1}{6}\right) + \frac{2}{3} \stackrel{?}{=} \frac{3}{4}$$

$$\frac{1}{12} + \frac{2}{3} \stackrel{?}{=} \frac{3}{4}$$

$$\frac{1}{12} + \frac{8}{12} \stackrel{?}{=} \frac{9}{12} \quad \text{Change each fraction to an equivalent fraction with 12 as the denominator}$$

$$\frac{9}{12} = \frac{9}{12}$$

**Classroom Example**
Solve $\frac{m}{3} - \frac{m}{5} = 2$.

## EXAMPLE 2

Solve $\frac{x}{2} + \frac{x}{3} = 10$.

### Solution

$$\frac{x}{2} + \frac{x}{3} = 10$$

$$6\left(\frac{x}{2} + \frac{x}{3}\right) = 6(10) \quad \text{Multiply both sides by the LCD}$$

$$6\left(\frac{x}{2}\right) + 6\left(\frac{x}{3}\right) = 6(10) \quad \text{Apply the distributive property to the left side}$$

$$3x + 2x = 60$$

$$5x = 60$$

$$x = 12$$

The solution set is $\{12\}$.

As you study the examples in this section, pay special attention to the steps shown in the solutions. There are no hard and fast rules as to which steps should be performed mentally; this is an individual decision. When you solve problems, show enough steps to allow the flow of the process to be understood and to minimize the chances of making careless computational errors.

**Classroom Example**
Solve $\frac{a-3}{2} - \frac{a+4}{9} = \frac{7}{6}$.

## EXAMPLE 3

Solve $\frac{x-2}{3} + \frac{x+1}{8} = \frac{5}{6}$.

### Solution

$$\frac{x-2}{3} + \frac{x+1}{8} = \frac{5}{6}$$

$$24\left(\frac{x-2}{3} + \frac{x+1}{8}\right) = 24\left(\frac{5}{6}\right) \quad \text{Multiply both sides by the LCD}$$

$$24\left(\frac{x-2}{3}\right) + 24\left(\frac{x+1}{8}\right) = 24\left(\frac{5}{6}\right) \quad \text{Apply the distributive property to the left side}$$

$$8(x-2) + 3(x+1) = 20 \quad \text{Divide the denominators into 24}$$

$$8x - 16 + 3x + 3 = 20$$

$$11x - 13 = 20$$

$$11x = 33$$

$$x = 3$$

The solution set is $\{3\}$.

**Classroom Example**
Solve $\frac{4x+7}{3} - \frac{x+3}{2} = 1$.

## EXAMPLE 4

Solve $\frac{3t-1}{5} - \frac{t-4}{3} = 1$.

### Solution

$$\frac{3t-1}{5} - \frac{t-4}{3} = 1$$

$$15\left(\frac{3t-1}{5} - \frac{t-4}{3}\right) = 15(1) \quad \text{Multiply both sides by the LCD}$$

$$15\left(\frac{3t-1}{5}\right) - 15\left(\frac{t-4}{3}\right) = 15(1) \quad \text{Apply the distributive property to the left side}$$

$$3(3t - 1) - 5(t - 4) = 15 \quad \text{Divide the denominators into 15}$$
$$9t - 3 - 5t + 20 = 15$$
$$4t + 17 = 15$$
$$4t = -2$$
$$t = -\frac{2}{4} = -\frac{1}{2} \quad \text{Reduce!}$$

The solution set is $\left\{-\frac{1}{2}\right\}$.

## Distinguish Between an Expression and an Equation

After students learn to multiply by the least common denominator to clear the fractions to solve an equation, they mistakenly do this with algebraic expressions. This confusion happens when fractions are involved.

An algebraic expression such as $\frac{1}{3}x + \frac{5}{6}x + \frac{1}{18}x$ does not have an equal sign. The properties of equality do not apply to expressions. Hence you cannot multiply by the LCD to clear the fractions. To simplify this expression you must find a common denominator and change each fraction to an equivalent fraction with the common denominator as the new denominator. The work could be done as follows.

$$\frac{1}{3}x + \frac{5}{6}x + \frac{1}{18}x \quad \text{The LCD is 18}$$
$$= \frac{1}{3} \cdot \frac{6}{6}x + \frac{5}{6} \cdot \frac{3}{3}x + \frac{1}{18}x$$
$$= \frac{6}{18}x + \frac{15}{18}x + \frac{1}{18}x$$
$$= \frac{22}{18}x$$
$$= \frac{11}{9}x \quad \text{Reduce}$$

An equation such as $\frac{3}{5}x + \frac{1}{2}x = \frac{1}{10}$ does have an equal sign. The properties of equality can be applied. Hence you can multiply both sides of the equation by the LCD to clear the fractions.

Remember the key is to determine if the given statement has an equal sign. If there is not an equal sign, then simplify the expression. If there is an equal sign, then solve the equation.

## Solving Word Problems

As we expand our skills for solving equations, we also expand our capabilities for solving word problems. There is no one definite procedure that will ensure success at solving word problems, but the following suggestions should be helpful.

### Suggestions for Solving Word Problems

1. Read the problem carefully and make certain that you understand the meanings of all of the words. Be especially alert for any technical terms used in the statement of the problem.
2. Read the problem a second time (perhaps even a third time) to get an overview of the situation being described. Determine the known facts as well as what is to be found.

3. Sketch any figure, diagram, or chart that might be helpful in analyzing the problem.
4. Choose a meaningful variable to represent an unknown quantity in the problem (perhaps $t$, if time is an unknown quantity) and represent any other unknowns in terms of that variable.
5. Look for a guideline that you can use to set up an equation. A guideline might be a formula, such as *distance equals rate times time*, or a statement of a relationship, such as "The sum of the two numbers is 28."
6. Form an equation that contains the variable and that translates the conditions of the guideline from English to algebra.
7. Solve the equation, and use the solution to determine all facts requested in the problem.
8. Check all answers back into the **original statement of the problem**.

Keep these suggestions in mind as we continue to solve problems. We will elaborate on some of these suggestions at different times throughout the text. Now let's consider some examples.

**Classroom Example**
The width of a rectangular parking lot is 4 meters less than two-thirds of the length. The perimeter of the lot is 192 meters. Find the length and width of the lot.

### EXAMPLE 5 Apply Your Skill

The width of a rectangular parking lot is 8 feet less than three-fifths of the length. The perimeter of the lot is 400 feet. Find the length and width of the lot.

### Solution

Let $l$ represent the length of the lot. Then $\frac{3}{5}l - 8$ represents the width (Figure 2.2).

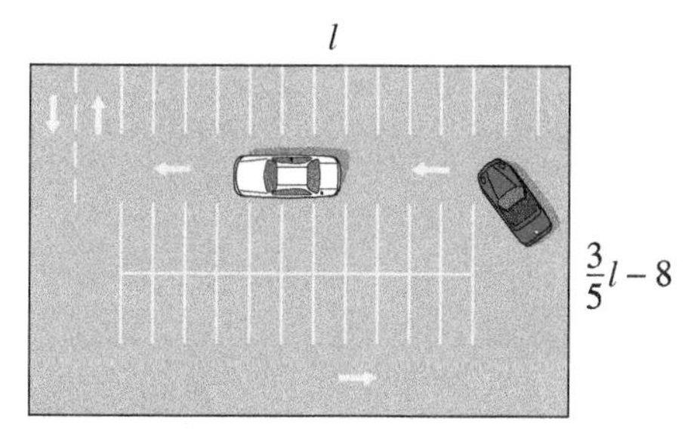


**Figure 2.2**

A guideline for this problem is the formula, *the perimeter of a rectangle equals twice the length plus twice the width* ($P = 2l + 2w$). Use this formula to form the following equation.

$$\begin{array}{ccccc} P & = & 2l & + & 2w \\ \downarrow & & \downarrow & & \downarrow \\ 400 & = & 2l & + & 2\left(\frac{3}{5}l - 8\right) \end{array}$$

Solving this equation, we obtain

$$400 = 2l + \frac{6l}{5} - 16$$

$$5(400) = 5\left(2l + \frac{6l}{5} - 16\right)$$

$$2000 = 10l + 6l - 80$$

$$2000 = 16l - 80$$
$$2080 = 16l$$
$$130 = l$$

The length of the lot is 130 feet, and the width is $\frac{3}{5}(130) - 8 = 70$ feet.

In Example 5, note the use of different letters as variables. It is helpful to choose a variable that has significance for the problem you are working on. For example, in Example 5, the choice of $l$ to represent the length seems natural and meaningful. (Certainly this is another matter of personal preference, but you might consider it.)

In Example 5 a geometric relationship, $(P = 2l + 2w)$, serves as a guideline for setting up the equation. The following geometric relationships pertaining to angle measure may also serve as guidelines.

1. Complementary angles are two angles that together measure 90°.
2. Supplementary angles are two angles that together measure 180°.
3. The sum of the measures of the three angles of a triangle is 180°.

**Classroom Example**
One of two supplementary angles is 15° larger than one-fourth of the other angle. Find the measure of each of the angles.

### EXAMPLE 6 Apply Your Skill

One of two complementary angles is 6° larger than one-half of the other angle. Find the measure of each of the angles.

#### Solution

Let $a$ represent the measure of one of the angles. Then $\frac{1}{2}a + 6$ represents the measure of the other angle. Because they are complementary angles, the sum of their measures is 90°.

$$a + \left(\frac{1}{2}a + 6\right) = 90$$
$$2a + a + 12 = 180$$
$$3a + 12 = 180$$
$$3a = 168$$
$$a = 56$$

If $a = 56$, then $\frac{1}{2}a + 6$ becomes $\frac{1}{2}(56) + 6 = 34$. The angles have measures of 34° and 56°.

Keep in mind that the problem-solving suggestions offered in this section simply outline a general algebraic approach to solving problems. You will add to this list throughout this course and in any subsequent mathematics courses that you take. Furthermore, you will be able to pick up additional problem-solving ideas from your instructor and from fellow classmates as you discuss problems in class. Always be on the alert for any ideas that might help you become a better problem solver.

## Concept Quiz 2.2

For Problems 1–8, answer true or false.

1. When solving an equation that involves fractions, the equation can be cleared of all the fractions by multiplying both sides of the equation by the least common multiple of all the denominators in the problem.

2. The least common multiple of a set of denominators is referred to as the least common denominator.
3. The least common multiple of 4, 6, and 9 is 36.
4. The least common multiple of 3, 9, and 18 is 36.
5. Answers for word problems need to be checked back into the original statement of the problem.
6. In a right triangle, the two acute angles are complementary angles.
7. A triangle can have two supplementary angles.
8. The sum of the measures of the three angles in a triangle is 100°.

## Problem Set 2.2

For Problems 1–48, determine whether the given problem is an equation or an expression. If it is an equation, then solve. If it is an expression, then simplify. (Objectives 1 and 2)

1. $\frac{3}{4}x = 9$
2. $\frac{2}{3}x = -14$
3. $\frac{-2x}{3} = \frac{2}{5}$
4. $\frac{-5x}{4} = \frac{7}{2}$
5. $\frac{n}{2} - \frac{2}{3} = \frac{5}{6}$
6. $\frac{n}{4} - \frac{5}{6} = \frac{5}{12}$
7. $\frac{3}{8}x - \frac{1}{4}x + \frac{5}{8}x$
8. $\frac{5}{6}x - \frac{5}{12}x + \frac{7}{24}x$
9. $\frac{5n}{6} - \frac{n}{8} = \frac{-17}{12}$
10. $\frac{2n}{5} - \frac{n}{6} = \frac{-7}{10}$
11. $\frac{a}{4} - 1 = \frac{a}{3} + 2$
12. $\frac{3a}{7} - 1 = \frac{a}{3}$
13. $\frac{h}{4} + \frac{h}{5} = 1$
14. $\frac{h}{6} + \frac{3h}{8} = 1$
15. $\frac{y}{4} + \frac{2y}{5} - \frac{3y}{10}$
16. $-\frac{2x}{7} - \frac{x}{21} - \frac{5x}{42}$
17. $\frac{h}{2} - \frac{h}{3} + \frac{h}{6} = 1$
18. $\frac{3h}{4} + \frac{2h}{5} = 1$
19. $\frac{x-2}{3} + \frac{x+2}{4} = -\frac{1}{6}$
20. $\frac{x+4}{5} + \frac{x-1}{4} = \frac{37}{10}$
21. $\frac{x+2}{2} - \frac{x-1}{5} = \frac{3}{5}$
22. $\frac{2x+1}{3} - \frac{x+1}{7} = -\frac{1}{3}$
23. $\frac{n+2}{4} - \frac{2n-1}{3} = \frac{1}{6}$
24. $\frac{n}{9} - \frac{n+3}{6} = -\frac{1}{2}$
25. $\frac{y}{3} + \frac{y-5}{10} = \frac{4y+3}{5}$
26. $\frac{y}{3} + \frac{y-2}{8} = \frac{6y-1}{12}$
27. $\frac{2x+1}{3} + x + \frac{5x+7}{9}$
28. $\frac{3x+1}{2} + \frac{x+4}{5} + x$
29. $\frac{4x-1}{10} - \frac{5x+2}{4} = -3$
30. $\frac{2x-1}{2} - \frac{3x+1}{4} = \frac{3}{10}$
31. $\frac{2x-1}{8} - 1 = \frac{x+5}{7}$
32. $\frac{3x+1}{9} + 2 = \frac{x-1}{4}$
33. $\frac{2a-3}{6} + \frac{3a-2}{4} + \frac{5a+6}{12} = 4$
34. $\frac{3a-1}{4} + \frac{a-2}{3} - \frac{a-1}{5} = \frac{21}{20}$
35. $x + \frac{3x-1}{9} - 4 = \frac{3x+1}{3}$
36. $\frac{2x+7}{8} + x - 2 = \frac{x-1}{2}$
37. $\frac{x+3}{2} + \frac{x+4}{5} = \frac{3}{10}$
38. $\frac{x-2}{5} - \frac{x-3}{4} = -\frac{1}{20}$

**39.** $n + \frac{2n - 3}{9} - 2 = \frac{2n + 1}{3}$

**40.** $n - \frac{3n + 1}{6} - 1 = \frac{2n + 4}{12}$

**41.** $\frac{3}{4}(t - 2) - \frac{2}{5}(2t - 3) = \frac{1}{5}$

**42.** $\frac{2}{3}(2t + 1) - \frac{1}{2}(3t - 2) = 2$

**43.** $\frac{1}{4}(x + 3) - \frac{2}{3}(x - 1)$

**44.** $\frac{2}{5}(x + 4) + \frac{1}{3}(x - 2)$

**45.** $\frac{1}{2}(2x - 1) - \frac{1}{3}(5x + 2) = 3$

**46.** $\frac{2}{5}(4x - 1) + \frac{1}{4}(5x + 2) = -1$

**47.** $3x - 1 + \frac{2}{7}(7x - 2) = -\frac{11}{7}$

**48.** $2x + 5 + \frac{1}{2}(6x - 1) = -\frac{1}{2}$

For Problems 49–62, use an algebraic approach to solve each problem. (Objective 3)

**49.** Find a number such that one-half of the number is 3 less than two-thirds of the number.

**50.** One-half of a number plus three-fourths of the number is 2 more than four-thirds of the number. Find the number.

**51.** Suppose that the width of a certain rectangle is 1 inch more than one-fourth of its length. The perimeter of the rectangle is 42 inches. Find the length and width of the rectangle.

**52.** Suppose that the width of a rectangle is 3 centimeters less than two-thirds of its length. The perimeter of the rectangle is 114 centimeters. Find the length and width of the rectangle.

**53.** Find three consecutive integers such that the sum of the first plus one-third of the second plus three-eighths of the third is 25.

**54.** Lou is paid $1\frac{1}{2}$ times his normal hourly rate for each hour he works over 40 hours in a week. Last week he worked 44 hours and earned \$483. What is his normal hourly rate?

**55.** A coaxial cable 20 feet long is cut into two pieces such that the length of one piece is two-thirds of the length of the other piece. Find the length of the shorter piece of cable.

**56.** Jody has a collection of 116 coins consisting of dimes, quarters, and silver dollars. The number of quarters is 5 less than three-fourths the number of dimes. The number of silver dollars is 7 more than five-eighths the number of dimes. How many coins of each kind are in her collection?

**57.** Aura took three biology exams and has an average score of 88. Her second exam score was 10 points better than her first, and her third exam score was 4 points better than her second exam. What were her three exam scores?

**58.** The average of the salaries of Tim, Maida, and Aaron is \$34,000 per year. Maida earns \$10,000 more than Tim, and Aaron's salary is \$8000 less than twice Tim's salary. Find the salary of each person.

**59.** One of two supplementary angles is 4° more than one-third of the other angle. Find the measure of each of the angles.

**60.** If one-half of the complement of an angle plus three-fourths of the supplement of the angle equals 110°, find the measure of the angle.

**61.** If the complement of an angle is 5° less than one-sixth of its supplement, find the measure of the angle.

**62.** In $\triangle ABC$, angle $B$ is 8° less than one-half of angle $A$, and angle $C$ is 28° larger than angle $A$. Find the measures of the three angles of the triangle.

## Thoughts Into Words

**63.** The fractions in the equation $\frac{x - 4}{3} + \frac{2x + 5}{4} = \frac{5}{6}$ could be cleared by multiplying by 12 or by 24. Which number would you choose and explain your choice.

**64.** Suppose that your friend shows you the following solution to an equation.

$$17 = 4 - 2x$$
$$17 + 2x = 4 - 2x + 2x$$
$$17 + 2x = 4$$
$$17 + 2x - 17 = 4 - 17$$
$$2x = -13$$
$$x = \frac{-13}{2}$$

Is this a correct solution? What suggestions would you have in terms of the method used to solve the equation?

### Answers to the Concept Quiz

**1.** True **2.** True **3.** True **4.** False **5.** True **6.** True **7.** False **8.** False

## 2.3 Equations Involving Decimals and Problem Solving

OBJECTIVES

1. Solve equations involving decimals
2. Solve word problems including those involving discount and selling price

In solving equations that involve fractions, usually the procedure is to clear the equation of all fractions. To solve equations that involve decimals, there are two commonly used procedures. One procedure is to keep the numbers in decimal form and solve the equation by applying the properties. Another procedure is to multiply both sides of the equation by an appropriate power of 10 to clear the equation of all decimals. Which technique to use depends on your personal preference and on the complexity of the equation. The following examples demonstrate both techniques.

Classroom Example
Solve $0.3t - 0.17 = 0.08t + 1.15$.

**EXAMPLE 1** Solve $0.2x + 0.24 = 0.08x + 0.72$.

### Solution

Let's clear the decimals by multiplying both sides of the equation by 100.

$$0.2x + 0.24 = 0.08x + 0.72$$
$$100(0.2x + 0.24) = 100(0.08x + 0.72)$$
$$100(0.2x) + 100(0.24) = 100(0.08x) + 100(0.72) \quad \text{Apply the distributive property}$$
$$20x + 24 = 8x + 72$$
$$12x + 24 = 72$$
$$12x = 48$$
$$x = 4$$

✔ Check

$$0.2x + 0.24 = 0.08x + 0.72$$
$$0.2(4) + 0.24 \stackrel{?}{=} 0.08(4) + 0.72$$
$$0.8 + 0.24 \stackrel{?}{=} 0.32 + 0.72$$
$$1.04 = 1.04$$

The solution set is $\{4\}$.

Classroom Example
Solve $0.04m + 0.08m = 4.8$.

**EXAMPLE 2** Solve $0.07x + 0.11x = 3.6$.

### Solution

Let's keep this problem in decimal form.

$$0.07x + 0.11x = 3.6$$
$$0.18x = 3.6 \quad \text{Add like terms}$$
$$x = \frac{3.6}{0.18}$$
$$x = 20$$

✔ **Check**

$$0.07x + 0.11x = 3.6$$
$$0.07(20) + 0.11(20) \stackrel{?}{=} 3.6$$
$$1.4 + 2.2 \stackrel{?}{=} 3.6$$
$$3.6 = 3.6$$

The solution set is $\{20\}$.

**Classroom Example**
Solve $y = 2.16 + 0.73y$.

**EXAMPLE 3** Solve $s = 1.95 + 0.35s$.

### Solution

Let's keep this problem in decimal form.

$$s = 1.95 + 0.35s$$
$$s + (-0.35s) = 1.95 + 0.35s + (-0.35s)$$
$$0.65s = 1.95 \qquad \text{Remember, } s = 1.00s$$
$$s = \frac{1.95}{0.65}$$
$$s = 3$$

The solution set is $\{3\}$. Check it!

**Classroom Example**
Solve
$0.6(x - 0.1) - 0.5(x + 4) = 0.34$.

**EXAMPLE 4** Solve $0.4(x - 0.6) - 0.3(x + 2) = 2.76$.

### Solution

Let's clear the decimals by multiplying both sides of the equation by 100.

$$0.4(x - 0.6) - 0.3(x + 2) = 2.76$$
$$100[0.4(x - 0.6) - 0.3(x + 2)] = 100(2.76) \qquad \text{Multiply both sides by 100}$$
$$100[0.4(x - 0.6)] - 100[0.3(x + 2)] = 276$$

The term $100[0.4(x - 0.6)]$ is the product of three factors; 100, 0.4, and $(x - 0.6)$. Find the product of 100 and 0.4, which is 40. Then multiply 40 by the factor $(x - 0.6)$. Do likewise for the term $100[0.3(x + 2)]$.

$$40(x - 0.6) - 30(x + 2) = 276$$
$$40x - 24 - 30x - 60 = 276$$
$$10x - 84 = 276$$
$$10x = 360$$
$$x = 36$$

The solution set is $\{36\}$.

## Solving Word Problems, Including Discount and Selling Price Problems

We can solve many consumer problems with an algebraic approach. For example, let's consider some discount sale problems involving the relationship, *original selling price minus discount equals discount sale price.*

Original selling price − Discount = Discount sale price

**Classroom Example**
Karyn bought a coat at a 25% discount sale for \$97.50. What was the original price of the coat?

## EXAMPLE 5 Apply Your Skill

Karyl bought a dress at a 35% discount sale for \$32.50. What was the original price of the dress?

### Solution

Let $p$ represent the original price of the dress. Using the discount sale relationship as a guideline, we find that the problem translates into an equation as follows:

Switching this equation to decimal form and solving the equation, we obtain

$$p - (35\%)(p) = 32.50$$
$$p - 0.35p = 32.50$$
$$0.65p = 32.50 \quad \text{Remember, } p = 1.00p$$
$$p = 50$$

The original price of the dress was \$50.

**Classroom Example**
John received a coupon from an electronic store that offered 15% off any item. If he uses the coupon, how much will he have to pay for a sound system that is priced at \$699?

## EXAMPLE 6 Apply Your Skill

Jason received a private mailing coupon from an electronic store that offered 12% off any item. If he uses the coupon, how much will he have to pay for a laptop computer that is priced at \$980?

### Solution

Let $s$ represent the discount sale price.

| Original price | Minus | Discount | Equals | Sale price |
|---|---|---|---|---|
| ↓ | | ↓ | | ↓ |
| \$980 | − | (12%)(\$980) | = | $s$ |

Solving this equation we obtain

$$980 - (12\%)(980) = s$$
$$980 - (0.12)(980) = s$$
$$980 - 117.60 = s$$
$$862.40 = s$$

With the coupon, Jason will pay \$862.40 for the laptop computer.

**Remark:** Keep in mind that if an item is on sale for 35% off, then the purchaser will pay $100\% - 35\% = 65\%$ of the original price. Thus in Example 5 you could begin with the equation $0.65p = 32.50$. Likewise in Example 6 you could start with the equation $s = 0.88(980)$.

Another basic relationship that pertains to consumer problems is *selling price equals cost plus profit*. We can state profit (also called markup, markon, and margin of profit) in different ways. Profit may be stated as a percent of the selling price, as a percent of the cost, or simply in terms of dollars and cents. We shall consider some problems for which the profit is calculated either as a percent of the cost or as a percent of the selling price.

$$\text{Selling price} = \text{Cost} + \text{Profit}$$

**Classroom Example**
Shauna bought antique bowls for \$270. She wants to resell the bowls and make a profit of 30% of the cost. What price should Shauna list to make her profit?

## EXAMPLE 7 Apply Your Skill

Heather bought some artwork at an online auction for \$400. She wants to resell the artwork on line and make a profit of 40% of the cost. What price should Heather list on line to make her profit?

### Solution

Let $s$ represent the selling price. Use the relationship *selling price equals cost plus profit* as a guideline.

| Selling price | Equals | Cost | Plus | Profit |
|---|---|---|---|---|
| ↓ | | ↓ | | ↓ |
| $s$ | $=$ | \$400 | $+$ | (40%)(\$400) |

Solving this equation yields

$$s = 400 + (40\%)(400)$$
$$s = 400 + (0.4)(400)$$
$$s = 400 + 160$$
$$s = 560$$

The selling price should be \$560.

**Remark:** A profit of 40% of the cost means that the selling price is 100% of the cost plus 40% of the cost, or 140% of the cost. Thus in Example 7 we could solve the equation $s = 1.4(400)$.

**Classroom Example**
A college bookstore bought official collegiate sweatshirts for \$12 each. At what price should the bookstore sell the sweatshirts to make a profit of 70% of the selling price?

## EXAMPLE 8 Apply Your Skill

A college bookstore purchased math textbooks for \$54 each. At what price should the bookstore sell the books if the bookstore wants to make a profit of 60% of the selling price?

### Solution

Let $s$ represent the selling price.

| Selling price | Equals | Cost | Plus | Profit |
|---|---|---|---|---|
| ↓ | | ↓ | | ↓ |
| $s$ | $=$ | \$54 | $+$ | (60%)($s$) |

Solving this equation yields

$$s = 54 + (60\%)(s)$$
$$s = 54 + 0.6s$$
$$0.4s = 54$$
$$s = 135$$

The selling price should be \$135.

Janusz Wrobel/Alamy

**Classroom Example**
If an antique desk cost a collector \$75, and she sells it for \$120, what is her rate of profit based on the cost?

## EXAMPLE 9 Apply Your Skill

If a maple tree costs a landscaper \$40.00, and he sells it for \$75.00, what is his rate of profit based on the cost? Express the rate to the nearest tenth of a percent.

### Solution

Let $r$ represent the rate of profit, and use the following guideline.

| Selling price | Equals | Cost | Plus | Profit (rate × cost) |
|---|---|---|---|---|
| ↓ | | ↓ | | ↓ |
| 75.00 | = | 40.00 | + | $r(40.00)$ |

Solving this equation yields

$$75.00 = 40 + r(40.00)$$
$$35.00 = r(40.00)$$
$$\frac{35.00}{40.00} = r$$
$$0.875 = r$$

To change the answer to a percent, multiply 0.875 by 100 and attach the % sign. Thus his rate of profit is 87.5%.

We can solve certain types of investment and money problems by using an algebraic approach. Consider the following examples.

Burke/Triolo Productions/Brand X Pictures/Jupiter Images

**Classroom Example**
Erin has 28 coins, consisting only of dimes and quarters, worth \$4.60. How many dimes and how many quarters does she have?

## EXAMPLE 10 Apply Your Skill

Erick has 40 coins, consisting only of dimes and nickels, worth \$3.35. How many dimes and how many nickels does he have?

### Solution

Let $x$ represent the number of dimes. Then the number of nickels can be represented by the total number of coins minus the number of dimes. Hence $40 - x$ represents the number of nickels. Because we know the amount of money Erick has, we need to multiply the number of each coin by its value. Use the following guideline.

| Money from the dimes | Plus | Money from the nickels | Equals | Total money |
|---|---|---|---|---|
| ↓ | | ↓ | | ↓ |
| $0.10x$ | + | $0.05(40 - x)$ | = | 3.35 |

Solving this equation yields

$$0.10x + 0.05(40 - x) = 3.35$$
$$10x + 5(40 - x) = 335 \quad \text{Multiply both sides by 100}$$
$$10x + 200 - 5x = 335$$
$$5x + 200 = 335$$
$$5x = 135$$
$$x = 27$$

The number of dimes is 27, and the number of nickels is $40 - x = 13$. So, Erick has 27 dimes and 13 nickels.

Steve Allen/Brand X Pictures/Jupiter Images

**Classroom Example**
A woman invests \$12,000, part of it at 2% and the remainder at 3%. Her total yearly interest from the two investments is \$304. How much did she invest at each rate?

**EXAMPLE 11** Apply Your Skill

A man invests \$8000, part of it at 6% and the remainder at 8%. His total yearly interest from the two investments is \$580. How much did he invest at each rate?

### Solution

Let $x$ represent the amount he invested at 6%. Then $8000 - x$ represents the amount he invested at 8%. Use the following guideline.

| Interest earned from 6% investment | + | Interest earned from 8% investment | = | Total amount of interest earned |
|---|---|---|---|---|
| ↓ | | ↓ | | ↓ |
| $(6\%)(x)$ | + | $(8\%)(8000 - x)$ | = | \$580 |

Solving this equation yields

$$(6\%)(x) + (8\%)(8000 - x) = 580$$
$$0.06x + 0.08(8000 - x) = 580$$
$$6x + 8(8000 - x) = 58{,}000 \quad \text{Multiply both sides by 100}$$
$$6x + 64{,}000 - 8x = 58{,}000$$
$$-2x + 64{,}000 = 58{,}000$$
$$-2x = -6000$$
$$x = 3000$$

Therefore, \$3000 was invested at 6%, and \$8000 − \$3000 = \$5000 was invested at 8%.

Don't forget to check word problems; determine whether the answers satisfy the conditions stated in the *original* problem. A check for Example 11 follows.

### ✔ Check

We claim that \$3000 is invested at 6% and \$5000 at 8%, and this satisfies the condition that \$8000 is invested. The \$3000 at 6% produces \$180 of interest, and the \$5000 at 8% produces \$400. Therefore, the interest from the investments is \$580. The conditions of the problem are satisfied, and our answers are correct.

As you tackle word problems throughout this text, keep in mind that our primary objective is to expand your repertoire of problem-solving techniques. We have chosen problems that provide you with the opportunity to use a variety of approaches to solving problems. Don't fall into the trap of thinking "I will never be faced with this kind of problem." That is not the issue; the goal is to develop problem-solving techniques. In the examples we are sharing some of our ideas for solving problems, but don't hesitate to use your own ingenuity. Furthermore, don't become discouraged—all of us have difficulty with some problems. Give each your best shot!

## Concept Quiz 2.3

For Problems 1–10, answer true or false.

1. To solve an equation involving decimals, you must first multiply both sides of the equation by a power of 10.
2. When using the formula "selling price = cost + profit" the profit is always a percentage of the cost.
3. If Kim bought a putter for \$50 and then sold it to a friend for \$60, her rate of profit based on the cost was 10%.

4. To determine the selling price when the profit is a percent of the selling price, you can subtract the percent of profit from 100% and then divide the cost by that result.
5. If an item is bought for \$30, then it should be sold for \$37.50 in order to obtain a profit of 20% based on the selling price.
6. A discount of 10% followed by a discount of 20% is the same as a discount of 30%.
7. If an item is bought for \$25, then it should be sold for \$30 in order to obtain a profit of 20% based on the cost.
8. To solve the equation $0.4x + 0.15 = 0.06x + 0.71$, you can start by multiplying both sides of the equation by 100.
9. A 10% discount followed by a 40% discount is the same as a 40% discount followed by a 10% discount.
10. Multiplying both sides of the equation $0.4(x - 1.2) = 0.6$ by 10 produces the equivalent equation $4(x - 12) = 6$.

## Problem Set 2.3

For Problems 1–28, solve each equation. (Objective 1)

1. $0.14x = 2.8$
2. $1.6x = 8$
3. $0.09y = 4.5$
4. $0.07y = 0.42$
5. $n + 0.4n = 56$
6. $n - 0.5n = 12$
7. $s = 9 + 0.25s$
8. $s = 15 + 0.4s$
9. $s = 3.3 + 0.45s$
10. $s = 2.1 + 0.6s$
11. $0.11x + 0.12(900 - x) = 104$
12. $0.09x + 0.11(500 - x) = 51$
13. $0.08(x + 200) = 0.07x + 20$
14. $0.07x = 152 - 0.08(2000 - x)$
15. $0.12t - 2.1 = 0.07t - 0.2$
16. $0.13t - 3.4 = 0.08t - 0.4$
17. $0.92 + 0.9(x - 0.3) = 2x - 5.95$
18. $0.3(2n - 5) = 11 - 0.65n$
19. $0.1d + 0.11(d + 1500) = 795$
20. $0.8x + 0.9(850 - x) = 715$
21. $0.12x + 0.1(5000 - x) = 560$
22. $0.10t + 0.12(t + 1000) = 560$
23. $0.09(x + 200) = 0.08x + 22$
24. $0.09x = 1650 - 0.12(x + 5000)$
25. $0.3(2t + 0.1) = 8.43$
26. $0.5(3t + 0.7) = 20.6$
27. $0.1(x - 0.1) - 0.4(x + 2) = -5.31$
28. $0.2(x + 0.2) + 0.5(x - 0.4) = 5.44$

For Problems 29–50, use an algebraic approach to solve each problem. (Objective 2)

29. Judy bought a coat at a 20% discount sale for \$72. What was the original price of the coat?
30. Jim bought a pair of jeans at a 25% discount sale for \$45. What was the original price of the jeans?
31. Find the discount sale price of a \$64 item that is on sale for 15% off.
32. Find the discount sale price of a \$72 item that is on sale for 35% off.
33. A retailer has some skirts that cost \$30 each. She wants to sell them at a profit of 60% of the cost. What price should she charge for the skirts?
34. The owner of a pizza parlor wants to make a profit of 70% of the cost for each pizza sold. If it costs \$7.50 to make a pizza, at what price should each pizza be sold?
35. If a ring costs a jeweler \$1200, at what price should it be sold to yield a profit of 50% on the selling price?
36. If a head of lettuce costs a retailer \$0.68, at what price should it be sold to yield a profit of 60% on the selling price?
37. If a pair of shoes costs a retailer \$24, and he sells them for \$39.60, what is his rate of profit based on the cost?
38. A retailer has some jackets that cost her \$45 each. If she sells them for \$83.25 per jacket, find her rate of profit based on the cost.
39. If a computer costs an electronics dealer \$300, and she sells them for \$800, what is her rate of profit based on the selling price?

**40.** A textbook costs a bookstore \$75, and the store sells it for \$93.75. Find the rate of profit based on the selling price.

**41.** Mitsuko's salary for next year is \$44,940. This represents a 7% increase over this year's salary. Find Mitsuko's present salary.

**42.** Don bought a used car for \$15,794, with 6% tax included. What was the price of the car without the tax?

**43.** Eva invested a certain amount of money at 4% interest and \$1500 more than that amount at 6%. Her total yearly interest was \$390. How much did she invest at each rate?

**44.** A total of \$4000 was invested, part of it at 5% interest and the remainder at 6%. If the total yearly interest amounted to \$230, how much was invested at each rate?

**45.** A sum of \$95,000 is split between two investments, one paying 3% and the other 5%. If the total yearly interest amounted to \$3910, how much was invested at 5%?

**46.** If \$1500 is invested at 2% interest, how much money must be invested at 4% so that the total return for both investments is \$100?

**47.** Suppose that Javier has a handful of coins, consisting of pennies, nickels, and dimes, worth \$2.63. The number of nickels is 1 less than twice the number of pennies, and the number of dimes is 3 more than the number of nickels. How many coins of each kind does he have?

**48.** Sarah has a collection of nickels, dimes, and quarters worth \$15.75. She has 10 more dimes than nickels and twice as many quarters as dimes. How many coins of each kind does she have?

**49.** A collection of 70 coins consisting of dimes, quarters, and half-dollars has a value of \$17.75. There are three times as many quarters as dimes. Find the number of each kind of coin.

**50.** Abby has 37 coins, consisting only of dimes and quarters, worth \$7.45. How many dimes and how many quarters does she have?

## Thoughts Into Words

**51.** Go to Problem 39 and calculate the rate of profit based on cost. Compare the rate of profit based on cost to the rate of profit based on selling price. From a consumer's viewpoint, would you prefer that a retailer figure his profit on the basis of the cost of an item or on the basis of its selling price? Explain your answer.

**52.** Is a 10% discount followed by a 30% discount the same as a 30% discount followed by a 10% discount? Justify your answer.

**53.** What is wrong with the following solution and how should it be done?

$$\begin{aligned} 1.2x + 2 &= 3.8 \\ 10(1.2x) + 2 &= 10(3.8) \\ 12x + 2 &= 38 \\ 12x &= 36 \\ x &= 3 \end{aligned}$$

## Further Investigations

For Problems 54–63, solve each equation and express the solutions in decimal form. Be sure to check your solutions. Use your calculator whenever it seems helpful.

**54.** $1.2x + 3.4 = 5.2$

**55.** $0.12x - 0.24 = 0.66$

**56.** $0.12x + 0.14(550 - x) = 72.5$

**57.** $0.14t + 0.13(890 - t) = 67.95$

**58.** $0.7n + 1.4 = 3.92$

**59.** $0.14n - 0.26 = 0.958$

**60.** $0.3(d + 1.8) = 4.86$

**61.** $0.6(d - 4.8) = 7.38$

**62.** $0.8(2x - 1.4) = 19.52$

**63.** $0.5(3x + 0.7) = 20.6$

**64.** The following formula can be used to determine the selling price of an item when the profit is based on a percent of the selling price.

$$\text{Selling price} = \frac{\text{Cost}}{100\% - \text{Percent of profit}}$$

Show how this formula is developed.

**65.** A retailer buys an item for \$90, resells it for \$100, and claims that she is making only a 10% profit. Is this claim correct?

**66.** Is a 10% discount followed by a 20% discount equal to a 30% discount? Defend your answer.

## Answers to the Concept Quiz

**1.** False **2.** False **3.** False **4.** True **5.** True **6.** False **7.** True **8.** True **9.** True **10.** False

# 2.4 Formulas

OBJECTIVES

1. Evaluate formulas for given values
2. Solve formulas for a specified variable
3. Use formulas to solve problems

To find the distance traveled in 4 hours at a rate of 55 miles per hour, we multiply the rate times the time; thus the distance is $55(4) = 220$ miles. We can state the rule *distance equals rate times time* as a formula, $d = rt$. **Formulas** are rules we state in symbolic form, usually as equations.

Formulas are typically used in two different ways. At times a formula is solved for a specific variable when we are given the numerical values for the other variables. This is much like evaluating an algebraic expression. At other times we need to change the form of an equation by solving for one variable in terms of the other variables. Throughout our work on formulas, we will use the properties of equality and the techniques we have previously learned for solving equations. Let's consider some examples.

**Classroom Example**
If we invest $P$ dollars at $r$ percent for $t$ years, the amount of simple interest $i$ is given by the formula $i = Prt$. Find the amount of interest earned by \$400 at 3% for 3 years.

### EXAMPLE 1

If we invest $P$ dollars at $r$ percent for $t$ years, the amount of simple interest $i$ is given by the formula $i = Prt$. Find the amount of interest earned by \$5000 invested at 4% for 2 years.

### Solution

By substituting \$5000 for $P$, 4% for $r$, and 2 for $t$, we obtain

$$i = Prt$$
$$i = (5000)(4\%)(2)$$
$$i = (5000)(0.04)(2)$$
$$i = 400$$

Thus we earn \$400 in interest.

**Classroom Example**
If we invest $P$ dollars at a simple rate of $r$ percent, then the amount $A$ accumulated after $t$ years is given by the formula $A = P + Prt$. If we invest \$2500 at 4%, how many years will it take to accumulate \$3000?

### EXAMPLE 2

If we invest $P$ dollars at a simple rate of $r$ percent, then the amount $A$ accumulated after $t$ years is given by the formula $A = P + Prt$. If we invest \$5000 at 5%, how many years will it take to accumulate \$6000?

### Solution

Substituting \$5000 for $P$, 5% for $r$, and \$6000 for $A$, we obtain

$$A = P + Prt$$
$$6000 = 5000 + 5000(5\%)(t)$$

Solving this equation for $t$ yields

$$6000 = 5000 + 5000(0.05)(t)$$
$$6000 = 5000 + 250t$$
$$1000 = 250t$$
$$4 = t$$

It will take 4 years to accumulate \$6000.

## Solving Formulas for a Specified Variable

When we are using a formula, it is sometimes necessary to change its form. If we wanted to use a calculator or a spreadsheet to complete the following chart, we would have to solve the perimeter formula for a rectangle ($P = 2l + 2w$) for $w$.

| Perimeter (P) | 32 | 24 | 36 | 18 | 56 | 80 |
|---|---|---|---|---|---|---|
| Length (l) | 10 | 7 | 14 | 5 | 15 | 22 |
| Width (w) | ? | ? | ? | ? | ? | ? |

All in centimeters

To perform the computational work or enter the formula into a spreadsheet, we would first solve the formula for $w$.

$$P = 2l + 2w$$
$$P - 2l = 2w \qquad \text{Add } -2l \text{ to both sides}$$
$$\frac{P - 2l}{2} = \frac{2w}{2} \qquad \text{Divide both sides by 2}$$
$$\frac{P - 2l}{2} = w$$
$$w = \frac{P - 2l}{2} \qquad \text{Apply the symmetric property of equality}$$

Now for each value for $P$ and $l$, we can easily determine the corresponding value for $w$. Be sure you agree with the following values for $w$: 6, 5, 4, 4, 13, and 18. Likewise we can also solve the formula $P = 2l + 2w$ for $l$ in terms of $P$ and $w$. The result would be $l = \frac{P - 2w}{2}$.

Let's consider some other often-used formulas and see how we can use the properties of equality to alter their forms. Here we will be solving a formula for a specified variable in terms of the other variables. The key is to isolate the term that contains the variable being solved for. Then, by appropriately applying the multiplication property of equality, we will solve the formula for the specified variable. Throughout this section, we will identify formulas when we first use them. (Some geometric formulas are also given on the endsheets.)

**Classroom Example**
Solve $V = \frac{1}{3}Bh$ for $B$ (volume of a pyramid).

**EXAMPLE 3** Solve $A = \frac{1}{2}bh$ for $h$ (area of a triangle).

### Solution

$$A = \frac{1}{2}bh$$
$$2A = bh \qquad \text{Multiply both sides by 2}$$
$$\frac{2A}{b} = \frac{bh}{b} \qquad \text{Divide both sides by } b$$

$$\frac{2A}{b} = h$$

$$h = \frac{2A}{b} \quad \text{Apply the symmetric property of equality}$$

Classroom Example
Solve $P = S - Sdt$ for $d$.

**EXAMPLE 4** Solve $A = P + Prt$ for $t$.

## Solution

$$A = P + Prt$$

$$A - P = Prt \quad \text{Add } -P \text{ to both sides}$$

$$\frac{A - P}{Pr} = \frac{Prt}{Pr} \quad \text{Divide both sides by Pr}$$

$$\frac{A - P}{Pr} = t$$

$$t = \frac{A - P}{Pr} \quad \text{Apply the symmetric property of equality}$$

Classroom Example
Solve $P = S - Sdt$ for $S$.

**EXAMPLE 5** Solve $A = P + Prt$ for $P$.

## Solution

$$A = P + Prt$$

$$A = P(1 + rt) \quad \text{Apply the distributive property to the right side}$$

$$\frac{A}{1 + rt} = \frac{P(1 + rt)}{(1 + rt)} \quad \text{Divide both sides by } (1 + rt)$$

$$\frac{A}{1 + rt} = P$$

$$P = \frac{A}{1 + rt} \quad \text{Apply the symmetric property of equality}$$

Classroom Example
Solve $A = \frac{1}{2}h(b_1 + b_2)$ for $b_2$.

**EXAMPLE 6** Solve $A = \frac{1}{2}h(b_1 + b_2)$ for $b_1$ (area of a trapezoid).

## Solution

Note the use of subscripts to identify the two bases of a trapezoid. Subscripts enable us to use the same letter $b$ to identify the bases, but $b_1$ represents one base and $b_2$ the other.

$$A = \frac{1}{2}h(b_1 + b_2)$$

$$2A = h(b_1 + b_2) \quad \text{Multiply both sides by 2}$$

$$2A = hb_1 + hb_2 \quad \text{Apply the distributive property to right side}$$

$$2A - hb_2 = hb_1 \quad \text{Add } -hb_2 \text{ to both sides}$$

$$\frac{2A - hb_2}{h} = \frac{hb_1}{h} \quad \text{Divide both sides by } h$$

$$\frac{2A - hb_2}{h} = b_1$$

$$b_1 = \frac{2A - hb_2}{h} \quad \text{Apply the symmetric property of equality}$$

In order to isolate the term containing the variable being solved for, we will apply the distributive property in different ways. In Example 5 you *must* use the distributive property to change from the form $P + Prt$ to $P(1 + rt)$. However, in Example 6 we used the distributive property to change $h(b_1 + b_2)$ to $hb_1 + hb_2$. In both problems the key is to isolate the term that contains the variable being solved for, so that an appropriate application of the multiplication property of equality will produce the desired result.

Sometimes we are faced with equations such as $ax + b = c$, where $x$ is the variable and $a$, $b$, and $c$ are referred to as *arbitrary constants*. Again we can use the properties of equality to solve the equation for $x$ as follows:

$$ax + b = c$$

$$ax = c - b \quad \text{Add } -b \text{ to both sides}$$

$$x = \frac{c - b}{a} \quad \text{Divide both sides by } a$$

In Chapter 7, we will be working with equations such as $2x - 5y = 7$, which are called equations of two variables in $x$ and $y$. Often we need to change the form of such equations by solving for one variable in terms of the other variable. The properties of equality provide the basis for doing this.

**Classroom Example**
Solve $2x - 5y = 7$ for $x$ in terms of $y$.

**EXAMPLE 7** Solve $2x - 5y = 7$ for $y$ in terms of $x$.

### Solution

$$2x - 5y = 7$$

$$-5y = 7 - 2x \quad \text{Add } -2x \text{ to both sides}$$

$$y = \frac{7 - 2x}{-5} \quad \text{Divide both sides by } -5$$

$$y = \frac{2x - 7}{5}$$

Multiply the numerator and denominator of the fraction on the right by $-1$ (This final step is not absolutely necessary, but usually we prefer to have a positive number as a denominator)

Equations of two variables may also contain arbitrary constants. For example, the equation $\frac{x}{a} + \frac{y}{b} = 1$ contains the variables $x$ and $y$ and the arbitrary constants $a$ and $b$.

**Classroom Example**
Solve the equation $\frac{x}{a} - \frac{y}{b} = 1$ for $y$.

**EXAMPLE 8** Solve the equation $\frac{x}{a} + \frac{y}{b} = 1$ for $x$.

### Solution

$$\frac{x}{a} + \frac{y}{b} = 1$$

$$ab\left(\frac{x}{a} + \frac{y}{b}\right) = ab(1) \quad \text{Multiply both sides by } ab$$

$$bx + ay = ab$$

$$bx = ab - ay \quad \text{Add } -ay \text{ to both sides}$$

$$x = \frac{ab - ay}{b} \quad \text{Divide both sides by } b$$

**Remark:** Traditionally, equations that contain more than one variable, such as those in Examples 3–8, are called **literal equations**. As illustrated, it is sometimes necessary to solve a literal equation for one variable in terms of the other variable(s).

## Using Formulas to Solve Problems

We often use formulas as guidelines for setting up an appropriate algebraic equation when solving a word problem. Let's consider an example to illustrate this point.



**Classroom Example**
How long will it take \$400 to double itself if we invest it at 4% simple interest?

### EXAMPLE 9 Apply Your Skill

How long will it take \$1000 to double itself if we invest it at 5% simple interest?

### Solution

For \$1000 to grow into \$2000 (double itself), it must earn \$1000 in interest. Thus we let $t$ represent the number of years it will take \$1000 to earn \$1000 in interest. Now we can use the formula $i = Prt$ as a guideline.

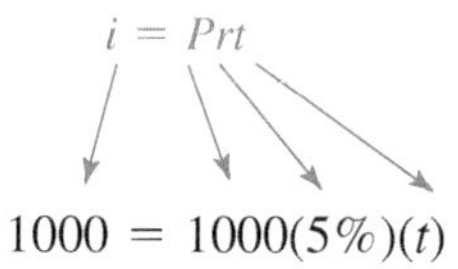


$$1000 = 1000(5\%)(t)$$

Solving this equation, we obtain

$$1000 = 1000(0.05)(t)$$
$$1000 = 50t$$
$$20 = t \qquad \text{Divided both sides by 50}$$

It will take 20 years.

Sometimes we use formulas in the analysis of a problem but not as the main guideline for setting up the equation. For example, although uniform-motion problems involve the formula $d = rt$, the main guideline for setting up an equation for such problems is usually a statement about times, rates, or distances. Let's consider an example to demonstrate.

**Classroom Example**
Latesha starts jogging at 3 miles per hour. Twenty minutes later, Sean starts jogging on the same route at 5 miles per hour. How long will it take Sean to catch Latesha?

### EXAMPLE 10 Apply Your Skill

Mercedes starts jogging at 5 miles per hour. One-half hour later, Karen starts jogging on the same route at 7 miles per hour. How long will it take Karen to catch Mercedes?

### Solution

First, let's sketch a diagram and record some information (Figure 2.3).

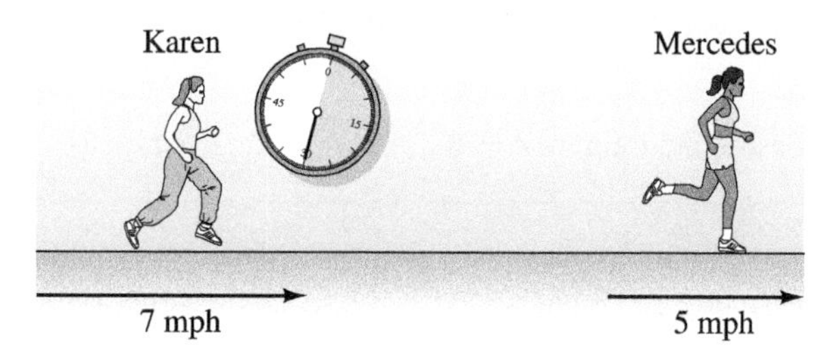


**Figure 2.3**

If we let $t$ represent Karen's time, then $t + \frac{1}{2}$ represents Mercedes' time. We can use the statement *Karen's distance equals Mercedes' distance* as a guideline.

$$7t = 5\left(t + \frac{1}{2}\right)$$

Solving this equation, we obtain

$$7t = 5t + \frac{5}{2}$$

$$2t = \frac{5}{2}$$

$$\frac{1}{2}(2t) = \frac{1}{2}\left(\frac{5}{2}\right) \qquad \text{Multiply both sides by } \frac{1}{2}$$

$$t = \frac{5}{4}$$

Karen should catch Mercedes in $1\frac{1}{4}$ hours.

**Remark:** An important tool for problem solving is sketching a meaningful figure that can be used to record the given information and help in the analysis of the problem. Our sketches were done by professional artists for aesthetic purposes. Your sketches can be very roughly drawn as long as they depict the situation in a way that helps you analyze the problem.

Note that in the solution of Example 10 we used a figure and a simple arrow diagram to record and organize the information pertinent to the problem. Some people find it helpful to use a chart for that purpose. We shall use a chart in Example 11. Keep in mind that we are not trying to dictate a particular approach; you decide what works best for you.

Goestock/Photodisc/Getty Images

**Classroom Example**
Two buses leave a city at the same time, one traveling east and the other traveling west. At the end of $5\frac{1}{2}$ hours, they are 671 miles apart. If the rate of the bus traveling east is 6 miles slower than the rate of the other bus, find their rates.

## EXAMPLE 11 Apply Your Skill

Two trains leave a city at the same time, one traveling east and the other traveling west. At the end of $9\frac{1}{2}$ hours, they are 1292 miles apart. If the rate of the train traveling east is 8 miles per hour faster than the rate of the other train, find their rates.

### Solution

If we let $r$ represent the rate of the westbound train, then $r + 8$ represents the rate of the eastbound train. Now we can record the times and rates in a chart and then use the distance formula ($d = rt$) to represent the distances.

| | Rate | Time | Distance ($d = rt$) |
|---|---|---|---|
| Westbound train | $r$ | $9\frac{1}{2}$ | $\frac{19}{2}r$ |
| Eastbound train | $r + 8$ | $9\frac{1}{2}$ | $\frac{19}{2}(r + 8)$ |

Because the distance that the westbound train travels plus the distance that the eastbound train travels equals 1292 miles, we can set up and solve the following equation.

$$\text{Eastbound distance} + \text{Westbound distance} = \text{Miles apart}$$

$$\frac{19r}{2} + \frac{19(r + 8)}{2} = 1292$$

$$19r + 19(r + 8) = 2584$$

$$19r + 19r + 152 = 2584$$

$$38r = 2432$$

$$r = 64$$

The westbound train travels at a rate of 64 miles per hour, and the eastbound train travels at a rate of $64 + 8 = 72$ miles per hour.

Now let's consider a problem that is often referred to as a mixture problem. There is no basic formula that applies to all of these problems, but we suggest that you think in terms of a pure substance, which is often helpful in setting up a guideline. Also keep in mind that the phrase "a 40% solution of some substance" means that the solution contains 40% of that particular substance and 60% of something else mixed with it. For example, a 40% salt solution contains 40% salt, and the other 60% is something else, probably water. Now let's illustrate what we mean by suggesting that you think in terms of a pure substance.

Don Mason/Brand X Pictures/ Jupiter Images

**Classroom Example**
Larson's Nursery stocks a 10% solution of herbicide and a 22% solution of herbicide. How many liters of each should be mixed to produce 20 liters of an 18% solution of herbicide?

## EXAMPLE 12 Apply Your Skill

Bryan's Pest Control stocks a 7% solution of insecticide for lawns and also a 15% solution. How many gallons of each should be mixed to produce 40 gallons that is 12% insecticide?

### Solution

The key idea in solving such a problem is to recognize the following guideline.

$$\begin{pmatrix}\text{Amount of insecticide}\\ \text{in the 7\% solution}\end{pmatrix} + \begin{pmatrix}\text{Amount of insecticide}\\ \text{in the 15\% solution}\end{pmatrix} = \begin{pmatrix}\text{Amount of insecticide in}\\ \text{40 gallons of 12\% solution}\end{pmatrix}$$

Let $x$ represent the gallons of 7% solution. Then $40 - x$ represents the gallons of 15% solution. The guideline translates into the following equation.

$$(7\%)(x) + (15\%)(40 - x) = 12\%(40)$$

Solving this equation yields

$$\begin{aligned} 0.07x + 0.15(40 - x) &= 0.12(40) \\ 7x + 15(40 - x) &= 12(40) \quad \text{Multiply both sides by 100} \\ 7x + 600 - 15x &= 480 \\ -8x + 600 &= 480 \\ -8x &= -120 \\ x &= 15 \end{aligned}$$

Thus 15 gallons of 7% solution and $40 - x = 25$ gallons of 15% solution need to be mixed to obtain 40 gallons of 12% solution.

Yuri Arcurs/Shutterstock.com

**Classroom Example**
How many gallons of pure antifreeze must be added to 12 gallons of a 30% solution to obtain a 70% solution?

## EXAMPLE 13 Apply Your Skill

How many liters of pure alcohol must we add to 20 liters of a 40% solution to obtain a 60% solution?

### Solution

The key idea in solving such a problem is to recognize the following guideline.

$$\begin{pmatrix}\text{Amount of pure}\\ \text{alcohol in the}\\ \text{original solution}\end{pmatrix} + \begin{pmatrix}\text{Amount of}\\ \text{pure alcohol}\\ \text{to be added}\end{pmatrix} = \begin{pmatrix}\text{Amount of pure}\\ \text{alcohol in the}\\ \text{final solution}\end{pmatrix}$$

Let $l$ represent the number of liters of pure alcohol to be added, and the guideline translates into the following equation.

$$(40\%)(20) + l = 60\%(20 + l)$$

Solving this equation yields

$$0.4(20) + l = 0.6(20 + l)$$
$$8 + l = 12 + 0.6l$$
$$0.4l = 4$$
$$l = 10$$

We need to add 10 liters of pure alcohol. (Remember to check this answer back into the original statement of the problem.)

## Concept Quiz 2.4

For Problems 1–10, answer true or false.

1. Formulas are rules stated in symbolic form, usually as algebraic expressions.
2. The properties of equality that apply to solving equations also apply to solving formulas.
3. The formula $A = P + Prt$ can be solved for $r$ or $t$ but not for $P$.
4. The formula $i = Prt$ is equivalent to $P = \frac{i}{rt}$.
5. The equation $y = mx + b$ is equivalent to $x = \frac{y - b}{m}$.
6. The formula $\text{F} = \frac{9}{5}\text{C} + 32$ is equivalent to $\text{C} = \frac{5}{9}(\text{F} - 32)$.
7. Using the formula $\text{F} = \frac{9}{5}\text{C} + 32$, a temperature of 30° Celsius is equal to 86° Fahrenheit.
8. Using the formula $\text{C} = \frac{5}{9}(\text{F} - 32)$, a temperature of 32° Fahrenheit is equal to 0° Celsius.
9. The amount of pure acid in 30 ounces of a 20% acid solution is 10 ounces.
10. For an equation such as $ax + b = c$, where $x$ is the variable, $a$, $b$, and $c$ are referred to as arbitrary constants.

## Problem Set 2.4

For Problems 1–16, use the formula to solve for the given variable. (Objective 1)

1. Solve $i = Prt$ for $i$, given that $P = \$3000$, $r = 4\%$, and $t = 5$ years.
2. Solve $i = Prt$ for $i$, given that $P = \$5000$, $r = 6\%$, and $t = 3\frac{1}{2}$ years.
3. Solve $i = Prt$ for $t$, given that $P = \$4000$, $r = 5\%$, and $i = \$600$.
4. Solve $i = Prt$ for $t$, given that $P = \$1250$, $r = 3\%$, and $i = \$150$.
5. Solve $i = Prt$ for $r$, given that $P = \$600$, $t = 2\frac{1}{2}$ years, and $i = \$90$. Express $r$ as a percent.
6. Solve $i = Prt$ for $r$, given that $P = \$700$, $t = 2$ years, and $i = \$84$. Express $r$ as a percent.
7. Solve $i = Prt$ for $P$, given that $r = 9\%$, $t = 3$ years, and $i = \$216$.
8. Solve $i = Prt$ for $P$, given that $r = 8\frac{1}{2}\%$, $t = 2$ years, and $i = \$204$.
9. Solve $A = P + Prt$ for $A$, given that $P = \$1000$, $r = 7\%$, and $t = 5$ years.
10. Solve $A = P + Prt$ for $A$, given that $P = \$850$, $r = 4\frac{1}{2}\%$, and $t = 10$ years.

**11.** Solve $A = P + Prt$ for $r$, given that $A = \$1372$, $P = \$700$, and $t = 12$ years. Express $r$ as a percent.

**12.** Solve $A = P + Prt$ for $r$, given that $A = \$516$, $P = \$300$, and $t = 8$ years. Express $r$ as a percent.

**13.** Solve $A = P + Prt$ for $P$, given that $A = \$326$, $r = 7\%$, and $t = 9$ years.

**14.** Solve $A = P + Prt$ for $P$, given that $A = \$720$, $r = 8\%$, and $t = 10$ years.

**15.** Solve the formula $A = \frac{1}{2}h(b_1 + b_2)$ for $b_2$ and complete the following chart.

| $A$ | 98 | 104 | 49 | 162 | $16\frac{1}{2}$ | $38\frac{1}{2}$ | square feet |
|---|---|---|---|---|---|---|---|
| $h$ | 14 | 8 | 7 | 9 | 3 | 11 | feet |
| $b_1$ | 8 | 12 | 4 | 16 | 4 | 5 | feet |
| $b_2$ | ? | ? | ? | ? | ? | ? | feet |

$A$ = area, $h$ = height, $b_1$ = one base, $b_2$ = other base

**16.** Solve the formula $P = 2l + 2w$ for $l$ and complete the following chart.

| $P$ | 28 | 18 | 12 | 34 | 68 | centimeters |
|---|---|---|---|---|---|---|
| $w$ | 6 | 3 | 2 | 7 | 14 | centimeters |
| $l$ | ? | ? | ? | ? | ? | centimeters |

$P$ = perimeter, $w$ = width, $l$ = length

For Problems 17–26, solve each of the following for the indicated variable. (Objective 2)

**17.** $V = Bh$ for $h$ (Volume of a prism)

**18.** $A = lw$ for $l$ (Area of a rectangle)

**19.** $V = \pi r^2 h$ for $h$ (Volume of a circular cylinder)

**20.** $V = \frac{1}{3}Bh$ for $B$ (Volume of a pyramid)

**21.** $C = 2\pi r$ for $r$ (Circumference of a circle)

**22.** $A = 2\pi r^2 + 2\pi rh$ for $h$ (Surface area of a circular cylinder)

**23.** $\text{I} = \frac{100M}{C}$ for $C$ (Intelligence quotient)

**24.** $A = \frac{1}{2}h(b_1 + b_2)$ for $h$ (Area of a trapezoid)

**25.** $\text{F} = \frac{9}{5}\text{C} + 32$ for C (Celsius to Fahrenheit)

**26.** $\text{C} = \frac{5}{9}(\text{F} - 32)$ for F (Fahrenheit to Celsius)

For Problems 27–36, solve each equation for $x$. (Objective 2)

**27.** $y = mx + b$

**28.** $\frac{x}{a} + \frac{y}{b} = 1$

**29.** $y - y_1 = m(x - x_1)$

**30.** $a(x + b) = c$

**31.** $a(x + b) = b(x - c)$

**32.** $x(a - b) = m(x - c)$

**33.** $\frac{x - a}{b} = c$

**34.** $\frac{x}{a} - 1 = b$

**35.** $\frac{1}{3}x + a = \frac{1}{2}b$

**36.** $\frac{2}{3}x - \frac{1}{4}a = b$

For Problems 37–46, solve each equation for the indicated variable. (Objective 2)

**37.** $2x - 5y = 7$ for $x$

**38.** $5x - 6y = 12$ for $x$

**39.** $-7x - y = 4$ for $y$

**40.** $3x - 2y = -1$ for $y$

**41.** $3(x - 2y) = 4$ for $x$

**42.** $7(2x + 5y) = 6$ for $y$

**43.** $\frac{y - a}{b} = \frac{x + b}{c}$ for $x$

**44.** $\frac{x - a}{b} = \frac{y - a}{c}$ for $y$

**45.** $(y + 1)(a - 3) = x - 2$ for $y$

**46.** $(y - 2)(a + 1) = x$ for $y$

Solve each of Problems 47–62 by setting up and solving an appropriate algebraic equation. (Objective 3)

**47.** Suppose that the length of a certain rectangle is 2 meters less than four times its width. The perimeter of the rectangle is 56 meters. Find the length and width of the rectangle.

**48.** The perimeter of a triangle is 42 inches. The second side is 1 inch more than twice the first side, and the third side is 1 inch less than three times the first side. Find the lengths of the three sides of the triangle.

**49.** How long will it take \$500 to double itself at 6% simple interest?

**50.** How long will it take \$700 to triple itself at 5% simple interest?

**51.** How long will it take $P$ dollars to double itself at 6% simple interest?

**52.** How long will it take $P$ dollars to triple itself at 5% simple interest?

**53.** Two airplanes leave Chicago at the same time and fly in opposite directions. If one travels at 450 miles per hour and the other at 550 miles per hour, how long will it take for them to be 4000 miles apart?

**54.** Look at Figure 2.4. Tyrone leaves city $A$ on a moped traveling toward city $B$ at 18 miles per hour. At the same time, Tina leaves city $B$ on a bicycle traveling toward city $A$ at 14 miles per hour. The distance between the two cities is 112 miles. How long will it take before Tyrone and Tina meet?

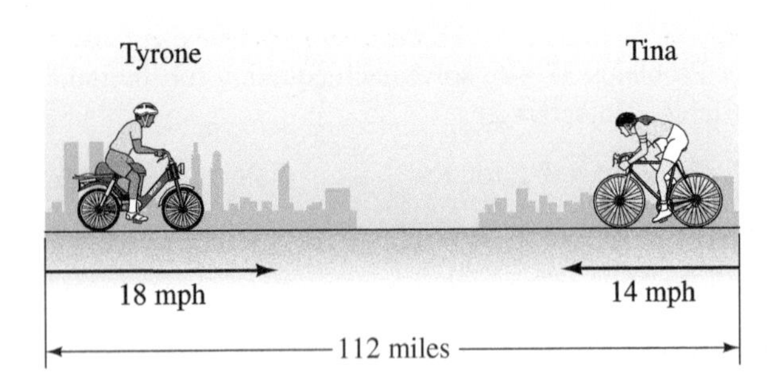


**Figure 2.4**

**55.** Juan starts walking at 4 miles per hour. An hour and a half later, Cathy starts jogging along the same route at 6 miles per hour. How long will it take Cathy to catch up with Juan?

**56.** A car leaves a town at 60 kilometers per hour. How long will it take a second car, traveling at 75 kilometers per hour, to catch the first car if it leaves 1 hour later?

**57.** Bret started on a 70-mile bicycle ride at 20 miles per hour. After a time he became a little tired and slowed down to 12 miles per hour for the rest of the trip. The entire trip of 70 miles took $4\frac{1}{2}$ hours. How far had Bret ridden when he reduced his speed to 12 miles per hour?

**58.** How many gallons of a 12%-salt solution must be mixed with 6 gallons of a 20%-salt solution to obtain a 15%-salt solution?

**59.** A pharmacist has a 6% solution of cough syrup and a 14% solution of the same cough syrup. How many ounces of each must be mixed to make 16 ounces of a 10% solution of cough syrup?

**60.** Suppose that you have a supply of a 30% solution of alcohol and a 70% solution of alcohol. How many quarts of each should be mixed to produce 20 quarts that is 40% alcohol?

**61.** How many milliliters of pure acid must be added to 150 milliliters of a 30% solution of acid to obtain a 40% solution?

**62.** How many cups of grapefruit juice must be added to 40 cups of punch that is 5% grapefruit juice to obtain a punch that is 10% grapefruit juice?

## Thoughts Into Words

**63.** Some people subtract 32 and then divide by 2 to estimate the change from a Fahrenheit reading to a Celsius reading. Why does this give an estimate, and how good is the estimate?

**64.** One of your classmates analyzes Problem 56 as follows: "The first car has traveled 60 kilometers before the second car starts. Because the second car travels 15 kilometers per hour faster, it will take $\frac{60}{15} = 4$ hours for the second car to overtake the first car." How would you react to this analysis of the problem?

**65.** Summarize the new ideas that you have learned thus far in this course that relate to problem solving.

## Further Investigations

For Problems 66–73, use your calculator to help solve each formula for the indicated variable.

**66.** Solve $i = Prt$ for $i$, given that $P = \$875$, $r = 3\frac{1}{2}\%$, and $t = 4$ years.

**67.** Solve $i = Prt$ for $i$, given that $P = \$1125$, $r = 6\frac{1}{4}\%$, and $t = 4$ years.

**68.** Solve $i = Prt$ for $t$, given that $i = \$129.50$, $P = \$925$, and $r = 4\%$.

**69.** Solve $i = Prt$ for $t$, given that $i = \$56.25$, $P = \$1250$, and $r = 3\%$.

**70.** Solve $i = Prt$ for $r$, given that $i = \$232.50$, $P = \$1550$, and $t = 2$ years. Express $r$ as a percent.

**71.** Solve $i = Prt$ for $r$, given that $i = \$88.00$, $P = \$2200$, and $t = 0.5$ of a year. Express $r$ as a percent.

**72.** Solve $A = P + Prt$ for $P$, given that $A = \$1358.50$, $r = 4\frac{1}{2}\%$, and $t = 1$ year.

**73.** Solve $A = P + Prt$ for $P$, given that $A = \$2173.75$, $r = 8\frac{3}{4}\%$, and $t = 2$ years.

**74.** If you have access to computer software that includes spreadsheets, go to Problems 15 and 16. You should be able to enter the given information in rows. Then, when you enter a formula in a cell below the information and drag that formula across the columns, the software should produce all the answers.

### Answers to the Concept Quiz

**1.** False **2.** True **3.** False **4.** True **5.** True **6.** True **7.** True **8.** True **9.** False **10.** True

## 2.5 Inequalities

OBJECTIVES

1. Write solution sets in interval notation
2. Solve inequalities

We listed the basic inequality symbols in Section 1.2. With these symbols we can make various **statements of inequality**:

$a < b$ means $a$ is less than $b$

$a \leq b$ means $a$ is less than or equal to $b$

$a > b$ means $a$ is greater than $b$

$a \geq b$ means $a$ is greater than or equal to $b$

Here are some examples of **numerical statements of inequality**:

| | |
|---|---|
| $7 + 8 > 10$ | $-4 + (-6) \geq -10$ |
| $-4 > -6$ | $7 - 9 \leq -2$ |
| $7 - 1 < 20$ | $3 + 4 > 12$ |
| $8(-3) < 5(-3)$ | $7 - 1 < 0$ |

Note that only $3 + 4 > 12$ and $7 - 1 < 0$ are *false*; the other six are *true* numerical statements.

**Algebraic inequalities** contain one or more variables. The following are examples of algebraic inequalities.

| | |
|---|---|
| $x + 4 > 8$ | $3x + 2y \leq 4$ |
| $3x - 1 < 15$ | $x^2 + y^2 + z^2 \geq 7$ |
| $y^2 + 2y - 4 \geq 0$ | |

An algebraic inequality such as $x + 4 > 8$ is neither true nor false as it stands, and we call it an **open sentence**. For each numerical value we substitute for $x$, the algebraic inequality $x + 4 > 8$ becomes a numerical statement of inequality that is true or false. For example, if $x = -3$, then $x + 4 > 8$ becomes $-3 + 4 > 8$, which is false. If $x = 5$, then $x + 4 > 8$ becomes $5 + 4 > 8$, which is true. **Solving an inequality** is the process of finding the numbers that make an algebraic inequality a true numerical statement. We call such numbers the *solutions* of the inequality; the solutions *satisfy* the inequality.

There are various ways to display the solution set of an inequality. The three most common ways to show the solution set are set builder notation, a line graph of the solution, or interval notation. The examples in Figure 2.5 contain some simple algebraic inequalities, their solution sets, graphs of the solution sets, and the solution sets written in interval notation. Look them over carefully to be sure you understand the symbols.

| Algebraic inequality | Solution set | Graph of solution set | Interval notation |
|---|---|---|---|
| $x < 2$ | $\{x \mid x < 2\}$ | | $(-\infty, 2)$ |
| $x > -1$ | $\{x \mid x > -1\}$ | | $(-1, \infty)$ |
| $3 < x$ | $\{x \mid x > 3\}$ | | $(3, \infty)$ |
| $x \geq 1$ ($\geq$ is read "greater than or equal to") | $\{x \mid x \geq 1\}$ | | $[1, \infty)$ |
| $x \leq 2$ ($\leq$ is read "less than or equal to") | $\{x \mid x \leq 2\}$ | | $(-\infty, 2]$ |
| $1 \geq x$ | $\{x \mid x \leq 1\}$ | | $(-\infty, 1]$ |

**Figure 2.5**

**Classroom Example**
Express the given inequalities in interval notation and graph the interval on a number line:
(a) $x > -1$
(b) $x \leq -2$
(c) $x < 2$
(d) $x \geq 1$

## EXAMPLE 1

Express the given inequalities in interval notation and graph the interval on a number line:

**(a)** $x > -2$ **(b)** $x \leq -1$ **(c)** $x < 3$ **(d)** $x \geq 2$

### Solution

**(a)** For the solution set of the inequality $x > -2$, we want all the numbers greater than $-2$ but not including $-2$. In interval notation, the solution set is written as $(-2, \infty)$; the parentheses are used to indicate exclusion of the endpoints. The use of a parenthesis carries over to the graph of the solution set. In Figure 2.6, the left-hand parenthesis at $-2$ indicates that $-2$ is *not* a solution, and the red part of the line to the right of $-2$ indicates that all real numbers greater than $-2$ are solutions. We refer to the red portion of the number line as the *graph* of the solution set.

| Inequality | Interval notation | Graph |
|---|---|---|
| $x > -2$ | $(-2, \infty)$ | |

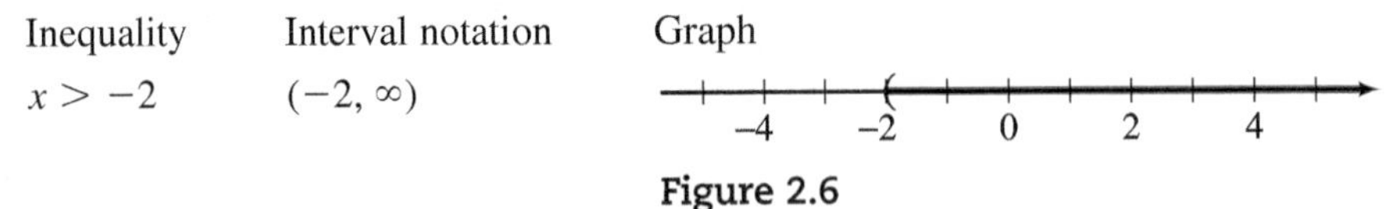

**Figure 2.6**

**(b)** For the solution set of the inequality $x \leq -1$, we want all the numbers less than or equal to $-1$. In interval notation, the solution set is written as $(-\infty, -1]$, where a square bracket is used to indicate inclusion of the endpoint. The use of a square bracket carries over to the graph of the solution set. In Figure 2.7, the right-hand square bracket at $-1$ indicates that $-1$ is part of the solution, and the red part of the line to the left of $-1$ indicates that all real numbers less than $-1$ are solutions.

| Inequality | Interval notation | Graph |
|---|---|---|
| $x \leq -1$ | $(-\infty, -1]$ | |

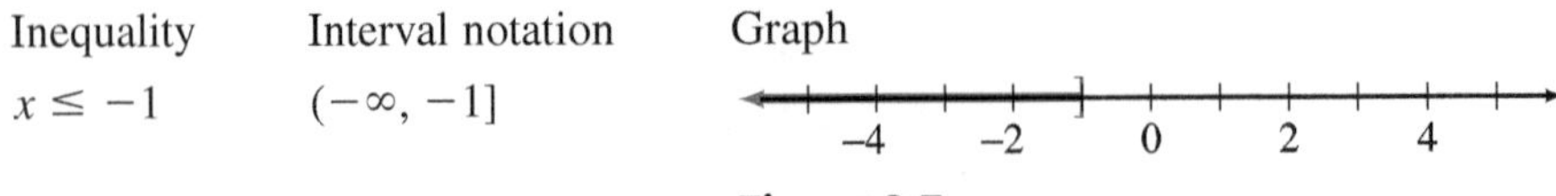

**Figure 2.7**

**(c)** For the solution set of the inequality $x < 3$, we want all the numbers less than 3 but not including 3. In interval notation, the solution set is written as $(-\infty, 3)$; see Figure 2.8.

| Inequality | Interval notation | Graph |
|---|---|---|
| $x < 3$ | $(-\infty, 3)$ | |

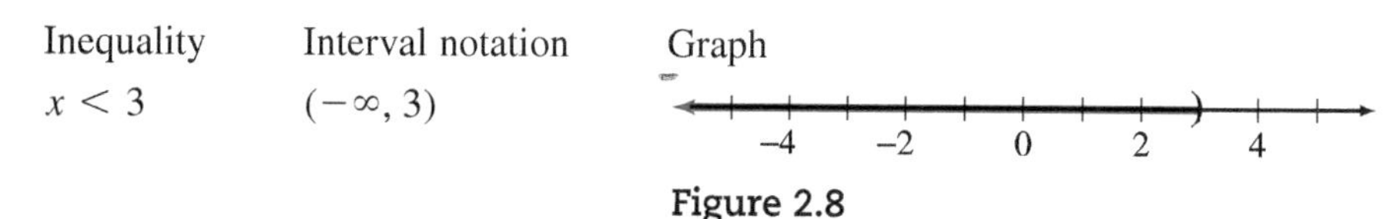


**Figure 2.8**

**(d)** For the solution set of the inequality $x \geq 2$, we want all the numbers greater than or equal to 2. In interval notation, the solution set is written as $[2, \infty)$; see Figure 2.9.

| Inequality | Interval notation | Graph |
|---|---|---|
| $x \geq 2$ | $[2, \infty)$ | |

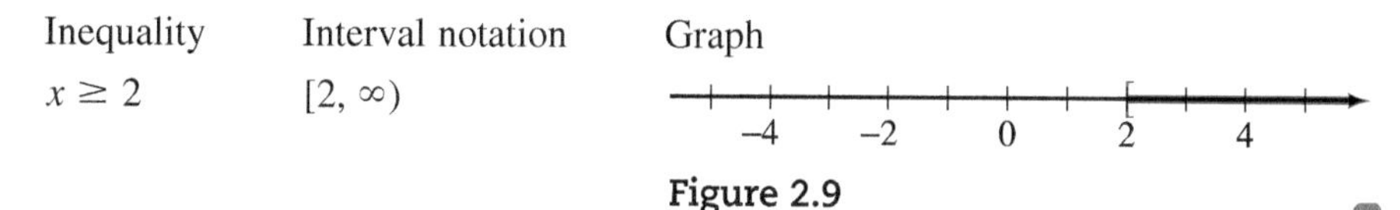


**Figure 2.9**

**Remark:** Note that the infinity symbol always has a parenthesis next to it because no actual endpoint could be included.

## Solving Inequalities

The general process for solving inequalities closely parallels the process for solving equations. We continue to replace the given inequality with equivalent, but simpler, inequalities. For example,

$$3x + 4 > 10 \quad (1)$$

$$3x > 6 \quad (2)$$

$$x > 2 \quad (3)$$

are all equivalent inequalities; that is, they all have the same solutions. By inspection we see that the solutions for (3) are all numbers greater than 2. Thus (1) has the same solutions.

The exact procedure for simplifying inequalities so that we can determine the solutions is based primarily on two properties. The first of these is the addition property of inequality.

### Addition Property of Inequality

For all real numbers $a$, $b$, and $c$,

$$a > b \quad \text{if and only if } a + c > b + c$$

The addition property of inequality states that we can add any number to both sides of an inequality to produce an equivalent inequality. We have stated the property in terms of $>$, but analogous properties exist for $<$, $\geq$, and $\leq$.

Before we state the multiplication property of inequality, let's look at some numerical examples.

| | | |
|---|---|---|
| $2 < 5$ | Multiply both sides by 4 | $8 < 20$ |
| $-3 > -7$ | Multiply both sides by 2 | $-6 > -14$ |
| $-4 < 6$ | Multiply both sides by 10 | $-40 < 60$ |
| $4 < 8$ | Multiply both sides by $-3$ | $-12 > -24$ |
| $3 > -2$ | Multiply both sides by $-4$ | $-12 < 8$ |
| $-4 < -1$ | Multiply both sides by $-2$ | $8 > 2$ |

Notice in the first three examples that when we multiply both sides of an inequality by a *positive number*, we get an inequality of the *same sense*. That means that if the original inequality is *less than*, then the new inequality is *less than*; and if the original inequality is *greater than*, then the new inequality is *greater than*. The last three examples illustrate that when we multiply both sides of an inequality by a *negative number* we get an inequality of the *opposite sense*.

We can state the multiplication property of inequality as follows.

**Multiplication Property of Inequality**

**(a)** For all real numbers $a$, $b$, and $c$, with $c > 0$,

$$a > b \quad \text{if and only if } ac > bc$$

**(b)** For all real numbers $a$, $b$, and $c$, with $c < 0$,

$$a > b \quad \text{if and only if } ac < bc$$

Similar properties hold if we reverse each inequality or if we replace $>$ with $\geq$ and $<$ with $\leq$. For example, if $a \leq b$ and $c < 0$, then $ac \geq bc$.

Let's clarify another point. We stated the properties of inequalities in terms of only two operations, addition and multiplication. We could also include the operations of subtraction and division. In terms of subtraction, if the same number is subtracted from both sides of an inequality, the sense of the inequality remains the same. In terms of division, if both sides of an inequality are divided by a positive number, the sense of the inequality remains the same. However, if both sides of an inequality are divided by a negative number, the sense of the inequality is reversed.

Classroom Example
Solve $2x + 5 > -1$, and graph the solutions.

**EXAMPLE 2** Solve $3x - 4 > 8$ and graph the solutions.

**Solution**

$$3x - 4 > 8$$

$$3x - 4 + 4 > 8 + 4 \qquad \text{Add 4 to both sides}$$

$$3x > 12$$

$$\frac{3x}{3} > \frac{12}{3} \qquad \text{Divide both sides by 3}$$

$$x > 4$$

The solution set is $(4, \infty)$. Figure 2.10 shows the graph of the solution set.

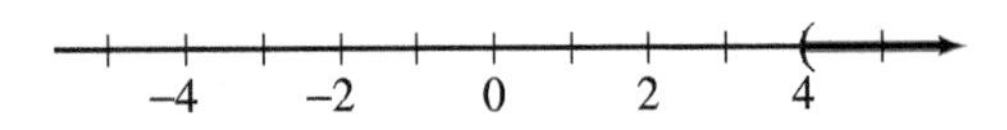


**Figure 2.10**

Classroom Example
Solve $-5x + 4 > 9$, and graph the solutions.

**EXAMPLE 3** Solve $-2x + 1 > 5$ and graph the solutions.

**Solution**

$$-2x + 1 > 5$$

$$-2x + 1 - 1 > 5 - 1 \qquad \text{Subtract 1 from both sides}$$

$$-2x > 4$$

$$\frac{-2x}{-2} < \frac{4}{-2} \qquad \text{Divide both sides by } -2$$

Note that the sense of the inequality has been reversed

$$x < -2$$

The solution set is $(-\infty, -2)$, which can be illustrated on a number line as in Figure 2.11.

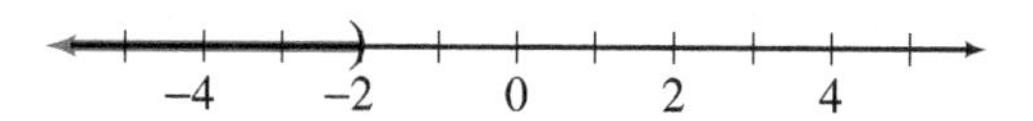


**Figure 2.11**

Checking solutions for an inequality presents a problem. Obviously, we cannot check all of the infinitely many solutions for a particular inequality. However, by checking at least one solution, especially when the multiplication property has been used, we might catch the common mistake of forgetting to change the sense of an inequality. In Example 3 we are claiming that all numbers less than $-2$ will satisfy the original inequality. Let's check one such number, say $-4$.

$$-2x + 1 > 5$$
$$-2(-4) + 1 \overset{?}{>} 5 \quad \text{when } x = -4$$
$$8 + 1 \overset{?}{>} 5$$
$$9 > 5$$

Thus $-4$ satisfies the original inequality. Had we forgotten to switch the sense of the inequality when both sides were divided by $-2$ our answer would have been $x > -2$, and we would have detected such an error by the check.

Many of the same techniques used to solve equations, such as removing parentheses and combining similar terms, may be used to solve inequalities. However, we must be extremely careful when using the multiplication property of inequality. Study each of the following examples very carefully. The format we used highlights the major steps of a solution.

**Classroom Example**
Solve
$-4x + 7x + 3 \leq 5x - 4 - x$.

**EXAMPLE 4** Solve $-3x + 5x - 2 \geq 8x - 7 - 9x$.

### Solution

$$-3x + 5x - 2 \geq 8x - 7 - 9x$$
$$2x - 2 \geq -x - 7 \quad \text{Combine similar terms on both sides}$$
$$3x - 2 \geq -7 \quad \text{Add } x \text{ to both sides}$$
$$3x \geq -5 \quad \text{Add 2 to both sides}$$
$$x \geq -\frac{5}{3} \quad \text{Divide both sides by 3}$$

The solution set is $\left[-\frac{5}{3}, \infty\right)$.

**Classroom Example**
Solve $-2(x + 3) \geq 4$, and graph the solutions.

**EXAMPLE 5** Solve $-5(x - 1) \leq 10$ and graph the solutions.

### Solution

$$-5(x - 1) \leq 10$$
$$-5x + 5 \leq 10 \quad \text{Apply the distributive property on the left}$$
$$-5x \leq 5 \quad \text{Add } -5 \text{ to both sides}$$
$$\frac{-5x}{-5} \geq \frac{5}{-5} \quad \text{Divide both sides by } -5\text{, which reverses the inequality}$$
$$x \geq -1$$

The solution set is $[-1, \infty)$, and it can be graphed as in Figure 2.12.

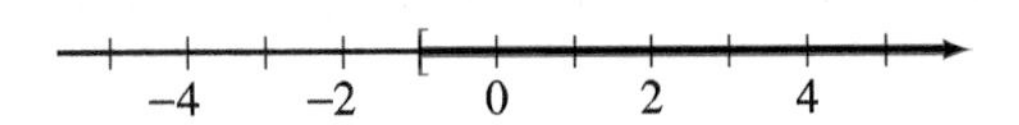


**Figure 2.12**

Classroom Example
Solve $3(x + 1) < 5(x - 2)$.

**EXAMPLE 6** Solve $4(x - 3) > 9(x + 1)$.

### Solution

$$4(x - 3) > 9(x + 1)$$

$$4x - 12 > 9x + 9 \quad \text{Apply the distributive property}$$

$$-5x - 12 > 9 \quad \text{Add } -9x \text{ to both sides}$$

$$-5x > 21 \quad \text{Add 12 to both sides}$$

$$\frac{-5x}{-5} < \frac{21}{-5} \quad \text{Divide both sides by } -5\text{, which reverses the inequality}$$

$$x < -\frac{21}{5}$$

The solution set is $\left(-\infty, -\frac{21}{5}\right)$.

The next example will solve the inequality without indicating the justification for each step. Be sure that you can supply the reasons for the steps.

Classroom Example
Solve $4(3x - 5) + 7(2x + 3) > 5(7x + 3)$.

**EXAMPLE 7** Solve $3(2x + 1) - 2(2x + 5) < 5(3x - 2)$.

### Solution

$$3(2x + 1) - 2(2x + 5) < 5(3x - 2)$$

$$6x + 3 - 4x - 10 < 15x - 10$$

$$2x - 7 < 15x - 10$$

$$-13x - 7 < -10$$

$$-13x < -3$$

$$-\frac{1}{13}(-13x) > -\frac{1}{13}(-3)$$

$$x > \frac{3}{13}$$

The solution set is $\left(\frac{3}{13}, \infty\right)$.

## Concept Quiz 2.5

For Problems 1–10, answer true or false.

1. Numerical statements of inequality are always true.
2. The algebraic statement $x + 4 > 6$ is called an open sentence.
3. The algebraic inequality $2x > 10$ has one solution.
4. The algebraic inequality $x < 3$ has an infinite number of solutions.

5. The solution set for the inequality $-3x - 1 > 2$ is $(-1, \infty)$.
6. When graphing the solution set of an inequality, a square bracket is used to include the endpoint.
7. The solution set of the inequality $x \geq 4$ is written $(4, \infty)$.
8. The solution set of the inequality $x < -5$ is written $(-\infty, -5)$.
9. When multiplying both sides of an inequality by a negative number, the sense of the inequality stays the same.
10. When adding a negative number to both sides of an inequality, the sense of the inequality stays the same.

## Problem Set 2.5

For Problems 1–8, express the given inequality in interval notation and sketch a graph of the interval. (Objective 1)

**1.** $x > 1$ **2.** $x > -2$

**3.** $x \geq -1$ **4.** $x \geq 3$

**5.** $x < -2$ **6.** $x < 1$

**7.** $x \leq 2$ **8.** $x \leq 0$

For Problems 9–16, express each interval as an inequality using the variable $x$. For example, we can express the interval $[5, \infty)$ as $x \geq 5$. (Objective 1)

**9.** $(-\infty, 4)$ **10.** $(-\infty, -2)$

**11.** $(-\infty, -7]$ **12.** $(-\infty, 9]$

**13.** $(8, \infty)$ **14.** $(-5, \infty)$

**15.** $[-7, \infty)$ **16.** $[10, \infty)$

For Problems 17–40, solve each of the inequalities and graph the solution set on a number line. (Objective 2)

**17.** $x - 3 > -2$ **18.** $x + 2 < 1$

**19.** $-2x \geq 8$ **20.** $-3x \leq -9$

**21.** $5x \leq -10$ **22.** $4x \geq -4$

**23.** $2x + 1 < 5$ **24.** $2x + 2 > 4$

**25.** $3x - 2 > -5$ **26.** $5x - 3 < -3$

**27.** $-7x - 3 \leq 4$ **28.** $-3x - 1 \geq 8$

**29.** $2 + 6x > -10$ **30.** $1 + 6x > -17$

**31.** $5 - 3x < 11$ **32.** $4 - 2x < 12$

**33.** $15 < 1 - 7x$ **34.** $12 < 2 - 5x$

**35.** $-10 \leq 2 + 4x$ **36.** $-9 \leq 1 + 2x$

**37.** $3(x + 2) > 6$ **38.** $2(x - 1) < -4$

**39.** $5x + 2 \geq 4x + 6$ **40.** $6x - 4 \leq 5x - 4$

For Problems 41–70, solve each inequality and express the solution set using interval notation. (Objective 2)

**41.** $2x - 1 > 6$

**42.** $3x - 2 < 12$

**43.** $-5x - 2 < -14$

**44.** $-2 - 4x > -2$

**45.** $-3(2x + 1) \geq 12$

**46.** $-2(3x + 2) \leq 18$

**47.** $4(3x - 2) \geq -8$

**48.** $3(4x - 3) \leq -11$

**49.** $6x - 2 > 4x - 14$

**50.** $9x + 5 < 6x - 10$

**51.** $2x - 7 < 6x + 13$

**52.** $2x - 3 > 7x + 22$

**53.** $4(x - 3) \leq -2(x + 1)$

**54.** $3(x - 1) \geq -(x + 4)$

**55.** $5(x - 4) - 6(x + 2) < 4$

**56.** $3(x + 2) - 4(x - 1) < 6$

**57.** $-3(3x + 2) - 2(4x + 1) \geq 0$

**58.** $-4(2x - 1) - 3(x + 2) \geq 0$

**59.** $-(x - 3) + 2(x - 1) < 3(x + 4)$

**60.** $3(x - 1) - (x - 2) > -2(x + 4)$

**61.** $7(x + 1) - 8(x - 2) < 0$

**62.** $5(x - 6) - 6(x + 2) < 0$

**63.** $-5(x - 1) + 3 > 3x - 4 - 4x$

**64.** $3(x + 2) + 4 < -2x + 14 + x$

**65.** $3(x - 2) - 5(2x - 1) \geq 0$

**66.** $4(2x - 1) - 3(3x + 4) \geq 0$

**67.** $-5(3x + 4) < -2(7x - 1)$

**68.** $-3(2x + 1) > -2(x + 4)$

**69.** $-3(x + 2) > 2(x - 6)$

**70.** $-2(x - 4) < 5(x - 1)$

## Thoughts Into Words

**71.** Do the *less than* and *greater than* relations possess a symmetric property similar to the symmetric property of equality? Defend your answer.

**72.** Give a step-by-step description of how you would solve the inequality $-3 > 5 - 2x$.

**73.** How would you explain to someone why it is necessary to reverse the inequality symbol when multiplying both sides of an inequality by a negative number?

## Further Investigations

**74.** Solve each of the following inequalities.

**(a)** $5x - 2 > 5x + 3$

**(b)** $3x - 4 < 3x + 7$

**(c)** $4(x + 1) < 2(2x + 5)$

**(d)** $-2(x - 1) > 2(x + 7)$

**(e)** $3(x - 2) < -3(x + 1)$

**(f)** $2(x + 1) + 3(x + 2) < 5(x - 3)$

### Answers to the Concept Quiz

**1.** False **2.** True **3.** False **4.** True **5.** False **6.** True **7.** False **8.** True **9.** False **10.** True

# 2.6 More on Inequalities and Problem Solving

OBJECTIVES

1. Solve inequalities involving fractions or decimals
2. Solve inequalities that are compound statements
3. Use inequalities to solve word problems

When we discussed solving equations that involve fractions, we found that *clearing the equation of all fractions* is frequently an effective technique. To accomplish this, we multiply both sides of the equation by the least common denominator (LCD) of all the fractions in the equation. This same basic approach also works very well with inequalities that involve fractions, as the next examples demonstrate.

Classroom Example
Solve $\frac{1}{2}m + \frac{3}{4}m < \frac{3}{8}$.

**EXAMPLE 1** Solve $\frac{2}{3}x - \frac{1}{2}x > \frac{3}{4}$.

### Solution

$$\frac{2}{3}x - \frac{1}{2}x > \frac{3}{4}$$

$$12\left(\frac{2}{3}x - \frac{1}{2}x\right) > 12\left(\frac{3}{4}\right) \quad \text{Multiply both sides by 12, which is the LCD of 3, 2, and 4}$$

$$12\left(\frac{2}{3}x\right) - 12\left(\frac{1}{2}x\right) > 12\left(\frac{3}{4}\right)$$ Apply the distributive property

$$8x - 6x > 9$$
$$2x > 9$$
$$x > \frac{9}{2}$$

The solution set is $\left(\frac{9}{2}, \infty\right)$.

**Classroom Example**
Solve $\frac{t+5}{3} + \frac{t-2}{9} > 3$.

## EXAMPLE 2

Solve $\frac{x+2}{4} + \frac{x-3}{8} < 1$.

### Solution

$$\frac{x+2}{4} + \frac{x-3}{8} < 1$$

$$8\left(\frac{x+2}{4} + \frac{x-3}{8}\right) < 8(1)$$ Multiply both sides by 8, which is the LCD of 4 and 8

$$8\left(\frac{x+2}{4}\right) + 8\left(\frac{x-3}{8}\right) < 8(1)$$

$$2(x+2) + (x-3) < 8$$ Divide the denominators into 8

$$2x + 4 + x - 3 < 8$$
$$3x + 1 < 8$$
$$3x < 7$$
$$x < \frac{7}{3}$$

The solution set is $\left(-\infty, \frac{7}{3}\right)$.

**Classroom Example**
Solve $\frac{d}{3} - \frac{d+3}{7} \leq \frac{d-3}{21} - 1$.

## EXAMPLE 3

Solve $\frac{x}{2} - \frac{x-1}{5} \geq \frac{x+2}{10} - 4$.

### Solution

$$\frac{x}{2} - \frac{x-1}{5} \geq \frac{x+2}{10} - 4$$

$$10\left(\frac{x}{2} - \frac{x-1}{5}\right) \geq 10\left(\frac{x+2}{10} - 4\right)$$

$$10\left(\frac{x}{2}\right) - 10\left(\frac{x-1}{5}\right) \geq 10\left(\frac{x+2}{10}\right) - 10(4)$$

$$5x - 2(x-1) \geq x + 2 - 40$$ Divide the denominators into 10

$$5x - 2x + 2 \geq x - 38$$
$$3x + 2 \geq x - 38$$
$$2x + 2 \geq -38$$
$$2x \geq -40$$
$$x \geq -20$$

The solution set is $[-20, \infty)$.

The idea of *clearing all decimals* also works with inequalities in much the same way as it does with equations. We can multiply both sides of an inequality by an appropriate power of 10 and then proceed in the usual way. The next two examples illustrate this procedure.

**Classroom Example**
Solve $m \leq 3.2 + 0.6m$.

**EXAMPLE 4** Solve $x \geq 1.6 + 0.2x$.

### Solution

$$x \geq 1.6 + 0.2x$$
$$10(x) \geq 10(1.6 + 0.2x) \quad \text{Multiply both sides by 10}$$
$$10x \geq 16 + 2x$$
$$8x \geq 16$$
$$x \geq 2$$

The solution set is $[2,\infty)$.

**Classroom Example**
Solve $0.03n + 0.05(n + 20) \leq 43$.

**EXAMPLE 5** Solve $0.08x + 0.09(x + 100) \geq 43$.

### Solution

$$0.08x + 0.09(x + 100) \geq 43$$
$$100[(0.08)x + 0.09(x + 100)] \geq 100(43) \quad \text{Multiply both sides by 100}$$
$$8x + 9(x + 100) \geq 4300$$
$$8x + 9x + 900 \geq 4300$$
$$17x + 900 \geq 4300$$
$$17x \geq 3400$$
$$x \geq 200$$

The solution set is $[200,\infty)$.

## Solving Inequalities That Are Compound Statements

We use the words "and" and "or" in mathematics to form **compound statements**. The following are examples of compound numerical statements that use "and." We call such statements **conjunctions**. We agree to call a conjunction true only if all of its component parts are true. Statements 1 and 2 below are true, but statements 3, 4, and 5 are false.

| | | | |
|---|---|---|---|
| **1.** $3 + 4 = 7$ | and | $-4 < -3$ | True |
| **2.** $-3 < -2$ | and | $-6 > -10$ | True |
| **3.** $6 > 5$ | and | $-4 < -8$ | False |
| **4.** $4 < 2$ | and | $0 < 10$ | False |
| **5.** $-3 + 2 = 1$ | and | $5 + 4 = 8$ | False |

We call compound statements that use "or" **disjunctions**. The following are examples of disjunctions that involve numerical statements.

| | | | |
|---|---|---|---|
| **6.** $0.14 > 0.13$ | or | $0.235 < 0.237$ | True |
| **7.** $\frac{3}{4} > \frac{1}{2}$ | or | $-4 + (-3) = 10$ | True |
| **8.** $-\frac{2}{3} > \frac{1}{3}$ | or | $(0.4)(0.3) = 0.12$ | True |
| **9.** $\frac{2}{5} < -\frac{2}{5}$ | or | $7 + (-9) = 16$ | False |

A disjunction is true if at least one of its component parts is true. In other words, disjunctions are false only if all of the component parts are false. Thus statements 6, 7, and 8 are true, but statement 9 is false.

Now let's consider finding solutions for some compound statements that involve algebraic inequalities. Keep in mind that our previous agreements for labeling conjunctions and disjunctions true or false form the basis for our reasoning.

**Classroom Example**
Graph the solution set for the conjunction $x > -2$ and $x < 1$.

**EXAMPLE 6** Graph the solution set for the conjunction $x > -1$ and $x < 3$.

## Solution

The key word is "and," so we need to satisfy both inequalities. Thus all numbers between $-1$ and 3 are solutions, and we can indicate this on a number line as in Figure 2.13.

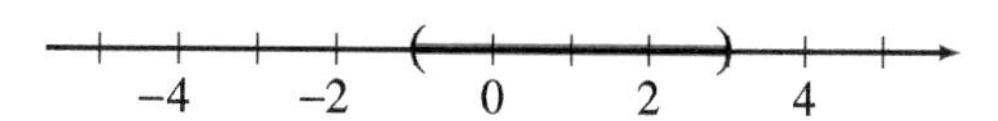


**Figure 2.13**

Using interval notation, we can represent the interval enclosed in parentheses in Figure 2.13 by $(-1, 3)$. Using set builder notation we can express the same interval as $\{x|-1 < x < 3\}$. The statement $-1 < x < 3$ can be read "Negative one is less than $x$, and $x$ is less than three." In other words, $x$ is between $-1$ and 3.

Example 6 represents another concept that pertains to sets. The set of all elements common to two sets is called the *intersection* of the two sets. Thus in Example 6, we found the intersection of the two sets $\{x|x > -1\}$ and $\{x|x < 3\}$ to be the set $\{x|-1 < x < 3\}$. In general, we define the intersection of two sets as follows:

**Definition 2.1**

The **intersection** of two sets $A$ and $B$ (written $A \cap B$) is the set of all elements that are in both $A$ and in $B$. Using set builder notation, we can write

$$A \cap B = \{x|x \in A \text{ and } x \in B\}$$

**Classroom Example**
Solve the conjunction $5x - 6 \geq 9$ *and* $4x - 5 \geq 3$, and graph the solution set on a number line.

**EXAMPLE 7**

Solve the conjunction $3x + 1 > -5$ *and* $2x + 5 > 7$, and graph its solution set on a number line.

## Solution

First, let's simplify both inequalities.

$$\begin{aligned} 3x + 1 &> -5 \quad \text{and} \quad & 2x + 5 &> 7 \\ 3x &> -6 \quad \text{and} \quad & 2x &> 2 \\ x &> -2 \quad \text{and} \quad & x &> 1 \end{aligned}$$

Because this is a conjunction, we must satisfy both inequalities. Thus all numbers greater than 1 are solutions, and the solution set is $(1, \infty)$. We show the graph of the solution set in Figure 2.14.

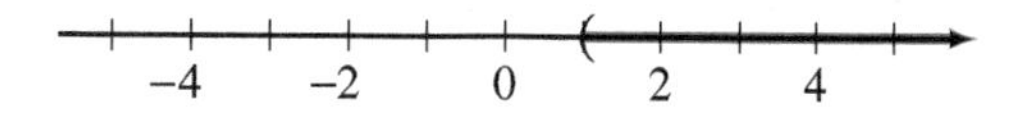


**Figure 2.14**

We can solve a conjunction such as $3x + 1 > -3$ and $3x + 1 < 7$, in which the same algebraic expression (in this case $3x + 1$) is contained in both inequalities, by using the *compact form* $-3 < 3x + 1 < 7$ as follows:

$$-3 < 3x + 1 < 7$$

$$-4 < 3x < 6 \qquad \text{Add } -1 \text{ to the left side, middle, and right side}$$

$$\frac{-4}{3} < \frac{3x}{3} < \frac{6}{3} \qquad \text{Divide each part by 3}$$

$$-\frac{4}{3} < x < 2$$

The solution set is $\left(-\frac{4}{3}, 2\right)$.

The word *and* ties the concept of a conjunction to the set concept of intersection. In a like manner, the word *or* links the idea of a disjunction to the set concept of *union*. We define the union of two sets as follows:

### Definition 2.2

The **union** of two sets $A$ and $B$ (written $A \cup B$) is the set of all elements that are in $A$ or in $B$, or in both. Using set builder notation, we can write

$$A \cup B = \{x | x \in A \text{ or } x \in B\}$$

**Classroom Example**
Graph the solution set for the disjunction $x < 0$ or $x > 3$, and express it using interval notation.

## EXAMPLE 8

Graph the solution set for the disjunction $x < -1$ *or* $x > 2$, and express it using interval notation.

### Solution

The key word is "or," so all numbers that satisfy either inequality (or both) are solutions. Thus all numbers less than $-1$, along with all numbers greater than 2, are the solutions. The graph of the solution set is shown in Figure 2.15.

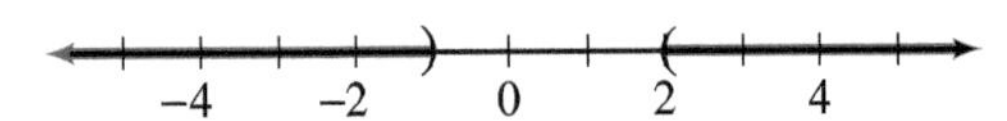


**Figure 2.15**

Using interval notation and the set concept of union, we can express the solution set as $(-\infty, -1) \cup (2, \infty)$.

Example 8 illustrates that in terms of set vocabulary, the solution set of a disjunction is the union of the solution sets of the component parts of the disjunction. Note that there is no compact form for writing $x < -1$ or $x > 2$.

**Classroom Example**
Solve the disjunction $3x + 2 \leq -1$ or $6x - 5 > 7$, and graph its solution set on a number line.

## EXAMPLE 9

Solve the disjunction $2x - 5 < -11$ or $5x + 1 \geq 6$, and graph its solution set on a number line.

### Solution

First, let's simplify both inequalities.

$$\begin{aligned} 2x - 5 < -11 \quad &\text{or} \quad 5x + 1 \geq 6 \\ 2x < -6 \quad &\text{or} \quad 5x \geq 5 \\ x < -3 \quad &\text{or} \quad x \geq 1 \end{aligned}$$

This is a disjunction, and all numbers less than $-3$, along with all numbers greater than or equal to 1, will satisfy it. Thus the solution set is $(-\infty, -3) \cup [1, \infty)$. Its graph is shown in Figure 2.16.

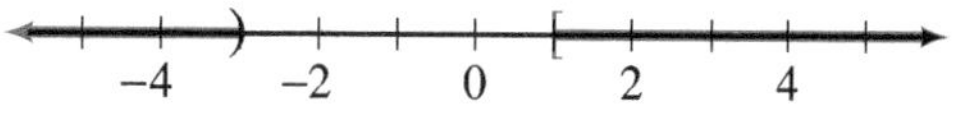


**Figure 2.16**

In summary, to solve a compound sentence involving an inequality, proceed as follows:

1. Solve separately each inequality in the compound sentence.
2. If it is a conjunction, the solution set is the intersection of the solution sets of each inequality.
3. If it is a disjunction, the solution set is the union of the solution sets of each inequality.

The following agreements on the use of interval notation (Figure 2.17) should be added to the list in Figure 2.5.

| Set | Graph | Interval notation |
|---|---|---|
| $\{x \vert a < x < b\}$ | 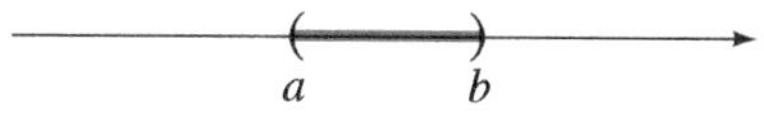 | $(a, b)$ |
| $\{x \vert a \leq x < b\}$ | 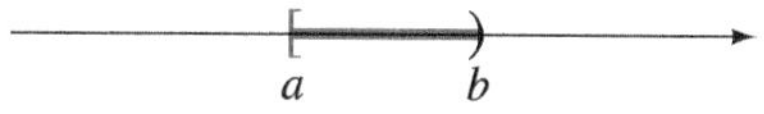 | $[a, b)$ |
| $\{x \vert a < x \leq b\}$ | 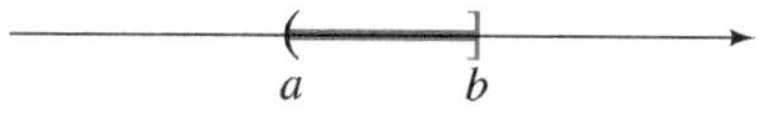 | $(a, b]$ |
| $\{x \vert a \leq x \leq b\}$ | 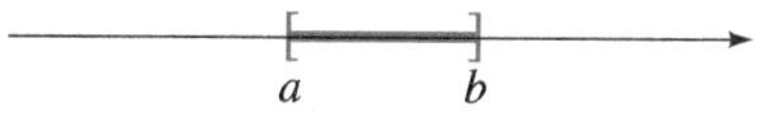 | $[a, b]$ |

**Figure 2.17**

## Using Inequalities to Solve Word Problems

We will conclude this section with some word problems that contain inequality statements.

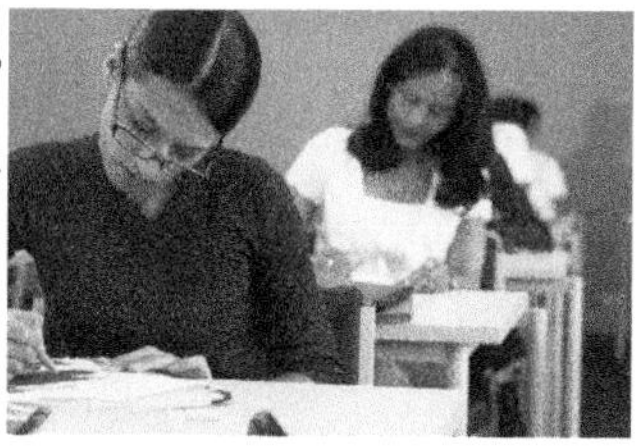
Dave & Les Jacobs/Jupiter Images

### EXAMPLE 10 Apply Your Skill

Sari had scores of 94, 84, 86, and 88 on her first four exams of the semester. What score must she obtain on the fifth exam to have an average of 90 or better for the five exams?

### Solution

Let $s$ represent the score Sari needs on the fifth exam. Because the average is computed by adding all scores and dividing by the number of scores, we have the following inequality to solve.

$$\frac{94 + 84 + 86 + 88 + s}{5} \geq 90$$

**Classroom Example**
Rebekah had scores of 92, 96, and 89 on her first three quizzes of the quarter. What score must she obtain on the fourth quiz to have an average of 93 or better for the four quizzes?

Solving this inequality, we obtain

$$\frac{352 + s}{5} \geq 90$$

$$5\left(\frac{352 + s}{5}\right) \geq 5(90) \quad \text{Multiply both sides by 5}$$

$$352 + s \geq 450$$

$$s \geq 98$$

Sari must receive a score of 98 or better.

Vorobyeva/Shutterstock.com

**Classroom Example**
An investor has $2500 to invest. Suppose he invests $1500 at 5% interest. At what rate must he invest the rest so that the two investments together yield more than $109 of yearly interest?

## EXAMPLE 11 Apply Your Skill

An investor has $1000 to invest. Suppose she invests $500 at 8% interest. At what rate must she invest the other $500 so that the two investments together yield more than $100 of yearly interest?

### Solution

Let $r$ represent the unknown rate of interest. We can use the following guideline to set up an inequality.

| Interest from 8% investment | + | Interest from $r$ percent investment | > | $100 |
|---|---|---|---|---|
| ↓ | | ↓ | | ↓ |
| (8%)($500) | + | $r$($500) | > | $100 |

Solving this inequality yields

$$40 + 500r > 100$$

$$500r > 60$$

$$r > \frac{60}{500}$$

$$r > 0.12 \quad \text{Change to a decimal}$$

She must invest the other $500 at a rate greater than 12%.

Perry Mastrovito/Creatas/ Jupiter Images

**Classroom Example**
If the temperature for a 24-hour period ranged between 41°F and 59°F, inclusive, what was the range in Celsius degrees?

## EXAMPLE 12 Apply Your Skill

A nursery advertises that a particular plant only thrives when the temperature is between 50°F and 86°F, inclusive. The nursery wants to display this information in both Fahrenheit and Celsius scales on an international website. What temperature range in Celsius should the nursery display for this particular plant?

### Solution

Use the formula $F = \frac{9}{5}C + 32$ to solve the following compound inequality.

$$50 \leq \frac{9}{5}C + 32 \leq 86$$

Solving this yields

$$18 \leq \frac{9}{5}C \leq 54 \quad \text{Add } -32 \text{ to all three parts}$$

$$\frac{5}{9}(18) \le \frac{5}{9}\left(\frac{9}{5}C\right) \le \frac{5}{9}(54) \qquad \text{Multiply all three parts by } \frac{5}{9}$$

$$10 \le C \le 30$$

The range is between 10°C and 30°C, inclusive.

## Concept Quiz 2.6

For Problems 1–5, answer true or false.

**1.** The solution set of a compound inequality formed by the word "and" is an intersection of the solution sets of the two inequalities.

**2.** The solution set of any compound inequality is the union of the solution sets of the two inequalities.

**3.** The intersection of two sets contains the elements that are common to both sets.

**4.** The union of two sets contains all the elements in both sets.

**5.** The intersection of set $A$ and set $B$ is denoted by $A \cap B$.

For Problems 6–10, match the compound statement with the graph of its solution set.

**6.** $x > 4$ or $x < -1$

**7.** $x > 4$ and $x > -1$

**8.** $x > 4$ or $x > -1$

**9.** $x \le 4$ and $x \ge -1$

**10.** $x > 4$ or $x \ge -1$

**A.** 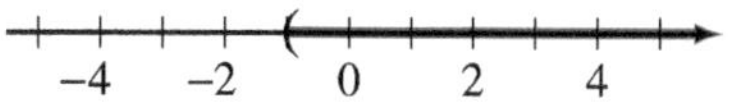


**B.** 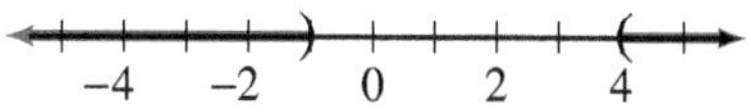


**C.** 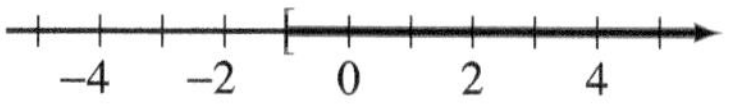


**D.** 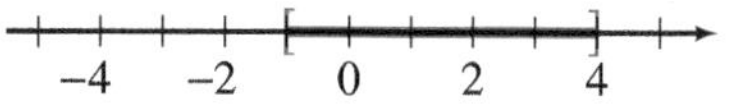


**E.** 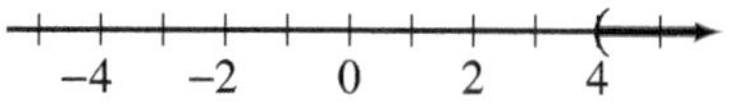


## Problem Set 2.6

For Problems 1–18, solve each of the inequalities and express the solution sets in interval notation. (Objective 1)

**1.** $\frac{2}{5}x + \frac{1}{3}x > \frac{44}{15}$

**2.** $\frac{1}{4}x - \frac{4}{3}x < -13$

**3.** $x - \frac{5}{6} < \frac{x}{2} + 3$

**4.** $x + \frac{2}{7} > \frac{x}{2} - 5$

**5.** $\frac{x-2}{3} + \frac{x+1}{4} \ge \frac{5}{2}$

**6.** $\frac{x-1}{3} + \frac{x+2}{5} \le \frac{3}{5}$

**7.** $\frac{3-x}{6} + \frac{x+2}{7} \le 1$

**8.** $\frac{4-x}{5} + \frac{x+1}{6} \ge 2$

**9.** $\frac{x+3}{8} - \frac{x+5}{5} \ge \frac{3}{10}$

**10.** $\frac{x-4}{6} - \frac{x-2}{9} \le \frac{5}{18}$

**11.** $\frac{4x-3}{6} - \frac{2x-1}{12} < -2$

**12.** $\frac{3x+2}{9} - \frac{2x+1}{3} > -1$

**13.** $0.06x + 0.08(250 - x) \ge 19$

**14.** $0.08x + 0.09(2x) \ge 130$

**15.** $0.09x + 0.1(x + 200) > 77$

**16.** $0.07x + 0.08(x + 100) > 38$

**17.** $x \ge 3.4 + 0.15x$

**18.** $x \ge 2.1 + 0.3x$

For Problems 19–34, graph the solution set for each compound inequality, and express the solution sets in interval notation. (Objective 2)

**19.** $x > -1$ and $x < 2$

**20.** $x > 1$ and $x < 4$

**21.** $x \le 2$ and $x > -1$

**22.** $x \le 4$ and $x \ge -2$

**23.** $x > 2$ or $x < -1$

**24.** $x > 1$ or $x < -4$

**25.** $x \le 1$ or $x > 3$

**26.** $x < -2$ or $x \ge 1$

**27.** $x > 0$ and $x > -1$

**28.** $x > -2$ and $x > 2$

**29.** $x < 0$ and $x > 4$

**30.** $x > 1$ or $x < 2$

**31.** $x > -2$ or $x < 3$

**32.** $x > 3$ and $x < -1$

**33.** $x > -1$ or $x > 2$

**34.** $x < -2$ or $x < 1$

For Problems 35–44, solve each compound inequality and graph the solution sets. Express the solution sets in interval notation. (Objective 2)

**35.** $x - 2 > -1$ and $x - 2 < 1$

**36.** $x + 3 > -2$ and $x + 3 < 2$

**37.** $x + 2 < -3$ or $x + 2 > 3$

**38.** $x - 4 < -2$ or $x - 4 > 2$

**39.** $2x - 1 \geq 5$ and $x > 0$

**40.** $3x + 2 > 17$ and $x \geq 0$

**41.** $5x - 2 < 0$ and $3x - 1 > 0$

**42.** $x + 1 > 0$ and $3x - 4 < 0$

**43.** $3x + 2 < -1$ or $3x + 2 > 1$

**44.** $5x - 2 < -2$ or $5x - 2 > 2$

For Problems 45–56, solve each compound inequality using the compact form. Express the solution sets in interval notation. (Objective 2)

**45.** $-3 < 2x + 1 < 5$

**46.** $-7 < 3x - 1 < 8$

**47.** $-17 \leq 3x - 2 \leq 10$

**48.** $-25 \leq 4x + 3 \leq 19$

**49.** $1 < 4x + 3 < 9$

**50.** $0 < 2x + 5 < 12$

**51.** $-6 < 4x - 5 < 6$

**52.** $-2 < 3x + 4 < 2$

**53.** $-4 \leq \dfrac{x - 1}{3} \leq 4$

**54.** $-1 \leq \dfrac{x + 2}{4} \leq 1$

**55.** $-3 < 2 - x < 3$

**56.** $-4 < 3 - x < 4$

For Problems 57–67, solve each problem by setting up and solving an appropriate inequality. (Objective 3)

**57.** Suppose that Lance has \$5000 to invest. If he invests \$3000 at 5% interest, at what rate must he invest the remaining \$2000 so that the two investments yield more than \$300 in yearly interest?

**58.** Mona invests \$1000 at 8% yearly interest. How much does she have to invest at 6% so that the total yearly interest from the two investments exceeds \$170?

**59.** The average height of the two forwards and the center of a basketball team is 6 feet and 8 inches. What must the average height of the two guards be so that the team average is at least 6 feet and 4 inches?

**60.** Thanh has scores of 52, 84, 65, and 74 on his first four math exams. What score must he make on the fifth exam to have an average of 70 or better for the five exams?

**61.** Marsha bowled 142 and 170 in her first two games. What must she bowl in the third game to have an average of at least 160 for the three games?

**62.** Candace had scores of 95, 82, 93, and 84 on her first four exams of the semester. What score must she obtain on the fifth exam to have an average of 90 or better for the five exams?

**63.** Suppose that Derwin shot rounds of 82, 84, 78, and 79 on the first four days of a golf tournament. What must he shoot on the fifth day of the tournament to average 80 or less for the five days?

**64.** The temperatures for a 24-hour period ranged between −4°F and 23°F, inclusive. What was the range in Celsius degrees? $\left(\text{Use } F = \dfrac{9}{5}C + 32.\right)$

**65.** Oven temperatures for baking various foods usually range between 325°F and 425°F, inclusive. Express this range in Celsius degrees. (Round answers to the nearest degree.)

**66.** A person's intelligence quotient (I) is found by dividing mental age ($M$), as indicated by standard tests, by chronological age ($C$) and then multiplying this ratio by 100. The formula $I = \dfrac{100M}{C}$ can be used. If the I range of a group of 11-year-olds is given by $80 \leq I \leq 140$, find the range of the mental age of this group.

**67.** Repeat Problem 66 for an I range of 70 to 125, inclusive, for a group of 9-year-olds.

## Thoughts Into Words

**68.** Explain the difference between a conjunction and a disjunction. Give an example of each (outside the field of mathematics).

**69.** How do you know by inspection that the solution set of the inequality $x + 3 > x + 2$ is the entire set of real numbers?

**70.** Find the solution set for each of the following compound statements, and in each case explain your reasoning.

**(a)** $x < 3$ and $5 > 2$

**(b)** $x < 3$ or $5 > 2$

**(c)** $x < 3$ and $6 < 4$

**(d)** $x < 3$ or $6 < 4$

### Answers to the Concept Quiz

**1.** True **2.** False **3.** True **4.** True **5.** True **6.** B **7.** E **8.** A **9.** D **10.** C

## 2.7 Equations and Inequalities Involving Absolute Value

OBJECTIVES

1. Solve equations that involve absolute value
2. Solve inequalities that involve absolute value

In Section 1.2, we defined the absolute value of a real number by

$$|a| = \begin{cases} a, & \text{if } a \geq 0 \\ -a, & \text{if } a < 0 \end{cases}$$

We also interpreted the absolute value of any real number to be the distance between the number and zero on a number line. For example, $|6| = 6$ translates to 6 units between 6 and 0. Likewise, $|-8| = 8$ translates to 8 units between $-8$ and 0.

The interpretation of absolute value as distance on a number line provides a straightforward approach to solving a variety of equations and inequalities involving absolute value. First, let's consider some equations.

Classroom Example
Solve $|x| = 6$.

**EXAMPLE 1** Solve $|x| = 2$.

### Solution

Think in terms of distance between the number and zero, and you will see that $x$ must be 2 or $-2$. That is, the equation $|x| = 2$ is equivalent to

$$x = -2 \quad \text{or} \quad x = 2$$

The solution set is $\{-2, 2\}$.

Classroom Example
Solve $|m - 3| = 4$.

**EXAMPLE 2** Solve $|x + 2| = 5$.

### Solution

The quantity, $x + 2$, must be $-5$ or 5. Thus $|x + 2| = 5$ is equivalent to

$$x + 2 = -5 \quad \text{or} \quad x + 2 = 5$$

Solving each equation of the disjunction yields

$$x + 2 = -5 \quad \text{or} \quad x + 2 = 5$$
$$x = -7 \quad \text{or} \quad x = 3$$

✔ **Check**

**When $x = -7$** | **When $x = 3$**

$$|x + 2| = 5 \qquad |x + 2| = 5$$
$$|-7 + 2| \stackrel{?}{=} 5 \qquad |3 + 2| \stackrel{?}{=} 5$$
$$|-5| \stackrel{?}{=} 5 \qquad |5| \stackrel{?}{=} 5$$
$$5 = 5 \qquad 5 = 5$$

The solution set is $\{-7, 3\}$.

The following general property should seem reasonable from the distance interpretation of absolute value.

**Property 2.1**

$|x| = k$ is equivalent to $x = -k$ or $x = k$, where $k$ is a positive number.

Example 3 demonstrates our format for solving equations of the form $|x| = k$.

Classroom Example
Solve $|2w - 5| = 4$.

**EXAMPLE 3** Solve $|5x + 3| = 7$.

**Solution**

$$|5x + 3| = 7$$
$$5x + 3 = -7 \quad \text{or} \quad 5x + 3 = 7$$
$$5x = -10 \quad \text{or} \quad 5x = 4$$
$$x = -2 \quad \text{or} \quad x = \frac{4}{5}$$

The solution set is $\left\{-2, \frac{4}{5}\right\}$. Check these solutions!

Classroom Example
Solve $|3x - 4| + 2 = 9$.

**EXAMPLE 4** Solve $|2x + 5| - 3 = 8$.

**Solution**

First isolate the absolute value expression by adding 3 to both sides of the equation.

$$|2x + 5| - 3 = 8$$
$$|2x + 5| - 3 + 3 = 8 + 3$$
$$|2x + 5| = 11$$
$$2x + 5 = 11 \quad \text{or} \quad 2x + 5 = -11$$
$$2x = 6 \quad \text{or} \quad 2x = -16$$
$$x = 3 \quad \text{or} \quad x = -8$$

The solution set is $\{-8, 3\}$. Check these solutions.

## Solving Inequalities That Involve Absolute Value

The distance interpretation for absolute value also provides a good basis for solving some inequalities that involve absolute value. Consider the following examples.

Classroom Example
Solve $|m| < 4$ and graph the solution set.

**EXAMPLE 5** Solve $|x| < 2$ and graph the solution set.

### Solution

The number, $x$, must be less than two units away from zero. Thus $|x| < 2$ is equivalent to

$$x > -2 \quad \text{and} \quad x < 2$$

The solution set is $(-2, 2)$, and its graph is shown in Figure 2.18.

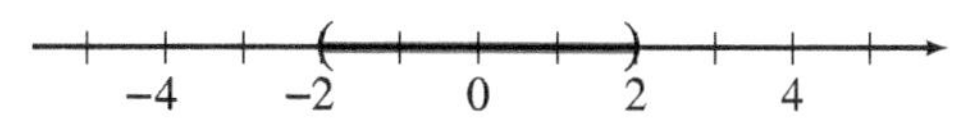


**Figure 2.18**

Classroom Example
Solve $|t - 2| \leq 3$ and graph the solution set.

**EXAMPLE 6** Solve $|x + 3| < 1$ and graph the solutions.

### Solution

Let's continue to think in terms of distance on a number line. The number, $x + 3$, must be less than one unit away from zero. Thus $|x + 3| < 1$ is equivalent to

$$x + 3 > -1 \quad \text{and} \quad x + 3 < 1$$

Solving this conjunction yields

$$x + 3 > -1 \quad \text{and} \quad x + 3 < 1$$
$$x > -4 \quad \text{and} \quad x < -2$$

The solution set is $(-4, -2)$, and its graph is shown in Figure 2.19.

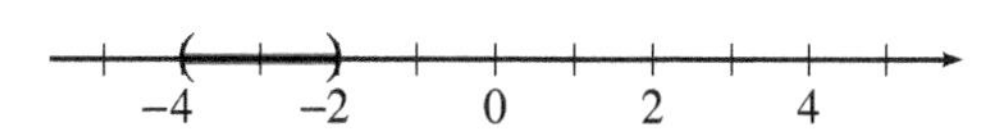


**Figure 2.19**

Take another look at Examples 5 and 6. The following general property should seem reasonable.

**Property 2.2**

$|x| < k$ is equivalent to $x > -k$ and $x < k$, where $k$ is a positive number.

Remember that we can write a conjunction such as $x > -k$ and $x < k$ in the compact form $-k < x < k$. The compact form provides a very convenient format for solving inequalities such as $|3x - 1| < 8$, as Example 7 illustrates.

Classroom Example
Solve $|4x - 7| < 9$ and graph the solutions.

**EXAMPLE 7** Solve $|3x - 1| < 8$ and graph the solutions.

### Solution

$$|3x - 1| < 8$$
$$-8 < 3x - 1 < 8$$
$$-7 < 3x < 9 \qquad \text{Add 1 to left side, middle, and right side}$$

$$\frac{-7}{3} < \frac{3x}{3} < \frac{9}{3} \qquad \text{Divide each part by 3}$$

$$-\frac{7}{3} < x < 3$$

The solution set is $\left(-\frac{7}{3}, 3\right)$, and its graph is shown in Figure 2.20.

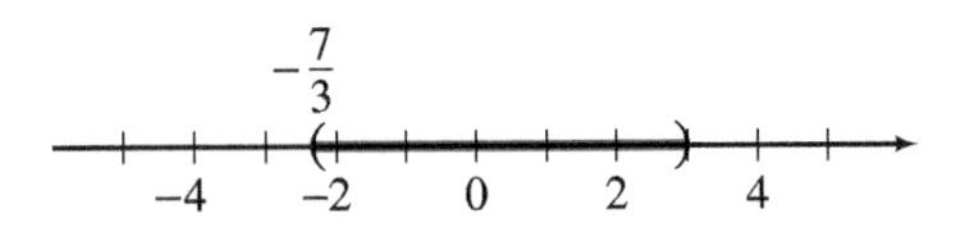


**Figure 2.20**

The distance interpretation also clarifies a property that pertains to *greater than* situations involving absolute value. Consider the following examples.

**Classroom Example**
Solve $|x| \geq \frac{3}{2}$ and graph the solutions.

**EXAMPLE 8** Solve $|x| > 1$ and graph the solutions.

### Solution

The number, $x$, must be more than one unit away from zero. Thus $|x| > 1$ is equivalent to

$$x < -1 \qquad \text{or} \qquad x > 1$$

The solution set is $(-\infty, -1) \cup (1, \infty)$, and its graph is shown in Figure 2.21.

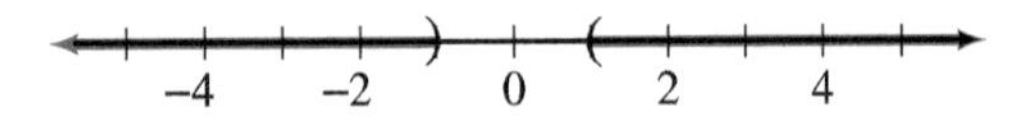


**Figure 2.21**

**Classroom Example**
Solve $|x + 2| \geq 4$ and graph the solutions.

**EXAMPLE 9** Solve $|x - 1| > 3$ and graph the solutions.

### Solution

The number, $x - 1$, must be more than three units away from zero. Thus $|x - 1| > 3$ is equivalent to

$$x - 1 < -3 \qquad \text{or} \qquad x - 1 > 3$$

Solving this disjunction yields

$$x - 1 < -3 \qquad \text{or} \qquad x - 1 > 3$$
$$x < -2 \qquad \text{or} \qquad x > 4$$

The solution set is $(-\infty, -2) \cup (4, \infty)$, and its graph is shown in Figure 2.22.

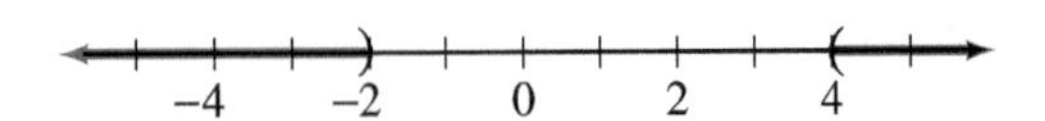


**Figure 2.22**

Examples 8 and 9 illustrate the following general property.

**Property 2.3**

$|x| > k$ is equivalent to $x < -k$ or $x > k$, where $k$ is a positive number.

Therefore, solving inequalities of the form $|x| > k$ can take the format shown in Example 10.

**Classroom Example**
Solve $|5x - 1| + 2 > 6$ and graph the solutions.

**EXAMPLE 10** Solve $|3x - 1| + 4 > 6$ and graph the solutions.

### Solution

First isolate the absolute value expression by subtracting 4 from both sides of the equation.

$$|3x - 1| + 4 > 6$$
$$|3x - 1| + 4 - 4 > 6 - 4 \quad \text{Subtract 4 from both sides}$$
$$|3x - 1| > 2$$
$$3x - 1 < -2 \quad \text{or} \quad 3x - 1 > 2$$
$$3x < -1 \quad \text{or} \quad 3x > 3$$
$$x < -\frac{1}{3} \quad \text{or} \quad x > 1$$

The solution set is $\left(-\infty, -\frac{1}{3}\right) \cup (1, \infty)$, and its graph is shown in Figure 2.23.

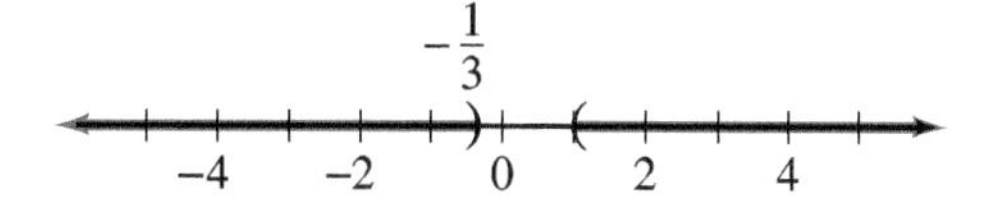


**Figure 2.23**

Properties 2.1, 2.2, and 2.3 provide the basis for solving a variety of equations and inequalities that involve absolute value. However, if at any time you become doubtful about what property applies, don't forget the distance interpretation. Furthermore, note that in each of the properties, $k$ is a positive number. If $k$ is a nonpositive number, we can determine the solution sets by inspection, as indicated by the following examples.

$|x + 3| = 0$ has a solution of $x = -3$, because the number $x + 3$ has to be 0. The solution set of $|x + 3| = 0$ is $\{-3\}$.

$|2x - 5| = -3$ has no solutions, because the absolute value (distance) cannot be negative. The solution set is $\varnothing$, the null set.

$|x - 7| < -4$ has no solutions, because we cannot obtain an absolute value less than $-4$. The solution set is $\varnothing$.

$|2x - 1| > -1$ is satisfied by all real numbers because the absolute value of $(2x - 1)$, regardless of what number is substituted for $x$, will always be greater than $-1$. The solution set is the set of all real numbers, which we can express in interval notation as $(-\infty, \infty)$.

## Concept Quiz 2.7

For Problems 1–10, answer true or false.

1. The absolute value of a negative number is the opposite of the number.
2. The absolute value of a number is always positive or zero.

3. The absolute value of a number is equal to the absolute value of its opposite.
4. The compound statement $x < 1$ or $x > 3$ can be written in compact form $3 < x < 1$.
5. The solution set for the equation $|x + 5| = 0$ is the null set, $\varnothing$.
6. The solution set for $|x - 2| \geq -6$ is all real numbers.
7. The solution set for $|x + 1| < -3$ is all real numbers.
8. The solution set for $|x - 4| \leq 0$ is $\{4\}$.
9. If a solution set in interval notation is $(-4, -2)$, then it can be expressed as $\{x|-4 < x < -2\}$ in set builder notation.
10. If a solution set in interval notation is $(-\infty, -2) \cup (4, \infty)$, then it can be expressed as $\{x|x < -2 \text{ or } x > 4\}$ in set builder notation.

## Problem Set 2.7

For Problems 1–16, solve each equation. (Objective 1)

1. $|x - 1| = 8$
2. $|x + 2| = 9$
3. $|2x - 4| = 6$
4. $|3x - 4| = 14$
5. $|3x + 4| = 11$
6. $|5x - 7| = 14$
7. $|4 - 2x| = 6$
8. $|3 - 4x| = 8$
9. $\left|x - \frac{3}{4}\right| = \frac{2}{3}$
10. $\left|x + \frac{1}{2}\right| = \frac{3}{5}$
11. $|2x - 3| + 2 = 5$
12. $|3x - 1| - 1 = 9$
13. $|x + 2| - 6 = -2$
14. $|x - 3| - 4 = -1$
15. $|4x - 3| + 2 = 2$
16. $|5x + 1| + 4 = 4$

For Problems 17–30, solve each inequality and graph the solution. (Objective 2)

17. $|x| < 5$
18. $|x| < 1$
19. $|x| \leq 2$
20. $|x| \leq 4$
21. $|x| > 2$
22. $|x| > 3$
23. $|x - 1| < 2$
24. $|x - 2| < 4$
25. $|x + 2| \leq 4$
26. $|x + 1| \leq 1$
27. $|x + 2| > 1$
28. $|x + 1| > 3$
29. $|x - 3| \geq 2$
30. $|x - 2| \geq 1$

For Problems 31–54, solve each inequality. (Objective 2)

31. $|x - 2| > 6$
32. $|x - 3| > 9$
33. $|x + 3| < 5$
34. $|x + 1| < 8$
35. $|2x - 1| \leq 9$
36. $|3x + 1| \leq 13$
37. $|4x + 2| \geq 12$
38. $|5x - 2| \geq 10$
39. $|2 - x| > 4$
40. $|4 - x| > 3$
41. $|1 - 2x| < 2$
42. $|2 - 3x| < 5$
43. $|5x + 9| \leq 16$
44. $|7x - 6| \geq 22$
45. $|-2x + 7| \leq 13$
46. $|-3x - 4| \leq 15$
47. $\left|\frac{x - 3}{4}\right| < 2$
48. $\left|\frac{x + 2}{3}\right| < 1$
49. $\left|\frac{2x + 1}{2}\right| > 1$
50. $\left|\frac{3x - 1}{4}\right| > 3$
51. $|x + 7| - 3 \geq 4$
52. $|x - 2| + 4 \geq 10$
53. $|2x - 1| + 1 \leq 6$
54. $|4x + 3| - 2 \leq 5$

For Problems 55–64, solve each equation and inequality *by inspection*. (Objectives 1 and 2)

55. $|2x + 1| = -4$
56. $|5x - 1| = -2$
57. $|3x - 1| > -2$
58. $|4x + 3| < -4$
59. $|5x - 2| = 0$
60. $|3x - 1| = 0$
61. $|4x - 6| < -1$
62. $|x + 9| > -6$
63. $|x + 4| < 0$
64. $|x + 6| > 0$

## Thoughts Into Words

**65.** Explain how you would solve the inequality $|2x + 5| > -3$.

**66.** Why is 2 the only solution for $|x - 2| \leq 0$?

**67.** Explain how you would solve the equation $|2x - 3| = 0$.

## Further Investigations

Consider the equation $|x| = |y|$. This equation will be a true statement if $x$ is equal to $y$ or if $x$ is equal to the opposite of $y$. Use the following format, $x = y$ or $x = -y$, to solve the equations in Problems 68–73.

For Problems 68–73, solve each equation.

**68.** $|3x + 1| = |2x + 3|$

**69.** $|-2x - 3| = |x + 1|$

**70.** $|2x - 1| = |x - 3|$

**71.** $|x - 2| = |x + 6|$

**72.** $|x + 1| = |x - 4|$

**73.** $|x + 1| = |x - 1|$

**74.** Use the definition of absolute value to help prove Property 2.1.

**75.** Use the definition of absolute value to help prove Property 2.2.

**76.** Use the definition of absolute value to help prove Property 2.3.

### Answers to the Concept Quiz

**1.** True **2.** True **3.** True **4.** False **5.** False **6.** True **7.** False **8.** True **9.** True **10.** True

| OBJECTIVE | SUMMARY | EXAMPLE |
|---|---|---|
| Evaluate formulas for given values. (Section 2.4/Objective 1) | A formula can be solved for a specific variable when we are given the numerical values for the other variables. | Solve $i = Prt$ for $r$, given that $P = \$1200$, $t = 4$ years, and $i = \$360$. **Solution** $i = Prt$ <br> $360 = (1200)(r)(4)$ <br> $360 = 4800r$ <br> $r = \frac{360}{4800}$ <br> $= 0.075$ <br> The rate, $r$, would be 0.075 or 7.5%. **Sample Problem 6** Solve $i = Prt$ for $t$ given that $P = \$600$, $r = 2\%$, and $i = \$36$. |
| Solve formulas for a specified variable. (Section 2.4/Objective 2) | We can change the form of an equation by solving for one variable in terms of the other variables. | Solve $A = \frac{1}{2}bh$ for $b$. **Solution** $A = \frac{1}{2}bh$ <br> $2A = 2\left(\frac{1}{2}bh\right)$ <br> $2A = bh$ <br> $\frac{2A}{h} = b$ <br> **Sample Problem 7** Solve $P = 2l + 2w$ for $l$. |
| Use formulas to solve problems. (Section 2.4/Objective 3) | Formulas are often used as guidelines for setting up an algebraic equation when solving a word problem. Sometimes formulas are used in the analysis of a problem but not as the main guideline. For example, uniform-motion problems use the formula $d = rt$, but the guideline is usually a statement about times, rates, or distances. | How long will it take \$400 to triple if it is invested at 8% simple interest? **Solution** Use the formula $i = Prt$. For \$400 to triple (to be worth \$1200), it must earn \$800 in interest. <br> $800 = 400(8\%)(t)$ <br> $800 = 400(0.08)(t)$ <br> $2 = 0.08t$ Divide by 400 <br> $t = \frac{2}{0.08} = 25$ <br> It will take 25 years to triple. **Sample Problem 8** What interest rate would you need to get to double an investment of \$200 in eight years? |

| OBJECTIVE | SUMMARY | EXAMPLE |
|---|---|---|
| Write solution sets in interval notation. (Section 2.5/Objective 1) | The solution set for an algebraic inequality can be written in interval notation. See the table below for examples of various algebraic inequalities and how their solution sets would be written in interval notation. | Express the solution set for $x \le 4$ in interval notation.<br>**Solution**<br>For the solution set we want all numbers less than or equal to 4. In interval notation, the solution set is written $(-\infty, 4]$.<br>**Sample Problem 9**<br>Express the solution set for $x > 5$ in interval notation. |

| Solution set | Graph | Interval notation |
|---|---|---|
| $\{x\|x > 1\}$ | –4 –2 0 2 4 | $(1, \infty)$ |
| $\{x\|x \ge 2\}$ | –4 –2 0 2 4 | $[2, \infty)$ |
| $\{x\|x < 0\}$ | –4 –2 0 2 4 | $(-\infty, 0)$ |
| $\{x\|x \le -1\}$ | –4 –2 0 2 4 | $(-\infty, -1]$ |
| $\{x\|-2 < x \le 2\}$ | –4 –2 0 2 4 | $(-2, 2]$ |
| $\{x\|x \le -1 \text{ or } x > 1\}$ | –4 –2 0 2 4 | $(-\infty, -1] \cup (1, \infty)$ |

| OBJECTIVE | SUMMARY | EXAMPLE |
|---|---|---|
| Solve inequalities. (Section 2.5/Objective 2) | The addition property of inequality states that any number can be added to each side of an inequality to produce an equivalent inequality. The multiplication property of inequality states that both sides of an inequality can be multiplied by a positive number to produce an equivalent inequality. If both sides of an inequality are multiplied by a negative number, then an inequality of the *opposite sense* is produced. When multiplying or dividing both sides of an inequality by a negative number, be sure to reverse the inequality symbol. | Solve $-8x + 2(x - 7) < 40$.<br>**Solution**<br>$-8x + 2(x - 7) < 40$<br>$-8x + 2x - 14 < 40$<br>$-6x - 14 < 40$<br>$-6x < 54$<br>$\frac{-6x}{-6} > \frac{54}{-6}$<br>$x > -9$<br>The solution set is $(-9, \infty)$.<br>**Sample Problem 10**<br>Solve $3x - 5(x + 7) \le 15$. |

(*continued*)

| OBJECTIVE | SUMMARY | EXAMPLE |
|---|---|---|
| Solve inequalities involving fractions or decimals.<br>(Section 2.6/Objective 1) | When solving inequalities that involve fractions, multiply the inequality by the least common multiple of all the denominators to clear the equation of fractions. The same technique can be used for inequalities involving decimals. | Solve $\frac{x+5}{3} - \frac{x+1}{2} < \frac{5}{6}$.<br>**Solution**<br>Multiply both sides of the inequality by 6.<br>$6\left(\frac{x+5}{3} - \frac{x+1}{2}\right) < 6\left(\frac{5}{6}\right)$<br>$2(x+5) - 3(x+1) < 5$<br>$2x + 10 - 3x - 3 < 5$<br>$-x + 7 < 5$<br>$-x < -2$<br>$-1(-x) > -1(-2)$<br>$x > 2$<br>The solution set is $(2, \infty)$.<br>**Sample Problem 11**<br>Solve $\frac{x-3}{2} - \frac{x-2}{5} \geq \frac{3}{10}$. |
| Solve inequalities that are compound statements.<br>(Section 2.6/Objective 2) | Inequalities connected with the words "and" form a compound statement called a conjunction. A conjunction is true only if all of its component parts are true. The solution set of a conjunction is the *intersection* of the solution sets of each inequality.<br><br>Inequalities connected with the words "or" form a compound statement called a disjunction. A disjunction is true if at least one of its component parts is true. The solution set of a disjunction is the *union* of the solution sets of each inequality. We define the intersection and union of two sets as follows.<br><br>**Intersection**<br>$A \cap B = \{x \mid x \in A \text{ and } x \in B\}$<br><br>**Union**<br>$A \cup B = \{x \mid x \in A \text{ or } x \in B\}$ | Solve the compound statement $x + 4 \leq -10$ or $x - 2 \geq 1$.<br>**Solution**<br>Simplify each inequality.<br>$x + 4 \leq -10$ or $x - 2 \geq 1$<br>$x \leq -14$ or $x \geq 3$<br>The solution set is<br>$(-\infty, -14] \cup [3, \infty)$.<br>**Sample Problem 12**<br>Solve the compound statement $x - 3 < 2$ and $x - 1 \geq 2$. |

| OBJECTIVE | SUMMARY | EXAMPLE |
| --- | --- | --- |
| Use inequalities to solve word problems.<br>(Section 2.6/Objective 3) | To solve word problems involving inequalities, use the same suggestions given for solving word problems; however, the guideline will translate into an inequality rather than an equation. | Cheryl bowled 156 and 180 in her first two games. What must she bowl in the third game to have an average of at least 170 for the three games?<br>**Solution**<br>Let $s$ represent the score in the third game.<br>$\frac{156 + 180 + s}{3} \geq 170$<br>$156 + 180 + s \geq 510$<br>$336 + s \geq 510$<br>$s \geq 174$<br>She must bowl 174 or greater.<br>**Sample Problem 13**<br>Kayla was playing in a 4-game basketball tournament. She scored 15 points, 8 points, and 12 points in the first three games. How many points must she score in the fourth game to have an average of at least 12 points for the 4-game tournament? |
| Solve absolute value equations.<br>(Section 2.7/Objective 1) | Property 2.1 states that $\lvert x \rvert = k$ is equivalent to $x = k$ or $x = -k$, where $k$ is a positive number. This property is applied to solve absolute value equations. | Solve $\lvert 2x - 5 \rvert = 9$.<br>**Solution**<br>$\lvert 2x - 5 \rvert = 9$<br>$2x - 5 = 9$ or $2x - 5 = -9$<br>$2x = 14$ or $2x = -4$<br>$x = 7$ or $x = -2$<br>The solution set is $\{-2, 7\}$.<br>**Sample Problem 14**<br>Solve $\lvert 3x + 7 \rvert = 16$. |
| Solve absolute value inequalities.<br>(Section 2.7/Objective 2) | Property 2.2 states that $\lvert x \rvert < k$ is equivalent to $x > -k$ and $x < k$, where $k$ is a positive number. This conjunction can be written in compact form as $-k < x < k$. For example, $\lvert x + 3 \rvert < 7$ can be written as $-7 < x + 3 < 7$ to begin the process of solving the inequality.<br><br>Property 2.3 states that $\lvert x \rvert > k$ is equivalent to $x < -k$ or $x > k$, where $k$ is a positive number. This disjunction cannot be written in a compact form. | Solve $\lvert x + 5 \rvert > 8$.<br>**Solution**<br>$\lvert x + 5 \rvert > 8$<br>$x + 5 < -8$ or $x + 5 > 8$<br>$x < -13$ or $x > 3$<br>The solution set is $(-\infty, -13) \cup (3, \infty)$.<br>**Sample Problem 15**<br>Solve $\lvert x - 8 \rvert \leq 10$. |

## Chapter 2 Review Problem Set

For Problems 1–28, determine if the given problem is an equation or an expression. If it is an equation, then solve. If it is an expression, then simplify.

1. $5(x - 6) = 3(x + 2)$
2. $2(2x + 1) - (x - 4) = 4(x + 5)$
3. $-(2n - 1) + 3(n + 2) = 7$
4. $2(3n - 4) + 3(2n - 3) = -2(n + 5)$
5. $\frac{3t - 2}{4} = \frac{2t + 1}{3}$
6. $\frac{x + 6}{5} + \frac{x - 1}{4} = 2$
7. $1 - \frac{2x - 1}{6} = \frac{3x}{8}$
8. $\frac{2x + 1}{3} + \frac{x - 4}{6}$
9. $\frac{3n - 1}{2} - \frac{2n + 3}{7} = 1$
10. $\frac{5x + 6}{2} - \frac{x - 4}{3} = \frac{5}{6}$
11. $\frac{3x + 4}{3} - \frac{2x + 6}{2} = \frac{1}{6}$
12. $\frac{2x + 3}{6} = \frac{4x + 6}{12}$
13. $\frac{n}{4} + \frac{n + 3}{2} = \frac{2n - 5}{3}$
14. $y + \frac{y - 4}{5} = \frac{3y + 5}{2}$
15. $\frac{1}{3}(5x + 1) - \frac{4}{3}(x + 2)$
16. $\frac{2x + 1}{3} + \frac{3x - 1}{5} = \frac{1}{10}$
17. $\frac{1}{2}(3x + 1) - \frac{4}{3}(x + 2) = 5$
18. $\frac{3}{5}(2x - 1) - \frac{3}{4}(x - 2) = -2$
19. $\frac{2x + 2}{4} = \frac{3x + 3}{6}$
20. $\frac{5x + 2}{5} - \frac{3x + 1}{3} = \frac{4}{15}$
21. $2x - 1 + \frac{2}{5}(3x - 2) = -\frac{9}{5}$
22. $3x + 2 + \frac{1}{2}(5x - 2) = -\frac{9}{2}$
23. $0.04(x + 1500) + 0.03x$
24. $0.03(2x + 8000) + 0.05(3x + 2000)$
25. $0.06x + 0.08(x + 100) = 15$
26. $0.4(t - 6) = 0.3(2t + 5)$
27. $0.1(n + 300) = 0.09n + 32$
28. $0.2(x - 0.5) - 0.3(x + 1) = 0.4$

Solve each of Problems 29–42 by setting up and solving an appropriate equation.

29. The width of a rectangle is 2 meters more than one-half of the length. The perimeter of the rectangle is 46 meters. Find the length and width of the rectangle.
30. Find three consecutive integers such that the sum of one-half of the smallest and one-third of the largest is one less than the other integer.
31. Pat is paid time-and-a-half for each hour he works over 36 hours in a week. Last week he worked 42 hours for a total of \$472.50. What is his normal hourly rate?
32. Marcela has a collection of nickels, dimes, and quarters worth \$24.75. The number of dimes is 10 more than twice the number of nickels, and the number of quarters is 25 more than the numbers of dimes. How many coins of each kind does she have?
33. If the complement of an angle is one-tenth of the supplement of the angle, find the measure of the angle.
34. A total of \$500 was invested, part of it at 7% interest and the remainder at 8%. If the total yearly interest from both investments amounted to \$38, how much was invested at each rate?
35. A retailer has some sweaters that cost her \$38 each. She wants to sell them at a profit of 20% of her cost. What price should she charge for each sweater?
36. If a necklace cost a jeweler \$60, at what price should it be sold to yield a profit of 80% based on the selling price?
37. If a DVD player costs a retailer \$40 and it sells for \$100, what is the rate of profit on the selling price?
38. Yuri bought a pair of running shoes at a 25% discount sale for \$78. What was the original price of the running shoes?
39. Solve $i = Prt$ for $P$, given that $r = 6\%$, $t = 3$ years, and $i = \$1440$.
40. Solve $A = P + Prt$ for $r$, given that $A = \$3706$, $P = \$3400$, and $t = 2$ years. Express $r$ as a percent.

41. Solve $P = 2w + 2l$ for $w$, given that $P = 86$ meters and $l = 32$ meters.

42. Solve $C = \frac{5}{9}(F - 32)$ for C, given that $F = -4°$.

For Problems 43–47, solve each equation for $x$.

43. $ax - b = b + 2$

44. $ax = bx + c$

45. $m(x + a) = p(x + b)$

46. $5x - 7y = 11$

47. $\frac{x - a}{b} = \frac{y + 1}{c}$

For Problems 48–52, solve each of the formulas for the indicated variable.

48. $A = \pi r^2 + \pi rs$ for $s$

49. $A = \frac{1}{2}h(b_1 + b_2)$ for $b_2$

50. $S_n = \frac{n(a_1 + a_2)}{2}$ for $n$

51. $\frac{1}{R} = \frac{1}{R_1} + \frac{1}{R_2}$ for $R$

52. $ax + by = c$ for $y$

53. How many pints of a 1% hydrogen peroxide solution should be mixed with a 4% hydrogen peroxide solution to obtain 10 pints of a 2% hydrogen peroxide solution?

54. Gladys leaves a town driving at a rate of 40 miles per hour. Two hours later, Reena leaves from the same place traveling the same route. She catches Gladys in 5 hours and 20 minutes. How fast was Reena traveling?

55. In $1\frac{1}{4}$ hours more time, Rita, riding her bicycle at 12 miles per hour, rode 2 miles farther than Sonya, who was riding her bicycle at 16 miles per hour. How long did each girl ride?

56. How many cups of orange juice must be added to 50 cups of a punch that is 10% orange juice to obtain a punch that is 20% orange juice?

For Problems 57–60, express the given inequality in interval notation.

57. $x \geq -2$

58. $x > 6$

59. $x < -1$

60. $x \leq 0$

For Problems 61–70, solve each of the inequalities.

61. $5x - 2 \geq 4x - 7$

62. $3 - 2x < -5$

63. $2(3x - 1) - 3(x - 3) > 0$

64. $3(x + 4) \leq 5(x - 1)$

65. $-3(2t - 1) - (t + 2) > -6(t - 3)$

66. $\frac{5}{6}n - \frac{1}{3}n < \frac{1}{6}$

67. $\frac{n - 4}{5} + \frac{n - 3}{6} > \frac{7}{15}$

68. $\frac{2}{3}(x - 1) + \frac{1}{4}(2x + 1) < \frac{5}{6}(x - 2)$

69. $s \geq 4.5 + 0.25s$

70. $0.07x + 0.09(500 - x) \geq 43$

For Problems 71–78, graph the solutions of each compound inequality.

71. $x > -1$ and $x < 1$

72. $x > 2$ or $x \leq -3$

73. $x > 2$ and $x > 3$

74. $x < 2$ or $x > -1$

75. $2x + 1 > 3$ or $2x + 1 < -3$

76. $2 < x + 4 < 5$

77. $-1 < 4x - 3 \leq 9$

78. $x + 1 > 3$ and $x - 3 < -5$

79. Susan's average score for her first three psychology exams is 84. What must she get on the fourth exam so that her average for the four exams is 85 or better?

80. Marci invests \$3000 at 6% yearly interest. How much does she have to invest at 8% so that the yearly interest from the two investments exceeds \$500?

For Problems 81–84, solve each of the equations.

81. $|3x - 1| = 11$

82. $|2n + 3| = 4$

83. $|3x + 1| - 8 = 2$

84. $\left|\frac{1}{2}x + 3\right| - 1 = 5$

For Problems 85–88, solve each of the inequalities.

85. $|2x - 1| < 11$

86. $|3x + 1| > 10$

87. $|5x - 4| \geq 8$

88. $\left|\frac{1}{4}x + 1\right| \leq 6$

# Chapter 2 Test

For Problems 1–10, solve each equation.

1. $5x - 2 = 2x - 11$

2. $6(n - 2) - 4(n + 3) = -14$

3. $-3(x + 4) = 3(x - 5)$

4. $3(2x - 1) - 2(x + 5) = -(x - 3)$

5. $\frac{3t - 2}{4} = \frac{5t + 1}{5}$

6. $\frac{5x + 2}{3} - \frac{2x + 4}{6} = -\frac{4}{3}$

7. $|4x - 3| = 9$

8. $\frac{1 - 3x}{4} + \frac{2x + 3}{3} = 1$

9. $2 - \frac{3x - 1}{5} = -4$

10. $0.05x + 0.06(1500 - x) = 83.5$

11. Solve $\frac{2}{3}x - \frac{3}{4}y = 2$ for $y$

12. Solve $S = 2\pi r(r + h)$ for $h$

For Problems 13–20, solve each inequality and express the solution set using interval notation.

13. $7x - 4 > 5x - 8$

14. $-3x - 4 \leq x + 12$

15. $2(x - 1) - 3(3x + 1) \geq -6(x - 5)$

16. $\frac{3}{5}x - \frac{1}{2}x < 1$

17. $\frac{x - 2}{6} - \frac{x + 3}{9} > -\frac{1}{2}$

18. $0.05x + 0.07(800 - x) \geq 52$

19. $|6x - 4| < 10$

20. $|4x + 5| \geq 6$

For Problems 21–25, solve each problem by setting up and solving an appropriate equation or inequality.

21. Dela bought a dress at a 20% discount sale for \$57.60. Find the original price of the dress.

22. The length of a rectangle is one centimeter more than three times its width. If the perimeter of the rectangle is 50 centimeters, find the length of the rectangle.

23. How many cups of grapefruit juice must be added to 30 cups of a punch that is 8% grapefruit juice to obtain a punch that is 10% grapefruit juice?

24. Rex has scores of 85, 92, 87, 88, and 91 on the first five exams. What score must he get on the sixth exam to have an average of 90 or better for all six exams?

25. If the complement of an angle is $\frac{2}{11}$ of the supplement of the angle, find the measure of the angle.

# Chapters 1–2 Cumulative Review Problem Set

**1.** Place a check mark in the table to identify all the sets that the identified number belongs to.

| Identified numbers | Natural numbers | Whole numbers | Integers | Rational numbers | Irrational numbers | Real numbers |
|---|---|---|---|---|---|---|
| $9$ | | | | | | |
| $-\frac{1}{2}$ | | | | | | |
| $-\sqrt{7}$ | | | | | | |
| $0.\overline{3}$ | | | | | | |
| $\frac{8}{3}$ | | | | | | |
| $-2$ | | | | | | |
| $0$ | | | | | | |

**2.** State the property of equality or the property of real numbers that justifies the statements.

**a.** $c(x) = x(c)$

**b.** $4(23 + 2) = 4(23) + 4(2)$

**c.** If $10 = a + 3$, then $a + 3 = 10$.

For Problems 3–9, simplify each numerical expression.

**3.** $20 \div 10 \cdot 2 - 6 \div 1 + 5$

**4.** $15 + 9\left(\frac{8 - 2}{4 - 1}\right) - 18 \div 2$

**5.** $(30 - 18)(16 \div 2 - 4 \div 4)$

**6.** $\left(\frac{30 - 3 \cdot 8 + 12 \div 2}{20 - 4 \cdot 4}\right) - 16 \div 8$

**7.** $4\left(-\frac{2}{3}\right) - 2\left(\frac{3}{5}\right) + 3\left(-\frac{1}{5}\right)$

**8.** $(-2)^2 - (-1)^3 - 5^2$

**9.** $\frac{4}{5}(10 - 15) - \frac{1}{3}(12 - 18)$

For Problems 10–12, simplify each algebraic expression by combining similar terms.

**10.** $3c^2 - 7 - 10c^2 + 8 + c^2$

**11.** $11(2a - 1) + 6(a + 3) - (3a - 2)$

**12.** $-\frac{1}{4}cd^2 + \frac{5}{6}cd^2 + \frac{11}{12}cd^2 - \frac{2}{3}cd^2$

For Problems 13–15, evaluate each algebraic expression for the given values of the variables.

**13.** $3x - 7y$ for $x = -5$ and $y = -2$

**14.** $5(x - 7) - 2(x - 18) - (x + 4)$ for $x = 5.2$

**15.** $6a^2 - b^2$ for $a = \frac{1}{2}$ and $b = -\frac{2}{3}$

**16.** Translate the following sentence into an algebraic expression. Use $x$ to represent the unknown number: The quotient of twice the number and the quantity two less than three times the number.

For Problems 17–23, solve each equation for $x$.

**17.** $4(x - 7) - (x + 5) = -7 - 8(x - 3)$

**18.** $\frac{2x - 1}{6} - 3 = \frac{6x + 13}{3}$

**19.** $0.05x + 0.04(x - 400) = 92$

**20.** $\frac{2}{3}x - \frac{1}{2}y = z$

**21.** $|3x - 4| - 7 = 22$

**22.** $|4x - 1| = -5$

**23.** $|9x - 6| = 0$

For Problems 24–28, solve each inequality, expressing the solution set in interval notation, and graph the solution set.

**24.** $7(2x - 4) \geq 2(6x - 11) - 10$

**25.** $\frac{x}{5} - \frac{x-6}{2} > \frac{x+5}{5}$

**26.** $|3x - 4| \leq 5$

**27.** $|5 - 2x| > 5$

**28.** $|10x - 1| < -4$

Solve each of Problems 29–36 by setting up and solving an appropriate equation.

**29.** Last week Kari worked 52 hours and earned \$1044. When she works more than 40 hours per week, she is paid time and a half for overtime hours. What is Kari's hourly rate?

**30.** Becky sells dog leashes and collars at agility shows every weekend. One weekend she sold a total of 34 items. The number of leashes she sold was two less than three times the number of collars. How many of each did Becky sell?

**31.** Carolyn has 19 bills consisting of ten-dollar bills, twenty-dollar bills, and fifty-dollar bills. The number of twenties is three times the number of tens, and the number of fifties is one less than the number of tens. How many of each bill does Carolyn have? How much money does she have?

**32.** A Florida beach house rents at a 30% discount during the month of January. If the usual weekly rental amount is \$3750, what would be the rent for one week in January?

**33.** Twice the complement of an angle plus one half the supplement of the angle equals 60°. Find the angle.

**34.** Uta is driving from Florida to Tennessee for a family reunion. Her brother, Sven, is driving the same route, but he is leaving one-half hour later. Uta drives at an average speed of 65 miles per hour, and her brother drives at an average speed of 70 miles per hour. How many hours will Sven drive before he catches up to Uta?

**35.** Glenn invests \$4000 at 5% annual interest. How much more must he invest at that rate if he wants to earn \$500 in annual interest?

**36.** An automobile dealership is advertising their hybrid vehicle for \$24,900. If the county sales tax is 6.5%, what will the vehicle actually cost with the tax included?

# 3 Polynomials

kristian sekulic/the Agency Collection/Getty Images

*"There is only one corner of the universe you can be certain of improving, and that's your own self."*
ALDOUS HUXLEY

## Study Skill Tip

Students often focus on improving homework scores but rarely take the time to learn how to become better test takers. Because course grades are based mostly on test scores, learning how to become a better test taker is well worth the time.

Here are some suggestions to improve your test taking.

- When you begin the test, do a *data dump*. Write down any formulas or rules that will be necessary for the problems on the test. This relieves the pressure of trying to keep those formulas in your memory while working on problems.
- Preview the test. Note which problems appear easy for you and do those problems first. A word of caution here is to work carefully on the easy problems. Every student is expected to miss some of the hard problems, but when you combine that with losing points on the easy problems, your grade drops significantly.
- Skip difficult problems after reading them over twice and go on to the next problem. When you have finished the other problems, return to those problems you skipped. A new idea might have become apparent, giving you the ability to now solve that problem.
- At the end, review the test. Analyze where you miss the most points on your tests. If you miss the most points at the beginning, perhaps from being nervous, then your review should concentrate on the beginning of the test. If you miss more questions at the end, perhaps from becoming tired, then focus on reviewing the end of the test.

*How important do you feel good test-taking skills are to your success in your math class?*

## Chapter Preview

Chapter 3 introduces polynomials and factoring. Just like with arithmetic, you will want to learn all the operations for polynomials. The chapter has sections on adding, subtracting, and multiplying polynomials. Division of monomials is covered in this chapter, but division of polynomials is covered in a different chapter.

Most students do very well learning addition, subtraction, and multiplication of polynomials. The calculations follow orderly and structured rules. Factoring is the reverse operation of multiplying polynomials and is a little more demanding. Factoring is a skill that can only be learned by a lot of practice. It is crucial that you learn factoring because from this chapter on, about 70% of the problems involve factoring.

## 3.1 Polynomials: Sums and Differences

OBJECTIVES

1. Find the degree of a polynomial
2. Add and subtract polynomials
3. Simplify polynomial expressions
4. Use polynomials in geometry problems

Recall that algebraic expressions such as $5x$, $-6y^2$, $7xy$, $14a^2b$, and $-17ab^2c^3$ are called terms. A **term** is an indicated product and may contain any number of factors. The variables in a term are called **literal factors**, and the numerical factor is called the **numerical coefficient**. Thus for $7xy$, the $x$ and $y$ are literal factors, 7 is the numerical coefficient, and the term is in two variables ($x$ and $y$).

Terms that contain variables with only whole numbers as exponents are called **monomials**. The terms previously listed, $5x$, $-6y^2$, $7xy$, $14a^2b$, and $-17ab^2c^3$, are all monomials. (We shall work later with some algebraic expressions, such as $7x^{-1}y^{-1}$ and $6a^{-2}b^{-3}$, which are not monomials.)

The **degree** of a monomial is the sum of the exponents of the literal factors.

$7xy$ is of degree 2

$14a^2b$ is of degree 3

$-17ab^2c^3$ is of degree 6

$5x$ is of degree 1

$-6y^2$ is of degree 2

If the monomial contains only one variable, then the exponent of the variable is the degree of the monomial. The last two examples illustrate this point. We say that any nonzero constant term is of degree zero.

A **polynomial** is a monomial or a finite sum (or difference) of monomials. Thus

$$4x^2, \qquad 3x^2 - 2x - 4, \qquad 7x^4 - 6x^3 + 4x^2 + x - 1,$$

$$3x^2y - 2xy^2, \qquad \frac{1}{5}a^2 - \frac{2}{3}b^2, \qquad \text{and} \qquad 14$$

are examples of polynomials. In addition to calling a polynomial with one term a **monomial**, we also classify polynomials with two terms as **binomials**, and those with three terms as **trinomials**.

The **degree of a polynomial** is the degree of the term with the highest degree in the polynomial. The following examples illustrate some of this terminology.

The polynomial $4x^3y^4$ is a monomial in two variables of degree 7.

The polynomial $4x^2y - 2xy$ is a binomial in two variables of degree 3.

The polynomial $9x^2 - 7x + 1$ is a trinomial in one variable of degree 2.

## Adding and Subtracting Polynomials

Remember that *similar terms,* or *like terms,* are terms that have the same literal factors. In the preceding chapters, we have frequently simplified algebraic expressions by combining similar terms, as the next examples illustrate.

$$\begin{aligned} 2x + 3y + 7x + 8y &= 2x + 7x + 3y + 8y \\ &= (2 + 7)x + (3 + 8)y \\ &= 9x + 11y \end{aligned}$$

Steps in dashed boxes are usually done mentally

$$\begin{aligned} 4a - 7 - 9a + 10 &= 4a + (-7) + (-9a) + 10 \\ &= 4a + (-9a) + (-7) + 10 \\ &= (4 + (-9))a + (-7) + 10 \\ &= -5a + 3 \end{aligned}$$

Steps in dashed boxes are usually done mentally

Both addition and subtraction of polynomials rely on basically the same ideas. The commutative, associative, and distributive properties provide the basis for rearranging, regrouping, and combining similar terms. Let's consider some examples.

**Classroom Example**
Add $3x^2 - 4x + 1$ and $5x^2 + 3x - 6$.

**EXAMPLE 1** Add $4x^2 + 5x + 1$ and $7x^2 - 9x + 4$.

### Solution

We generally use the horizontal format for such work. Thus

$$\begin{aligned} (4x^2 + 5x + 1) + (7x^2 - 9x + 4) &= (4x^2 + 7x^2) + (5x - 9x) + (1 + 4) \\ &= 11x^2 - 4x + 5 \end{aligned}$$

**Classroom Example**
Add $2m + 9$, $5m - 2$, and $10m - 6$.

**EXAMPLE 2** Add $5x - 3$, $3x + 2$, and $8x + 6$.

### Solution

$$\begin{aligned} (5x - 3) + (3x + 2) + (8x + 6) &= (5x + 3x + 8x) + (-3 + 2 + 6) \\ &= 16x + 5 \end{aligned}$$

**Classroom Example**
Find the indicated sum:
$(3a^2b - 2ab^2) + (-6a^2b + 9ab^2) + (7a^2b - 5ab^2)$

**EXAMPLE 3**

Find the indicated sum: $(-4x^2y + xy^2) + (7x^2y - 9xy^2) + (5x^2y - 4xy^2)$

### Solution

$$\begin{aligned} &(-4x^2y + xy^2) + (7x^2y - 9xy^2) + (5x^2y - 4xy^2) \\ &\quad = (-4x^2y + 7x^2y + 5x^2y) + (xy^2 - 9xy^2 - 4xy^2) \\ &\quad = 8x^2y - 12xy^2 \end{aligned}$$

The concept of subtraction as adding the opposite extends to polynomials in general. Hence the expression $a - b$ is equivalent to $a + (-b)$. We can form the opposite of a polynomial by taking the opposite of each term. For example, the opposite of $3x^2 - 7x + 1$ is $-3x^2 + 7x - 1$. We express this in symbols as

$$-(3x^2 - 7x + 1) = -3x^2 + 7x - 1$$

Now consider the following subtraction problems.

**Classroom Example**
Subtract $2x^2 - 5x + 4$ from $6x^2 + 7x - 3$.

**EXAMPLE 4** Subtract $3x^2 + 7x - 1$ from $7x^2 - 2x - 4$.

## Solution

Use the horizontal format to obtain

$$\begin{aligned}(7x^2 - 2x - 4) - (3x^2 + 7x - 1) &= (7x^2 - 2x - 4) + (-3x^2 - 7x + 1)\\ &= (7x^2 - 3x^2) + (-2x - 7x) + (-4 + 1)\\ &= 4x^2 - 9x - 3\end{aligned}$$

**Classroom Example**
Subtract $4m^2 - 9m - 7$ from $10m^2 + 3$.

**EXAMPLE 5** Subtract $-3y^2 + y - 2$ from $4y^2 + 7$.

## Solution

Because subtraction is not a commutative operation, be sure to perform the subtraction in the correct order.

$$\begin{aligned}(4y^2 + 7) - (-3y^2 + y - 2) &= (4y^2 + 7) + (3y^2 - y + 2)\\ &= (4y^2 + 3y^2) + (-y) + (7 + 2)\\ &= 7y^2 - y + 9\end{aligned}$$

The following example demonstrates the use of the vertical format for this work.

**Classroom Example**
Subtract $5a^2 - 4ab + 11$ from $2a^2 + 3ab + 7$.

**EXAMPLE 6** Subtract $4x^2 - 7xy + 5y^2$ from $3x^2 - 2xy + y^2$.

## Solution

$$\begin{array}{r}3x^2 - 2xy + y^2\\ \underline{4x^2 - 7xy + 5y^2}\end{array}$$

Note which polynomial goes on the bottom and how the similar terms are aligned

Now we can form the opposite of the bottom polynomial and add.

$$\begin{array}{r}3x^2 - 2xy + y^2\\ \underline{-4x^2 + 7xy - 5y^2}\\ -x^2 + 5xy - 4y^2\end{array}$$

The opposite of $4x^2 - 7xy + 5y^2$ is $-4x^2 + 7xy - 5y^2$

We can also use the distributive property and the properties $a = 1(a)$ and $-a = -1(a)$ when adding and subtracting polynomials. The next example illustrates this approach.

**Classroom Example**
Perform the indicated operations: $(12t + 3) - (4t - 5) + (7t + 1)$

**EXAMPLE 7**

Perform the indicated operations: $(5x - 2) + (2x - 1) - (3x + 4)$

## Solution

$$\begin{aligned}(5x - 2) + (2x - 1) - (3x + 4) &= 1(5x - 2) + 1(2x - 1) - 1(3x + 4)\\ &= 1(5x) - 1(2) + 1(2x) - 1(1) - 1(3x) - 1(4)\\ &= 5x - 2 + 2x - 1 - 3x - 4\\ &= 5x + 2x - 3x - 2 - 1 - 4\\ &= 4x - 7\end{aligned}$$

We can do some of the steps mentally and simplify our format, as shown in the next two examples.

**Classroom Example**
Perform the indicated operations: $(9x^2 - 4y) - (2x^2 - 3) + (-3y + 6)$

## EXAMPLE 8

Perform the indicated operations: $(5a^2 - 2b) - (2a^2 + 4) + (-7b - 3)$

### Solution

$$\begin{aligned}(5a^2 - 2b) - (2a^2 + 4) + (-7b - 3) &= 5a^2 - 2b - 2a^2 - 4 - 7b - 3 \\ &= 3a^2 - 9b - 7\end{aligned}$$

**Classroom Example**
Simplify $(8x^2 + 3x - 7) - (3x^2 - x - 2)$.

## EXAMPLE 9

Simplify $(4t^2 - 7t - 1) - (t^2 + 2t - 6)$.

### Solution

Remember that a polynomial in parentheses preceded by a negative sign can be written without the parentheses by replacing each term with its opposite. Thus $-(t^2 + 2t - 6) = -t^2 - 2t + 6$.

$$\begin{aligned}(4t^2 - 7t - 1) - (t^2 + 2t - 6) &= 4t^2 - 7t - 1 - t^2 - 2t + 6 \\ &= 3t^2 - 9t + 5\end{aligned}$$

Finally, let's consider a simplification problem that contains grouping symbols within grouping symbols.

**Classroom Example**
Simplify $12m + [5m - (m - 6)]$.

## EXAMPLE 10

Simplify $7x + [3x - (2x + 7)]$.

### Solution

$$\begin{aligned}7x + [3x - (2x + 7)] &= 7x + [3x - 2x - 7] \quad \text{Remove the innermost parentheses first} \\ &= 7x + [x - 7] \\ &= 7x + x - 7 \\ &= 8x - 7\end{aligned}$$

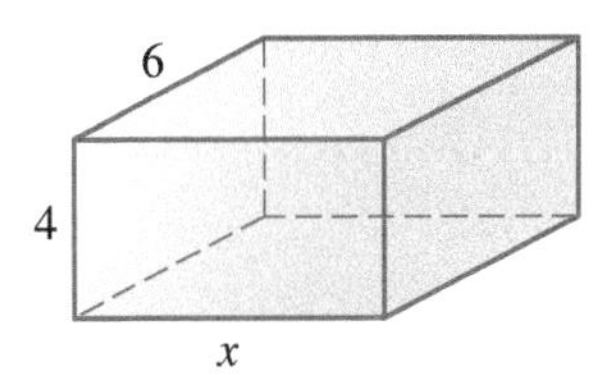

**Figure 3.1**

Sometimes we encounter polynomials in a geometric setting. For example, we can find a polynomial that represents the total surface area of the rectangular solid in Figure 3.1 as follows:

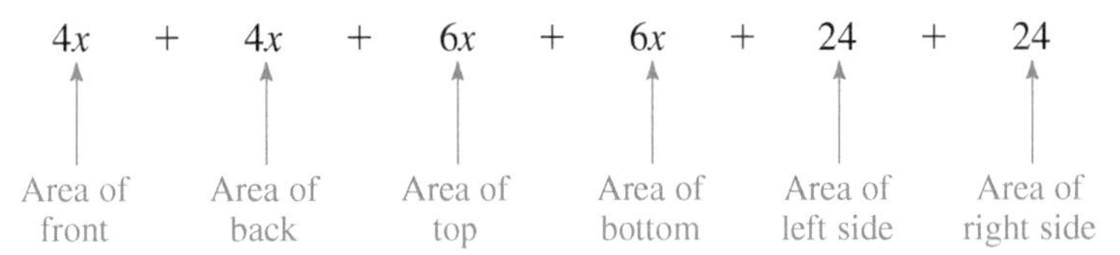


Simplifying $4x + 4x + 6x + 6x + 24 + 24$, we obtain the polynomial $20x + 48$, which represents the total surface area of the rectangular solid. Furthermore, by evaluating the polynomial $20x + 48$ for different positive values of $x$, we can determine the total surface area of any rectangular solid for which two dimensions are 4 and 6. The following chart contains some specific rectangular solids.

| x | 4 by 6 by x Rectangular solid | Total surface area (20x + 48) |
|---|---|---|
| 2 | 4 by 6 by 2 | 20(2) + 48 = 88 |
| 4 | 4 by 6 by 4 | 20(4) + 48 = 128 |
| 5 | 4 by 6 by 5 | 20(5) + 48 = 148 |
| 7 | 4 by 6 by 7 | 20(7) + 48 = 188 |
| 12 | 4 by 6 by 12 | 20(12) + 48 = 288 |

## Concept Quiz 3.1

For Problems 1–10, answer true or false.

1. The degree of the monomial $4x^2y$ is 3.
2. The degree of the polynomial $2x^4 - 5x^3 + 7x^2 - 4x + 6$ is 10.
3. A three-term polynomial is called a binomial.
4. A polynomial is a monomial or a finite sum of monomials.
5. Monomial terms must have whole number exponents for each variable.
6. The sum of $-2x - 1$, $-x + 4$, and $5x - 7$ is $8x - 4$.
7. If $3x - 4$ is subtracted from $-7x + 2$, the result is $-10x + 6$.
8. Polynomials must be of the same degree if they are to be added.
9. If $-x - 1$ is subtracted from the sum of $2x - 1$ and $-4x - 6$, the result is $-x - 6$.
10. We can form the opposite of a polynomial by taking the opposite of each term.

## Problem Set 3.1

For Problems 1–10, determine the degree of the given polynomials. (Objective 1)

1. $7xy + 6y$
2. $-5x^2y^2 - 6xy^2 + x$
3. $-x^2y + 2xy^2 - xy$
4. $5x^3y^2 - 6x^3y^3$
5. $5x^2 - 7x - 2$
6. $7x^3 - 2x + 4$
7. $8x^6 + 9$
8. $5y^6 + y^4 - 2y^2 - 8$
9. $-12$
10. $7x - 2y$

For Problems 11–20, add the given polynomials. (Objective 2)

11. $3x - 7$ and $7x + 4$
12. $9x + 6$ and $5x - 3$
13. $-5t - 4$ and $-6t + 9$
14. $-7t + 14$ and $-3t - 6$
15. $3x^2 - 5x - 1$ and $-4x^2 + 7x - 1$
16. $6x^2 + 8x + 4$ and $-7x^2 - 7x - 10$
17. $12a^2b^2 - 9ab$ and $5a^2b^2 + 4ab$
18. $15a^2b^2 - ab$ and $-20a^2b^2 - 6ab$
19. $2x - 4$, $-7x + 2$, and $-4x + 9$
20. $-x^2 - x - 4$, $2x^2 - 7x + 9$, and $-3x^2 + 6x - 10$

For Problems 21–30, subtract the polynomials using the horizontal format. (Objective 2)

21. $5x - 2$ from $3x + 4$
22. $7x + 5$ from $2x - 1$
23. $-4a - 5$ from $6a + 2$
24. $5a + 7$ from $-a - 4$
25. $3x^2 - x + 2$ from $7x^2 + 9x + 8$
26. $5x^2 + 4x - 7$ from $3x^2 + 2x - 9$
27. $2a^2 - 6a - 4$ from $-4a^2 + 6a + 10$
28. $-3a^2 - 6a + 3$ from $3a^2 + 6a - 11$
29. $2x^3 + x^2 - 7x - 2$ from $5x^3 + 2x^2 + 6x - 13$
30. $6x^3 + x^2 + 4$ from $9x^3 - x - 2$

For Problems 31–40, subtract the polynomials using the vertical format. (Objective 2)

31. $5x - 2$ from $12x + 6$
32. $3x - 7$ from $2x + 1$
33. $-4x + 7$ from $-7x - 9$
34. $-6x - 2$ from $5x + 6$
35. $2x^2 + x + 6$ from $4x^2 - x - 2$
36. $4x^2 - 3x - 7$ from $-x^2 - 6x + 9$
37. $x^3 + x^2 - x - 1$ from $-2x^3 + 6x^2 - 3x + 8$
38. $2x^3 - x + 6$ from $x^3 + 4x^2 + 1$
39. $-5x^2 + 6x - 12$ from $2x - 1$
40. $2x^2 - 7x - 10$ from $-x^3 - 12$

For Problems 41–46, perform the operations as described. (Objective 2)

41. Subtract $2x^2 - 7x - 1$ from the sum of $x^2 + 9x - 4$ and $-5x^2 - 7x + 10$.
42. Subtract $4x^2 + 6x + 9$ from the sum of $-3x^2 - 9x + 6$ and $-2x^2 + 6x - 4$.
43. Subtract $-x^2 - 7x - 1$ from the sum of $4x^2 + 3$ and $-7x^2 + 2x$.

**44.** Subtract $-4x^2 + 6x - 3$ from the sum of $-3x + 4$ and $9x^2 - 6$.

**45.** Subtract the sum of $5n^2 - 3n - 2$ and $-7n^2 + n + 2$ from $-12n^2 - n + 9$.

**46.** Subtract the sum of $-6n^2 + 2n - 4$ and $4n^2 - 2n + 4$ from $-n^2 - n + 1$.

For Problems 47–56, perform the indicated operations. (Objective 2)

**47.** $(5x + 2) + (7x - 1) + (-4x - 3)$

**48.** $(-3x + 1) + (6x - 2) + (9x - 4)$

**49.** $(12x - 9) - (-3x + 4) - (7x + 1)$

**50.** $(6x + 4) - (4x - 2) - (-x - 1)$

**51.** $(2x^2 - 7x - 1) + (-4x^2 - x + 6) + (-7x^2 - 4x - 1)$

**52.** $(5x^2 + x + 4) + (-x^2 + 2x + 4) + (-14x^2 - x + 6)$

**53.** $(7x^2 - x - 4) - (9x^2 - 10x + 8) + (12x^2 + 4x - 6)$

**54.** $(-6x^2 + 2x + 5) - (4x^2 + 4x - 1) + (7x^2 + 4)$

**55.** $(n^2 - 7n - 9) - (-3n + 4) - (2n^2 - 9)$

**56.** $(6n^2 - 4) - (5n^2 + 9) - (6n + 4)$

For Problems 57–70, simplify by removing the inner parentheses first and working outward. (Objective 3)

**57.** $3x - [5x - (x + 6)]$

**58.** $7x - [2x - (-x - 4)]$

**59.** $2x^2 - [-3x^2 - (x^2 - 4)]$

**60.** $4x^2 - [-x^2 - (5x^2 - 6)]$

**61.** $-2n^2 - [n^2 - (-4n^2 + n + 6)]$

**62.** $-7n^2 - [3n^2 - (-n^2 - n + 4)]$

**63.** $[4t^2 - (2t + 1) + 3] - [3t^2 + (2t - 1) - 5]$

**64.** $-(3n^2 - 2n + 4) - [2n^2 - (n^2 + n + 3)]$

**65.** $[2n^2 - (2n^2 - n + 5)] + [3n^2 + (n^2 - 2n - 7)]$

**66.** $3x^2 - [4x^2 - 2x - (x^2 - 2x + 6)]$

**67.** $[7xy - (2x - 3xy + y)] - [3x - (x - 10xy - y)]$

**68.** $[9xy - (4x + xy - y)] - [4y - (2x - xy + 6y)]$

**69.** $[4x^3 - (2x^2 - x - 1)] - [5x^3 - (x^2 + 2x - 1)]$

**70.** $[x^3 - (x^2 - x + 1)] - [-x^3 + (7x^2 - x + 10)]$

For Problems 71–73, use geometry to solve the problems. (Objective 4)

**71.** Find a polynomial that represents the perimeter of each of the following figures (Figures 3.2, 3.3, and 3.4).

**(a)**

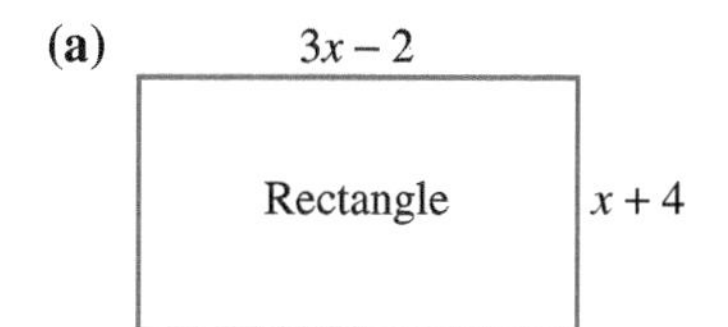


Figure 3.2

**(b)**

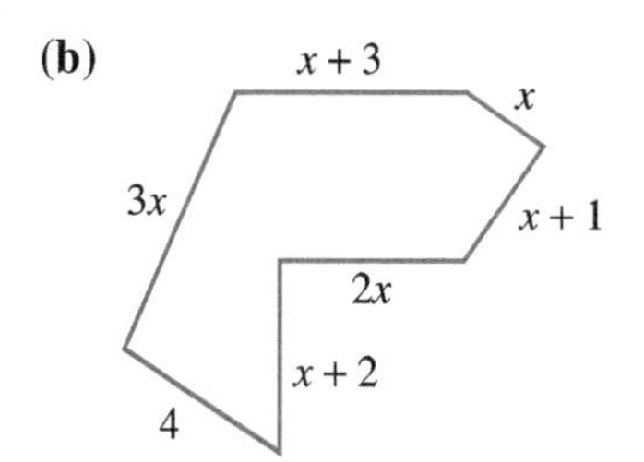


Figure 3.3

**(c)**

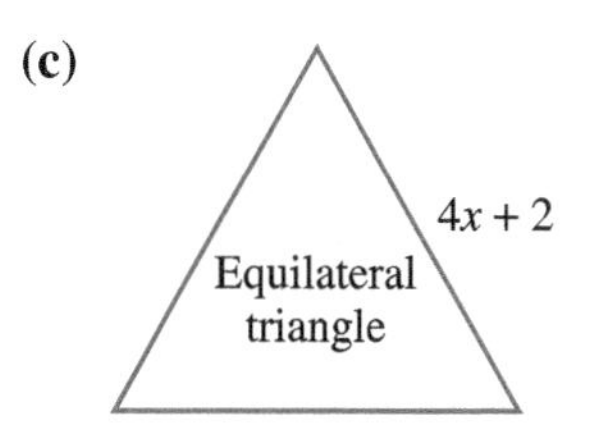


Figure 3.4

**72.** Find a polynomial that represents the total surface area of the rectangular solid in Figure 3.5.

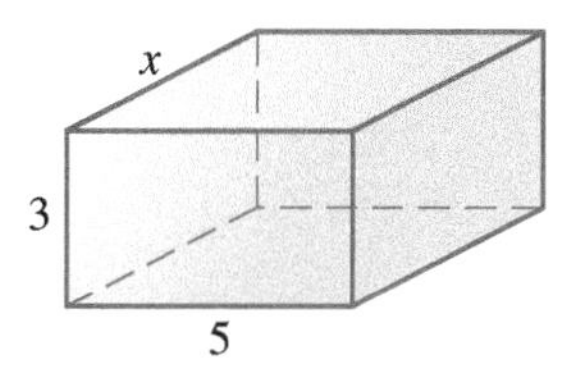


Figure 3.5

Now use that polynomial to determine the total surface area of each of the following rectangular solids.

**(a)** 3 by 5 by 4 **(b)** 3 by 5 by 7
**(c)** 3 by 5 by 11 **(d)** 3 by 5 by 13

**73.** Find a polynomial that represents the total surface area of the right circular cylinder in Figure 3.6. Now use that polynomial to determine the total surface area of each of the following right circular cylinders that have a base with a radius of 4. Use 3.14 for $\pi$, and express the answers to the nearest tenth.

**(a)** $h = 5$ **(b)** $h = 7$
**(c)** $h = 14$ **(d)** $h = 18$

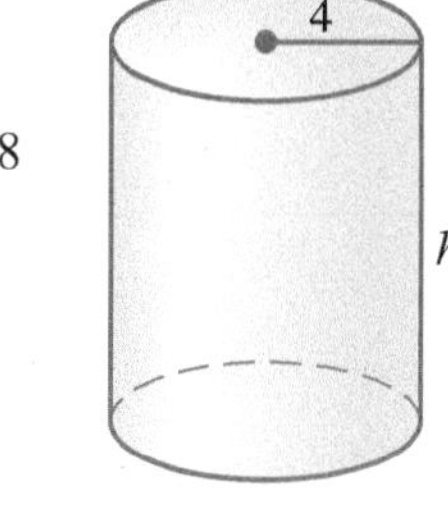


Figure 3.6

## Thoughts Into Words

**74.** Explain how to subtract the polynomial $-3x^2 + 2x - 4$ from $4x^2 + 6$.

**75.** Is the sum of two binomials always another binomial? Defend your answer.

**76.** Explain how to simplify the expression $7x - [3x - (2x - 4) + 2] - x$

### Answers to the Concept Quiz

**1.** True **2.** False **3.** False **4.** True **5.** True **6.** False **7.** True **8.** False **9.** True **10.** True

## 3.2 Products and Quotients of Monomials

OBJECTIVES

1. Multiply monomials
2. Raise a monomial to an exponent
3. Divide monomials
4. Use polynomials in geometry problems

Suppose that we want to find the product of two monomials such as $3x^2y$ and $4x^3y^2$. To proceed, use the properties of real numbers, and keep in mind that exponents indicate repeated multiplication.

$$\begin{aligned}(3x^2y)(4x^3y^2) &= (3 \cdot x \cdot x \cdot y)(4 \cdot x \cdot x \cdot x \cdot y \cdot y)\\ &= 3 \cdot 4 \cdot x \cdot x \cdot x \cdot x \cdot x \cdot y \cdot y \cdot y\\ &= 12x^5y^3\end{aligned}$$

You can use such an approach to find the product of any two monomials. However, there are some basic properties of exponents that make the process of multiplying monomials a much easier task. Let's consider each of these properties and illustrate its use when multiplying monomials. The following examples demonstrate the first property.

$$x^2 \cdot x^3 = (x \cdot x)(x \cdot x \cdot x) = x^5$$
$$a^4 \cdot a^2 = (a \cdot a \cdot a \cdot a)(a \cdot a) = a^6$$
$$b^3 \cdot b^4 = (b \cdot b \cdot b)(b \cdot b \cdot b \cdot b) = b^7$$

In general,

$$\begin{aligned}b^n \cdot b^m &= \underbrace{(b \cdot b \cdot b \cdot \ldots \cdot b)}_{n \text{ factors of } b}\underbrace{(b \cdot b \cdot b \cdot \ldots \cdot b)}_{m \text{ factors of } b}\\ &= \underbrace{b \cdot b \cdot b \cdot \ldots \cdot b}_{(n+m) \text{ factors of } b}\\ &= b^{n+m}\end{aligned}$$

We can state the first property as follows:

**Property 3.1 Product of the Same Base with Integer Exponents**

If $b$ is any real number, and $n$ and $m$ are positive integers, then

$$b^n \cdot b^m = b^{n+m}$$

Property 3.1 says that to find the product of two positive integral powers of the same base, we add the exponents and use this sum as the exponent of the common base.

$$x^7 \cdot x^8 = x^{7+8} = x^{15} \qquad y^6 \cdot y^4 = y^{6+4} = y^{10}$$

$$2^3 \cdot 2^8 = 2^{3+8} = 2^{11} \qquad (-3)^4 \cdot (-3)^5 = (-3)^{4+5} = (-3)^9$$

$$\left(\frac{2}{3}\right)^7 \cdot \left(\frac{2}{3}\right)^5 = \left(\frac{2}{3}\right)^{5+7} = \left(\frac{2}{3}\right)^{12}$$

The following examples illustrate the use of Property 3.1, along with the commutative and associative properties of multiplication, to form the basis for multiplying monomials. The steps enclosed in the dashed boxes could be performed mentally.

Classroom Example
$(2x^3y^4)(5xy^2)$

**EXAMPLE 1**

$$\begin{aligned}(3x^2y)(4x^3y^2) &= 3 \cdot 4 \cdot x^2 \cdot x^3 \cdot y \cdot y^2 \\ &= 12x^{2+3}y^{1+2} \\ &= 12x^5y^3\end{aligned}$$

Classroom Example
$(3m^2n^5)(-7m^2n^2)$

**EXAMPLE 2**

$$\begin{aligned}(-5a^3b^4)(7a^2b^5) &= -5 \cdot 7 \cdot a^3 \cdot a^2 \cdot b^4 \cdot b^5 \\ &= -35a^{3+2}b^{4+5} \\ &= -35a^5b^9\end{aligned}$$

Classroom Example
$\left(\frac{1}{3}x^5y^4\right)\left(\frac{3}{8}x^2y^3\right)$

**EXAMPLE 3**

$$\begin{aligned}\left(\frac{3}{4}xy\right)\left(\frac{1}{2}x^5y^6\right) &= \frac{3}{4} \cdot \frac{1}{2} \cdot x \cdot x^5 \cdot y \cdot y^6 \\ &= \frac{3}{8}x^{1+5}y^{1+6} \\ &= \frac{3}{8}x^6y^7\end{aligned}$$

Classroom Example
$(-4m^3n^2)(-m^2n)$

**EXAMPLE 4**

$$\begin{aligned}(-ab^2)(-5a^2b) &= (-1)(-5)(a)(a^2)(b^2)(b) \\ &= 5a^{1+2}b^{2+1} \\ &= 5a^3b^3\end{aligned}$$

Classroom Example
$(6xy^4)(2x^4)(3x^3y^2)$

**EXAMPLE 5**

$$\begin{aligned}(2x^2y^2)(3x^2y)(4y^3) &= 2 \cdot 3 \cdot 4 \cdot x^2 \cdot x^2 \cdot y^2 \cdot y \cdot y^3 \\ &= 24x^{2+2}y^{2+1+3} \\ &= 24x^4y^6\end{aligned}$$

**Classroom Example**
Find the product $(4x^{n+1})(3x^{n-2})(x^{2n+3})$.

**EXAMPLE 6** Find the product $(x^{2n})(3x^{n-1})(5x^6)$.

## Solution

For this problem there are variables in the exponents. The assumption will be made that the variables in the exponents represent positive numbers. We will apply Property 3.1 and add the exponents.

$$\begin{aligned}(x^{2n})(3x^{n-1})(5x^6) &= 3 \cdot 5 \cdot x^{2n} \cdot x^{n-1} \cdot x^6 \\ &= 15x^{2n+n-1+6} \\ &= 15x^{3n+5}\end{aligned}$$

The following examples demonstrate another useful property of exponents.

$$(x^2)^3 = x^2 \cdot x^2 \cdot x^2 = x^{2+2+2} = x^6$$
$$(a^3)^2 = a^3 \cdot a^3 = a^{3+3} = a^6$$
$$(b^4)^3 = b^4 \cdot b^4 \cdot b^4 = b^{4+4+4} = b^{12}$$

In general,

$$\begin{aligned}(b^n)^m &= \underbrace{b^n \cdot b^n \cdot b^n \cdot \ldots \cdot b^n}_{m \text{ factors of } b^n} \\ &= b^{\overbrace{n+n+n+\cdots+n}^{\text{adding } m \text{ of these}}} \\ &= b^{mn}\end{aligned}$$

We can state this property as follows:

**Property 3.2 Power Raised to a Power**

If $b$ is any real number, and $m$ and $n$ are positive integers, then

$$(b^n)^m = b^{mn}$$

The following examples show how Property 3.2 is used to find "the power of a power."

$$(x^4)^5 = x^{5(4)} = x^{20} \qquad (y^6)^3 = y^{3(6)} = y^{18}$$
$$(2^3)^7 = 2^{7(3)} = 2^{21}$$

A third property of exponents pertains to raising a monomial to a power. Consider the following examples, which we use to introduce the property.

$$(3x)^2 = (3x)(3x) = 3 \cdot 3 \cdot x \cdot x = 3^2 \cdot x^2$$
$$(4y^2)^3 = (4y^2)(4y^2)(4y^2) = 4 \cdot 4 \cdot 4 \cdot y^2 \cdot y^2 \cdot y^2 = (4)^3(y^2)^3$$
$$\begin{aligned}(-2a^3b^4)^2 = (-2a^3b^4)(-2a^3b^4) &= (-2)(-2)(a^3)(a^3)(b^4)(b^4) \\ &= (-2)^2(a^3)^2(b^4)^2\end{aligned}$$

In general,

$$\begin{aligned}(ab)^n &= \underbrace{(ab)(ab)(ab) \cdot \ldots \cdot (ab)}_{n \text{ factors of } ab} \\ &= \underbrace{(a \cdot a \cdot a \cdot \ldots \cdot a)}_{n \text{ factors of } a}\underbrace{(b \cdot b \cdot b \cdot \ldots \cdot b)}_{n \text{ factors of } b} \\ &= a^n b^n\end{aligned}$$

We can formally state Property 3.3 as follows:

**Property 3.3 Power of a Product**

If $a$ and $b$ are real numbers, and $n$ is a positive integer, then

$$(ab)^n = a^n b^n$$

Properties 3.2 and 3.3 form the basis for raising a monomial to a power, as in the next examples.

**Classroom Example**
$(m^4n^2)^7$

**EXAMPLE 7**

$$(x^2y^3)^4 = (x^2)^4(y^3)^4 \quad \text{Use } (ab)^n = a^nb^n$$
$$= x^8y^{12} \quad \text{Use } (b^n)^m = b^{mn}$$

**Classroom Example**
$(2r^3)^4$

**EXAMPLE 8**

$$(3a^5)^3 = (3)^3(a^5)^3$$
$$= 27a^{15}$$

**Classroom Example**
$(-3m^5n)^3$

**EXAMPLE 9**

$$(-2xy^4)^5 = (-2)^5(x)^5(y^4)^5$$
$$= -32x^5y^{20}$$

## Dividing Monomials

To develop an effective process for dividing by a monomial, we need yet another property of exponents. This property is a direct consequence of the definition of an exponent. Study the following examples.

$$\frac{x^4}{x^3} = \frac{\cancel{x} \cdot \cancel{x} \cdot \cancel{x} \cdot x}{\cancel{x} \cdot \cancel{x} \cdot \cancel{x}} = x \qquad \frac{x^3}{x^3} = \frac{\cancel{x} \cdot \cancel{x} \cdot \cancel{x}}{\cancel{x} \cdot \cancel{x} \cdot \cancel{x}} = 1$$

$$\frac{a^5}{a^2} = \frac{\cancel{a} \cdot \cancel{a} \cdot a \cdot a \cdot a}{\cancel{a} \cdot \cancel{a}} = a^3 \qquad \frac{y^5}{y^5} = \frac{\cancel{y} \cdot \cancel{y} \cdot \cancel{y} \cdot \cancel{y} \cdot \cancel{y}}{\cancel{y} \cdot \cancel{y} \cdot \cancel{y} \cdot \cancel{y} \cdot \cancel{y}} = 1$$

$$\frac{y^8}{y^4} = \frac{\cancel{y} \cdot \cancel{y} \cdot \cancel{y} \cdot \cancel{y} \cdot y \cdot y \cdot y \cdot y}{\cancel{y} \cdot \cancel{y} \cdot \cancel{y} \cdot \cancel{y}} = y^4$$

We can state the general property as follows:

**Property 3.4 Quotient of Same Base with Integer Exponents**

If $b$ is any nonzero real number, and $m$ and $n$ are positive integers, then

**1.** $\dfrac{b^n}{b^m} = b^{n-m}$ when $n > m$ **2.** $\dfrac{b^n}{b^m} = 1$ when $n = m$

Applying Property 3.4 to the previous examples yields

$$\frac{x^4}{x^3} = x^{4-3} = x^1 = x \qquad \frac{x^3}{x^3} = 1$$

$$\frac{a^5}{a^2} = a^{5-2} = a^3 \qquad \frac{y^5}{y^5} = 1$$

$$\frac{y^8}{y^4} = y^{8-4} = y^4$$

(We will discuss the situation when $n < m$ in a later chapter.)

Property 3.4, along with our knowledge of dividing integers, provides the basis for dividing monomials. The following example demonstrates the process.

**Classroom Example**
Simplify the following:
(a) $\frac{32y^6}{4y^3}$ (b) $\frac{-42m^{11}}{-14m^6}$
(c) $\frac{-48t^7}{6t^2}$ (d) $\frac{54a^4}{6a^4}$
(e) $\frac{56y^6}{-14y}$ (f) $\frac{16x^6y^9}{4x^2y^5}$

**EXAMPLE 10** Simplify the following:

(a) $\frac{24x^5}{3x^2}$ (b) $\frac{-36a^{13}}{-12a^5}$ (c) $\frac{-56x^9}{7x^4}$ (d) $\frac{72b^5}{8b^5}$ (e) $\frac{48y^7}{-12y}$ (f) $\frac{12x^4y^7}{2x^2y^4}$

### Solution

(a) $\frac{24x^5}{3x^2} = 8x^{5-2} = 8x^3$

(b) $\frac{-36a^{13}}{-12a^5} = 3a^{13-5} = 3a^8$

(c) $\frac{-56x^9}{7x^4} = -8x^{9-4} = -8x^5$

(d) $\frac{72b^5}{8b^5} = 9 \qquad \frac{b^5}{b^5} = 1$

(e) $\frac{48y^7}{-12y} = -4y^{7-1} = -4y^6$

(f) $\frac{12x^4y^7}{2x^2y^4} = 6x^{4-2}y^{7-4} = 6x^2y^3$

## Concept Quiz 3.2

For Problems 1–10, answer true or false.

1. When multiplying factors with the same base, add the exponents.
2. $3^2 \cdot 3^2 = 9^4$
3. $2x^2 \cdot 3x^3 = 6x^6$
4. $(x^2)^3 = x^5$
5. $(-4x^3)^2 = -4x^6$
6. To simplify $(3x^2y)(2x^3y^2)^4$ according to the order of operations, first raise $2x^3y^2$ to the fourth power and then multiply the monomials.
7. $\frac{-8x^6}{2x^2} = -4x^3$
8. $\frac{24x^3y^2}{-xy} = -24x^2y$
9. $\frac{-14xy^3}{-7xy^3} = 2$
10. $\frac{36a^2b^3c}{-18ab^2} = -2abc$

## Problem Set 3.2

For Problems 1–36, find each product. (Objective 1)

1. $(4x^3)(9x)$
2. $(6x^3)(7x^2)$
3. $(-2x^2)(6x^3)$
4. $(2xy)(-4x^2y)$
5. $(-a^2b)(-4ab^3)$
6. $(-8a^2b^2)(-3ab^3)$
7. $(x^2yz^2)(-3xyz^4)$
8. $(-2xy^2z^2)(-x^2y^3z)$

9. $(5xy)(-6y^3)$

10. $(-7xy)(4x^4)$

11. $(3a^2b)(9a^2b^4)$

12. $(-8a^2b^2)(-12ab^5)$

13. $(m^2n)(-mn^2)$

14. $(-x^3y^2)(xy^3)$

15. $\left(\frac{2}{5}xy^2\right)\left(\frac{3}{4}x^2y^4\right)$

16. $\left(\frac{1}{2}x^2y^6\right)\left(\frac{2}{3}xy\right)$

17. $\left(-\frac{3}{4}ab\right)\left(\frac{1}{5}a^2b^3\right)$

18. $\left(-\frac{2}{7}a^2\right)\left(\frac{3}{5}ab^3\right)$

19. $\left(-\frac{1}{2}xy\right)\left(\frac{1}{3}x^2y^3\right)$

20. $\left(\frac{3}{4}x^4y^5\right)(-x^2y)$

21. $(3x)(-2x^2)(-5x^3)$

22. $(-2x)(-6x^3)(x^2)$

23. $(-6x^2)(3x^3)(x^4)$

24. $(-7x^2)(3x)(4x^3)$

25. $(x^2y)(-3xy^2)(x^3y^3)$

26. $(xy^2)(-5xy)(x^2y^4)$

27. $(-3y^2)(-2y^2)(-4y^5)$

28. $(-y^3)(-6y)(-8y^4)$

29. $(4ab)(-2a^2b)(7a)$

30. $(3b)(-2ab^2)(7a)$

31. $(-ab)(-3ab)(-6ab)$

32. $(-3a^2b)(-ab^2)(-7a)$

33. $\left(\frac{2}{3}xy\right)(-3x^2y)(5x^4y^5)$

34. $\left(\frac{3}{4}x\right)(-4x^2y^2)(9y^3)$

35. $(12y)(-5x)\left(-\frac{5}{6}x^4y\right)$

36. $(-12x)(3y)\left(-\frac{3}{4}xy^6\right)$

For Problems 37–58, raise each monomial to the indicated power. (Objective 2)

37. $(3xy^2)^3$

38. $(4x^2y^3)^3$

39. $(-2x^2y)^5$

40. $(-3xy^4)^3$

41. $(-x^4y^5)^4$

42. $(-x^5y^2)^4$

43. $(ab^2c^3)^6$

44. $(a^2b^3c^5)^5$

45. $(2a^2b^3)^6$

46. $(2a^3b^2)^6$

47. $(9xy^4)^2$

48. $(8x^2y^5)^2$

49. $(-3ab^3)^4$

50. $(-2a^2b^4)^4$

51. $-(2ab)^4$

52. $-(3ab)^4$

53. $-(xy^2z^3)^6$

54. $-(xy^2z^3)^8$

55. $(-5a^2b^2c)^3$

56. $(-4abc^4)^3$

57. $(-xy^4z^2)^7$

58. $(-x^2y^4z^5)^5$

For Problems 59–74, find each quotient. (Objective 3)

59. $\frac{9x^4y^5}{3xy^2}$

60. $\frac{12x^2y^7}{6x^2y^3}$

61. $\frac{25x^5y^6}{-5x^2y^4}$

62. $\frac{56x^6y^4}{-7x^2y^3}$

63. $\frac{-54ab^2c^3}{-6abc}$

64. $\frac{-48a^3bc^5}{-6a^2c^4}$

65. $\frac{-18x^2y^2z^6}{xyz^2}$

66. $\frac{-32x^4y^5z^8}{x^2yz^3}$

67. $\frac{a^3b^4c^7}{-abc^5}$

68. $\frac{-a^4b^5c}{a^2b^4c}$

69. $\frac{-72x^2y^4}{-8x^2y^4}$

70. $\frac{-96x^4y^5}{12x^4y^4}$

71. $\frac{14ab^3}{-14ab}$

72. $\frac{-12abc^2}{12bc}$

73. $\frac{-36x^3y^5}{2y^5}$

74. $\frac{-48xyz^2}{2xz}$

For Problems 75–90, find each product. Assume that the variables in the exponents represent positive integers. (Objective 1) For example, $(x^{2n})(x^{3n}) = x^{2n+3n} = x^{5n}$.

75. $(2x^n)(3x^{2n})$

76. $(3x^{2n})(x^{3n-1})$

77. $(a^{2n-1})(a^{3n+4})$

78. $(a^{5n-1})(a^{5n+1})$

79. $(x^{3n-2})(x^{n+2})$

80. $(x^{n-1})(x^{4n+3})$

81. $(a^{5n-2})(a^3)$

82. $(x^{3n-4})(x^4)$

83. $(2x^n)(-5x^n)$

84. $(4x^{2n-1})(-3x^{n+1})$

85. $(-3a^2)(-4a^{n+2})$

86. $(-5x^{n-1})(-6x^{2n+4})$

87. $(x^n)(2x^{2n})(3x^2)$

88. $(2x^n)(3x^{3n-1})(-4x^{2n+5})$

89. $(3x^{n-1})(x^{n+1})(4x^{2-n})$

90. $(-5x^{n+2})(x^{n-2})(4x^{3-2n})$

For Problems 91–93, use geometry to solve the problems. (Objective 4)

91. Find a polynomial that represents the total surface area of the rectangular solid in Figure 3.7. Also find a polynomial that represents the volume.

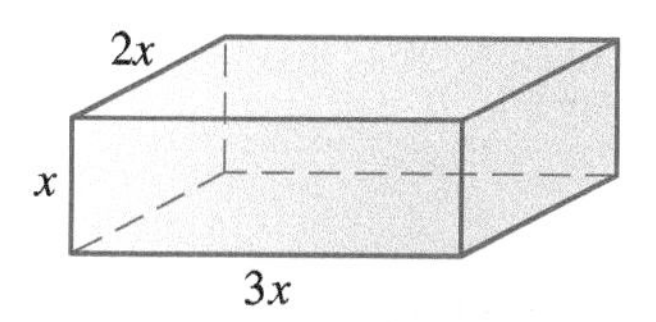


**Figure 3.7**

**92.** Find a polynomial that represents the total surface area of the rectangular solid in Figure 3.8. Also find a polynomial that represents the volume.

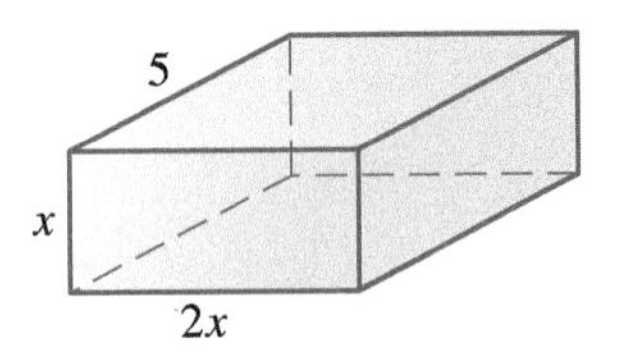


**Figure 3.8**

**93.** Find a polynomial that represents the area of the shaded region in Figure 3.9. The length of a radius of the larger circle is $r$ units, and the length of a radius of the smaller circle is 6 units.

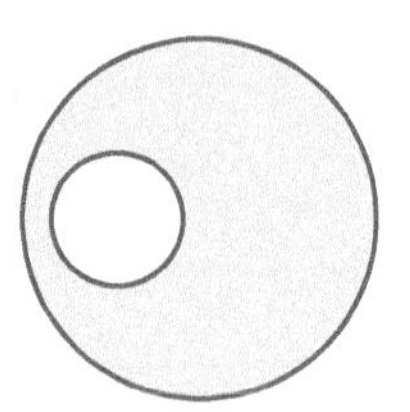

**Figure 3.9**

## Thoughts Into Words

**94.** How would you convince someone that $\frac{x^6}{x^2}$ is $x^4$ and not $x^3$?

**95.** Your friend simplifies $2^3 \cdot 2^2$ as follows:

$$2^3 \cdot 2^2 = 4^{3+2} = 4^5 = 1024$$

What has she done incorrectly and how would you help her?

### Answers to the Concept Quiz

**1.** True **2.** False **3.** False **4.** False **5.** False **6.** True **7.** False **8.** True **9.** True **10.** True

## 3.3 Multiplying Polynomials

OBJECTIVES

1. Multiply polynomials
2. Multiply two binomials
3. Use a pattern to find the square of a binomial and the difference of two squares
4. Find the cube of a binomial
5. Use polynomials in geometry problems

We usually state the distributive property as $a(b + c) = ab + ac$; however, we can extend it as follows:

$$a(b + c + d) = ab + ac + ad$$

$$a(b + c + d + e) = ab + ac + ad + ae, \text{ etc.}$$

We apply the commutative and associative properties, the properties of exponents, and the distributive property together to find the product of a monomial and a polynomial. The following examples illustrate this idea.

Classroom Example
$4n^3(3n^2 + 2n - 3)$

**EXAMPLE 1**

$$3x^2(2x^2 + 5x + 3) = 3x^2(2x^2) + 3x^2(5x) + 3x^2(3)$$
$$= 6x^4 + 15x^3 + 9x^2$$

Classroom Example
$-5mn(2m^3 + 6m^2n - 4mn^2 - 3n^3)$

**EXAMPLE 2**

$$\begin{aligned}&-2xy(3x^3 - 4x^2y - 5xy^2 + y^3)\\&= -2xy(3x^3) - (-2xy)(4x^2y) - (-2xy)(5xy^2) + (-2xy)(y^3)\\&= -6x^4y + 8x^3y^2 + 10x^2y^3 - 2xy^4\end{aligned}$$

Now let's consider the product of two polynomials, neither of which is a monomial. Consider the following examples.

Classroom Example
$(x + 4)(y + 8)$

**EXAMPLE 3**

$$\begin{aligned}(x + 2)(y + 5) &= x(y + 5) + 2(y + 5)\\&= x(y) + x(5) + 2(y) + 2(5)\\&= xy + 5x + 2y + 10\end{aligned}$$

Note that each term of the first polynomial is multiplied by each term of the second polynomial.

Classroom Example
$(r - 5)(s - t + 4)$

**EXAMPLE 4**

$$\begin{aligned}(x - 3)(y + z + 3) &= x(y + z + 3) - 3(y + z + 3)\\&= xy + xz + 3x - 3y - 3z - 9\end{aligned}$$

Multiplying polynomials often produces similar terms that can be combined to simplify the resulting polynomial.

Classroom Example
$(x + 3)(x + 9)$

**EXAMPLE 5**

$$\begin{aligned}(x + 5)(x + 7) &= x(x + 7) + 5(x + 7)\\&= x^2 + 7x + 5x + 35\\&= x^2 + 12x + 35\end{aligned}$$

Classroom Example
$(x - 3)(x^2 + 2x - 7)$

**EXAMPLE 6**

$$\begin{aligned}(x - 2)(x^2 - 3x + 4) &= x(x^2 - 3x + 4) - 2(x^2 - 3x + 4)\\&= x^3 - 3x^2 + 4x - 2x^2 + 6x - 8\\&= x^3 - 5x^2 + 10x - 8\end{aligned}$$

In Example 6, we are claiming that

$$(x - 2)(x^2 - 3x + 4) = x^3 - 5x^2 + 10x - 8$$

*for all real numbers*. In addition to going back over our work, how can we verify such a claim? Obviously, we cannot try all real numbers, but trying at least one number gives us a partial check. Let's try the number 4 in the left side of the equation and the right side.

When $x = 4$

$$\begin{aligned}&(x - 2)(x^2 - 3x + 4)\\&= (4 - 2)(4^2 - 3(4) + 4)\\&= (2)(16 - 12 + 4)\\&= 2(8)\\&= 16\end{aligned}$$

When $x = 4$

$$\begin{aligned}&x^3 - 5x^2 + 10x - 8\\&= 4^3 - 5(4)^2 + 10(4) - 8\\&= 64 - 80 + 40 - 8\\&= 16\end{aligned}$$

Classroom Example
$(5m - 3n)(m^2 - 3mn + n^2)$

**EXAMPLE 7**

$$\begin{aligned}&(3x - 2y)(x^2 + xy - y^2)\\&= 3x(x^2 + xy - y^2) - 2y(x^2 + xy - y^2)\\&= 3x^3 + 3x^2y - 3xy^2 - 2x^2y - 2xy^2 + 2y^3\\&= 3x^3 + x^2y - 5xy^2 + 2y^3\end{aligned}$$

It helps to be able to find the product of two binomials without showing all of the intermediate steps. This is quite easy to do with the *three-step shortcut pattern* demonstrated by Figures 3.10 and 3.11 in the following examples.

**Classroom Example**
Multiply $(x + 2)(x + 7)$.

**EXAMPLE 8** Multiply $(x + 3)(x + 8)$.

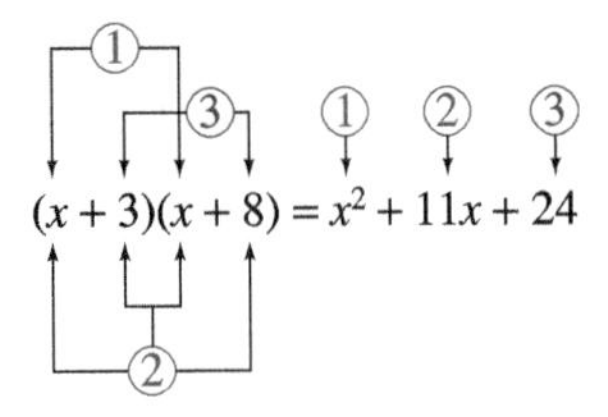


**Figure 3.10**

**Step 1** Multiply $x \cdot x$.
**Step 2** Multiply $3 \cdot x$ and $8 \cdot x$ and combine.
**Step 3** Multiply $3 \cdot 8$.

**Classroom Example**
Multiply $(4d + 1)(3d - 2)$.

**EXAMPLE 9** Multiply $(3x + 2)(2x - 1)$.

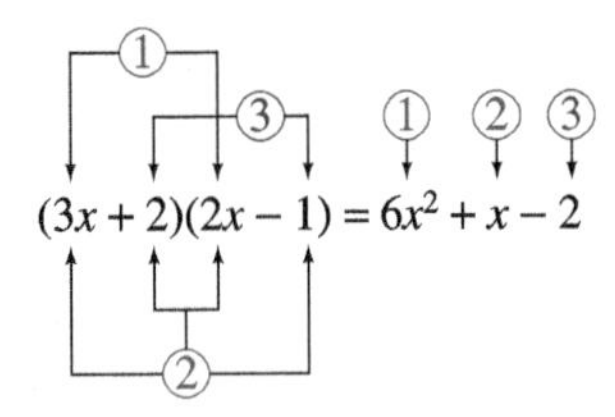


**Figure 3.11**

The acronym FOIL is often used to remember the pattern for multiplying binomials. The letters in FOIL represent, First, Outside, Inside, and Last. If you look back at Examples 8 and 9, step 1 is to find the product of the first terms in the binomial; step 2 is to find the sum of the product of the outside terms and the product of the inside terms; and step 3 is to find the product of the last terms in each binomial.

Now see if you can use the pattern to find the following products.

$(x + 2)(x + 6) = ?$
$(x - 3)(x + 5) = ?$
$(2x + 5)(3x + 7) = ?$
$(3x - 1)(4x - 3) = ?$

Your answers should be $x^2 + 8x + 12$, $x^2 + 2x - 15$, $6x^2 + 29x + 35$, and $12x^2 - 13x + 3$. Keep in mind that this shortcut pattern applies only to finding the product of two binomials.

We can use exponents to indicate repeated multiplication of polynomials. For example, $(x + 3)^2$ means $(x + 3)(x + 3)$, and $(x + 4)^3$ means $(x + 4)(x + 4)(x + 4)$. To square a binomial, we simply write it as the product of two equal binomials and apply the shortcut pattern. Thus

$$(x + 3)^2 = (x + 3)(x + 3) = x^2 + 6x + 9$$
$$(x - 6)^2 = (x - 6)(x - 6) = x^2 - 12x + 36 \quad \text{and}$$
$$(3x - 4)^2 = (3x - 4)(3x - 4) = 9x^2 - 24x + 16$$

When squaring binomials, be careful not to forget the middle term. That is to say, $(x + 3)^2 \neq x^2 + 3^2$; instead, $(x + 3)^2 = x^2 + 6x + 9$.

When multiplying binomials, there are some special patterns that you should recognize. We can use these patterns to find products, and later we will use some of them when factoring polynomials.

**Pattern 1** $(a + b)^2 = (a + b)(a + b) = a^2 + 2ab + b^2$

Square of first term of binomial + Twice the product of the two terms of binomial + Square of second term of binomial

**Classroom Example**
Expand the following squares of binomials:
(a) $(x + 6)^2$
(b) $(3x + y)^2$
(c) $(4x + 5y)^2$

**EXAMPLE 10** Expand the following squares of binomials:

**(a)** $(x + 4)^2$ **(b)** $(2x + 3y)^2$ **(c)** $(5a + 7b)^2$

### Solution

Square of the first term of binomial + Twice the product of the terms of binomial + Square of second term of binomial

**(a)** $(x + 4)^2 = x^2 + 8x + 16$

**(b)** $(2x + 3y)^2 = 4x^2 + 12xy + 9y^2$

**(c)** $(5a + 7b)^2 = 25a^2 + 70ab + 49b^2$

**Pattern 2** $(a - b)^2 = (a - b)(a - b) = a^2 - 2ab + b^2$

Square of first term of binomial − Twice the product of the two terms of binomial + Square of second term of binomial

**Classroom Example**
Expand the following squares of binomials:
(a) $(m - 4)^2$
(b) $(2x - 5y)^2$
(c) $(3x - 7y)^2$

**EXAMPLE 11** Expand the following squares of binomials:

**(a)** $(x - 8)^2$ **(b)** $(3x - 4y)^2$ **(c)** $(4a - 9b)^2$

### Solution

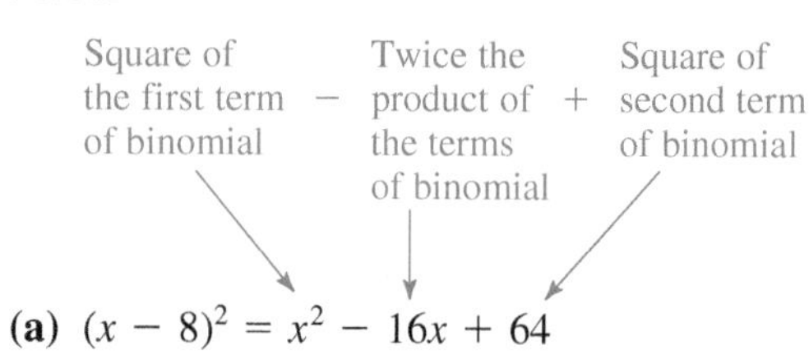


**(a)** $(x - 8)^2 = x^2 - 16x + 64$

**(b)** $(3x - 4y)^2 = 9x^2 - 24xy + 16y^2$

**(c)** $(4a - 9b)^2 = 16a^2 - 72ab + 81b^2$

**Pattern 3** $(a + b)(a - b) = a^2 - b^2$

Square of first term of binomials − Square of second term of binomials

Classroom Example
Find the product of the following:
(a) $(x + 8)(x - 8)$
(b) $(x - 4y)(x + 4y)$
(c) $(6x + 5y)(6x - 5y)$

**EXAMPLE 12** Find the product for the following:

**(a)** $(x + 7)(x - 7)$ **(b)** $(2x + y)(2x - y)$ **(c)** $(3a - 2b)(3a + 2b)$

### Solution

Square of the first term of binomial − Square of second term of binomial

**(a)** $(x + 7)(x - 7) = x^2 - 49$
**(b)** $(2x + y)(2x - y) = 4x^2 - y^2$
**(c)** $(3a - 2b)(3a + 2b) = 9a^2 - 4b^2$

Now suppose that we want to cube a binomial. One approach is as follows:

$$
\begin{aligned}
(x + 4)^3 &= (x + 4)(x + 4)(x + 4) \\
&= (x + 4)(x^2 + 8x + 16) \\
&= x(x^2 + 8x + 16) + 4(x^2 + 8x + 16) \\
&= x^3 + 8x^2 + 16x + 4x^2 + 32x + 64 \\
&= x^3 + 12x^2 + 48x + 64
\end{aligned}
$$

Another approach is to cube a general binomial and then use the resulting pattern.

**Pattern 4**
$$
\begin{aligned}
(a + b)^3 &= (a + b)(a + b)(a + b) \\
&= (a + b)(a^2 + 2ab + b^2) \\
&= a(a^2 + 2ab + b^2) + b(a^2 + 2ab + b^2) \\
&= a^3 + 2a^2b + ab^2 + a^2b + 2ab^2 + b^3 \\
&= a^3 + 3a^2b + 3ab^2 + b^3
\end{aligned}
$$

Classroom Example
Expand $(x + 5)^3$.

**EXAMPLE 13** Expand $(x + 4)^3$.

### Solution

Let's use the pattern $(a + b)^3 = a^3 + 3a^2b + 3ab^2 + b^3$ to cube the binomial $x + 4$.

$$
\begin{aligned}
(x + 4)^3 &= x^3 + 3x^2(4) + 3x(4)^2 + 4^3 \\
&= x^3 + 12x^2 + 48x + 64
\end{aligned}
$$

Because $a - b = a + (-b)$, we can easily develop a pattern for cubing $a - b$.

**Pattern 5**
$$
\begin{aligned}
(a - b)^3 &= [a + (-b)]^3 \\
&= a^3 + 3a^2(-b) + 3a(-b)^2 + (-b)^3 \\
&= a^3 - 3a^2b + 3ab^2 - b^3
\end{aligned}
$$

**Classroom Example**
Expand $(2x - 3y)^3$.

**EXAMPLE 14** Expand $(3x - 2y)^3$.

### Solution

Now let's use the pattern $(a - b)^3 = a^3 - 3a^2b + 3ab^2 - b^3$ to cube the binomial $3x - 2y$.

$$\begin{aligned}(3x - 2y)^3 &= (3x)^3 - 3(3x)^2(2y) + 3(3x)(2y)^2 - (2y)^3 \\ &= 27x^3 - 54x^2y + 36xy^2 - 8y^3\end{aligned}$$

Finally, we need to realize that if the patterns are forgotten or do not apply, then we can revert to applying the distributive property.

$$\begin{aligned}(2x - 1)(x^2 - 4x + 6) &= 2x(x^2 - 4x + 6) - 1(x^2 - 4x + 6) \\ &= 2x^3 - 8x^2 + 12x - x^2 + 4x - 6 \\ &= 2x^3 - 9x^2 + 16x - 6\end{aligned}$$

## Back to the Geometry Connection

As you might expect, there are geometric interpretations for many of the algebraic concepts we present in this section. We will give you the opportunity to make some of these connections between algebra and geometry in the next problem set. Let's conclude this section with a problem that allows us to use some algebra and geometry.

**Classroom Example**
A rectangular piece of steel is 20 centimeters long and 8 centimeters wide. From each corner a square piece $x$ centimeters on a side is cut out. The flaps are then turned up to form an open box. Find polynomials that represent the volume and outside surface area of the box.

**EXAMPLE 15** Apply Your Skill

A rectangular piece of tin is 16 inches long and 12 inches wide as shown in Figure 3.12. From each corner a square piece $x$ inches on a side is cut out. The flaps are then turned up to form an open box. Find polynomials that represent the volume and outside surface area of the box.

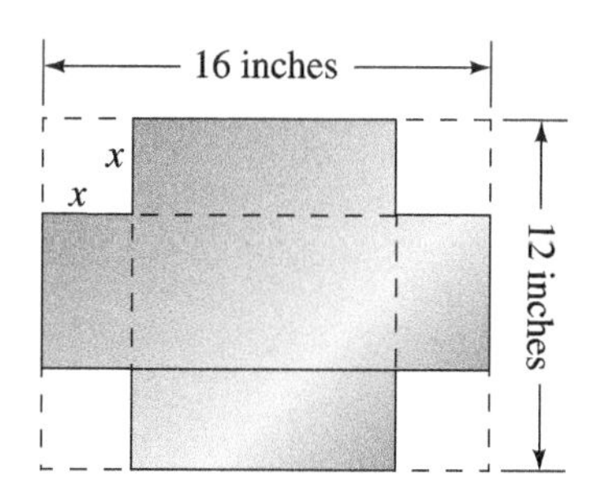


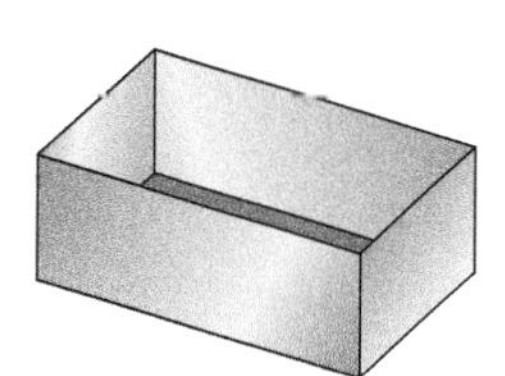

**Figure 3.12**

### Solution

The length of the box will be $16 - 2x$, the width $12 - 2x$, and the height $x$. With the volume formula $V = lwh$, the polynomial $(16 - 2x)(12 - 2x)(x)$, which simplifies to $4x^3 - 56x^2 + 192x$, represents the volume.

The outside surface area of the box is the area of the original piece of tin, minus the four corners that were cut off. Therefore, the polynomial $16(12) - 4x^2$, or $192 - 4x^2$, represents the outside surface area of the box.

**Remark:** Recall that in Section 3.1 we found the total surface area of a rectangular solid by adding the areas of the sides, top, and bottom. Use this approach for the open box in Example 15 to check our answer of $192 - 4x^2$. Keep in mind that the box has no top.

## Concept Quiz 3.3

For Problems 1–10, answer true or false.

1. The algebraic expression $(x + y)^2$ is called the square of a binomial.
2. The algebraic expression $(x + y)(x + 2xy + y)$ is called the product of two binomials.
3. The acronym FOIL stands for first, outside, inside, and last.
4. Although the distributive property is usually stated as $a(b + c) = ab + ac$, it can be extended, as in $a(b + c + d + e) = ab + ac + ad + ae$, when multiplying polynomials.
5. Multiplying polynomials often produces similar terms that can be combined to simplify the resulting product.
6. The pattern for $(a + b)^2$ is $a^2 + b^2$.
7. The pattern for $(a - b)^2$ is $a^2 - 2ab - b^2$.
8. The pattern for $(a + b)(a - b)$ is $a^2 - b^2$.
9. The pattern for $(a + b)^3$ is $a^3 + 3ab + b^3$.
10. The pattern for $(a - b)^3$ is $a^3 + 3a^2b - 3ab^2 - b^3$.

## Problem Set 3.3

For Problems 1–74, find each indicated product. Remember the shortcut for multiplying binomials and the other special patterns we discussed in this section. (Objectives 1–4)

1. $2xy(5xy^2 + 3x^2y^3)$
2. $3x^2y(6y^2 - 5x^2y^4)$
3. $-3a^2b(4ab^2 - 5a^3)$
4. $-7ab^2(2b^3 - 3a^2)$
5. $8a^3b^4(3ab - 2ab^2 + 4a^2b^2)$
6. $9a^3b(2a - 3b + 7ab)$
7. $-x^2y(6xy^2 + 3x^2y^3 - x^3y)$
8. $-ab^2(5a + 3b - 6a^2b^3)$
9. $(a + 2b)(x + y)$
10. $(t - s)(x + y)$
11. $(a - 3b)(c + 4d)$
12. $(a - 4b)(c - d)$
13. $(x + 6)(x + 10)$
14. $(x + 2)(x + 10)$
15. $(y - 5)(y + 11)$
16. $(y - 3)(y + 9)$
17. $(n + 2)(n - 7)$
18. $(n + 3)(n - 12)$
19. $(x + 6)(x - 6)$
20. $(t + 8)(t - 8)$
21. $(x - 6)^2$
22. $(x - 2)^2$
23. $(x - 6)(x - 8)$
24. $(x - 3)(x - 13)$
25. $(x + 1)(x - 2)(x - 3)$
26. $(x - 1)(x + 4)(x - 6)$
27. $(x - 3)(x + 3)(x - 1)$
28. $(x - 5)(x + 5)(x - 8)$
29. $(t + 9)^2$
30. $(t + 13)^2$
31. $(y - 7)^2$
32. $(y - 4)^2$
33. $(4x + 5)(x + 7)$
34. $(6x + 5)(x + 3)$
35. $(3y - 1)(3y + 1)$
36. $(5y - 2)(5y + 2)$
37. $(7x - 2)(2x + 1)$
38. $(6x - 1)(3x + 2)$
39. $(1 + t)(5 - 2t)$
40. $(3 - t)(2 + 4t)$
41. $(3t + 7)^2$
42. $(4t + 6)^2$
43. $(2 - 5x)(2 + 5x)$
44. $(6 - 3x)(6 + 3x)$
45. $(7x - 4)^2$
46. $(5x - 7)^2$
47. $(6x + 7)(3x - 10)$
48. $(4x - 7)(7x + 4)$
49. $(2x - 5y)(x + 3y)$
50. $(x - 4y)(3x + 7y)$
51. $(5x - 2a)(5x + 2a)$
52. $(9x - 2y)(9x + 2y)$
53. $(t + 3)(t^2 - 3t - 5)$
54. $(t - 2)(t^2 + 7t + 2)$
55. $(x - 4)(x^2 + 5x - 4)$
56. $(x + 6)(2x^2 - x - 7)$
57. $(2x - 3)(x^2 + 6x + 10)$
58. $(3x + 4)(2x^2 - 2x - 6)$
59. $(4x - 1)(3x^2 - x + 6)$
60. $(5x - 2)(6x^2 + 2x - 1)$
61. $(x^2 + 2x + 1)(x^2 + 3x + 4)$
62. $(x^2 - x + 6)(x^2 - 5x - 8)$
63. $(2x^2 + 3x - 4)(x^2 - 2x - 1)$
64. $(3x^2 - 2x + 1)(2x^2 + x - 2)$
65. $(x + 2)^3$
66. $(x + 1)^3$
67. $(x - 4)^3$
68. $(x - 5)^3$
69. $(2x + 3)^3$
70. $(3x + 1)^3$
71. $(4x - 1)^3$
72. $(3x - 2)^3$
73. $(5x + 2)^3$
74. $(4x - 5)^3$

For Problems 75–84, find the indicated products. Assume all variables that appear as exponents represent positive integers. (Objectives 2 and 3)

75. $(x^n - 4)(x^n + 4)$

76. $(x^{3a} - 1)(x^{3a} + 1)$

77. $(x^a + 6)(x^a - 2)$

78. $(x^a + 4)(x^a - 9)$

79. $(2x^n + 5)(3x^n - 7)$

80. $(3x^n + 5)(4x^n - 9)$

81. $(x^{2a} - 7)(x^{2a} - 3)$

82. $(x^{2a} + 6)(x^{2a} - 4)$

83. $(2x^n + 5)^2$

84. $(3x^n - 7)^2$

For Problems 85–89, use geometry to solve the problems. (Objective 5)

85. Explain how Figure 3.13 can be used to demonstrate geometrically that $(x + 2)(x + 6) = x^2 + 8x + 12$.

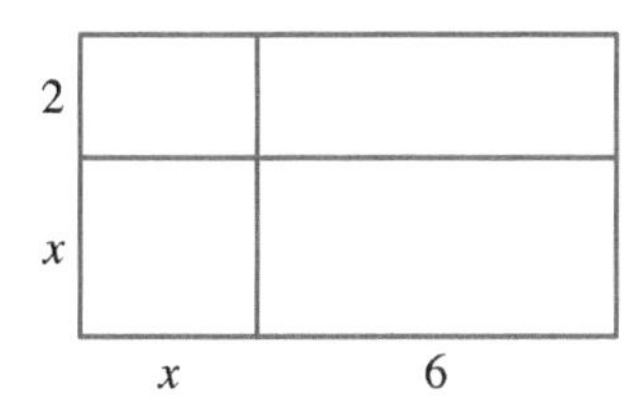


**Figure 3.13**

86. Find a polynomial that represents the sum of the areas of the two rectangles shown in Figure 3.14.

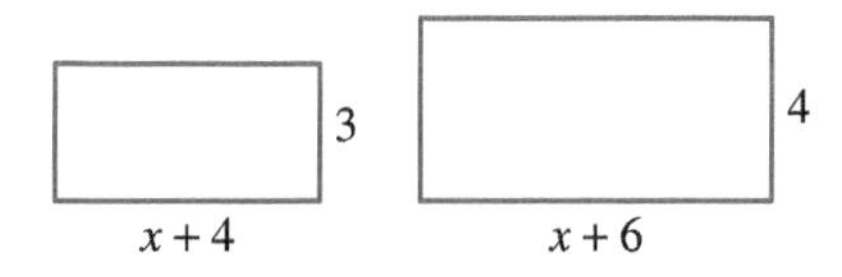


**Figure 3.14**

87. Find a polynomial that represents the area of the shaded region in Figure 3.15.

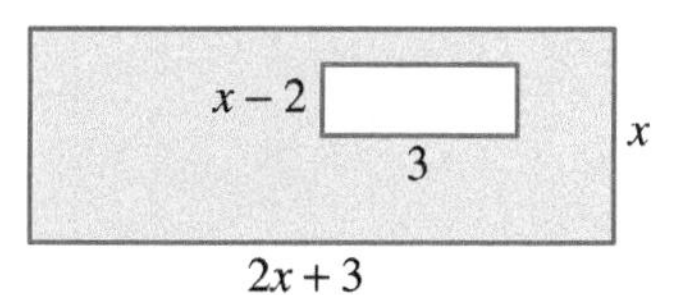


**Figure 3.15**

88. Explain how Figure 3.16 can be used to demonstrate geometrically that $(x + 7)(x - 3) = x^2 + 4x - 21$.

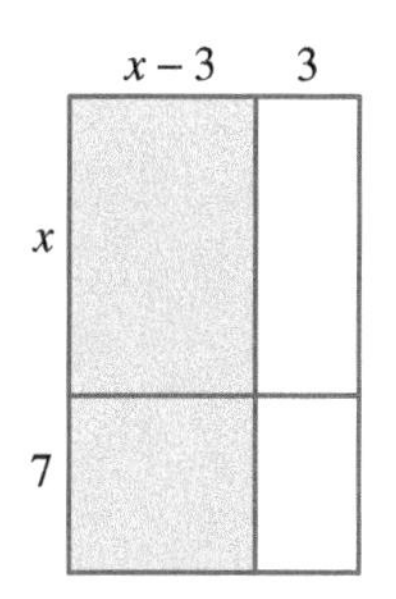


**Figure 3.16**

89. A square piece of cardboard is 16 inches on a side. A square piece $x$ inches on a side is cut out from each corner. The flaps are then turned up to form an open box. Find polynomials that represent the volume and the outside surface area of the box.

## Thoughts Into Words

90. How would you simplify $(2^3 + 2^2)^2$? Explain your reasoning.

91. Describe the process of multiplying two polynomials.

92. Determine the number of terms in the product of $(x + y)$ and $(a + b + c + d)$ without doing the multiplication. Explain how you arrived at your answer.

## Further Investigations

93. We have used the following two multiplication patterns.

$$(a + b)^2 = a^2 + 2ab + b^2$$

$$(a + b)^3 = a^3 + 3a^2b + 3ab^2 + b^3$$

By multiplying, we can extend these patterns as follows:

$$(a + b)^4 = a^4 + 4a^3b + 6a^2b^2 + 4ab^3 + b^4$$

$$(a + b)^5 = a^5 + 5a^4b + 10a^3b^2 + 10a^2b^3 + 5ab^4 + b^5$$

On the basis of these results, see if you can determine a pattern that will enable you to complete each of the following without using the long-multiplication process.

**(a)** $(a + b)^6$ **(b)** $(a + b)^7$

**(c)** $(a + b)^8$ **(d)** $(a + b)^9$

94. Find each of the following indicated products. These patterns will be used again in Section 3.5.

**(a)** $(x - 1)(x^2 + x + 1)$

**(b)** $(x + 1)(x^2 - x + 1)$

**(c)** $(x + 3)(x^2 - 3x + 9)$

**(d)** $(x - 4)(x^2 + 4x + 16)$

**(e)** $(2x - 3)(4x^2 + 6x + 9)$

**(f)** $(3x + 5)(9x^2 - 15x + 25)$

**95.** Some of the product patterns can be used to do arithmetic computations mentally. For example, let's use the pattern $(a + b)^2 = a^2 + 2ab + b^2$ to compute $31^2$ mentally. Your thought process should be "$31^2 = (30 + 1)^2 = 30^2 + 2(30)(1) + 1^2 = 961$." Compute each of the following numbers mentally, and then check your answers.

**(a)** $21^2$ **(b)** $41^2$ **(c)** $71^2$

**(d)** $32^2$ **(e)** $52^2$ **(f)** $82^2$

**96.** Use the pattern $(a - b)^2 = a^2 - 2ab + b^2$ to compute each of the following numbers mentally, and then check your answers.

**(a)** $19^2$ **(b)** $29^2$ **(c)** $49^2$

**(d)** $79^2$ **(e)** $38^2$ **(f)** $58^2$

**97.** Every whole number with a units digit of 5 can be represented by the expression $10x + 5$, where $x$ is a whole number. For example, $35 = 10(3) + 5$ and $145 = 10(14) + 5$. Now let's observe the following pattern when squaring such a number.

$$(10x + 5)^2 = 100x^2 + 100x + 25$$
$$= \boxed{100x(x + 1) + 25}$$

The pattern inside the dashed box can be stated as "add 25 to the product of $x$, $x + 1$, and 100." Thus, to compute $35^2$ mentally, we can figure "$35^2 = 3(4)(100) + 25 = 1225$." Compute each of the following numbers mentally, and then check your answers.

**(a)** $15^2$ **(b)** $25^2$ **(c)** $45^2$

**(d)** $55^2$ **(e)** $65^2$ **(f)** $75^2$

**(g)** $85^2$ **(h)** $95^2$ **(i)** $105^2$

### Answers to the Concept Quiz

**1.** True **2.** False **3.** True **4.** True **5.** True **6.** False **7.** False **8.** True **9.** False **10.** False

## 3.4 Factoring: Greatest Common Factor and Common Binomial Factor

OBJECTIVES

1. Understand the rules about completely factored form
2. Factor out the highest common monomial factor
3. Factor out a common binomial factor
4. Factor by grouping
5. Use factoring to solve equations
6. Solve word problems that involve factoring

Recall that 2 and 3 are said to be *factors* of 6 because the product of 2 and 3 is 6. Likewise, in an indicated product such as $7ab$, the 7, $a$, and $b$ are called factors of the product. In general, factoring is the reverse of multiplication. Previously, we have used the distributive property to find the product of a monomial and a polynomial, as in the Table 3.1.

**Table 3.1** Use the Distributive Property to Find a Product

| Expression | Apply the distributive property | Product |
|---|---|---|
| $3(x + 2)$ | $3(x) + 3(2)$ | $3x + 6$ |
| $5(2x - 1)$ | $5(2x) + 5(-1)$ | $10x - 5$ |
| $x(x^2 + 6x - 4)$ | $x(x^2) + x(6x) + x(-4)$ | $x^3 + 6x^2 - 4x$ |

We shall also use the distributive property [in the form $ab + ac = a(b + c)$] to reverse the process—that is, to factor a given polynomial. Consider the examples in Table 3.2.

**Table 3.2** Use the Distributive Property to Factor

| Expression | Rewrite the expression | Factored form when the distributive property is applied |
|---|---|---|
| $3x + 6$ | $3(x) + 3(2)$ | $3(x + 2)$ |
| $10x - 5$ | $5(2x) + 5(-1)$ | $5(2x - 1)$ |
| $x^3 + 6x^2 - 4x$ | $x(x^2) + x(6x) + x(-4)$ | $x(x^2 + 6x - 4)$ |

Note that in each example a given polynomial has been factored into the product of a monomial and a polynomial. Obviously, polynomials could be factored in a variety of ways. Consider some factorizations of $3x^2 + 12x$.

$$3x^2 + 12x = 3x(x + 4) \quad \text{or} \quad 3x^2 + 12x = 3(x^2 + 4x) \quad \text{or}$$

$$3x^2 + 12x = x(3x + 12) \quad \text{or} \quad 3x^2 + 12x = \frac{1}{2}x(6x + 24)$$

We are, however, primarily interested in the first of the previous factorization forms, which we refer to as the **completely factored form**. A polynomial with integral coefficients is in completely factored form if

1. It is expressed as a product of polynomials with *integral coefficients*, and
2. No polynomial, other than a monomial, within the factored form can be further factored into polynomials with integral coefficients.

Do you see why only the first of the above factored forms of $3x^2 + 12x$ is said to be in completely factored form? In each of the other three forms, the polynomial inside the parentheses can be factored further. Moreover, in the last form, $\frac{1}{2}x(6x + 24)$, the condition of using only integral coefficients is violated.

Classroom Example
For each of the following, determine if the factorization is in completely factored form. If it is not in completely factored form, state which rule is violated.
(a) $5x^6 + 15x^7y = 5x^5(x + 3x^2y)$
(b) $12m^2n^3 + 4mn^2 = 4mn^2(3mn + 1)$
(c) $12p^7q^3 + 3p^3q^4 = 12p^3q^3(p^4 + 0.25q)$
(d) $24a^4b + 12ab^3 = 3ab(8a^3 + 4b^2)$

EXAMPLE 1

For each of the following, determine if the factorization is in completely factored form. If it is not in completely factored form, state which rule has been violated.

**(a)** $4m^3 + 8m^4n = 4m^2(m + 2m^2n)$  **(b)** $32p^2q^4 + 8pq = 8pq(4pq^3 + 1)$
**(c)** $8x^2y^5 + 4x^3y^2 = 8x^2y^2(y^3 + 0.5x)$  **(d)** $10ab^3 + 20a^4b = 2ab(5b^2 + 10a^3)$

### Solution

**(a)** No, it is not completely factored. The polynomial inside the parentheses can be factored further.

**(b)** Yes, it is completely factored.

**(c)** No, it is not completely factored. The coefficient of 0.5 is not an integer.

**(d)** No, it is not completely factored. The polynomial inside the parentheses can be factored further.

## Factoring Out the Highest Common Monomial Factor

The factoring process that we discuss in this section, $ab + ac = a(b + c)$, is often referred to as **factoring out the highest common monomial factor**. The key idea in this process is to recognize the monomial factor that is common to all terms. For example, we observe that each term of the polynomial $2x^3 + 4x^2 + 6x$ has a factor of $2x$. Thus we write

$$2x^3 + 4x^2 + 6x = 2x(\quad)$$

and insert within the parentheses the appropriate polynomial factor. We determine the terms of this polynomial factor by dividing each term of the original polynomial by the factor of $2x$. The final, completely factored form is

$$2x^3 + 4x^2 + 6x = 2x(x^2 + 2x + 3)$$

The following examples further demonstrate this process of factoring out the highest common monomial factor.

$$12x^3 + 16x^2 = 4x^2(3x + 4) \qquad 6x^2y^3 + 27xy^4 = 3xy^3(2x + 9y)$$
$$8ab - 18b = 2b(4a - 9) \qquad 8y^3 + 4y^2 = 4y^2(2y + 1)$$
$$30x^3 + 42x^4 - 24x^5 = 6x^3(5 + 7x - 4x^2)$$

Note that in each example, the common monomial factor itself is not in a completely factored form. For example, $4x^2(3x + 4)$ is not written as $2 \cdot 2 \cdot x \cdot x \cdot (3x + 4)$.

**Classroom Example**
Factor out the highest common factor for each of the following:
**(a)** $4x^5 - 12x^4 + 32x^3$
**(b)** $18x^2y^3 + 6xy^4 - 24x^3y^2$

**EXAMPLE 2** Factor out the highest common factor for each of the following:

**(a)** $3x^4 + 15x^3 - 21x^2$ **(b)** $8x^3y^2 - 2x^4y - 12xy^2$

**Solution**

**(a)** Each term of the polynomial has a common factor of $3x^2$.

$$3x^4 + 15x^3 - 21x^2 = 3x^2(x^2 + 5x - 7)$$

**(b)** Each term of the polynomial has a common factor of $2xy$.

$$8x^3y^2 - 2x^4y - 12xy^2 = 2xy(4x^2y - x^3 - 6y)$$

## Factoring Out a Common Binomial Factor

Sometimes there may be a common binomial factor rather than a common monomial factor. For example, each of the two terms of the expression $x(y + 2) + z(y + 2)$ has a binomial factor of $(y + 2)$. Thus we can factor $(y + 2)$ from each term, and our result is

$$x(y + 2) + z(y + 2) = (y + 2)(x + z)$$

Consider an example that involves a common binomial factor.

**Classroom Example**
For each of the following, factor out the common binomial factor:
**(a)** $n^3(m + 2) + 4(m + 2)$
**(b)** $a(3b + 4) - b(3b + 4)$
**(c)** $y(y - 3) + 5(y - 3)$

**EXAMPLE 3** For each of the following, factor out the common binomial factor:

**(a)** $a^2(b + 1) + 2(b + 1)$ **(b)** $x(2y - 1) - y(2y - 1)$ **(c)** $x(x + 2) + 3(x + 2)$

**Solution**

**(a)** $a^2(b + 1) + 2(b + 1) = (b + 1)(a^2 + 2)$

**(b)** $x(2y - 1) - y(2y - 1) = (2y - 1)(x - y)$

**(c)** $x(x + 2) + 3(x + 2) = (x + 2)(x + 3)$

## Factoring by Grouping

It may be that the original polynomial exhibits no apparent common monomial or binomial factor, which is the case with $ab + 3a + bc + 3c$. However, there is a common factor for the first two terms and a common factor for the last two terms. Proceeding by factoring $a$ from the first two terms and $c$ from the last two terms, we get

$$ab + 3a + bc + 3c = a(b + 3) + c(b + 3)$$

Now we observe that we have a common binomial factor, $(b + 3)$, in each of the two terms, and we can proceed as before.

$$a(b + 3) + c(b + 3) = (b + 3)(a + c)$$

We refer to this factoring process as **factoring by grouping**. Let's consider a few examples of this type.

Classroom Example
Factor $x^2 - 5x + 2x - 10$.

**EXAMPLE 4** Factor $x^2 - x + 5x - 5$.

### Solution

$$x^2 - x + 5x - 5 = x(x - 1) + 5(x - 1)$$ Factor $x$ from the first two terms and 5 from the last two terms

$$= (x - 1)(x + 5)$$ Factor $(x - 1)$ from both terms

Classroom Example
Factor $m^2 - 2m + 4m - 8$.

**EXAMPLE 5** Factor $x^2 + 2x - 3x - 6$.

### Solution

$$x^2 + 2x - 3x - 6 = x(x + 2) - 3(x + 2)$$ Factor $x$ from the first two terms and $-3$ from the last two terms

$$= (x + 2)(x - 3)$$ Factor $(x + 2)$ from both terms

Classroom Example
Factor $x^2y + 4x^2 - 3y^2 - 12y$.

**EXAMPLE 6** Factor $ab^2 - 4b^2 + 3a - 12$.

### Solution

$$ab^2 - 4b^2 + 3a - 12 = b^2(a - 4) + 3(a - 4)$$ Factor $b^2$ from the first two terms and 3 from the last two terms

$$= (a - 4)(b^2 + 3)$$ Factor $(a - 4)$ from both terms

Classroom Example
Factor $3x^2 + yz^2 + 3y + x^2z^2$.

**EXAMPLE 7** Factor $4a^2 - bc^2 - a^2b + 4c^2$.

### Solution

First we determine that there is not a factor common to all four terms. Therefore, we will have to use factoring by grouping.

Terms that contain common factors need to be grouped together. So in order to apply factoring by grouping, the terms need to be rearranged. This can be done in more than one way. Two different ways will be shown.

**Method 1**

$$4a^2 - bc^2 - a^2b + 4c^2 = 4a^2 - a^2b + 4c^2 - bc^2$$ Group the terms with $a^2$ and $c^2$ together

$$= a^2(4 - b) + c^2(4 - b)$$

$$= (4 - b)(a^2 + c^2)$$ Factor $(4 - b)$ from both terms

**Method 2**

$$4a^2 - bc^2 - a^2b + 4c^2 = 4a^2 + 4c^2 - bc^2 - a^2b$$ Group the terms with 4 and $b$ together

$$= 4(a^2 + c^2) - b(c^2 + a^2)$$

$$= 4(a^2 + c^2) - b(a^2 + c^2)$$ Apply commutative property

$$= (a^2 + c^2)(4 - b)$$ Factor $(a^2 + c^2)$ from both terms

## Using Factoring to Solve Equations

One reason that factoring is an important algebraic skill is that it extends our techniques for solving equations. Each time we examine a factoring technique, we will then use it to help solve certain types of equations.

We need another property of equality before we consider some equations for which the highest-common-factor technique is useful. Suppose that the product of two numbers is zero. Can we conclude that at least one of these numbers must itself be zero? Yes. Let's state a property that formalizes this idea. Property 3.5, along with the highest-common-factor pattern, provides us with another technique for solving equations.

### Property 3.5

Let $a$ and $b$ be real numbers. Then

$$ab = 0 \quad \text{if and only if } a = 0 \text{ or } b = 0$$

**Classroom Example**
Solve $a^2 + 7a = 0$.

**EXAMPLE 8** Solve $x^2 + 6x = 0$.

### Solution

$$x^2 + 6x = 0$$

$$x(x + 6) = 0 \qquad \text{Factor the left side}$$

$$x = 0 \quad \text{or} \quad x + 6 = 0 \qquad ab = 0 \text{ if and only if } a = 0 \text{ or } b = 0$$

$$x = 0 \quad \text{or} \quad x = -6$$

Thus both 0 and $-6$ will satisfy the original equation, and the solution set is $\{-6, 0\}$.

**Classroom Example**
Solve $x^2 = 10x$.

**EXAMPLE 9** Solve $a^2 = 11a$.

### Solution

$$a^2 = 11a$$

$$a^2 - 11a = 0 \qquad \text{Add } -11a \text{ to both sides}$$

$$a(a - 11) = 0 \qquad \text{Factor the left side}$$

$$a = 0 \quad \text{or} \quad a - 11 = 0 \qquad ab = 0 \text{ if and only if } a = 0 \text{ or } b = 0$$

$$a = 0 \quad \text{or} \quad a = 11$$

The solution set is $\{0, 11\}$.

**Remark:** Note that in Example 9 we did *not* divide both sides of the equation by $a$. This would cause us to lose the solution of 0.

**Classroom Example**
Solve $7n^2 - 8n = 0$.

**EXAMPLE 10** Solve $3n^2 - 5n = 0$.

### Solution

$$3n^2 - 5n = 0$$

$$n(3n - 5) = 0$$

$$n = 0 \quad \text{or} \quad 3n - 5 = 0$$

$$n = 0 \quad \text{or} \quad 3n = 5$$

$$n = 0 \quad \text{or} \quad n = \frac{5}{3}$$

The solution set is $\left\{0, \frac{5}{3}\right\}$.

**Classroom Example**
Solve $2ax^2 - bx = 0$ for $x$.

**EXAMPLE 11** Solve $3ax^2 + bx = 0$ for $x$.

### Solution

$$3ax^2 + bx = 0$$
$$x(3ax + b) = 0$$
$$x = 0 \quad \text{or} \quad 3ax + b = 0$$
$$x = 0 \quad \text{or} \quad 3ax = -b$$
$$x = 0 \quad \text{or} \quad x = -\frac{b}{3a}$$

The solution set is $\left\{0, -\frac{b}{3a}\right\}$.

## Solving Word Problems That Involve Factoring

Many of the problems that we solve in the next few sections have a geometric setting. Some basic geometric figures, along with appropriate formulas, are listed in the inside front cover of this text. You may need to refer to them to refresh your memory.

**Classroom Example**
The area of a square is four times its perimeter. Find the length of a side of the square.

**EXAMPLE 12** Apply Your Skill

The area of a square is three times its perimeter. Find the length of a side of the square.

### Solution

Let $s$ represent the length of a side of the square (Figure 3.17). The area is represented by $s^2$ and the perimeter by $4s$. Thus

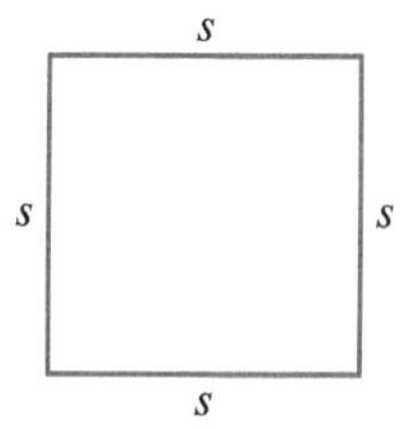

$$s^2 = 3(4s) \qquad \text{The area is to be three times the perimeter}$$
$$s^2 = 12s$$
$$s^2 - 12s = 0$$
$$s(s - 12) = 0$$
$$s = 0 \quad \text{or} \quad s = 12$$

s
s s
s

**Figure 3.17**

Because 0 is not a reasonable solution, it must be a 12-by-12 square. (Be sure to check this answer in the original statement of the problem!)

**Classroom Example**
Suppose that the volume of a right circular cylinder is numerically equal to three-fourths the total surface area of the cylinder. If the height of the cylinder is equal to the length of a radius of the base, find the height.

**EXAMPLE 13** Apply Your Skill

Suppose that the volume of a right circular cylinder is numerically equal to the total surface area of the cylinder. If the height of the cylinder is equal to the length of a radius of the base, find the height.

### Solution

Because $r = h$, the formula for volume $V = \pi r^2 h$ becomes $V = \pi r^3$, and the formula for the total surface area $S = 2\pi r^2 + 2\pi rh$ becomes $S = 2\pi r^2 + 2\pi r^2$, or $S = 4\pi r^2$. Therefore, we can set up and solve the following equation.

$$\pi r^3 = 4\pi r^2 \qquad \text{Volume is to be equal to the surface area}$$
$$\pi r^3 - 4\pi r^2 = 0$$
$$\pi r^2(r - 4) = 0$$
$$\pi r^2 = 0 \quad \text{or} \quad r - 4 = 0$$
$$r = 0 \quad \text{or} \quad r = 4$$

Zero is not a reasonable answer, therefore the height must be 4 units.

## Concept Quiz 3.4

For Problems 1–10, answer true or false.

1. Factoring is the reverse of multiplication.
2. The distributive property in the form $ab + ac = a(b + c)$ is applied to factor polynomials.
3. A polynomial could have many factored forms but only one completely factored form.
4. The greatest common factor of $6x^2y^3 - 12x^3y^2 + 18x^4y$ is $2x^2y$.
5. If the factored form of a polynomial can be factored further, then it has not met the conditions to be considered "factored completely."
6. Common factors are always monomials.
7. If the product of $x$ and $y$ is zero, then $x$ is zero or $y$ is zero.
8. The factored form, $3a(2a^2 + 4)$, is factored completely.
9. The solutions for the equation $x(x + 2) = 7$ are 7 and 5.
10. The solution set for $x^2 = 7x$ is $\{7\}$.

## Problem Set 3.4

For Problems 1–4, state if the polynomial is factored completely. (Objective 1)

1. $6x^2y + 12xy^2 = 2xy(3x + 6y)$
2. $2a^3b^2 + 4a^2b^2 = 4a^2b^2\left(\frac{1}{2}a + 1\right)$
3. $10m^2n^3 + 15m^4n^2 = 5m^2n(2n^2 + 3m^2n)$
4. $24ab + 12bc - 18bd = 6b(4a + 2c - 3d)$

For Problems 5–42, factor completely. (Objectives 2 and 3)

5. $6x + 3y$
6. $12x + 8y$
7. $6x^2 + 14x$
8. $15x^2 + 6x$
9. $28y^2 - 4y$
10. $42y^2 - 6y$
11. $20xy - 15x$
12. $27xy - 36y$
13. $15xy - 6yz$
14. $3ab - 39a^2b$
15. $7m^5n + 16mn^3$
16. $4xy^3 - 9x^4y$
17. $7x^3 + 10x^2$
18. $12x^3 - 10x^2$
19. $18a^2b + 27ab^2$
20. $24a^3b^2 + 36a^2b$
21. $12x^3y^4 - 39x^4y^3$
22. $15x^4y^2 - 45x^5y^4$
23. $12m^3 - 6m^2 + 3m$
24. $24n^5 + 6n^4 - 2n^3$
25. $8x^4 + 12x^3 - 24x^2$
26. $6x^5 - 18x^3 + 24x$
27. $5x + 7x^2 + 9x^4$
28. $9x^2 - 17x^4 + 21x^5$
29. $5ab^3 + 10a^2b^4 - 15a^3b^5$
30. $18a^3b^3 - 6ab + 12a^2b^2$
31. $15x^2y^3 + 20xy^2 + 35x^3y^4$
32. $8x^5y^3 - 6x^4y^5 + 12x^2y^3$
33. $x(y + 2) + 3(y + 2)$
34. $x(y - 1) + 5(y - 1)$
35. $4a(x + 8) - 5b(x + 8)$
36. $3x(a - 7) - y(a - 7)$
37. $6a^2(b^3 - c^5) - x^2y(b^3 - c^5)$
38. $3x^2y(a^2 - b^5) - 8(a^2 - b^5)$
39. $3x(2a + b) - 2y(2a + b)$
40. $5x(a - b) + y(a - b)$
41. $x(x + 2) + 5(x + 2)$
42. $x(x - 1) - 3(x - 1)$

For Problems 43–64, factor by grouping. (Objective 4)

43. $ax + 4x + ay + 4y$
44. $ax - 2x + ay - 2y$
45. $ax - 2bx + ay - 2by$
46. $2ax - bx + 2ay - by$
47. $3ax - 3bx - ay + by$
48. $5ax - 5bx - 2ay + 2by$
49. $2ax + 2x + ay + y$
50. $3bx + 3x + by + y$
51. $ax^2 - x^2 + 2a - 2$
52. $ax^2 - 2x^2 + 3a - 6$
53. $2ac + 3bd + 2bc + 3ad$
54. $2bx + cy + cx + 2by$
55. $ax - by + bx - ay$
56. $2a^2 - 3bc - 2ab + 3ac$
57. $x^2 + 9x + 6x + 54$
58. $x^2 - 2x + 5x - 10$

**59.** $2x^2 + 8x + x + 4$

**60.** $3x^2 + 18x - 2x - 12$

**61.** $a^2b + 18x + 3b + 6a^2x$

**62.** $2m^2 - 10np - 4n + 5m^2p$

**63.** $4xz + 2y^2 + 8y^2z + x$

**64.** $3x^2y + 10z + 15y + 2x^2z$

For Problems 65–80, solve each of the equations. (Objective 7)

**65.** $x^2 + 7x = 0$

**66.** $x^2 + 9x = 0$

**67.** $x^2 - x = 0$

**68.** $x^2 - 14x = 0$

**69.** $a^2 = 5a$

**70.** $b^2 = -7b$

**71.** $-2y = 4y^2$

**72.** $-6x = 2x^2$

**73.** $3x^2 + 7x = 0$

**74.** $-4x^2 + 9x = 0$

**75.** $4x^2 = 5x$

**76.** $3x = 11x^2$

**77.** $x - 4x^2 = 0$

**78.** $x - 6x^2 = 0$

**79.** $12a = -a^2$

**80.** $-5a = -a^2$

For Problems 81–86, solve each equation for the indicated variable. (Objective 7)

**81.** $5bx^2 - 3ax = 0$ for $x$

**82.** $ax^2 + bx = 0$ for $x$

**83.** $2by^2 = -3ay$ for $y$

**84.** $3ay^2 = by$ for $y$

**85.** $y^2 - ay + 2by - 2ab = 0$ for $y$

**86.** $x^2 + ax + bx + ab = 0$ for $x$

For Problems 87–96, set up an equation and solve each of the following problems. (Objective 8)

**87.** The square of a number equals seven times the number. Find the number.

**88.** Suppose that the area of a square is six times its perimeter. Find the length of a side of the square.

**89.** The area of a circular region is numerically equal to three times the circumference of the circle. Find the length of a radius of the circle.

**90.** Find the length of a radius of a circle such that the circumference of the circle is numerically equal to the area of the circle.

**91.** Suppose that the area of a circle is numerically equal to the perimeter of a square and that the length of a radius of the circle is equal to the length of a side of the square. Find the length of a side of the square. Express your answer in terms of $\pi$.

**92.** Find the length of a radius of a sphere such that the surface area of the sphere is numerically equal to the volume of the sphere.

**93.** Suppose that the area of a square lot is twice the area of an adjoining rectangular plot of ground. If the rectangular plot is 50 feet wide, and its length is the same as the length of a side of the square lot, find the dimensions of both the square and the rectangle.

**94.** The area of a square is one-fourth as large as the area of a triangle. One side of the triangle is 16 inches long, and the altitude to that side is the same length as a side of the square. Find the length of a side of the square.

**95.** Suppose that the volume of a sphere is numerically equal to twice the surface area of the sphere. Find the length of a radius of the sphere.

**96.** Suppose that a radius of a sphere is equal in length to a radius of a circle. If the volume of the sphere is numerically equal to four times the area of the circle, find the length of a radius for both the sphere and the circle.

## Thoughts Into Words

**97.** Your classmate solves the equation $3ax + bx = 0$ for $x$ as follows:

$$3ax + bx = 0$$
$$3ax = -bx$$
$$x = \frac{-bx}{3a}$$

How should he know that the solution is incorrect? How would you help him obtain the correct solution?

**98.** Suppose that your friend factors $36x^2y + 48xy^2$ as follows:

$$\begin{aligned} 36x^2y + 48xy^2 &= (4xy)(9x + 12y) \\ &= (4xy)(3)(3x + 4y) \\ &= 12xy(3x + 4y) \end{aligned}$$

Is this a correct approach? Would you have any suggestion to offer your friend?

## Further Investigations

The total surface area of a right circular cylinder is given by the formula $A = 2\pi r^2 + 2\pi rh$, where $r$ represents the radius of a base, and $h$ represents the height of the cylinder. For computational purposes, it may be more convenient to change the form of the right side of the formula by factoring it.

$$A = 2\pi r^2 + 2\pi rh$$
$$= 2\pi r(r + h)$$

Use $A = 2\pi r(r + h)$ to find the total surface area of each of the following cylinders. Also, use $\frac{22}{7}$ as an approximation for $\pi$.

**99.** $r = 7$ centimeters and $h = 12$ centimeters

**100.** $r = 14$ meters and $h = 20$ meters

**101.** $r = 3$ feet and $h = 4$ feet

**102.** $r = 5$ yards and $h = 9$ yards

For Problems 103–108, factor each expression. Assume that all variables that appear as exponents represent positive integers.

**103.** $2x^{2a} - 3x^a$

**104.** $6x^{2a} + 8x^a$

**105.** $y^{3m} + 5y^{2m}$

**106.** $3y^{5m} - y^{4m} - y^{3m}$

**107.** $2x^{6a} - 3x^{5a} + 7x^{4a}$

**108.** $6x^{3a} - 10x^{2a}$

### Answers to the Concept Quiz

**1.** True **2.** True **3.** True **4.** False **5.** True **6.** False **7.** True **8.** False **9.** False **10.** False

## 3.5 Factoring: Difference of Two Squares and Sum or Difference of Two Cubes

OBJECTIVES

1. Factor the difference of two squares
2. Factor the sum or difference of two cubes
3. Use factoring to solve equations
4. Solve word problems that involve factoring

In Section 3.3, we examined some special multiplication patterns. One of these patterns was

$$(a + b)(a - b) = a^2 - b^2$$

This same pattern, viewed as a factoring pattern, is referred to as the difference of two squares.

**Difference of Two Squares**

$$a^2 - b^2 = (a + b)(a - b)$$

Applying the pattern is fairly simple, as the next example demonstrates.

**Classroom Example**
Factor each of the following:
**(a)** $x^2 - 49$
**(b)** $9x^2 - 16$
**(c)** $25x^2 - 4y^2$
**(d)** $1 - y^2$

**EXAMPLE 1** Factor each of the following:

**(a)** $x^2 - 16$ **(b)** $4x^2 - 25$

**(c)** $16x^2 - 9y^2$ **(d)** $1 - a^2$

## Solution

**(a)** $x^2 - 16 = (x)^2 - (4)^2 = (x + 4)(x - 4)$

**(b)** $4x^2 - 25 = (2x)^2 - (5)^2 = (2x + 5)(2x - 5)$

**(c)** $16x^2 - 9y^2 = (4x)^2 - (3y)^2 = (4x + 3y)(4x - 3y)$

**(d)** $1 - a^2 = (1)^2 - (a)^2 = (1 + a)(1 - a)$

Multiplication is commutative, so the order of writing the factors is not important. For example, $(x + 4)(x - 4)$ can also be written as $(x - 4)(x + 4)$.

You must be careful not to assume an analogous factoring pattern for the *sum* of two squares; *it does not exist*. For example, $x^2 + 4 \neq (x + 2)(x + 2)$ because $(x + 2)(x + 2) = x^2 + 4x + 4$. We say that a polynomial such as $x^2 + 4$ is a **prime polynomial** or that it is not factorable using integers.

Sometimes the difference-of-two-squares pattern can be applied more than once, as the next example illustrates.

Classroom Example
Completely factor each of the following:
**(a)** $x^4 - 625y^4$
**(b)** $81x^4 - 16y^4$

### EXAMPLE 2

Completely factor each of the following:

**(a)** $x^4 - y^4$ **(b)** $16x^4 - 81y^4$

## Solution

**(a)** $x^4 - y^4 = (x^2 + y^2)(x^2 - y^2) = (x^2 + y^2)(x + y)(x - y)$

**(b)** $16x^4 - 81y^4 = (4x^2 + 9y^2)(4x^2 - 9y^2) = (4x^2 + 9y^2)(2x + 3y)(2x - 3y)$

It may also be that the squares are other than simple monomial squares, as in the next example.

Classroom Example
Completely factor each of the following:
**(a)** $(x + 5)^2 - 4y^2$
**(b)** $9m^2 - (3n + 7)^2$
**(c)** $(a + 2)^2 - (a - 1)^2$

### EXAMPLE 3

Completely factor each of the following:

**(a)** $(x + 3)^2 - y^2$ **(b)** $4x^2 - (2y + 1)^2$ **(c)** $(x - 1)^2 - (x + 4)^2$

## Solution

**(a)** $(x + 3)^2 - y^2 = ((x + 3) + y)((x + 3) - y) = (x + 3 + y)(x + 3 - y)$

**(b)** $4x^2 - (2y + 1)^2 = (2x + (2y + 1))(2x - (2y + 1))$
$= (2x + 2y + 1)(2x - 2y - 1)$

**(c)** $(x - 1)^2 - (x + 4)^2 = ((x - 1) + (x + 4))((x - 1) - (x + 4))$
$= (x - 1 + x + 4)(x - 1 - x - 4)$
$= (2x + 3)(-5)$

It is possible to apply both the technique of factoring out a common monomial factor and the pattern of the difference of two squares to the same problem. In general, it is best to look first for a common monomial factor. Consider the following example.

Classroom Example
Completely factor each of the following:
**(a)** $3x^2 - 12$
**(b)** $4x^2 - 36$
**(c)** $18y^3 - 8y$

### EXAMPLE 4

Completely factor each of the following:

**(a)** $2x^2 - 50$ **(b)** $9x^2 - 36$ **(c)** $48y^3 - 27y$

## Solution

**(a)** $2x^2 - 50 = 2(x^2 - 25) = 2(x + 5)(x - 5)$

**(b)** $9x^2 - 36 = 9(x^2 - 4) = 9(x + 2)(x - 2)$

**(c)** $48y^3 - 27y = 3y(16y^2 - 9) = 3y(4y + 3)(4y - 3)$

**Word of Caution** The polynomial $9x^2 - 36$ can be factored as follows:

$$\begin{aligned} 9x^2 - 36 &= (3x + 6)(3x - 6) \\ &= 3(x + 2)(3)(x - 2) \\ &= 9(x + 2)(x - 2) \end{aligned}$$

However, when one takes this approach, there seems to be a tendency to stop at the step $(3x + 6)(3x - 6)$. Therefore, remember the suggestion to *look first for a common monomial factor.*

The following examples should help you summarize all of the factoring techniques we have considered thus far.

| | |
|---|---|
| $7x^2 + 28 = 7(x^2 + 4)$ | Common factor |
| $x^2 - 4 = (x + 2)(x - 2)$ | Difference of two squares |
| $18 - 2x^2 = 2(9 - x^2) = 2(3 + x)(3 - x)$ | Common factor and the difference of two squares |
| $y^2 + 9$ is not factorable using integers | |
| $x^4 - 16 = (x^2 + 4)(x^2 - 4) = (x^2 + 4)(x + 2)(x - 2)$ | Difference of two squares applied twice |

## Factoring the Sum and Difference of Two Cubes

As we pointed out before, there exists no sum-of-squares pattern analogous to the difference-of-squares factoring pattern. That is, a polynomial such as $x^2 + 9$ is not factorable using integers. However, patterns do exist for both the sum and the difference of two cubes. These patterns are as follows:

**Sum and Difference of Two Cubes**

$$a^3 + b^3 = (a + b)(a^2 - ab + b^2)$$

$$a^3 - b^3 = (a - b)(a^2 + ab + b^2)$$

Note how we apply these patterns in the next example.

**Classroom Example**
Factor each of the following:
**(a)** $x^3 + 64$
**(b)** $27m^3 + 1000n^3$
**(c)** $1 - y^3$
**(d)** $8x^3 - 27y^3$

**EXAMPLE 5** Factor each of the following:

**(a)** $x^3 + 27$ **(b)** $8a^3 + 125b^3$ **(c)** $x^3 - 1$ **(d)** $27y^3 - 64x^3$

### Solution

**(a)** $x^3 + 27 = (x)^3 + (3)^3 = (x + 3)(x^2 - 3x + 9)$

**(b)** $8a^3 + 125b^3 = (2a)^3 + (5b)^3 = (2a + 5b)(4a^2 - 10ab + 25b^2)$

**(c)** $x^3 - 1 = (x)^3 - (1)^3 = (x - 1)(x^2 + x + 1)$

**(d)** $27y^3 - 64x^3 = (3y)^3 - (4x)^3 = (3y - 4x)(9y^2 + 12xy + 16x^2)$

## Using Factoring to Solve Equations

Remember that each time we pick up a new factoring technique we also develop more power for solving equations. Let's consider how we can use the difference-of-two-squares factoring pattern to help solve certain types of equations.

**Classroom Example**
Solve $n^2 = 49$.

**EXAMPLE 6** Solve $x^2 = 16$.

### Solution

$$x^2 = 16$$
$$x^2 - 16 = 0 \quad \text{Set the polynomial equal to zero}$$
$$(x + 4)(x - 4) = 0$$
$$x + 4 = 0 \quad \text{or} \quad x - 4 = 0$$
$$x = -4 \quad \text{or} \quad x = 4$$

The solution set is $\{-4, 4\}$. (Be sure to check these solutions in the original equation!)

**Classroom Example**
Solve $16m^2 = 81$.

**EXAMPLE 7** Solve $9x^2 = 64$.

### Solution

$$9x^2 = 64$$
$$9x^2 - 64 = 0 \quad \text{Set the polynomial equal to zero}$$
$$(3x + 8)(3x - 8) = 0$$
$$3x + 8 = 0 \quad \text{or} \quad 3x - 8 = 0$$
$$3x = -8 \quad \text{or} \quad 3x = 8$$
$$x = -\frac{8}{3} \quad \text{or} \quad x = \frac{8}{3}$$

The solution set is $\left\{-\frac{8}{3}, \frac{8}{3}\right\}$.

**Classroom Example**
Solve $3x^2 - 12 = 0$.

**EXAMPLE 8** Solve $7x^2 - 7 = 0$.

### Solution

$$7x^2 - 7 = 0$$
$$7(x^2 - 1) = 0$$
$$x^2 - 1 = 0 \quad \text{Divide both sides by 7}$$
$$(x + 1)(x - 1) = 0$$
$$x + 1 = 0 \quad \text{or} \quad x - 1 = 0$$
$$x = -1 \quad \text{or} \quad x = 1$$

The solution set is $\{-1, 1\}$.

In the previous examples we have been using the property $ab = 0$ if and only if $a = 0$ or $b = 0$. This property can be extended to any number of factors whose product is zero. Thus for three factors, the property could be stated $abc = 0$ if and only if $a = 0$ or $b = 0$ or $c = 0$. The next two examples illustrate this idea.

**Classroom Example**
Solve $a^4 - 81 = 0$.

**EXAMPLE 9** Solve $x^4 - 16 = 0$.

### Solution

$$x^4 - 16 = 0$$
$$(x^2 + 4)(x^2 - 4) = 0$$
$$(x^2 + 4)(x + 2)(x - 2) = 0$$

$$x^2 + 4 = 0 \quad \text{or} \quad x + 2 = 0 \quad \text{or} \quad x - 2 = 0$$
$$x^2 = -4 \quad \text{or} \quad x = -2 \quad \text{or} \quad x = 2$$

The solution set is $\{-2, 2\}$. (Because no real numbers, when squared, will produce $-4$, the equation $x^2 = -4$ yields no additional real number solutions.)

Classroom Example
Solve $t^3 - 25t = 0$.

**EXAMPLE 10** Solve $x^3 - 49x = 0$.

### Solution

$$x^3 - 49x = 0$$
$$x(x^2 - 49) = 0$$
$$x(x + 7)(x - 7) = 0$$
$$x = 0 \quad \text{or} \quad x + 7 = 0 \quad \text{or} \quad x - 7 = 0$$
$$x = 0 \quad \text{or} \quad x = -7 \quad \text{or} \quad x = 7$$

The solution set is $\{-7, 0, 7\}$.

## Solving Word Problems That Involve Factoring

The more we know about solving equations, the more resources we have for solving word problems.

Classroom Example
The combined area of two squares is 2600 square inches. Each side of one square is five times as long as a side of the other square. Find the dimensions of each of the squares.

**EXAMPLE 11** Apply Your Skill

The combined area of two squares is 40 square centimeters. Each side of one square is three times as long as a side of the other square. Find the dimensions of each of the squares.

### Solution

Let $s$ represent the length of a side of the smaller square. Then $3s$ represents the length of a side of the larger square (Figure 3.18).

$$s^2 + (3s)^2 = 40$$
$$s^2 + 9s^2 = 40$$
$$10s^2 = 40$$
$$s^2 = 4$$
$$s^2 - 4 = 0$$
$$(s + 2)(s - 2) = 0$$
$$s + 2 = 0 \quad \text{or} \quad s - 2 = 0$$
$$s = -2 \quad \text{or} \quad s = 2$$

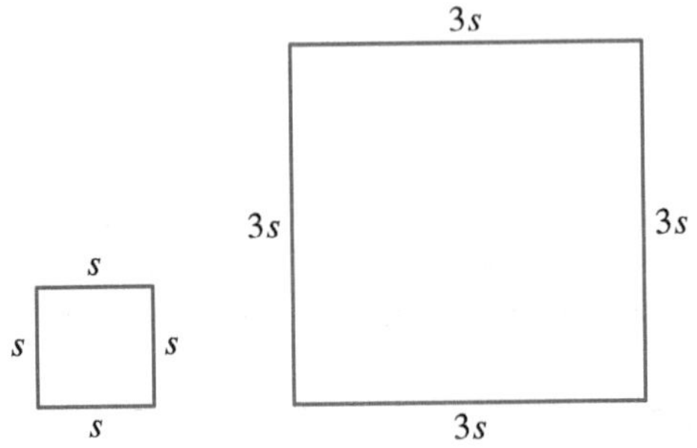

**Figure 3.18**

Because $s$ represents the length of a side of a square, the solution $-2$ has to be disregarded. Thus the length of a side of the small square is 2 centimeters, and the large square has sides of length $3(2) = 6$ centimeters.

## Concept Quiz 3.5

For Problems 1–10, answer true or false.

1. A binomial that has two perfect square terms that are subtracted is called the difference of two squares.
2. The sum of two squares is factorable using integers.
3. The sum of two cubes is factorable using integers.

4. The difference of two squares is factorable using integers.
5. The difference of two cubes is factorable using integers.
6. When factoring it is usually best to look for a common factor first.
7. The polynomial $4x^2 + y^2$ factors into $(2x + y)(2x + y)$.
8. The completely factored form of $y^4 - 81$ is $(y^2 + 9)(y^2 - 9)$.
9. The equation $x^2 = -9$ does not have any real number solutions.
10. The equation $abc = 0$ if and only if $a = 0$.

## Problem Set 3.5

For Problems 1–26, use the difference-of-squares pattern to factor each of the following. (Objective 1)

1. $x^2 - 1$
2. $x^2 - 9$
3. $16x^2 - 25$
4. $4x^2 - 49$
5. $9x^2 - 25y^2$
6. $x^2 - 64y^2$
7. $25a^2 - b^2$
8. $49a^2 - 36b^2$
9. $4m^2 - 9n^2$
10. $16b^2 - 25a^2$
11. $1 - 16y^2$
12. $1 - 4c^2$
13. $25x^2y^2 - 36$
14. $x^2y^2 - a^2b^2$
15. $4x^2 - y^4$
16. $x^6 - 9y^2$
17. $1 - 144n^2$
18. $25 - 49n^2$
19. $(x + 2)^2 - y^2$
20. $(3x + 5)^2 - y^2$
21. $4x^2 - (y + 1)^2$
22. $x^2 - (y - 5)^2$
23. $9a^2 - (2b + 3)^2$
24. $16s^2 - (3t + 1)^2$
25. $(x + 2)^2 - (x + 7)^2$
26. $(x - 1)^2 - (x - 8)^2$

For Problems 27–50, factor each of the following polynomials completely. Indicate any that are not factorable using integers. Don't forget to look first for a common monomial factor. (Objective 1)

27. $9x^2 - 36$
28. $8x^2 - 72$
29. $5x^2 + 5$
30. $7x^2 + 28$
31. $8y^2 - 32$
32. $5y^2 - 80$
33. $a^3b - 9ab$
34. $x^3y^2 - xy^2$
35. $16x^2 + 25$
36. $x^4 - 16$
37. $n^4 - 81$
38. $4x^2 + 9$
39. $3x^3 + 27x$
40. $20x^3 + 45x$
41. $4x^3y - 64xy^3$
42. $12x^3 - 27xy^2$
43. $6x - 6x^3$
44. $1 - 16x^4$
45. $1 - x^4y^4$
46. $20x - 5x^3$
47. $4x^2 - 64y^2$
48. $9x^2 - 81y^2$
49. $3x^4 - 48$
50. $2x^5 - 162x$

For Problems 51–68, use the sum-of-two-cubes or the difference-of-two-cubes pattern to factor each of the following. (Objective 2)

51. $a^3 - 64$
52. $a^3 - 27$
53. $x^3 + 1$
54. $x^3 + 8$
55. $8a^3 - 1$
56. $27a^3 - 8$
57. $64m^3 + 125$
58. $125y^3 + 8$
59. $8a^3 - 125b^3$
60. $125x^3 - y^3$
61. $27x^3 + 64y^3$
62. $8x^3 + 27y^3$
63. $1 - 27a^3$
64. $1 - 8x^3$
65. $x^3y^3 - 1$
66. $125x^3 + 27y^3$
67. $x^6 - y^6$
68. $x^6 + y^6$

For Problems 69–82, find all real number solutions for each equation. (Objective 3)

69. $x^2 - 25 = 0$
70. $x^2 - 1 = 0$
71. $9x^2 - 49 = 0$
72. $4y^2 = 25$
73. $8x^2 - 32 = 0$
74. $3x^2 - 108 = 0$
75. $3x^3 = 3x$
76. $4x^3 = 64x$
77. $20 - 5x^2 = 0$
78. $54 - 6x^2 = 0$
79. $x^4 - 81 = 0$
80. $x^5 - x = 0$
81. $6x^3 + 24x = 0$
82. $4x^3 + 12x = 0$

For Problems 83–92, set up an equation and solve each of the following problems. (Objective 4)

83. The cube of a number equals nine times the same number. Find the number.

84. The cube of a number equals the square of the same number. Find the number.

**85.** The combined area of two circles is $80\pi$ square centimeters. The length of a radius of one circle is twice the length of a radius of the other circle. Find the length of the radius of each circle.

**86.** The combined area of two squares is 26 square meters. The sides of the larger square are five times as long as the sides of the smaller square. Find the dimensions of each of the squares.

**87.** A rectangle is twice as long as it is wide, and its area is 50 square meters. Find the length and the width of the rectangle.

**88.** Suppose that the length of a rectangle is one and one-third times as long as its width. The area of the rectangle is 48 square centimeters. Find the length and width of the rectangle.

**89.** The total surface area of a right circular cylinder is $54\pi$ square inches. If the altitude of the cylinder is twice the length of a radius, find the altitude of the cylinder.

**90.** The total surface area of a right circular cone is $108\pi$ square feet. If the slant height of the cone is twice the length of a radius of the base, find the length of a radius.

**91.** The sum, in square yards, of the areas of a circle and a square is $(16\pi + 64)$. If a side of the square is twice the length of a radius of the circle, find the length of a side of the square.

**92.** The length of an altitude of a triangle is one-third the length of the side to which it is drawn. If the area of the triangle is 6 square centimeters, find the length of that altitude.

## Thoughts Into Words

**93.** Explain how you would solve the equation $4x^3 = 64x$.

**94.** What is wrong with the following factoring process?

$$25x^2 - 100 = (5x + 10)(5x - 10)$$

How would you correct the error?

**95.** Consider the following solution:

$$6x^2 - 24 = 0$$
$$6(x^2 - 4) = 0$$
$$6(x + 2)(x - 2) = 0$$
$$6 = 0 \quad \text{or} \quad x + 2 = 0 \quad \text{or} \quad x - 2 = 0$$
$$6 = 0 \quad \text{or} \quad x = -2 \quad \text{or} \quad x = 2$$

The solution set is $\{-2, 2\}$.

Is this a correct solution? Would you have any suggestion to offer the person who used this approach?

**Answers to the Concept Quiz**

**1.** True **2.** False **3.** True **4.** True **5.** True **6.** True **7.** False **8.** False **9.** True **10.** False

# 3.6 Factoring Trinomials

OBJECTIVES

1. Factor trinomials of the form $x^2 + bx + c$
2. Factor trinomials of the form $ax^2 + bx + c$
3. Factor perfect-square trinomials
4. Summary of factoring techniques

One of the most common types of factoring used in algebra is the expression of a trinomial as the product of two binomials. To develop a factoring technique, we first look at some multiplication ideas. Let's consider the product $(x + a)(x + b)$ and use the distributive property to show how each term of the resulting trinomial is formed.

$$\begin{aligned}(x + a)(x + b) &= x(x + b) + a(x + b) \\ &= x(x) + x(b) + a(x) + a(b) \\ &= x^2 + (a + b)x + ab\end{aligned}$$

Note that the coefficient of the middle term is the sum of $a$ and $b$ and that the last term is the product of $a$ and $b$. These two relationships can be used to factor trinomials. Let's consider some examples.

**Classroom Example**
Factor $x^2 + 13x + 40$.

**EXAMPLE 1** Factor $x^2 + 8x + 12$.

### Solution

We need to complete the following with two integers whose sum is 8 and whose product is 12.

$$x^2 + 8x + 12 = (x + \_\_\_)(x + \_\_\_)$$

The possible pairs of factors of 12 are 1(12), 2(6), and 3(4). Because $6 + 2 = 8$, we can complete the factoring as follows:

$$x^2 + 8x + 12 = (x + 6)(x + 2)$$

To check our answer, we find the product of $(x + 6)$ and $(x + 2)$.

**Classroom Example**
Factor $m^2 - 8m + 15$.

**EXAMPLE 2** Factor $x^2 - 10x + 24$.

### Solution

We need two integers whose product is 24 and whose sum is $-10$. Let's use a small table to organize our thinking.

| Factors | Product of the factors | Sum of the factors |
|---|---|---|
| $(-1)(-24)$ | 24 | $-25$ |
| $(-2)(-12)$ | 24 | $-14$ |
| $(-3)(-8)$ | 24 | $-11$ |
| $(-4)(-6)$ | 24 | $-10$ |

The bottom line contains the numbers that we need. Thus

$$x^2 - 10x + 24 = (x - 4)(x - 6)$$

**Classroom Example**
Factor $a^2 + 7a - 44$.

**EXAMPLE 3** Factor $x^2 + 7x - 30$.

### Solution

We need two integers whose product is $-30$ and whose sum is 7.

| Factors | Product of the factors | Sum of the factors |
|---|---|---|
| $(-1)(30)$ | $-30$ | 29 |
| $(1)(-30)$ | $-30$ | $-29$ |
| $(2)(-15)$ | $-30$ | $-13$ |
| $(-2)(15)$ | $-30$ | 13 |
| $(-3)(10)$ | $-30$ | 7 |

No need to search any further

The numbers that we need are $-3$ and 10, and we can complete the factoring.

$$x^2 + 7x - 30 = (x + 10)(x - 3)$$

Classroom Example
Factor $y^2 + 5y + 12$.

**EXAMPLE 4** Factor $x^2 + 7x + 16$.

### Solution

We need two integers whose product is 16 and whose sum is 7.

| Factors | Product of the factors | Sum of the factors |
|---|---|---|
| (1)(16) | 16 | 17 |
| (2)(8) | 16 | 10 |
| (4)(4) | 16 | 8 |

We have exhausted all possible pairs of factors of 16 and no two factors have a sum of 7, so we conclude that $x^2 + 7x + 16$ *is not factorable using integers.*

The tables in Examples 2, 3, and 4 were used to illustrate one way of organizing your thoughts for such problems. Normally you would probably factor such problems mentally without taking the time to formulate a table. Note, however, that in Example 4 the table helped us to be absolutely sure that we tried all the possibilities. Whether or not you use the table, keep in mind that the key ideas are the product and sum relationships.

Classroom Example
Factor $x^2 - x - 12$.

**EXAMPLE 5** Factor $n^2 - n - 72$.

### Solution

Note that the coefficient of the middle term is $-1$. Hence we are looking for two integers whose product is $-72$, and because their sum is $-1$, the absolute value of the negative number must be 1 larger than the positive number. The numbers are $-9$ and 8, and we can complete the factoring.

$$n^2 - n - 72 = (n - 9)(n + 8)$$

Classroom Example
Factor $m^2 + 4m - 117$.

**EXAMPLE 6** Factor $t^2 + 2t - 168$.

### Solution

We need two integers whose product is $-168$ and whose sum is 2. Because the absolute value of the constant term is rather large, it might help to look at it in prime factored form.

$$168 = 2 \cdot 2 \cdot 2 \cdot 3 \cdot 7$$

Now we can mentally form two numbers by using all of these factors in different combinations. Using two 2s and a 3 in one number and the other 2 and the 7 in the second number produces $2 \cdot 2 \cdot 3 = 12$ and $2 \cdot 7 = 14$. The coefficient of the middle term of the trinomial is 2, so we know that we must use 14 and $-12$. Thus we obtain

$$t^2 + 2t - 168 = (t + 14)(t - 12)$$

## Factoring Trinomials of the Form $ax^2 + bx + c$

We have been factoring trinomials of the form $x^2 + bx + c$, that is, trinomials where the coefficient of the squared term is 1. Now let's consider factoring trinomials where the coefficient of the squared term is not 1. First, let's illustrate an informal trial-and-error technique that works quite well for certain types of trinomials. This technique is based on our knowledge of multiplication of binomials.

**Classroom Example**
Factor $3x^2 + 11x + 6$.

**EXAMPLE 7** Factor $2x^2 + 11x + 5$.

### Solution

By looking at the first term, $2x^2$, and the positive signs of the other two terms, we know that the binomials are of the form

$$(x + \_\_\_)(2x + \_\_\_)$$

Because the factors of the last term, 5, are 1 and 5, we have only the following two possibilities to try.

$$(x + 1)(2x + 5) \qquad \text{or} \qquad (x + 5)(2x + 1)$$

By checking the middle term formed in each of these products, we find that the second possibility yields the correct middle term of $11x$. Therefore,

$$2x^2 + 11x + 5 = (x + 5)(2x + 1)$$

**Classroom Example**
Factor $15x^2 - 17x - 4$.

**EXAMPLE 8** Factor $10x^2 - 17x + 3$.

### Solution

First, observe that $10x^2$ can be written as $x \cdot 10x$ or $2x \cdot 5x$. Second, because the middle term of the trinomial is negative, and the last term is positive, we know that the binomials are of the form

$$(x - \_\_\_)(10x - \_\_\_) \qquad \text{or} \qquad (2x - \_\_\_)(5x - \_\_\_)$$

The factors of the last term, 3, are 1 and 3, so the following possibilities exist.

$$(x - 1)(10x - 3) \qquad (2x - 1)(5x - 3)$$
$$(x - 3)(10x - 1) \qquad (2x - 3)(5x - 1)$$

By checking the middle term formed in each of these products, we find that the product $(2x - 3)(5x - 1)$ yields the desired middle term of $-17x$. Therefore,

$$10x^2 - 17x + 3 = (2x - 3)(5x - 1)$$

**Classroom Example**
Factor $9x^2 + 14x + 16$.

**EXAMPLE 9** Factor $4x^2 + 6x + 9$.

### Solution

The first term, $4x^2$, and the positive signs of the middle and last terms indicate that the binomials are of the form

$$(x + \_\_\_)(4x + \_\_\_) \qquad \text{or} \qquad (2x + \_\_\_)(2x + \_\_\_)$$

Because the factors of 9 are 1 and 9 or 3 and 3, we have the following five possibilities to try.

$$(x + 1)(4x + 9) \qquad (2x + 1)(2x + 9)$$
$$(x + 9)(4x + 1) \qquad (2x + 3)(2x + 3)$$
$$(x + 3)(4x + 3)$$

When we try all of these possibilities we find that none of them yields a middle term of $6x$. Therefore, $4x^2 + 6x + 9$ is not factorable using integers.

## Another Method of Factoring the Form $ax^2 + bx + c$

By now it is obvious that factoring trinomials of the form $ax^2 + bx + c$ can be tedious. The key idea is to organize your work so that you consider all possibilities. We suggested one possible format in the previous three examples. As you practice such problems, you may come across a format of your own. Whatever works best for you is the right approach.

There is another, more systematic technique that you may wish to use with some trinomials. It is an extension of the technique we used at the beginning of this section. To see the basis of this technique, let's look at the following product.

$$(px + r)(qx + s) = px(qx) + px(s) + r(qx) + r(s)$$
$$= (pq)x^2 + (ps + rq)x + rs$$

Note that the product of the coefficient of the $x^2$ term and the constant term is $pqrs$. Likewise, the product of the two coefficients of $x$, $ps$ and $rq$, is also $pqrs$. Therefore, when we are factoring the trinomial $(pq)x^2 + (ps + rq)x + rs$, the two coefficients of $x$ must have a sum of $(ps) + (rq)$ and a product of $pqrs$. Let's see how this works in some examples.

Classroom Example
Factor $8x^2 - 2x - 15$.

**EXAMPLE 10** Factor $6x^2 - 11x - 10$.

**Solution**

**Step 1** Multiply the coefficient of the $x^2$ term, 6, and the constant term, $-10$.

$$(6)(-10) = -60$$

**Step 2** Find two integers whose sum is $-11$ and whose product is $-60$. It will be helpful to make a listing of the factor pairs for 60.

(1)(60)
(2)(30)
(3)(20)
(4)(15)
(5)(12)
(6)(10)

Because the product from step 1 is $-60$, we want a pair of factors for which the absolute value of their difference is 11. The factors are 4 and 15. For the sum to be $-11$ and the product to be $-60$, we will assign the signs so that we have $+4$ and $-15$.

**Step 3** Rewrite the original problem and express the middle term as a sum of terms using the factors in step 2 as the coefficients of the terms.

| **Original problem** | **Problem rewritten** |
|---|---|
| $6x^2 - 11x - 10$ | $6x^2 - 15x + 4x - 10$ |

**Step 4** Now use factoring by grouping to factor the rewritten problem.

$$6x^2 - 15x + 4x - 10 = 3x(2x - 5) + 2(2x - 5)$$
$$= (2x - 5)(3x + 2)$$

Thus $6x^2 - 11x - 10 = (2x - 5)(3x + 2)$.

Classroom Example
Factor $5x^2 + 38x - 16$.

**EXAMPLE 11** Factor $4x^2 - 29x + 30$.

**Solution**

**Step 1** Multiply the coefficient of the $x^2$ term, 4, and the constant term, 30.

$$(4)(30) = 120$$

**Step 2** Find two integers whose sum is $-29$ and whose product is 120. It will be helpful to make a listing of the factor pairs for 120.

| | |
|---|---|
| (1)(120) | (5)(24) |
| (2)(60) | (6)(20) |
| (3)(40) | (8)(15) |
| (4)(30) | (10)(12) |

Because our product from step 1 is $+120$, we want a pair of factors for which the absolute value of their sum is 29. The factors are 5 and 24. For the sum to be $-29$ and the product to be $+120$, we will assign the signs so that we have $-5$ and $-24$.

**Step 3** Rewrite the original problem and express the middle term as a sum of terms using the factors in step 2 as the coefficients of the terms.

| **Original problem** | **Problem rewritten** |
|---|---|
| $4x^2 - 29x + 30$ | $4x^2 - 5x - 24x + 30$ |

**Step 4** Now use factoring by grouping to factor the rewritten problem.

$$4x^2 - 5x - 24x + 30 = x(4x - 5) - 6(4x - 5)$$
$$= (4x - 5)(x - 6)$$

Thus $4x^2 - 29x + 30 = (4x - 5)(x - 6)$.

The technique presented in Examples 10 and 11 has concrete steps to follow. Examples 7 through 9 were factored by trial-and-error. Both of the techniques we used have their strengths and weaknesses. Which technique to use depends on the complexity of the problem and on your personal preference. The more that you work with both techniques, the more comfortable you will feel using them.

## Factoring Perfect-Square Trinomials

Before we summarize our work with factoring techniques, let's look at two more special factoring patterns. In Section 3.3 we used the following two patterns to square binomials.

$$(a + b)^2 = a^2 + 2ab + b^2 \quad \text{and} \quad (a - b)^2 = a^2 - 2ab + b^2$$

These patterns can also be used for factoring purposes.

$$a^2 + 2ab + b^2 = (a + b)^2 \quad \text{and} \quad a^2 - 2ab + b^2 = (a - b)^2$$

The trinomials on the left sides are called **perfect-square trinomials**; they are the result of squaring a binomial. We can always factor perfect-square trinomials using the usual techniques for factoring trinomials. However, they are easily recognized by the nature of their terms. For example, $4x^2 + 12x + 9$ is a perfect-square trinomial because

1. The first term is a perfect square $(2x)^2$
2. The last term is a perfect square $(3)^2$
3. The middle term is twice the product of the quantities being squared in the first and last terms $2(2x)(3)$

Likewise, $9x^2 - 30x + 25$ is a perfect-square trinomial because

1. The first term is a perfect square. $(3x)^2$
2. The last term is a perfect square. $(5)^2$
3. The middle term is the negative of twice the product of the quantities being squared in the first and last terms. $-2(3x)(5)$

Once we know that we have a perfect-square trinomial, the factors follow immediately from the two basic patterns. Thus

$$4x^2 + 12x + 9 = (2x + 3)^2 \qquad 9x^2 - 30x + 25 = (3x - 5)^2$$

The next example illustrates perfect-square trinomials and their factored forms.

**Classroom Example**
Factor each of the following:
**(a)** $x^2 + 18x + 81$
**(b)** $n^2 - 14n + 49$
**(c)** $4a^2 - 28ab + 49b^2$
**(d)** $9x^2 + 6xy + y^2$

**EXAMPLE 12** Factor each of the following:

**(a)** $x^2 + 14x + 49$ **(b)** $n^2 - 16n + 64$
**(c)** $36a^2 + 60ab + 25b^2$ **(d)** $16x^2 - 8xy + y^2$

**Solution**

**(a)** $x^2 + 14x + 49 = (x)^2 + 2(x)(7) + (7)^2 = (x + 7)^2$

**(b)** $n^2 - 16n + 64 = (n)^2 - 2(n)(8) + (8)^2 = (n - 8)^2$

**(c)** $36a^2 + 60ab + 25b^2 = (6a)^2 + 2(6a)(5b) + (5b)^2 = (6a + 5b)^2$

**(d)** $16x^2 - 8xy + y^2 = (4x)^2 - 2(4x)(y) + (y)^2 = (4x - y)^2$

## Summary of Factoring Techniques

As we have indicated, factoring is an important algebraic skill. We learned some basic factoring techniques one at a time, but you must be able to apply whichever is (or are) appropriate to the situation. Let's review the techniques and consider examples that demonstrate their use.

1. As a general guideline, always look for a common factor first. The common factor could be a binomial factor.

   $3x^2y^3 + 27xy = 3xy(xy^2 + 9)$ $\quad$ $x(y + 2) + 5(y + 2) = (y + 2)(x + 5)$

2. If the polynomial has two terms, then the pattern could be the difference-of-squares pattern or the sum or difference-of-two cubes pattern.

   $9a^2 - 25 = (3a + 5)(3a - 5)$ $\quad$ $8x^3 + 125 = (2x + 5)(4x^2 - 10x + 25)$

3. If the polynomial has three terms, then the polynomial may factor into the product of two binomials. Examples 10 and 11 presented concrete steps for factoring trinomials. Examples 7 through 9 were factored by trial-and-error. The perfect-square-trinomial pattern is a special case of the technique.

   $30n^2 - 31n + 5 = (5n - 1)(6n - 5)$ $\quad$ $t^4 + 3t^2 + 2 = (t^2 + 2)(t^2 + 1)$

4. If the polynomial has four or more terms, then factoring by grouping may apply. It may be necessary to rearrange the terms before factoring.

   $ab + ac + 4b + 4c = a(b + c) + 4(b + c) = (b + c)(a + 4)$

5. If none of the mentioned patterns or techniques work, then the polynomial may not be factorable using integers.

   $x^2 + 5x + 12$ $\quad$ Not factorable using integers

## Concept Quiz 3.6

For Problems 1–10, answer true or false.

1. To factor $x^2 - 4x - 60$ we look for two numbers whose product is $-60$ and whose sum is $-4$.
2. To factor $2x^2 - x - 3$ we look for two numbers whose product is $-3$ and whose sum is $-1$.
3. A trinomial of the form $x^2 + bx + c$ will never have a common factor other than 1.
4. A trinomial of the form $ax^2 + bx + c$ will never have a common factor other than 1.
5. The polynomial $x^2 + 25x + 72$ is not factorable using integers.
6. The polynomial $x^2 + 27x + 72$ is not factorable using integers.
7. The polynomial $2x^2 + 5x - 3$ is not factorable using integers.

8. The trinomial $49x^2 - 42x + 9$ is a perfect-square trinomial.
9. The trinomial $25x^2 + 80x - 64$ is a perfect-square trinomial.
10. To factor $12x^2 - 38x + 30$ one technique is to rewrite the problem as $12x^2 - 20x - 18x + 30$ and to factor by grouping.

## Problem Set 3.6

For Problems 1–30, factor completely each of the trinomials and indicate any that are not factorable using integers. (Objective 1)

1. $x^2 + 9x + 20$
2. $x^2 + 11x + 24$
3. $x^2 - 11x + 28$
4. $x^2 - 8x + 12$
5. $a^2 + 5a - 36$
6. $a^2 + 6a - 40$
7. $y^2 + 20y + 84$
8. $y^2 + 21y + 98$
9. $x^2 - 5x - 14$
10. $x^2 - 3x - 54$
11. $x^2 + 9x + 12$
12. $35 - 2x - x^2$
13. $6 + 5x - x^2$
14. $x^2 + 8x - 24$
15. $x^2 + 15xy + 36y^2$
16. $x^2 - 14xy + 40y^2$
17. $a^2 - ab - 56b^2$
18. $a^2 + 2ab - 63b^2$
19. $x^2 + 25x + 150$
20. $x^2 + 21x + 108$
21. $n^2 - 36n + 320$
22. $n^2 - 26n + 168$
23. $t^2 + 3t - 180$
24. $t^2 - 2t - 143$
25. $t^4 - 5t^2 + 6$
26. $t^4 + 10t^2 + 24$
27. $x^4 - 9x^2 + 8$
28. $x^4 - x^2 - 12$
29. $x^4 - 17x^2 + 16$
30. $x^4 - 13x^2 + 36$

For Problems 31–56, factor completely each of the trinomials and indicate any that are not factorable using integers. (Objective 2)

31. $15x^2 + 23x + 6$
32. $9x^2 + 30x + 16$
33. $12x^2 - x - 6$
34. $20x^2 - 11x - 3$
35. $4a^2 + 3a - 27$
36. $12a^2 + 4a - 5$
37. $3n^2 - 7n - 20$
38. $4n^2 + 7n - 15$
39. $3x^2 + 10x + 4$
40. $4n^2 - 19n + 21$
41. $10n^2 - 29n - 21$
42. $4x^2 - x + 6$
43. $8x^2 + 26x - 45$
44. $6x^2 + 13x - 33$
45. $6 - 35x - 6x^2$
46. $4 - 4x - 15x^2$
47. $20y^2 + 31y - 9$
48. $8y^2 + 22y - 21$
49. $24n^2 - 2n - 5$
50. $3n^2 - 16n - 35$
51. $5n^2 + 33n + 18$
52. $7n^2 + 31n + 12$
53. $10x^4 + 3x^2 - 4$
54. $3x^4 + 7x^2 - 6$
55. $18n^4 + 25n^2 - 3$
56. $4n^4 + 3n^2 - 27$

For Problems 57–62, factor completely each of the perfect-square trinomials. (Objective 3)

57. $y^2 - 16y + 64$
58. $a^2 + 30a + 225$
59. $4x^2 + 12xy + 9y^2$
60. $25x^2 - 60xy + 36y^2$
61. $8y^2 - 8y + 2$
62. $12x^2 + 36x + 27$

Problems 63–112 should help you pull together all of the factoring techniques of this chapter. Factor completely each polynomial, and indicate any that are not factorable using integers. (Objective 4)

63. $x^2 - 14x + 24$
64. $x^2 - 4x - 21$
65. $a^2b^2 + 6b^2 + 3a^2 + 18$
66. $6ab - 15b + 2a - 5$
67. $27y^3 - 8$
68. $1000x^3 + 27y^3$
69. $2t^2 - 8$
70. $14w^2 - 29w - 15$
71. $12x^2 + 7xy - 10y^2$
72. $8x^2 + 2xy - y^2$
73. $18n^3 + 39n^2 - 15n$
74. $n^2 + 18n + 77$
75. $n^2 - 17n + 60$
76. $(x + 5)^2 - y^2$
77. $36a^2 - 12a + 1$
78. $2n^2 - n - 5$
79. $6x^2 + 54$
80. $x^5 - x$
81. $3x^2 + x - 5$
82. $5x^2 + 42x - 27$
83. $x^2 - 11x + 18$
84. $x^2 + 10x - 56$
85. $20mn + 28m + 5n + 7$
86. $a^2x + b^2x + a^2y + b^2y$
87. $125a^3 - 64b^3$
88. $8x^3 - 125y^3$
89. $x^2 - (y - 7)^2$
90. $2n^3 + 6n^2 + 10n$
91. $1 - 16x^4$
92. $9a^2 - 30a + 25$
93. $4n^2 + 25n + 36$
94. $x^3 - 9x$
95. $n^3 - 49n$
96. $4x^2 + 16$
97. $x^2 - 7x - 8$
98. $x^2 + 3x - 54$
99. $3x^4 - 81x$
100. $x^3 + 125$

**101.** $x^4 + 6x^2 + 9$

**102.** $18x^2 - 12x + 2$

**103.** $x^4 - 5x^2 - 36$

**104.** $6x^4 - 5x^2 - 21$

**105.** $6w^2 - 11w - 35$

**106.** $10x^3 + 15x^2 + 20x$

**107.** $25n^2 + 64$

**108.** $4x^2 - 37x + 40$

**109.** $2n^3 + 14n^2 - 20n$

**110.** $25t^2 - 100$

**111.** $2xy + 6x + y + 3$

**112.** $3xy + 15x - 2y - 10$

## Thoughts Into Words

**113.** How can you determine that $x^2 + 5x + 12$ is not factorable using integers?

**114.** Explain your thought process when factoring $30x^2 + 13x - 56$.

**115.** Consider the following approach to factoring $12x^2 + 54x + 60$:

$$\begin{aligned} 12x^2 + 54x + 60 &= (3x + 6)(4x + 10) \\ &= 3(x + 2)(2)(2x + 5) \\ &= 6(x + 2)(2x + 5) \end{aligned}$$

Is this a correct factoring process? Do you have any suggestion for the person using this approach?

## Further Investigations

For Problems 116–121, factor each trinomial and assume that all variables that appear as exponents represent positive integers.

**116.** $x^{2a} + 2x^a - 24$

**117.** $x^{2a} + 10x^a + 21$

**118.** $6x^{2a} - 7x^a + 2$

**119.** $4x^{2a} + 20x^a + 25$

**120.** $12x^{2n} + 7x^n - 12$

**121.** $20x^{2n} + 21x^n - 5$

Consider the following approach to factoring the problem $(x - 2)^2 + 3(x - 2) - 10$.

$$\begin{aligned} &(x - 2)^2 + 3(x - 2) - 10 \\ &= y^2 + 3y - 10 && \text{Replace } x - 2 \text{ with } y \\ &= (y + 5)(y - 2) && \text{Factor} \\ &= (x - 2 + 5)(x - 2 - 2) && \text{Replace } y \text{ with } x - 2 \\ &= (x + 3)(x - 4) \end{aligned}$$

Use this approach to factor Problems 122–127.

**122.** $(x - 3)^2 + 10(x - 3) + 24$

**123.** $(x + 1)^2 - 8(x + 1) + 15$

**124.** $(2x + 1)^2 + 3(2x + 1) - 28$

**125.** $(3x - 2)^2 - 5(3x - 2) - 36$

**126.** $6(x - 4)^2 + 7(x - 4) - 3$

**127.** $15(x + 2)^2 - 13(x + 2) + 2$

### Answers to the Concept Quiz

**1.** True **2.** False **3.** True **4.** False **5.** True **6.** False **7.** False **8.** True **9.** False **10.** True

# 3.7 Equations and Problem Solving

OBJECTIVES

1. Solve equations by factoring
2. Solve word problems that involve factoring

The techniques for factoring trinomials that were presented in the previous section provide us with more power to solve equations. That is, the property "$ab = 0$ if and only if $a = 0$ or $b = 0$" continues to play an important role as we solve equations that contain factorable trinomials. Let's consider some examples.

Classroom Example
Solve $m^2 + 5m - 36 = 0$.

**EXAMPLE 1** Solve $x^2 - 11x - 12 = 0$.

### Solution

$$x^2 - 11x - 12 = 0$$
$$(x - 12)(x + 1) = 0$$
$$x - 12 = 0 \quad \text{or} \quad x + 1 = 0$$
$$x = 12 \quad \text{or} \quad x = -1$$

The solution set is $\{-1, 12\}$.

Classroom Example
Solve $21x^2 + x - 2 = 0$.

**EXAMPLE 2** Solve $20x^2 + 7x - 3 = 0$.

### Solution

$$20x^2 + 7x - 3 = 0$$
$$(4x - 1)(5x + 3) = 0$$
$$4x - 1 = 0 \quad \text{or} \quad 5x + 3 = 0$$
$$4x = 1 \quad \text{or} \quad 5x = -3$$
$$x = \frac{1}{4} \quad \text{or} \quad x = -\frac{3}{5}$$

The solution set is $\left\{-\frac{3}{5}, \frac{1}{4}\right\}$.

Classroom Example
Solve $-3t^2 + 15t + 72 = 0$.

**EXAMPLE 3** Solve $-2n^2 - 10n + 12 = 0$.

### Solution

$$-2n^2 - 10n + 12 = 0$$
$$-2(n^2 + 5n - 6) = 0$$
$$n^2 + 5n - 6 = 0 \qquad \text{Multiply both sides by } -\frac{1}{2}$$
$$(n + 6)(n - 1) = 0$$
$$n + 6 = 0 \quad \text{or} \quad n - 1 = 0$$
$$n = -6 \quad \text{or} \quad n = 1$$

The solution set is $\{-6, 1\}$.

Classroom Example
Solve $9x^2 + 48x + 64 = 0$.

**EXAMPLE 4** Solve $16x^2 - 56x + 49 = 0$.

### Solution

$$16x^2 - 56x + 49 = 0$$
$$(4x - 7)^2 = 0$$
$$(4x - 7)(4x - 7) = 0$$
$$4x - 7 = 0 \quad \text{or} \quad 4x - 7 = 0$$
$$4x = 7 \quad \text{or} \quad 4x = 7$$
$$x = \frac{7}{4} \quad \text{or} \quad x = \frac{7}{4}$$

The only solution is $\frac{7}{4}$; thus the solution set is $\left\{\frac{7}{4}\right\}$.

Classroom Example
Solve $x(4x + 4) = 15$.

**EXAMPLE 5** Solve $9a(a + 1) = 4$.

### Solution

$$\begin{aligned} 9a(a + 1) &= 4 \\ 9a^2 + 9a &= 4 && \text{Distributive property} \\ 9a^2 + 9a - 4 &= 0 && \text{Set the polynomial equal to zero} \\ (3a + 4)(3a - 1) &= 0 \\ 3a + 4 = 0 \quad &\text{or} \quad 3a - 1 = 0 \\ 3a = -4 \quad &\text{or} \quad 3a = 1 \\ a = -\frac{4}{3} \quad &\text{or} \quad a = \frac{1}{3} \end{aligned}$$

The solution set is $\left\{-\frac{4}{3}, \frac{1}{3}\right\}$.

Classroom Example
Solve $(x - 6)(x + 1) = 8$.

**EXAMPLE 6** Solve $(x - 1)(x + 9) = 11$.

### Solution

$$\begin{aligned} (x - 1)(x + 9) &= 11 \\ x^2 + 8x - 9 &= 11 && \text{Multiply } (x - 1) \text{ by } (x + 9) \\ x^2 + 8x - 20 &= 0 && \text{Set the polynomial equal to zero} \\ (x + 10)(x - 2) &= 0 \\ x + 10 = 0 \quad &\text{or} \quad x - 2 = 0 \\ x = -10 \quad &\text{or} \quad x = 2 \end{aligned}$$

The solution set is $\{-10, 2\}$.

## Solving Word Problems

As you might expect, the increase in our power to solve equations broadens our base for solving problems. Now we are ready to tackle some problems using equations of the types presented in this section.

Classroom Example
An accounting spreadsheet contains 78 cells. The number of columns is one more than twice the number of rows. Find the number of rows and the number of columns.

**EXAMPLE 7** Apply Your Skill

A cryptographer needs to arrange 60 numbers in a rectangular array in which the number of columns is two more than twice the number of rows. Find the number of rows and the number of columns.

### Solution

Let $r$ represent the numbers of rows. Then $2r + 2$ represents the number of columns.

$$\begin{aligned} r(2r + 2) &= 60 && \text{The number of rows times the number of columns yields the total amount of numbers in the array} \\ 2r^2 + 2r &= 60 \\ 2r^2 + 2r - 60 &= 0 \\ 2(r^2 + r - 30) &= 0 \\ 2(r + 6)(r - 5) &= 0 \\ r + 6 = 0 \quad &\text{or} \quad r - 5 = 0 \\ r = -6 \quad &\text{or} \quad r = 5 \end{aligned}$$

The solution $-6$ must be discarded, so there are 5 rows and $2r + 2$ or $2(5) + 2 = 12$ columns.

**Classroom Example**
A strip of uniform width cut from both sides and both ends of an 8-inch by 11-inch sheet of paper reduces the size of the paper to an area of 40 square inches. Find the width of the strip.

## EXAMPLE 8 Apply Your Skill

A strip of uniform width cut from both sides and both ends of a 5-inch by 7-inch photograph reduces the size of the photo to an area of 15 square inches. Find the width of the strip.

### Solution

Let $x$ represent the width of the strip, as indicated in Figure 3.19.

The length of the photograph after the strips of width $x$ are cut from both ends and both sides will be $7 - 2x$, and the width of the newly cropped photo will be $5 - 2x$. Because the area ($A = lw$) is to be 15 square inches, we can set up and solve the following equation.

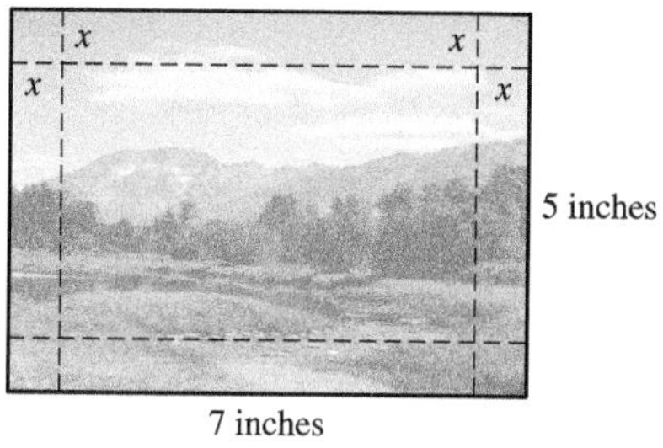

x, x, x, x, 5 inches, 7 inches

**Figure 3.19**

$$(7 - 2x)(5 - 2x) = 15$$
$$35 - 24x + 4x^2 = 15$$
$$4x^2 - 24x + 20 = 0$$
$$4(x^2 - 6x + 5) = 0$$
$$4(x - 5)(x - 1) = 0$$
$$x - 5 = 0 \quad \text{or} \quad x - 1 = 0$$
$$x = 5 \quad \text{or} \quad x = 1$$

The solution of 5 must be discarded because the width of the original photograph is only 5 inches. Therefore, the strip to be cropped from all four sides must be 1 inch wide. (Check this answer!)

The Pythagorean theorem, an important theorem pertaining to right triangles, can sometimes serve as a guideline for solving problems that deal with right triangles (see Figure 3.20). The Pythagorean theorem states that "in any right triangle, the square of the longest side (called the hypotenuse) is equal to the sum of the squares of the other two sides (called legs)." Let's use this relationship to help solve a problem.

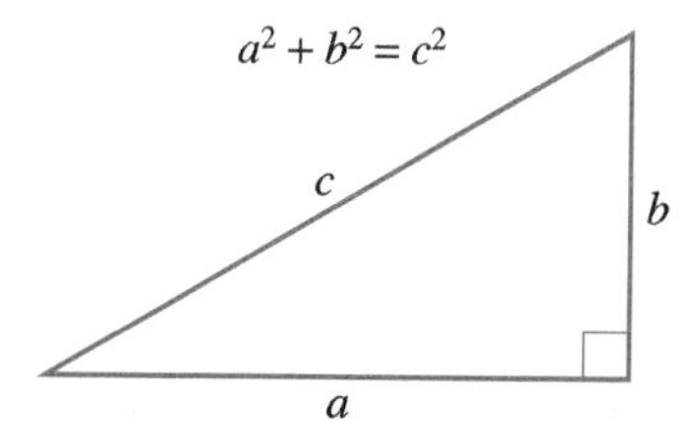


**Figure 3.20**

**Classroom Example**
The longer leg of a right triangle is 6 centimeters less than twice the shorter leg. The hypotenuse is 3 centimeters more than the longer leg. Find the length of the three sides of the right triangle.

## EXAMPLE 9 Apply Your Skill

One leg of a right triangle is 2 centimeters more than twice as long as the other leg. The hypotenuse is 1 centimeter longer than the longer of the two legs. Find the lengths of the three sides of the right triangle.

### Solution

Let $l$ represent the length of the shortest leg. Then $2l + 2$ represents the length of the other leg, and $2l + 3$ represents the length of the hypotenuse. Use the Pythagorean theorem as a guideline to set up and solve the following equation.

$$l^2 + (2l + 2)^2 = (2l + 3)^2$$
$$l^2 + 4l^2 + 8l + 4 = 4l^2 + 12l + 9$$
$$l^2 - 4l - 5 = 0$$
$$(l - 5)(l + 1) = 0$$
$$l - 5 = 0 \quad \text{or} \quad l + 1 = 0$$
$$l = 5 \quad \text{or} \quad l = -1$$

The negative solution must be discarded, so the length of one leg is 5 centimeters; the other leg is $2(5) + 2 = 12$ centimeters long, and the hypotenuse is $2(5) + 3 = 13$ centimeters long.

## Concept Quiz 3.7

For Problems 1–5, answer true or false.

1. If $xy = 0$, then $x = 0$ or $y = 0$.
2. If the product of three numbers is zero, then at least one of the numbers must be zero.
3. The Pythagorean theorem is true for all triangles.
4. The longest side of a right triangle is called the hypotenuse.
5. If we know the length of any two sides of a right triangle, the third can be determined by using the Pythagorean theorem.

## Problem Set 3.7

For Problems 1–54, solve each equation. You will need to use the factoring techniques that we discussed throughout this chapter. (Objective 1)

**1.** $x^2 + 4x + 3 = 0$

**2.** $x^2 + 7x + 10 = 0$

**3.** $x^2 + 18x + 72 = 0$

**4.** $n^2 + 20n + 91 = 0$

**5.** $n^2 - 13n + 36 = 0$

**6.** $n^2 - 10n + 16 = 0$

**7.** $x^2 + 4x - 12 = 0$

**8.** $x^2 + 7x - 30 = 0$

**9.** $w^2 - 4w = 5$

**10.** $s^2 - 4s = 21$

**11.** $n^2 + 25n + 156 = 0$

**12.** $n(n - 24) = -128$

**13.** $3t^2 + 14t - 5 = 0$

**14.** $4t^2 - 19t - 30 = 0$

**15.** $6x^2 + 25x + 14 = 0$

**16.** $25x^2 + 30x + 8 = 0$

**17.** $3t(t - 4) = 0$

**18.** $1 - x^2 = 0$

**19.** $-6n^2 + 13n - 2 = 0$

**20.** $(x + 1)^2 - 4 = 0$

**21.** $2n^3 = 72n$

**22.** $a(a - 1) = 2$

**23.** $(x - 5)(x + 3) = 9$

**24.** $3w^3 - 24w^2 + 36w = 0$

**25.** $16 - x^2 = 0$

**26.** $16t^2 - 72t + 81 = 0$

**27.** $n^2 + 7n - 44 = 0$

**28.** $2x^3 = 50x$

**29.** $3x^2 = 75$

**30.** $x^2 + x - 2 = 0$

**31.** $15x^2 + 34x + 15 = 0$

**32.** $20x^2 + 41x + 20 = 0$

**33.** $8n^2 - 47n - 6 = 0$

**34.** $7x^2 + 62x - 9 = 0$

**35.** $28n^2 - 47n + 15 = 0$

**36.** $24n^2 - 38n + 15 = 0$

**37.** $35n^2 - 18n - 8 = 0$

**38.** $8n^2 - 6n - 5 = 0$

**39.** $-3x^2 - 19x + 14 = 0$

**40.** $5x^2 = 43x - 24$

**41.** $n(n + 2) = 360$

**42.** $n(n + 1) = 182$

**43.** $9x^4 - 37x^2 + 4 = 0$

**44.** $4x^4 - 13x^2 + 9 = 0$

**45.** $3x^2 - 46x - 32 = 0$

**46.** $x^4 - 9x^2 = 0$

47. $2x^2 + x - 3 = 0$

48. $x^3 + 5x^2 - 36x = 0$

49. $12x^3 + 46x^2 + 40x = 0$

50. $5x(3x - 2) = 0$

51. $(3x - 1)^2 - 16 = 0$

52. $(x + 8)(x - 6) = -24$

53. $4a(a + 1) = 3$

54. $-18n^2 - 15n + 7 = 0$

For Problems 55–70, set up an equation and solve each problem. (Objective 2)

55. Find two consecutive integers whose product is 72.

56. Find two consecutive even whole numbers whose product is 224.

57. Find two integers whose product is 105 such that one of the integers is one more than twice the other integer.

58. Find two integers whose product is 104 such that one of the integers is three less than twice the other integer.

59. The perimeter of a rectangle is 32 inches, and the area is 60 square inches. Find the length and width of the rectangle.

60. Suppose that the length of a certain rectangle is two centimeters more than three times its width. If the area of the rectangle is 56 square centimeters, find its length and width.

61. The sum of the squares of two consecutive integers is 85. Find the integers.

62. The sum of the areas of two circles is $65\pi$ square feet. The length of a radius of the larger circle is 1 foot less than twice the length of a radius of the smaller circle. Find the length of a radius of each circle.

63. The combined area of a square and a rectangle is 64 square centimeters. The width of the rectangle is 2 centimeters more than the length of a side of the square, and the length of the rectangle is 2 centimeters more than its width. Find the dimensions of the square and the rectangle.

64. The Ortegas have an apple orchard that contains 90 trees. The number of trees in each row is 3 more than twice the number of rows. Find the number of rows and the number of trees per row.

65. The lengths of the three sides of a right triangle are represented by consecutive whole numbers. Find the lengths of the three sides.

66. The area of the floor of the rectangular room shown in Figure 3.21 is 175 square feet. The length of the room is $1\frac{1}{2}$ feet longer than the width. Find the length of the room.

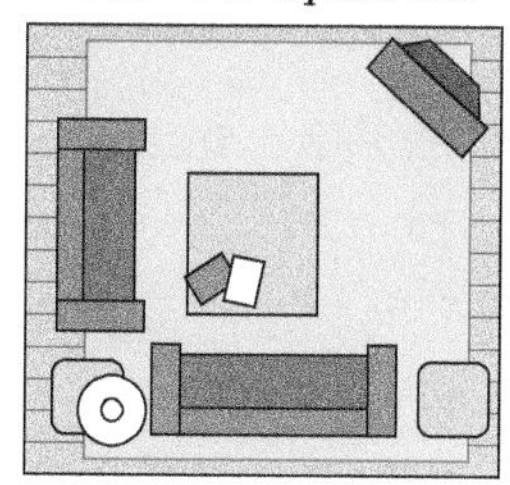


Figure 3.21

67. Suppose that the length of one leg of a right triangle is 3 inches more than the length of the other leg. If the length of the hypotenuse is 15 inches, find the lengths of the two legs.

68. The lengths of the three sides of a right triangle are represented by consecutive even whole numbers. Find the lengths of the three sides.

69. The area of a triangular sheet of paper is 28 square inches. One side of the triangle is 2 inches more than three times the length of the altitude to that side. Find the length of that side and the altitude to the side.

70. A strip of uniform width is shaded along both sides and both ends of a rectangular poster that measures 12 inches by 16 inches (see Figure 3.22). How wide is the shaded strip if one-half of the poster is shaded?

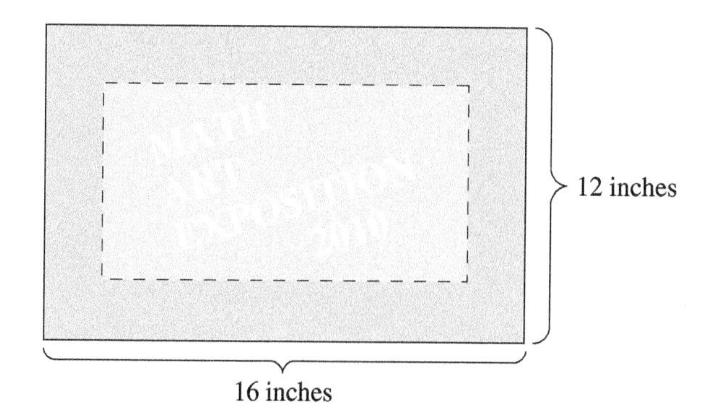


Figure 3.22

## Thoughts Into Words

71. Discuss the role that factoring plays in solving equations.

72. Explain how you would solve the equation $(x + 6)(x - 4) = 0$ and also how you would solve $(x + 6)(x - 4) = -16$.

73. Explain how you would solve the equation $3(x - 1)(x + 2) = 0$ and also how you would solve the equation $x(x - 1)(x + 2) = 0$.

**74.** Consider the following two solutions for the equation $(x + 3)(x - 4) = (x + 3)(2x - 1)$.

**Solution A**

$$(x + 3)(x - 4) = (x + 3)(2x - 1)$$
$$(x + 3)(x - 4) - (x + 3)(2x - 1) = 0$$
$$(x + 3)[x - 4 - (2x - 1)] = 0$$
$$(x + 3)(x - 4 - 2x + 1) = 0$$
$$(x + 3)(-x - 3) = 0$$
$$x + 3 = 0 \quad \text{or} \quad -x - 3 = 0$$
$$x = -3 \quad \text{or} \quad -x = 3$$
$$x = -3 \quad \text{or} \quad x = -3$$

The solution set is $\{-3\}$.

**Solution B**

$$(x + 3)(x - 4) = (x + 3)(2x - 1)$$
$$x^2 - x - 12 = 2x^2 + 5x - 3$$
$$0 = x^2 + 6x + 9$$
$$0 = (x + 3)^2$$
$$x + 3 = 0$$
$$x = -3$$

The solution set is $\{-3\}$.

Are both approaches correct? Which approach would you use, and why?

**Answers to the Concept Quiz**

**1.** True **2.** True **3.** False **4.** True **5.** True

# Chapter 3 Summary

| OBJECTIVE | SUMMARY | EXAMPLE |
| --- | --- | --- |
| Find the degree of a polynomial. (Section 3.1/Objective 1) | A polynomial is a monomial or a finite sum (or difference) of monomials. We classify polynomials as follows:<br><br>Polynomial with one term: Monomial<br>Polynomial with two terms: Binomial<br>Polynomial with three terms: Trinomial<br><br>The degree of a monomial is the sum of the exponents of the literal factors. The degree of a polynomial is the degree of the term with the highest degree in the polynomial. | Find the degree of the given polynomial:<br>$6x^4 - 7x^3 + 8x^2 + 2x - 10$<br><br>**Solution**<br>The degree of the polynomial is 4, because the term with the highest degree, $6x^4$, has a degree of 4.<br><br>**Sample Problem 1**<br>Find the degree of the given polynomial:<br>$3x^5 + 2x^2 + 7x - 8$ |
| Add, subtract, and simplify polynomial expressions. (Section 3.1/Objectives 2 and 3) | Similar (or like) terms have the same literal factors. The commutative, associative, and distributive properties provide the basis for rearranging, regrouping, and combining similar terms. | Perform the indicated operations:<br>$4x - [9x^2 - 2(7x - 3x^2)]$<br><br>**Solution**<br>$4x - [9x^2 - 2(7x - 3x^2)]$<br>$= 4x - [9x^2 - 14x + 6x^2]$<br>$= 4x - [15x^2 - 14x]$<br>$= 4x - 15x^2 + 14x$<br>$= -15x^2 + 18x$<br><br>**Sample Problem 2**<br>Perform the indicated operations:<br>$5y - [6y^2 - 3(7y + 6)]$ |
| Multiply monomials and raise a monomial to an exponent. (Section 3.2/Objectives 1 and 2) | The following properties provide the basis for multiplying monomials:<br><br>**1.** $b^n \cdot b^m = b^{n+m}$<br>**2.** $(b^n)^m = b^{mn}$<br>**3.** $(ab)^n = a^n b^n$ | Simplify each of the following:<br>**(a)** $(-5a^4b)(2a^2b^3)$<br>**(b)** $(-3x^3y)^2$<br><br>**Solution**<br>**(a)** $(-5a^4b)(2a^2b^3) = -10a^6b^4$<br>**(b)** $(-3x^3y)^2 = (-3)^2(x^3)^2(y)^2$<br>$= 9x^6y^2$<br><br>**Sample Problem 3**<br>Simplify each of the following:<br>**(a)** $(2x^2y)(-3x^5y)$<br>**(b)** $(-2xy^3)^3$ |

Answers to **Sample Problems** are located in the back of the book.

*(continued)*

| OBJECTIVE | SUMMARY | EXAMPLE |
| --- | --- | --- |
| Divide monomials. (Section 3.2/Objective 3) | The following properties provide the basis for dividing monomials: **1.** $\frac{b^n}{b^m} = b^{n-m}$ if $n > m$ **2.** $\frac{b^n}{b^m} = 1$ if $n = m$ | Find the quotient: $\frac{8x^5y^4}{-8xy^2}$ **Solution** $\frac{8x^5y^4}{-8xy^2} = -x^4y^2$ **Sample Problem 4** Find the quotient: $\frac{-24x^6yz^5}{3xyz^3}$ |
| Multiply polynomials. (Section 3.3/Objective 1) | To multiply two polynomials, every term of the first polynomial is multiplied by each term of the second polynomial. Multiplying polynomials often produces similar terms that can be combined to simplify the resulting polynomial. | Find the indicated product: $(3x + 4)(x^2 + 6x - 5)$ **Solution** $(3x + 4)(x^2 + 6x - 5)$ $= 3x(x^2 + 6x - 5)$ $+ 4(x^2 + 6x - 5)$ $= 3x^3 + 18x^2 - 15x + 4x^2 + 24x - 20$ $= 3x^3 + 22x^2 + 9x - 20$ **Sample Problem 5** Find the indicated product: $(3y - 2)(y^2 + 4y - 7)$ |
| Multiply two binomials using a shortcut pattern. (Section 3.3/Objective 2) | A three-step shortcut pattern, often referred to as FOIL, is used to find the product of two binomials. | Find the indicated product: $(3x + 5)(x - 4)$ **Solution** $(3x + 5)(x - 4)$ $= 3x^2 + (-12x + 5x) - 20$ $= 3x^2 - 7x - 20$ **Sample Problem 6** Find the indicated product: $(7x + 6)(3x - 5)$ |
| Find the square of a binomial using a shortcut pattern. (Section 3.3/Objective 3) | The patterns for squaring a binomial are: $(a + b)^2 = a^2 + 2ab + b^2$ and $(a - b)^2 = a^2 - 2ab + b^2$ | Expand $(4x - 3)^2$. **Solution** $(4x - 3)^2 = (4x)^2 - 2(4x)(3) + (-3)^2$ $= 16x^2 - 24x + 9$ **Sample Problem 7** Expand $(3x + 2)^2$. |

| OBJECTIVE | SUMMARY | EXAMPLE |
|---|---|---|
| Use a pattern to find the product of $(a + b)(a - b)$.<br>(Section 3.3/Objective 3) | The pattern is<br>$(a + b)(a - b) = a^2 - b^2$. | Find the product:<br>$(x - 3y)(x + 3y)$<br>**Solution**<br>$(x - 3y)(x + 3y) = (x)^2 - (3y)^2$<br>$= x^2 - 9y^2$<br>**Sample Problem 8**<br>Find the product: $(4x - y)(4x + y)$ |
| Find the cube of a binomial.<br>(Section 3.3/Objective 4) | The patterns for cubing a binomial are:<br>$(a + b)^3 = a^3 + 3a^2b + 3ab^2 + b^3$<br>and<br>$(a - b)^3 = a^3 - 3a^2b + 3ab^2 - b^3$ | Expand $(2a + 5)^3$.<br>**Solution**<br>$(2a + 5)^3$<br>$= (2a)^3 + 3(2a)^2(5) + 3(2a)(5)^2 + (5)^3$<br>$= 8a^3 + 60a^2 + 150a + 125$<br>**Sample Problem 9**<br>Expand: $(4x - 1)^3$ |
| Use polynomials in geometry problems.<br>(Section 3.1/Objective 4; Section 3.2/Objective 4; Section 3.3/Objective 5) | Sometimes polynomials are encountered in a geometric setting. A polynomial may be used to represent area or volume. | A rectangular piece of cardboard is 20 inches long and 10 inches wide. From each corner a square piece $x$ inches on a side is cut out. The flaps are turned up to form an open box. Find a polynomial that represents the volume.<br>**Solution**<br>The length of the box will be $20 - 2x$, the width of the box will be $10 - 2x$, and the height will be $x$, so<br>$V = (20 - 2x)(10 - 2x)(x)$.<br>Simplifying the polynomial gives<br>$V = 4x^3 - 60x^2 + 200x$.<br>**Sample Problem 10**<br>A square piece of cardboard is 12 inches on each side. From each corner a square piece $x$ inches on a side is cut out. The flaps are turned up to form an open box. Find a polynomial to represent the volume. |

*(continued)*

| OBJECTIVE | SUMMARY | EXAMPLE |
|---|---|---|
| Understand the rules about completely factored form. (Section 3.4/Objective 1) | A polynomial with integral coefficients is completely factored if:<br>**1.** It is expressed as a product of polynomials with integral coefficients; and<br>**2.** No polynomial, other than a monomial, within the factored form can be further factored into polynomials with integral coefficients. | Which of the following is the completely factored form of $2x^3y + 6x^2y^2$?<br>**(a)** $2x^3y + 6x^2y^2 = x^2y(2x + 6y)$<br>**(b)** $2x^3y + 6x^2y^2 = 6x^2y\left(\frac{1}{3}x + y\right)$<br>**(c)** $2x^3y + 6x^2y^2 = 2x^2y(x + 3y)$<br>**(d)** $2x^3y + 6x^2y^2 = 2xy(x^2 + 3xy)$<br>**Solution**<br>Only (c) is completely factored. For parts (a) and (d), the polynomial inside the parentheses can be factored further. For part (b), the coefficients are not integers.<br>**Sample Problem 11**<br>Which of the following is the completely factored form of $6x^2y + 12xy^3$?<br>**(a)** $2xy(3x + 6y^2)$<br>**(b)** $6xy(x + 2y^2)$<br>**(c)** $12xy\left(\frac{1}{2}x + y^2\right)$<br>**(d)** $xy(6x + 12y^2)$ |
| Factor out the greatest common monomial factor. (Section 3.4/Objective 2) | The distributive property in the form $ab + ac = a(b + c)$ is the basis for factoring out the greatest common monomial factor. | Factor $-4x^3y^4 - 2x^4y^3 - 6x^5y^2$.<br>**Solution**<br>$-4x^3y^4 - 2x^4y^3 - 6x^5y^2$<br>$= -2x^3y^2(2y^2 + xy + 3x^2)$<br>**Sample Problem 12**<br>Factor $-6x^3y + 3x^2y - 9x^2y^2$. |
| Factor out a common binomial factor. (Section 3.4/Objective 3) | The common factor can be a binomial factor. | Factor $y(x - 4) + 6(x - 4)$.<br>**Solution**<br>$y(x - 4) + 6(x - 4)$<br>$= (x - 4)(y + 6)$<br>**Sample Problem 13**<br>Factor $x(2y + 3) - 7(2y + 3)$. |
| Factor by grouping. (Section 3.4/Objective 4) | It may be that the polynomial exhibits no common monomial or binomial factor. However, after factoring common factors from groups of terms, a common factor may be evident. | Factor $2xz + 6x + yz + 3y$.<br>**Solution**<br>$2xz + 6x + yz + 3y$<br>$= 2x(z + 3) + y(z + 3)$<br>$= (z + 3)(2x + y)$<br>**Sample Problem 14**<br>Factor $6ax - 9ay - 8bx + 12by$. |

| OBJECTIVE | SUMMARY | EXAMPLE |
|---|---|---|
| Factor the difference of two squares. (Section 3.5/Objective 1) | The factoring pattern for the difference of two squares is: $a^2 - b^2 = (a + b)(a - b)$ | Factor $36a^2 - 25b^2$. **Solution** $36a^2 - 25b^2$ $= (6a - 5b)(6a + 5b)$ **Sample Problem 15** Factor $49x^2 - 81y^2$. |
| Factor the sum or difference of two cubes. (Section 3.5/Objective 2) | The factoring patterns $a^3 + b^3 = (a + b)(a^2 - ab + b^2)$ and $a^3 - b^3 = (a - b)(a^2 + ab + b^2)$ are called the sum of two cubes and the difference of two cubes, respectively. | Factor $8x^3 + 27y^3$. **Solution** $8x^3 + 27y^3$ $= (2x + 3y)(4x^2 - 6xy + 9y^2)$ **Sample Problem 16** Factor $x^3 - 64y^3$. |
| Factor trinomials of the form $x^2 + bx + c$. (Section 3.6/Objective 1) | Expressing a trinomial (for which the coefficient of the squared term is 1) as a product of two binomials is based on the relationship $(x + a)(x + b) = x^2 + (a + b)x + ab$. The coefficient of the middle term is the sum of $a$ and $b$, and the last term is the product of $a$ and $b$. | Factor $x^2 - 2x - 35$. **Solution** $x^2 - 2x - 35 = (x - 7)(x + 5)$ **Sample Problem 17** Factor $x^2 + 6x - 27$. |
| Factor trinomials of the form $ax^2 + bx + c$. (Section 3.6/Objective 2) | Two methods were presented for factoring trinomials of the form $ax^2 + bx + c$. One technique is to try the various possibilities of factors and check by multiplying. This method is referred to as trial-and-error. The other method is a structured technique that is shown in Examples 10 and 11 of Section 3.6. | Factor $4x^2 + 16x + 15$. **Solution** Multiply 4 times 15 to get 60. The factors of 60 that add to 16 are 6 and 10. Rewrite the problem and factor by grouping: $4x^2 + 16x + 15$ $= 4x^2 + 10x + 6x + 15$ $= 2x(2x + 5) + 3(2x + 5)$ $= (2x + 5)(2x + 3)$ **Sample Problem 18** Factor $6x^2 - 7x - 5$. |
| Factor perfect-square trinomials. (Section 3.6/Objective 3) | A perfect-square trinomial is the result of squaring a binomial. There are two basic perfect-square trinomial factoring patterns, $a^2 + 2ab + b^2 = (a + b)^2$ and $a^2 - 2ab + b^2 = (a - b)^2$ | Factor $16x^2 + 40x + 25$. **Solution** $16x^2 + 40x + 25 = (4x + 5)^2$ **Sample Problem 19** Factor $25x^2 - 20x + 4$. |

*(continued)*

| OBJECTIVE | SUMMARY | EXAMPLE |
| --- | --- | --- |
| Summarize the factoring techniques.<br>(Section 3.6/Objective 4) | **1.** As a general guideline, always look for a common factor first. The common factor could be a binomial term.<br>**2.** If the polynomial has two terms, then its pattern could be the difference of squares or the sum or difference of two cubes.<br>**3.** If the polynomial has three terms, then the polynomial may factor into the product of two binomials.<br>**4.** If the polynomial has four or more terms, then factoring by grouping may apply. It may be necessary to rearrange the terms before factoring.<br>**5.** If none of the mentioned patterns or techniques work, then the polynomial may not be factorable using integers. | Factor $18x^2 - 50$.<br>**Solution**<br>First factor out a common factor of 2:<br>$18x^2 - 50 = 2(9x^2 - 25)$<br>Now factor the difference of squares:<br>$18x^2 - 50$<br>$= 2(9x^2 - 25)$<br>$= 2(3x - 5)(3x + 5)$<br>**Sample Problem 20**<br>Factor $20x^2 - 45$. |
| Solve equations.<br>(Section 3.4/Objective 5; Section 3.5/Objective 3; Section 3.7/Objective 1) | The factoring techniques in this chapter, along with the property $ab = 0$, provide the basis for some additional equation-solving skills. | Solve $x^2 - 11x + 28 = 0$.<br>**Solution**<br>$x^2 - 11x + 28 = 0$<br>$(x - 7)(x - 4) = 0$<br>$x - 7 = 0$ or $x - 4 = 0$<br>$x = 7$ or $x = 4$<br>The solution set is $\{4, 7\}$.<br>**Sample Problem 21**<br>Solve $x^2 + x - 12 = 0$. |
| Solve word problems.<br>(Section 3.4/Objective 6; Section 3.5/Objective 4; Section 3.7/Objective 2) | The ability to solve more types of equations increased our capabilities to solve word problems. | Suppose that the area of a square is numerically equal to three times its perimeter. Find the length of a side of the square.<br>**Solution**<br>Let $x$ represent the length of a side of the square. The area is $x^2$ and the perimeter is $4x$. Because the area is numerically equal to three times the perimeter, we have the equation $x^2 = 3(4x)$. By solving this equation, we can determine that the length of a side of the square is 12 units.<br>**Sample Problem 22**<br>Suppose that the area of a square is numerically equal to one-half its perimeter. Find the length of a side of the square. |

# Chapter 3 Review Problem Set

For Problems 1–4, find the degree of the polynomial.

1. $-2x^3 + 4x^2 - 8x + 10$
2. $x^4 + 11x^2 - 15$
3. $5x^3y + 4x^4y^2 - 3x^3y^2$
4. $5xy^3 + 2x^2y^2 - 3x^3y^2$

For Problems 5–40, perform the indicated operations and then simplify.

5. $(3x - 2) + (4x - 6) + (-2x + 5)$
6. $(8x^2 + 9x - 3) - (5x^2 - 3x - 1)$
7. $(6x^2 - 2x - 1) + (4x^2 + 2x + 5) - (-2x^2 + x - 1)$
8. $(-3x^2 - 4x + 8) + (5x^2 + 7x + 2) - (-9x^2 + x + 6)$
9. $[3x - (2x - 3y + 1)] - [2y - (x - 1)]$
10. $[8x - (5x - y + 3)] - [-4y - (2x + 1)]$
11. $(-5x^2y^3)(4x^3y^4)$
12. $(-2a^2)(3ab^2)(a^2b^3)$
13. $\left(\frac{1}{2}ab\right)(8a^3b^2)(-2a^3)$
14. $\left(\frac{3}{4}x^2y^3\right)(12x^3y^2)(3y^3)$
15. $(4x^2y^3)^4$
16. $(-2x^2y^3z)^3$
17. $-(3ab)(2a^2b^3)^2$
18. $(3x^{n+1})(2x^{3n-1})$
19. $\dfrac{-39x^3y^4}{3xy^3}$
20. $\dfrac{30x^5y^4}{15x^2y}$
21. $\dfrac{12a^2b^5}{-3a^2b^3}$
22. $\dfrac{20a^4b^6}{5ab^3}$
23. $5a^2(3a^2 - 2a - 1)$
24. $-2x^3(4x^2 - 3x - 5)$
25. $(x + 4)(3x^2 - 5x - 1)$
26. $(3x + 2)(2x^2 - 5x + 1)$
27. $(x^2 - 2x - 5)(x^2 + 3x - 7)$
28. $(3x^2 - x - 4)(x^2 + 2x - 5)$
29. $(4x - 3y)(6x + 5y)$
30. $(7x - 9)(x + 4)$
31. $(7 - 3x)(3 + 5x)$
32. $(x^2 - 3)(x^2 + 8)$
33. $(2x - 3)^2$
34. $(5x - 1)^2$
35. $(4x + 3y)^2$
36. $(2x + 5y)^2$
37. $(2x - 7)(2x + 7)$
38. $(3x - 1)(3x + 1)$
39. $(x - 2)^3$
40. $(2x + 5)^3$
41. Find a polynomial that represents the area of the shaded region in Figure 3.23.

$x - 1$
$x - 2$
$x$
$3x + 4$

**Figure 3.23**

42. Find a polynomial that represents the volume of the rectangular solid in Figure 3.24.

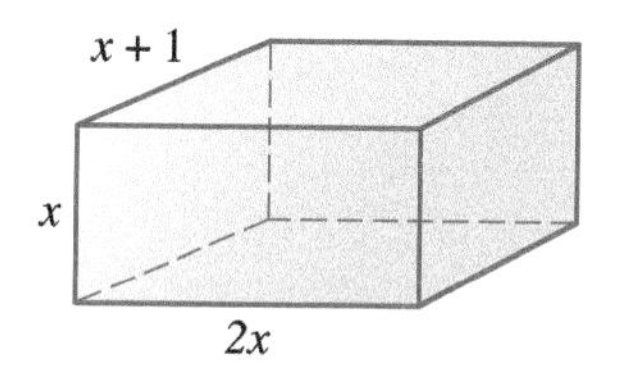


**Figure 3.24**

For Problems 43–62, factor each polynomial.

43. $10a^2b - 5ab^3 - 15a^3b^2$
44. $3xy - 5x^2y^2 - 15x^3y^3$
45. $a(x + 4) + b(x + 4)$
46. $y(3x - 1) + 7(3x - 1)$
47. $6x^3 + 3x^2y + 2xz^2 + yz^2$
48. $mn + 5n^2 - 4m - 20n$
49. $49a^2 - 25b^2$
50. $36x^2 - y^2$
51. $125a^3 - 8$
52. $27x^3 + 64y^3$
53. $x^2 - 9x + 18$
54. $x^2 + 11x + 28$

55. $x^2 - 4x - 21$

56. $x^2 + 6x - 16$

57. $2x^2 + 9x + 4$

58. $6x^2 - 11x + 4$

59. $12x^2 - 5x - 2$

60. $8x^2 - 10x - 3$

61. $4x^2 - 12xy + 9y^2$

62. $x^2 + 16xy + 64y^2$

For Problems 63–84, factor each polynomial completely. Indicate any that are not factorable using integers.

63. $x^2 + 3x - 28$

64. $2t^2 - 18$

65. $4n^2 + 9$

66. $12n^2 - 7n + 1$

67. $x^6 - x^2$

68. $x^3 - 6x^2 - 72x$

69. $6a^3b + 4a^2b^2 - 2a^2bc$

70. $x^2 - (y - 1)^2$

71. $8x^2 + 12$

72. $12x^2 + x - 35$

73. $16n^2 - 40n + 25$

74. $4n^2 - 8n$

75. $3w^3 + 18w^2 - 24w$

76. $20x^2 + 3xy - 2y^2$

77. $16a^2 - 64a$

78. $3x^3 - 15x^2 - 18x$

79. $n^2 - 8n - 128$

80. $t^4 - 22t^2 - 75$

81. $35x^2 - 11x - 6$

82. $15 - 14x + 3x^2$

83. $64n^3 - 27$

84. $16x^3 + 250$

For Problems 85–104, solve each equation.

85. $4x^2 - 36 = 0$

86. $x^2 + 5x - 6 = 0$

87. $49n^2 - 28n + 4 = 0$

88. $(3x - 1)(5x + 2) = 0$

89. $(3x - 4)^2 - 25 = 0$

90. $6a^3 = 54a$

91. $x^5 = x$

92. $-n^2 + 2n + 63 = 0$

93. $7n(7n + 2) = 8$

94. $30w^2 - w - 20 = 0$

95. $5x^4 - 19x^2 - 4 = 0$

96. $9n^2 - 30n + 25 = 0$

97. $n(2n + 4) = 96$

98. $7x^2 + 33x - 10 = 0$

99. $(x + 1)(x + 2) = 42$

100. $x^2 + 12x - x - 12 = 0$

101. $2x^4 + 9x^2 + 4 = 0$

102. $30 - 19x - 5x^2 = 0$

103. $3t^3 - 27t^2 + 24t = 0$

104. $-4n^2 - 39n + 10 = 0$

For Problems 105–114, set up an equation and solve each problem.

105. Find three consecutive integers such that the product of the smallest and the largest is one less than 9 times the middle integer.

106. Find two integers whose sum is 2 and whose product is $-48$.

107. Find two consecutive odd whole numbers whose product is 195.

108. Two cars leave an intersection at the same time, one traveling north and the other traveling east. Some time later, they are 20 miles apart, and the car going east has traveled 4 miles farther than the other car. How far has each car traveled?

109. The perimeter of a rectangle is 32 meters, and its area is 48 square meters. Find the length and width of the rectangle.

110. A room contains 144 chairs. The number of chairs per row is two less than twice the number of rows. Find the number of rows and the number of chairs per row.

111. The area of a triangle is 39 square feet. The length of one side is 1 foot more than twice the altitude to that side. Find the length of that side and the altitude to the side.

112. A rectangular-shaped pool 20 feet by 30 feet has a sidewalk of uniform width around the pool (see Figure 3.25). The area of the sidewalk is 336 square feet. Find the width of the sidewalk.

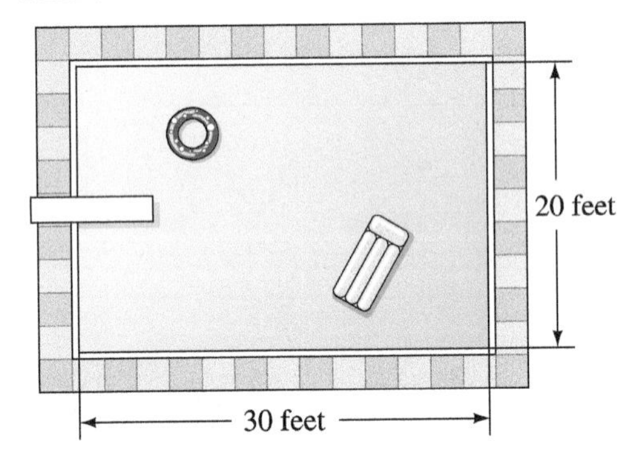


**Figure 3.25**

113. The sum of the areas of two squares is 89 square centimeters. The length of a side of the larger square is 3 centimeters more than the length of a side of the smaller square. Find the dimensions of each square.

114. The total surface area of a right circular cylinder is $32\pi$ square inches. If the altitude of the cylinder is three times the length of a radius, find the altitude of the cylinder.

# Chapter 3 Test

For Problems 1–8, perform the indicated operations and simplify each expression.

1. $(-3x - 1) + (9x - 2) - (4x + 8)$

2. $(-6xy^2)(8x^3y^2)$

3. $(-3x^2y^4)^3$

4. $(5x - 7)(4x + 9)$

5. $(3n - 2)(2n - 3)$

6. $(x - 4y)^3$

7. $(x + 6)(2x^2 - x - 5)$

8. $\dfrac{-70x^4y^3}{5xy^2}$

For Problems 9–14, factor each expression completely.

9. $6x^2 + 19x - 20$

10. $12x^2 - 3$

11. $64 + t^3$

12. $30x + 4x^2 - 16x^3$

13. $x^2 - xy + 4x - 4y$

14. $24n^2 + 55n - 24$

For Problems 15–22, solve each equation.

15. $x^2 + 8x - 48 = 0$

16. $4n^2 = n$

17. $4x^2 - 12x + 9 = 0$

18. $(n - 2)(n + 7) = -18$

19. $3x^3 + 21x^2 - 54x = 0$

20. $12 + 13x - 35x^2 = 0$

21. $n(3n - 5) = 2$

22. $9x^2 - 36 = 0$

For Problems 23–25, set up an equation and solve each problem.

23. The perimeter of a rectangle is 30 inches, and its area is 54 square inches. Find the length of the longest side of the rectangle.

24. A room contains 105 chairs arranged in rows. The number of rows is one more than twice the number of chairs per row. Find the number of rows.

25. The combined area of a square and a rectangle is 57 square feet. The width of the rectangle is 3 feet more than the length of a side of the square, and the length of the rectangle is 5 feet more than the length of a side of the square. Find the length of the rectangle.

# 4 Rational Expressions

*"Success does not consist in never making mistakes but in never making the same mistake a second time."*
GEORGE BERNARD SHAW

## Study Skill Tip

It is very important that you review the problems you did incorrectly on a test and determine whether the errors were "simple" mistakes or "concept" errors.

An example of a simple mistake might be not following directions or an arithmetic mistake. Simple errors occur regardless of your proficiency in the math material.

A concept error is where you did not understand *how* to do the problem. For example to solve the inequality $-2x > 8$, you did not know to reverse the inequality sign when dividing both sides of the inequality by a negative number.

It's hard to avoid simple mistakes, but checking your answers on a test can minimize simple mistakes. Though simple errors affect your grade, they don't affect your ability to do well on the next test.

Concept errors do affect your ability to be successful in the course. If you miss 10% on the first test because of concept errors and don't review those concepts, you will miss that 10% again on the next test. This has a snowball effect and by the third or fourth test, students are so far behind they give up.

Go through your test and mark every error as "simple" or "concept." After reviewing the errors, you should be confident that you could get those problems correct on the next test. Doing this will definitely improve your grade on the next test.

***How do you plan to learn from the mistakes you made on your tests?***

## Chapter Preview

Rational expressions are to algebra what rational numbers are to arithmetic. Most of the work we will do with rational expressions in this chapter parallels the work you have previously done with arithmetic fractions. The same basic properties we use to explain reducing, adding, subtracting, multiplying, and dividing arithmetic fractions will serve as a basis for our work with rational expressions. The techniques of factoring that we studied in Chapter 3 will also play an important role in our discussions. At the end of this chapter, we will work with some fractional equations that contain rational expressions.

## 4.1 Simplifying Rational Expressions

OBJECTIVES
1 Simplify rational numbers
2 Simplify rational expressions

We reviewed the basic operations with rational numbers in Chapter 1. In this review, we relied primarily on your knowledge of arithmetic. At this time, we want to become a little more formal with our review so that we can use the work with rational numbers as a basis for operating with rational expressions. We will define a rational expression shortly.

You will recall that any number that can be written in the form $\frac{a}{b}$, where $a$ and $b$ are integers and $b \neq 0$, is called a rational number. The following are examples of rational numbers.

$$\frac{1}{2} \qquad \frac{3}{4} \qquad \frac{15}{7} \qquad \frac{-5}{6} \qquad \frac{7}{-8} \qquad \frac{-12}{-17}$$

Numbers such as 6, $-4$, 0, $4\frac{1}{2}$, 0.7, and 0.21 are also rational because we can express them as the indicated quotient of two integers. For example,

$$6 = \frac{6}{1} = \frac{12}{2} = \frac{18}{3} \quad \text{and so on} \qquad 4\frac{1}{2} = \frac{9}{2}$$

$$-4 = \frac{4}{-1} = \frac{-4}{1} = \frac{8}{-2} \quad \text{and so on} \qquad 0.7 = \frac{7}{10}$$

$$0 = \frac{0}{1} = \frac{0}{2} = \frac{0}{3} \quad \text{and so on} \qquad 0.21 = \frac{21}{100}$$

Because a rational number is the quotient of two integers, our previous work with division of integers can help us understand the various forms of rational numbers. If the signs of the numerator and denominator are different, then the rational number is negative. If the signs of the numerator and denominator are the same, then the rational number is positive. The next examples and Property 4.1 show the equivalent forms of rational numbers. Generally, it is preferred to express the denominator of a rational number as a positive integer.

$$\frac{8}{-2} = \frac{-8}{2} = -\frac{8}{2} = -4 \qquad \frac{12}{3} = \frac{-12}{-3} = 4$$

Observe the following general properties.

**Property 4.1**

1. $\frac{-a}{b} = \frac{a}{-b} = -\frac{a}{b}$ where $b \neq 0$
2. $\frac{-a}{-b} = \frac{a}{b}$ where $b \neq 0$

Therefore, a rational number such as $\frac{-2}{5}$ can also be written as $\frac{2}{-5}$ or $-\frac{2}{5}$.

We use the following property, often referred to as the **fundamental principle of fractions**, to reduce fractions to lowest terms or to express fractions in simplest form.

**Property 4.2 Fundamental Principle of Fractions**

If $b$ and $k$ are nonzero integers and $a$ is any integer, then

$$\frac{a \cdot k}{b \cdot k} = \frac{a}{b}$$

Let's apply Properties 4.1 and 4.2 to the following examples.

**Classroom Example**
Reduce $\frac{14}{21}$ to lowest terms.

**EXAMPLE 1** Simplify $\frac{18}{24}$ to lowest terms.

**Solution**

$$\frac{18}{24} = \frac{3 \cdot 6}{4 \cdot 6} = \frac{3}{4}$$

**Classroom Example**
Change $\frac{32}{56}$ to simplest form.

**EXAMPLE 2** Change $\frac{40}{48}$ to simplest form.

**Solution**

$$\frac{\overset{5}{\cancel{40}}}{\underset{6}{\cancel{48}}} = \frac{5}{6}$$ A common factor of 8 was divided out of both numerator and denominator

**Classroom Example**
Express $\frac{-28}{44}$ in reduced form.

**EXAMPLE 3** Express $\frac{-36}{63}$ in simplest form.

**Solution**

$$\frac{-36}{63} = -\frac{36}{63} = -\frac{4 \cdot 9}{7 \cdot 9} = -\frac{4}{7}$$

**Classroom Example**
Reduce $\frac{36}{-84}$ to simplest form.

**EXAMPLE 4** Reduce $\frac{72}{-90}$ to simplest form.

**Solution**

$$\frac{72}{-90} = -\frac{72}{90} = -\frac{\cancel{2} \cdot 2 \cdot 2 \cdot \cancel{3} \cdot \cancel{3}}{\cancel{2} \cdot \cancel{3} \cdot \cancel{3} \cdot 5} = -\frac{4}{5}$$

In Examples 1–4, keep in mind that the number is not being changed, but the form of the numeral representing the number is being changed. In Example 1, $\frac{18}{24}$ and $\frac{3}{4}$ are equivalent fractions; they name the same number. Also note the use of prime factors in Example 4.

## Simplifying Rational Expressions

A **rational expression** is the indicated quotient of two polynomials. The following are examples of rational expressions.

$$\frac{3x^2}{5} \qquad \frac{x-2}{x+3} \qquad \frac{x^2+5x-1}{x^2-9} \qquad \frac{xy^2+x^2y}{xy} \qquad \frac{a^3-3a^2-5a-1}{a^4+a^3+6}$$

Because we must avoid division by zero, no values that create a denominator of zero can be assigned to variables. Thus the rational expression $\frac{x-2}{x+3}$ is meaningful for all values of $x$ except $x = -3$. Rather than making restrictions for each individual expression, we will merely assume that all denominators represent nonzero real numbers.

Property 4.2 $\left(\frac{a \cdot k}{b \cdot k} = \frac{a}{b}\right)$ serves as the basis for simplifying rational expressions, as the next examples illustrate.

Classroom Example
Simplify $\frac{18mn}{45m}$.

### EXAMPLE 5

Simplify $\frac{15xy}{25y}$.

**Solution**

$$\frac{15xy}{25y} = \frac{3 \cdot \cancel{5} \cdot x \cdot \cancel{y}}{\cancel{5} \cdot 5 \cdot \cancel{y}} = \frac{3x}{5}$$

Classroom Example
Simplify $\frac{-12}{36ab^2}$.

### EXAMPLE 6

Simplify $\frac{-9}{18x^2y}$.

**Solution**

$$\frac{-9}{18x^2y} = -\frac{\overset{1}{\cancel{9}}}{\underset{2}{\cancel{18}}x^2y} = -\frac{1}{2x^2y}$$

A common factor of 9 was divided out of numerator and denominator

Classroom Example
Simplify $\frac{-42x^3y^2}{-54x^2y^2}$.

### EXAMPLE 7

Simplify $\frac{-28a^2b^2}{-63a^2b^3}$.

**Solution**

$$\frac{-28a^2b^2}{-63a^2b^3} = \frac{4 \cdot \cancel{7} \cdot \cancel{a^2} \cdot \cancel{b^2}}{9 \cdot \cancel{7} \cdot \cancel{a^2} \cdot \underset{b}{\cancel{b^3}}} = \frac{4}{9b}$$

The factoring techniques from Chapter 3 can be used to factor numerators and/or denominators so that we can apply the property $\frac{a \cdot k}{b \cdot k} = \frac{a}{b}$. Examples 8–12 should clarify this process.

Classroom Example
Simplify $\frac{x^2 - 7x}{x^2 - 49}$.

**EXAMPLE 8** Simplify $\frac{x^2 + 4x}{x^2 - 16}$.

### Solution

$$\frac{x^2 + 4x}{x^2 - 16} = \frac{x(x + 4)}{(x - 4)(x + 4)} = \frac{x}{x - 4}$$

Classroom Example
Simplify $\frac{9x^2 + 6x + 1}{3x + 1}$.

**EXAMPLE 9** Simplify $\frac{4a^2 + 12a + 9}{2a + 3}$.

### Solution

$$\frac{4a^2 + 12a + 9}{2a + 3} = \frac{(2a + 3)(2a + 3)}{1(2a + 3)} = \frac{2a + 3}{1} = 2a + 3$$

Classroom Example
Simplify $\frac{7n^2 + 23n + 6}{21n^2 - n - 2}$.

**EXAMPLE 10** Simplify $\frac{5n^2 + 6n - 8}{10n^2 - 3n - 4}$.

### Solution

$$\frac{5n^2 + 6n - 8}{10n^2 - 3n - 4} = \frac{(5n - 4)(n + 2)}{(5n - 4)(2n + 1)} = \frac{n + 2}{2n + 1}$$

Classroom Example
Simplify $\frac{3x^3y - 12xy}{x^2y - xy - 6y}$.

**EXAMPLE 11** Simplify $\frac{6x^3y - 6xy}{x^3 + 5x^2 + 4x}$.

### Solution

$$\frac{6x^3y - 6xy}{x^3 + 5x^2 + 4x} = \frac{6xy(x^2 - 1)}{x(x^2 + 5x + 4)} = \frac{6xy(x + 1)(x - 1)}{x(x + 1)(x + 4)} = \frac{6y(x - 1)}{x + 4}$$

Note that in Example 11 we left the numerator of the final fraction in factored form. This is often done if expressions other than monomials are involved. Either $\frac{6y(x - 1)}{x + 4}$ or $\frac{6xy - 6y}{x + 4}$ is an acceptable answer.

Remember that the quotient of any nonzero real number and its opposite is $-1$. For example, $\frac{6}{-6} = -1$ and $\frac{-8}{8} = -1$. Likewise, the indicated quotient of any polynomial and its opposite is equal to $-1$; that is,

$\frac{a}{-a} = -1$ because $a$ and $-a$ are opposites

$\frac{a - b}{b - a} = -1$ because $a - b$ and $b - a$ are opposites

$\frac{x^2 - 4}{4 - x^2} = -1$ because $x^2 - 4$ and $4 - x^2$ are opposites

Example 12 shows how we use this idea when simplifying rational expressions.

**Classroom Example**
Simplify $\frac{6a^2 - 17a + 5}{15a - 6a^2}$.

**EXAMPLE 12** Simplify $\frac{6a^2 - 7a + 2}{10a - 15a^2}$.

**Solution**

$$\frac{6a^2 - 7a + 2}{10a - 15a^2} = \frac{(2a - 1)(3a - 2)}{5a(2 - 3a)} \qquad \frac{3a - 2}{2 - 3a} = -1$$

$$= (-1)\left(\frac{2a - 1}{5a}\right)$$

$$= -\frac{2a - 1}{5a} \quad \text{or} \quad \frac{1 - 2a}{5a}$$

## Concept Quiz 4.1

For Problems 1–10, answer true or false.

1. When a rational number is being reduced, the form of the numeral is being changed but not the number it represents.
2. A rational number is the ratio of two integers where the denominator is not zero.
3. $-3$ is a rational number.
4. The rational expression $\frac{x + 2}{x + 3}$ is meaningful for all values of $x$ except when $x = -2$ and $x = 3$.
5. The binomials $x - y$ and $y - x$ are opposites.
6. The binomials $x + 3$ and $x - 3$ are opposites.
7. The rational expression $\frac{2 - x}{x + 2}$ reduces to $-1$.
8. The rational expression $\frac{x - y}{y - x}$ reduces to $-1$.
9. $\frac{x^2 + 5x - 14}{x^2 + 2x + 1} = \frac{5x - 14}{2x + 1}$
10. The rational expression $\frac{2x - x^2}{x^2 - 4}$ reduces to $\frac{x}{x + 2}$.

## Problem Set 4.1

For Problems 1–8, express each rational number in simplest form. (Objective 1)

1. $\frac{27}{36}$
2. $\frac{14}{21}$
3. $\frac{45}{54}$
4. $\frac{-14}{42}$
5. $\frac{24}{-60}$
6. $\frac{45}{-75}$
7. $\frac{-16}{-56}$
8. $\frac{-30}{-42}$

For Problems 9–50, simplify each rational expression. (Objective 2)

9. $\frac{12xy}{42y}$
10. $\frac{21xy}{35x}$
11. $\frac{18a^2}{45ab}$
12. $\frac{48ab}{84b^2}$
13. $\frac{-14y^3}{56xy^2}$
14. $\frac{-14x^2y^3}{63xy^2}$
15. $\frac{54c^2d}{-78cd^2}$
16. $\frac{60x^3z}{-64xyz^2}$
17. $\frac{-40x^3y}{-24xy^4}$
18. $\frac{-30x^2y^2z^2}{-35xz^3}$
19. $\frac{x^2 - 4}{x^2 + 2x}$
20. $\frac{xy + y^2}{x^2 - y^2}$

**21.** $\dfrac{18x + 12}{12x - 6}$

**22.** $\dfrac{20x + 50}{15x - 30}$

**23.** $\dfrac{a^2 + 7a + 10}{a^2 - 7a - 18}$

**24.** $\dfrac{a^2 + 4a - 32}{3a^2 + 26a + 16}$

**25.** $\dfrac{2n^2 + n - 21}{10n^2 + 33n - 7}$

**26.** $\dfrac{4n^2 - 15n - 4}{7n^2 - 30n + 8}$

**27.** $\dfrac{5x^2 + 7}{10x}$

**28.** $\dfrac{12x^2 + 11x - 15}{20x^2 - 23x + 6}$

**29.** $\dfrac{6x^2 + x - 15}{8x^2 - 10x - 3}$

**30.** $\dfrac{4x^2 + 8x}{x^3 + 8}$

**31.** $\dfrac{3x^2 - 12x}{x^3 - 64}$

**32.** $\dfrac{x^2 - 14x + 49}{6x^2 - 37x - 35}$

**33.** $\dfrac{3x^2 + 17x - 6}{9x^2 - 6x + 1}$

**34.** $\dfrac{9y^2 - 1}{3y^2 + 11y - 4}$

**35.** $\dfrac{2x^3 + 3x^2 - 14x}{x^2y + 7xy - 18y}$

**36.** $\dfrac{3x^3 + 12x}{9x^2 + 18x}$

**37.** $\dfrac{5y^2 + 22y + 8}{25y^2 - 4}$

**38.** $\dfrac{16x^3y + 24x^2y^2 - 16xy^3}{24x^2y + 12xy^2 - 12y^3}$

**39.** $\dfrac{15x^3 - 15x^2}{5x^3 + 5x}$

**40.** $\dfrac{5n^2 + 18n - 8}{3n^2 + 13n + 4}$

**41.** $\dfrac{4x^2y + 8xy^2 - 12y^3}{18x^3y - 12x^2y^2 - 6xy^3}$

**42.** $\dfrac{3 + x - 2x^2}{2 + x - x^2}$

**43.** $\dfrac{3n^2 + 16n - 12}{7n^2 + 44n + 12}$

**44.** $\dfrac{x^4 - 2x^2 - 15}{2x^4 + 9x^2 + 9}$

**45.** $\dfrac{8 + 18x - 5x^2}{10 + 31x + 15x^2}$

**46.** $\dfrac{6x^4 - 11x^2 + 4}{2x^4 + 17x^2 - 9}$

**47.** $\dfrac{27x^4 - x}{6x^3 + 10x^2 - 4x}$

**48.** $\dfrac{64x^4 + 27x}{12x^3 - 27x^2 - 27x}$

**49.** $\dfrac{-40x^3 + 24x^2 + 16x}{20x^3 + 28x^2 + 8x}$

**50.** $\dfrac{-6x^3 - 21x^2 + 12x}{-18x^3 - 42x^2 + 120x}$

For Problems 51–58, simplify each rational expression. You will need to use factoring by grouping. (Objective 2)

**51.** $\dfrac{xy + ay + bx + ab}{xy + ay + cx + ac}$

**52.** $\dfrac{xy + 2y + 3x + 6}{xy + 2y + 4x + 8}$

**53.** $\dfrac{ax - 3x + 2ay - 6y}{2ax - 6x + ay - 3y}$

**54.** $\dfrac{x^2 - 2x + ax - 2a}{x^2 - 2x + 3ax - 6a}$

**55.** $\dfrac{5x^2 + 5x + 3x + 3}{5x^2 + 3x - 30x - 18}$

**56.** $\dfrac{x^2 + 3x + 4x + 12}{2x^2 + 6x - x - 3}$

**57.** $\dfrac{2st - 30 - 12s + 5t}{3st - 6 - 18s + t}$

**58.** $\dfrac{nr - 6 - 3n + 2r}{nr + 10 + 2r + 5n}$

For Problems 59–68, simplify each rational expression. You may want to refer to Example 12 of this section. (Objective 2)

**59.** $\dfrac{5x - 7}{7 - 5x}$

**60.** $\dfrac{4a - 9}{9 - 4a}$

**61.** $\dfrac{n^2 - 49}{7 - n}$

**62.** $\dfrac{9 - y}{y^2 - 81}$

**63.** $\dfrac{2y - 2xy}{x^2y - y}$

**64.** $\dfrac{3x - x^2}{x^2 - 9}$

**65.** $\dfrac{2x^3 - 8x}{4x - x^3}$

**66.** $\dfrac{x^2 - (y - 1)^2}{(y - 1)^2 - x^2}$

**67.** $\dfrac{n^2 - 5n - 24}{40 + 3n - n^2}$

**68.** $\dfrac{x^2 + 2x - 24}{20 - x - x^2}$

## Thoughts Into Words

**69.** Compare the concept of a rational number in arithmetic to the concept of a rational expression in algebra.

**70.** What role does factoring play in the simplifying of rational expressions?

**71.** Why is the rational expression $\dfrac{x + 3}{x^2 - 4}$ undefined for $x = 2$ and $x = -2$ but defined for $x = -3$?

**72.** How would you convince someone that $\dfrac{x - 4}{4 - x} = -1$ for all real numbers except 4?

**Answers to the Concept Quiz**

**1.** True **2.** True **3.** True **4.** False **5.** True **6.** False **7.** False **8.** True **9.** False **10.** False

## 4.2 Multiplying and Dividing Rational Expressions

OBJECTIVES
1. Multiply rational numbers
2. Multiply rational expressions
3. Divide rational numbers
4. Divide rational expressions
5. Simplify problems that involve both multiplication and division of rational expressions

We define multiplication of rational numbers in common-fraction form as follows:

**Definition 4.1 Multiplication of Fractions**

If $a$, $b$, $c$, and $d$ are integers, and $b$ and $d$ are not equal to zero, then

$$\frac{a}{b}\cdot\frac{c}{d}=\frac{a\cdot c}{b\cdot d}=\frac{ac}{bd}$$

To multiply rational numbers in common-fraction form, we *multiply numerators and multiply denominators,* as the following examples demonstrate. (The steps in the dashed boxes are usually done mentally.)

$$\frac{2}{3}\cdot\frac{4}{5}=\boxed{\frac{2\cdot 4}{3\cdot 5}}=\frac{8}{15}$$

$$\frac{-3}{4}\cdot\frac{5}{7}=\boxed{\frac{-3\cdot 5}{4\cdot 7}=\frac{-15}{28}}=-\frac{15}{28}$$

$$-\frac{5}{6}\cdot\frac{13}{3}=\boxed{\frac{-5}{6}\cdot\frac{13}{3}=\frac{-5\cdot 13}{6\cdot 3}=\frac{-65}{18}}=-\frac{65}{18}$$

We also agree, when multiplying rational numbers, to express the final product in simplest form. The following examples show some different formats used to multiply and simplify rational numbers.

$$\frac{3}{4}\cdot\frac{4}{7}=\frac{3\cdot \cancel{4}}{\cancel{4}\cdot 7}=\frac{3}{7}$$

$$\frac{\overset{1}{\cancel{8}}}{\underset{1}{\cancel{9}}}\cdot\frac{\overset{3}{\cancel{27}}}{\underset{4}{\cancel{32}}}=\frac{3}{4}$$

A common factor of 9 was divided out of 9 and 27, and a common factor of 8 was divided out of 8 and 32

$$\left(-\frac{28}{25}\right)\left(-\frac{65}{78}\right)=\frac{\cancel{2}\cdot 2\cdot 7\cdot \cancel{5}\cdot \cancel{13}}{\cancel{5}\cdot 5\cdot \cancel{2}\cdot 3\cdot \cancel{13}}=\frac{14}{15}$$

We should recognize that a negative times a negative is positive; also, note the use of prime factors to help us recognize common factors

### Multiplying Rational Expressions

Multiplication of rational expressions follows the same basic pattern as multiplication of rational numbers in common-fraction form. That is to say, we multiply numerators and multiply denominators and express the final product in simplified or reduced form. Let's consider some examples.

$$\frac{3x}{4y}\cdot\frac{8y^2}{9x}=\frac{\cancel{3}\cdot \overset{2}{\cancel{8}}\cdot \cancel{x}\cdot \overset{y}{\cancel{y^2}}}{\cancel{4}\cdot \underset{3}{\cancel{9}}\cdot \cancel{x}\cdot \cancel{y}}=\frac{2y}{3}$$

Note that we use the commutative property of multiplication to rearrange the factors in a form that allows us to identify common factors of the numerator and denominator

$$\frac{-4a}{6a^2b^2} \cdot \frac{9ab}{12a^2} = -\frac{4 \cdot 9 \cdot a^2 \cdot b}{6 \cdot 12 \cdot a^4 \cdot b^2} = -\frac{1}{2a^2b}$$

$$\frac{12x^2y}{-18xy} \cdot \frac{-24xy^2}{56y^3} = \frac{12 \cdot 24 \cdot x^3 \cdot y^3}{18 \cdot 56 \cdot x \cdot y^4} = \frac{2x^2}{7y}$$

You should recognize that the first fraction is negative, and the second fraction is negative. Thus the product is positive.

If the rational expressions contain polynomials (other than monomials) that are factorable, then the first step will be to factor.

**Classroom Example**
Multiply and simplify $\frac{m}{n^2 - 9} \cdot \frac{n-3}{m^3}$.

**EXAMPLE 1** Multiply and simplify $\frac{y}{x^2 - 4} \cdot \frac{x+2}{y^2}$.

### Solution

$$\frac{y}{x^2 - 4} \cdot \frac{x+2}{y^2} = \frac{y}{(x+2)(x-2)} \cdot \frac{x+2}{y^2} \quad \text{Factor } x^2 - 4$$

$$= \frac{y(x+2)}{y^2(x+2)(x-2)} = \frac{1}{y(x-2)}$$

Note that we left the final answer in factored form. Either $\frac{1}{y(x-2)}$ or $\frac{1}{xy - 2y}$ would be an acceptable answer.

**Classroom Example**
Multiply and simplify:
$\frac{m^2 + m}{m + 4} \cdot \frac{m^2 - 4m + 3}{m^4 - m^2}$

**EXAMPLE 2** Multiply and simplify $\frac{x^2 - x}{x + 5} \cdot \frac{x^2 + 5x + 4}{x^4 - x^2}$.

### Solution

$$\frac{x^2 - x}{x + 5} \cdot \frac{x^2 + 5x + 4}{x^4 - x^2} = \frac{x(x-1)}{x+5} \cdot \frac{(x+1)(x+4)}{x^2(x-1)(x+1)} \quad \text{Factor}$$

$$= \frac{x(x-1)(x+1)(x+4)}{(x+5)(x^2)(x-1)(x+1)} = \frac{x+4}{x(x+5)}$$

**Classroom Example**
Multiply and simplify:
$\frac{8x^2 + 10x - 3}{6x^2 + 7x - 3} \cdot \frac{3x^2 + 20x - 7}{8x^2 + 18x - 5}$

**EXAMPLE 3** Multiply and simplify $\frac{6n^2 + 7n - 5}{n^2 + 2n - 24} \cdot \frac{4n^2 + 21n - 18}{12n^2 + 11n - 15}$.

### Solution

$$\frac{6n^2 + 7n - 5}{n^2 + 2n - 24} \cdot \frac{4n^2 + 21n - 18}{12n^2 + 11n - 15} = \frac{(3n+5)(2n-1)}{(n+6)(n-4)} \cdot \frac{(4n-3)(n+6)}{(3n+5)(4n-3)}$$

$$= \frac{(3n+5)(2n-1)(4n-3)(n+6)}{(n+6)(n-4)(3n+5)(4n-3)}$$

$$= \frac{2n-1}{n-4}$$

## Dividing Rational Numbers

We define division of rational numbers in common-fraction form as follows:

**Definition 4.2 Division of Fractions**

If $a$, $b$, $c$, and $d$ are integers, and $b$, $c$, and $d$ are not equal to zero, then

$$\frac{a}{b} \div \frac{c}{d} = \frac{a}{b} \cdot \frac{d}{c} = \frac{ad}{bc}$$

Definition 4.2 states that to divide two rational numbers in fraction form, we **invert the divisor and multiply**. We call the numbers $\frac{c}{d}$ and $\frac{d}{c}$ "reciprocals" or "multiplicative inverses" of each other, because their product is 1. Thus we can describe division by saying "to divide by a fraction, multiply by its reciprocal." The following examples demonstrate the use of Definition 4.2.

$$\frac{7}{8} \div \frac{5}{6} = \frac{7}{8} \cdot \frac{6}{5} = \frac{7}{\underset{4}{\cancel{8}}} \cdot \frac{\overset{3}{\cancel{6}}}{5} = \frac{21}{20}$$

$$\frac{-5}{9} \div \frac{15}{18} = -\frac{5}{9} \cdot \frac{18}{15} = -\frac{\cancel{5}}{\cancel{9}} \cdot \frac{\overset{2}{\cancel{18}}}{\underset{3}{\cancel{15}}} = -\frac{2}{3}$$

$$\frac{14}{-19} \div \frac{21}{-38} = \left(-\frac{14}{19}\right) \div \left(-\frac{21}{38}\right) = \left(-\frac{14}{19}\right)\left(-\frac{38}{21}\right) = \left(-\frac{\overset{2}{\cancel{14}}}{\cancel{19}}\right)\left(-\frac{\overset{2}{\cancel{38}}}{\underset{3}{\cancel{21}}}\right) = \frac{4}{3}$$

## Dividing Rational Expressions

We define division of algebraic rational expressions in the same way that we define division of rational numbers. That is, the quotient of two rational expressions is the product we obtain when we multiply the first expression by the reciprocal of the second. Consider the following examples.

**Classroom Example**
Divide and simplify:

$$\frac{18mn^3}{32m^3n^2} \div \frac{9m^3n^2}{12m^2n^2}$$

**EXAMPLE 4** Divide and simplify $\frac{16x^2y}{24xy^3} \div \frac{9xy}{8x^2y^2}$.

### Solution

$$\frac{16x^2y}{24xy^3} \div \frac{9xy}{8x^2y^2} = \frac{16x^2y}{24xy^3} \cdot \frac{8x^2y^2}{9xy} = \frac{16 \cdot \cancel{8} \cdot \overset{x^2}{\cancel{x^4}} \cdot \cancel{y^3}}{\underset{3}{\cancel{24}} \cdot 9 \cdot \cancel{x^2} \cdot \underset{y}{\cancel{y^4}}} = \frac{16x^2}{27y}$$

**Classroom Example**
Divide and simplify:

$$\frac{4x^2 + 36}{8x^2 + 4x} \div \frac{x^4 - 81}{2x^2 - 5x - 3}$$

**EXAMPLE 5** Divide and simplify $\frac{3a^2 + 12}{3a^2 - 15a} \div \frac{a^4 - 16}{a^2 - 3a - 10}$.

### Solution

$$\frac{3a^2 + 12}{3a^2 - 15a} \div \frac{a^4 - 16}{a^2 - 3a - 10} = \frac{3a^2 + 12}{3a^2 - 15a} \cdot \frac{a^2 - 3a - 10}{a^4 - 16}$$

$$= \frac{3(a^2 + 4)}{3a(a - 5)} \cdot \frac{(a - 5)(a + 2)}{(a^2 + 4)(a + 2)(a - 2)} \quad \text{Factor}$$

$$= \frac{\overset{1}{\cancel{3}}\cancel{(a^2 + 4)}\cancel{(a - 5)}\cancel{(a + 2)}}{\underset{1}{\cancel{3}}a\cancel{(a - 5)}\cancel{(a^2 + 4)}\cancel{(a + 2)}(a - 2)}$$

$$= \frac{1}{a(a - 2)}$$

**Classroom Example**
Divide and simplify:

$$\frac{35x^3 - 8x^2 - 3x}{45x^2 - x - 2} \div (7x - 3)$$

**EXAMPLE 6** Divide and simplify $\frac{28t^3 - 51t^2 - 27t}{49t^2 + 42t + 9} \div (4t - 9)$.

### Solution

$$\frac{28t^3 - 51t^2 - 27t}{49t^2 + 42t + 9} \div \frac{4t - 9}{1} = \frac{28t^3 - 51t^2 - 27t}{49t^2 + 42t + 9} \cdot \frac{1}{4t - 9}$$

$$= \frac{t(7t+3)(4t-9)}{(7t+3)(7t+3)} \cdot \frac{1}{(4t-9)} \quad \text{Factor}$$

$$= \frac{t\cancel{(7t+3)}\cancel{(4t-9)}}{\cancel{(7t+3)}(7t+3)\cancel{(4t-9)}}$$

$$= \frac{t}{7t+3}$$

In a problem such as Example 6, it may be helpful to write the divisor with a denominator of 1. Thus we write $4t - 9$ as $\frac{4t-9}{1}$; its reciprocal is obviously $\frac{1}{4t-9}$.

Let's consider one final example that involves both multiplication and division.

**Classroom Example**
Perform the indicated operations and simplify:

$$\frac{5x^2+13x-6}{2xy^2-3y^2} \cdot \frac{2x^2+5x-12}{x^3+3x^2} \div \frac{2x^2+13x+20}{x^2y}$$

### EXAMPLE 7

Perform the indicated operations and simplify:

$$\frac{x^2+5x}{3x^2-4x-20} \cdot \frac{x^2y+y}{2x^2+11x+5} \div \frac{xy^2}{6x^2-17x-10}$$

**Solution**

$$\frac{x^2+5x}{3x^2-4x-20} \cdot \frac{x^2y+y}{2x^2+11x+5} \div \frac{xy^2}{6x^2-17x-10}$$

$$= \frac{x^2+5x}{3x^2-4x-20} \cdot \frac{x^2y+y}{2x^2+11x+5} \cdot \frac{6x^2-17x-10}{xy^2}$$

$$= \frac{x(x+5)}{(3x-10)(x+2)} \cdot \frac{y(x^2+1)}{(2x+1)(x+5)} \cdot \frac{(2x+1)(3x-10)}{xy^2} \quad \text{Factor}$$

$$= \frac{\cancel{x}\cancel{(x+5)}\cancel{(y)}(x^2+1)\cancel{(2x+1)}\cancel{(3x-10)}}{\cancel{(3x-10)}(x+2)\cancel{(2x+1)}\cancel{(x+5)}\cancel{(x)}\cancel{(y^2)}_{y}} = \frac{x^2+1}{y(x+2)}$$

## Concept Quiz 4.2

For Problems 1–10, answer true or false.

1. To multiply two rational numbers in fraction form, we need to change to equivalent fractions with a common denominator.
2. When multiplying rational expressions that contain polynomials, the polynomials are factored so that common factors can be divided out.
3. In the division problem $\frac{2x^2y}{3z} \div \frac{4x^3}{5y^2}$, the fraction $\frac{4x^3}{5y^2}$ is the divisor.
4. The numbers $-\frac{2}{3}$ and $\frac{3}{2}$ are multiplicative inverses.
5. To divide two numbers in fraction form, we invert the divisor and multiply.
6. If $x \neq 0$, then $\left(\frac{4xy}{x}\right)\left(\frac{3y}{2x}\right) = \frac{6y^2}{x}$.
7. $\frac{3}{4} \div \frac{4}{3} = 1$.
8. If $x \neq 0$ and $y \neq 0$, then $\frac{5x^2y}{2y} \div \frac{10x^2}{3y} = \frac{3}{4}$.

9. If $x \neq 0$ and $y \neq 0$, then $\frac{1}{x} \div \frac{1}{y} = xy$.

10. If $x \neq y$, then $\frac{1}{x-y} \div \frac{1}{y-x} = -1$.

## Problem Set 4.2

For Problems 1–12, perform the indicated operations involving rational numbers. Express final answers in reduced form. (Objectives 1 and 3)

1. $\frac{7}{12} \cdot \frac{6}{35}$

2. $\frac{5}{8} \cdot \frac{12}{20}$

3. $\frac{-4}{9} \cdot \frac{18}{30}$

4. $\frac{-6}{9} \cdot \frac{36}{48}$

5. $\frac{3}{-8} \cdot \frac{-6}{12}$

6. $\frac{-12}{16} \cdot \frac{18}{-32}$

7. $\left(-\frac{5}{7}\right) \div \frac{6}{7}$

8. $\left(-\frac{5}{9}\right) \div \frac{10}{3}$

9. $\frac{-9}{5} \div \frac{27}{10}$

10. $\frac{4}{7} \div \frac{16}{-21}$

11. $\frac{4}{9} \cdot \frac{6}{11} \div \frac{4}{15}$

12. $\frac{2}{3} \cdot \frac{6}{7} \div \frac{8}{3}$

For Problems 13–50, perform the indicated operations involving rational expressions. Express final answers in simplest form. (Objectives 2, 4, and 5)

13. $\frac{6xy}{9y^4} \cdot \frac{30x^3y}{-48x}$

14. $\frac{-14xy^4}{18y^2} \cdot \frac{24x^2y^3}{35y^2}$

15. $\frac{5a^2b^2}{11ab} \cdot \frac{22a^3}{15ab^2}$

16. $\frac{10a^2}{5b^2} \cdot \frac{15b^3}{2a^4}$

17. $\frac{5xy}{8y^2} \cdot \frac{18x^2y}{15}$

18. $\frac{4x^2}{5y^2} \cdot \frac{15xy}{24x^2y^2}$

19. $\frac{5x^4}{12x^2y^3} \div \frac{9}{5xy}$

20. $\frac{7x^2y}{9xy^3} \div \frac{3x^4}{2x^2y^2}$

21. $\frac{9a^2c}{12bc^2} \div \frac{21ab}{14c^3}$

22. $\frac{3ab^3}{4c} \div \frac{21ac}{12bc^3}$

23. $\frac{9x^2y^3}{14x} \cdot \frac{21y}{15xy^2} \cdot \frac{10x}{12y^3}$

24. $\frac{5xy}{7a} \cdot \frac{14a^2}{15x} \cdot \frac{3a}{8y}$

25. $\frac{3x+6}{5y} \cdot \frac{x^2+4}{x^2+10x+16}$

26. $\frac{5xy}{x+6} \cdot \frac{x^2-36}{x^2-6x}$

27. $\frac{5a^2+20a}{a^3-2a^2} \cdot \frac{a^2-a-12}{a^2-16}$

28. $\frac{2a^2+6}{a^2-a} \cdot \frac{a^3-a^2}{8a-4}$

29. $\frac{3n^2+15n-18}{3n^2+10n-48} \cdot \frac{6n^2-n-40}{4n^2+6n-10}$

30. $\frac{6n^2+11n-10}{3n^2+19n-14} \cdot \frac{2n^2+6n-56}{2n^2-3n-20}$

31. $\frac{9y^2}{x^2+12x+36} \div \frac{12y}{x^2+6x}$

32. $\frac{7xy}{x^2-4x+4} \div \frac{14y}{x^2-4}$

33. $\frac{x^2-4xy+4y^2}{7xy^2} \div \frac{4x^2-3xy-10y^2}{20x^2y+25xy^2}$

34. $\frac{x^2+5xy-6y^2}{xy^2-y^3} \cdot \frac{2x^2+15xy+18y^2}{xy+4y^2}$

35. $\frac{5-14n-3n^2}{1-2n-3n^2} \cdot \frac{9+7n-2n^2}{27-15n+2n^2}$

36. $\frac{6-n-2n^2}{12-11n+2n^2} \cdot \frac{24-26n+5n^2}{2+3n+n^2}$

37. $\frac{3x^4+2x^2-1}{3x^4+14x^2-5} \cdot \frac{x^4-2x^2-35}{x^4-17x^2+70}$

38. $\frac{2x^4+x^2-3}{2x^4+5x^2+2} \cdot \frac{3x^4+10x^2+8}{3x^4+x^2-4}$

39. $\frac{3x^2-20x+25}{2x^2-7x-15} \div \frac{9x^2-3x-20}{12x^2+28x+15}$

40. $\frac{21t^2+t-2}{2t^2-17t-9} \div \frac{12t^2-5t-3}{8t^2-2t-3}$

41. $\frac{10t^3+25t}{20t+10} \cdot \frac{2t^2-t-1}{t^5-t}$

42. $\frac{t^4-81}{t^2-6t+9} \cdot \frac{6t^2-11t-21}{5t^2+8t-21}$

43. $\frac{4t^2+t-5}{t^3-t^2} \cdot \frac{t^4+6t^3}{16t^2+40t+25}$

44. $\frac{9n^2-12n+4}{n^2-4n-32} \cdot \frac{n^2+4n}{3n^3-2n^2}$

45. $\frac{nr+3n+2r+6}{nr+3n-3r-9} \cdot \frac{n^2-9}{n^3-4n}$

**46.** $\dfrac{xy + xc + ay + ac}{xy - 2xc + ay - 2ac} \cdot \dfrac{2x^3 - 8x}{12x^3 + 20x^2 - 8x}$

**47.** $\dfrac{x^2 - x}{4y} \cdot \dfrac{10xy^2}{2x - 2} \div \dfrac{3x^2 + 3x}{15x^2y^2}$

**48.** $\dfrac{4xy^2}{7x} \cdot \dfrac{14x^3y}{12y} \div \dfrac{7y}{9x^3}$

**49.** $\dfrac{a^2 - 4ab + 4b^2}{6a^2 - 4ab} \cdot \dfrac{3a^2 + 5ab - 2b^2}{6a^2 + ab - b^2} \div \dfrac{a^2 - 4b^2}{8a + 4b}$

**50.** $\dfrac{2x^2 + 3x}{2x^3 - 10x^2} \cdot \dfrac{x^2 - 8x + 15}{3x^3 - 27x} \div \dfrac{14x + 21}{x^2 - 6x - 27}$

## Thoughts Into Words

**51.** Explain in your own words how to divide two rational expressions.

**52.** Suppose that your friend missed class the day the material in this section was discussed. How could you draw on her background in arithmetic to explain to her how to multiply and divide rational expressions?

**53.** Give a step-by-step description of how to do the following multiplication problem.

$$\frac{x^2 + 5x + 6}{x^2 - 2x - 8} \cdot \frac{x^2 - 16}{16 - x^2}$$

### Answers to the Concept Quiz

**1.** False **2.** True **3.** True **4.** False **5.** True **6.** True **7.** False **8.** False **9.** False **10.** True

# 4.3 Adding and Subtracting Rational Expressions

OBJECTIVES

1. Add and subtract rational numbers
2. Add and subtract rational expressions

We can define addition and subtraction of rational numbers as follows:

**Definition 4.3 Addition and Subtraction of Fractions**

If $a$, $b$, and $c$ are integers, and $b$ is not zero, then

$$\frac{a}{b} + \frac{c}{b} = \frac{a + c}{b} \quad \text{Addition}$$

$$\frac{a}{b} - \frac{c}{b} = \frac{a - c}{b} \quad \text{Subtraction}$$

*We can add or subtract rational numbers with a common denominator by adding or subtracting the numerators and placing the result over the common denominator.* The following examples illustrate Definition 4.3.

$$\frac{2}{9} + \frac{3}{9} = \frac{2 + 3}{9} = \frac{5}{9}$$

$$\frac{7}{8} - \frac{3}{8} = \frac{7 - 3}{8} = \frac{4}{8} = \frac{1}{2} \quad \text{Don't forget to simplify!}$$

$$\frac{4}{6} + \frac{-5}{6} = \frac{4 + (-5)}{6} = \frac{-1}{6} = -\frac{1}{6}$$

$$\frac{7}{10} + \frac{4}{-10} = \frac{7}{10} + \frac{-4}{10} = \frac{7 + (-4)}{10} = \frac{3}{10}$$

We use this same *common denominator* approach when adding or subtracting rational expressions, as in these next examples.

$$\frac{3}{x} + \frac{9}{x} = \frac{3+9}{x} = \frac{12}{x}$$

$$\frac{8}{x-2} - \frac{3}{x-2} = \frac{8-3}{x-2} = \frac{5}{x-2}$$

$$\frac{9}{4y} + \frac{5}{4y} = \frac{9+5}{4y} = \frac{14}{4y} = \frac{7}{2y}$$ Don't forget to simplify the final answer!

$$\frac{n^2}{n-1} - \frac{1}{n-1} = \frac{n^2-1}{n-1} = \frac{(n+1)\cancel{(n-1)}}{\cancel{n-1}} = n+1$$

$$\frac{6a^2}{2a+1} + \frac{13a+5}{2a+1} = \frac{6a^2+13a+5}{2a+1} = \frac{\cancel{(2a+1)}(3a+5)}{\cancel{2a+1}} = 3a+5$$

In each of the previous examples that involve rational expressions, we should technically restrict the variables to exclude division by zero. For example, $\frac{3}{x} + \frac{9}{x} = \frac{12}{x}$ is true for all real number values for $x$, except $x = 0$. Likewise, $\frac{8}{x-2} - \frac{3}{x-2} = \frac{5}{x-2}$ as long as $x$ does not equal 2. Rather than taking the time and space to write down restrictions for each problem, we will merely assume that such restrictions exist.

*If rational numbers that do not have a common denominator are to be added or subtracted, then we apply the fundamental principle of fractions* $\left(\frac{a}{b} = \frac{ak}{bk}\right)$ *to obtain equivalent fractions with a common denominator.* Equivalent fractions are fractions such as $\frac{1}{2}$ and $\frac{2}{4}$ that name the same number. Consider the following example.

$$\frac{1}{2} + \frac{1}{3} = \frac{3}{6} + \frac{2}{6} = \frac{3+2}{6} = \frac{5}{6}$$

($\frac{1}{2}$ and $\frac{3}{6}$ are equivalent fractions.) ($\frac{1}{3}$ and $\frac{2}{6}$ are equivalent fractions.)

Note that we chose 6 as our common denominator, and 6 is the *least common multiple* of the original denominators 2 and 3. The least common multiple of a set of whole numbers is the smallest nonzero whole number divisible by each of the numbers. In general, we use the least common multiple of the denominators of the fractions to be added or subtracted as a *least common denominator* (LCD). A least common denominator may be found by inspection or by using the prime factored forms of the numbers.

When studying Examples 1–10, note that all the examples follow the same basic procedure for addition or subtraction of rational expressions.

**Step 1** If necessary, factor the denominators.

**Step 2** Find the least common denominator (LCD).

**Step 3** Change each fraction to an equivalent fraction that has the LCD as its denominator.

**Step 4** Combine the numerators and place over the LCD.

**Step 5** Simplify the numerator by performing the addition or subtraction.

**Step 6** If necessary, simplify the resulting fraction.

Classroom Example
Subtract $\frac{7}{9} - \frac{1}{6}$.

**EXAMPLE 1** Subtract $\frac{5}{6} - \frac{3}{8}$.

### Solution

By inspection, we can see that the LCD is 24. Thus both fractions can be changed to equivalent fractions, each with a denominator of 24.

$$\frac{5}{6} - \frac{3}{8} = \frac{5}{6} \cdot \frac{4}{4} - \frac{3}{8} \cdot \frac{3}{3} \quad \text{Change to equivalent fractions}$$
$$= \frac{20}{24} - \frac{9}{24}$$
$$= \frac{11}{24}$$

Classroom Example
Perform the indicated operations:
$\frac{1}{4} + \frac{3}{7} - \frac{5}{28}$

**EXAMPLE 2** Perform the indicated operations: $\frac{3}{5} + \frac{1}{6} - \frac{13}{15}$

### Solution

Again by inspection, we can determine that the LCD is 30. Thus we can proceed as follows:

$$\frac{3}{5} + \frac{1}{6} - \frac{13}{15} = \frac{3}{5} \cdot \frac{6}{6} + \frac{1}{6} \cdot \frac{5}{5} - \frac{13}{15} \cdot \frac{2}{2} \quad \text{Change to equivalent fractions}$$
$$= \frac{18}{30} + \frac{5}{30} - \frac{26}{30}$$
$$= \frac{18 + 5 - 26}{30}$$
$$= \frac{-3}{30} = -\frac{1}{10} \quad \text{Don't forget to simplify!}$$

Classroom Example
Add $\frac{4}{9} + \frac{7}{15}$.

**EXAMPLE 3** Add $\frac{7}{18} + \frac{11}{24}$.

### Solution

Let's use the prime factored forms of the denominators to help find the LCD.

$$18 = 2 \cdot 3 \cdot 3 \qquad 24 = 2 \cdot 2 \cdot 2 \cdot 3$$

The LCD must contain three factors of 2 because 24 contains three 2s. The LCD must also contain two factors of 3 because 18 has two 3s. Thus the LCD $= 2 \cdot 2 \cdot 2 \cdot 3 \cdot 3 = 72$. Now we can proceed as usual.

$$\frac{7}{18} + \frac{11}{24} = \left(\frac{7}{18}\right)\left(\frac{4}{4}\right) + \left(\frac{11}{24}\right)\left(\frac{3}{3}\right) = \frac{28}{72} + \frac{33}{72} = \frac{61}{72}$$

To add and subtract rational expressions with different denominators, follow the same basic routine that you follow when you add or subtract rational numbers with different denominators. Study the following examples carefully and note the similarity to our previous work with rational numbers.

**Classroom Example**
Add $\frac{2x+3}{5}+\frac{x+4}{2}$.

**EXAMPLE 4** Add $\frac{x+2}{4}+\frac{3x+1}{3}$.

### Solution

By inspection, we see that the LCD is 12.

$$\frac{x+2}{4}+\frac{3x+1}{3}=\left(\frac{x+2}{4}\right)\left(\frac{3}{3}\right)+\left(\frac{3x+1}{3}\right)\left(\frac{4}{4}\right)$$ Change to equivalent fractions

$$=\frac{3(x+2)}{12}+\frac{4(3x+1)}{12}$$

$$=\frac{3(x+2)+4(3x+1)}{12}$$ Combine numerators over the LCD

$$=\frac{3x+6+12x+4}{12}$$ Distributive property

$$=\frac{15x+10}{12}$$

Note the final result in Example 4. The numerator, $15x + 10$, could be factored as $5(3x + 2)$. However, because this produces no common factors with the denominator, the fraction cannot be simplified. Thus the final answer can be left as $\frac{15x+10}{12}$. It would also be acceptable to express it as $\frac{5(3x+2)}{12}$.

**Classroom Example**
Subtract $\frac{x-3}{3}-\frac{x+12}{12}$.

**EXAMPLE 5** Subtract $\frac{a-2}{2}-\frac{a-6}{6}$.

### Solution

By inspection, we see that the LCD is 6.

$$\frac{a-2}{2}-\frac{a-6}{6}=\left(\frac{a-2}{2}\right)\left(\frac{3}{3}\right)-\frac{a-6}{6}$$

$$=\frac{3(a-2)}{6}-\frac{a-6}{6}$$

$$=\frac{3(a-2)-(a-6)}{6}$$ Combine numerators over the LCD

$$=\frac{3a-6-a+6}{6}$$ Distributive property

$$=\frac{2a}{6}=\frac{a}{3}$$ Don't forget to simplify

**Classroom Example**
Perform the indicated operations:
$\frac{x+2}{12}-\frac{x-4}{6}+\frac{3x-5}{20}$

**EXAMPLE 6** Perform the indicated operations: $\frac{x+3}{10}+\frac{2x+1}{15}-\frac{x-2}{18}$

### Solution

If you cannot determine the LCD by inspection, then use the prime factored forms of the denominators.

$$10 = 2 \cdot 5 \qquad 15 = 3 \cdot 5 \qquad 18 = 2 \cdot 3 \cdot 3$$

The LCD must contain one factor of 2, two factors of 3, and one factor of 5. Thus the LCD is $2 \cdot 3 \cdot 3 \cdot 5 = 90$.

$$\frac{x+3}{10}+\frac{2x+1}{15}-\frac{x-2}{18}=\left(\frac{x+3}{10}\right)\left(\frac{9}{9}\right)+\left(\frac{2x+1}{15}\right)\left(\frac{6}{6}\right)-\left(\frac{x-2}{18}\right)\left(\frac{5}{5}\right)$$

$$=\frac{9(x+3)}{90}+\frac{6(2x+1)}{90}-\frac{5(x-2)}{90}$$

$$=\frac{9(x+3)+6(2x+1)-5(x-2)}{90}$$ Combine numerators over the LCD

$$=\frac{9x+27+12x+6-5x+10}{90}$$ Distributive property

$$=\frac{16x+43}{90}$$

A denominator that contains variables does not create any serious difficulties; our approach remains basically the same.

Classroom Example
Add $\frac{4}{3a}+\frac{2}{7b}$.

**EXAMPLE 7** Add $\frac{3}{2x}+\frac{5}{3y}$.

### Solution

Using an LCD of $6xy$, we can proceed as follows:

$$\frac{3}{2x}+\frac{5}{3y}=\left(\frac{3}{2x}\right)\left(\frac{3y}{3y}\right)+\left(\frac{5}{3y}\right)\left(\frac{2x}{2x}\right)$$ Change to equivalent fractions

$$=\frac{9y}{6xy}+\frac{10x}{6xy}$$

$$=\frac{9y+10x}{6xy}$$

Classroom Example
Subtract $\frac{3}{14x^2}-\frac{5}{21xy}$.

**EXAMPLE 8** Subtract $\frac{7}{12ab}-\frac{11}{15a^2}$.

### Solution

We can prime factor the numerical coefficients of the denominators to help find the LCD.

$$\left.\begin{aligned}12ab&=2\cdot2\cdot3\cdot a\cdot b\\15a^2&=3\cdot5\cdot a^2\end{aligned}\right\}\longrightarrow \text{LCD}=2\cdot2\cdot3\cdot5\cdot a^2\cdot b=60a^2b$$

$$\frac{7}{12ab}-\frac{11}{15a^2}=\left(\frac{7}{12ab}\right)\left(\frac{5a}{5a}\right)-\left(\frac{11}{15a^2}\right)\left(\frac{4b}{4b}\right)$$ Change to equivalent fractions

$$=\frac{35a}{60a^2b}-\frac{44b}{60a^2b}$$

$$=\frac{35a-44b}{60a^2b}$$

Classroom Example
Add $\frac{x}{x-4}+\frac{2}{x}$.

**EXAMPLE 9** Add $\frac{x}{x-3}+\frac{4}{x}$.

### Solution

By inspection, the LCD is $x(x-3)$.

$$\frac{x}{x-3}+\frac{4}{x}=\left(\frac{x}{x-3}\right)\left(\frac{x}{x}\right)+\left(\frac{4}{x}\right)\left(\frac{x-3}{x-3}\right)$$ Change to equivalent fractions

$$= \frac{x^2}{x(x-3)} + \frac{4(x-3)}{x(x-3)}$$

$$= \frac{x^2 + 4(x-3)}{x(x-3)}$$ Combine numerators over the LCD

$$= \frac{x^2 + 4x - 12}{x(x-3)} \quad \text{or} \quad \frac{(x+6)(x-2)}{x(x-3)}$$

**Classroom Example**

Subtract $\frac{4x}{x+3} - 5$.

**EXAMPLE 10** Subtract $\frac{2x}{x+1} - 3$.

### Solution

$$\frac{2x}{x+1} - 3 = \frac{2x}{x+1} - 3\left(\frac{x+1}{x+1}\right)$$ Change to equivalent fractions

$$= \frac{2x}{x+1} - \frac{3(x+1)}{x+1}$$

$$= \frac{2x - 3(x+1)}{x+1}$$ Combine numerators over the LCD

$$= \frac{2x - 3x - 3}{x+1}$$ Distributive property

$$= \frac{-x - 3}{x+1}$$

## Concept Quiz 4.3

For Problems 1–10, answer true or false.

1. The addition problem $\frac{2x}{x+4} + \frac{1}{x+4}$ is equal to $\frac{2x+1}{x+4}$ for all values of $x$ except $x = -\frac{1}{2}$ and $x = -4$.
2. Any common denominator can be used to add rational expressions, but typically we use the least common denominator.
3. The fractions $\frac{2x^2}{3y}$ and $\frac{10x^2z}{15yz}$ are equivalent fractions.
4. The least common multiple of the denominators is always the least common denominator.
5. To simplify the expression $\frac{5}{2x-1} + \frac{3}{1-2x}$, we could use $2x - 1$ for the common denominator.
6. If $x \neq \frac{1}{2}$, then $\frac{5}{2x-1} + \frac{3}{1-2x} = \frac{2}{2x-1}$.
7. $\frac{3}{-4} - \frac{-2}{3} = \frac{17}{12}$
8. $\frac{4x-1}{5} + \frac{2x+1}{6} = \frac{x}{5}$
9. $\frac{x}{4} - \frac{3x}{2} + \frac{5x}{3} = \frac{5x}{12}$
10. If $x \neq 0$, then $\frac{2}{3x} - \frac{3}{2x} - 1 = \frac{-5 - 6x}{6x}$.

## Problem Set 4.3

For Problems 1–12, perform the indicated operations involving rational numbers. Be sure to express your answers in reduced form. (Objective 1)

**1.** $\frac{1}{4} + \frac{5}{6}$

**2.** $\frac{3}{5} + \frac{1}{6}$

**3.** $\frac{7}{8} - \frac{3}{5}$

**4.** $\frac{7}{9} - \frac{1}{6}$

**5.** $\frac{6}{5} + \frac{1}{-4}$

**6.** $\frac{7}{8} + \frac{5}{-12}$

**7.** $\frac{8}{15} + \frac{3}{25}$

**8.** $\frac{5}{9} - \frac{11}{12}$

**9.** $\frac{1}{5} + \frac{5}{6} - \frac{7}{15}$

**10.** $\frac{2}{3} - \frac{7}{8} + \frac{1}{4}$

**11.** $\frac{1}{3} - \frac{1}{4} - \frac{3}{14}$

**12.** $\frac{5}{6} - \frac{7}{9} - \frac{3}{10}$

For Problems 13–66, add or subtract the rational expressions as indicated. Be sure to express your answers in simplest form. (Objective 2)

**13.** $\frac{2x}{x-1} + \frac{4}{x-1}$

**14.** $\frac{3x}{2x+1} - \frac{5}{2x+1}$

**15.** $\frac{4a}{a+2} + \frac{8}{a+2}$

**16.** $\frac{6a}{a-3} - \frac{18}{a-3}$

**17.** $\frac{3(y-2)}{7y} + \frac{4(y-1)}{7y}$

**18.** $\frac{2x-1}{4x^2} + \frac{3(x-2)}{4x^2}$

**19.** $\frac{x-1}{2} + \frac{x+3}{3}$

**20.** $\frac{x-2}{4} + \frac{x+6}{5}$

**21.** $\frac{2a-1}{4} + \frac{3a+2}{6}$

**22.** $\frac{a-4}{6} + \frac{4a-1}{8}$

**23.** $\frac{n+2}{6} - \frac{n-4}{9}$

**24.** $\frac{2n+1}{9} - \frac{n+3}{12}$

**25.** $\frac{3x-1}{3} - \frac{5x+2}{5}$

**26.** $\frac{4x-3}{6} - \frac{8x-2}{12}$

**27.** $\frac{x-2}{5} - \frac{x+3}{6} + \frac{x+1}{15}$

**28.** $\frac{x+1}{4} + \frac{x-3}{6} - \frac{x-2}{8}$

**29.** $\frac{3}{8x} + \frac{7}{10x}$

**30.** $\frac{5}{6x} - \frac{3}{10x}$

**31.** $\frac{5}{7x} - \frac{11}{4y}$

**32.** $\frac{5}{12x} - \frac{9}{8y}$

**33.** $\frac{4}{3x} + \frac{5}{4y} - 1$

**34.** $\frac{7}{3x} - \frac{8}{7y} - 2$

**35.** $\frac{7}{10x^2} + \frac{11}{15x}$

**36.** $\frac{7}{12a^2} - \frac{5}{16a}$

**37.** $\frac{10}{7n} - \frac{12}{4n^2}$

**38.** $\frac{6}{8n^2} - \frac{3}{5n}$

**39.** $\frac{3}{n^2} - \frac{2}{5n} + \frac{4}{3}$

**40.** $\frac{1}{n^2} + \frac{3}{4n} - \frac{5}{6}$

**41.** $\frac{3}{x} - \frac{5}{3x^2} - \frac{7}{6x}$

**42.** $\frac{7}{3x^2} - \frac{9}{4x} - \frac{5}{2x}$

**43.** $\frac{6}{5t^2} - \frac{4}{7t^3} + \frac{9}{5t^3}$

**44.** $\frac{5}{7t} + \frac{3}{4t^2} + \frac{1}{14t}$

**45.** $\frac{5b}{24a^2} - \frac{11a}{32b}$

**46.** $\frac{9}{14x^2y} - \frac{4x}{7y^2}$

**47.** $\frac{7}{9xy^3} - \frac{4}{3x} + \frac{5}{2y^2}$

**48.** $\frac{7}{16a^2b} + \frac{3a}{20b^2}$

**49.** $\frac{2x}{x-1} + \frac{3}{x}$

**50.** $\frac{3x}{x-4} - \frac{2}{x}$

**51.** $\frac{a-2}{a} - \frac{3}{a+4}$

**52.** $\frac{a+1}{a} - \frac{2}{a+1}$

**53.** $\frac{-3}{4n+5} - \frac{8}{3n+5}$

**54.** $\frac{-2}{n-6} - \frac{6}{2n+3}$

**55.** $\frac{-1}{x+4} + \frac{4}{7x-1}$

**56.** $\frac{-3}{4x+3} + \frac{5}{2x-5}$

**57.** $\frac{7}{3x-5} - \frac{5}{2x+7}$

**58.** $\frac{5}{x-1} - \frac{3}{2x-3}$

**59.** $\frac{5}{3x-2} + \frac{6}{4x+5}$

**60.** $\frac{3}{2x+1} + \frac{2}{3x+4}$

**61.** $\frac{3x}{2x+5} + 1$

**62.** $2 + \frac{4x}{3x-1}$

**63.** $\frac{4x}{x-5} - 3$

**64.** $\frac{7x}{x+4} - 2$

**65.** $-1 - \frac{3}{2x+1}$

**66.** $-2 - \frac{5}{4x-3}$

**67.** Recall that the indicated quotient of a polynomial and its opposite is $-1$. For example, $\frac{x-2}{2-x}$ simplifies to $-1$. Keep this idea in mind as you add or subtract the following rational expressions.

**(a)** $\frac{1}{x-1} - \frac{x}{x-1}$ **(b)** $\frac{3}{2x-3} - \frac{2x}{2x-3}$

**(c)** $\frac{4}{x-4} - \frac{x}{x-4} + 1$ **(d)** $-1 + \frac{2}{x-2} - \frac{x}{x-2}$

**68.** Consider the addition problem $\frac{8}{x-2} + \frac{5}{2-x}$. Note that the denominators are opposites of each other. If the property $\frac{a}{-b} = -\frac{a}{b}$ is applied to the second fraction, we have $\frac{5}{2-x} = -\frac{5}{x-2}$. Thus we proceed as follows:

$$\frac{8}{x-2} + \frac{5}{2-x} = \frac{8}{x-2} - \frac{5}{x-2} = \frac{8-5}{x-2} = \frac{3}{x-2}$$

Use this approach to do the following problems.

**(a)** $\frac{7}{x-1} + \frac{2}{1-x}$ **(b)** $\frac{5}{2x-1} + \frac{8}{1-2x}$

**(c)** $\frac{4}{a-3} - \frac{1}{3-a}$ **(d)** $\frac{10}{a-9} - \frac{5}{9-a}$

**(e)** $\frac{x^2}{x-1} - \frac{2x-3}{1-x}$ **(f)** $\frac{x^2}{x-4} - \frac{3x-28}{4-x}$

## Thoughts Into Words

**69.** What is the difference between the concept of least common multiple and the concept of least common denominator?

**70.** A classmate tells you that she finds the least common multiple of two counting numbers by listing the multiples of each number and then choosing the smallest number that appears in both lists. Is this a correct procedure? What is the weakness of this procedure?

**71.** For which real numbers does $\frac{x}{x-3} + \frac{4}{x}$ equal $\frac{(x+6)(x-2)}{x(x-3)}$? Explain your answer.

**72.** Suppose that your friend does an addition problem as follows:

$$\frac{5}{8} + \frac{7}{12} = \frac{5(12) + 8(7)}{8(12)} = \frac{60 + 56}{96} = \frac{116}{96} = \frac{29}{24}$$

Is this answer correct? If not, what advice would you offer your friend?

**Answers to the Concept Quiz**

**1.** False **2.** True **3.** True **4.** True **5.** True **6.** True **7.** False **8.** False **9.** True **10.** True

## 4.4 More on Rational Expressions and Complex Fractions

OBJECTIVES

1. Add and subtract rational expressions
2. Simplify complex fractions

In this section, we expand our work with adding and subtracting rational expressions, and we discuss the process of simplifying complex fractions. Before we begin, however, this seems like an appropriate time to offer a bit of advice regarding your study of algebra. Success in algebra depends on having a good understanding of the concepts and being able to perform the various computations. As for the computational work, you should adopt a carefully organized format that shows as many steps as you need in order to minimize the chances of making careless errors. Don't be eager to find shortcuts for certain computations before you have a thorough understanding of the steps involved in the process. This advice is especially appropriate at the beginning of this section.

Study Examples 1–4 very carefully. Again we have included the steps to follow in solving each problem:

**Step 1** Factor the denominators.

**Step 2** Find the LCD.

**Step 3** Change each fraction to an equivalent fraction that has the LCD as its denominator.

**Step 4** Combine the numerators and place over the LCD.

**Step 5** Simplify by performing the addition or subtraction.

**Step 6** Look for ways to reduce the resulting fraction.

Classroom Example

Add $\frac{8}{a^2 - 2a} + \frac{4}{a}$.

**EXAMPLE 1** Add $\frac{8}{x^2 - 4x} + \frac{2}{x}$.

### Solution

$$\frac{8}{x^2 - 4x} + \frac{2}{x} = \frac{8}{x(x-4)} + \frac{2}{x}$$ Factor the denominators

The LCD is $x(x - 4)$. Find the LCD

$$= \frac{8}{x(x-4)} + \left(\frac{2}{x}\right)\left(\frac{x-4}{x-4}\right)$$ Change each fraction to an equivalent fraction that has the LCD as its denominator

$$= \frac{8 + 2(x-4)}{x(x-4)}$$ Combine numerators and place over the LCD

$$= \frac{8 + 2x - 8}{x(x-4)}$$

$$= \frac{2x}{x(x-4)}$$ Simplify by performing the addition or subtraction

$$= \frac{2}{x-4}$$ Simplify

Classroom Example

Subtract $\frac{x}{x^2 - 9} - \frac{7}{x+3}$.

**EXAMPLE 2** Subtract $\frac{a}{a^2 - 4} - \frac{3}{a+2}$.

### Solution

$$\frac{a}{a^2 - 4} - \frac{3}{a+2} = \frac{a}{(a+2)(a-2)} - \frac{3}{a+2}$$ Factor the denominators

The LCD is $(a + 2)(a - 2)$. Find the LCD

$$= \frac{a}{(a+2)(a-2)} - \left(\frac{3}{a+2}\right)\left(\frac{a-2}{a-2}\right)$$ Change each fraction to an equivalent fraction that has the LCD as its denominator

$$= \frac{a - 3(a-2)}{(a+2)(a-2)}$$ Combine numerators and place over the LCD

$$= \frac{a - 3a + 6}{(a+2)(a-2)}$$

$$= \frac{-2a + 6}{(a+2)(a-2)} \quad \text{or} \quad \frac{-2(a-3)}{(a+2)(a-2)}$$ Simplify by performing the addition or subtraction

**Classroom Example**
Add:

$$\frac{2x}{x^2+5x+6}+\frac{5}{x^2-5x-14}$$

**EXAMPLE 3** Add $\dfrac{3n}{n^2+6n+5}+\dfrac{4}{n^2-7n-8}$.

### Solution

$$\frac{3n}{n^2+6n+5}+\frac{4}{n^2-7n-8}$$

$$=\frac{3n}{(n+5)(n+1)}+\frac{4}{(n-8)(n+1)}$$ Factor the denominators

The LCD is $(n+5)(n+1)(n-8)$. Find the LCD

$$=\left(\frac{3n}{(n+5)(n+1)}\right)\left(\frac{n-8}{n-8}\right)+\left(\frac{4}{(n-8)(n+1)}\right)\left(\frac{n+5}{n+5}\right)$$ Change each fraction to an equivalent fraction that has the LCD as its denominator

$$=\frac{3n(n-8)+4(n+5)}{(n+5)(n+1)(n-8)}$$ Combine numerators and place over the LCD

$$=\frac{3n^2-24n+4n+20}{(n+5)(n+1)(n-8)}$$

$$=\frac{3n^2-20n+20}{(n+5)(n+1)(n-8)}$$ Simplify by performing the addition or subtraction

**Classroom Example**
Perform the indicated operations:

$$\frac{4x^2}{x^4-16}+\frac{x}{x^2-4}-\frac{1}{x-2}$$

**EXAMPLE 4** Perform the indicated operations:

$$\frac{2x^2}{x^4-1}+\frac{x}{x^2-1}-\frac{1}{x-1}$$

### Solution

$$\frac{2x^2}{x^4-1}+\frac{x}{x^2-1}-\frac{1}{x-1}$$

$$=\frac{2x^2}{(x^2+1)(x+1)(x-1)}+\frac{x}{(x+1)(x-1)}-\frac{1}{x-1}$$ Factor the denominators

The LCD is $(x^2+1)(x+1)(x-1)$. Find the LCD

$$=\frac{2x^2}{(x^2+1)(x+1)(x-1)}+\left(\frac{x}{(x+1)(x-1)}\right)\left(\frac{x^2+1}{x^2+1}\right)-\left(\frac{1}{x-1}\right)\frac{(x^2+1)(x+1)}{(x^2+1)(x+1)}$$ Change each fraction to an equivalent fraction that has the LCD as its denominator

$$=\frac{2x^2+x(x^2+1)-(x^2+1)(x+1)}{(x^2+1)(x+1)(x-1)}$$ Combine numerators and place over the LCD

$$=\frac{2x^2+x^3+x-x^3-x^2-x-1}{(x^2+1)(x+1)(x-1)}$$ Distributive property

$$=\frac{x^2-1}{(x^2+1)(x+1)(x-1)}$$ Simplify by performing the addition or subtraction

$$=\frac{(x+1)(x-1)}{(x^2+1)(x+1)(x-1)}$$

$$=\frac{1}{x^2+1}$$ Simplify

## Simplifying Complex Fractions

**Complex fractions** are fractional forms that contain rational numbers or rational expressions in the numerators and/or denominators. The following are examples of complex fractions.

$$\frac{\frac{4}{x}}{\frac{2}{xy}} \qquad \frac{\frac{1}{2}+\frac{3}{4}}{\frac{5}{6}-\frac{3}{8}} \qquad \frac{\frac{3}{x}+\frac{2}{y}}{\frac{5}{x}-\frac{6}{y^2}} \qquad \frac{\frac{1}{x}+\frac{1}{y}}{2} \qquad \frac{-3}{\frac{2}{x}-\frac{3}{y}}$$

It is often necessary to **simplify** a complex fraction. We will take each of these five examples and examine some techniques for simplifying complex fractions.

Classroom Example
Simplify $\frac{\frac{6}{m}}{\frac{3}{m^2n}}$.

### EXAMPLE 5

Simplify $\frac{\frac{4}{x}}{\frac{2}{xy}}$.

### Solution

This type of problem is a simple division problem. Let's rewrite in a division format. Then invert the divisor and multiply.

$$\frac{\frac{4}{x}}{\frac{2}{xy}} = \frac{4}{x} \div \frac{2}{xy}$$

$$= \frac{\overset{2}{\cancel{4}}}{\cancel{x}} \cdot \frac{\cancel{x}y}{\cancel{2}} = 2y$$

Classroom Example
Simplify $\frac{\frac{3}{4}-\frac{1}{3}}{\frac{5}{6}+\frac{2}{9}}$.

### EXAMPLE 6

Simplify $\frac{\frac{1}{2}+\frac{3}{4}}{\frac{5}{6}-\frac{3}{8}}$.

Let's look at two possible ways to simplify such a problem.

### Solution A

Here we will simplify the numerator by performing the addition and simplify the denominator by performing the subtraction. Then the problem is a simple division problem as in Example 5.

$$\frac{\frac{1}{2}+\frac{3}{4}}{\frac{5}{6}-\frac{3}{8}} = \frac{\frac{2}{4}+\frac{3}{4}}{\frac{20}{24}-\frac{9}{24}}$$

$$= \frac{\frac{5}{4}}{\frac{11}{24}} = \frac{5}{4} \div \frac{11}{24} = \frac{5}{4} \cdot \frac{24}{11} = \frac{5}{\cancel{4}} \cdot \frac{\overset{6}{\cancel{24}}}{11}$$

$$= \frac{30}{11}$$

### Solution B

Here we find the LCD of all four denominators (2, 4, 6, and 8). The LCD is 24. Use this LCD to multiply the entire complex fraction by a form of 1, specifically $\frac{24}{24}$.

$$\frac{\frac{1}{2}+\frac{3}{4}}{\frac{5}{6}-\frac{3}{8}} = \left(\frac{24}{24}\right)\left(\frac{\frac{1}{2}+\frac{3}{4}}{\frac{5}{6}-\frac{3}{8}}\right)$$

$$= \frac{24\left(\frac{1}{2}+\frac{3}{4}\right)}{24\left(\frac{5}{6}-\frac{3}{8}\right)}$$

$$= \frac{24\left(\frac{1}{2}\right)+24\left(\frac{3}{4}\right)}{24\left(\frac{5}{6}\right)-24\left(\frac{3}{8}\right)} \quad \text{Distributive property}$$

$$= \frac{12+18}{20-9} = \frac{30}{11}$$

Classroom Example

Simplify $\frac{\frac{1}{x}+\frac{3}{y}}{\frac{4}{x}-\frac{2}{y^2}}$.

**EXAMPLE 7**

Simplify $\frac{\frac{3}{x}+\frac{2}{y}}{\frac{5}{x}-\frac{6}{y^2}}$.

### Solution A

Simplify the numerator and the denominator. Then the problem becomes a division problem.

$$\frac{\frac{3}{x}+\frac{2}{y}}{\frac{5}{x}-\frac{6}{y^2}} = \frac{\left(\frac{3}{x}\right)\left(\frac{y}{y}\right)+\left(\frac{2}{y}\right)\left(\frac{x}{x}\right)}{\left(\frac{5}{x}\right)\left(\frac{y^2}{y^2}\right)-\left(\frac{6}{y^2}\right)\left(\frac{x}{x}\right)}$$

$$= \frac{\frac{3y}{xy}+\frac{2x}{xy}}{\frac{5y^2}{xy^2}-\frac{6x}{xy^2}}$$

$$= \frac{\frac{3y+2x}{xy}}{\frac{5y^2-6x}{xy^2}}$$

$$= \frac{3y+2x}{xy} \div \frac{5y^2-6x}{xy^2} = \frac{3y+2x}{xy} \cdot \frac{xy^2}{5y^2-6x}$$

$$= \frac{3y+2x}{\cancel{xy}} \cdot \frac{\cancel{xy^2}^{\,y}}{5y^2-6x}$$

$$= \frac{y(3y+2x)}{5y^2-6x}$$

### Solution B

Here we find the LCD of all four denominators ($x$, $y$, $x$, and $y^2$). The LCD is $xy^2$. Use this LCD to multiply the entire complex fraction by a form of 1, specifically $\frac{xy^2}{xy^2}$.

$$\frac{\frac{3}{x}+\frac{2}{y}}{\frac{5}{x}-\frac{6}{y^2}} = \left(\frac{xy^2}{xy^2}\right)\left(\frac{\frac{3}{x}+\frac{2}{y}}{\frac{5}{x}-\frac{6}{y^2}}\right)$$

$$= \frac{xy^2\left(\frac{3}{x}+\frac{2}{y}\right)}{xy^2\left(\frac{5}{x}-\frac{6}{y^2}\right)}$$

$$= \frac{xy^2\left(\frac{3}{x}\right)+xy^2\left(\frac{2}{y}\right)}{xy^2\left(\frac{5}{x}\right)-xy^2\left(\frac{6}{y^2}\right)} \quad \text{Distributive property}$$

$$= \frac{3y^2 + 2\ldots}{5y^2 - \ldots} \ldots \frac{y(3y+2x)}{5y^2-6x}$$

Certainly either ap[...]lution A or Solution B) will work with problems such as Examples 6 and 7. Exami[...]3 in both examples carefully. This approach works effectively with complex fraction[...]CD of all the denominators is easy to find. (Don't be misled by the length of Soluti[...]ple 6; we were especially careful to show every step.)

Classroom Example
Simplify $\frac{\frac{1}{m}-\frac{1}{n}}{3}$.

**EXAMPLE 8** [...] $\frac{\ldots+\frac{1}{y}}{2}$.

### Solution

The number 2 can be [...] LCD of all three denominators ($x$, $y$, and 1) is $xy$.

Therefore, let's multip[...] fraction by a form of 1, specifically $\frac{xy}{xy}$.

$$\left(\frac{xy}{xy}\right)\left(\frac{\frac{1}{x}+\frac{1}{y}}{\frac{2}{1}}\right) = \ldots$$

$$= \ldots$$

Classroom Example
Simplify $\frac{-5}{\frac{4}{x}-\frac{8}{y}}$.

**EXAMPLE 9** Simplify $\frac{-3}{\frac{2}{x}-\frac{3}{y}}$.

### Solution

$$\left(\frac{xy}{xy}\right)\left(\frac{\frac{-3}{1}}{\frac{2}{x}-\frac{3}{y}}\right) = \frac{-3(xy)}{xy\left(\frac{2}{x}\right)-xy\left(\frac{3}{y}\right)}$$

$$= \frac{-3xy}{2y-3x}$$

Let's conclude this section with an example that has a complex fraction as part of an algebraic expression.

Classroom Example

Simplify $1 + \dfrac{x}{1 + \dfrac{1}{x}}$.

**EXAMPLE 10**

Simplify $1 - \dfrac{n}{1 - \dfrac{1}{n}}$.

## Solution

First simplify the complex fraction $\dfrac{n}{1 - \dfrac{1}{n}}$ by multiplying by $\dfrac{n}{n}$.

$$\left(\frac{n}{n}\right)\left(\frac{n}{1 - \dfrac{1}{n}}\right) = \frac{n(n)}{n\left(1 - \dfrac{1}{n}\right)} = \frac{n^2}{n - 1}$$

Now we can perform the subtraction.

$$1 - \frac{n^2}{n - 1} = \left(\frac{n - 1}{n - 1}\right)\left(\frac{1}{1}\right) - \frac{n^2}{n - 1}$$

$$= \frac{n - 1}{n - 1} - \frac{n^2}{n - 1}$$

$$= \frac{n - 1 - n^2}{n - 1} \quad \text{or} \quad \frac{-n^2 + n - 1}{n - 1}$$

## Concept Quiz 4.4

For Problems 1–7, answer true or false.

1. A complex fraction can be described as a fraction within a fraction.
2. Division can simplify the complex fraction $\dfrac{\dfrac{2y}{x}}{\dfrac{6}{x^2}}$.
3. The complex fraction $\dfrac{\dfrac{3}{x - 2} + \dfrac{2}{x + 2}}{\dfrac{7x}{(x + 2)(x - 2)}}$ is defined for all values of $x$ except $x = 0$.
4. The complex fraction $\dfrac{\dfrac{1}{3} - \dfrac{5}{6}}{\dfrac{1}{6} + \dfrac{5}{9}}$ simplifies to $-\dfrac{9}{13}$.
5. One method for simplifying a complex fraction is to multiply the entire fraction by a form of 1.
6. The complex fraction $\dfrac{\dfrac{3}{4} - \dfrac{1}{2}}{\dfrac{2}{3}}$ simplifies to $\dfrac{3}{8}$.
7. The complex fraction $\dfrac{\dfrac{7}{8} - \dfrac{1}{18}}{\dfrac{5}{6} + \dfrac{4}{15}}$ simplifies to $\dfrac{59}{33}$.

**8.** Arrange in order the following steps for adding rational expressions.

**A.** Combine numerators and place over the LCD.

**B.** Find the LCD.

**C.** Reduce.

**D.** Factor the denominators.

**E.** Simplify by performing the addition or subtraction.

**F.** Change each fraction to an equivalent fraction that has the LCD as its denominator.

## Problem Set 4.4

For Problems 1–40, perform the indicated operations, and express your answers in simplest form. (Objective 1)

**1.** $\frac{2x}{x^2+4x}+\frac{5}{x}$

**2.** $\frac{3x}{x^2-6x}+\frac{4}{x}$

**3.** $\frac{4}{x^2+7x}-\frac{1}{x}$

**4.** $\frac{-10}{x^2-9x}-\frac{2}{x}$

**5.** $\frac{x}{x^2-1}+\frac{5}{x+1}$

**6.** $\frac{2x}{x^2-16}+\frac{7}{x-4}$

**7.** $\frac{6a+4}{a^2-1}-\frac{5}{a-1}$

**8.** $\frac{4a-4}{a^2-4}-\frac{3}{a+2}$

**9.** $\frac{2n}{n^2-25}-\frac{3}{4n+20}$

**10.** $\frac{3n}{n^2-36}-\frac{2}{5n+30}$

**11.** $\frac{5}{x}-\frac{5x-30}{x^2+6x}+\frac{x}{x+6}$

**12.** $\frac{3}{x+1}+\frac{x+5}{x^2-1}-\frac{3}{x-1}$

**13.** $\frac{3}{x^2+9x+14}+\frac{5}{2x^2+15x+7}$

**14.** $\frac{6}{x^2+11x+24}+\frac{4}{3x^2+13x+12}$

**15.** $\frac{1}{a^2-3a-10}-\frac{4}{a^2+4a-45}$

**16.** $\frac{6}{a^2-3a-54}-\frac{10}{a^2+5a-6}$

**17.** $\frac{3a}{8a^2-2a-3}+\frac{1}{4a^2+13a-12}$

**18.** $\frac{2a}{6a^2+13a-5}+\frac{a}{2a^2+a-10}$

**19.** $\frac{5}{x^2+3}-\frac{2}{x^2+4x-21}$

**20.** $\frac{7}{x^2+1}-\frac{3}{x^2+7x-60}$

**21.** $\frac{3x}{x^2-6x+9}-\frac{2}{x-3}$

**22.** $\frac{3}{x+4}+\frac{2x}{x^2+8x+16}$

**23.** $\frac{5}{x^2-1}+\frac{9}{x^2+2x+1}$

**24.** $\frac{6}{x^2-9}-\frac{9}{x^2-6x+9}$

**25.** $\frac{2}{y^2+6y-16}-\frac{4}{y+8}-\frac{3}{y-2}$

**26.** $\frac{7}{y-6}-\frac{10}{y+12}+\frac{4}{y^2+6y-72}$

**27.** $x-\frac{x^2}{x-2}+\frac{3}{x^2-4}$

**28.** $x+\frac{5}{x^2-25}-\frac{x^2}{x+5}$

**29.** $\frac{x+3}{x+10}+\frac{4x-3}{x^2+8x-20}+\frac{x-1}{x-2}$

**30.** $\frac{2x-1}{x+3}+\frac{x+4}{x-6}+\frac{3x-1}{x^2-3x-18}$

**31.** $\frac{n}{n-6}+\frac{n+3}{n+8}+\frac{12n+26}{n^2+2n-48}$

**32.** $\frac{n-1}{n+4}+\frac{n}{n+6}+\frac{2n+18}{n^2+10n+24}$

**33.** $\frac{4x-3}{2x^2+x-1}-\frac{2x+7}{3x^2+x-2}-\frac{3}{3x-2}$

**34.** $\frac{2x+5}{x^2+3x-18}-\frac{3x-1}{x^2+4x-12}+\frac{5}{x-2}$

**35.** $\frac{n}{n^2+1}+\frac{n^2+3n}{n^4-1}-\frac{1}{n-1}$

**36.** $\frac{2n^2}{n^4-16}-\frac{n}{n^2-4}+\frac{1}{n+2}$

**37.** $\frac{15x^2-10}{5x^2-7x+2}-\frac{3x+4}{x-1}-\frac{2}{5x-2}$

38. $\dfrac{32x + 9}{12x^2 + x - 6} - \dfrac{3}{4x + 3} - \dfrac{x + 5}{3x - 2}$

39. $\dfrac{t + 3}{3t - 1} + \dfrac{8t^2 + 8t + 2}{3t^2 - 7t + 2} - \dfrac{2t + 3}{t - 2}$

40. $\dfrac{t - 3}{2t + 1} + \dfrac{2t^2 + 19t - 46}{2t^2 - 9t - 5} - \dfrac{t + 4}{t - 5}$

For Problems 41–64, simplify each complex fraction. (Objective 2)

41. $\dfrac{\frac{1}{2} - \frac{1}{4}}{\frac{5}{8} + \frac{3}{4}}$

42. $\dfrac{\frac{3}{8} + \frac{3}{4}}{\frac{5}{8} - \frac{7}{12}}$

43. $\dfrac{\frac{3}{28} - \frac{5}{14}}{\frac{5}{7} + \frac{1}{4}}$

44. $\dfrac{\frac{5}{9} + \frac{7}{36}}{\frac{3}{18} - \frac{5}{12}}$

45. $\dfrac{\frac{5}{6y}}{\frac{10}{3xy}}$

46. $\dfrac{\frac{9}{8xy^2}}{\frac{5}{4x^2}}$

47. $\dfrac{\frac{3}{x} - \frac{2}{y}}{\frac{4}{y} - \frac{7}{xy}}$

48. $\dfrac{\frac{9}{x} + \frac{7}{x^2}}{\frac{5}{y} + \frac{3}{y^2}}$

49. $\dfrac{\frac{6}{a} - \frac{5}{b^2}}{\frac{12}{a^2} + \frac{2}{b}}$

50. $\dfrac{\frac{4}{ab} - \frac{3}{b^2}}{\frac{1}{a} + \frac{3}{b}}$

51. $\dfrac{\frac{2}{x} - 3}{\frac{3}{y} + 4}$

52. $\dfrac{1 + \frac{3}{x}}{1 - \frac{6}{x}}$

53. $\dfrac{3 + \frac{2}{n + 4}}{5 - \frac{1}{n + 4}}$

54. $\dfrac{4 + \frac{6}{n - 1}}{7 - \frac{4}{n - 1}}$

55. $\dfrac{5 - \frac{2}{n - 3}}{4 - \frac{1}{n - 3}}$

56. $\dfrac{\frac{3}{n - 5} - 2}{1 - \frac{4}{n - 5}}$

57. $\dfrac{\frac{-1}{y - 2} + \frac{5}{x}}{\frac{3}{x} - \frac{4}{xy - 2x}}$

58. $\dfrac{\frac{-2}{x} - \frac{4}{x + 2}}{\frac{3}{x^2 + 2x} + \frac{3}{x}}$

59. $\dfrac{\frac{2}{x - 3} - \frac{3}{x + 3}}{\frac{5}{x^2 - 9} - \frac{2}{x - 3}}$

60. $\dfrac{\frac{2}{x - y} + \frac{3}{x + y}}{\frac{5}{x + y} - \frac{1}{x^2 - y^2}}$

61. $\dfrac{3a}{2 - \frac{1}{a}} - 1$

62. $\dfrac{a}{\frac{1}{a} + 4} + 1$

63. $2 - \dfrac{x}{3 - \frac{2}{x}}$

64. $1 + \dfrac{x}{1 + \frac{1}{x}}$

## Thoughts Into Words

... Which of the two techniques presented in the text would ... use to simplify $\dfrac{\frac{1}{4} + \frac{1}{3}}{\frac{3}{4} - \frac{1}{6}}$? Which technique would you ... plify $\dfrac{\frac{3}{8} - \frac{5}{7}}{\frac{7}{9} + \frac{6}{25}}$? Explain your choice for each

66. Give a step-by-step description of how to do the following addition problem.

$$\frac{3x + 4}{8} + \frac{5x - 2}{12}$$

... the Concept Quiz

... True **3.** False **4.** True **5.** True **6.** True **7.** False **8.** D, B, F, A, E, C

## 4.5 Dividing Polynomials

OBJECTIVES

1. Divide polynomials
2. Use synthetic division to divide polynomials

In Chapter 3, we saw how the property $\frac{b^n}{b^m} = b^{n-m}$, along with our knowledge of dividing integers, is used to divide monomials. For example,

$$\frac{12x^3}{3x} = 4x^2 \qquad \frac{-36x^4y^5}{4xy^2} = -9x^3y^3$$

In Section 4.3, we used $\frac{a}{b} + \frac{c}{b} = \frac{a+c}{b}$ and $\frac{a}{b} - \frac{c}{b} = \frac{a-c}{b}$ as the basis for adding and subtracting rational expressions. These same equalities, viewed as $\frac{a+c}{b} = \frac{a}{b} + \frac{c}{b}$ and $\frac{a-c}{b} = \frac{a}{b} - \frac{c}{b}$, along with our knowledge of dividing monomials, provide the basis for dividing polynomials by monomials. Consider the following examples.

$$\frac{18x^3 + 24x^2}{6x} = \frac{18x^3}{6x} + \frac{24x^2}{6x} = 3x^2 + 4x$$

$$\frac{35x^2y^3 - 55x^3y^4}{5xy^2} = \frac{35x^2y^3}{5xy^2} - \frac{55x^3y^4}{5xy^2} = 7xy - 11x^2y^2$$

To divide a polynomial by a monomial, we divide each term of the polynomial by the monomial. As with many skills, once you feel comfortable with the process, you may then want to perform some of the steps mentally. Your work could take on the following format.

$$\frac{40x^4y^5 + 72x^5y^7}{8x^2y} = 5x^2y^4 + 9x^3y^6 \qquad \frac{36a^3b^4 - 45a^4b^6}{-9a^2b^3} = -4ab + 5a^2b^3$$

### Dividing a Polynomial by a Binomial

In Section 4.1, we saw that a fraction like $\frac{3x^2 + 11x - 4}{x + 4}$ can be simplified by factoring and reducing as follows:

$$\frac{3x^2 + 11x - 4}{x + 4} = \frac{(3x - 1)\cancel{(x + 4)}}{\cancel{x + 4}} = 3x - 1$$

We can obtain the same result by using a dividing process similar to long division in arithmetic.

**Step 1** Use the conventional long-division format, and arrange both the dividend and the divisor in descending powers of the variable.

$$x + 4\overline{)3x^2 + 11x - 4}$$

**Step 2** Find the first term of the quotient by dividing the first term of the dividend by the first term of the divisor.

$$\begin{array}{r} 3x \phantom{+ 11x - 4} \\ x + 4\overline{)3x^2 + 11x - 4} \end{array}$$

**Step 3** Multiply the entire divisor by the term of the quotient found in step 2, and position the product to be subtracted from the dividend.

$$\begin{array}{r} 3x \phantom{+ 11x - 4} \\ x + 4\overline{)3x^2 + 11x - 4} \\ \underline{3x^2 + 12x} \phantom{- 4} \end{array}$$

**Step 4** Subtract.

Remember to add the opposite!

$(3x^2 + 11x - 4) - (3x^2 + 12x) = -x - 4$

$$\begin{array}{r} 3x \phantom{+11x-4} \\ x + 4\overline{)3x^2 + 11x - 4} \\ \underline{3x^2 + 12x} \phantom{-4} \\ -x - 4 \end{array}$$

**Step 5** Repeat the process beginning with step 2; use the polynomial that resulted from the subtraction in step 4 as a new dividend.

$$\begin{array}{r} 3x - 1 \phantom{1x-4} \\ x + 4\overline{)3x^2 + 11x - 4} \\ \underline{3x^2 + 12x} \phantom{-4} \\ -x - 4 \\ \underline{-x - 4} \end{array}$$

In the next example, let's *think* in terms of the previous step-by-step procedure but arrange our work in a more compact form.

Classroom Example
Divide $3x^2 - 5x - 28$ by $x - 4$.

**EXAMPLE 1** Divide $5x^2 + 6x - 8$ by $x + 2$.

**Solution**

$$\begin{array}{r} 5x - 4 \phantom{x-8} \\ x + 2\overline{)5x^2 + 6x - 8} \\ \underline{5x^2 + 10x} \phantom{-8} \\ -4x - 8 \\ \underline{-4x - 8} \\ 0 \end{array}$$

**Think Steps**

1. $\dfrac{5x^2}{x} = 5x$
2. $5x(x + 2) = 5x^2 + 10x$
3. $(5x^2 + 6x - 8) - (5x^2 + 10x) = -4x - 8$
4. $\dfrac{-4x}{x} = -4$
5. $-4(x + 2) = -4x - 8$

Recall that to check a division problem, we can multiply the divisor by the quotient and add the remainder. In other words,

Dividend = (Divisor)(Quotient) + (Remainder)

Sometimes the remainder is expressed as a fractional part of the divisor. The relationship then becomes

$$\frac{\text{Dividend}}{\text{Divisor}} = \text{Quotient} + \frac{\text{Remainder}}{\text{Divisor}}$$

Classroom Example
Divide $2x^2 + 11x + 20$ by $x + 3$.

**EXAMPLE 2** Divide $2x^2 - 3x + 1$ by $x - 5$.

**Solution**

$$\begin{array}{r} 2x + 7 \phantom{x+1} \\ x - 5\overline{)2x^2 - 3x + 1} \\ \underline{2x^2 - 10x} \phantom{+1} \\ 7x + 1 \\ \underline{7x - 35} \\ 36 \end{array}$$

← Remainder

Thus

$$\frac{2x^2 - 3x + 1}{x - 5} = 2x + 7 + \frac{36}{x - 5}, \qquad x \neq 5$$

✔ **Check**

$$(x - 5)(2x + 7) + 36 \stackrel{?}{=} 2x^2 - 3x + 1$$
$$2x^2 - 3x - 35 + 36 \stackrel{?}{=} 2x^2 - 3x + 1$$
$$2x^2 - 3x + 1 = 2x^2 - 3x + 1$$

Each of the next two examples illustrates another point regarding the division process. Study them carefully, and then you should be ready to work the exercises in the next problem set.

**Classroom Example**
Divide $t^3 - 1$ by $t - 1$.

**EXAMPLE 3** Divide $t^3 - 8$ by $t - 2$.

### Solution

$$\begin{array}{r}
t^2 + 2t + 4 \\
t - 2\overline{)t^3 + 0t^2 + 0t - 8} \\
\underline{t^3 - 2t^2} \qquad\qquad \\
2t^2 + 0t - 8 \\
\underline{2t^2 - 4t} \quad\; \\
4t - 8 \\
\underline{4t - 8} \\
0
\end{array}$$

← Note the insertion of a "$t$-squared" term and a "$t$ term" with zero coefficients

Check this result!

**Classroom Example**
Divide $x^3 + x^2 - 7x - 2$ by $x^2 - 3x$.

**EXAMPLE 4** Divide $y^3 + 3y^2 - 2y - 1$ by $y^2 + 2y$.

### Solution

$$\begin{array}{r}
y + 1 \\
y^2 + 2y\overline{)y^3 + 3y^2 - 2y - 1} \\
\underline{y^3 + 2y^2} \qquad\qquad \\
y^2 - 2y - 1 \\
\underline{y^2 + 2y} \quad\; \\
-4y - 1
\end{array}$$

← Remainder of $-4y - 1$

The division process is complete when the degree of the remainder is less than the degree of the divisor. Thus

$$\frac{y^3 + 3y^2 - 2y - 1}{y^2 + 2y} = y + 1 + \frac{-4y - 1}{y^2 + 2y}$$

## Synthetic Division

If the divisor is of the form $x - k$, where the coefficient of the $x$ term is 1, then the format of the division process described in this section can be simplified by a procedure called **synthetic division**. This procedure is a shortcut for this type of polynomial division. If you are continuing on to study college algebra, then you will want to know synthetic division. If you are not continuing on to college algebra, then you probably will not need a shortcut, and the long-division process will be sufficient.

First, let's consider an example and use the usual division process. Then, in step-by-step fashion, we can observe some shortcuts that will lead us into the synthetic-division procedure. Consider the division problem $(2x^4 + x^3 - 17x^2 + 13x + 2) \div (x - 2)$. We can apply the synthetic-division procedure to this problem because the coefficient of the $x$ term in the divisor is 1.

$$\begin{array}{r}
2x^3 + 5x^2 - 7x - 1 \\
x - 2\overline{)2x^4 + x^3 - 17x^2 + 13x + 2} \\
\underline{2x^4 - 4x^3} \qquad\qquad\qquad\quad \\
5x^3 - 17x^2 \qquad\qquad \\
\underline{5x^3 - 10x^2} \qquad\qquad \\
-7x^2 + 13x \qquad \\
\underline{-7x^2 + 14x} \qquad \\
-x + 2 \\
\underline{-x + 2}
\end{array}$$

Note that because the dividend $(2x^4 + x^3 - 17x^2 + 13x + 2)$ is written in descending powers of $x$, the quotient $(2x^3 + 5x^2 - 7x - 1)$ is produced, also in descending powers of $x$. In other words, the numerical coefficients are the important numbers. Thus let's rewrite this problem in terms of its coefficients.

$$
\begin{array}{r}
2 + 5 - \phantom{1}7 - \phantom{1}1\phantom{\,+\,2} \\
1 - 2\overline{)2 + 1 - 17 + 13 + 2} \\
\underline{\textcircled{2} - 4}\phantom{\,-\,17 + 13 + 2} \\
5 \textcircled{$-17$}\phantom{\,+\,13 + 2} \\
\underline{\textcircled{5} - 10}\phantom{\,+\,13 + 2} \\
-7 + \textcircled{13}\phantom{\,+\,2} \\
\underline{\textcircled{$-7$} + 14}\phantom{\,+\,2} \\
-1 + \textcircled{2} \\
\underline{\textcircled{$-1$} + 2}
\end{array}
$$

Now observe that the numbers that are circled are simply repetitions of the numbers directly above them in the format. Therefore, by removing the circled numbers, we can write the process in a more compact form as

$$
\begin{array}{rrrrrr}
 & 2 & 5 & -7 & -1 & \qquad (1) \\
-2) & 2 & 1 & -17 & 13 \quad 2 & \qquad (2) \\
 & & -4 & -10 & 14 \quad 2 & \qquad (3) \\
\hline
 & & 5 & -7 & -1 \quad 0 & \qquad (4)
\end{array}
$$

where the repetitions are omitted and where 1, the coefficient of $x$ in the divisor, is omitted.

Note that line (4) reveals all of the coefficients of the quotient, line (1), except for the first coefficient of 2. Thus we can begin line (4) with the first coefficient and then use the following form.

$$
\begin{array}{rrrrrrl}
-2) & 2 & 1 & -17 & 13 & 2 & \qquad (5) \\
 & & -4 & -10 & 14 & 2 & \qquad (6) \\
\hline
 & 2 & 5 & -7 & -1 & 0 & \qquad (7)
\end{array}
$$

Line (7) contains the coefficients of the quotient, where the 0 indicates the remainder.

Finally, by changing the constant in the divisor to 2 (instead of $-2$), we can add the corresponding entries in lines (5) and (6) rather than subtract. Thus the final synthetic division form for this problem is

$$
\begin{array}{rrrrrr}
2) & 2 & 1 & -17 & 13 & 2 \\
 & & 4 & 10 & -14 & -2 \\
\hline
 & 2 & 5 & -7 & -1 & 0
\end{array}
$$

Now let's consider another problem that illustrates a step-by-step procedure for carrying out the synthetic-division process. Suppose that we want to divide $3x^3 - 2x^2 + 6x - 5$ by $x + 4$.

**Step 1** Write the coefficients of the dividend as follows:

$$
\overline{)3 \quad -2 \quad 6 \quad -5}
$$

**Step 2** In the divisor, $(x + 4)$, use $-4$ instead of 4 so that later we can add rather than subtract.

$$
-4\overline{)3 \quad -2 \quad 6 \quad -5}
$$

**Step 3** Bring down the first coefficient of the dividend (3).

$$
\begin{array}{rrrrr}
-4) & 3 & -2 & 6 & -5 \\
 & & & & \\
\hline
 & 3 & & &
\end{array}
$$

**Step 4** Multiply $(3)(-4)$, which yields $-12$; this result is to be added to the second coefficient of the dividend $(-2)$.

$$\begin{array}{r|rrrr} -4) & 3 & -2 & 6 & -5 \\ & & -12 & & \\ \hline & 3 & -14 & & \end{array}$$

**Step 5** Multiply $(-14)(-4)$, which yields 56; this result is to be added to the third coefficient of the dividend (6).

$$\begin{array}{r|rrrr} -4) & 3 & -2 & 6 & -5 \\ & & -12 & 56 & \\ \hline & 3 & -14 & 62 & \end{array}$$

**Step 6** Multiply $(62)(-4)$, which yields $-248$; this result is added to the last term of the dividend $(-5)$.

$$\begin{array}{r|rrrr} -4) & 3 & -2 & 6 & -5 \\ & & -12 & 56 & -248 \\ \hline & 3 & -14 & 62 & -253 \end{array}$$

The last row indicates a quotient of $3x^2 - 14x + 62$ and a remainder of $-253$. Thus we have

$$\frac{3x^3 - 2x^2 + 6x - 5}{x + 4} = 3x^2 - 14x + 62 - \frac{253}{x + 4}$$

We will consider one more example, which shows only the final compact form for synthetic division.

**Classroom Example**
Find the quotient and remainder for $(2x^4 - 11x^3 + 17x^2 + 2x - 9) \div (x - 3)$.

## EXAMPLE 5

Find the quotient and remainder for $(4x^4 - 2x^3 + 6x - 1) \div (x - 1)$.

### Solution

We can use synthetic division to find the quotient because the coefficient of the $x$ term in the divisor is 1.

$$\begin{array}{r|rrrrr} 1) & 4 & -2 & 0 & 6 & -1 \\ & & 4 & 2 & 2 & 8 \\ \hline & 4 & 2 & 2 & 8 & 7 \end{array}$$

Note that a zero has been inserted as the coefficient of the missing $x^2$ term

Therefore,

$$\frac{4x^4 - 2x^3 + 6x - 1}{x - 1} = 4x^3 + 2x^2 + 2x + 8 + \frac{7}{x - 1}$$

The quotient is $4x^3 + 2x^2 + 2x + 8$, and the remainder is $\frac{7}{x - 1}$.

## Concept Quiz 4.5

For Problems 1–10, answer true or false.

1. A division problem written as $(x^2 - x - 6) \div (x - 1)$ could also be written as $\frac{x^2 - x - 6}{x - 1}$.

2. The division of $\frac{x^2 + 7x + 12}{x + 3} = x + 4$ could be checked by multiplying $(x + 4)$ by $(x + 3)$.

3. For the division problem $(2x^2 + 5x + 9) \div (2x + 1)$, the remainder is 7. The remainder for the division problem can be expressed as $\frac{7}{2x+1}$.

4. In general, to check a division problem we can multiply the divisor by the quotient and subtract the remainder.

5. If a term is inserted to act as a placeholder, then the coefficient of the term must be zero.

6. When performing division, the process ends when the degree of the remainder is less than the degree of the divisor.

7. Synthetic division is a shortcut process for polynomial division.

8. Synthetic division can be used when the divisor is of the form $x - k$.

9. The fraction $\frac{x^2 - x - 6}{x - 3}$ can only be simplified by using synthetic division.

10. Synthetic division cannot be used for the problem $(6x^3 + x - 4) \div (x + 2)$ because there is no $x^2$ term in the dividend.

## Problem Set 4.5

For Problems 1–10, perform the indicated divisions of polynomials by monomials. (Objective 1)

1. $\frac{9x^4 + 18x^3}{3x}$
2. $\frac{12x^3 - 24x^2}{6x^2}$
3. $\frac{-24x^6 + 36x^8}{4x^2}$
4. $\frac{-35x^5 - 42x^3}{-7x^2}$
5. $\frac{15a^3 - 25a^2 - 40a}{5a}$
6. $\frac{-16a^4 + 32a^3 - 56a^2}{-8a}$
7. $\frac{13x^3 - 17x^2 + 28x}{-x}$
8. $\frac{14xy - 16x^2y^2 - 20x^3y^4}{-xy}$
9. $\frac{-18x^2y^2 + 24x^3y^2 - 48x^2y^3}{6xy}$
10. $\frac{-27a^3b^4 - 36a^2b^3 + 72a^2b^5}{9a^2b^2}$

For Problems 11–52, perform the indicated divisions. (Objective 1)

11. $\frac{x^2 - 7x - 78}{x + 6}$
12. $\frac{x^2 + 11x - 60}{x - 4}$
13. $(x^2 + 12x - 160) \div (x - 8)$
14. $(x^2 - 18x - 175) \div (x + 7)$
15. $\frac{2x^2 - x - 4}{x - 1}$
16. $\frac{3x^2 - 2x - 7}{x + 2}$
17. $\frac{15x^2 + 22x - 5}{3x + 5}$
18. $\frac{12x^2 - 32x - 35}{2x - 7}$
19. $\frac{3x^3 + 7x^2 - 13x - 21}{x + 3}$
20. $\frac{4x^3 - 21x^2 + 3x + 10}{x - 5}$
21. $(2x^3 + 9x^2 - 17x + 6) \div (2x - 1)$
22. $(3x^3 - 5x^2 - 23x - 7) \div (3x + 1)$
23. $(4x^3 - x^2 - 2x + 6) \div (x - 2)$
24. $(6x^3 - 2x^2 + 4x - 3) \div (x + 1)$
25. $(x^4 - 10x^3 + 19x^2 + 33x - 18) \div (x - 6)$
26. $(x^4 + 2x^3 - 16x^2 + x + 6) \div (x - 3)$
27. $\frac{x^3 - 125}{x - 5}$
28. $\frac{x^3 + 64}{x + 4}$
29. $(x^3 + 64) \div (x + 1)$
30. $(x^3 - 8) \div (x - 4)$
31. $(2x^3 - x - 6) \div (x + 2)$
32. $(5x^3 + 2x - 3) \div (x - 2)$
33. $\frac{4a^2 - 8ab + 4b^2}{a - b}$
34. $\frac{3x^2 - 2xy - 8y^2}{x - 2y}$
35. $\frac{4x^3 - 5x^2 + 2x - 6}{x^2 - 3x}$
36. $\frac{3x^3 + 2x^2 - 5x - 1}{x^2 + 2x}$
37. $\frac{8y^3 - y^2 - y + 5}{y^2 + y}$
38. $\frac{5y^3 - 6y^2 - 7y - 2}{y^2 - y}$
39. $(2x^3 + x^2 - 3x + 1) \div (x^2 + x - 1)$
40. $(3x^3 - 4x^2 + 8x + 8) \div (x^2 - 2x + 4)$
41. $(4x^3 - 13x^2 + 8x - 15) \div (4x^2 - x + 5)$
42. $(5x^3 + 8x^2 - 5x - 2) \div (5x^2 - 2x - 1)$
43. $(5a^3 + 7a^2 - 2a - 9) \div (a^2 + 3a - 4)$
44. $(4a^3 - 2a^2 + 7a - 1) \div (a^2 - 2a + 3)$

45. $(2n^4 + 3n^3 - 2n^2 + 3n - 4) \div (n^2 + 1)$

46. $(3n^4 + n^3 - 7n^2 - 2n + 2) \div (n^2 - 2)$

47. $(x^5 - 1) \div (x - 1)$

48. $(x^5 + 1) \div (x + 1)$

49. $(x^4 - 1) \div (x + 1)$

50. $(x^4 - 1) \div (x - 1)$

51. $(3x^4 + x^3 - 2x^2 - x + 6) \div (x^2 - 1)$

52. $(4x^3 - 2x^2 + 7x - 5) \div (x^2 + 2)$

For Problems 53–64, use synthetic division to determine the quotient and remainder. (Objective 2)

53. $(x^2 - 8x + 12) \div (x - 2)$

54. $(x^2 + 9x + 18) \div (x + 3)$

55. $(x^2 + 2x - 10) \div (x - 4)$

56. $(x^2 - 10x + 15) \div (x - 8)$

57. $(x^3 - 2x^2 - x + 2) \div (x - 2)$

58. $(x^3 - 5x^2 + 2x + 8) \div (x + 1)$

59. $(x^3 - 7x - 6) \div (x + 2)$

60. $(x^3 + 6x^2 - 5x - 1) \div (x - 1)$

61. $(2x^3 - 5x^2 - 4x + 6) \div (x - 2)$

62. $(3x^4 - x^3 + 2x^2 - 7x - 1) \div (x + 1)$

63. $(x^4 + 4x^3 - 7x - 1) \div (x - 3)$

64. $(2x^4 + 3x^2 + 3) \div (x + 2)$

## Thoughts Into Words

65. Describe the process of long division of polynomials.

66. Give a step-by-step description of how you would do the following division problem.

$$(4 - 3x - 7x^3) \div (x + 6)$$

67. How do you know by inspection that $3x^2 + 5x + 1$ cannot be the correct answer for the division problem $(3x^3 - 7x^2 - 22x + 8) \div (x - 4)$?

### Answers to the Concept Quiz

**1.** True **2.** True **3.** True **4.** False **5.** True **6.** True **7.** True **8.** True **9.** False **10.** False

# 4.6 Fractional Equations

OBJECTIVES

1. Solve rational equations
2. Solve proportions
3. Solve word problems involving ratios

The fractional equations used in this text are of two basic types. One type has only constants as denominators, and the other type contains variables in the denominators.

In Chapter 2, we considered fractional equations that involved only constants in the denominators. Let's briefly review our approach to solving such equations because we will be using that same basic technique to solve any type of fractional equation.

**Classroom Example**

Solve $\frac{x + 5}{2} + \frac{x - 3}{6} = \frac{2}{3}$.

**EXAMPLE 1** Solve $\frac{x - 2}{3} + \frac{x + 1}{4} = \frac{1}{6}$.

### Solution

$$\frac{x - 2}{3} + \frac{x + 1}{4} = \frac{1}{6}$$

$$12\left(\frac{x - 2}{3} + \frac{x + 1}{4}\right) = 12\left(\frac{1}{6}\right)$$ Multiply both sides by 12, which is the LCD of all of the denominators

$$12\left(\frac{x-2}{3}\right)+12\left(\frac{x+1}{4}\right)=12\left(\frac{1}{6}\right)$$
$$4(x-2)+3(x+1)=2$$
$$4x-8+3x+3=2$$
$$7x-5=2$$
$$7x=7$$
$$x=1$$

The solution set is $\{1\}$. Check it!

If an equation contains a variable (or variables) in one or more denominators, then we proceed in essentially the same way as in Example 1 *except that we must avoid any value of the variable that makes a denominator zero.* Consider the following examples.

Classroom Example
Solve $\frac{3}{n}+\frac{1}{4}=\frac{5}{n}$.

**EXAMPLE 2** Solve $\frac{5}{n}+\frac{1}{2}=\frac{9}{n}$.

### Solution

First, we need to realize that $n$ cannot equal zero. (Let's indicate this restriction so that it is not forgotten!) Then we can proceed.

$$\frac{5}{n}+\frac{1}{2}=\frac{9}{n}, \quad n \neq 0$$
$$2n\left(\frac{5}{n}+\frac{1}{2}\right)=2n\left(\frac{9}{n}\right) \quad \text{Multiply both sides by the LCD, which is } 2n$$
$$2n\left(\frac{5}{n}\right)+2n\left(\frac{1}{2}\right)=2n\left(\frac{9}{n}\right)$$
$$10+n=18$$
$$n=8$$

The solution set is $\{8\}$. Check it!

Classroom Example
Solve $\frac{27-x}{x}=9-\frac{3}{x}$.

**EXAMPLE 3** Solve $\frac{35-x}{x}=7+\frac{3}{x}$.

### Solution

$$\frac{35-x}{x}=7+\frac{3}{x}, \quad x \neq 0$$
$$x\left(\frac{35-x}{x}\right)=x\left(7+\frac{3}{x}\right) \quad \text{Multiply both sides by } x$$
$$35-x=x(7)+x\left(\frac{3}{x}\right)$$
$$35-x=7x+3$$
$$32=8x$$
$$4=x$$

The solution set is $\{4\}$.

Classroom Example
Solve $\frac{5}{x-3}=\frac{6}{x+2}$.

**EXAMPLE 4** Solve $\frac{3}{a-2}=\frac{4}{a+1}$.

### Solution

$$\frac{3}{a-2} = \frac{4}{a+1}, \qquad a \neq 2 \text{ and } a \neq -1$$

$$(a-2)(a+1)\left(\frac{3}{a-2}\right) = (a-2)(a+1)\left(\frac{4}{a+1}\right) \qquad \text{Multiply both sides by } (a-2)(a+1)$$

$$3(a+1) = 4(a-2)$$

$$3a + 3 = 4a - 8$$

$$11 = a$$

The solution set is $\{11\}$.

Keep in mind that listing the restrictions at the beginning of a problem does not replace checking the potential solutions. In Example 4, the answer 11 needs to be checked in the original equation.

**Classroom Example**
Solve $\frac{x}{x+3} + \frac{3}{2} = \frac{-3}{x+3}$.

**EXAMPLE 5**

Solve $\frac{a}{a-2} + \frac{2}{3} = \frac{2}{a-2}$.

### Solution

$$\frac{a}{a-2} + \frac{2}{3} = \frac{2}{a-2}, \qquad a \neq 2$$

$$3(a-2)\left(\frac{a}{a-2} + \frac{2}{3}\right) = 3(a-2)\left(\frac{2}{a-2}\right) \qquad \text{Multiply both sides by } 3(a-2)$$

$$3(a-2)\left(\frac{a}{a-2}\right) + 3(a-2)\left(\frac{2}{3}\right) = 3(a-2)\left(\frac{2}{a-2}\right)$$

$$3(a) + 2(a-2) = 3(2)$$

$$3a + 2a - 4 = 6$$

$$5a = 10$$

$$a = 2$$

Because our initial restriction was $a \neq 2$, we conclude that this equation has no solution. Thus the solution set is $\varnothing$.

## Solving Proportions

A **ratio** is the comparison of two numbers by division. We often use the fractional form to express ratios. For example, we can write the ratio of $a$ to $b$ as $\frac{a}{b}$. A statement of equality between two ratios is called a **proportion**. Thus if $\frac{a}{b}$ and $\frac{c}{d}$ are two equal ratios, we can form the proportion $\frac{a}{b} = \frac{c}{d}$ ($b \neq 0$ and $d \neq 0$). We deduce an important property of proportions as follows:

$$\frac{a}{b} = \frac{c}{d}, \qquad b \neq 0 \text{ and } d \neq 0$$

$$bd\left(\frac{a}{b}\right) = bd\left(\frac{c}{d}\right) \qquad \text{Multiply both sides by } bd$$

$$ad = bc$$

**Cross-Multiplication Property of Proportions**

If $\frac{a}{b} = \frac{c}{d}$ ($b \neq 0$ and $d \neq 0$), then $ad = bc$.

We can treat some fractional equations as proportions and solve them by using the cross-multiplication idea, as in the next examples.

**Classroom Example**
Solve $\frac{4}{x+3} = \frac{9}{x-4}$.

**EXAMPLE 6** Solve $\frac{5}{x+6} = \frac{7}{x-5}$.

### Solution

$$\frac{5}{x+6} = \frac{7}{x-5}, \quad x \neq -6 \text{ and } x \neq 5$$

$$5(x-5) = 7(x+6) \quad \text{Apply the cross-multiplication property}$$

$$5x - 25 = 7x + 42$$

$$-67 = 2x$$

$$-\frac{67}{2} = x$$

The solution set is $\left\{-\frac{67}{2}\right\}$.

**Classroom Example**
Solve $\frac{x}{9} = \frac{3}{x-6}$.

**EXAMPLE 7** Solve $\frac{x}{7} = \frac{4}{x+3}$.

### Solution

$$\frac{x}{7} = \frac{4}{x+3}, \quad x \neq -3$$

$$x(x+3) = 7(4) \quad \text{Cross-multiplication property}$$

$$x^2 + 3x = 28$$

$$x^2 + 3x - 28 = 0$$

$$(x+7)(x-4) = 0$$

$$x + 7 = 0 \quad \text{or} \quad x - 4 = 0$$

$$x = -7 \quad \text{or} \quad x = 4$$

The solution set is $\{-7, 4\}$. Check these solutions in the original equation.

## Solving Word Problems Involving Ratios

We can conveniently set up some problems and solve them using the concepts of ratio and proportion. Let's conclude this section with two such examples.

**Classroom Example**
On a drawing, $\frac{5}{8}$ inch represents 10 feet. If two sewer pipes are $4\frac{1}{2}$ inches apart on the drawing, find the number of feet between the pipes.

**EXAMPLE 8** Apply Your Skill

On a blueprint for a construction job, $1\frac{1}{2}$ inches represents 15 feet. If two columns are 2 inches apart on the blueprint, find the number of feet between the two columns.

### Solution

Let $d$ represent the number of feet between the two columns. To set up the proportion, we will use a ratio of inches on the blueprint to feet. Be sure to keep the ratio "inches on the blueprint" the same for both sides of the proportion.

Before we set up the proportion, let's change $1\frac{1}{2}$ to $\frac{3}{2}$.

$$\frac{\frac{3}{2}}{15} = \frac{2}{d}$$

$$\frac{3}{2}d = 15(2) \qquad \text{Cross-multiplication property}$$

$$\frac{2}{3}\left(\frac{3}{2}d\right) = \frac{2}{3}(15)(2) \qquad \text{Multiply both sides by } \frac{2}{3}$$

$$d = \frac{2}{3}(15)(2)$$

$$d = 20$$

The distance between the two columns is 20 feet.

**Classroom Example**
A sum of \$3600 is to be divided between two people in the ratio of 3 to 5. How much does each person receive?

### EXAMPLE 9 Apply Your Skill

A sum of \$750 is to be divided between two people in the ratio of 2 to 3. How much does each person receive?

#### Solution

Let $d$ represent the amount of money that one person receives. Then $750 - d$ represents the amount for the other person.

$$\frac{d}{750 - d} = \frac{2}{3}, \qquad d \neq 750$$

$$3d = 2(750 - d)$$

$$3d = 1500 - 2d$$

$$5d = 1500$$

$$d = 300$$

If $d = 300$, then $750 - d$ equals 450. Therefore, one person receives \$300 and the other person receives \$450.

## Concept Quiz 4.6

For Problems 1–3, answer true or false.

1. In solving rational equations, any value of the variable that makes a denominator zero cannot be a solution of the equation.
2. One method of solving rational equations is to multiply both sides of the equation by the least common denominator of the fractions in the equation.
3. In solving a rational equation that is a proportion, cross products can be set equal to each other.
4. Identify the following equations as a proportion or not a proportion.

   **(a)** $\frac{2x}{x + 1} + x = \frac{7}{x + 1}$ **(b)** $\frac{x - 8}{2x + 5} = \frac{7}{9}$ **(c)** $5 + \frac{2x}{x + 6} = \frac{x - 3}{x + 4}$

5. Select all the equations that could represent the following problem: John bought three bottles of energy drink for \$5.07. If the price remains the same, what will eight bottles of the energy drink cost?

   **(a)** $\frac{3}{5.07} = \frac{x}{8}$ **(b)** $\frac{5.07}{8} = \frac{x}{3}$ **(c)** $\frac{3}{8} = \frac{5.07}{x}$ **(d)** $\frac{5.07}{3} = \frac{x}{8}$

For Problems 6–10, match each equation with its solution set.

**Equations**

6. $\frac{3}{x+1} = \frac{3}{x-1}$

7. $\frac{x}{5} = \frac{3x}{15}$

8. $\frac{2x+1}{7} = \frac{3x}{7}$

9. $\frac{-x+9}{x-4} = \frac{5}{x-4}$

10. $\frac{4}{x+2} = \frac{4}{2x-1}$

**Solution Sets**

**A.** {All real numbers}

**B.** $\varnothing$

**C.** {3}

**D.** {1}

## Problem Set 4.6

For Problems 1–44, solve each equation. (Objectives 1 and 2)

**1.** $\frac{x+1}{4} + \frac{x-2}{6} = \frac{3}{4}$

**2.** $\frac{x+2}{5} + \frac{x-1}{6} = \frac{3}{5}$

**3.** $\frac{x+3}{2} - \frac{x-4}{7} = 1$

**4.** $\frac{x+4}{3} - \frac{x-5}{9} = 1$

**5.** $\frac{5}{n} + \frac{1}{3} = \frac{7}{n}$

**6.** $\frac{3}{n} + \frac{1}{6} = \frac{11}{3n}$

**7.** $\frac{7}{2x} + \frac{3}{5} = \frac{2}{3x}$

**8.** $\frac{9}{4x} + \frac{1}{3} = \frac{5}{2x}$

**9.** $\frac{3}{4x} + \frac{5}{6} = \frac{4}{3x}$

**10.** $\frac{5}{7x} - \frac{5}{6} = \frac{1}{6x}$

**11.** $\frac{47-n}{n} = 8 + \frac{2}{n}$

**12.** $\frac{45-n}{n} = 6 + \frac{3}{n}$

**13.** $\frac{n}{65-n} = 8 + \frac{2}{65-n}$

**14.** $\frac{n}{70-n} = 7 + \frac{6}{70-n}$

**15.** $n + \frac{1}{n} = \frac{17}{4}$

**16.** $n + \frac{1}{n} = \frac{37}{6}$

**17.** $n - \frac{2}{n} = \frac{23}{5}$

**18.** $n - \frac{3}{n} = \frac{26}{3}$

**19.** $\frac{5}{7x-3} = \frac{3}{4x-5}$

**20.** $\frac{3}{2x-1} = \frac{5}{3x+2}$

**21.** $\frac{-2}{x-5} = \frac{1}{x+9}$

**22.** $\frac{5}{2a-1} = \frac{-6}{3a+2}$

**23.** $\frac{x}{x+1} - 2 = \frac{3}{x-3}$

**24.** $\frac{x}{x-2} + 1 = \frac{8}{x-1}$

**25.** $\frac{a}{a+5} - 2 = \frac{3a}{a+5}$

**26.** $\frac{a}{a-3} - \frac{3}{2} = \frac{3}{a-3}$

**27.** $\frac{5}{x+6} = \frac{6}{x-3}$

**28.** $\frac{3}{x-1} = \frac{4}{x+2}$

**29.** $\frac{3x-7}{10} = \frac{2}{x}$

**30.** $\frac{x}{-4} = \frac{3}{12x-25}$

**31.** $\frac{x}{x-6} - 3 = \frac{6}{x-6}$

**32.** $\frac{x}{x+1} + 3 = \frac{4}{x+1}$

**33.** $\frac{3s}{s+2} + 1 = \frac{35}{2(3s+1)}$

**34.** $\frac{s}{2s-1} - 3 = \frac{-32}{3(s+5)}$

**35.** $2 - \frac{3x}{x-4} = \frac{14}{x+7}$

**36.** $-1 + \frac{2x}{x+3} = \frac{-4}{x+4}$

**37.** $\frac{n+6}{27} = \frac{1}{n}$

**38.** $\frac{n}{5} = \frac{10}{n-5}$

**39.** $\frac{3n}{n-1} - \frac{1}{3} = \frac{-40}{3n-18}$

**40.** $\frac{n}{n+1} + \frac{1}{2} = \frac{-2}{n+2}$

**41.** $\frac{-3}{4x+5} = \frac{2}{5x-7}$

**42.** $\frac{7}{x+4} = \frac{3}{x-8}$

**43.** $\frac{2x}{x-2} + \frac{15}{x^2-7x+10} = \frac{3}{x-5}$

**44.** $\frac{x}{x-4} - \frac{2}{x+3} = \frac{20}{x^2-x-12}$

For Problems 45–56, set up an algebraic equation and solve each problem. (Objective 3)

**45.** A sum of $1750 is to be divided between two people in the ratio of 3 to 4. How much does each person receive?

**46.** A blueprint has a scale in which 1 inch represents 5 feet. Find the dimensions of a rectangular room that measures $3\frac{1}{2}$ inches by $5\frac{3}{4}$ inches on the blueprint.

**47.** One angle of a triangle has a measure of 60°, and the measures of the other two angles are in the ratio of 2 to 3. Find the measures of the other two angles.

48. The ratio of the complement of an angle to its supplement is 1 to 4. Find the measure of the angle.

49. If a home valued at \$150,000 is assessed \$2500 in real estate taxes, then what are the taxes on a home valued at \$210,000 if assessed at the same rate?

50. The ratio of male students to female students at a certain university is 5 to 7. If there is a total of 16,200 students, find the number of male students and the number of female students.

51. Employees of an accounting firm were given a health assessment. Of the 540 employees, it was determined the ratio of the number of overweight employees to the employees that were not overweight was 5 to 4. Find the number of overweight employees.

52. The total value of a house and a lot is \$168,000. If the ratio of the value of the house to the value of the lot is 7 to 1, find the value of the house.

53. A 20-foot board is to be cut into two pieces whose lengths are in the ratio of 7 to 3. Find the lengths of the two pieces.

54. An inheritance of \$300,000 is to be divided between a son and the local heart fund in the ratio of 3 to 1. How much money will the son receive?

55. Suppose that in a certain precinct, 1150 people voted in the last presidential election. If the ratio of female voters to male voters was 3 to 2, how many females and how many males voted?

56. The perimeter of a rectangle is 114 centimeters. If the ratio of its width to its length is 7 to 12, find the dimensions of the rectangle.

## Thoughts Into Words

57. How could you do Problem 53 without using algebra?

58. How can you tell by inspection that the equation $\frac{x}{x+2} = \frac{-2}{x+2}$ has no solution?

59. How would you help someone solve the equation $\frac{3}{x} - \frac{4}{x} = \frac{-1}{x}$?

### Answers to the Concept Quiz

**1.** True **2.** True **3.** True **4. (a)** Not a proportion **(b)** Proportion **(c)** Not a proportion **5.** c, d
**6.** B **7.** A **8.** D **9.** B **10.** C

# 4.7 More Fractional Equations and Applications

OBJECTIVES

1. Solve rational equations with denominators that require factoring
2. Solve formulas that involve fractional forms
3. Solve rate-time word problems

Let's begin this section by considering a few more fractional equations. We will continue to solve them using the same basic techniques as in the previous section. That is, we will multiply both sides of the equation by the least common denominator of all of the denominators in the equation, with the necessary restrictions to avoid division by zero. Some of the denominators in these problems will require factoring before we can determine a least common denominator.

Classroom Example
Solve $\frac{x}{3x+9} + \frac{9}{x^2-9} = \frac{1}{3}$.

**EXAMPLE 1** Solve $\frac{x}{2x-8} + \frac{16}{x^2-16} = \frac{1}{2}$.

### Solution

To determine the necessary restrictions, first factor the denominators.

$$\frac{x}{2x-8}+\frac{16}{x^2-16}=\frac{1}{2}$$

$$\frac{x}{2(x-4)}+\frac{16}{(x+4)(x-4)}=\frac{1}{2}, \quad x \neq 4 \text{ and } x \neq -4$$

$$2(x-4)(x+4)\left(\frac{x}{2(x-4)}+\frac{16}{(x+4)(x-4)}\right)=2(x+4)(x-4)\left(\frac{1}{2}\right)$$

Multiply both sides by the LCD, $2(x-4)(x+4)$

$$x(x+4)+2(16)=(x+4)(x-4)$$

$$x^2+4x+32=x^2-16$$

$$4x=-48$$

$$x=-12$$

The solution set is $\{-12\}$. Perhaps you should check it!

Classroom Example
Solve $\frac{4}{x+5}+\frac{3}{3x-2}=\frac{x+12}{3x^2+13x-10}$.

**EXAMPLE 2** Solve $\frac{3}{n-5}-\frac{2}{2n+1}=\frac{n+3}{2n^2-9n-5}$.

### Solution

$$\frac{3}{n-5}-\frac{2}{2n+1}=\frac{n+3}{2n^2-9n-5}$$

$$\frac{3}{n-5}-\frac{2}{2n+1}=\frac{n+3}{(2n+1)(n-5)}, \quad n \neq -\frac{1}{2} \text{ and } n \neq 5$$

$$(2n+1)(n-5)\left(\frac{3}{n-5}-\frac{2}{2n+1}\right)=(2n+1)(n-5)\left(\frac{n+3}{(2n+1)(n-5)}\right)$$

Multiply both sides by the LCD, $(2n+1)\cdot(n-5)$

$$3(2n+1)-2(n-5)=n+3$$

$$6n+3-2n+10=n+3$$

$$4n+13=n+3$$

$$3n=-10$$

$$n=-\frac{10}{3}$$

The solution set is $\left\{-\frac{10}{3}\right\}$.

Classroom Example
Solve $3-\frac{9}{x+3}=\frac{27}{x^2+3x}$.

**EXAMPLE 3** Solve $2+\frac{4}{x-2}=\frac{8}{x^2-2x}$.

### Solution

$$2+\frac{4}{x-2}=\frac{8}{x^2-2x}$$

$$2+\frac{4}{x-2}=\frac{8}{x(x-2)}, \quad x \neq 0 \text{ and } x \neq 2$$

$$x(x-2)\left(2+\frac{4}{x-2}\right)=x(x-2)\left(\frac{8}{x(x-2)}\right)$$

Multiply both sides by the LCD, $x(x-2)$

$$2x(x-2)+4x=8$$

$$2x^2-4x+4x=8$$

$$2x^2=8$$

$$x^2=4$$

$$x^2 - 4 = 0$$
$$(x + 2)(x - 2) = 0$$
$$x + 2 = 0 \quad \text{or} \quad x - 2 = 0$$
$$x = -2 \quad \text{or} \quad x = 2$$

Because our initial restriction indicated that $x \neq 2$, the only solution is $-2$. Thus the solution set is $\{-2\}$.

## Solving Formulas that Involve Fractional Forms

In Section 2.4, we discussed using the properties of equality to change the form of various formulas. For example, we considered the simple interest formula $A = P + Prt$ and changed its form by solving for $P$ as follows:

$$A = P + Prt$$
$$A = P(1 + rt) \qquad \text{Factor out the common factor of } P$$
$$\frac{A}{1 + rt} = P \qquad \text{Divide both sides by } (1 + rt)$$

If the formula is in the form of a fractional equation, then the techniques of these last two sections are applicable. Consider the following example.

**Classroom Example**
Solve the future value formula for $r$:
$A = P\left(1 + \frac{r}{n}\right)$

**EXAMPLE 4**

If the original cost of some business property is $C$ dollars and it is depreciated linearly over $N$ years, then its value, $V$, at the end of $T$ years is given by

$$V = C\left(1 - \frac{T}{N}\right)$$

Solve this formula for $N$ in terms of $V$, $C$, and $T$.

### Solution

$$V = C\left(1 - \frac{T}{N}\right)$$
$$V = C - \frac{CT}{N} \qquad \text{Distributive property}$$
$$N(V) = N\left(C - \frac{CT}{N}\right) \qquad \text{Multiply both sides by } N$$
$$NV = NC - CT$$
$$NV - NC = -CT$$
$$N(V - C) = -CT \qquad \text{Factor out a common factor of } N$$
$$N = \frac{-CT}{V - C} \qquad \text{Divide both sides by } (V - C)$$
$$N = -\frac{CT}{V - C}$$

## Solving Rate-Time Word Problems

In Section 2.4 we solved some uniform motion problems. The formula $d = rt$ was used in the analysis of the problems, and we used guidelines that involve distance relationships. Now let's consider some uniform motion problems for which guidelines involving either times or rates are appropriate. These problems will generate fractional equations to solve.

olly/Shutterstock.com

**Classroom Example**
An airplane travels 2852 miles in the same time that a car travels 299 miles. If the rate of the plane is 555 miles per hour greater than the rate of the car, find the rate of each.

## EXAMPLE 5 Apply Your Skill

An airplane travels 2050 miles in the same time that a car travels 260 miles. If the rate of the plane is 358 miles per hour greater than the rate of the car, find the rate of each.

### Solution

Let $r$ represent the rate of the car. Then $r + 358$ represents the rate of the plane. The fact that the times are equal can be a guideline. Remember from the basic formula, $d = rt$, that $t = \frac{d}{r}$.

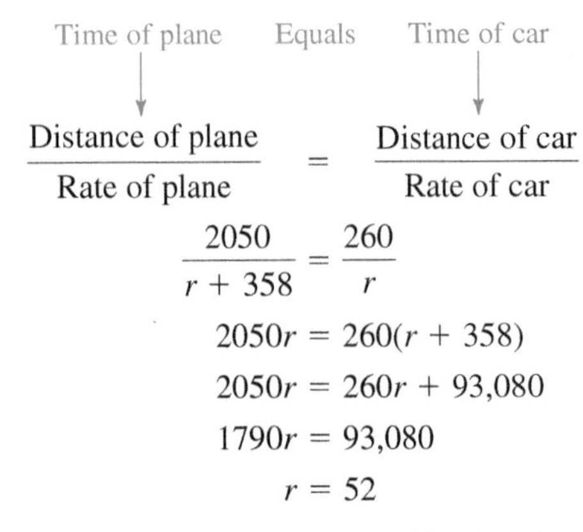


$$\frac{\text{Distance of plane}}{\text{Rate of plane}} = \frac{\text{Distance of car}}{\text{Rate of car}}$$

$$\frac{2050}{r + 358} = \frac{260}{r}$$

$$2050r = 260(r + 358)$$

$$2050r = 260r + 93{,}080$$

$$1790r = 93{,}080$$

$$r = 52$$

If $r = 52$, then $r + 358$ equals 410. Thus the rate of the car is 52 miles per hour, and the rate of the plane is 410 miles per hour.

Darren Hedges/Shutterstock.com

**Classroom Example**
It takes a freight train 1 hour longer to travel 180 miles than it takes an express train to travel 195 miles. The rate of the express train is 20 miles per hour greater than the rate of the freight train. Find the times and rates of both trains.

## EXAMPLE 6 Apply Your Skill

It takes a freight train 2 hours longer to travel 300 miles than it takes an express train to travel 280 miles. The rate of the express train is 20 miles per hour greater than the rate of the freight train. Find the times and rates of both trains.

### Solution

Let $t$ represent the time of the express train. Then $t + 2$ represents the time of the freight train. Let's record the information of this problem in a table.

| | Distance | Time | Rate $= \frac{\text{distance}}{\text{time}}$ |
|---|---|---|---|
| Express train | 280 | $t$ | $\frac{280}{t}$ |
| Freight train | 300 | $t + 2$ | $\frac{300}{t + 2}$ |

The fact that the rate of the express train is 20 miles per hour greater than the rate of the freight train can be a guideline.

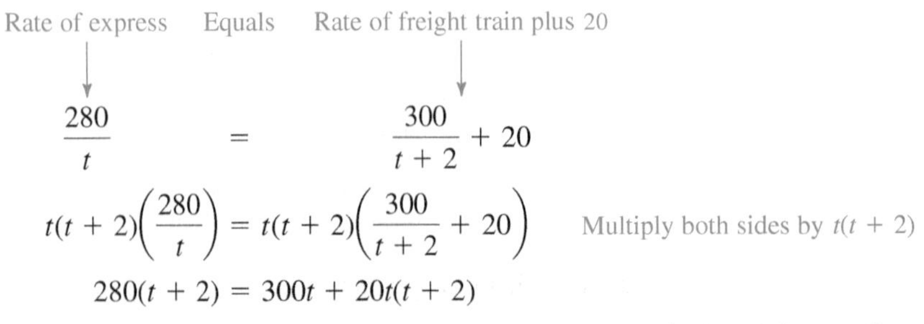


$$\frac{280}{t} = \frac{300}{t + 2} + 20$$

$$t(t + 2)\left(\frac{280}{t}\right) = t(t + 2)\left(\frac{300}{t + 2} + 20\right) \quad \text{Multiply both sides by } t(t + 2)$$

$$280(t + 2) = 300t + 20t(t + 2)$$

$$
\begin{aligned}
280t + 560 &= 300t + 20t^2 + 40t \\
280t + 560 &= 340t + 20t^2 \\
0 &= 20t^2 + 60t - 560 \\
0 &= t^2 + 3t - 28 \\
0 &= (t + 7)(t - 4)
\end{aligned}
$$

$$t + 7 = 0 \quad \text{or} \quad t - 4 = 0$$

$$t = -7 \quad \text{or} \quad t = 4$$

The negative solution must be discarded, so the time of the express train ($t$) is 4 hours, and the time of the freight train ($t + 2$) is 6 hours. The rate of the express train $\left(\frac{280}{t}\right)$ is $\frac{280}{4} = 70$ miles per hour, and the rate of the freight train $\left(\frac{300}{t+2}\right)$ is $\frac{300}{6} = 50$ miles per hour.

**Remark:** Note that to solve Example 5 we went directly to a guideline without the use of a table, but for Example 6 we used a table. Remember that this is a personal preference; we are merely acquainting you with a variety of techniques.

Uniform-motion problems are a special case of a larger group of problems we refer to as **rate-time problems**. For example, if a certain machine can produce 150 items in 10 minutes, then we say that the machine is producing at a rate of $\frac{150}{10} = 15$ items per minute. Likewise, if a person can do a certain job in 3 hours, then, assuming a constant rate of work, we say that the person is working at a rate of $\frac{1}{3}$ of the job per hour. In general, if $Q$ is the quantity of something done in $t$ units of time, then the rate, $r$, is given by $r = \frac{Q}{t}$. We state the rate in terms of *so much quantity per unit of time.* (In uniform-motion problems the "quantity" is distance.) Let's consider some examples of rate-time problems.

David R. Frazier Photolibrary, Inc./ Alamy

**Classroom Example**
If Shayla can paint a chair in 45 minutes, and her sister Jamie can paint a similar chair in 60 minutes, how long will it take them to paint a chair if they work together?

## EXAMPLE 7 Apply Your Skill

If Jim can mow a lawn in 50 minutes, and his son, Todd, can mow the same lawn in 40 minutes, how long will it take them to mow the lawn if they work together?

### Solution

Jim's rate is $\frac{1}{50}$ of the lawn per minute, and Todd's rate is $\frac{1}{40}$ of the lawn per minute. If we let $m$ represent the number of minutes that they work together, then $\frac{1}{m}$ represents their rate when working together. Therefore, because the sum of the individual rates must equal the rate working together, we can set up and solve the following equation.

Jim's rate → Todd's rate → Combined rate

$$\frac{1}{50} + \frac{1}{40} = \frac{1}{m}$$

$$200m\left(\frac{1}{50} + \frac{1}{40}\right) = 200m\left(\frac{1}{m}\right) \qquad \text{Multiply both sides by the LCD, } 200m$$

$$
\begin{aligned}
4m + 5m &= 200 \\
9m &= 200 \\
m &= \frac{200}{9} = 22\frac{2}{9}
\end{aligned}
$$

It should take them $22\frac{2}{9}$ minutes.

Creatas/Super Stock

**Classroom Example**
Working together, Kevin and Casey can wash the windows in $3\frac{1}{2}$ hours. Kevin can wash the windows by himself in $6\frac{1}{2}$ hours. How long would it take Casey to wash the windows by herself?

## EXAMPLE 8 Apply Your Skill

Working together, Linda and Kathy can type a term paper in $3\frac{3}{5}$ hours. Linda can type the paper by herself in 6 hours. How long would it take Kathy to type the paper by herself?

### Solution

Their rate working together is $\frac{1}{3\frac{3}{5}} = \frac{1}{\frac{18}{5}} = \frac{5}{18}$ of the job per hour, and Linda's rate is $\frac{1}{6}$ of the job per hour. If we let $h$ represent the number of hours that it would take Kathy to do the job by herself, then her rate is $\frac{1}{h}$ of the job per hour. Thus we have

$$\underset{\text{Linda's rate}}{\frac{1}{6}} + \underset{\text{Kathy's rate}}{\frac{1}{h}} = \underset{\text{Combined rate}}{\frac{5}{18}}$$

Solving this equation yields

$$18h\left(\frac{1}{6}+\frac{1}{h}\right) = 18h\left(\frac{5}{18}\right) \quad \text{Multiply both sides by the LCD, } 18h$$

$$3h + 18 = 5h$$

$$18 = 2h$$

$$9 = h$$

It would take Kathy 9 hours to type the paper by herself.

Our final example of this section illustrates another approach that some people find meaningful for rate-time problems. For this approach, think in terms of fractional parts of the job. For example, if a person can do a certain job in 5 hours, then at the end of 2 hours, he or she has done $\frac{2}{5}$ of the job. (Again, assume a constant rate of work.) At the end of 4 hours, he or she has finished $\frac{4}{5}$ of the job; and, in general, at the end of $h$ hours, he or she has done $\frac{h}{5}$ of the job. Just as for the motion problems in which distance equals rate times the time, here the fractional part done equals the working rate times the time. Let's see how this works in a problem.

Ugur Anahtarci/Shutterstock.com

**Classroom Example**
It takes Wayne 9 hours to tile a backsplash. After he had been working for 2 hours, he was joined by Greg, and together they finished the task in 4 hours. How long would it take Greg to do the job by himself?

## EXAMPLE 9 Apply Your Skill

It takes Pat 12 hours to detail a boat. After he had been working for 3 hours, he was joined by his brother Mike, and together they finished the detailing in 5 hours. How long would it take Mike to detail the boat by himself?

### Solution

Let $h$ represent the number of hours that it would take Mike to do the detailing by himself. The fractional part of the job that Pat does equals his working rate times his time. Because it takes Pat 12 hours to do the entire job, his working rate is $\frac{1}{12}$. He works for 8 hours (3 hours before Mike and then 5 hours with Mike). Therefore, Pat's part of the job is $\frac{1}{12}(8) = \frac{8}{12}$. The fractional part of the job that Mike does equals his working rate times his time. Because $h$ represents Mike's time to do the entire job, his working rate is $\frac{1}{h}$; he works for 5 hours. Therefore, Mike's part of the job is $\frac{1}{h}(5) = \frac{5}{h}$. Adding the two fractional parts together results in 1 entire job being done. Let's also show this information in chart form and set up our guideline. Then we can set up and solve the equation.

| | Time to do entire job | Working rate | Time working | Fractional part of the job done |
|---|---|---|---|---|
| Pat | 12 | $\frac{1}{12}$ | 8 | $\frac{8}{12}$ |
| Mike | $h$ | $\frac{1}{h}$ | 5 | $\frac{5}{h}$ |

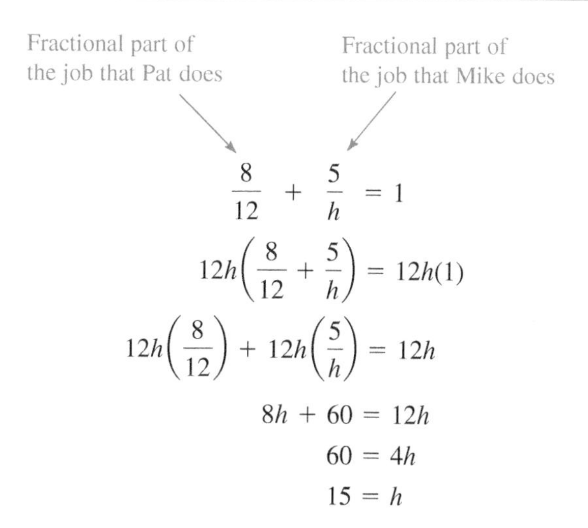


$$\frac{8}{12} + \frac{5}{h} = 1$$

$$12h\left(\frac{8}{12} + \frac{5}{h}\right) = 12h(1)$$

$$12h\left(\frac{8}{12}\right) + 12h\left(\frac{5}{h}\right) = 12h$$

$$8h + 60 = 12h$$

$$60 = 4h$$

$$15 = h$$

It would take Mike 15 hours to detail the boat by himself.

## Concept Quiz 4.7

For Problems 1–10, answer true or false.

1. Assuming uniform motion, the rate at which a car travels is equal to the time traveled divided by the distance traveled.
2. If a worker can lay 640 square feet of tile in 8 hours, we can say his rate of work is 80 square feet per hour.
3. If a person can complete two jobs in 5 hours, then the person is working at the rate of $\frac{5}{2}$ of the job per hour.
4. In a time-rate problem involving two workers, the sum of their individual rates must equal the rate working together.
5. If a person works at the rate of $\frac{2}{15}$ of the job per hour, then at the end of 3 hours the job would be $\frac{6}{15}$ completed.
6. If a person can do a job in 7 hours, then at the end of 5 hours he or she will have completed $\frac{5}{7}$ of the job.
7. If a person can do a job in $h$ hours, then at the end of 3 hours he or she will have completed $\frac{h}{3}$ of the job.

**8.** The equation $A = P + Prt$ cannot be solved for $P$, because $P$ occurs in two different terms.

**9.** If Zorka can complete a certain task in 5 hours and Mitzie can complete the same task in 9 hours, then working together they should be able to complete the task in 7 hours.

**10.** Uniform-motion problems are one type of rate-time problem.

## Problem Set 4.7

For Problems 1–30, solve each equation. (Objective 1)

**1.** $\frac{x}{4x-4} + \frac{5}{x^2-1} = \frac{1}{4}$

**2.** $\frac{x}{3x-6} + \frac{4}{x^2-4} = \frac{1}{3}$

**3.** $3 + \frac{6}{t-3} = \frac{6}{t^2-3t}$

**4.** $2 + \frac{4}{t-1} = \frac{4}{t^2-t}$

**5.** $\frac{3}{n-5} + \frac{4}{n+7} = \frac{2n+11}{n^2+2n-35}$

**6.** $\frac{2}{n+3} + \frac{3}{n-4} = \frac{2n-1}{n^2-n-12}$

**7.** $\frac{5x}{2x+6} - \frac{4}{x^2-9} = \frac{5}{2}$

**8.** $\frac{3x}{5x+5} - \frac{2}{x^2-1} = \frac{3}{5}$

**9.** $1 + \frac{1}{n-1} = \frac{1}{n^2-n}$

**10.** $3 + \frac{9}{n-3} = \frac{27}{n^2-3n}$

**11.** $\frac{2}{n-2} - \frac{n}{n+5} = \frac{10n+15}{n^2+3n-10}$

**12.** $\frac{n}{n+3} + \frac{1}{n-4} = \frac{11-n}{n^2-n-12}$

**13.** $\frac{2}{2x-3} - \frac{2}{10x^2-13x-3} = \frac{x}{5x+1}$

**14.** $\frac{1}{3x+4} + \frac{6}{6x^2+5x-4} = \frac{x}{2x-1}$

**15.** $\frac{2x}{x+3} - \frac{3}{x-6} = \frac{29}{x^2-3x-18}$

**16.** $\frac{x}{x-4} - \frac{2}{x+8} = \frac{63}{x^2+4x-32}$

**17.** $\frac{a}{a-5} + \frac{2}{a-6} = \frac{2}{a^2-11a+30}$

**18.** $\frac{a}{a+2} + \frac{3}{a+4} = \frac{14}{a^2+6a+8}$

**19.** $\frac{-1}{2x-5} + \frac{2x-4}{4x^2-25} = \frac{5}{6x+15}$

**20.** $\frac{-2}{3x+2} + \frac{x-1}{9x^2-4} = \frac{3}{12x-8}$

**21.** $\frac{7y+2}{12y^2+11y-15} - \frac{1}{3y+5} = \frac{2}{4y-3}$

**22.** $\frac{5y-4}{6y^2+y-12} - \frac{2}{2y+3} = \frac{5}{3y-4}$

**23.** $\frac{2n}{6n^2+7n-3} - \frac{n-3}{3n^2+11n-4} = \frac{5}{2n^2+11n+12}$

**24.** $\frac{x+1}{2x^2+7x-4} - \frac{x}{2x^2-7x+3} = \frac{1}{x^2+x-12}$

**25.** $\frac{1}{2x^2-x-1} + \frac{3}{2x^2+x} = \frac{2}{x^2-1}$

**26.** $\frac{2}{n^2+4n} + \frac{3}{n^2-3n-28} = \frac{5}{n^2-6n-7}$

**27.** $\frac{x+1}{x^3-9x} - \frac{1}{2x^2+x-21} = \frac{1}{2x^2+13x+21}$

**28.** $\frac{x}{2x^2+5x} - \frac{x}{2x^2+7x+5} = \frac{2}{x^2+x}$

**29.** $\frac{4t}{4t^2-t-3} + \frac{2-3t}{3t^2-t-2} = \frac{1}{12t^2+17t+6}$

**30.** $\frac{2t}{2t^2+9t+10} + \frac{1-3t}{3t^2+4t-4} = \frac{4}{6t^2+11t-10}$

For Problems 31–44, solve each equation for the indicated variable. (Objective 2)

**31.** $y = \frac{5}{6}x + \frac{2}{9}$ for $x$

**32.** $y = \frac{3}{4}x - \frac{2}{3}$ for $x$

**33.** $\frac{-2}{x-4} = \frac{5}{y-1}$ for $y$

**34.** $\frac{7}{y-3} = \frac{3}{x+1}$ for $y$

**35.** $I = \frac{100M}{C}$ for $M$

**36.** $V = C\left(1 - \frac{T}{N}\right)$ for $T$

**37.** $\frac{R}{S} = \frac{T}{S+T}$ for $R$

**38.** $\frac{1}{R} = \frac{1}{S} + \frac{1}{T}$ for $R$

**39.** $\frac{y-1}{x-3} = \frac{b-1}{a-3}$ for $y$

**40.** $y = -\frac{a}{b}x + \frac{c}{d}$ for $x$

**41.** $\frac{x}{a} + \frac{y}{b} = 1$ for $y$

**42.** $\frac{y - b}{x} = m$ for $y$

**43.** $\frac{y - 1}{x + 6} = \frac{-2}{3}$ for $y$

**44.** $\frac{y + 5}{x - 2} = \frac{3}{7}$ for $y$

Set up an equation and solve each of the following problems. (Objective 3)

**45.** Kent drives 270 miles in the same time that it takes Dave to drive 250 miles. If Kent averages 4 miles per hour faster than Dave, find their rates.

**46.** Suppose that Wendy rides her bicycle 30 miles in the same time that it takes Kim to ride her bicycle 20 miles. If Wendy rides 5 miles per hour faster than Kim, find the rate of each.

**47.** An inlet pipe can fill a tank (see Figure 4.1) in 10 minutes. A drain can empty the tank in 12 minutes. If the tank is empty, and both the pipe and drain are open, how long will it take before the tank overflows?

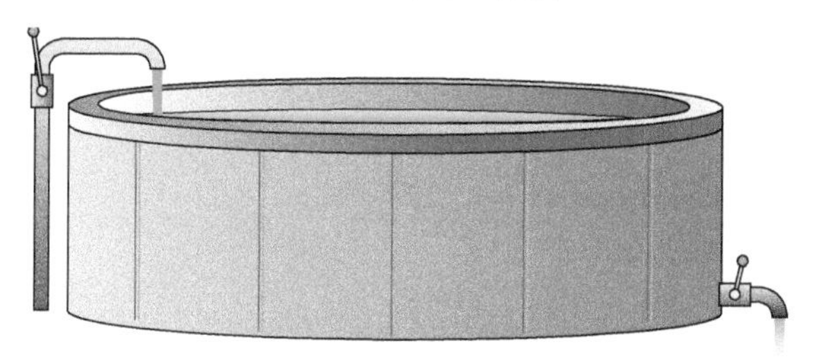

**Figure 4.1**

**48.** Barry can do a certain job in 3 hours, whereas it takes Sanchez 5 hours to do the same job. How long would it take them to do the job working together?

**49.** Connie can type 600 words in 5 minutes less time than it takes Katie to type 600 words. If Connie types at a rate of 20 words per minute faster than Katie types, find the typing rate of each woman.

**50.** Walt can mow a lawn in 1 hour, and his son, Malik, can mow the same lawn in 50 minutes. One day Malik started mowing the lawn by himself and worked for 30 minutes. Then Walt joined him and they finished the lawn. How long did it take them to finish mowing the lawn after Walt started to help?

**51.** Plane A can travel 1400 miles in 1 hour less time than it takes plane B to travel 2000 miles. The rate of plane B is 50 miles per hour greater than the rate of plane A. Find the times and rates of both planes.

**52.** To travel 60 miles, it takes Sue, riding a moped, 2 hours less time than it takes Doreen to travel 50 miles riding a bicycle. Sue travels 10 miles per hour faster than Doreen. Find the times and rates of both girls.

**53.** It takes Amy twice as long to deliver papers as it does Nancy. How long would it take each girl to deliver the papers by herself if they can deliver the papers together in 40 minutes?

**54.** If two inlet pipes are both open, they can fill a pool in 1 hour and 12 minutes. One of the pipes can fill the pool by itself in 2 hours. How long would it take the other pipe to fill the pool by itself?

**55.** Debbie rode her bicycle out into the country for a distance of 24 miles. On the way back, she took a much shorter route of 12 miles and made the return trip in one-half hour less time. If her rate out into the country was 4 miles per hour greater than her rate on the return trip, find both rates.

**56.** Felipe jogs for 10 miles and then walks another 10 miles. He jogs $2\frac{1}{2}$ miles per hour faster than he walks, and the entire distance of 20 miles takes 6 hours. Find the rate at which he walks and the rate at which he jogs.

## Thoughts Into Words

**57.** Why is it important to consider more than one way to do a problem?

**58.** Write a paragraph or two summarizing the new ideas about problem solving you have acquired thus far in this course.

### Answers to the Concept Quiz

**1.** False **2.** True **3.** False **4.** True **5.** True **6.** True **7.** False **8.** False **9.** False **10.** True

# Chapter 4 Summary

| OBJECTIVE | SUMMARY | EXAMPLE |
|---|---|---|
| Reduce rational numbers and rational expressions. (Section 4.1/Objectives 1 and 2) | Any number that can be written in the form $\frac{a}{b}$, where $a$ and $b$ are integers and $b \neq 0$, is a rational number. A rational expression is defined as the indicated quotient of two polynomials. The fundamental principle of fractions, $\frac{a \cdot k}{b \cdot k} = \frac{a}{b}$, is used when reducing rational numbers or rational expressions. | Simplify $\frac{x^2 - 2x - 15}{x^2 + x - 6}$.<br>**Solution**<br>$\frac{x^2 - 2x - 15}{x^2 + x - 6}$<br>$= \frac{(x + 3)(x - 5)}{(x + 3)(x - 2)} = \frac{x - 5}{x - 2}$<br>**Sample Problem 1**<br>Simplify $\frac{x^2 - 3x - 10}{x^2 + 9x + 14}$. |
| Multiply rational numbers and rational expressions. (Section 4.2/Objectives 1 and 2) | Multiplication of rational expressions is based on the following definition:<br>$\frac{a}{b} \cdot \frac{c}{d} = \frac{ac}{bd}$,<br>where $b \neq 0$ and $d \neq 0$ | Find the product:<br>$\frac{3y^2 + 12y}{y^3 - 2y^2} \cdot \frac{y^2 - 3y + 2}{y^2 + 7y + 12}$<br>**Solution**<br>$\frac{3y^2 + 12y}{y^3 - 2y^2} \cdot \frac{y^2 - 3y + 2}{y^2 + 7y + 12}$<br>$= \frac{3y(y + 4)}{y^2(y - 2)} \cdot \frac{(y - 2)(y - 1)}{(y + 3)(y + 4)}$<br>$= \frac{3\cancel{y}\cancel{(y + 4)}}{\cancel{y^2}_{y}\cancel{(y - 2)}} \cdot \frac{\cancel{(y - 2)}(y - 1)}{(y + 3)\cancel{(y + 4)}}$<br>$= \frac{3(y - 1)}{y(y + 3)}$<br>**Sample Problem 2**<br>Find the product: $\frac{x^2 - 3x - 18}{2x^2 - 12x} \cdot \frac{x^2 - 7x}{x^2 + 10x + 21}$ |
| Divide rational numbers and rational expressions. (Section 4.2/Objectives 3 and 4) | Division of rational expressions is based on the following definition:<br>$\frac{a}{b} \div \frac{c}{d} = \frac{a}{b} \cdot \frac{d}{c} = \frac{ad}{bc}$,<br>where $b \neq 0$, $c \neq 0$, and $d \neq 0$ | Find the quotient:<br>$\frac{6xy}{x^2 - 6x + 9} \div \frac{18x}{x^2 - 9}$<br>**Solution**<br>$\frac{6xy}{x^2 - 6x + 9} \div \frac{18x}{x^2 - 9}$<br>$= \frac{6xy}{x^2 - 6x + 9} \cdot \frac{x^2 - 9}{18x}$<br>$= \frac{6xy}{(x - 3)(x - 3)} \cdot \frac{(x + 3)(x - 3)}{18x}$<br>$= \frac{\cancel{6x}y}{\cancel{(x - 3)}(x - 3)} \cdot \frac{(x + 3)\cancel{(x - 3)}}{\cancel{18x}_{3}}$<br>$= \frac{y(x + 3)}{3(x - 3)}$ |

Answers to **Sample Problems** are located in the back of the book.

| OBJECTIVE | SUMMARY | EXAMPLE |
| --- | --- | --- |
| | | **Sample Problem 3**<br>Find the quotient: $\frac{4x^2 - 4x + 1}{12x^2y} \div \frac{4x^2 - 1}{3xy}$ |
| Simplify problems that involve both multiplication and division of rational expressions.<br>(Section 4.2/Objective 5) | Perform the multiplications and divisions from left to right according to the order of operations. You can change division to multiplication by multiplying by the reciprocal of the divisor and then finding the product. | Perform the indicated operations:<br>$\frac{6xy^3}{5x} \div \frac{3xy}{10} \cdot \frac{y}{7x^2}$<br>**Solution**<br>$\frac{6xy^3}{5x} \div \frac{3xy}{10} \cdot \frac{y}{7x^2}$<br>$= \frac{6xy^3}{5x} \cdot \frac{10}{3xy} \cdot \frac{y}{7x^2}$<br>$= \frac{\overset{2}{\cancel{6}}\cancel{x}y^3}{\cancel{5}x} \cdot \frac{\overset{2}{\cancel{10}}}{\cancel{3}\cancel{x}\cancel{y}} \cdot \frac{\cancel{y}}{7x^2}$<br>$= \frac{4y^3}{7x^3}$<br>**Sample Problem 4**<br>Perform the indicated operations: $\frac{4xy}{3x^2y} \div \frac{24xy}{15x} \cdot \frac{2x^2}{y}$ |
| Add and subtract rational numbers or rational expressions.<br>(Section 4.3/Objectives 1 and 2; Section 4.4/Objective 1) | Addition and subtraction of rational expressions are based on the following definitions.<br>$\frac{a}{b} + \frac{c}{b} = \frac{a+c}{b}$ Addition<br>$\frac{a}{b} - \frac{c}{b} = \frac{a-c}{b}$ Subtraction<br>The following basic procedure is used to add or subtract rational expressions.<br>**1.** Factor the denominators.<br>**2.** Find the LCD.<br>**3.** Change each fraction to an equivalent fraction that has the LCD as the denominator.<br>**4.** Combine the numerators and place over the LCD.<br>**5.** Simplify by performing the addition or subtraction in the numerator.<br>**6.** If possible, reduce the resulting fraction. | Subtract:<br>$\frac{2}{x^2 - 2x - 3} - \frac{5}{x^2 + 5x + 4}$<br>**Solution**<br>$\frac{2}{x^2 - 2x - 3} - \frac{5}{x^2 + 5x + 4}$<br>$= \frac{2}{(x-3)(x+1)} - \frac{5}{(x+1)(x+4)}$<br>The LCD is $(x-3)(x+1)(x+4)$.<br>$= \frac{2(x+4)}{(x-3)(x+1)(x+4)}$<br>$- \frac{5(x-3)}{(x+1)(x+4)(x-3)}$<br>$= \frac{2(x+4) - 5(x-3)}{(x-3)(x+1)(x+4)}$<br>$= \frac{2x + 8 - 5x + 15}{(x-3)(x+1)(x+4)}$<br>$= \frac{-3x + 23}{(x-3)(x+1)(x+4)}$<br>**Sample Problem 5**<br>Subtract: $\frac{6}{x^2 + 4x - 5} - \frac{2}{x^2 + 7x + 10}$ |

*(continued)*

| OBJECTIVE | SUMMARY | EXAMPLE |
| --- | --- | --- |
| Simplify complex fractions. (Section 4.4/Objective 2) | Fractions that contain rational numbers or rational expressions in the numerators or denominators are called complex fractions. In Section 4.4, two methods were shown for simplifying complex fractions. | Simplify $\dfrac{\dfrac{2}{x}-\dfrac{3}{y}}{\dfrac{4}{x^2}+\dfrac{5}{y}}$.<br>**Solution**<br>$\dfrac{\dfrac{2}{x}-\dfrac{3}{y}}{\dfrac{4}{x^2}+\dfrac{5}{y}}$<br>Multiply the numerator and denominator by $x^2y$:<br>$\dfrac{x^2y\left(\dfrac{2}{x}-\dfrac{3}{y}\right)}{x^2y\left(\dfrac{4}{x^2}+\dfrac{5}{y}\right)}$<br>$=\dfrac{x^2y\left(\dfrac{2}{x}\right)+x^2y\left(-\dfrac{3}{y}\right)}{x^2y\left(\dfrac{4}{x^2}\right)+x^2y\left(\dfrac{5}{y}\right)}$<br>$=\dfrac{2xy-3x^2}{4y+5x^2}$<br>**Sample Problem 6**<br>Simplify $\dfrac{\dfrac{1}{x}+\dfrac{6}{y^2}}{\dfrac{5}{x}-\dfrac{3}{y}}$. |
| Divide polynomials. (Section 4.5/Objective 1) | **1.** To divide a polynomial by a monomial, divide each term of the polynomial by the monomial.<br>**2.** The procedure for dividing a polynomial by a polynomial resembles the long-division process. | Divide $(2x^2+11x+19)$ by $(x+3)$.<br>**Solution**<br>$\begin{array}{r} 2x+5 \\ x+3\overline{)2x^2+11x+19} \\ \underline{2x^2+\ \ 6x} \\ 5x+19 \\ \underline{5x+15} \\ 4 \end{array}$<br>Thus $\dfrac{2x^2+11x+19}{x+3}$<br>$=2x+5+\dfrac{4}{x+3}$.<br>**Sample Problem 7**<br>Divide $(3x^2+x-2)$ by $(x+2)$. |

| OBJECTIVE | SUMMARY | EXAMPLE |
|---|---|---|
| Use synthetic division to divide polynomials. (Section 4.5/Objective 2) | Synthetic division is a shortcut to the long-division process when the divisor is of the form $x - k$. In other words, the coefficient of the $x$ term in the divisor has to be 1. | Divide $(x^4 - 3x^2 + 5x + 6)$ by $(x + 2)$.<br>**Solution**<br>$\begin{array}{r\|rrrrr} -2 & 1 & 0 & -3 & 5 & 6 \\ & & -2 & 4 & -2 & -6 \\ \hline & 1 & -2 & 1 & 3 & 0 \end{array}$<br>Thus $\dfrac{x^4 - 3x^2 + 5x + 6}{x + 2}$<br>$= x^3 - 2x^2 + x + 3$.<br>**Sample Problem 8**<br>Divide $(x^3 - 3x^2 + 4)$ by $(x - 2)$. |
| Solve rational equations. (Section 4.6/Objective 1) | To solve a rational equation, it is often easiest to begin by multiplying both sides of the equation by the LCD of all the denominators in the equation. Recall that any value of the variable that makes the denominators zero cannot be a solution to the equation. | Solve $\dfrac{2}{3x} + \dfrac{5}{12} = \dfrac{1}{4x}$.<br>**Solution**<br>$\dfrac{2}{3x} + \dfrac{5}{12} = \dfrac{1}{4x}, \quad x \neq 0$<br>Multiply both sides by $12x$:<br>$12x\left(\dfrac{2}{3x} + \dfrac{5}{12}\right) = 12x\left(\dfrac{1}{4x}\right)$<br>$12x\left(\dfrac{2}{3x}\right) + 12x\left(\dfrac{5}{12}\right) = 12x\left(\dfrac{1}{4x}\right)$<br>$8 + 5x = 3$<br>$5x = -5$<br>$x = -1$<br>The solution set is $\{-1\}$.<br>**Sample Problem 9**<br>Solve $\dfrac{3}{5} + \dfrac{1}{x} = \dfrac{2}{3}$. |
| Solve proportions. (Section 4.6/Objective 2) | A ratio is the comparison of two numbers by division. A proportion is a statement of equality between two ratios. Proportions can be solved using the cross-multiplication property of proportions. | Solve $\dfrac{5}{2x - 1} = \dfrac{3}{x + 4}$.<br>**Solution**<br>$\dfrac{5}{2x - 1} = \dfrac{3}{x + 4}, \quad x \neq -4, x \neq \dfrac{1}{2}$<br>$3(2x - 1) = 5(x + 4)$<br>$6x - 3 = 5x + 20$<br>$x = 23$<br>The solution set is $\{23\}$.<br>**Sample Problem 10**<br>Solve $\dfrac{x - 1}{4} = \dfrac{x + 3}{3}$. |

*(continued)*

| OBJECTIVE | SUMMARY | EXAMPLE |
|---|---|---|
| Solve rational equations where the denominators require factoring.<br>(Section 4.7/Objective 1) | It may be necessary to factor the denominators in a rational equation in order to determine the LCD of all the denominators. | Solve<br>$\frac{7x}{3x+12} - \frac{2}{x^2-16} = \frac{7}{3}$.<br>**Solution**<br>$\frac{7x}{3x+12} - \frac{2}{x^2-16} = \frac{7}{3}, \quad x \neq -4, x \neq 4$<br>$\frac{7x}{3(x+4)} - \frac{2}{(x-4)(x+4)} = \frac{7}{3}$<br>Multiply both sides by $3(x+4)(x-4)$:<br>$7x(x-4) - 2(3) = 7(x+4)(x-4)$<br>$7x^2 - 28x - 6 = 7x^2 - 112$<br>$-28x = -106$<br>$x = \frac{-106}{-28} = \frac{53}{14}$<br>The solution set is $\left\{\frac{53}{14}\right\}$.<br>**Sample Problem 11**<br>Solve $\frac{6}{5x+10} - \frac{1}{x-5} = \frac{4}{x^2-3x-10}$. |
| Solve formulas that involve fractional forms.<br>(Section 4.7/Objective 2) | The techniques that are used for solving rational equations can also be used to change the form of formulas. | Solve $\frac{x}{2a} - \frac{y}{2b} = 1$ for $y$.<br>**Solution**<br>$\frac{x}{2a} - \frac{y}{2b} = 1$<br>Multiply both sides by $2ab$:<br>$2ab\left(\frac{x}{2a} - \frac{y}{2b}\right) = 2ab(1)$<br>$bx - ay = 2ab$<br>$-ay = 2ab - bx$<br>$y = \frac{2ab - bx}{-a}$<br>$y = \frac{-2ab + bx}{a}$<br>**Sample Problem 12**<br>Solve $\frac{a}{3x} - \frac{b}{2y} = 4$ for $a$. |

| OBJECTIVE | SUMMARY | EXAMPLE |
|---|---|---|
| Solve word problems involving ratios. (Section 4.6/Objective 3) | Many real-world situations can be solved by using ratios and setting up a proportion to be solved. | At a law firm, the ratio of female attorneys to male attorneys is 1 to 4. If the firm has a total of 125 attorneys, find the number of female attorneys.<br>**Solution**<br>Let $x$ represent the number of female attorneys. Then $125 - x$ represents the numbers of male attorneys. The following proportion can be set up.<br>$\frac{x}{125 - x} = \frac{1}{4}$<br>Solve by cross-multiplication:<br>$\frac{x}{125 - x} = \frac{1}{4}$<br>$4x = 1(125 - x)$<br>$4x = 125 - x$<br>$5x = 125$<br>$x = 25$<br>There are 25 female attorneys.<br>**Sample Problem 13**<br>The ratio of male to female firefighters in a certain town is 7 to 1. If the town has 48 firefighters, find the number of female firefighters. |
| Solve rate-time word problems. (Section 4.7/Objective 3) | Uniform-motion problems are a special case of rate-time problems. In general, if $Q$ is the quantity of some job done in $t$ time units, then the rate, $r$, is given by $r = \frac{Q}{t}$. | At a veterinarian clinic, it takes Laurie twice as long to feed the animals as it does Janet. How long would it take each person to feed the animals by herself if they can feed the animals together in 60 minutes?<br>**Solution**<br>Let $t$ represent the time it takes Janet to feed the animals. Then $2t$ represents the time it would take Laurie to feed the animals. Laurie's rate plus Janet's rate equals the rate working together.<br>$\frac{1}{2t} + \frac{1}{t} = \frac{1}{60}$<br>Multiply both sides by $60t$:<br>$60t\left(\frac{1}{2t} + \frac{1}{t}\right) = 60t\left(\frac{1}{60}\right)$<br>$30 + 60 = t$<br>$90 = t$<br>It would take Janet 90 minutes working alone to feed the animals, and it would take Laurie 180 minutes working alone to feed the animals.<br>**Sample Problem 14**<br>One painter can paint a room in four hours. Another painter can paint the same room in six hours. How long would it take the two painters if they worked together? |

## Chapter 4 Review Problem Set

For Problems 1–6, simplify each rational expression.

1. $\dfrac{26x^2y^3}{39x^4y^2}$

2. $\dfrac{a^2 - 9}{a^2 + 3a}$

3. $\dfrac{n^2 - 3n - 10}{n^2 + n - 2}$

4. $\dfrac{x^4 - 1}{x^3 - x}$

5. $\dfrac{8x^3 - 2x^2 - 3x}{12x^2 - 9x}$

6. $\dfrac{x^4 - 7x^2 - 30}{2x^4 + 7x^2 + 3}$

For Problems 7–10, simplify each complex fraction.

7. $\dfrac{\frac{5}{8} - \frac{1}{2}}{\frac{1}{6} + \frac{3}{4}}$

8. $\dfrac{\frac{3}{2x} + \frac{5}{3y}}{\frac{4}{x} - \frac{3}{4y}}$

9. $\dfrac{\frac{3}{x-2} - \frac{4}{x^2-4}}{\frac{2}{x+2} + \frac{1}{x-2}}$

10. $1 - \dfrac{1}{2 - \frac{1}{x}}$

For Problems 11–24, perform the indicated operations, and express your answers in simplest form.

11. $\dfrac{6xy^2}{7y^3} \div \dfrac{15x^2y}{5x^2}$

12. $\dfrac{9ab}{3a + 6} \cdot \dfrac{a^2 - 4a - 12}{a^2 - 6a}$

13. $\dfrac{n^2 + 10n + 25}{n^2 - n} \cdot \dfrac{5n^3 - 3n^2}{5n^2 + 22n - 15}$

14. $\dfrac{x^2 - 2xy - 3y^2}{x^2 + 9y^2} \div \dfrac{2x^2 + xy - y^2}{2x^2 - xy}$

15. $\dfrac{2x + 1}{5} + \dfrac{3x - 2}{4}$

16. $\dfrac{3}{2n} + \dfrac{5}{3n} - \dfrac{1}{9}$

17. $\dfrac{3x}{x + 7} - \dfrac{2}{x}$

18. $\dfrac{10}{x^2 - 5x} + \dfrac{2}{x}$

19. $\dfrac{3}{n^2 - 5n - 36} + \dfrac{2}{n^2 + 3n - 4}$

20. $\dfrac{3}{2y + 3} + \dfrac{5y - 2}{2y^2 - 9y - 18} - \dfrac{1}{y - 6}$

21. $\dfrac{2x^2y}{3x} \cdot \dfrac{xy^2}{6} \div \dfrac{x}{9y}$

22. $\dfrac{10x^4y^3}{8x^2y} \div \dfrac{5}{xy^2} \cdot \dfrac{3y}{x}$

23. $\dfrac{8x}{2x - 6} \div \dfrac{2x - 1}{x^2 - 9} \cdot \dfrac{2x^2 + x - 1}{x^2 + 7x + 12}$

24. $\dfrac{2 - x}{6} \cdot \dfrac{x + 1}{x^2 - 4} \div \dfrac{x^2 + 2x + 1}{10}$

For Problems 25–26, perform the long division.

25. $(18x^2 + 9x - 2) \div (3x + 2)$

26. $(3x^3 + 5x^2 - 6x - 2) \div (x + 4)$

For Problems 27–30, divide using synthetic division.

27. Divide $(3x^4 - 14x^3 + 7x^2 + 6x - 8)$ by $(x - 4)$.

28. Divide $(2x^4 + x^2 - x + 3)$ by $(x + 1)$.

29. Divide $(5x^3 + 14x^2 + 9x - 6) \div (x + 2)$.

30. Divide $(2x^3 - 7x^2 + 2x + 3) \div (x - 3)$.

For Problems 31–42, solve each equation.

31. $\dfrac{4x + 5}{3} + \dfrac{2x - 1}{5} = 2$

32. $\dfrac{3}{4x} + \dfrac{4}{5} = \dfrac{9}{10x}$

33. $\dfrac{a}{a - 2} - \dfrac{3}{2} = \dfrac{2}{a - 2}$

34. $\dfrac{4}{5y - 3} = \dfrac{2}{3y + 7}$

35. $n + \dfrac{1}{n} = \dfrac{53}{14}$

36. $\dfrac{1}{2x - 7} + \dfrac{x - 5}{4x^2 - 49} = \dfrac{4}{6x - 21}$

37. $\dfrac{x}{2x + 1} - 1 = \dfrac{-4}{7(x - 2)}$

38. $\dfrac{2x}{-5} = \dfrac{3}{4x - 13}$

39. $\dfrac{2n}{2n^2 + 11n - 21} - \dfrac{n}{n^2 + 5n - 14} = \dfrac{3}{n^2 + 5n - 14}$

40. $\dfrac{2}{t^2 - t - 6} + \dfrac{t + 1}{t^2 + t - 12} = \dfrac{t}{t^2 + 6t + 8}$

41. Solve $\dfrac{y - 6}{x + 1} = \dfrac{3}{4}$ for $y$.

42. Solve $\dfrac{x}{a} - \dfrac{y}{b} = 1$ for $y$.

For Problems 43–54, set up an equation, and solve the problem.

43. A sum of \$1400 is to be divided between two people in the ratio of $\frac{3}{5}$. How much does each person receive?

44. At a restaurant the tips are split between the busboy and the waiter in the ratio of 2 to 7. Find the amount each received in tips if there was a total of \$162 in tips.

45. In a physician's practice the ratio of patients on Medicare insurance to patients insured by other means is 5 to 8. If there is a total of 3120 patients in the physician's practice, find the number of patients on Medicare insurance.

**46.** On a 40-question final exam in history the problems are split between true/false questions and multiple-choice problems in the ratio of 3 to 5. Find the number of each type of problem.

**47.** Working together, Dan and Julio can mow a lawn in 12 minutes. Julio can mow the lawn by himself in 10 minutes less time than it takes Dan by himself. How long does it take each of them to mow the lawn alone?

**48.** It takes Meredith 7 hours to clean an office building, whereas Otis can clean the same office building in 5 hours. How long would it take them to clean the office building working together?

**49.** Maria can assembly 200 tamales in half the time that it takes George to assemble the same number of tamales. How long does it take each person to make the 200 tamales, if they can assemble them in 60 minutes working together?

**50.** Mark can overhaul an engine in 20 hours, and Phil can do the same job by himself in 30 hours. If they both work together for a time and then Mark finishes the job by himself in 5 hours, how long did they work together?

**51.** In his speedboat Joey can travel the 12 miles between the Boat Tree Marina and the Swamp House Restaurant in half the time it takes him to travel the same distance in his jon boat. The rate of the speedboat is 8 miles per hour faster than the jon boat. Find the rate of both boats.

**52.** A mallard duck can migrate 500 miles in the same time a heron can migrate 150 miles. If the speed of the mallard duck is 35 miles per hour faster than the speed of the heron, find the speed for both birds.

**53.** Suppose that car A can travel 250 miles in 3 hours less time than it takes car B to travel 440 miles. The rate of car B is 5 miles per hour faster than that of car A. Find the rates of both cars.

**54.** Nasser rode his bicycle 66 miles in $4\frac{1}{2}$ hours. For the first 40 miles he averaged a certain rate, and then for the last 26 miles he reduced his rate by 3 miles per hour. Find his rate for the last 26 miles.

# Chapter 4 Test

For Problems 1–4, simplify each rational expression.

1. $\dfrac{39x^2y^3}{72x^3y}$

2. $\dfrac{3x^2 + 17x - 6}{x^3 - 36x}$

3. $\dfrac{6n^2 - 5n - 6}{3n^2 + 14n + 8}$

4. $\dfrac{2x - 2x^2}{x^2 - 1}$

For Problems 5–13, perform the indicated operations, and express your answers in simplest form.

5. $\dfrac{5x^2y}{8x} \cdot \dfrac{12y^2}{20xy}$

6. $\dfrac{5a + 5b}{20a + 10b} \cdot \dfrac{a^2 - ab}{2a^2 + 2ab}$

7. $\dfrac{3x^2 + 10x - 8}{5x^2 + 19x - 4} \div \dfrac{3x^2 - 23x + 14}{x^2 - 3x - 28}$

8. $\dfrac{3x - 1}{4} + \dfrac{2x + 5}{6}$

9. $\dfrac{5x - 6}{3} - \dfrac{x - 12}{6}$

10. $\dfrac{3}{5n} + \dfrac{2}{3} - \dfrac{7}{3n}$

11. $\dfrac{3x}{x - 6} + \dfrac{2}{x}$

12. $\dfrac{9}{x^2 - x} - \dfrac{2}{x}$

13. $\dfrac{3}{2n^2 + n - 10} + \dfrac{5}{n^2 + 5n - 14}$

14. Divide $3x^3 + 10x^2 - 9x - 4$ by $x + 4$.

15. Simplify the complex fraction $\dfrac{\dfrac{3}{2x} - \dfrac{1}{6}}{\dfrac{2}{3x} + \dfrac{3}{4}}$.

16. Solve $\dfrac{x + 2}{y - 4} = \dfrac{3}{4}$ for $y$.

For Problems 17–22, solve each equation.

17. $\dfrac{x - 1}{2} - \dfrac{x + 2}{5} = -\dfrac{3}{5}$

18. $\dfrac{5}{4x} + \dfrac{3}{2} = \dfrac{7}{5x}$

19. $\dfrac{-3}{4n - 1} = \dfrac{-2}{3n + 11}$

20. $n - \dfrac{5}{n} = 4$

21. $\dfrac{6}{x - 4} - \dfrac{4}{x + 3} = \dfrac{8}{x - 4}$

22. $\dfrac{1}{3x - 1} + \dfrac{x - 2}{9x^2 - 1} = \dfrac{7}{6x - 2}$

For Problems 23–25, set up an equation and then solve the problem.

23. The denominator of a rational number is 9 less than three times the numerator. The number in simplest form is $\dfrac{3}{8}$. Find the number.

24. It takes Jodi three times as long to deliver papers as it does Jannie. Together they can deliver the papers in 15 minutes. How long would it take Jodi by herself?

25. René can ride her bike 60 miles in 1 hour less time than it takes Sue to ride 60 miles. René's rate is 3 miles per hour faster than Sue's rate. Find René's rate.

# Chapters 1–4 Cumulative Review Problem Set

1. Simplify the numerical expression $16 \div 4(2) + 8$.

2. Simplify the numerical expression $(-2)^2 + (-2)^3 - 3^2$

3. Evaluate $-2xy + 5y^2$ for $x = -3$ and $y = 4$.

4. Evaluate $3(n - 2) + 4(n - 4) - 8(n - 3)$ for $n = -\frac{1}{2}$.

For Problems 5–14, perform the indicated operations and then simplify.

5. $(6a^2 + 3a - 4) + (8a + 6) + (a^2 - 1)$

6. $(x^2 + 5x + 2) - (3x^2 - 4x + 6)$

7. $(2x^2y)(-xy^4)$

8. $(4xy^3)^2$

9. $(-3a^3)^2(4ab^2)$

10. $(4a^2b)(-3a^3b^2)(2ab)$

11. $-3x^2(6x^2 - x + 4)$

12. $(5x + 3y)(2x - y)$

13. $(x + 4y)^2$

14. $(a + 3b)(a^2 - 4ab + b^2)$

For Problems 15–20, factor each polynomial completely.

15. $x^2 - 5x + 6$

16. $6x^2 - 5x - 4$

17. $2x^2 - 8x + 6$

18. $3x^2 + 18x - 48$

19. $9m^2 - 16n^2$

20. $27a^3 + 8$

21. Simplify $\frac{-28x^2y^5}{4x^4y}$.

22. Simplify $\frac{4x - x^2}{x - 4}$.

For Problems 23–28, perform the indicated operations and express the answer in simplest form.

23. $\frac{6xy}{2x + 4} \cdot \frac{x^2 - 3x - 10}{3xy - 3y}$

24. $\frac{x^2 - 3x - 4}{x^2 - 1} \div \frac{x^2 - x - 12}{x^2 + 6x - 7}$

25. $\frac{7n - 3}{5} - \frac{n + 4}{2}$

26. $\frac{3}{x^2 + x - 6} + \frac{5}{x^2 - 9}$

27. $\frac{\frac{2}{x} + \frac{3}{y}}{6}$

28. $\frac{\frac{1}{n^2} - \frac{1}{m^2}}{\frac{1}{m} + \frac{1}{n}}$

29. Divide $(6x^3 + 7x^2 + 5x + 12)$ by $(2x + 3)$.

30. Use synthetic division to divide $(2x^3 - 3x^2 - 23x + 14)$ by $(x - 4)$.

For Problems 31–40, solve the equation.

31. $8n - 3(n + 2) = 2n + 12$

32. $0.2(y - 6) = 0.02y + 3.12$

33. $\frac{x + 1}{4} + \frac{3x + 2}{2} = 5$

34. $\frac{5}{8}(x + 2) - \frac{1}{2}x = 2$

35. $|3x - 2| = 8$

36. $|x + 8| - 4 = 16$

37. $x^2 + 7x - 8 = 0$

38. $2x^2 + 13x + 15 = 0$

39. $n - \frac{3}{n} = \frac{26}{3}$

40. $\frac{3}{n - 7} + \frac{4}{n + 2} = \frac{27}{n^2 - 5n - 14}$

41. Solve the formula $A = P + Prt$ for $P$.

42. Solve the formula $V = \frac{1}{3}BH$ for $B$.

For Problems 43–56, solve the inequality and express the solution in interval notation.

43. $-3x + 2(x - 4) \geq -10$

44. $-10 < 3x + 2 < 8$

45. $|4x + 3| < 15$

46. $|2x + 6| \geq 20$

47. $|x + 4| - 6 > 0$

**48.** The owner of a local café wants to make a profit of 80% of the cost for each Caesar salad sold. If it costs \$5.00 to make a Caesar salad, at what price should each salad be sold?

**49.** Find the discount sale price of a \$920 television that is on sale for 25% off.

**50.** Suppose that the length of a rectangle is 8 inches less than twice the width. The perimeter of the rectangle is 122 inches. Find the length and width of the rectangle.

**51.** Two planes leave Kansas City at the same time and fly in opposite directions. If one travels at 450 miles per hour and the other travels at 400 miles per hour, how long will it take for them to be 3400 miles apart?

**52.** A sum of \$68,000 is to be divided between two partners in the ratio of $\frac{1}{4}$. How much does each person receive?

**53.** Victor can rake the lawn in 20 minutes, and his sister Lucia can rake the same lawn in 30 minutes. How long will it take them to rake the lawn if they work together?

**54.** One leg of a right triangle is 7 inches less than the other leg. The hypotenuse is 1 inch longer than the longer of the two legs. Find the length of the three sides of the right triangle.

**55.** How long will it take \$1500 to double itself at 6% simple interest?

**56.** A collection of 40 coins consisting of dimes and quarters has a value of \$5.95. Find the number of each kind of coin.

# 5 Exponents and Radicals

## Study Skill Tip

Homework is an important element of your math course. Here are some suggestions to help improve your homework grade.

If at all possible, never skip the homework. Skipping the homework between classes has the effect that you will not be prepared for the next class. You won't be prepared to ask about that group of problems, and quite possibly the instructor won't take questions on that group of problems again. Also, due to the sequential nature of math, you probably will need to understand that material to learn the new material presented in class. So skipping the homework puts you behind.

While doing your homework, break up your study periods. A good length of time for doing math homework is about 45 minutes to an hour. Don't spend too much time on one problem. If you are unsuccessful in solving the problem after 5 minutes, go on to the next problem. Be prepared to ask about that problem in the next class session.

When checking your answers in the back of the book, keep in mind that there are three possible explanations for wrong answers. One possibility is that you made a mistake in working the problem. Another possibility is that the answer in the back of the book is incorrect. It could be that both your version of the answer and the answer in the back of the book are equivalent forms and hence both are correct answers. If you think that is the case, be sure to ask the instructor.

***"One of life's most painful moments comes when we must admit that we didn't do our homework, that we are not prepared."***
MERLIN OLSEN, NFL PLAYER AND ACTOR

***What is your plan for being sure you have time to do your math homework?***

## Chapter Preview

This chapter begins with the use of negative exponents. The same laws of exponents you have been using for positive integers also hold for all integers. The inverse operation of raising a number to an exponent is finding the root of a number. Finding roots, simplifying radicals, and operations with radicals are covered in this chapter. Because many problems involve simplifying radicals, the suggestion is made in the chapter to make a reference list of the perfect squares, cubes, etc. Scientific notation, which you may have already seen in your science courses, is covered in the last section of the chapter.

# 5.1 Using Integers as Exponents

OBJECTIVES

1. Simplify numerical expressions that have integer exponents
2. Simplify algebraic expressions that have integer exponents
3. Multiply and divide algebraic expressions that have integer exponents
4. Simplify sums and differences of expressions involving integer exponents

Thus far in the text we have used only positive integers as exponents. In Chapter 1 the expression $b^n$, where $b$ is any real number and $n$ is a positive integer, was defined by

$$b^n = b \cdot b \cdot b \cdot \cdots \cdot b \qquad n \text{ factors of } b$$

Then, in Chapter 3, more properties of exponents served as a basis for manipulation with polynomials. We are now ready to extend the concept of an exponent to include the use of zero and the negative integers as exponents.

First, let's consider the use of zero as an exponent. We want to use zero in such a way that the previously listed properties continue to hold. If $b^n \cdot b^m = b^{n+m}$ is to hold, then $x^4 \cdot x^0 = x^{4+0} = x^4$. In other words, $x^0$ *acts like* 1 because $x^4 \cdot x^0 = x^4$. This line of reasoning suggests the following definition.

**Definition 5.1 Exponent of Zero**

If $b$ is a nonzero real number, then

$$b^0 = 1$$

According to Definition 5.1, the following statements are all true.

$$5^0 = 1 \qquad (-413)^0 = 1 \qquad (x^3y^4)^0 = 1, \quad x \neq 0, y \neq 0$$

$$\left(\frac{3}{11}\right)^0 = 1 \qquad n^0 = 1, \quad n \neq 0$$

We can use a similar line of reasoning to motivate a definition for the use of negative integers as exponents. Consider the example $x^4 \cdot x^{-4}$. If $b^n \cdot b^m = b^{n+m}$ is to hold, then $x^4 \cdot x^{-4} = x^{4+(-4)} = x^0 = 1$. Thus $x^{-4}$ must be the reciprocal of $x^4$, because their product is 1. That is,

$$x^{-4} = \frac{1}{x^4}$$

This suggests the following general definition.

**Definition 5.2 Negative Exponent**

If $n$ is a positive integer, and $b$ is a nonzero real number, then

$$b^{-n} = \frac{1}{b^n}$$

According to Definition 5.2, the following statements are all true.

$$x^{-5} = \frac{1}{x^5} \qquad 2^{-4} = \frac{1}{2^4} = \frac{1}{16} \qquad \left(\frac{3}{4}\right)^{-2} = \frac{1}{\left(\frac{3}{4}\right)^2} = \frac{1}{\frac{9}{16}} = \frac{16}{9}$$

$$10^{-2} = \frac{1}{10^2} = \frac{1}{100} \qquad \frac{2}{x^{-3}} = \frac{2}{\frac{1}{x^3}} = (2)\left(\frac{x^3}{1}\right) = 2x^3$$

It can be verified (although it is beyond the scope of this text) that all of the parts of Property 5.1 hold for *all integers*. We will now apply the following equality for finding the quotient of like bases raised to integer powers.

$$\frac{b^n}{b^m} = b^{n-m} \quad \text{for all integers } n \text{ and } m$$

Property 5.1 states the rules that hold for all integer exponents and includes a name tag for easy reference.

**Property 5.1 Properties for Integer Exponents**

If $m$ and $n$ are integers, and $a$ and $b$ are real numbers (and $b \neq 0$ whenever it appears in a denominator), then

1. $b^n \cdot b^m = b^{n+m}$ Product of two powers
2. $(b^n)^m = b^{mn}$ Power of a power
3. $(ab)^n = a^n b^n$ Power of a product
4. $\left(\frac{a}{b}\right)^n = \frac{a^n}{b^n}$ Power of a quotient
5. $\frac{b^n}{b^m} = b^{n-m}$ Quotient of two powers

Having the use of all integers as exponents enables us to work with a large variety of numerical and algebraic expressions. Let's consider some examples that illustrate the use of the various parts of Property 5.1.

**Classroom Example**
Simplify each of the following numerical expressions:
(a) $10^{-5} \cdot 10^2$ (b) $(3^{-2})^{-2}$
(c) $(3^{-1} \cdot 5^2)^{-1}$ (d) $\left(\frac{3^{-3}}{5^{-2}}\right)^{-1}$
(e) $\frac{10^{-6}}{10^{-9}}$

**EXAMPLE 1** Simplify each of the following numerical expressions:

**(a)** $10^{-3} \cdot 10^2$ **(b)** $(2^{-3})^{-2}$ **(c)** $(2^{-1} \cdot 3^2)^{-1}$

**(d)** $\left(\frac{2^{-3}}{3^{-2}}\right)^{-1}$ **(e)** $\frac{10^{-2}}{10^{-4}}$

## Solution

**(a)** $10^{-3} \cdot 10^{2} = 10^{-3+2}$ Product of two powers

$= 10^{-1}$

$= \frac{1}{10^{1}} = \frac{1}{10}$

**(b)** $(2^{-3})^{-2} = 2^{(-2)(-3)}$ Power of a power

$= 2^{6} = 64$

**(c)** $(2^{-1} \cdot 3^{2})^{-1} = (2^{-1})^{-1}(3^{2})^{-1}$ Power of a product

$= 2^{1} \cdot 3^{-2}$

$= 2 \cdot \frac{1}{3^{2}} = 2 \cdot \frac{1}{9} = \frac{2}{9}$

**(d)** $\left(\frac{2^{-3}}{3^{-2}}\right)^{-1} = \frac{(2^{-3})^{-1}}{(3^{-2})^{-1}}$ Power of a quotient

$= \frac{2^{3}}{3^{2}} = \frac{8}{9}$

**(e)** $\frac{10^{-2}}{10^{-4}} = 10^{-2-(-4)}$ Quotient of two powers

$= 10^{2} = 100$

**Classroom Example**
Simplify each of the following; express final results without using zero or negative integers as exponents:
**(a)** $x^{2} \cdot x^{-7}$
**(b)** $(x^{3})^{-2}$
**(c)** $(m^{4} \cdot n^{-2})^{-3}$
**(d)** $\left(\frac{x^{2}}{y^{-4}}\right)^{-3}$
**(e)** $\frac{x^{-5}}{x^{-1}}$

## EXAMPLE 2

Simplify each of the following; express final results without using zero or negative integers as exponents:

**(a)** $x^{2} \cdot x^{-5}$ **(b)** $(x^{-2})^{4}$ **(c)** $(x^{2}y^{-3})^{-4}$

**(d)** $\left(\frac{a^{3}}{b^{-5}}\right)^{-2}$ **(e)** $\frac{x^{-4}}{x^{-2}}$

## Solution

**(a)** $x^{2} \cdot x^{-5} = x^{2+(-5)}$ Product of two powers

$= x^{-3}$

$= \frac{1}{x^{3}}$

**(b)** $(x^{-2})^{4} = x^{4(-2)}$ Power of a power

$= x^{-8}$

$= \frac{1}{x^{8}}$

**(c)** $(x^{2}y^{-3})^{-4} = (x^{2})^{-4}(y^{-3})^{-4}$ Power of a product

$= x^{-4(2)}y^{-4(-3)}$

$= x^{-8}y^{12}$

$= \frac{y^{12}}{x^{8}}$

**(d)** $\left(\frac{a^{3}}{b^{-5}}\right)^{-2} = \frac{(a^{3})^{-2}}{(b^{-5})^{-2}}$ Power of a quotient

$= \frac{a^{-6}}{b^{10}}$

$$= \frac{1}{a^6b^{10}}$$

**(e)** $\dfrac{x^{-4}}{x^{-2}} = x^{-4-(-2)}$ Quotient of two powers

$$= x^{-2}$$

$$= \frac{1}{x^2}$$

**Classroom Example**
Find the indicated products and quotients; express your results using positive integral exponents only:

(a) $(5x^3y^{-2})(2x^{-5}y^3)$

(b) $\dfrac{14a^2b^5}{-2a^{-1}b^6}$

(c) $\left(\dfrac{24x^{-5}y^{-3}}{8x^2y^{-9}}\right)^{-1}$

## EXAMPLE 3

Find the indicated products and quotients; express your results using positive integral exponents only:

**(a)** $(3x^2y^{-4})(4x^{-3}y)$ **(b)** $\dfrac{12a^3b^2}{-3a^{-1}b^5}$ **(c)** $\left(\dfrac{15x^{-1}y^2}{5xy^{-4}}\right)^{-1}$

### Solution

**(a)** $(3x^2y^{-4})(4x^{-3}y) = 12x^{2+(-3)}y^{-4+1}$

$$= 12x^{-1}y^{-3}$$

$$= \frac{12}{xy^3}$$

**(b)** $\dfrac{12a^3b^2}{-3a^{-1}b^5} = -4a^{3-(-1)}b^{2-5}$

$$= -4a^4b^{-3}$$

$$= -\frac{4a^4}{b^3}$$

**(c)** $\left(\dfrac{15x^{-1}y^2}{5xy^{-4}}\right)^{-1} = (3x^{-1-1}y^{2-(-4)})^{-1}$ Note that we are first simplifying inside the parentheses

$$= (3x^{-2}y^6)^{-1}$$

$$= 3^{-1}x^2y^{-6}$$

$$= \frac{x^2}{3y^6}$$

The final examples of this section show the simplification of numerical and algebraic expressions that involve sums and differences. In such cases, we use Definition 5.2 to change from negative to positive exponents so that we can proceed in the usual way.

**Classroom Example**
Simplify $3^{-2} + 5^{-1}$.

## EXAMPLE 4

Simplify $2^{-3} + 3^{-1}$.

### Solution

$$2^{-3} + 3^{-1} = \frac{1}{2^3} + \frac{1}{3^1}$$

$$= \frac{1}{8} + \frac{1}{3}$$

$$= \frac{3}{24} + \frac{8}{24}$$ Use 24 as the LCD

$$= \frac{11}{24}$$

**Classroom Example**
Express $a^{-2} + b^{-1}$ as a single fraction involving positive exponents only.

### EXAMPLE 5

Express $a^{-1} + b^{-2}$ as a single fraction involving positive exponents only.

**Solution**

$$a^{-1} + b^{-2} = \frac{1}{a^1} + \frac{1}{b^2}$$ Use $ab^2$ as the common denominator

$$= \left(\frac{1}{a}\right)\left(\frac{b^2}{b^2}\right) + \left(\frac{1}{b^2}\right)\left(\frac{a}{a}\right)$$ Change to equivalent fractions with $ab^2$ as the common denominator

$$= \frac{b^2}{ab^2} + \frac{a}{ab^2}$$

$$= \frac{b^2 + a}{ab^2}$$

## Concept Quiz 5.1

For Problems 1–10, answer true or false.

1. $\left(\frac{2}{5}\right)^{-2} = \left(\frac{5}{2}\right)^2$
2. $(3)^0(3)^2 = 9^2$
3. $(2)^{-4}(2)^4 = 2$
4. $(4^{-2})^{-1} = 16$
5. $(2^{-2} \cdot 2^{-3})^{-1} = \frac{1}{16}$
6. $\left(\frac{3^{-2}}{3^{-1}}\right)^2 = \frac{1}{9}$
7. $\frac{1}{\left(\frac{2}{3}\right)^{-3}} = \frac{8}{27}$
8. $(10^4)(10^{-6}) = \frac{1}{100}$
9. $\frac{x^{-6}}{x^{-3}} = x^2$
10. $x^{-1} - x^{-2} = \frac{x - 1}{x^2}$

## Problem Set 5.1

For Problems 1–42, simplify each numerical expression. (Objective 1)

1. $3^{-3}$
2. $2^{-4}$
3. $-10^{-2}$
4. $10^{-3}$
5. $\frac{1}{3^{-4}}$
6. $\frac{1}{2^{-6}}$
7. $-\left(\frac{1}{3}\right)^{-3}$
8. $\left(\frac{1}{2}\right)^{-3}$
9. $\left(-\frac{1}{2}\right)^{-3}$
10. $\left(\frac{2}{7}\right)^{-2}$
11. $\left(-\frac{3}{4}\right)^0$
12. $\frac{1}{\left(\frac{4}{5}\right)^{-2}}$

13. $\dfrac{1}{\left(\dfrac{3}{7}\right)^{-2}}$

14. $-\left(\dfrac{5}{6}\right)^{0}$

15. $2^7 \cdot 2^{-3}$

16. $3^{-4} \cdot 3^6$

17. $10^{-5} \cdot 10^2$

18. $10^4 \cdot 10^{-6}$

19. $10^{-1} \cdot 10^{-2}$

20. $10^{-2} \cdot 10^{-2}$

21. $(3^{-1})^{-3}$

22. $(2^{-2})^{-4}$

23. $(5^3)^{-1}$

24. $(3^{-1})^3$

25. $(2^3 \cdot 3^{-2})^{-1}$

26. $(2^{-2} \cdot 3^{-1})^{-3}$

27. $(4^2 \cdot 5^{-1})^2$

28. $(2^{-3} \cdot 4^{-1})^{-1}$

29. $\left(\dfrac{2^{-1}}{5^{-2}}\right)^{-1}$

30. $\left(\dfrac{2^{-4}}{3^{-2}}\right)^{-2}$

31. $\left(\dfrac{2^{-1}}{3^{-2}}\right)^{2}$

32. $\left(\dfrac{3^2}{5^{-1}}\right)^{-1}$

33. $\dfrac{3^3}{3^{-1}}$

34. $\dfrac{2^{-2}}{2^3}$

35. $\dfrac{10^{-2}}{10^2}$

36. $\dfrac{10^{-2}}{10^{-5}}$

37. $2^{-2} + 3^{-2}$

38. $2^{-4} + 5^{-1}$

39. $\left(\dfrac{1}{3}\right)^{-1} - \left(\dfrac{2}{5}\right)^{-1}$

40. $\left(\dfrac{3}{2}\right)^{-1} - \left(\dfrac{1}{4}\right)^{-1}$

41. $(2^{-3} + 3^{-2})^{-1}$

42. $(5^{-1} - 2^{-3})^{-1}$

For Problems 43–62, simplify each expression. Express final results without using zero or negative integers as exponents. (Objective 2)

43. $x^2 \cdot x^{-8}$

44. $x^{-3} \cdot x^{-4}$

45. $a^3 \cdot a^{-5} \cdot a^{-1}$

46. $b^{-2} \cdot b^3 \cdot b^{-6}$

47. $(a^{-4})^2$

48. $(b^4)^{-3}$

49. $(x^2y^{-6})^{-1}$

50. $(x^5y^{-1})^{-3}$

51. $(ab^3c^{-2})^{-4}$

52. $(a^3b^{-3}c^{-2})^{-5}$

53. $(2x^3y^{-4})^{-3}$

54. $(4x^5y^{-2})^{-2}$

55. $\left(\dfrac{x^{-1}}{y^{-4}}\right)^{-3}$

56. $\left(\dfrac{y^3}{x^{-4}}\right)^{-2}$

57. $\left(\dfrac{3a^{-2}}{2b^{-1}}\right)^{-2}$

58. $\left(\dfrac{2xy^2}{5a^{-1}b^{-2}}\right)^{-1}$

59. $\dfrac{x^{-6}}{x^{-4}}$

60. $\dfrac{a^{-2}}{a^2}$

61. $\dfrac{a^3b^{-2}}{a^{-2}b^{-4}}$

62. $\dfrac{x^{-3}y^{-4}}{x^2y^{-1}}$

For Problems 63–74, find the indicated products and quotients. Express final results using positive integral exponents only. (Objective 3)

63. $(2xy^{-1})(3x^{-2}y^4)$

64. $(-4x^{-1}y^2)(6x^3y^{-4})$

65. $(-7a^2b^{-5})(-a^{-2}b^7)$

66. $(-9a^{-3}b^{-6})(-12a^{-1}b^4)$

67. $\dfrac{28x^{-2}y^{-3}}{4x^{-3}y^{-1}}$

68. $\dfrac{63x^2y^{-4}}{7xy^{-4}}$

69. $\dfrac{-72a^2b^{-4}}{6a^3b^{-7}}$

70. $\dfrac{108a^{-5}b^{-4}}{9a^{-2}b}$

71. $\left(\dfrac{35x^{-1}y^{-2}}{7x^4y^3}\right)^{-1}$

72. $\left(\dfrac{-48ab^2}{-6a^3b^5}\right)^{-2}$

73. $\left(\dfrac{-36a^{-1}b^{-6}}{4a^{-1}b^4}\right)^{-2}$

74. $\left(\dfrac{8xy^3}{-4x^4y}\right)^{-3}$

For Problems 75–84, express each of the following as a single fraction involving positive exponents only. (Objective 4)

75. $x^{-2} + x^{-3}$

76. $x^{-1} + x^{-5}$

77. $x^{-3} - y^{-1}$

78. $2x^{-1} - 3y^{-2}$

79. $3a^{-2} + 4b^{-1}$

80. $a^{-1} + a^{-1}b^{-3}$

81. $x^{-1}y^{-2} - xy^{-1}$

82. $x^2y^{-2} - x^{-1}y^{-3}$

83. $2x^{-1} - 3x^{-2}$

84. $5x^{-2}y + 6x^{-1}y^{-2}$

## Thoughts Into Words

85. Is the following simplification process correct?

$$(3^{-2})^{-1} = \left(\frac{1}{3^2}\right)^{-1} = \left(\frac{1}{9}\right)^{-1} = \frac{1}{\left(\frac{1}{9}\right)^1} = 9$$

Could you suggest a better way to do the problem?

86. Explain how to simplify $(2^{-1} \cdot 3^{-2})^{-1}$ and also how to simplify $(2^{-1} + 3^{-2})^{-1}$.

## Further Investigations

**87.** Use a calculator to check your answers for Problems 1–42.

**88.** Use a calculator to simplify each of the following numerical expressions. Express your answers to the nearest hundredth.

**(a)** $(2^{-3} + 3^{-3})^{-2}$

**(b)** $(4^{-3} - 2^{-1})^{-2}$

**(c)** $(5^{-3} - 3^{-5})^{-1}$

**(d)** $(6^{-2} + 7^{-4})^{-2}$

**(e)** $(7^{-3} - 2^{-4})^{-2}$

**(f)** $(3^{-4} + 2^{-3})^{-3}$

### Answers to the Concept Quiz

**1.** True **2.** False **3.** False **4.** True **5.** False **6.** True **7.** True **8.** True **9.** False **10.** True

# 5.2 Roots and Radicals

OBJECTIVES

1. Evaluate roots of numbers
2. Express radicals in simplest radical form
3. Rationalizing the denominator to simplify radicals
4. Applications of radicals

We will begin our discussion of roots and radicals with square roots. Then the discussion will continue on to cube roots and higher power roots such as fourth root. We already know that to square a number means to raise it to the second power—that is, to use the number as a factor twice, as shown.

$$4^2 = 4 \cdot 4 = 16 \qquad (-3)^2 = (-3)(-3) = 9$$

A **square root of a number** is one of its two equal factors. Thus 4 is a square root of 16 because $4 \cdot 4 = 16$. Likewise, $-4$ is also a square root of 16 because $(-4)(-4) = 16$. To avoid confusion between the two different square roots of a number, the nonnegative (positive or zero) root is called the **principal square root**. The definition of the principal square root is as follows.

### Definition 5.3 Principal Square Root

If $a \geq 0$ and $b \geq 0$, then $\sqrt{b} = a$ if and only if $a^2 = b$; $a$ is called the principal square root of $b$.

The symbol $\sqrt{\ }$ is called a **radical sign** and is used as noted in the definition to designate the nonnegative or principal square root. The number under the radical sign is called the **radicand**. The entire expression such as $\sqrt{16}$ is called a radical.

The following examples summarize some of the information about square roots.

$\sqrt{36} = 6$ — $\sqrt{36}$ indicates the principal square root of 36, which is 6.

$-\sqrt{36} = -6$ — $-\sqrt{36}$ indicates the negative square root of 36, which is $-6$.

$\sqrt{0} = 0$ — Zero only has one square root, which is zero.

$\sqrt{-4}$ The square root of $-4$ does not exist in the real number system. This is because there is no real number that when squared gives a negative value. Note that in Definition 5.3 the condition exists that the radicand, the number under the radical sign, must be positive or zero.

To cube a number means to raise it to the third power—that is, to use the number as a factor three times as shown.

$$2^3 = 2 \cdot 2 \cdot 2 = 8 \qquad (-4)^3 = (-4)(-4)(-4) = -64$$

A **cube root of a number** is one of its three equal factors. Thus 2 is a cube root of 8 because $2 \cdot 2 \cdot 2 = 8$. In fact, 2 is the only real number cube root of 8. Furthermore, $-4$ is a cube root of $-64$ because $(-4)(-4)(-4) = -64$. In fact, $-4$ is the only real number cube root of $-64$. The definition of the cube root of a number is as follows.

### Definition 5.4 Cube Root of a Number

$\sqrt[3]{b} = a$ if and only if $a^3 = b$.

The symbol $\sqrt[3]{\phantom{b}}$ designates the cube root of a number. The following examples summarize some of the information about cube roots.

$\sqrt[3]{125} = 5$ Five is the cube root of 125 because $5^3 = 125$. The cube root of a positive number is always a positive number.

$\sqrt[3]{-27} = -3$ Negative three is the cube root of $-27$ because $(-3)^3 = -27$. The cube root of a negative number is always a negative number.

$\sqrt[3]{0} = 0$ The cube root of zero is zero.

**Remark:** Technically, every nonzero real number has three cube roots, but only one of them is a real number. The other two roots are classified as imaginary numbers. We are restricting our work at this time to the set of real numbers.

The concept of root can be extended to fourth roots, fifth roots, sixth roots, and in general, $n$th roots. The fourth root of a number is one of its four equal factors. Thus 3 is a fourth root of 81 because $3^4 = 81$. The definition of the $n$th root of a number is as follows.

### Definition 5.5 nth Root of a Number

The $n$th root of $b$ is $a$ if and only if $a^n = b$.

The symbol $\sqrt[n]{\phantom{b}}$ designates the principal $n$th root. To complete our terminology, the $n$ in the radical $\sqrt[n]{b}$ is called the **index of the radical**. For square roots, when $n = 2$, we commonly write $\sqrt{b}$ rather than $\sqrt[2]{b}$. The following examples summarize some of the information about higher power roots.

$\sqrt[4]{16} = 2$ $\sqrt[4]{16}$ indicates the principal fourth root of 16, which is 2 because $2^4 = 16$.

$\sqrt[5]{-32} = -2$ The principal fifth root of $-32$ is $-2$ because $(-2)^5 = -32$.

$\sqrt[4]{-81}$ The fourth root of $-81$ does not exist in the real number system because there is no real number that when raised to the fourth power gives a negative value.

Because the rules are different for even-numbered roots and odd-numbered roots, let's make some generalizations about roots for each category.

For even-numbered roots (square roots, 4th roots, 6th roots, etc.):

1. Negative real numbers do not have real number even-numbered roots.

   For example, no real number root exists for $\sqrt{-25}$ because no real number, when squared, gives a negative value.
2. The even-numbered root of zero is zero.

   For example, $\sqrt[4]{0} = 0$.
3. For even-numbered roots, positive real numbers have exactly two real number roots.

   For example, $\sqrt[4]{81} = 3$ and $-\sqrt[4]{81} = -3$.
4. The symbol, $\sqrt[n]{\phantom{x}}$, indicates the principal root which is a positive number or zero.

   For example, $\sqrt{16} = 4$. Although $-4$ is a square root of 16, $\sqrt{16} \neq -4$ because the radical sign here means only the principal root, which is the positive root.

For odd-numbered roots (cube roots, 5th roots, 7th roots, etc.):

1. For odd-numbered roots, every number has exactly one real number $n$th root.
2. For odd-numbered roots, the $n$th root of a positive number is a positive number.

   For example, $\sqrt[5]{243} = 3$ because $3^5 = 243$.
3. For odd-numbered roots, the $n$th root of a negative number is a negative number.

   For example, $\sqrt[3]{-1000} = -10$ because $(-10)^3 = -1000$.

The following property is a direct consequence of Definition 5.5.

**Property 5.2 nth Root Raised to the nth Power**

For any positive integer $n$ greater than 1,

1. $\left(\sqrt[n]{b}\right)^n = b$ where $\sqrt[n]{b}$ is a real number
2. $\sqrt[n]{b^n} = b$ if $b \geq 0$
3. $\sqrt[n]{b^n} = b$ if $b < 0$ and $n$ is an odd number

We can use Property 5.2 to simplify certain rational expressions. Applying the transitive property where $\left(\sqrt[n]{b}\right)^n$ is equal to $b$ and $\sqrt[n]{b^n}$ is also equal to $b$, we can conclude that $\left(\sqrt[n]{b}\right)^n = \sqrt[n]{b^n}$. The arithmetic is usually easier to perform when we use the form $\left(\sqrt[n]{b}\right)^n$ where we take the root before applying the exponent. The following examples demonstrate the use of Property 5.2.

$\sqrt{144^2} = \left(\sqrt{144}\right)^2 = (12)^2 = 144$ Rewrite as $\left(\sqrt{144}\right)^2$, then evaluate the square root of 144. That result will then be raised to the second power.

$\sqrt[3]{(-8)^3} = \left(\sqrt[3]{-8}\right)^3 = (-2)^3 = -8$ Rewrite as $\left(\sqrt[3]{-8}\right)^3$, then evaluate the cube root of $-8$. That result will then be raised to the third power.

## Expressing a Radical in Simplest Radical Form

Let's apply the transitive property to some examples to lead into the next very useful property of radicals.

$\sqrt{36} = 6$ and $\sqrt{4} \cdot \sqrt{9} = 2 \cdot 3 = 6$ hence $\sqrt{36} = \sqrt{4} \cdot \sqrt{9}$

$\sqrt{100} = 10$ and $\sqrt{4} \cdot \sqrt{25} = 2 \cdot 5 = 10$ hence $\sqrt{100} = \sqrt{4} \cdot \sqrt{25}$

$\sqrt[3]{216} = 6$ and $\sqrt[3]{8} \cdot \sqrt[3]{27} = 2 \cdot 3 = 6$ hence $\sqrt[3]{216} = \sqrt[3]{8} \cdot \sqrt[3]{27}$

$\sqrt[3]{-216} = -6$ and $\sqrt[3]{-8} \cdot \sqrt[3]{27} = -2 \cdot 3 = -6$ hence $\sqrt[3]{-216} = \sqrt[3]{-8} \cdot \sqrt[3]{27}$

In general, we can state the following property.

**Property 5.3 Product of Two Radicals With the Same Index**

$\sqrt[n]{bc} = \sqrt[n]{b}\sqrt[n]{c}$, when $\sqrt[n]{b}$ and $\sqrt[n]{c}$ are real numbers

Property 5.3 states that the $n$th root of a product is equal to the product of the $n$th roots.

## Expressing a Radical in Simplest Radical Form

The definition of $n$th root, along with Property 5.3, provides the basis for changing radicals to simplest radical form. The concept of **simplest radical form** takes on additional meaning as we encounter more complicated expressions, but for now it simply means that the radicand is not to contain any perfect powers of the index. Let's consider some examples to clarify this idea.

The first step in each example is to express the radicand of the given radical as the product of two factors, one of which must be a perfect $n$th power other than 1.

Because one of the factors has to be a perfect $n$th power, we suggest that you keep a list of some of the perfect powers close by as a reference when simplifying radicals. Presented here is a list of some of the perfect powers. Most likely one of your factors will be a number on this list. You may want to expand the list.

| | | | |
|---|---|---|---|
| $2^2 = 4$ | $2^3 = 8$ | $2^4 = 16$ | $2^5 = 32$ |
| $3^2 = 9$ | $3^3 = 27$ | $3^4 = 81$ | $3^5 = 243$ |
| $4^2 = 16$ | $4^3 = 64$ | $4^4 = 256$ | |
| $5^2 = 25$ | $5^3 = 125$ | $5^4 = 625$ | |
| $6^2 = 36$ | $6^3 = 216$ | | |
| $7^2 = 49$ | $7^3 = 343$ | | |
| $8^2 = 64$ | | | |
| $9^2 = 81$ | | | |
| $10^2 = 100$ | | | |

**Classroom Example**
Express each of the following in simplest radical form:
(a) $\sqrt{12}$
(b) $\sqrt{32}$
(c) $\sqrt[3]{48}$
(d) $\sqrt[3]{40}$

**EXAMPLE 1** Express each of the following in simplest radical form:

**(a)** $\sqrt{8}$ **(b)** $\sqrt{45}$ **(c)** $\sqrt[3]{24}$ **(d)** $\sqrt[3]{54}$

### Solution

**(a)** $\sqrt{8} = \sqrt{4 \cdot 2} = \sqrt{4}\sqrt{2} = 2\sqrt{2}$ 4 is a perfect square

**(b)** $\sqrt{45} = \sqrt{9 \cdot 5} = \sqrt{9}\sqrt{5} = 3\sqrt{5}$ 9 is a perfect square

**(c)** $\sqrt[3]{24} = \sqrt[3]{8 \cdot 3} = \sqrt[3]{8}\sqrt[3]{3} = 2\sqrt[3]{3}$ 8 is a perfect cube

**(d)** $\sqrt[3]{54} = \sqrt[3]{27 \cdot 2} = \sqrt[3]{27}\sqrt[3]{2} = 3\sqrt[3]{2}$ 27 is a perfect cube

Observe the radicands of the final radicals. In each case, the radicand cannot have a factor that is a perfect $n$th power other than 1. We say that the final radicals $2\sqrt{2}$, $3\sqrt{5}$, $2\sqrt[3]{3}$, and $3\sqrt[3]{2}$ are in *simplest radical form.*

You may vary the steps somewhat in changing to simplest radical form, but the final result should be the same. Consider some different approaches to changing $\sqrt{72}$ to simplest form:

$\sqrt{72} = \sqrt{9}\sqrt{8} = 3\sqrt{8} = 3\sqrt{4}\sqrt{2} = 3 \cdot 2\sqrt{2} = 6\sqrt{2}$ or
$\sqrt{72} = \sqrt{4}\sqrt{18} = 2\sqrt{18} = 2\sqrt{9}\sqrt{2} = 2 \cdot 3\sqrt{2} = 6\sqrt{2}$ or
$\sqrt{72} = \sqrt{36}\sqrt{2} = 6\sqrt{2}$

Another variation of the technique for changing radicals to simplest form is to prime factor the radicand and then to look for perfect $n$th powers in exponential form. The following example illustrates the use of this technique.

**Classroom Example**
Express each of the following in simplest radical form:
(a) $\sqrt{48}$
(b) $5\sqrt{72}$
(c) $\sqrt[3]{200}$

**EXAMPLE 2** Express each of the following in simplest radical form:

**(a)** $\sqrt{50}$ **(b)** $3\sqrt{80}$ **(c)** $\sqrt[3]{108}$

### Solution

**(a)** $\sqrt{50} = \sqrt{2 \cdot 5 \cdot 5} = \sqrt{5^2}\sqrt{2} = 5\sqrt{2}$

**(b)** $3\sqrt{80} = 3\sqrt{2 \cdot 2 \cdot 2 \cdot 2 \cdot 5} = 3\sqrt{2^4}\sqrt{5} = 3 \cdot 2^2\sqrt{5} = 12\sqrt{5}$

**(c)** $\sqrt[3]{108} = \sqrt[3]{2 \cdot 2 \cdot 3 \cdot 3 \cdot 3} = \sqrt[3]{3^3}\sqrt[3]{4} = 3\sqrt[3]{4}$

Another property of $n$th roots is demonstrated by the following examples.

$$\sqrt{\frac{36}{9}} = \sqrt{4} = 2 \quad \text{and} \quad \frac{\sqrt{36}}{\sqrt{9}} = \frac{6}{3} = 2$$

$$\sqrt[3]{\frac{64}{8}} = \sqrt[3]{8} = 2 \quad \text{and} \quad \frac{\sqrt[3]{64}}{\sqrt[3]{8}} = \frac{4}{2} = 2$$

$$\sqrt[3]{\frac{-8}{64}} = \sqrt[3]{-\frac{1}{8}} = -\frac{1}{2} \quad \text{and} \quad \frac{\sqrt[3]{-8}}{\sqrt[3]{64}} = \frac{-2}{4} = -\frac{1}{2}$$

In general, we can state the following property.

**Property 5.4**

$\sqrt[n]{\frac{b}{c}} = \frac{\sqrt[n]{b}}{\sqrt[n]{c}}$, when $\sqrt[n]{b}$ and $\sqrt[n]{c}$ are real numbers, and $c \neq 0$

Property 5.4 states that the $n$th root of a quotient is equal to the quotient of the $n$th roots.

To evaluate radicals such as $\sqrt{\frac{4}{25}}$ and $\sqrt[3]{\frac{27}{8}}$, for which the numerator and denominator of the fractional radicand are perfect $n$th powers, you may use Property 5.4 or merely rely on the definition of $n$th root.

$$\sqrt{\frac{4}{25}} = \frac{\sqrt{4}}{\sqrt{25}} = \frac{2}{5} \quad \text{or} \quad \sqrt{\frac{4}{25}} = \frac{2}{5} \text{ because } \frac{2}{5} \cdot \frac{2}{5} = \frac{4}{25}$$

Property 5.4 ↕ Definition of $n$th root

$$\sqrt[3]{\frac{27}{8}} = \frac{\sqrt[3]{27}}{\sqrt[3]{8}} = \frac{3}{2} \quad \text{or} \quad \sqrt[3]{\frac{27}{8}} = \frac{3}{2} \text{ because } \frac{3}{2} \cdot \frac{3}{2} \cdot \frac{3}{2} = \frac{27}{8}$$

Radicals such as $\sqrt{\frac{28}{9}}$ and $\sqrt[3]{\frac{24}{27}}$, in which only the denominators of the radicand are perfect $n$th powers, can be simplified as follows:

$$\sqrt{\frac{28}{9}} = \frac{\sqrt{28}}{\sqrt{9}} = \frac{\sqrt{28}}{3} = \frac{\sqrt{4}\sqrt{7}}{3} = \frac{2\sqrt{7}}{3}$$

$$\sqrt[3]{\frac{24}{27}} = \frac{\sqrt[3]{24}}{\sqrt[3]{27}} = \frac{\sqrt[3]{24}}{3} = \frac{\sqrt[3]{8}\sqrt[3]{3}}{3} = \frac{2\sqrt[3]{3}}{3}$$

Before we consider more examples, let's summarize some ideas that pertain to the simplifying of radicals. A radical is said to be in *simplest radical form* if the following conditions are satisfied.

1. No fraction appears with a radical sign. $\sqrt{\frac{3}{4}}$ violates this condition
2. No radical appears in the denominator. $\frac{\sqrt{2}}{\sqrt{3}}$ violates this condition
3. No radicand, when expressed in prime factored form, contains a factor raised to a power equal to or greater than the index. $\sqrt{2^3 \cdot 5}$ violates this condition

## Rationalizing the Denominator to Simplify Radicals

Now let's consider an example in which neither the numerator nor the denominator of the radicand is a perfect $n$th power.

Classroom Example
Simplify $\sqrt{\frac{5}{7}}$.

**EXAMPLE 3** Simplify $\sqrt{\frac{2}{3}}$.

### Solution

$$\sqrt{\frac{2}{3}} = \frac{\sqrt{2}}{\sqrt{3}} = \frac{\sqrt{2}}{\sqrt{3}} \cdot \frac{\sqrt{3}}{\sqrt{3}} = \frac{\sqrt{6}}{3}$$

Form of 1 (pointing to $\frac{\sqrt{3}}{\sqrt{3}}$)

We refer to the process we used to simplify the radical in Example 3 as **rationalizing the denominator**. Note that the denominator becomes a rational number. The process of rationalizing the denominator can often be accomplished in more than one way, as we will see in the next example.

Classroom Example
Simplify $\frac{\sqrt{7}}{\sqrt{12}}$.

**EXAMPLE 4** Simplify $\frac{\sqrt{5}}{\sqrt{8}}$.

### Solution A

$$\frac{\sqrt{5}}{\sqrt{8}} = \frac{\sqrt{5}}{\sqrt{8}} \cdot \frac{\sqrt{8}}{\sqrt{8}} = \frac{\sqrt{40}}{\sqrt{64}} = \frac{\sqrt{4}\sqrt{10}}{8} = \frac{2\sqrt{10}}{8} = \frac{\sqrt{10}}{4}$$

### Solution B

$$\frac{\sqrt{5}}{\sqrt{8}} = \frac{\sqrt{5}}{\sqrt{8}} \cdot \frac{\sqrt{2}}{\sqrt{2}} = \frac{\sqrt{10}}{\sqrt{16}} = \frac{\sqrt{10}}{4}$$

### Solution C

$$\frac{\sqrt{5}}{\sqrt{8}} = \frac{\sqrt{5}}{\sqrt{4}\sqrt{2}} = \frac{\sqrt{5}}{2\sqrt{2}} = \frac{\sqrt{5}}{2\sqrt{2}} \cdot \frac{\sqrt{2}}{\sqrt{2}} = \frac{\sqrt{10}}{2\sqrt{4}} = \frac{\sqrt{10}}{2(2)} = \frac{\sqrt{10}}{4}$$

The three approaches to Example 4 again illustrate the need to think first and only then push the pencil. You may find one approach easier than another. To conclude this section, study the following examples and check the final radicals against the three conditions previously listed for simplest radical form.

**Classroom Example**
Simplify each of the following:

(a) $\frac{3\sqrt{5}}{4\sqrt{3}}$ (b) $\frac{6\sqrt{11}}{5\sqrt{27}}$

(c) $\sqrt[3]{\frac{7}{25}}$ (d) $\frac{\sqrt[3]{3}}{\sqrt[3]{36}}$

**EXAMPLE 5** Simplify each of the following:

**(a)** $\frac{3\sqrt{2}}{5\sqrt{3}}$ **(b)** $\frac{3\sqrt{7}}{2\sqrt{18}}$ **(c)** $\sqrt[3]{\frac{5}{9}}$ **(d)** $\frac{\sqrt[3]{5}}{\sqrt[3]{16}}$

### Solution

**(a)** $$\frac{3\sqrt{2}}{5\sqrt{3}} = \frac{3\sqrt{2}}{5\sqrt{3}} \cdot \frac{\sqrt{3}}{\sqrt{3}} = \frac{3\sqrt{6}}{5\sqrt{9}} = \frac{3\sqrt{6}}{5(3)} = \frac{\sqrt{6}}{5}$$

Form of 1

**(b)** $$\frac{3\sqrt{7}}{2\sqrt{18}} = \frac{3\sqrt{7}}{2\sqrt{18}} \cdot \frac{\sqrt{2}}{\sqrt{2}} = \frac{3\sqrt{14}}{2\sqrt{36}} = \frac{3\sqrt{14}}{2(6)} = \frac{3\sqrt{14}}{12} = \frac{\sqrt{14}}{4}$$

Form of 1

**(c)** $$\sqrt[3]{\frac{5}{9}} = \frac{\sqrt[3]{5}}{\sqrt[3]{9}} = \frac{\sqrt[3]{5}}{\sqrt[3]{9}} \cdot \frac{\sqrt[3]{3}}{\sqrt[3]{3}} = \frac{\sqrt[3]{15}}{\sqrt[3]{27}} = \frac{\sqrt[3]{15}}{3}$$

Form of 1

**(d)** $$\frac{\sqrt[3]{5}}{\sqrt[3]{16}} = \frac{\sqrt[3]{5}}{\sqrt[3]{16}} \cdot \frac{\sqrt[3]{4}}{\sqrt[3]{4}} = \frac{\sqrt[3]{20}}{\sqrt[3]{64}} = \frac{\sqrt[3]{20}}{4}$$

Form of 1

## Applications of Radicals

Many real-world applications involve radical expressions. For example, police often use the formula $S = \sqrt{30Df}$ to estimate the speed of a car on the basis of the length of the skid marks at the scene of an accident. In this formula, $S$ represents the speed of the car in miles per hour, $D$ represents the length of the skid marks in feet, and $f$ represents a coefficient of friction. For a particular situation, the coefficient of friction is a constant that depends on the type and condition of the road surface.

**Classroom Example**
Using 0.46 as a coefficient of friction, determine how fast a car was traveling if it skidded 275 feet.

**EXAMPLE 6** Apply Your Skill

Using 0.35 as a coefficient of friction, determine how fast a car was traveling if it skidded 325 feet.

### Solution

Substitute 0.35 for $f$ and 325 for $D$ in the formula.

$$S = \sqrt{30Df} = \sqrt{30(325)(0.35)} = 58 \quad \text{to the nearest whole number}$$

The car was traveling at approximately 58 miles per hour.

The **period** of a pendulum is the time it takes to swing from one side to the other side and back. The formula

$$T = 2\pi\sqrt{\frac{L}{32}}$$

where $T$ represents the time in seconds and $L$ the length in feet, can be used to determine the period of a pendulum (see Figure 5.1).

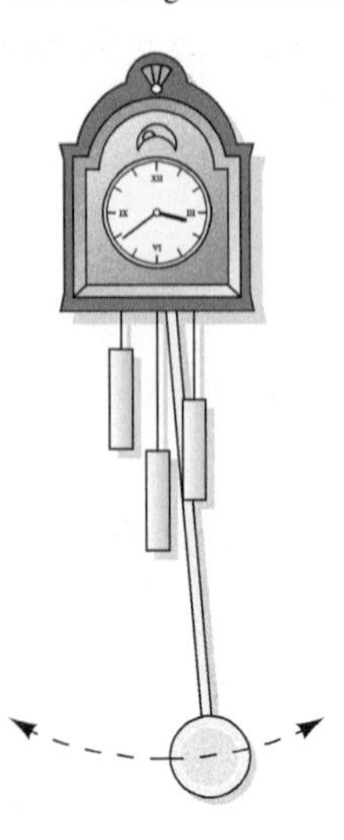

**Figure 5.1**

**Classroom Example**
Find, to the nearest tenth of a second, the period of a pendulum of length 2.1 feet.

**EXAMPLE 7** Apply Your Skill

Find, to the nearest tenth of a second, the period of a pendulum of length 3.5 feet.

### Solution

Let's use 3.14 as an approximation for $\pi$ and substitute 3.5 for $L$ in the formula.

$$T = 2\pi\sqrt{\frac{L}{32}} = 2(3.14)\sqrt{\frac{3.5}{32}} = 2.1 \quad \text{to the nearest tenth}$$

The period is approximately 2.1 seconds.

Radical expressions are also used in some geometric applications. For example, the area of a triangle can be found by using a formula that involves a square root. If $a$, $b$, and $c$ represent the lengths of the three sides of a triangle, the formula $K = \sqrt{s(s - a)(s - b)(s - c)}$, known as Heron's formula, can be used to determine the area ($K$) of the triangle. The letter $s$ represents the semiperimeter of the triangle; that is, $s = \frac{a + b + c}{2}$.

**Classroom Example**
Find the area of a triangular piece of sheet metal that has sides of lengths 34 cm, 32 cm, and 60 cm.

**EXAMPLE 8** Apply Your Skill

Find the area of a triangular piece of sheet metal that has sides of lengths 17 inches, 19 inches, and 26 inches.

### Solution

First, let's find the value of $s$, the semiperimeter of the triangle.

$$s = \frac{17 + 19 + 26}{2} = 31$$

Now we can use Heron's formula.

$$\begin{aligned} K = \sqrt{s(s - a)(s - b)(s - c)} &= \sqrt{31(31 - 17)(31 - 19)(31 - 26)} \\ &= \sqrt{31(14)(12)(5)} \\ &= \sqrt{20{,}640} \\ &= 161.4 \quad \text{to the nearest tenth} \end{aligned}$$

Thus the area of the piece of sheet metal is approximately 161.4 square inches.

**Remark:** Note that in Examples 6–8, we did not simplify the radicals. When one is using a calculator to approximate the square roots, there is no need to simplify first.

## Concept Quiz 5.2

For Problems 1–10, answer true or false.

1. The cube root of a number is one of its three equal factors.
2. Every positive real number has one positive real number square root.
3. The principal square root of a number is the nonnegative square root of the number.
4. The symbol $\sqrt{\phantom{x}}$ is called a radical.
5. The square root of 0 is not a real number.
6. The number under the radical sign is called the radicand.
7. Every positive real number has two square roots.
8. The $n$ in the radical $\sqrt[n]{a}$ is called the index of the radical.

**9.** If $n$ is an odd integer greater than 1 and $b$ is a negative real number, then $\sqrt[n]{b}$ is a negative real number.

**10.** $\dfrac{3\sqrt{24}}{8}$ is in simplest radical form.

## Problem Set 5.2

For Problems 1–20, evaluate each of the following. For example, $\sqrt{25} = 5$. (Objective 1)

1. $\sqrt{64}$
2. $\sqrt{49}$
3. $-\sqrt{100}$
4. $-\sqrt{81}$
5. $\sqrt[3]{27}$
6. $\sqrt[3]{216}$
7. $\sqrt[3]{-64}$
8. $\sqrt[3]{-125}$
9. $\sqrt[4]{81}$
10. $-\sqrt[4]{16}$
11. $\sqrt{\dfrac{16}{25}}$
12. $\sqrt{\dfrac{25}{64}}$
13. $-\sqrt{\dfrac{36}{49}}$
14. $\sqrt{\dfrac{16}{64}}$
15. $\sqrt{\dfrac{9}{36}}$
16. $\sqrt{\dfrac{144}{36}}$
17. $\sqrt[3]{\dfrac{27}{64}}$
18. $\sqrt[3]{-\dfrac{8}{27}}$
19. $\sqrt[3]{8^3}$
20. $\sqrt[4]{16^4}$

For Problems 21–74, change each radical to simplest radical form. (Objectives 2 and 3)

21. $\sqrt{27}$
22. $\sqrt{48}$
23. $\sqrt{32}$
24. $\sqrt{98}$
25. $\sqrt{80}$
26. $\sqrt{125}$
27. $\sqrt{160}$
28. $\sqrt{112}$
29. $4\sqrt{18}$
30. $5\sqrt{32}$
31. $-6\sqrt{20}$
32. $-4\sqrt{54}$
33. $\dfrac{2}{5}\sqrt{75}$
34. $\dfrac{1}{3}\sqrt{90}$
35. $\dfrac{3}{2}\sqrt{24}$
36. $\dfrac{3}{4}\sqrt{45}$
37. $-\dfrac{5}{6}\sqrt{28}$
38. $-\dfrac{2}{3}\sqrt{96}$
39. $\sqrt{\dfrac{19}{4}}$
40. $\sqrt{\dfrac{22}{9}}$
41. $\sqrt{\dfrac{27}{16}}$
42. $\sqrt{\dfrac{8}{25}}$
43. $\sqrt{\dfrac{75}{81}}$
44. $\sqrt{\dfrac{24}{49}}$
45. $\sqrt{\dfrac{2}{7}}$
46. $\sqrt{\dfrac{3}{8}}$
47. $\sqrt{\dfrac{2}{3}}$
48. $\sqrt{\dfrac{7}{12}}$
49. $\dfrac{\sqrt{5}}{\sqrt{12}}$
50. $\dfrac{\sqrt{3}}{\sqrt{7}}$
51. $\dfrac{\sqrt{11}}{\sqrt{24}}$
52. $\dfrac{\sqrt{5}}{\sqrt{48}}$
53. $\dfrac{\sqrt{18}}{\sqrt{27}}$
54. $\dfrac{\sqrt{10}}{\sqrt{20}}$
55. $\dfrac{\sqrt{35}}{\sqrt{7}}$
56. $\dfrac{\sqrt{42}}{\sqrt{6}}$
57. $\dfrac{2\sqrt{3}}{\sqrt{7}}$
58. $\dfrac{3\sqrt{2}}{\sqrt{6}}$
59. $-\dfrac{4\sqrt{12}}{\sqrt{5}}$
60. $\dfrac{-6\sqrt{5}}{\sqrt{18}}$
61. $\dfrac{3\sqrt{2}}{4\sqrt{3}}$
62. $\dfrac{6\sqrt{5}}{5\sqrt{12}}$
63. $\dfrac{-8\sqrt{18}}{10\sqrt{50}}$
64. $\dfrac{4\sqrt{45}}{-6\sqrt{20}}$
65. $\sqrt[3]{16}$
66. $\sqrt[3]{40}$
67. $2\sqrt[3]{81}$
68. $-3\sqrt[3]{54}$
69. $\dfrac{2}{\sqrt[3]{9}}$
70. $\dfrac{3}{\sqrt[3]{3}}$
71. $\dfrac{\sqrt[3]{27}}{\sqrt[3]{4}}$
72. $\dfrac{\sqrt[3]{8}}{\sqrt[3]{16}}$
73. $\dfrac{\sqrt[3]{6}}{\sqrt[3]{4}}$
74. $\dfrac{\sqrt[3]{4}}{\sqrt[3]{2}}$

For Problems 75–80, use radicals to solve the problems. (Objective 4)

75. Use a coefficient of friction of 0.4 in the formula from Example 6 and find the speeds of cars that left skid

marks of lengths 150 feet, 200 feet, and 350 feet. Express your answers to the nearest mile per hour.

**76.** Use the formula from Example 7, and find the periods of pendulums of lengths 2 feet, 3 feet, and 4.5 feet. Express your answers to the nearest tenth of a second.

**77.** Find, to the nearest square centimeter, the area of a triangle that measures 14 centimeters by 16 centimeters by 18 centimeters.

**78.** Find, to the nearest square yard, the area of a triangular plot of ground that measures 45 yards by 60 yards by 75 yards.

**79.** Find the area of an equilateral triangle, each of whose sides is 18 inches long. Express the area to the nearest square inch.

**80.** Find, to the nearest square inch, the area of the quadrilateral in Figure 5.2.

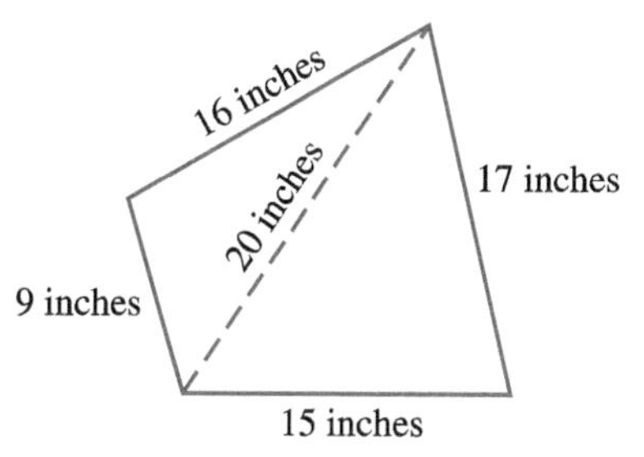


**Figure 5.2**

## Thoughts Into Words

**81.** Why is $\sqrt{-9}$ not a real number?

**82.** Why is it that we say 25 has two square roots (5 and $-5$), but we write $\sqrt{25} = 5$?

**83.** How is the multiplication property of 1 used when simplifying radicals?

**84.** How could you find a whole number approximation for $\sqrt{2750}$ if you did not have a calculator or table available?

## Further Investigations

**85.** Use your calculator to find a rational approximation, to the nearest thousandth, for (a) through (i).

**(a)** $\sqrt{2}$ **(b)** $\sqrt{75}$

**(c)** $\sqrt{156}$ **(d)** $\sqrt{691}$

**(e)** $\sqrt{3249}$ **(f)** $\sqrt{45{,}123}$

**(g)** $\sqrt{0.14}$ **(h)** $\sqrt{0.023}$

**(i)** $\sqrt{0.8649}$

**86.** Sometimes a fairly good estimate can be made of a radical expression by using whole number approximations. For example, $5\sqrt{35} + 7\sqrt{50}$ is approximately $5(6) + 7(7) = 79$. Using a calculator, we find that $5\sqrt{35} + 7\sqrt{50} = 79.1$, to the nearest tenth. In this case our whole number estimate is very good. For (a) through (f), first make a whole number estimate, and then use your calculator to see how well you estimated.

**(a)** $3\sqrt{10} - 4\sqrt{24} + 6\sqrt{65}$

**(b)** $9\sqrt{27} + 5\sqrt{37} - 3\sqrt{80}$

**(c)** $12\sqrt{5} + 13\sqrt{18} + 9\sqrt{47}$

**(d)** $3\sqrt{98} - 4\sqrt{83} - 7\sqrt{120}$

**(e)** $4\sqrt{170} + 2\sqrt{198} + 5\sqrt{227}$

**(f)** $-3\sqrt{256} - 6\sqrt{287} + 11\sqrt{321}$

### Answers to the Concept Quiz

**1.** True **2.** True **3.** True **4.** False **5.** False **6.** True **7.** True **8.** True **9.** True **10.** False

# 5.3 Combining Radicals and Simplifying Radicals That Contain Variables

OBJECTIVES

1. Simplify expressions by combining radicals
2. Simplify radicals that contain variables

Recall our use of the distributive property as the basis for combining similar terms. For example,

$$3x + 2x = (3 + 2)x = 5x$$
$$8y - 5y = (8 - 5)y = 3y$$
$$\frac{2}{3}a^2 + \frac{3}{4}a^2 = \left(\frac{2}{3} + \frac{3}{4}\right)a^2 = \left(\frac{8}{12} + \frac{9}{12}\right)a^2 = \frac{17}{12}a^2$$

In a like manner, expressions that contain radicals can often be simplified by using the distributive property, as follows:

$$3\sqrt{2} + 5\sqrt{2} = (3 + 5)\sqrt{2} = 8\sqrt{2}$$
$$7\sqrt[3]{5} - 3\sqrt[3]{5} = (7 - 3)\sqrt[3]{5} = 4\sqrt[3]{5}$$
$$4\sqrt{7} + 5\sqrt{7} + 6\sqrt{11} - 2\sqrt{11} = (4 + 5)\sqrt{7} + (6 - 2)\sqrt{11} = 9\sqrt{7} + 4\sqrt{11}$$

Note that *in order to be added or subtracted, radicals must have the same index and the same radicand.* Thus we cannot simplify an expression such as $5\sqrt{2} + 7\sqrt{11}$.

Simplifying by combining radicals sometimes requires that you first express the given radicals in simplest form and then apply the distributive property. The following examples illustrate this idea.

**Classroom Example**
Simplify $2\sqrt{12} + 5\sqrt{48} - 7\sqrt{3}$.

**EXAMPLE 1** Simplify $3\sqrt{8} + 2\sqrt{18} - 4\sqrt{2}$.

**Solution**

$$\begin{aligned}3\sqrt{8} + 2\sqrt{18} - 4\sqrt{2} &= 3\sqrt{4}\sqrt{2} + 2\sqrt{9}\sqrt{2} - 4\sqrt{2}\\ &= 3 \cdot 2 \cdot \sqrt{2} + 2 \cdot 3 \cdot \sqrt{2} - 4\sqrt{2}\\ &= 6\sqrt{2} + 6\sqrt{2} - 4\sqrt{2}\\ &= (6 + 6 - 4)\sqrt{2} = 8\sqrt{2}\end{aligned}$$

**Classroom Example**
Simplify $\frac{3}{4}\sqrt{98} - \frac{2}{5}\sqrt{50}$.

**EXAMPLE 2** Simplify $\frac{1}{4}\sqrt{45} + \frac{1}{3}\sqrt{20}$.

**Solution**

$$\begin{aligned}\frac{1}{4}\sqrt{45} + \frac{1}{3}\sqrt{20} &= \frac{1}{4}\sqrt{9}\sqrt{5} + \frac{1}{3}\sqrt{4}\sqrt{5}\\ &= \frac{1}{4} \cdot 3 \cdot \sqrt{5} + \frac{1}{3} \cdot 2 \cdot \sqrt{5}\\ &= \frac{3}{4}\sqrt{5} + \frac{2}{3}\sqrt{5} = \left(\frac{3}{4} + \frac{2}{3}\right)\sqrt{5}\\ &= \left(\frac{9}{12} + \frac{8}{12}\right)\sqrt{5} = \frac{17}{12}\sqrt{5}\end{aligned}$$

Classroom Example
Simplify $3\sqrt[3]{4} + 5\sqrt[3]{32} - 2\sqrt[3]{108}$.

**EXAMPLE 3** Simplify $5\sqrt[3]{2} - 2\sqrt[3]{16} - 6\sqrt[3]{54}$.

### Solution

$$\begin{aligned} 5\sqrt[3]{2} - 2\sqrt[3]{16} - 6\sqrt[3]{54} &= 5\sqrt[3]{2} - 2\sqrt[3]{8}\sqrt[3]{2} - 6\sqrt[3]{27}\sqrt[3]{2} \\ &= 5\sqrt[3]{2} - 2 \cdot 2 \cdot \sqrt[3]{2} - 6 \cdot 3 \cdot \sqrt[3]{2} \\ &= 5\sqrt[3]{2} - 4\sqrt[3]{2} - 18\sqrt[3]{2} \\ &= (5 - 4 - 18)\sqrt[3]{2} \\ &= -17\sqrt[3]{2} \end{aligned}$$

## Simplifying Radicals That Contain Variables

Before we discuss the process of simplifying radicals that contain variables, there is one technicality that we should call to your attention. Let's look at some examples to clarify the point. Consider the radical $\sqrt{x^2}$.

Let $x = 3$; then $\sqrt{x^2} = \sqrt{3^2} = \sqrt{9} = 3$.

Let $x = -3$; then $\sqrt{x^2} = \sqrt{(-3)^2} = \sqrt{9} = 3$.

Thus if $x \geq 0$, then $\sqrt{x^2} = x$, *but* if $x < 0$, then $\sqrt{x^2} = -x$. Using the concept of absolute value, we can state that for all real numbers, $\sqrt{x^2} = |x|$.

Now consider the radical $\sqrt{x^3}$. Because $x^3$ is negative when $x$ is negative, we need to restrict $x$ to the nonnegative real numbers when working with $\sqrt{x^3}$. Thus we can write, "if $x \geq 0$, then $\sqrt{x^3} = \sqrt{x^2}\sqrt{x} = x\sqrt{x}$," and no absolute-value sign is necessary. Finally, let's consider the radical $\sqrt[3]{x^3}$.

Let $x = 2$; then $\sqrt[3]{x^3} = \sqrt[3]{2^3} = \sqrt[3]{8} = 2$.

Let $x = -2$; then $\sqrt[3]{x^3} = \sqrt[3]{(-2)^3} = \sqrt[3]{-8} = -2$.

Thus it is correct to write, "$\sqrt[3]{x^3} = x$ for all real numbers," and again no absolute-value sign is necessary.

The previous discussion indicates that technically, every radical expression involving variables in the radicand needs to be analyzed individually in terms of any necessary restrictions imposed on the variables. To help you gain experience with this skill, examples and problems are discussed under *Further Investigations* in the problem set. For now, however, to avoid considering such restrictions on a problem-to-problem basis, we shall merely assume that all variables represent positive real numbers. Let's consider the process of simplifying radicals that contain variables in the radicand. Study the following examples, and note that the same basic approach we used in Section 5.2 is applied here.

Classroom Example
Simplify each of the following:
**(a)** $\sqrt{125m^3}$
**(b)** $\sqrt{28m^5n^9}$
**(c)** $\sqrt{147x^6y^7}$
**(d)** $\sqrt[3]{128m^{10}n^5}$

**EXAMPLE 4** Simplify each of the following:

**(a)** $\sqrt{8x^3}$ **(b)** $\sqrt{45x^3y^7}$ **(c)** $\sqrt{180a^4b^3}$ **(d)** $\sqrt[3]{40x^4y^8}$

### Solution

**(a)** $\sqrt{8x^3} = \sqrt{4x^2}\sqrt{2x} = 2x\sqrt{2x}$ — $4x^2$ is a perfect square

**(b)** $\sqrt{45x^3y^7} = \sqrt{9x^2y^6}\sqrt{5xy} = 3xy^3\sqrt{5xy}$ — $9x^2y^6$ is a perfect square

**(c)** If the numerical coefficient of the radicand is quite large, you may want to look at it in the prime factored form.

$$\begin{aligned} \sqrt{180a^4b^3} &= \sqrt{2 \cdot 2 \cdot 3 \cdot 3 \cdot 5 \cdot a^4 \cdot b^3} \\ &= \sqrt{36 \cdot 5 \cdot a^4 \cdot b^3} \end{aligned}$$

$$= \sqrt{36a^4b^2}\sqrt{5b}$$

$$= 6a^2b\sqrt{5b}$$

**(d)** $\sqrt[3]{40x^4y^8} = \sqrt[3]{8x^3y^6}\sqrt[3]{5xy^2} = 2xy^2\sqrt[3]{5xy^2}$ — $8x^3y^6$ is a perfect cube

Before we consider more examples, let's restate (in such a way that includes radicands containing variables) the conditions necessary for a radical to be in simplest radical form.

1. A radicand contains no polynomial factor raised to a power equal to or greater than the index of the radical. — $\sqrt{x^3}$ violates this condition
2. No fraction appears within a radical sign. — $\sqrt{\dfrac{2x}{3y}}$ violates this condition
3. No radical appears in the denominator. — $\dfrac{3}{\sqrt[3]{4x}}$ violates this condition

**Classroom Example**
Express each of the following in simplest radical form:

(a) $\sqrt{\dfrac{5m}{7n}}$ (b) $\dfrac{\sqrt{3}}{\sqrt{8x^5}}$

(c) $\dfrac{\sqrt{125m^7}}{\sqrt{48n^4}}$

(d) $\dfrac{7}{\sqrt[3]{9y}}$

(e) $\dfrac{\sqrt[3]{81a^7}}{\sqrt[3]{32b^2}}$

**EXAMPLE 5** Express each of the following in simplest radical form:

**(a)** $\sqrt{\dfrac{2x}{3y}}$ **(b)** $\dfrac{\sqrt{5}}{\sqrt{12a^3}}$ **(c)** $\dfrac{\sqrt{8x^2}}{\sqrt{27y^5}}$ **(d)** $\dfrac{3}{\sqrt[3]{4x}}$ **(e)** $\dfrac{\sqrt[3]{16x^2}}{\sqrt[3]{9y^5}}$

### Solution

**(a)** $$\sqrt{\frac{2x}{3y}} = \frac{\sqrt{2x}}{\sqrt{3y}} = \frac{\sqrt{2x}}{\sqrt{3y}} \cdot \frac{\sqrt{3y}}{\sqrt{3y}} = \frac{\sqrt{6xy}}{3y}$$

(↑ $\frac{\sqrt{3y}}{\sqrt{3y}}$: Form of 1)

**(b)** $$\frac{\sqrt{5}}{\sqrt{12a^3}} = \frac{\sqrt{5}}{\sqrt{12a^3}} \cdot \frac{\sqrt{3a}}{\sqrt{3a}} = \frac{\sqrt{15a}}{\sqrt{36a^4}} = \frac{\sqrt{15a}}{6a^2}$$

(↑ $\frac{\sqrt{3a}}{\sqrt{3a}}$: Form of 1)

**(c)** $$\frac{\sqrt{8x^2}}{\sqrt{27y^5}} = \frac{\sqrt{4x^2}\sqrt{2}}{\sqrt{9y^4}\sqrt{3y}} = \frac{2x\sqrt{2}}{3y^2\sqrt{3y}} = \frac{2x\sqrt{2}}{3y^2\sqrt{3y}} \cdot \frac{\sqrt{3y}}{\sqrt{3y}} = \frac{2x\sqrt{6y}}{(3y^2)(3y)} = \frac{2x\sqrt{6y}}{9y^3}$$

**(d)** $$\frac{3}{\sqrt[3]{4x}} = \frac{3}{\sqrt[3]{4x}} \cdot \frac{\sqrt[3]{2x^2}}{\sqrt[3]{2x^2}} = \frac{3\sqrt[3]{2x^2}}{\sqrt[3]{8x^3}} = \frac{3\sqrt[3]{2x^2}}{2x}$$

**(e)** $$\frac{\sqrt[3]{16x^2}}{\sqrt[3]{9y^5}} = \frac{\sqrt[3]{16x^2}}{\sqrt[3]{9y^5}} \cdot \frac{\sqrt[3]{3y}}{\sqrt[3]{3y}} = \frac{\sqrt[3]{48x^2y}}{\sqrt[3]{27y^6}} = \frac{\sqrt[3]{8}\sqrt[3]{6x^2y}}{3y^2} = \frac{2\sqrt[3]{6x^2y}}{3y^2}$$

Note that in part (c) we did some simplifying first before rationalizing the denominator, whereas in part (b) we proceeded immediately to rationalize the denominator. This is an individual choice, and you should probably do it both ways a few times to decide which you prefer.

## Concept Quiz 5.3

For Problems 1–10, answer true or false.

1. In order to be combined when adding, radicals must have the same index and the same radicand.
2. If $x \geq 0$, then $\sqrt{x^2} = x$.

3. For all real numbers, $\sqrt{x^2} = x$.
4. For all real numbers, $\sqrt[3]{x^3} = x$.
5. A radical is not in simplest radical form if it has a fraction within the radical sign.
6. If a radical contains a factor raised to a power that is equal to the index of the radical, then the radical is not in simplest radical form.
7. The radical $\dfrac{1}{\sqrt{x}}$ is in simplest radical form.
8. $3\sqrt{2} + 4\sqrt{3} = 7\sqrt{5}$.
9. If $x > 0$, then $\sqrt{45x^3} = 3x^2\sqrt{5x}$.
10. If $x > 0$, then $\dfrac{4\sqrt{x^5}}{3\sqrt{4x^2}} = \dfrac{2x\sqrt{x}}{3}$.

## Problem Set 5.3

For Problems 1–20, use the distributive property to help simplify each of the following. (Objective 1)
For example,

$$\begin{aligned} 3\sqrt{8} - \sqrt{32} &= 3\sqrt{4}\sqrt{2} - \sqrt{16}\sqrt{2} \\ &= 3(2)\sqrt{2} - 4\sqrt{2} \\ &= 6\sqrt{2} - 4\sqrt{2} \\ &= (6 - 4)\sqrt{2} = 2\sqrt{2} \end{aligned}$$

1. $5\sqrt{18} - 2\sqrt{2}$
2. $7\sqrt{12} + 4\sqrt{3}$
3. $7\sqrt{12} + 10\sqrt{48}$
4. $6\sqrt{8} - 5\sqrt{18}$
5. $-2\sqrt{50} - 5\sqrt{32}$
6. $-2\sqrt{20} - 7\sqrt{45}$
7. $3\sqrt{20} - \sqrt{5} - 2\sqrt{45}$
8. $6\sqrt{12} + \sqrt{3} - 2\sqrt{48}$
9. $-9\sqrt{24} + 3\sqrt{54} - 12\sqrt{6}$
10. $13\sqrt{28} - 2\sqrt{63} - 7\sqrt{7}$
11. $\frac{3}{4}\sqrt{7} - \frac{2}{3}\sqrt{28}$
12. $\frac{3}{5}\sqrt{5} - \frac{1}{4}\sqrt{80}$
13. $\frac{3}{5}\sqrt{40} + \frac{5}{6}\sqrt{90}$
14. $\frac{3}{8}\sqrt{96} - \frac{2}{3}\sqrt{54}$
15. $\dfrac{3\sqrt{18}}{5} - \dfrac{5\sqrt{72}}{6} + \dfrac{3\sqrt{98}}{4}$
16. $\dfrac{-2\sqrt{20}}{3} + \dfrac{3\sqrt{45}}{4} - \dfrac{5\sqrt{80}}{6}$
17. $5\sqrt[3]{3} + 2\sqrt[3]{24} - 6\sqrt[3]{81}$
18. $-3\sqrt[3]{2} - 2\sqrt[3]{16} + \sqrt[3]{54}$
19. $-\sqrt[3]{16} + 7\sqrt[3]{54} - 9\sqrt[3]{2}$
20. $4\sqrt[3]{24} - 6\sqrt[3]{3} + 13\sqrt[3]{81}$

For Problems 21–64, express each of the following in simplest radical form. All variables represent positive real numbers. (Objective 2)

21. $\sqrt{32x}$
22. $\sqrt{50y}$
23. $\sqrt{75x^2}$
24. $\sqrt{108y^2}$
25. $\sqrt{20x^2y}$
26. $\sqrt{80xy^2}$
27. $\sqrt{64x^3y^7}$
28. $\sqrt{36x^5y^6}$
29. $\sqrt{54a^4b^3}$
30. $\sqrt{96a^7b^8}$
31. $\sqrt{63x^6y^8}$
32. $\sqrt{28x^4y^{12}}$
33. $2\sqrt{40a^3}$
34. $4\sqrt{90a^5}$
35. $\frac{2}{3}\sqrt{96xy^3}$
36. $\frac{4}{5}\sqrt{125x^4y}$
37. $\sqrt{\dfrac{2x}{5y}}$
38. $\sqrt{\dfrac{3x}{2y}}$
39. $\sqrt{\dfrac{5}{12x^4}}$
40. $\sqrt{\dfrac{7}{8x^2}}$
41. $\dfrac{5}{\sqrt{18y}}$
42. $\dfrac{3}{\sqrt{12x}}$
43. $\dfrac{\sqrt{7x}}{\sqrt{8y^5}}$
44. $\dfrac{\sqrt{5y}}{\sqrt{18x^3}}$
45. $\dfrac{\sqrt{18y^3}}{\sqrt{16x}}$
46. $\dfrac{\sqrt{2x^3}}{\sqrt{9y}}$

**47.** $\dfrac{\sqrt{24a^2b^3}}{\sqrt{7ab^6}}$

**48.** $\dfrac{\sqrt{12a^2b}}{\sqrt{5a^3b^3}}$

**49.** $\sqrt[3]{24y}$

**50.** $\sqrt[3]{16x^2}$

**51.** $\sqrt[3]{16x^4}$

**52.** $\sqrt[3]{54x^3}$

**53.** $\sqrt[3]{56x^6y^8}$

**54.** $\sqrt[3]{81x^5y^6}$

**55.** $\sqrt[3]{\dfrac{7}{9x^2}}$

**56.** $\sqrt[3]{\dfrac{5}{2x}}$

**57.** $\dfrac{\sqrt[3]{3y}}{\sqrt[3]{16x^4}}$

**58.** $\dfrac{\sqrt[3]{2y}}{\sqrt[3]{3x}}$

**59.** $\dfrac{\sqrt[3]{12xy}}{\sqrt[3]{3x^2y^5}}$

**60.** $\dfrac{5}{\sqrt[3]{9xy^2}}$

**61.** $\sqrt{8x + 12y}$ [*Hint*: $\sqrt{8x + 12y} = \sqrt{4(2x + 3y)}$]

**62.** $\sqrt{4x + 4y}$

**63.** $\sqrt{16x + 48y}$

**64.** $\sqrt{27x + 18y}$

For Problems 65–74, use the distributive property to help simplify each of the following. All variables represent positive real numbers. (Objectives 1 and 2)

**65.** $-3\sqrt{4x} + 5\sqrt{9x} + 6\sqrt{16x}$

**66.** $-2\sqrt{25x} - 4\sqrt{36x} + 7\sqrt{64x}$

**67.** $2\sqrt{18x} - 3\sqrt{8x} - 6\sqrt{50x}$

**68.** $4\sqrt{20x} + 5\sqrt{45x} - 10\sqrt{80x}$

**69.** $5\sqrt{27n} - \sqrt{12n} - 6\sqrt{3n}$

**70.** $4\sqrt{8n} + 3\sqrt{18n} - 2\sqrt{72n}$

**71.** $7\sqrt{4ab} - \sqrt{16ab} - 10\sqrt{25ab}$

**72.** $4\sqrt{ab} - 9\sqrt{36ab} + 6\sqrt{49ab}$

**73.** $-3\sqrt{2x^3} + 4\sqrt{8x^3} - 3\sqrt{32x^3}$

**74.** $2\sqrt{40x^5} - 3\sqrt{90x^5} + 5\sqrt{160x^5}$

## Thoughts Into Words

**75.** Is the expression $3\sqrt{2} + \sqrt{50}$ in simplest radical form? Defend your answer.

**76.** Your friend simplified $\dfrac{\sqrt{6}}{\sqrt{8}}$ as follows:

$$\frac{\sqrt{6}}{\sqrt{8}} \cdot \frac{\sqrt{8}}{\sqrt{8}} = \frac{\sqrt{48}}{8} = \frac{\sqrt{16}\sqrt{3}}{8} = \frac{4\sqrt{3}}{8} = \frac{\sqrt{3}}{2}$$

Is this a correct procedure? Can you show her a better way to do this problem?

**77.** Does $\sqrt{x + y}$ equal $\sqrt{x} + \sqrt{y}$? Defend your answer.

## Further Investigations

**78.** Use your calculator and evaluate each expression in Problems 1–16. Then evaluate the simplified expression that you obtained when doing these problems. Your two results for each problem should be the same.

Consider these problems, in which the variables could represent any real number. However, we would still have the restriction that the radical would represent a real number. In other words, the radicand must be nonnegative.

$\sqrt{98x^2} = \sqrt{49x^2}\sqrt{2} = 7|x|\sqrt{2}$ An absolute-value sign is necessary to ensure that the principal root is nonnegative

$\sqrt{24x^4} = \sqrt{4x^4}\sqrt{6} = 2x^2\sqrt{6}$ Because $x^2$ is nonnegative, there is no need for an absolute-value sign to ensure that the principal root is nonnegative

$\sqrt{25x^3} = \sqrt{25x^2}\sqrt{x} = 5x\sqrt{x}$ Because the radicand is defined to be nonnegative, $x$ must be nonnegative, and there is no need for an absolute-value sign to ensure that the principal root is nonnegative

$\sqrt{18b^5} = \sqrt{9b^4}\sqrt{2b} = 3b^2\sqrt{2b}$ An absolute-value sign is not necessary to ensure that the principal root is nonnegative

$\sqrt{12y^6} = \sqrt{4y^6}\sqrt{3} = 2|y^3|\sqrt{3}$ An absolute-value sign is necessary to ensure that the principal root is nonnegative

**79.** Do the following problems, in which the variable could be any real number as long as the radical represents a real number. Use absolute-value signs in the answers as necessary.

(a) $\sqrt{125x^2}$ (b) $\sqrt{16x^4}$ (g) $\sqrt{128c^{10}}$ (h) $\sqrt{18d^7}$

(c) $\sqrt{8b^3}$ (d) $\sqrt{3y^5}$ (i) $\sqrt{49x^2}$ (j) $\sqrt{80n^{20}}$

(e) $\sqrt{288x^6}$ (f) $\sqrt{28m^8}$ (k) $\sqrt{81h^3}$

### Answers to the Concept Quiz

**1.** True **2.** True **3.** False **4.** True **5.** True **6.** True **7.** False **8.** False **9.** False **10.** True

## 5.4 Products and Quotients Involving Radicals

OBJECTIVES

1. Multiply two radicals
2. Use the distributive property to multiply radical expressions
3. Rationalize binomial denominators

As we have seen, Property 5.3 $\left(\sqrt[n]{bc} = \sqrt[n]{b}\sqrt[n]{c}\right)$ is used to express one radical as the product of two radicals and also to express the product of two radicals as one radical. In fact, we have used the property for both purposes within the framework of simplifying radicals. For example,

$$\frac{\sqrt{3}}{\sqrt{32}} = \frac{\sqrt{3}}{\sqrt{16}\sqrt{2}} = \frac{\sqrt{3}}{4\sqrt{2}} = \frac{\sqrt{3}}{4\sqrt{2}} \cdot \frac{\sqrt{2}}{\sqrt{2}} = \frac{\sqrt{6}}{8}$$

$\sqrt[n]{bc} = \sqrt[n]{b}\sqrt[n]{c}$ $\qquad$ $\sqrt[n]{b}\sqrt[n]{c} = \sqrt[n]{bc}$

The following examples demonstrate the use of Property 5.3 to multiply radicals and to express the product in simplest form.

Classroom Example
Multiply and simplify where possible:
(a) $\left(4\sqrt{2}\right)\left(6\sqrt{7}\right)$
(b) $\left(5\sqrt{18}\right)\left(2\sqrt{2}\right)$
(c) $\left(2\sqrt{5}\right)\left(5\sqrt{10}\right)$
(d) $\left(8\sqrt[3]{9}\right)\left(5\sqrt[3]{6}\right)$

**EXAMPLE 1** Multiply and simplify where possible:

(a) $\left(2\sqrt{3}\right)\left(3\sqrt{5}\right)$ (b) $\left(3\sqrt{8}\right)\left(5\sqrt{2}\right)$

(c) $\left(7\sqrt{6}\right)\left(3\sqrt{8}\right)$ (d) $\left(2\sqrt[3]{6}\right)\left(5\sqrt[3]{4}\right)$

### Solution

(a) $\left(2\sqrt{3}\right)\left(3\sqrt{5}\right) = 2 \cdot 3 \cdot \sqrt{3} \cdot \sqrt{5} = 6\sqrt{15}$

(b) $\left(3\sqrt{8}\right)\left(5\sqrt{2}\right) = 3 \cdot 5 \cdot \sqrt{8} \cdot \sqrt{2} = 15\sqrt{16} = 15 \cdot 4 = 60$

(c) $\left(7\sqrt{6}\right)\left(3\sqrt{8}\right) = 7 \cdot 3 \cdot \sqrt{6} \cdot \sqrt{8} = 21\sqrt{48} = 21\sqrt{16}\sqrt{3}$
$= 21 \cdot 4 \cdot \sqrt{3} = 84\sqrt{3}$

(d) $\left(2\sqrt[3]{6}\right)\left(5\sqrt[3]{4}\right) = 2 \cdot 5 \cdot \sqrt[3]{6} \cdot \sqrt[3]{4} = 10\sqrt[3]{24}$
$= 10\sqrt[3]{8}\sqrt[3]{3}$
$= 10 \cdot 2 \cdot \sqrt[3]{3}$
$= 20\sqrt[3]{3}$

## Using the Distributive Property to Multiply Radical Expressions

Recall the use of the distributive property when finding the product of a monomial and a polynomial. For example, $3x^2(2x + 7) = 3x^2(2x) + 3x^2(7) = 6x^3 + 21x^2$. In a similar manner, the distributive property and Property 5.3 provide the basis for finding certain special products that involve radicals. The following examples illustrate this idea.

**Classroom Example**
Multiply and simplify where possible:
(a) $\sqrt{2}(\sqrt{10} + \sqrt{8})$
(b) $3\sqrt{5}(\sqrt{10} + \sqrt{15})$
(c) $\sqrt{7a}(\sqrt{14a} - \sqrt{28ab})$
(d) $(3\sqrt[3]{12})(4\sqrt[3]{9})$

**EXAMPLE 2** Multiply and simplify where possible:

(a) $\sqrt{3}(\sqrt{6} + \sqrt{12})$ (b) $2\sqrt{2}(4\sqrt{3} - 5\sqrt{6})$
(c) $\sqrt{6x}(\sqrt{8x} + \sqrt{12xy})$ (d) $\sqrt[3]{2}(5\sqrt[3]{4} - 3\sqrt[3]{16})$

### Solution

(a)
$$\begin{aligned}\sqrt{3}(\sqrt{6} + \sqrt{12}) &= \sqrt{3}\sqrt{6} + \sqrt{3}\sqrt{12} \\ &= \sqrt{18} + \sqrt{36} \\ &= \sqrt{9}\sqrt{2} + 6 \\ &= 3\sqrt{2} + 6\end{aligned}$$

(b)
$$\begin{aligned}2\sqrt{2}(4\sqrt{3} - 5\sqrt{6}) &= (2\sqrt{2})(4\sqrt{3}) - (2\sqrt{2})(5\sqrt{6}) \\ &= 8\sqrt{6} - 10\sqrt{12} \\ &= 8\sqrt{6} - 10\sqrt{4}\sqrt{3} \\ &= 8\sqrt{6} - 20\sqrt{3}\end{aligned}$$

(c)
$$\begin{aligned}\sqrt{6x}(\sqrt{8x} + \sqrt{12xy}) &= (\sqrt{6x})(\sqrt{8x}) + (\sqrt{6x})(\sqrt{12xy}) \\ &= \sqrt{48x^2} + \sqrt{72x^2y} \\ &= \sqrt{16x^2}\sqrt{3} + \sqrt{36x^2}\sqrt{2y} \\ &= 4x\sqrt{3} + 6x\sqrt{2y}\end{aligned}$$

(d)
$$\begin{aligned}\sqrt[3]{2}(5\sqrt[3]{4} - 3\sqrt[3]{16}) &= (\sqrt[3]{2})(5\sqrt[3]{4}) - (\sqrt[3]{2})(3\sqrt[3]{16}) \\ &= 5\sqrt[3]{8} - 3\sqrt[3]{32} \\ &= 5 \cdot 2 - 3\sqrt[3]{8}\sqrt[3]{4} \\ &= 10 - 6\sqrt[3]{4}\end{aligned}$$

The distributive property also plays a central role in determining the product of two binomials. For example, $(x + 2)(x + 3) = x(x + 3) + 2(x + 3) = x^2 + 3x + 2x + 6 = x^2 + 5x + 6$. Finding the product of two binomial expressions that involve radicals can be handled in a similar fashion, as in the next examples.

**Classroom Example**
Find the following products and simplify:
(a) $(\sqrt{2} - \sqrt{7})(\sqrt{5} + \sqrt{3})$
(b) $(5\sqrt{5} + \sqrt{3})(4\sqrt{5} - 6\sqrt{3})$
(c) $(\sqrt{10} + \sqrt{3})(\sqrt{10} - \sqrt{3})$
(d) $(\sqrt{m} - \sqrt{n})(\sqrt{m} + \sqrt{n})$

**EXAMPLE 3** Find the following products and simplify:

(a) $(\sqrt{3} + \sqrt{5})(\sqrt{2} + \sqrt{6})$ (b) $(2\sqrt{2} - \sqrt{7})(3\sqrt{2} + 5\sqrt{7})$
(c) $(\sqrt{8} + \sqrt{6})(\sqrt{8} - \sqrt{6})$ (d) $(\sqrt{x} + \sqrt{y})(\sqrt{x} - \sqrt{y})$

### Solution

(a)
$$\begin{aligned}(\sqrt{3} + \sqrt{5})(\sqrt{2} + \sqrt{6}) &= \sqrt{3}(\sqrt{2} + \sqrt{6}) + \sqrt{5}(\sqrt{2} + \sqrt{6}) \\ &= \sqrt{3}\sqrt{2} + \sqrt{3}\sqrt{6} + \sqrt{5}\sqrt{2} + \sqrt{5}\sqrt{6} \\ &= \sqrt{6} + \sqrt{18} + \sqrt{10} + \sqrt{30} \\ &= \sqrt{6} + 3\sqrt{2} + \sqrt{10} + \sqrt{30}\end{aligned}$$

**(b)** $(2\sqrt{2} - \sqrt{7})(3\sqrt{2} + 5\sqrt{7}) = 2\sqrt{2}(3\sqrt{2} + 5\sqrt{7})$
$-\sqrt{7}(3\sqrt{2} + 5\sqrt{7})$
$= (2\sqrt{2})(3\sqrt{2}) + (2\sqrt{2})(5\sqrt{7})$
$-(\sqrt{7})(3\sqrt{2}) - (\sqrt{7})(5\sqrt{7})$
$= 12 + 10\sqrt{14} - 3\sqrt{14} - 35$
$= -23 + 7\sqrt{14}$

**(c)** $(\sqrt{8} + \sqrt{6})(\sqrt{8} - \sqrt{6}) = \sqrt{8}(\sqrt{8} - \sqrt{6}) + \sqrt{6}(\sqrt{8} - \sqrt{6})$
$= \sqrt{8}\sqrt{8} - \sqrt{8}\sqrt{6} + \sqrt{6}\sqrt{8} - \sqrt{6}\sqrt{6}$
$= 8 - \sqrt{48} + \sqrt{48} - 6$
$= 2$

**(d)** $(\sqrt{x} + \sqrt{y})(\sqrt{x} - \sqrt{y}) = \sqrt{x}(\sqrt{x} - \sqrt{y}) + \sqrt{y}(\sqrt{x} - \sqrt{y})$
$= \sqrt{x}\sqrt{x} - \sqrt{x}\sqrt{y} + \sqrt{y}\sqrt{x} - \sqrt{y}\sqrt{y}$
$= x - \sqrt{xy} + \sqrt{xy} - y$
$= x - y$

## Rationalizing Binomial Denominators

Note parts (c) and (d) of Example 3; they fit the special-product pattern $(a + b)(a - b) = a^2 - b^2$. Furthermore, in each case the final product is in rational form. The factors $a + b$ and $a - b$ are called **conjugates**. This suggests a way of rationalizing the denominator in an expression that contains a binomial denominator with radicals. We will multiply by the conjugate of the binomial denominator. Consider the following example.

**Classroom Example**
Simplify $\frac{2}{\sqrt{7} - \sqrt{3}}$ by rationalizing the denominator.

**EXAMPLE 4**

Simplify $\frac{4}{\sqrt{5} + \sqrt{2}}$ by rationalizing the denominator.

### Solution

$$\frac{4}{\sqrt{5} + \sqrt{2}} = \frac{4}{\sqrt{5} + \sqrt{2}} \cdot \left(\frac{\sqrt{5} - \sqrt{2}}{\sqrt{5} - \sqrt{2}}\right)$$

Form of 1

$$= \frac{4(\sqrt{5} - \sqrt{2})}{(\sqrt{5} + \sqrt{2})(\sqrt{5} - \sqrt{2})} = \frac{4(\sqrt{5} - \sqrt{2})}{5 - 2}$$

$$= \frac{4(\sqrt{5} - \sqrt{2})}{3} \quad \text{or} \quad \frac{4\sqrt{5} - 4\sqrt{2}}{3}$$

Either answer is acceptable

The next examples further illustrate the process of rationalizing and simplifying expressions that contain binomial denominators.

**EXAMPLE 5**

For each of the following, rationalize the denominator and simplify:

**(a)** $\frac{\sqrt{3}}{\sqrt{6} - 9}$ **(b)** $\frac{7}{3\sqrt{5} + 2\sqrt{3}}$ **(c)** $\frac{\sqrt{x} + 2}{\sqrt{x} - 3}$ **(d)** $\frac{2\sqrt{x} - 3\sqrt{y}}{\sqrt{x} + \sqrt{y}}$

**Classroom Example**
For each of the following, rationalize the denominator and simplify:

(a) $\dfrac{\sqrt{6}}{\sqrt{2}-8}$

(b) $\dfrac{5}{2\sqrt{5}-3\sqrt{2}}$

(c) $\dfrac{\sqrt{y}-6}{\sqrt{y}+2}$

(d) $\dfrac{8\sqrt{m}+5\sqrt{n}}{\sqrt{m}-\sqrt{n}}$

## Solution

**(a)**
$$\frac{\sqrt{3}}{\sqrt{6}-9} = \frac{\sqrt{3}}{\sqrt{6}-9}\cdot\frac{\sqrt{6}+9}{\sqrt{6}+9}$$
$$= \frac{\sqrt{3}(\sqrt{6}+9)}{(\sqrt{6}-9)(\sqrt{6}+9)}$$
$$= \frac{\sqrt{18}+9\sqrt{3}}{6-81}$$
$$= \frac{3\sqrt{2}+9\sqrt{3}}{-75}$$
$$= \frac{3(\sqrt{2}+3\sqrt{3})}{(-3)(25)}$$
$$= -\frac{\sqrt{2}+3\sqrt{3}}{25} \quad \text{or} \quad \frac{-\sqrt{2}-3\sqrt{3}}{25}$$

**(b)**
$$\frac{7}{3\sqrt{5}+2\sqrt{3}} = \frac{7}{3\sqrt{5}+2\sqrt{3}}\cdot\frac{3\sqrt{5}-2\sqrt{3}}{3\sqrt{5}-2\sqrt{3}}$$
$$= \frac{7(3\sqrt{5}-2\sqrt{3})}{(3\sqrt{5}+2\sqrt{3})(3\sqrt{5}-2\sqrt{3})}$$
$$= \frac{7(3\sqrt{5}-2\sqrt{3})}{45-12}$$
$$= \frac{7(3\sqrt{5}-2\sqrt{3})}{33} \quad \text{or} \quad \frac{21\sqrt{5}-14\sqrt{3}}{33}$$

**(c)**
$$\frac{\sqrt{x}+2}{\sqrt{x}-3} = \frac{\sqrt{x}+2}{\sqrt{x}-3}\cdot\frac{\sqrt{x}+3}{\sqrt{x}+3} = \frac{(\sqrt{x}+2)(\sqrt{x}+3)}{(\sqrt{x}-3)(\sqrt{x}+3)}$$
$$= \frac{x+3\sqrt{x}+2\sqrt{x}+6}{x-9}$$
$$= \frac{x+5\sqrt{x}+6}{x-9}$$

**(d)**
$$\frac{2\sqrt{x}-3\sqrt{y}}{\sqrt{x}+\sqrt{y}} = \frac{2\sqrt{x}-3\sqrt{y}}{\sqrt{x}+\sqrt{y}}\cdot\frac{\sqrt{x}-\sqrt{y}}{\sqrt{x}-\sqrt{y}}$$
$$= \frac{(2\sqrt{x}-3\sqrt{y})(\sqrt{x}-\sqrt{y})}{(\sqrt{x}+\sqrt{y})(\sqrt{x}-\sqrt{y})}$$
$$= \frac{2x-2\sqrt{xy}-3\sqrt{xy}+3y}{x-y}$$
$$= \frac{2x-5\sqrt{xy}+3y}{x-y}$$

## Concept Quiz 5.4

For Problems 1–10, answer true or false.

1. The property $\sqrt[n]{x}\sqrt[n]{y} = \sqrt[n]{xy}$ can be used to express the product of two radicals as one radical.

2. The product of two radicals always results in an expression that has a radical even after simplifying.
3. The conjugate of $5 + \sqrt{3}$ is $-5 - \sqrt{3}$.
4. The product of $2 - \sqrt{7}$ and $2 + \sqrt{7}$ is a rational number.
5. To rationalize the denominator for the expression $\frac{2\sqrt{5}}{4 - \sqrt{5}}$, we would multiply by $\frac{\sqrt{5}}{\sqrt{5}}$.
6. To rationalize the denominator for the expression $\frac{\sqrt{x} + 8}{\sqrt{x} - 4}$, we would multiply the numerator and denominator by $\sqrt{x} - 4$.
7. $\frac{\sqrt{8} + \sqrt{12}}{\sqrt{2}} = 2 + \sqrt{6}$
8. $\frac{\sqrt{2}}{\sqrt{8} + \sqrt{12}} = \frac{1}{2 + \sqrt{6}}$
9. The product of $5 + \sqrt{3}$ and $-5 - \sqrt{3}$ is $-28$.
10. The product of $\sqrt{5} - 1$ and $\sqrt{5} + 1$ is 24.

## Problem Set 5.4

For Problems 1–14, multiply and simplify where possible. (Objective 1)

1. $\sqrt{6}\sqrt{12}$
2. $\sqrt{8}\sqrt{6}$
3. $(3\sqrt{3})(2\sqrt{6})$
4. $(5\sqrt{2})(3\sqrt{12})$
5. $(4\sqrt{2})(-6\sqrt{5})$
6. $(-7\sqrt{3})(2\sqrt{5})$
7. $(-3\sqrt{3})(-4\sqrt{8})$
8. $(-5\sqrt{8})(-6\sqrt{7})$
9. $(5\sqrt{6})(4\sqrt{6})$
10. $(3\sqrt{7})(2\sqrt{7})$
11. $(2\sqrt[3]{4})(6\sqrt[3]{2})$
12. $(4\sqrt[3]{3})(5\sqrt[3]{9})$
13. $(4\sqrt[3]{6})(7\sqrt[3]{4})$
14. $(9\sqrt[3]{6})(2\sqrt[3]{9})$

For Problems 15–52, find the following products and express answers in simplest radical form. All variables represent nonnegative real numbers. (Objective 2)

15. $\sqrt{2}(\sqrt{3} + \sqrt{5})$
16. $\sqrt{3}(\sqrt{7} + \sqrt{10})$
17. $3\sqrt{5}(2\sqrt{2} - \sqrt{7})$
18. $5\sqrt{6}(2\sqrt{5} - 3\sqrt{11})$
19. $2\sqrt{6}(3\sqrt{8} - 5\sqrt{12})$
20. $4\sqrt{2}(3\sqrt{12} + 7\sqrt{6})$
21. $-4\sqrt{5}(2\sqrt{5} + 4\sqrt{12})$
22. $-5\sqrt{3}(3\sqrt{12} - 9\sqrt{8})$
23. $3\sqrt{x}(5\sqrt{2} + \sqrt{y})$
24. $\sqrt{2x}(3\sqrt{y} - 7\sqrt{5})$
25. $\sqrt{xy}(5\sqrt{xy} - 6\sqrt{x})$
26. $4\sqrt{x}(2\sqrt{xy} + 2\sqrt{x})$
27. $\sqrt{5y}(\sqrt{8x} + \sqrt{12y^2})$
28. $\sqrt{2x}(\sqrt{12xy} - \sqrt{8y})$
29. $5\sqrt{3}(2\sqrt{8} - 3\sqrt{18})$
30. $2\sqrt{2}(3\sqrt{12} - \sqrt{27})$
31. $(\sqrt{3} + 4)(\sqrt{3} - 7)$
32. $(\sqrt{2} + 6)(\sqrt{2} - 2)$
33. $(\sqrt{5} - 6)(\sqrt{5} - 3)$
34. $(\sqrt{7} - 2)(\sqrt{7} - 8)$
35. $(3\sqrt{5} - 2\sqrt{3})(2\sqrt{7} + \sqrt{2})$
36. $(\sqrt{2} + \sqrt{3})(\sqrt{5} - \sqrt{7})$
37. $(2\sqrt{6} + 3\sqrt{5})(\sqrt{8} - 3\sqrt{12})$
38. $(5\sqrt{2} - 4\sqrt{6})(2\sqrt{8} + \sqrt{6})$
39. $(2\sqrt{6} + 5\sqrt{5})(3\sqrt{6} - \sqrt{5})$
40. $(7\sqrt{3} - \sqrt{7})(2\sqrt{3} + 4\sqrt{7})$
41. $(3\sqrt{2} - 5\sqrt{3})(6\sqrt{2} - 7\sqrt{3})$
42. $(\sqrt{8} - 3\sqrt{10})(2\sqrt{8} - 6\sqrt{10})$
43. $(\sqrt{6} + 4)(\sqrt{6} - 4)$
44. $(\sqrt{7} - 2)(\sqrt{7} + 2)$
45. $(\sqrt{2} + \sqrt{10})(\sqrt{2} - \sqrt{10})$

46. $(2\sqrt{3} + \sqrt{11})(2\sqrt{3} - \sqrt{11})$

47. $(\sqrt{2x} + \sqrt{3y})(\sqrt{2x} - \sqrt{3y})$

48. $(2\sqrt{x} - 5\sqrt{y})(2\sqrt{x} + 5\sqrt{y})$

49. $2\sqrt[3]{3}(5\sqrt[3]{4} + \sqrt[3]{6})$

50. $2\sqrt[3]{2}(3\sqrt[3]{6} - 4\sqrt[3]{5})$

51. $3\sqrt[3]{4}(2\sqrt[3]{2} - 6\sqrt[3]{4})$

52. $3\sqrt[3]{3}(4\sqrt[3]{9} + 5\sqrt[3]{7})$

For Problems 53–76, rationalize the denominator and simplify. All variables represent positive real numbers. (Objective 3)

53. $\dfrac{2}{\sqrt{7} + 1}$

54. $\dfrac{6}{\sqrt{5} + 2}$

55. $\dfrac{3}{\sqrt{2} - 5}$

56. $\dfrac{-4}{\sqrt{6} - 3}$

57. $\dfrac{1}{\sqrt{2} + \sqrt{7}}$

58. $\dfrac{3}{\sqrt{3} + \sqrt{10}}$

59. $\dfrac{\sqrt{2}}{\sqrt{10} - \sqrt{3}}$

60. $\dfrac{\sqrt{3}}{\sqrt{7} - \sqrt{2}}$

61. $\dfrac{\sqrt{3}}{2\sqrt{5} + 4}$

62. $\dfrac{\sqrt{7}}{3\sqrt{2} - 5}$

63. $\dfrac{6}{3\sqrt{7} - 2\sqrt{6}}$

64. $\dfrac{5}{2\sqrt{5} + 3\sqrt{7}}$

65. $\dfrac{\sqrt{6}}{3\sqrt{2} + 2\sqrt{3}}$

66. $\dfrac{3\sqrt{6}}{5\sqrt{3} - 4\sqrt{2}}$

67. $\dfrac{2}{\sqrt{x} + 4}$

68. $\dfrac{3}{\sqrt{x} + 7}$

69. $\dfrac{\sqrt{x}}{\sqrt{x} - 5}$

70. $\dfrac{\sqrt{x}}{\sqrt{x} - 1}$

71. $\dfrac{\sqrt{x} - 2}{\sqrt{x} + 6}$

72. $\dfrac{\sqrt{x} + 1}{\sqrt{x} - 10}$

73. $\dfrac{\sqrt{x}}{\sqrt{x} + 2\sqrt{y}}$

74. $\dfrac{\sqrt{y}}{2\sqrt{x} - \sqrt{y}}$

75. $\dfrac{3\sqrt{y}}{2\sqrt{x} - 3\sqrt{y}}$

76. $\dfrac{2\sqrt{x}}{3\sqrt{x} + 5\sqrt{y}}$

## Thoughts Into Words

77. How would you help someone rationalize the denominator and simplify $\dfrac{4}{\sqrt{8} + \sqrt{12}}$?

78. Discuss how the distributive property has been used thus far in this chapter.

79. How would you simplify the expression $\dfrac{\sqrt{8} + \sqrt{12}}{\sqrt{2}}$?

## Further Investigations

80. Use your calculator to evaluate each expression in Problems 53–66. Then evaluate the results you obtained when you did the problems.

**Answers to the Concept Quiz**

**1.** True **2.** False **3.** False **4.** True **5.** False **6.** False **7.** True **8.** False **9.** False **10.** False

# 5.5 Equations Involving Radicals

OBJECTIVES

1. Solve radical equations
2. Solve radical equations for real-world problems

We often refer to equations that contain radicals with variables in a radicand as **radical equations**. In this section we discuss techniques for solving such equations that contain one or more radicals. To solve radical equations, we need the following property of equality.

**Property 5.5**

Let $a$ and $b$ be real numbers and $n$ be a positive integer.

If $a = b$ then $a^n = b^n$

Property 5.5 states that we can raise both sides of an equation to a positive integral power. However, raising both sides of an equation to a positive integral power sometimes produces results that do not satisfy the original equation. Let's consider two examples to illustrate this point.

**Classroom Example**
Solve $\sqrt{2x - 1} = 3$.

**EXAMPLE 1** Solve $\sqrt{2x - 5} = 7$.

**Solution**

$$\sqrt{2x - 5} = 7$$
$$(\sqrt{2x - 5})^2 = 7^2 \quad \text{Square both sides}$$
$$2x - 5 = 49$$
$$2x = 54$$
$$x = 27$$

✔ **Check**

$$\sqrt{2x - 5} = 7$$
$$\sqrt{2(27) - 5} \stackrel{?}{=} 7$$
$$\sqrt{49} \stackrel{?}{=} 7$$
$$7 = 7$$

The solution set for $\sqrt{2x - 5} = 7$ is $\{27\}$.

**Classroom Example**
Solve $\sqrt{5y + 9} = -8$.

**EXAMPLE 2** Solve $\sqrt{3a + 4} = -4$.

**Solution**

$$\sqrt{3a + 4} = -4$$
$$(\sqrt{3a + 4})^2 = (-4)^2 \quad \text{Square both sides}$$
$$3a + 4 = 16$$
$$3a = 12$$
$$a = 4$$

✔ **Check**

$$\begin{aligned}\sqrt{3a+4} &= -4\\ \sqrt{3(4)+4} &\stackrel{?}{=} -4\\ \sqrt{16} &\stackrel{?}{=} -4\\ 4 &\neq -4\end{aligned}$$

Because 4 does not check, the original equation has no real number solution. Thus the solution set is $\varnothing$.

In general, raising bo th sides of an equation to a positive integral power produces an equation that has all of the solutions of the original equation, but it may also have some extra solutions that do not satisfy the original equation. Such extra solutions are called **extraneous solutions**. Therefore, when using Property 5.5, you *must* check each potential solution in the original equation.

Let's consider some examples to illustrate different situations that arise when we are solving radical equations.

Classroom Example
Solve $\sqrt{2x+6} = x+3$.

**EXAMPLE 3** Solve $\sqrt{2t-4} = t-2$.

### Solution

$$\begin{aligned}\sqrt{2t-4} &= t-2\\ \left(\sqrt{2t-4}\right)^2 &= (t-2)^2 && \text{Square both sides}\\ 2t-4 &= t^2-4t+4\\ 0 &= t^2-6t+8\\ 0 &= (t-2)(t-4) && \text{Factor the right side}\end{aligned}$$

$t-2=0$ or $t-4=0$ Apply: $ab=0$ if and only if $a=0$ or $b=0$

$t=2$ or $t=4$

✔ **Check**

$$\begin{aligned}\sqrt{2t-4} &= t-2\\ \sqrt{2(2)-4} &\stackrel{?}{=} 2-2 \quad \text{when } t=2\\ \sqrt{0} &\stackrel{?}{=} 0\\ 0 &= 0\end{aligned}$$

or

$$\begin{aligned}\sqrt{2t-4} &= t-2\\ \sqrt{2(4)-4} &\stackrel{?}{=} 4-2 \quad \text{when } t=4\\ \sqrt{4} &\stackrel{?}{=} 2\\ 2 &= 2\end{aligned}$$

The solution set is $\{2, 4\}$.

Classroom Example
Solve $\sqrt{m}+2 = m$.

**EXAMPLE 4** Solve $\sqrt{y}+6 = y$.

### Solution

$$\begin{aligned}\sqrt{y}+6 &= y\\ \sqrt{y} &= y-6\\ \left(\sqrt{y}\right)^2 &= (y-6)^2 && \text{Square both sides}\\ y &= y^2-12y+36\\ 0 &= y^2-13y+36\\ 0 &= (y-4)(y-9) && \text{Factor the right side}\end{aligned}$$

$y-4=0$ or $y-9=0$ Apply: $ab=0$ if and only if $a=0$ or $b=0$

$y=4$ or $y=9$

✔ **Check**

$$\sqrt{y} + 6 = y \qquad\qquad \sqrt{y} + 6 = y$$

$$\sqrt{4} + 6 \stackrel{?}{=} 4 \quad \text{when } y = 4 \qquad \text{or} \qquad \sqrt{9} + 6 \stackrel{?}{=} 9 \quad \text{when } y = 9$$

$$2 + 6 \stackrel{?}{=} 4 \qquad\qquad 3 + 6 \stackrel{?}{=} 9$$

$$8 \neq 4 \qquad\qquad 9 = 9$$

The only solution is 9; the solution set is $\{9\}$.

In Example 4, note that we changed the form of the original equation $\sqrt{y} + 6 = y$ to $\sqrt{y} = y - 6$ before we squared both sides. Squaring both sides of $\sqrt{y} + 6 = y$ produces $y + 12\sqrt{y} + 36 = y^2$, which is a much more complex equation that still contains a radical. Here again, it pays to think ahead before carrying out all the steps. Now let's consider an example involving a cube root.

Classroom Example
Solve $\sqrt[3]{x^2 + 2} = 3$.

**EXAMPLE 5** Solve $\sqrt[3]{n^2 - 1} = 2$.

### Solution

$$\sqrt[3]{n^2 - 1} = 2$$

$$\left(\sqrt[3]{n^2 - 1}\right)^3 = 2^3 \qquad \text{Cube both sides}$$

$$n^2 - 1 = 8$$

$$n^2 - 9 = 0$$

$$(n + 3)(n - 3) = 0$$

$$n + 3 = 0 \quad \text{or} \quad n - 3 = 0$$

$$n = -3 \quad \text{or} \quad n = 3$$

✔ **Check**

$$\sqrt[3]{n^2 - 1} = 2 \qquad\qquad \sqrt[3]{n^2 - 1} = 2$$

$$\sqrt[3]{(-3)^2 - 1} \stackrel{?}{=} 2 \quad \text{when } n = -3 \qquad \text{or} \qquad \sqrt[3]{3^2 - 1} \stackrel{?}{=} 2 \quad \text{when } n = 3$$

$$\sqrt[3]{8} \stackrel{?}{=} 2 \qquad\qquad \sqrt[3]{8} \stackrel{?}{=} 2$$

$$2 = 2 \qquad\qquad 2 = 2$$

The solution set is $\{-3, 3\}$.

It may be necessary to square both sides of an equation, simplify the resulting equation, and then square both sides again. The next example illustrates this type of problem.

Classroom Example
Solve $\sqrt{x + 4} = 1 + \sqrt{x - 1}$.

**EXAMPLE 6** Solve $\sqrt{x + 2} = 7 - \sqrt{x + 9}$.

### Solution

$$\sqrt{x + 2} = 7 - \sqrt{x + 9}$$

$$\left(\sqrt{x + 2}\right)^2 = \left(7 - \sqrt{x + 9}\right)^2 \qquad \text{Square both sides}$$

$$x + 2 = 49 - 14\sqrt{x + 9} + x + 9$$

$$x + 2 = x + 58 - 14\sqrt{x + 9}$$

$$-56 = -14\sqrt{x + 9}$$

$$4 = \sqrt{x + 9}$$

$$(4)^2 = \left(\sqrt{x + 9}\right)^2 \qquad \text{Square both sides}$$

$$16 = x + 9$$

$$7 = x$$

✔ **Check**

$$\sqrt{x+2} = 7 - \sqrt{x+9}$$
$$\sqrt{7+2} \stackrel{?}{=} 7 - \sqrt{7+9} \quad \text{when } x = 7$$
$$\sqrt{9} \stackrel{?}{=} 7 - \sqrt{16}$$
$$3 \stackrel{?}{=} 7 - 4$$
$$3 = 3$$

The solution set is $\{7\}$.

## Solving Radical Equations for Real-World Problems

In Section 5.1 we used the formula $S = \sqrt{30Df}$ to approximate how fast a car was traveling on the basis of the length of skid marks. (Remember that $S$ represents the speed of the car in miles per hour, $D$ represents the length of the skid marks in feet, and $f$ represents a coefficient of friction.) This same formula can be used to estimate the length of skid marks that are produced by cars traveling at different rates on various types of road surfaces. To use the formula for this purpose, let's change the form of the equation by solving for $D$.

$$\sqrt{30Df} = S$$
$$30Df = S^2 \quad \text{The result of squaring both sides of the original equation}$$
$$D = \frac{S^2}{30f} \quad D,\ S, \text{ and } f \text{ are positive numbers, so this final equation and the original one are equivalent}$$

**Classroom Example**
Suppose that for a particular road surface, the coefficient of friction is 0.27. How far will a car skid when the brakes are applied at 65 miles per hour?

### EXAMPLE 7 Apply Your Skill

Suppose that for a particular road surface, the coefficient of friction is 0.35. How far will a car skid when the brakes are applied at 60 miles per hour?

### Solution

We can substitute 0.35 for $f$ and 60 for $S$ in the formula $D = \frac{S^2}{30f}$.

$$D = \frac{60^2}{30(0.35)} = 343 \quad \text{to the nearest whole number}$$

The car will skid approximately 343 feet.

**Remark:** Pause for a moment and think about the result in Example 7. The coefficient of friction 0.35 refers to a wet concrete road surface. Note that a car traveling at 60 miles per hour on such a surface will skid more than the length of a football field.

## Concept Quiz 5.5

For Problems 1–10, answer true or false.

1. To solve a radical equation, we can raise each side of the equation to a positive integer power.
2. Solving the equation that results from squaring each side of an original equation may not give all the solutions of the original equation.
3. The equation $\sqrt[3]{x-1} = -2$ has a solution.
4. Potential solutions that do not satisfy the original equation are called extraneous solutions.
5. The equation $\sqrt{x+1} = -2$ has no real number solutions.
6. The solution set for $\sqrt{x+2} = x$ is $\{1, 4\}$.
7. The solution set for $\sqrt{x+1} + \sqrt{x-2} = -3$ is the null set.

**8.** The solution set for $\sqrt[3]{x+2} = -2$ is the null set.

**9.** The solution set for the equation $\sqrt{x^2 - 2x + 1} = x - 3$ is $\{2\}$.

**10.** The solution set for the equation $\sqrt{5x+1} + \sqrt{x+4} = 3$ is $\{0\}$.

## Problem Set 5.5

For Problems 1–56, solve each equation. Don't forget to check each of your potential solutions. (Objective 1)

**1.** $\sqrt{5x} = 10$

**2.** $\sqrt{3x} = 9$

**3.** $\sqrt{2x} + 4 = 0$

**4.** $\sqrt{4x} + 5 = 0$

**5.** $2\sqrt{n} = 5$

**6.** $5\sqrt{n} = 3$

**7.** $3\sqrt{n} - 2 = 0$

**8.** $2\sqrt{n} - 7 = 0$

**9.** $\sqrt{3y+1} = 4$

**10.** $\sqrt{2y-3} = 5$

**11.** $\sqrt{4y-3} - 6 = 0$

**12.** $\sqrt{3y+5} - 2 = 0$

**13.** $\sqrt{3x-1} + 1 = 4$

**14.** $\sqrt{4x-1} - 3 = 2$

**15.** $\sqrt{2n+3} - 2 = -1$

**16.** $\sqrt{5n+1} - 6 = -4$

**17.** $\sqrt{2x-5} = -1$

**18.** $\sqrt{4x-3} = -4$

**19.** $\sqrt{5x+2} = \sqrt{6x+1}$

**20.** $\sqrt{4x+2} = \sqrt{3x+4}$

**21.** $\sqrt{3x+1} = \sqrt{7x-5}$

**22.** $\sqrt{6x+5} = \sqrt{2x+10}$

**23.** $\sqrt{3x-2} - \sqrt{x+4} = 0$

**24.** $\sqrt{7x-6} - \sqrt{5x+2} = 0$

**25.** $5\sqrt{t-1} = 6$

**26.** $4\sqrt{t+3} = 6$

**27.** $\sqrt{x^2+7} = 4$

**28.** $\sqrt{x^2+3} - 2 = 0$

**29.** $\sqrt{x^2+13x+37} = 1$

**30.** $\sqrt{x^2+5x-20} = 2$

**31.** $\sqrt{x^2-x+1} = x+1$

**32.** $\sqrt{n^2-2n-4} = n$

**33.** $\sqrt{x^2+3x+7} = x+2$

**34.** $\sqrt{x^2+2x+1} = x+3$

**35.** $\sqrt{-4x+17} = x-3$

**36.** $\sqrt{2x-1} = x-2$

**37.** $\sqrt{n+4} = n+4$

**38.** $\sqrt{n+6} = n+6$

**39.** $\sqrt{3y} = y-6$

**40.** $2\sqrt{n} = n-3$

**41.** $4\sqrt{x} + 5 = x$

**42.** $\sqrt{-x} - 6 = x$

**43.** $\sqrt[3]{x-2} = 3$

**44.** $\sqrt[3]{x+1} = 4$

**45.** $\sqrt[3]{2x+3} = -3$

**46.** $\sqrt[3]{3x-1} = -4$

**47.** $\sqrt[3]{2x+5} = \sqrt[3]{4-x}$

**48.** $\sqrt[3]{3x-1} = \sqrt[3]{2-5x}$

**49.** $\sqrt{x+19} - \sqrt{x+28} = -1$

**50.** $\sqrt{x+4} = \sqrt{x-1} + 1$

**51.** $\sqrt{3x+1} + \sqrt{2x+4} = 3$

**52.** $\sqrt{2x-1} - \sqrt{x+3} = 1$

**53.** $\sqrt{n-4} + \sqrt{n+4} = 2\sqrt{n-1}$

**54.** $\sqrt{n-3} + \sqrt{n+5} = 2\sqrt{n}$

**55.** $\sqrt{t+3} - \sqrt{t-2} = \sqrt{7-t}$

**56.** $\sqrt{t+7} - 2\sqrt{t-8} = \sqrt{t-5}$

For Problems 57–59, use the appropriate formula to solve the problems. (Objective 2)

**57.** Use the formula given in Example 7 with a coefficient of friction of 0.95. How far will a car skid at 40 miles per hour? at 55 miles per hour? at 65 miles per hour? Express the answers to the nearest foot.

**58.** Solve the formula $T = 2\pi\sqrt{\frac{L}{32}}$ for $L$. (Remember that in this formula, which was used in Section 5.2, $T$ represents the period of a pendulum expressed in seconds, and $L$ represents the length of the pendulum in feet.)

**59.** In Problem 58, you should have obtained the equation $L = \frac{8T^2}{\pi^2}$. What is the length of a pendulum that has a period of 2 seconds? of 2.5 seconds? of 3 seconds? Express your answers to the nearest tenth of a foot.

## Thoughts Into Words

**60.** Your friend makes an effort to solve the equation $3 + 2\sqrt{x} = x$ as follows:

$$\left(3 + 2\sqrt{x}\right)^2 = x^2$$
$$9 + 12\sqrt{x} + 4x = x^2$$

At this step he stops and doesn't know how to proceed. What help would you give him?

**61.** Explain why possible solutions for radical equations *must* be checked.

**62.** Explain the concept of extraneous solutions.

### Answers to the Concept Quiz

**1.** True **2.** False **3.** True **4.** True **5.** True **6.** False **7.** True **8.** False **9.** False **10.** True

# 5.6 Merging Exponents and Roots

OBJECTIVES

1. Evaluate a number raised to a rational exponent
2. Write expressions with rational exponents as radicals
3. Write radical expressions as expressions with rational exponents
4. Simplify algebraic expressions that have rational exponents
5. Multiply and divide radicals with different indexes

Recall that the basic properties of positive integral exponents led to a definition for the use of negative integers as exponents. In this section, the properties of integral exponents are used to form definitions for the use of rational numbers as exponents. These definitions will tie together the concepts of exponent and root.

Let's consider the following comparisons.

From our study of radicals, we know that

$$(\sqrt{5})^2 = 5$$
$$(\sqrt[3]{8})^3 = 8$$
$$(\sqrt[4]{21})^4 = 21$$

If $(b^n)^m = b^{mn}$ is to hold when $n$ equals a rational number of the form $\frac{1}{p}$, where $p$ is a positive integer greater than 1, then

$$(5^{\frac{1}{2}})^2 = 5^{2(\frac{1}{2})} = 5^1 = 5$$
$$(8^{\frac{1}{3}})^3 = 8^{3(\frac{1}{3})} = 8^1 = 8$$
$$(21^{\frac{1}{4}})^4 = 21^{4(\frac{1}{4})} = 21^1 = 21$$

It would seem reasonable to make the following definition.

**Definition 5.6**

If $b$ is a real number, $n$ is a positive integer greater than 1, and $\sqrt[n]{b}$ exists, then

$$b^{\frac{1}{n}} = \sqrt[n]{b}$$

Definition 5.6 states that $b^{\frac{1}{n}}$ means the $n$th root of $b$. We shall assume that $b$ and $n$ are chosen so that $\sqrt[n]{b}$ exists. For example, $(-25)^{\frac{1}{2}}$ is not meaningful at this time because $\sqrt{-25}$ is not a real number. Consider the following examples, which demonstrate the use of Definition 5.6.

$$25^{\frac{1}{2}} = \sqrt{25} = 5 \qquad 16^{\frac{1}{4}} = \sqrt[4]{16} = 2$$

$$8^{\frac{1}{3}} = \sqrt[3]{8} = 2 \qquad \left(\frac{36}{49}\right)^{\frac{1}{2}} = \sqrt{\frac{36}{49}} = \frac{6}{7}$$

$$(-27)^{\frac{1}{3}} = \sqrt[3]{-27} = -3$$

The following definition provides the basis for the use of *all* rational numbers as exponents.

**Definition 5.7**

If $\frac{m}{n}$ is a rational number, where $n$ is a positive integer greater than 1, and $b$ is a real number such that $\sqrt[n]{b}$ exists, then

$$b^{\frac{m}{n}} = \sqrt[n]{b^m} = (\sqrt[n]{b})^m$$

In Definition 5.7, note that the denominator of the exponent is the index of the radical and that the numerator of the exponent is either the exponent of the radicand or the exponent of the root.

Whether we use the form $\sqrt[n]{b^m}$ or the form $\left(\sqrt[n]{b}\right)^m$ for computational purposes depends somewhat on the magnitude of the problem. Let's use both forms on two problems to illustrate this point.

$$8^{\frac{2}{3}} = \sqrt[3]{8^2} = \sqrt[3]{64} = 4 \qquad \text{or} \qquad 8^{\frac{2}{3}} = \left(\sqrt[3]{8}\right)^2 = 2^2 = 4$$

$$27^{\frac{2}{3}} = \sqrt[3]{27^2} = \sqrt[3]{729} = 9 \qquad \text{or} \qquad 27^{\frac{2}{3}} = \left(\sqrt[3]{27}\right)^2 = 3^2 = 9$$

To compute $8^{\frac{2}{3}}$, either form seems to work about as well as the other one. However, to compute $27^{\frac{2}{3}}$, it should be obvious that $\left(\sqrt[3]{27}\right)^2$ is much easier to handle than $\sqrt[3]{27^2}$.

**Classroom Example**
Simplify each of the following numerical expressions:
(a) $49^{\frac{3}{2}}$ (b) $81^{\frac{3}{4}}$
(c) $8^{-\frac{5}{3}}$ (d) $(-27)^{\frac{4}{3}}$
(e) $-64^{\frac{1}{3}}$

**EXAMPLE 1** Simplify each of the following numerical expressions:

(a) $25^{\frac{3}{2}}$ (b) $16^{\frac{3}{4}}$ (c) $(32)^{-\frac{2}{5}}$ (d) $(-64)^{\frac{2}{3}}$ (e) $-8^{\frac{1}{3}}$

### Solution

(a) $25^{\frac{3}{2}} = \left(\sqrt{25}\right)^3 = 5^3 = 125$

(b) $16^{\frac{3}{4}} = \left(\sqrt[4]{16}\right)^3 = 2^3 = 8$

(c) $(32)^{-\frac{2}{5}} = \dfrac{1}{(32)^{\frac{2}{5}}} = \dfrac{1}{\left(\sqrt[5]{32}\right)^2} = \dfrac{1}{2^2} = \dfrac{1}{4}$

(d) $(-64)^{\frac{2}{3}} = \left(\sqrt[3]{-64}\right)^2 = (-4)^2 = 16$

(e) $-8^{\frac{1}{3}} = -\sqrt[3]{8} = -2$

The basic laws of exponents that we stated in Property 5.1 are true for all rational exponents. Therefore, from now on we will use Property 5.1 for rational as well as integral exponents.

Some problems can be handled better in exponential form and others in radical form. Thus we must be able to switch forms with a certain amount of ease. Let's consider some examples that require a switch from one form to the other.

**Classroom Example**
Write each of the following expressions in radical form:
(a) $m^{\frac{2}{5}}$ (b) $6a^{\frac{4}{7}}$
(c) $m^{\frac{1}{3}}n^{\frac{2}{3}}$ (d) $(a + b)^{\frac{3}{4}}$

**EXAMPLE 2** Write each of the following expressions in radical form:

(a) $x^{\frac{3}{4}}$ (b) $3y^{\frac{2}{5}}$ (c) $x^{\frac{1}{4}}y^{\frac{3}{4}}$ (d) $(x + y)^{\frac{2}{3}}$

### Solution

(a) $x^{\frac{3}{4}} = \sqrt[4]{x^3}$

(b) $3y^{\frac{2}{5}} = 3\sqrt[5]{y^2}$

(c) $x^{\frac{1}{4}}y^{\frac{3}{4}} = (xy^3)^{\frac{1}{4}} = \sqrt[4]{xy^3}$

(d) $(x + y)^{\frac{2}{3}} = \sqrt[3]{(x + y)^2}$

**Classroom Example**
Write each of the following using positive rational exponents:
(a) $\sqrt{ab}$ (b) $\sqrt[3]{m^2n}$
(c) $5\sqrt[4]{x^3}$ (d) $\sqrt[6]{(m + n)^5}$

**EXAMPLE 3** Write each of the following using positive rational exponents:

(a) $\sqrt{xy}$ (b) $\sqrt[4]{a^3b}$ (c) $4\sqrt[3]{x^2}$ (d) $\sqrt[5]{(x + y)^4}$

### Solution

(a) $\sqrt{xy} = (xy)^{\frac{1}{2}} = x^{\frac{1}{2}}y^{\frac{1}{2}}$

(b) $\sqrt[4]{a^3b} = (a^3b)^{\frac{1}{4}} = a^{\frac{3}{4}}b^{\frac{1}{4}}$

(c) $4\sqrt[3]{x^2} = 4x^{\frac{2}{3}}$

(d) $\sqrt[5]{(x + y)^4} = (x + y)^{\frac{4}{5}}$

The properties of exponents provide the basis for simplifying algebraic expressions that contain rational exponents, as these next examples illustrate.

**Classroom Example**
Simplify each of the following. Express final results using positive exponents only:
**(a)** $(5x^{\frac{1}{3}})(2x^{\frac{3}{5}})$ **(b)** $(3m^{\frac{1}{4}}n^{\frac{1}{6}})^3$
**(c)** $\dfrac{18y^{\frac{1}{6}}}{9y^{\frac{1}{4}}}$ **(d)** $\left(\dfrac{5x^{\frac{1}{7}}}{8y^{\frac{3}{5}}}\right)^3$

## EXAMPLE 4

Simplify each of the following. Express final results using positive exponents only:

**(a)** $(3x^{\frac{1}{2}})(4x^{\frac{2}{3}})$ **(b)** $(5a^{\frac{1}{3}}b^{\frac{1}{2}})^2$ **(c)** $\dfrac{12y^{\frac{1}{3}}}{6y^{\frac{1}{2}}}$ **(d)** $\left(\dfrac{3x^{\frac{2}{5}}}{2y^{\frac{2}{3}}}\right)^4$

### Solution

**(a)**
$$\begin{aligned}(3x^{\frac{1}{2}})(4x^{\frac{2}{3}}) &= 3 \cdot 4 \cdot x^{\frac{1}{2}} \cdot x^{\frac{2}{3}} \\ &= 12x^{\frac{1}{2}+\frac{2}{3}} && b^n \cdot b^m = b^{n+m} \\ &= 12x^{\frac{3}{6}+\frac{4}{6}} && \text{Use 6 as LCD} \\ &= 12x^{\frac{7}{6}}\end{aligned}$$

**(b)**
$$\begin{aligned}(5a^{\frac{1}{3}}b^{\frac{1}{2}})^2 &= 5^2 \cdot (a^{\frac{1}{3}})^2 \cdot (b^{\frac{1}{2}})^2 && (ab)^n = a^nb^n \\ &= 25a^{\frac{2}{3}}b && (b^n)^m = b^{mn}\end{aligned}$$

**(c)**
$$\begin{aligned}\frac{12y^{\frac{1}{3}}}{6y^{\frac{1}{2}}} &= 2y^{\frac{1}{3}-\frac{1}{2}} && \frac{b^n}{b^m} = b^{n-m} \\ &= 2y^{\frac{2}{6}-\frac{3}{6}} \\ &= 2y^{-\frac{1}{6}} \\ &= \frac{2}{y^{\frac{1}{6}}}\end{aligned}$$

**(d)**
$$\begin{aligned}\left(\frac{3x^{\frac{2}{5}}}{2y^{\frac{2}{3}}}\right)^4 &= \frac{(3x^{\frac{2}{5}})^4}{(2y^{\frac{2}{3}})^4} && \left(\frac{a}{b}\right)^n = \frac{a^n}{b^n} \\ &= \frac{3^4 \cdot (x^{\frac{2}{5}})^4}{2^4 \cdot (y^{\frac{2}{3}})^4} && (ab)^n = a^nb^n \\ &= \frac{81x^{\frac{8}{5}}}{16y^{\frac{8}{3}}} && (b^n)^m = b^{mn}\end{aligned}$$

## Multiplying and Dividing Radicals with Different Indexes

The link between exponents and roots also provides a basis for multiplying and dividing some radicals even if they have different indexes. The general procedure is as follows:

1. **Change from radical form to exponential form.**
2. **Apply the properties of exponents.**
3. **Then change back to radical form.**

The three parts of Example 5 illustrate this process.

**Classroom Example**
Perform the indicated operations and express the answers in simplest radical form:
**(a)** $\sqrt[3]{3} \cdot \sqrt[4]{3}$
**(b)** $\dfrac{\sqrt[3]{2}}{\sqrt{2}}$
**(c)** $\dfrac{\sqrt{9}}{\sqrt[3]{3}}$

## EXAMPLE 5

Perform the indicated operations and express the answers in simplest radical form:

**(a)** $\sqrt{2}\sqrt[3]{2}$ **(b)** $\dfrac{\sqrt{5}}{\sqrt[3]{5}}$ **(c)** $\dfrac{\sqrt{4}}{\sqrt[3]{2}}$

**Solution**

**(a)** $\sqrt{2}\sqrt[3]{2} = 2^{\frac{1}{2}} \cdot 2^{\frac{1}{3}}$

$= 2^{\frac{1}{2}+\frac{1}{3}}$

$= 2^{\frac{3}{6}+\frac{2}{6}}$ Use 6 as LCD

$= 2^{\frac{5}{6}}$

$= \sqrt[6]{2^5} = \sqrt[6]{32}$

**(b)** $\dfrac{\sqrt{5}}{\sqrt[3]{5}} = \dfrac{5^{\frac{1}{2}}}{5^{\frac{1}{3}}}$

$= 5^{\frac{1}{2}-\frac{1}{3}}$

$= 5^{\frac{3}{6}-\frac{2}{6}}$ Use 6 as LCD

$= 5^{\frac{1}{6}} = \sqrt[6]{5}$

**(c)** $\dfrac{\sqrt{4}}{\sqrt[3]{2}} = \dfrac{4^{\frac{1}{2}}}{2^{\frac{1}{3}}}$

$= \dfrac{(2^2)^{\frac{1}{2}}}{2^{\frac{1}{3}}}$

$= \dfrac{2^1}{2^{\frac{1}{3}}}$

$= 2^{1-\frac{1}{3}}$

$= 2^{\frac{2}{3}} = \sqrt[3]{2^2} = \sqrt[3]{4}$

## Concept Quiz 5.6

For Problems 1–10, answer true or false.

1. Assuming the $n$th root of $x$ exists, $\sqrt[n]{x}$ can be written as $x^{\frac{1}{n}}$.
2. An exponent of $\frac{1}{3}$ means that we need to find the cube root of the number.
3. To evaluate $16^{\frac{2}{3}}$ we would find the square root of 16 and then cube the result.
4. When an expression with a rational exponent is written as a radical expression, the denominator of the rational exponent is the index of the radical.
5. The expression $\sqrt[n]{x^m}$ is equivalent to $\left(\sqrt[n]{x}\right)^m$.
6. $-16^{-3} = \dfrac{1}{64}$
7. $\dfrac{\sqrt{7}}{\sqrt[3]{7}} = \sqrt[6]{7}$
8. $(16)^{-\frac{3}{4}} = \dfrac{1}{8}$
9. $\dfrac{\sqrt[3]{16}}{\sqrt{2}} = 2\sqrt{2}$
10. $\sqrt[3]{64^2} = 16$

## Problem Set 5.6

For Problems 1–30, evaluate each numerical expression. (Objective 1)

1. $81^{\frac{1}{2}}$
2. $64^{\frac{1}{2}}$
3. $27^{\frac{1}{3}}$
4. $(-32)^{\frac{1}{5}}$
5. $(-8)^{\frac{1}{3}}$
6. $\left(-\dfrac{27}{8}\right)^{\frac{1}{3}}$
7. $-25^{\frac{1}{2}}$
8. $-64^{\frac{1}{3}}$
9. $36^{-\frac{1}{2}}$
10. $81^{-\frac{1}{2}}$
11. $\left(\dfrac{1}{27}\right)^{-\frac{1}{3}}$
12. $\left(-\dfrac{8}{27}\right)^{-\frac{1}{3}}$
13. $4^{\frac{3}{2}}$
14. $64^{\frac{2}{3}}$
15. $27^{\frac{4}{3}}$
16. $4^{\frac{7}{2}}$
17. $(-1)^{\frac{7}{3}}$
18. $(-8)^{\frac{4}{3}}$

19. $-4^{\frac{5}{2}}$

20. $-16^{\frac{3}{2}}$

21. $\left(\frac{27}{8}\right)^{\frac{4}{3}}$

22. $\left(\frac{8}{125}\right)^{\frac{2}{3}}$

23. $\left(\frac{1}{8}\right)^{-\frac{2}{3}}$

24. $\left(-\frac{1}{27}\right)^{-\frac{2}{3}}$

25. $64^{-\frac{7}{6}}$

26. $32^{-\frac{4}{5}}$

27. $-25^{\frac{3}{2}}$

28. $-16^{\frac{3}{4}}$

29. $125^{\frac{4}{3}}$

30. $81^{\frac{5}{4}}$

For Problems 31–44, write each of the following in radical form. (Objective 2)
For example,

$$3x^{\frac{2}{3}} = 3\sqrt[3]{x^2}$$

31. $x^{\frac{4}{3}}$

32. $x^{\frac{2}{5}}$

33. $3x^{\frac{1}{2}}$

34. $5x^{\frac{1}{4}}$

35. $(2y)^{\frac{1}{3}}$

36. $(3xy)^{\frac{1}{2}}$

37. $(2x - 3y)^{\frac{1}{2}}$

38. $(5x + y)^{\frac{1}{3}}$

39. $(2a - 3b)^{\frac{2}{3}}$

40. $(5a + 7b)^{\frac{3}{5}}$

41. $x^{\frac{2}{3}}y^{\frac{1}{3}}$

42. $x^{\frac{3}{7}}y^{\frac{5}{7}}$

43. $-3x^{\frac{1}{5}}y^{\frac{2}{5}}$

44. $-4x^{\frac{3}{4}}y^{\frac{1}{4}}$

For Problems 45–58, write each of the following using positive rational exponents. (Objective 3)
For example,

$$\sqrt{ab} = (ab)^{\frac{1}{2}} = a^{\frac{1}{2}}b^{\frac{1}{2}}$$

45. $\sqrt{5y}$

46. $\sqrt{2xy}$

47. $3\sqrt{y}$

48. $5\sqrt{ab}$

49. $\sqrt[3]{xy^2}$

50. $\sqrt[5]{x^2y^4}$

51. $\sqrt[4]{a^2b^3}$

52. $\sqrt[6]{ab^5}$

53. $\sqrt[5]{(2x - y)^3}$

54. $\sqrt[7]{(3x - y)^4}$

55. $5x\sqrt{y}$

56. $4y\sqrt[3]{x}$

57. $-\sqrt[3]{x + y}$

58. $-\sqrt[5]{(x - y)^2}$

For Problems 59–80, simplify each of the following. Express final results using positive exponents only. (Objective 4)
For example,

$$\left(2x^{\frac{1}{2}}\right)\left(3x^{\frac{1}{3}}\right) = 6x^{\frac{5}{6}}$$

59. $\left(2x^{\frac{2}{5}}\right)\left(6x^{\frac{1}{4}}\right)$

60. $\left(3x^{\frac{1}{4}}\right)\left(5x^{\frac{1}{3}}\right)$

61. $\left(y^{\frac{2}{3}}\right)\left(y^{-\frac{1}{4}}\right)$

62. $\left(y^{\frac{3}{4}}\right)\left(y^{-\frac{1}{2}}\right)$

63. $\left(x^{\frac{2}{5}}\right)\left(4x^{-\frac{1}{2}}\right)$

64. $\left(2x^{\frac{1}{3}}\right)\left(x^{-\frac{1}{2}}\right)$

65. $\left(4x^{\frac{1}{2}}y\right)^2$

66. $\left(3x^{\frac{1}{4}}y^{\frac{1}{5}}\right)^3$

67. $\left(8x^6y^3\right)^{\frac{1}{3}}$

68. $\left(9x^2y^4\right)^{\frac{1}{2}}$

69. $\dfrac{24x^{\frac{3}{5}}}{6x^{\frac{1}{3}}}$

70. $\dfrac{18x^{\frac{1}{2}}}{9x^{\frac{1}{3}}}$

71. $\dfrac{48b^{\frac{1}{3}}}{12b^{\frac{3}{4}}}$

72. $\dfrac{56a^{\frac{1}{6}}}{8a^{\frac{1}{4}}}$

73. $\left(\dfrac{6x^{\frac{2}{5}}}{7y^{\frac{2}{3}}}\right)^2$

74. $\left(\dfrac{2x^{\frac{1}{3}}}{3y^{\frac{1}{4}}}\right)^4$

75. $\left(\dfrac{x^2}{y^3}\right)^{-\frac{1}{2}}$

76. $\left(\dfrac{a^3}{b^{-2}}\right)^{-\frac{1}{3}}$

77. $\left(\dfrac{18x^{\frac{1}{3}}}{9x^{\frac{1}{4}}}\right)^2$

78. $\left(\dfrac{72x^{\frac{3}{4}}}{6x^{\frac{1}{2}}}\right)^2$

79. $\left(\dfrac{60a^{\frac{1}{5}}}{15a^{\frac{3}{4}}}\right)^2$

80. $\left(\dfrac{64a^{\frac{1}{3}}}{16a^{\frac{5}{9}}}\right)^3$

For Problems 81–90, perform the indicated operations and express answers in simplest radical form. (Objective 5) (See Example 5.)

81. $\sqrt[3]{3}\sqrt{3}$

82. $\sqrt{2}\sqrt[4]{2}$

83. $\sqrt[4]{6}\sqrt{6}$

84. $\sqrt[3]{5}\sqrt{5}$

85. $\dfrac{\sqrt[3]{3}}{\sqrt[4]{3}}$

86. $\dfrac{\sqrt{2}}{\sqrt[3]{2}}$

87. $\dfrac{\sqrt[3]{8}}{\sqrt[4]{4}}$

88. $\dfrac{\sqrt{9}}{\sqrt[3]{3}}$

89. $\dfrac{\sqrt[4]{27}}{\sqrt{3}}$

90. $\dfrac{\sqrt[3]{16}}{\sqrt[6]{4}}$

## Thoughts Into Words

91. Your friend keeps getting an error message when evaluating $-4^{\frac{5}{2}}$ on his calculator. What error is he probably making?

92. Explain how you would evaluate $27^{\frac{2}{3}}$ without a calculator.

## Further Investigations

93. Use your calculator to evaluate each of the following.

(a) $\sqrt[3]{1728}$

(b) $\sqrt[3]{5832}$

(c) $\sqrt[4]{2401}$

(d) $\sqrt[4]{65{,}536}$

(e) $\sqrt[5]{161{,}051}$

(f) $\sqrt[5]{6{,}436{,}343}$

**94.** Definition 5.7 states that

$b^{\frac{m}{n}} = \sqrt[n]{b^m} = \left(\sqrt[n]{b}\right)^m$

Use your calculator to verify each of the following.

**(a)** $\sqrt[3]{27^2} = \left(\sqrt[3]{27}\right)^2$ **(b)** $\sqrt[3]{8^5} = \left(\sqrt[3]{8}\right)^5$

**(c)** $\sqrt[4]{16^3} = \left(\sqrt[4]{16}\right)^3$ **(d)** $\sqrt[3]{16^2} = \left(\sqrt[3]{16}\right)^2$

**(e)** $\sqrt[5]{9^4} = \left(\sqrt[5]{9}\right)^4$ **(f)** $\sqrt[3]{12^4} = \left(\sqrt[3]{12}\right)^4$

**95.** Use your calculator to evaluate each of the following.

**(a)** $16^{\frac{5}{2}}$ **(b)** $25^{\frac{7}{2}}$ **(c)** $16^{\frac{9}{4}}$

**(d)** $27^{\frac{5}{3}}$ **(e)** $343^{\frac{2}{3}}$ **(f)** $512^{\frac{4}{3}}$

**96.** Use your calculator to estimate each of the following to the nearest one-thousandth.

**(a)** $7^{\frac{4}{3}}$ **(b)** $10^{\frac{4}{5}}$

**(c)** $12^{\frac{3}{5}}$ **(d)** $19^{\frac{2}{5}}$

**(e)** $7^{\frac{3}{4}}$ **(f)** $10^{\frac{5}{4}}$

**97. (a)** Because $\frac{4}{5} = 0.8$, we can evaluate $10^{\frac{4}{5}}$ by evaluating $10^{0.8}$, which involves a shorter sequence of "calculator steps." Evaluate parts (b), (c), (d), (e), and (f) of Problem 96 and take advantage of decimal exponents.

**(b)** What problem is created when we try to evaluate $7^{\frac{4}{3}}$ by changing the exponent to decimal form?

### Answers to the Concept Quiz

**1.** True **2.** True **3.** False **4.** True **5.** True **6.** False **7.** True **8.** True **9.** False **10.** True

## 5.7 Scientific Notation

OBJECTIVES

1. Write numbers in scientific notation
2. Convert numbers from scientific notation to ordinary decimal notation
3. Perform calculations with numbers using scientific notation

Many applications of mathematics involve the use of very large or very small numbers:

1. The speed of light is approximately 29,979,200,000 centimeters per second.
2. A light year—the distance that light travels in 1 year—is approximately 5,865,696,000,000 miles.
3. A millimicron equals 0.000000001 of a meter.

Working with numbers of this type in standard decimal form is quite cumbersome. It is much more convenient to represent very small and very large numbers in *scientific notation.* Although negative numbers can be written in scientific form, we will restrict our discussion to positive numbers. The expression $(N)(10^k)$, where $N$ is a number greater than or equal to 1 and less than 10, written in decimal form, and $k$ is any integer, is commonly called **scientific notation** or the scientific form of a number. Consider the following examples, which show a comparison between ordinary decimal notation and scientific notation.

| Ordinary notation | Scientific notation |
|---|---|
| 2.14 | $(2.14)(10^0)$ |
| 31.78 | $(3.178)(10^1)$ |
| 412.9 | $(4.129)(10^2)$ |
| 8,000,000 | $(8)(10^6)$ |
| 0.14 | $(1.4)(10^{-1})$ |
| 0.0379 | $(3.79)(10^{-2})$ |
| 0.00000049 | $(4.9)(10^{-7})$ |

To switch from ordinary notation to scientific notation, you can use the following procedure.

> Write the given number as the product of a number greater than or equal to 1 and less than 10, and a power of 10. The exponent of 10 is determined by counting the number of places that the decimal point was moved when going from the original number to the number greater than or equal to 1 and less than 10. This exponent is (a) negative if the original number is less than 1, (b) positive if the original number is greater than 10, and (c) 0 if the original number itself is between 1 and 10.

Thus we can write

$0.00467 = (4.67)(10^{-3})$

$87{,}000 = (8.7)(10^{4})$

$3.1416 = (3.1416)(10^{0})$

We can express the applications given earlier in scientific notation as follows:

**Speed of light** $29{,}979{,}200{,}000 = (2.99792)(10^{10})$ centimeters per second

**Light year** $5{,}865{,}696{,}000{,}000 = (5.865696)(10^{12})$ miles

**Metric units** A millimicron is $0.000000001 = (1)(10^{-9})$ meter

To switch from scientific notation to ordinary decimal notation, you can use the following procedure.

> Move the decimal point the number of places indicated by the exponent of 10. The decimal point is moved to the right if the exponent is positive and to the left if the exponent is negative.

Thus we can write

$(4.78)(10^{4}) = 47{,}800$

$(8.4)(10^{-3}) = 0.0084$

Scientific notation can frequently be used to simplify numerical calculations. We merely change the numbers to scientific notation and use the appropriate properties of exponents. Consider the following examples.

**Classroom Example**
Convert each number to scientific notation and perform the indicated operations:
**(a)** $(0.00051)(4000)$
**(b)** $\dfrac{8{,}600{,}000}{0.00043}$
**(c)** $\dfrac{(0.000052)(0.032)}{(0.000016)(0.00104)}$
**(d)** $\sqrt{0.000025}$

## EXAMPLE 1

Convert each number to scientific notation and perform the indicated operations. Express the result in ordinary decimal notation:

**(a)** $(0.00024)(20{,}000)$ **(b)** $\dfrac{7{,}800{,}000}{0.0039}$

**(c)** $\dfrac{(0.00069)(0.0034)}{(0.0000017)(0.023)}$ **(d)** $\sqrt{0.000004}$

### Solution

**(a)**
$$\begin{aligned}(0.00024)(20{,}000) &= (2.4)(10^{-4})(2)(10^{4})\\ &= (2.4)(2)(10^{-4})(10^{4})\\ &= (4.8)(10^{0})\\ &= (4.8)(1)\\ &= 4.8\end{aligned}$$

(b) $\dfrac{7{,}800{,}000}{0.0039} = \dfrac{(7.8)(10^{6})}{(3.9)(10^{-3})}$
$= (2)(10^{9})$
$= 2{,}000{,}000{,}000$

(c) $\dfrac{(0.00069)(0.0034)}{(0.0000017)(0.023)} = \dfrac{(6.9)(10^{-4})(3.4)(10^{-3})}{(1.7)(10^{-6})(2.3)(10^{-2})}$

$= \dfrac{\overset{3}{(\cancel{6.9})}\overset{2}{(\cancel{3.4})}(10^{-7})}{(\cancel{1.7})(\cancel{2.3})(10^{-8})}$
$= (6)(10^{1})$
$= 60$

(d) $\sqrt{0.000004} = \sqrt{(4)(10^{-6})}$
$= ((4)(10^{-6}))^{\frac{1}{2}}$
$= (4)^{\frac{1}{2}}(10^{-6})^{\frac{1}{2}}$
$= (2)(10^{-3})$
$= 0.002$

**Classroom Example**
The speed of light is approximately $(1.86)(10^{5})$ miles per second. When Saturn is $(8.9)(10^{8})$ miles away from the sun, how long does it take light from the sun to reach Saturn?

### EXAMPLE 2 Apply Your Skill

The speed of light is approximately $(1.86)(10^{5})$ miles per second. When Earth is $(9.3)(10^{7})$ miles away from the sun, how long does it take light from the sun to reach Earth?

### Solution

We will use the formula $t = \dfrac{d}{r}$.

$t = \dfrac{(9.3)(10^{7})}{(1.86)(10^{5})}$

$t = \dfrac{(9.3)}{(1.86)}(10^{2})$ Subtract exponents

$t = (5)(10^{2}) = 500$ seconds

At this distance it takes light about 500 seconds to travel from the sun to Earth. To find the answer in minutes, divide 500 seconds by 60 seconds/minute. That gives a result of approximately 8.33 minutes.

Many calculators are equipped to display numbers in scientific notation. The display panel shows the number between 1 and 10 and the appropriate exponent of 10. For example, evaluating $3{,}800{,}000^2$ yields

| 1.444E13 |
|---|

Thus $3{,}800{,}000^2 = (1.444)(10^{13}) = 14{,}440{,}000{,}000{,}000$.
Similarly, the answer for $0.000168^2$ is displayed as

| 2.8224E-8 |
|---|

Thus $0.000168^2 = (2.8224)(10^{-8}) = 0.000000028224$.
Calculators vary as to the number of digits displayed in the number between 1 and 10 when scientific notation is used. For example, we used two different calculators to estimate $6729^6$ and obtained the following results.

| 9.2833E22 |
|---|

| 9.283316768E22 |
|---|

Obviously, you need to know the capabilities of your calculator when working with problems in scientific notation. Many calculators also allow the entry of a number in scientific notation. Such calculators are equipped with an enter-the-exponent key (often labeled as EE or EEX). Thus a number such as $(3.14)(10^8)$ might be entered as follows:

| Enter | Press | Display |
|---|---|---|
| 3.14 | EE | 3.14E |
| 8 | | 3.14E8 |

or

| Enter | Press | Display |
|---|---|---|
| 3.14 | EE | $3.14^{00}$ |
| 8 | | $3.14^{08}$ |

A MODE key is often used on calculators to let you choose normal decimal notation, scientific notation, or engineering notation. (The abbreviations Norm, Sci, and Eng are commonly used.) If the calculator is in scientific mode, then a number can be entered and changed to scientific form by pressing the ENTER key. For example, when we enter 589 and press the ENTER key, the display will show 5.89E2. Likewise, when the calculator is in scientific mode, the answers to computational problems are given in scientific form. For example, the answer for (76)(533) is given as 4.0508E4.

It should be evident from this brief discussion that even when you are using a calculator, you need to have a thorough understanding of scientific notation.

## Concept Quiz 5.7

For Problems 1–10, answer true or false.

1. A positive number written in scientific notation has the form $(N)(10^k)$, where $1 \leq N < 10$ and $k$ is an integer.
2. A number is less than zero if the exponent is negative when the number is written in scientific notation.
3. $(3.11)(10^{-2}) = 311$
4. $(5.24)(10^{-1}) = 0.524$
5. $(8.91)(10^2) = 89.1$
6. $(4.163)(10^{-5}) = 0.00004163$
7. $(0.00715) = (7.15)(10^{-3})$
8. Scientific notation provides a way of working with numbers that are very large or very small in magnitude.
9. $(0.0012)(5000) = 60$
10. $\dfrac{6{,}200{,}000}{0.0031} = 2{,}000{,}000{,}000$

## Problem Set 5.7

For Problems 1–18, write each of the following in scientific notation. (Objective 1)

For example,

$$27{,}800 = (2.78)(10^4)$$

1. 89
2. 117
3. 4290
4. 812,000
5. 6,120,000
6. 72,400,000
7. 40,000,000
8. 500,000,000
9. 376.4
10. 9126.21
11. 0.347
12. 0.2165
13. 0.0214
14. 0.0037
15. 0.00005
16. 0.00000082
17. 0.00000000194
18. 0.00000000003

For Problems 19–32, write each of the following in ordinary decimal notation. (Objective 2)
For example,

$(3.18)(10^2) = 318$

**19.** $(2.3)(10^1)$

**20.** $(1.62)(10^2)$

**21.** $(4.19)(10^3)$

**22.** $(7.631)(10^4)$

**23.** $(5)(10^8)$

**24.** $(7)(10^9)$

**25.** $(3.14)(10^{10})$

**26.** $(2.04)(10^{12})$

**27.** $(4.3)(10^{-1})$

**28.** $(5.2)(10^{-2})$

**29.** $(9.14)(10^{-4})$

**30.** $(8.76)(10^{-5})$

**31.** $(5.123)(10^{-8})$

**32.** $(6)(10^{-9})$

For Problems 33–50, convert each number to scientific notation and perform the indicated operations. Express the result in ordinary decimal notation. (Objective 3)

**33.** $(0.0037)(0.00002)$

**34.** $(0.00003)(0.00025)$

**35.** $(0.00007)(11{,}000)$

**36.** $(0.000004)(120{,}000)$

**37.** $\dfrac{360{,}000{,}000}{0.0012}$

**38.** $\dfrac{66{,}000{,}000{,}000}{0.022}$

**39.** $\dfrac{0.000064}{16{,}000}$

**40.** $\dfrac{0.00072}{0.0000024}$

**41.** $\dfrac{(60{,}000)(0.006)}{(0.0009)(400)}$

**42.** $\dfrac{(0.00063)(960{,}000)}{(3{,}200)(0.0000021)}$

**43.** $\dfrac{(0.0045)(60{,}000)}{(1800)(0.00015)}$

**44.** $\dfrac{(0.00016)(300)(0.028)}{0.064}$

**45.** $\sqrt{9{,}000{,}000}$

**46.** $\sqrt{0.00000009}$

**47.** $\sqrt[3]{8000}$

**48.** $\sqrt[3]{0.001}$

**49.** $90{,}000^{\frac{3}{2}}$

**50.** $8000^{\frac{2}{3}}$

**51.** Avogadro's number, 602,000,000,000,000,000,000,000, is the number of atoms in 1 mole of a substance. Express this number in scientific notation.

**52.** The Social Security program paid out approximately 821,000,000,000 dollars in benefits in 2013. Express this number in scientific notation.

**53.** Carlos' first computer had a processing speed of $(1.6)(10^6)$ hertz. He recently purchased a laptop computer with a processing speed of $(1.33)(10^9)$ hertz. Approximately how many times faster is the processing speed of his laptop than that of his first computer? Express the result in decimal form.

**54.** Alaska has an area of approximately $(6.15)(10^5)$ square miles. In 2012 the state had a population of approximately 730,000 people. Compute the population density to the nearest hundredth. Population density is the number of people per square mile. Express the result in decimal form rounded to the nearest hundredth.

**55.** In the year 2013 the public debt of the United States was approximately \$11,600,000,000,000. For July 2012, the census reported that 313,000,000 people lived in the United States. Convert these figures to scientific notation, and compute the average debt per person.

**56.** The space shuttle can travel at approximately 410,000 miles per day. If the shuttle could travel to Mars, and Mars was 140,000,000 miles away, how many days would it take the shuttle to travel to Mars? Express the result in decimal form.

**57.** A square pixel on a computer screen has a side of length $(1.17)(10^{-2})$ inch. Find the approximate area of the pixel in inches. Express the result in decimal form.

**58.** The field of view of a microscope is $(4)(10^{-4})$ meter. If a single cell organism occupies $\frac{1}{5}$ of the field of view, find the length of the organism in meters. Express the result in scientific notation.

**59.** The mass of an electron is $(9.11)(10^{-31})$ kilogram, and the mass of a proton is $(1.67)(10^{-27})$ kilogram. Approximately how many times more is the mass of a proton than is the mass of an electron? Express the result in decimal form.

**60.** Atomic masses are measured in atomic mass units (amu). The amu, $(1.66)(10^{-27})$ kilogram, is defined as $\frac{1}{12}$ the mass of a common carbon atom. Find the mass of a carbon atom in kilograms. Express the result in scientific notation.

## Thoughts Into Words

**61.** Explain the importance of scientific notation.

**62.** Why do we need scientific notation even when using calculators and computers?

## Further Investigations

**63.** Sometimes it is more convenient to express a number as a product of a power of 10 and a number that is not between 1 and 10. For example, suppose that we want to calculate $\sqrt{640{,}000}$. We can proceed as follows:

$$\begin{aligned}\sqrt{640{,}000} &= \sqrt{(64)(10^4)}\\ &= ((64)(10^4))^{\frac{1}{2}}\\ &= (64)^{\frac{1}{2}}(10^4)^{\frac{1}{2}}\\ &= (8)(10^2)\\ &= 8(100) = 800\end{aligned}$$

Compute each of the following without a calculator, and then use a calculator to check your answers.

**(a)** $\sqrt{49{,}000{,}000}$ **(b)** $\sqrt{0.0025}$

**(c)** $\sqrt{14{,}400}$ **(d)** $\sqrt{0.000121}$

**(e)** $\sqrt[3]{27{,}000}$ **(f)** $\sqrt[3]{0.000064}$

**64.** Use your calculator to evaluate each of the following. Express final answers in ordinary notation.

**(a)** $27{,}000^2$ **(b)** $450{,}000^2$

**(c)** $14{,}800^2$ **(d)** $1700^3$

**(e)** $900^4$ **(f)** $60^5$

**(g)** $0.0213^2$ **(h)** $0.000213^2$

**(i)** $0.000198^2$ **(j)** $0.000009^3$

**65.** Use your calculator to estimate each of the following. Express final answers in scientific notation with the number between 1 and 10 rounded to the nearest one-thousandth.

**(a)** $4576^4$ **(b)** $719^{10}$

**(c)** $28^{12}$ **(d)** $8619^6$

**(e)** $314^5$ **(f)** $145{,}723^2$

**66.** Use your calculator to estimate each of the following. Express final answers in ordinary decimal notation rounded to the nearest one-thousandth.

**(a)** $1.09^5$ **(b)** $1.08^{10}$

**(c)** $1.14^7$ **(d)** $1.12^{20}$

**(e)** $0.785^4$ **(f)** $0.492^5$

### Answers to the Concept Quiz

**1.** True **2.** False **3.** False **4.** True **5.** False **6.** True **7.** True **8.** True **9.** False **10.** True

# Chapter 5 Summary

| OBJECTIVE | SUMMARY | EXAMPLE |
|---|---|---|
| Simplify numerical expressions that have integer exponents. (Section 5.1/Objective 1) | The concept of exponent is expanded to include negative exponents and exponents of zero. If $b$ is a nonzero number, then $b^0 = 1$. If $n$ is a positive integer and $b$ is a nonzero number, then $b^{-n} = \frac{1}{b^n}$. | Simplify $\left(\frac{2}{5}\right)^{-2}$. **Solution** $\left(\frac{2}{5}\right)^{-2} = \frac{2^{-2}}{5^{-2}} = \frac{5^2}{2^2} = \frac{25}{4}$ **Sample Problem 1** Simplify $\left(\frac{1}{3}\right)^{-3}$. |
| Simplify algebraic expressions that have integer exponents. (Section 5.1/Objective 2) | The properties for integer exponents listed on page 245 form the basis for manipulating with integer exponents. These properties, along with Definition 5.2; that is, $b^{-n} = \frac{1}{b^n}$, enable us to simplify algebraic expressions and express the results with positive exponents. | Simplify $(2x^{-3}y)^{-2}$ and express the final result using positive exponents. **Solution** $(2x^{-3}y)^{-2} = 2^{-2}x^6y^{-2}$ $= \frac{x^6}{2^2y^2} = \frac{x^6}{4y^2}$ **Sample Problem 2** Simplify $(4xy^{-1})^{-2}$ and express the final result using positive exponents. |
| Multiply and divide algebraic expressions that have integer exponents. (Section 5.1/Objective 3) | The previous remark also applies to simplifying multiplication and division problems that involve integer exponents. | Simplify $(-3x^5y^{-2})(4x^{-1}y^{-1})$ and express the final result using positive exponents. **Solution** $(-3x^5y^{-2})(4x^{-1}y^{-1}) = -12x^4y^{-3}$ $= -\frac{12x^4}{y^3}$ **Sample Problem 3** Simplify $(2x^3y^{-1})(-3x^{-2}y^{-1})$ and express the final result using positive exponents. |
| Simplify sums and differences of expressions involving integer exponents. (Section 5.1/Objective 4) | To find the sum or difference of expressions involving integer exponents, change all expressions having negative or zero exponents to equivalent expressions with positive exponents only. To find the sum or difference, it may be necessary to find a common denominator. | Simplify $5x^{-2} + 6y^{-1}$ and express the result as a single fraction involving positive exponents only. **Solution** $5x^{-2} + 6y^{-1} = \frac{5}{x^2} + \frac{6}{y}$ $= \frac{5}{x^2} \cdot \frac{y}{y} + \frac{6}{y} \cdot \frac{x^2}{x^2}$ $= \frac{5y + 6x^2}{x^2y}$ **Sample Problem 4** Simplify $3x^{-1} - 4y^{-2}$ and express the result as a single fraction involving positive exponents only. |

Answers to **Sample Problems** are located in the back of the book.

*(continued)*

| OBJECTIVE | SUMMARY | EXAMPLE |
|---|---|---|
| Express a radical in simplest radical form. (Section 5.2/Objective 2) | The principal $n$th root of $b$ is designated by $\sqrt[n]{b}$, where $n$ is the index and $b$ is the radicand.<br>A radical expression is in simplest form if:<br>**1.** A radicand contains no polynomial factor raised to a power equal to or greater than the index of the radical;<br>**2.** No fraction appears within a radical sign; and<br>**3.** No radical appears in the denominator.<br>The following properties are used to express radicals in simplest form:<br>$\sqrt[n]{bc} = \sqrt[n]{b}\sqrt[n]{c}$ and $\sqrt[n]{\frac{b}{c}} = \frac{\sqrt[n]{b}}{\sqrt[n]{c}}$ | Simplify $\sqrt{150a^3b^2}$. Assume all variables represent nonnegative values.<br>**Solution**<br>$\sqrt{150a^3b^2} = \sqrt{25a^2b^2}\sqrt{6a}$<br>$= 5ab\sqrt{6a}$<br>**Sample Problem 5**<br>Simplify $\sqrt{20x^8y^5}$. |
| Rationalize the denominator to simplify radicals. (Section 5.2/Objective 3) | If a radical appears in the denominator, it will be necessary to rationalize the denominator for the expression to be in simplest form. | Simplify $\frac{2\sqrt{18}}{\sqrt{5}}$.<br>**Solution**<br>$\frac{2\sqrt{18}}{\sqrt{5}} = \frac{2\sqrt{9}\sqrt{2}}{\sqrt{5}}$<br>$= \frac{2(3)\sqrt{2}}{\sqrt{5}} = \frac{6\sqrt{2}}{\sqrt{5}}$<br>$= \frac{6\sqrt{2}}{\sqrt{5}} \cdot \frac{\sqrt{5}}{\sqrt{5}} = \frac{6\sqrt{10}}{\sqrt{25}}$<br>$= \frac{6\sqrt{10}}{5}$<br>**Sample Problem 6**<br>Simplify $\frac{3\sqrt{12}}{\sqrt{7}}$. |
| Simplify expressions by combining radicals. (Section 5.3/Objective 1) | Simplifying by combining radicals sometimes requires that we first express the given radicals in simplest form. | Simplify $\sqrt{24} - \sqrt{54} + 8\sqrt{6}$.<br>**Solution**<br>$\sqrt{24} - \sqrt{54} + 8\sqrt{6}$<br>$= \sqrt{4}\sqrt{6} - \sqrt{9}\sqrt{6} + 8\sqrt{6}$<br>$= 2\sqrt{6} - 3\sqrt{6} + 8\sqrt{6}$<br>$= (2 - 3 + 8)(\sqrt{6})$<br>$= 7\sqrt{6}$<br>**Sample Problem 7**<br>Simplify $4\sqrt{18} + \sqrt{50} - \sqrt{8}$. |

| OBJECTIVE | SUMMARY | EXAMPLE |
|---|---|---|
| Multiply two radicals. (Section 5.4/Objective 1) | The property $\sqrt[n]{b}\sqrt[n]{c} = \sqrt[n]{bc}$ is used to find the product of two radicals. | Multiply $\sqrt[3]{4x^2y}\sqrt[3]{6x^2y^2}$.<br>**Solution**<br>$\sqrt[3]{4x^2y}\sqrt[3]{6x^2y^2} = \sqrt[3]{24x^4y^3}$<br>$= \sqrt[3]{8x^3y^3}\sqrt[3]{3x}$<br>$= 2xy\sqrt[3]{3x}$<br>**Sample Problem 8**<br>Multiply $\sqrt[3]{2xy^2}\sqrt[3]{4x^4y}$. |
| Use the distributive property to multiply radical expressions. (Section 5.4/Objective 2) | The distributive property and the property $\sqrt[n]{b}\sqrt[n]{c} = \sqrt[n]{bc}$ are used to find products of radical expressions. | Multiply $\sqrt{2x}(\sqrt{6x} + \sqrt{18xy})$ and simplify where possible.<br>**Solution**<br>$\sqrt{2x}(\sqrt{6x} + \sqrt{18xy})$<br>$= \sqrt{12x^2} + \sqrt{36x^2y}$<br>$= \sqrt{4x^2}\sqrt{3} + \sqrt{36x^2}\sqrt{y}$<br>$= 2x\sqrt{3} + 6x\sqrt{y}$<br>**Sample Problem 9**<br>Multiply $\sqrt{3x}(\sqrt{12x} - \sqrt{6xy})$ and simplify where possible. |
| Rationalize binomial denominators. (Section 5.4/Objective 3) | The factors $(a - b)$ and $(a + b)$ are called conjugates. To rationalize a binomial denominator involving radicals, multiply the numerator or denominator by the conjugate of the denominator. | Simplify $\frac{3}{\sqrt{7} - \sqrt{5}}$ by rationalizing the denominator.<br>**Solution**<br>$\frac{3}{\sqrt{7} - \sqrt{5}}$<br>$= \frac{3}{(\sqrt{7} - \sqrt{5})} \cdot \frac{(\sqrt{7} + \sqrt{5})}{(\sqrt{7} + \sqrt{5})}$<br>$= \frac{3(\sqrt{7} + \sqrt{5})}{\sqrt{49} - \sqrt{25}} = \frac{3(\sqrt{7} + \sqrt{5})}{7 - 5}$<br>$= \frac{3(\sqrt{7} + \sqrt{5})}{2}$<br>**Sample Problem 10**<br>Simplify $\frac{5}{\sqrt{6} - \sqrt{3}}$ by rationalizing the denominator. |

*(continued)*

| OBJECTIVE | SUMMARY | EXAMPLE |
|---|---|---|
| Solve radical equations. (Section 5.5/Objective 1) | Equations with variables in a radicand are called radical equations. Radical equations are solved by raising each side of the equation to the appropriate power. However, raising both sides of the equation to a power may produce extraneous roots. Therefore, you must check each potential solution. | Solve $\sqrt{x} + 20 = x$.<br>**Solution**<br>$\sqrt{x} + 20 = x$<br>$\sqrt{x} = x - 20$ Isolate the radical<br>$(\sqrt{x})^2 = (x - 20)^2$<br>$x = x^2 - 40x + 400$<br>$0 = x^2 - 41x + 400$<br>$0 = (x - 25)(x - 16)$<br>$x = 25$ or $x = 16$<br>✔ **Check**<br>$\sqrt{x} + 20 = x$<br>**If $x = 25$** **If $x = 16$**<br>$\sqrt{25} + 20 \stackrel{?}{=} 25$ $\sqrt{16} + 20 \stackrel{?}{=} 16$<br>$25 = 25$ $24 \neq 16$<br>The solution set is $\{25\}$.<br>**Sample Problem 11**<br>Solve $\sqrt{2x - 1} = x - 2$. |
| Solve radical equations for real-world problems. (Section 5.5/Objective 2) | Various formulas involve radical equations. These formulas are solved in the same manner as radical equations. | Use the formula $\sqrt{30Df} = S$ (given in Section 5.5) to determine the coefficient of friction, to the nearest hundredth, if a car traveling at 50 miles per hour skidded 300 feet.<br>**Solution**<br>Solve $\sqrt{30Df} = S$ for $f$.<br>$(\sqrt{30Df})^2 = S^2$<br>$30Df = S^2$<br>$f = \frac{S^2}{30D}$<br>Substituting the values for $S$ and $D$ gives<br>$f = \frac{50^2}{30(300)}$<br>$= 0.28$<br>**Sample Problem 12**<br>Use the formula $\sqrt{30Df} = S$ to determine the coefficient of friction, to the nearest hundredth, if a car traveling at 40 miles per hour skidded 200 feet. |
| Evaluate a number raised to a rational exponent. (Section 5.6/Objective 1) | To simplify a number raised to a rational exponent, we apply either the property $b^{\frac{1}{n}} = \sqrt[n]{b}$ or the property $b^{\frac{m}{n}} = \sqrt[n]{b^m} = (\sqrt[n]{b})^m$. When simplifying $b^{\frac{m}{n}}$, the arithmetic computations are usually easiest using the form $(\sqrt[n]{b})^m$, where the $n$th root is taken first, and that result is raised to the $m$ power. | Simplify $16^{\frac{3}{2}}$.<br>**Solution**<br>$16^{\frac{3}{2}} = (16^{\frac{1}{2}})^3$<br>$= 4^3$<br>$= 64$<br>**Sample Problem 13**<br>Simplify $8^{\frac{5}{3}}$. |

| OBJECTIVE | SUMMARY | EXAMPLE |
|---|---|---|
| Write an expression with rational exponents as a radical. (Section 5.6/Objective 2) | If $\frac{m}{n}$ is a rational number, $n$ is a positive integer greater than 1, and $b$ is a real number such that $\sqrt[n]{b}$ exists, then $b^{\frac{m}{n}} = \sqrt[n]{b^m} = \left(\sqrt[n]{b}\right)^m$. | Write $x^{\frac{3}{5}}$ in radical form. <br> **Solution** <br> $x^{\frac{3}{5}} = \sqrt[5]{x^3}$ <br> **Sample Problem 14** <br> Write $y^{\frac{2}{3}}$ in radical form. |
| Write radical expressions as expressions with rational exponents. (Section 5.6/Objective 3) | The index of the radical will be the denominator of the rational exponent. | Write $\sqrt[4]{x^3y}$ using positive rational exponents. <br> **Solution** <br> $\sqrt[4]{x^3y} = x^{\frac{3}{4}}y^{\frac{1}{4}}$ <br> **Sample Problem 15** <br> Write $\sqrt[5]{xy^3}$ using positive rational exponents. |
| Simplify algebraic expressions that have rational exponents. (Section 5.6/Objective 4 ) | Properties of exponents are used to simplify products and quotients involving rational exponents. | Simplify $\left(4x^{\frac{1}{3}}\right)\left(-3x^{-\frac{3}{4}}\right)$ and express the result with positive exponents only. <br> **Solution** <br> $\left(4x^{\frac{1}{3}}\right)\left(-3x^{-\frac{3}{4}}\right) = -12x^{\frac{1}{3}-\frac{3}{4}}$ <br> $= -12x^{-\frac{5}{12}}$ <br> $= \frac{-12}{x^{\frac{5}{12}}}$ <br> **Sample Problem 16** <br> Simplify $\left(2x^{\frac{1}{2}}\right)\left(-5x^{-\frac{1}{4}}\right)$ and express the result with positive exponents only. |
| Multiply and divide radicals with different indexes. (Section 5.6/Objective 5) | The link between rational exponents and roots provides the basis for multiplying and dividing radicals with different indexes. | Multiply $\sqrt[3]{y^2}\sqrt{y}$ and express in simplest radical form. <br> **Solution** <br> $\sqrt[3]{y^2}\sqrt{y} = y^{\frac{2}{3}}y^{\frac{1}{2}}$ <br> $= y^{\frac{2}{3}+\frac{1}{2}} = y^{\frac{7}{6}}$ <br> $= \sqrt[6]{y^7} = y\sqrt[6]{y}$ <br> **Sample Problem 17** <br> Multiply $\sqrt{x^3}\sqrt[3]{x}$ and express in simplest radical form. |
| Write numbers in scientific notation. (Section 5.7/Objective 1) | Scientific notation is often used to write numbers that are very small or very large in magnitude. The scientific form of a number is expressed as $(N)(10^k)$, where the absolute value of $N$ is a number greater than or equal to 1 and less than 10, written in decimal form, and $k$ is an integer. | Write each of the following in scientific notation: <br> **(a)** 0.000000843 <br> **(b)** 456,000,000,000 <br> **Solution** <br> **(a)** $0.000000843 = (8.43)(10^{-7})$ <br> **(b)** $456{,}000{,}000{,}000 = (4.56)(10^{11})$ <br> **Sample Problem 18** <br> Write each of the following in scientific notation: <br> **(a)** 0.000702 <br> **(b)** 398,100,000 |

*(continued)*

| OBJECTIVE | SUMMARY | EXAMPLE |
|---|---|---|
| Convert numbers from scientific notation to ordinary decimal notation. (Section 5.7/Objective 2) | To switch from scientific notation to ordinary notation, move the decimal point the number of places indicated by the exponent of 10. The decimal point is moved to the right if the exponent is positive and to the left if the exponent is negative. | Write each of the following in ordinary decimal notation: **(a)** $(8.5)(10^{-5})$ **(b)** $(3.4)(10^{6})$ <br> **Solution** <br> **(a)** $(8.5)(10^{-5}) = 0.000085$ <br> **(b)** $(3.4)(10^{6}) = 3{,}400{,}000$ <br> **Sample Problem 19** <br> Write each of the following in ordinary decimal notation: **(a)** $(3.24)(10^{-1})$ **(b)** $(9.23)(10^{5})$ |
| Perform calculations with numbers using scientific notation. (Section 5.7/Objective 3) | Scientific notation can often be used to simplify numerical calculations. | Use scientific notation and the properties of exponents to simplify $\frac{0.0000084}{0.002}$. <br> **Solution** <br> Change the numbers to scientific notation and use the appropriate properties of exponents. Express the result in ordinary decimal notation. <br> $\frac{0.0000084}{0.002} = \frac{(8.4)(10^{-6})}{(2)(10^{-3})}$ <br> $= (4.2)(10^{-3}) = 0.0042$ <br> **Sample Problem 20** <br> Use scientific notation and the properties of exponents to simplify $\frac{0.000048}{0.03}$. |

## Chapter 5 Review Problem Set

For Problems 1–6, evaluate the numerical expression.

1. $4^{-3}$
2. $\left(\frac{2}{3}\right)^{-2}$
3. $(3^2 \cdot 3^{-3})^{-1}$
4. $(4^{-2} \cdot 4^2)^{-1}$
5. $\left(\frac{3^{-1}}{3^2}\right)^{-1}$
6. $\left(\frac{5^2}{5^{-1}}\right)^{-1}$

For Problems 7–18, simplify and express the final result using positive exponents.

7. $(x^{-3}y^4)^{-2}$
8. $\left(\frac{2a^{-1}}{3b^4}\right)^{-3}$
9. $\left(\frac{4a^{-2}}{3b^{-2}}\right)^{-2}$
10. $(5x^3y^{-2})^{-3}$
11. $\left(\frac{6x^{-2}}{2x^4}\right)^{-2}$
12. $\left(\frac{8y^2}{2y^{-1}}\right)^{-1}$
13. $(-5x^{-3})(2x^6)$
14. $(a^{-4}b^3)(3ab^2)$
15. $\frac{a^{-1}b^{-2}}{a^4b^{-5}}$
16. $\frac{x^3y^5}{x^{-1}y^6}$
17. $\frac{-12x^3}{6x^5}$
18. $\frac{10a^2b^3}{-5ab^4}$

For Problems 19–22, express as a single fraction involving positive exponents only.

19. $x^{-2} + y^{-1}$
20. $a^{-2} - 2a^{-1}b^{-1}$
21. $2x^{-1} + 3y^{-2}$
22. $(2x)^{-1} + 3y^{-2}$

For Problems 23–34, express the radical in simplest radical form. Assume the variables represent positive real numbers.

23. $\sqrt{54}$
24. $\sqrt{48x^3y}$
25. $\sqrt[3]{56}$
26. $\sqrt[3]{108x^4y^8}$
27. $\frac{3}{4}\sqrt{150}$
28. $\frac{2}{3}\sqrt{45xy^3}$
29. $\frac{4\sqrt{3}}{\sqrt{6}}$
30. $\sqrt{\frac{5}{12x^3}}$
31. $\frac{\sqrt[3]{2}}{\sqrt[3]{9}}$
32. $\sqrt{\frac{9}{5}}$
33. $\sqrt{\frac{3x^3}{7}}$
34. $\frac{\sqrt{8x^2}}{\sqrt{2x}}$

For Problems 35–38, use the distributive property to help simplify the expression.

**35.** $3\sqrt{45} - 2\sqrt{20} - \sqrt{80}$

**36.** $4\sqrt[3]{24} + 3\sqrt[3]{3} - 2\sqrt[3]{81}$

**37.** $3\sqrt{24} - \frac{2\sqrt{54}}{5} + \frac{\sqrt{96}}{4}$

**38.** $-2\sqrt{12x} + 3\sqrt{27x} - 5\sqrt{48x}$

For Problems 39–48, multiply and simplify. Assume the variables represent nonnegative real numbers.

**39.** $(3\sqrt{8})(4\sqrt{5})$

**40.** $(5\sqrt[3]{2})(6\sqrt[3]{4})$

**41.** $(\sqrt{6xy})(\sqrt{10x})$

**42.** $(-3\sqrt{6xy^3})(\sqrt{6y})$

**43.** $3\sqrt{2}(4\sqrt{6} - 2\sqrt{7})$

**44.** $(\sqrt{x} + 3)(\sqrt{x} - 5)$

**45.** $(2\sqrt{5} - \sqrt{3})(2\sqrt{5} + \sqrt{3})$

**46.** $(3\sqrt{2} + \sqrt{6})(5\sqrt{2} - 3\sqrt{6})$

**47.** $(2\sqrt{a} + \sqrt{b})(3\sqrt{a} - 4\sqrt{b})$

**48.** $(4\sqrt{8} - \sqrt{2})(\sqrt{8} + 3\sqrt{2})$

For Problems 49–52, rationalize the denominator and simplify.

**49.** $\frac{4}{\sqrt{7} - 1}$

**50.** $\frac{\sqrt{3}}{\sqrt{8} + \sqrt{5}}$

**51.** $\frac{3}{2\sqrt{3} + 3\sqrt{5}}$

**52.** $\frac{3\sqrt{2}}{2\sqrt{6} - \sqrt{10}}$

For Problems 53–60, solve the equation.

**53.** $\sqrt{7x - 3} = 4$

**54.** $\sqrt{2y + 1} = \sqrt{5y - 11}$

**55.** $\sqrt{2x} = x - 4$

**56.** $\sqrt{n^2 - 4n - 4} = n$

**57.** $\sqrt[3]{2x - 1} = 3$

**58.** $\sqrt{t^2 + 9t - 1} = 3$

**59.** $\sqrt{x^2 + 3x - 6} = x$

**60.** $\sqrt{x + 1} - \sqrt{2x} = -1$

**61.** The formula $S = \sqrt{30Df}$ is used to approximate the speed $S$, where $D$ represents the length of the skid marks in feet and $f$ represents the coefficient of friction for the road surface. Suppose that the coefficient of friction is 0.38. How far will a car skid, to the nearest foot, when the brakes are applied at 75 miles per hour?

**62.** The formula $T = 2\pi\sqrt{\frac{L}{32}}$ is used for pendulum motion, where $T$ represents the period of the pendulum in seconds, and $L$ represents the length of the pendulum in feet. Find the length of a pendulum, to the nearest tenth of a foot, if the period is 2.4 seconds.

For Problems 63–70, simplify.

**63.** $4^{\frac{5}{2}}$

**64.** $(-1)^{\frac{2}{3}}$

**65.** $\left(\frac{8}{27}\right)^{\frac{2}{3}}$

**66.** $-16^{\frac{3}{2}}$

**67.** $(27)^{-\frac{2}{3}}$

**68.** $(32)^{-\frac{2}{5}}$

**69.** $9^{\frac{3}{2}}$

**70.** $16^{\frac{3}{4}}$

For Problems 71–74, write the expression in radical form.

**71.** $x^{\frac{1}{3}}y^{\frac{2}{3}}$

**72.** $a^{\frac{3}{4}}$

**73.** $4y^{\frac{1}{2}}$

**74.** $(x + 5y)^{\frac{2}{3}}$

For Problems 75–78, write the expression using positive rational exponents.

**75.** $\sqrt[5]{x^3y}$

**76.** $\sqrt[3]{4a^2}$

**77.** $6\sqrt[4]{y^2}$

**78.** $\sqrt[3]{(3a + b)^5}$

For Problems 79–84, simplify and express the final result using positive exponents.

**79.** $(4x^{\frac{1}{2}})(5x^{\frac{1}{5}})$

**80.** $\frac{42a^{\frac{3}{4}}}{6a^{\frac{1}{3}}}$

**81.** $\left(\frac{x^3}{y^4}\right)^{-\frac{1}{3}}$

**82.** $(-3a^{\frac{1}{4}})(2a^{-\frac{1}{2}})$

**83.** $(x^{\frac{4}{5}})^{-\frac{1}{2}}$

**84.** $\frac{-24y^{\frac{2}{3}}}{4y^{\frac{1}{4}}}$

For Problems 85–88, perform the indicated operation and express the answer in simplest radical form.

**85.** $\sqrt[4]{3}\sqrt{3}$

**86.** $\sqrt[3]{9}\sqrt{3}$

**87.** $\frac{\sqrt[3]{5}}{\sqrt[4]{5}}$

**88.** $\frac{\sqrt[3]{16}}{\sqrt{2}}$

For Problems 89–92, write the number in scientific notation.

**89.** 540,000,000

**90.** 84,000

**91.** 0.000000032

**92.** 0.000768

For Problems 93–96, write the number in ordinary decimal notation.

**93.** $(1.4)(10^{-6})$

**94.** $(6.38)(10^{-4})$

**95.** $(4.12)(10^{7})$

**96.** $(1.25)(10^{5})$

For Problems 97–104, use scientific notation and the properties of exponents to help perform the calculations.

**97.** $(0.00002)(0.0003)$

**98.** $(120{,}000)(300{,}000)$

**99.** $(0.000015)(400{,}000)$

**100.** $\frac{0.000045}{0.0003}$

**101.** $\frac{(0.00042)(0.0004)}{0.006}$

**102.** $\sqrt{0.000004}$

**103.** $\sqrt[3]{0.000000008}$

**104.** $4{,}000{,}000^{\frac{3}{2}}$

# Chapter 5 Test

For Problems 1–4, simplify each of the numerical expressions.

**1.** $(4)^{-\frac{5}{2}}$

**2.** $-16^{\frac{5}{4}}$

**3.** $\left(\frac{2}{3}\right)^{-4}$

**4.** $\left(\frac{2^{-1}}{2^{-2}}\right)^{-2}$

For Problems 5–9, express each radical expression in simplest radical form. Assume the variables represent positive real numbers.

**5.** $\sqrt{63}$

**6.** $\sqrt[3]{108}$

**7.** $\sqrt{52x^4y^3}$

**8.** $\frac{5\sqrt{18}}{3\sqrt{12}}$

**9.** $\sqrt{\frac{7}{24x^3}}$

**10.** Multiply and simplify: $(4\sqrt{6})(3\sqrt{12})$

**11.** Multiply and simplify: $(3\sqrt{2} + \sqrt{3})(\sqrt{2} - 2\sqrt{3})$

**12.** Simplify by combining similar radicals:

$2\sqrt{50} - 4\sqrt{18} - 9\sqrt{32}$

**13.** Rationalize the denominator and simplify:

$\frac{3\sqrt{2}}{4\sqrt{3} - \sqrt{8}}$

**14.** Simplify and express the answer using positive exponents: $\left(\frac{2x^{-1}}{3y}\right)^{-2}$

**15.** Simplify and express the answer using positive exponents: $\frac{-84a^{\frac{1}{2}}}{7a^{\frac{4}{5}}}$

**16.** Express $x^{-1} + y^{-3}$ as a single fraction involving positive exponents.

**17.** Multiply and express the answer using positive exponents: $(3x^{-\frac{1}{2}})(-4x^{-\frac{3}{4}})$

**18.** Multiply and simplify:

$(3\sqrt{5} - 2\sqrt{3})(3\sqrt{5} + 2\sqrt{3})$

For Problems 19 and 20, use scientific notation and the properties of exponents to help with the calculations.

**19.** $\frac{(0.00004)(300)}{0.00002}$

**20.** $\sqrt{0.000009}$

For Problems 21–25, solve each equation.

**21.** $\sqrt{3x + 1} = 3$

**22.** $\sqrt[3]{3x + 2} = 2$

**23.** $\sqrt{x} = x - 2$

**24.** $\sqrt{5x - 2} = \sqrt{3x + 8}$

**25.** $\sqrt{x^2 - 10x + 28} = 2$

# 6 Quadratic Equations and Inequalities

Zigy Kaluzny-Charles Thatcher/Stone/Getty Images

*"When you change the way you look at things, the things you look at change."*
WAYNE DYER

## Study Skill Tip

Understanding some things about how memory works in the learning process can help you identify ways to improve your learning. The memory process includes the following stages.

- **Sensory input and sensory register.** We receive information through hearing, seeing, tasting, touching, and smelling. We hold the sensory information in a registry until it can be processed.
- **Short-term memory.** Short-term memory is the ability to recall information immediately after receiving it. If we don't put this information in notes or learn it immediately, then we lose this information. This is why it is important to be focused during your class period.
- **Long-term memory.** Information already stored in long-term memory along with the new information just processed is used to learn math.
- **Memory output.** The ability to recall the information in long-term memory and express this knowledge is memory output.

If there is interference with any of these steps, then learning does not take place. If you do not input or register the input, the process ends there. You need to get the information into short-term memory for a sufficient time to get the information into your notes. Otherwise you will have to relearn the material. Spending enough quality time on your homework will get the information into your long-term memory.

***What step(s) in the memory process are encountering interference when you are learning math?***

## Chapter Preview

Chapter 6 is about solving quadratic equations and inequalities. This chapter could be considered the heart of Intermediate Algebra. You will be presented with four methods for solving a quadratic equation. The method you choose to apply often depends on the form of the quadratic equation.

# 6.1 Complex Numbers

OBJECTIVES

1. Know about the set of complex numbers
2. Add and subtract complex numbers
3. Simplify radicals involving negative numbers
4. Perform operations on radicals involving negative numbers
5. Multiply complex numbers
6. Divide complex numbers

Because the square of any real number is nonnegative, a simple equation such as $x^2 = -4$ has no solutions in the set of real numbers. To handle this situation, we can expand the set of real numbers into a larger set called the *complex numbers*. In this section we will instruct you on how to manipulate complex numbers.

To provide a solution for the equation $x^2 + 1 = 0$, we use the number $i$, such that

$$i^2 = -1$$

The number $i$ is not a real number and is often called the **imaginary unit**, but the number $i^2$ is the real number $-1$. The imaginary unit $i$ is used to define a complex number as follows:

### Definition 6.1

A **complex number** is any number that can be expressed in the form

$$a + bi$$

where $a$ and $b$ are real numbers.

The form $a + bi$ is called the **standard form** of a complex number. The real number $a$ is called the **real part** of the complex number, and $b$ is called the **imaginary part**. (Note that $b$ is a real number even though it is called the imaginary part.) The following list exemplifies this terminology.

1. The number $7 + 5i$ is a complex number that has a real part of 7 and an imaginary part of 5.
2. The number $\frac{2}{3} + i\sqrt{2}$ is a complex number that has a real part of $\frac{2}{3}$ and an imaginary part of $\sqrt{2}$. (It is easy to mistake $\sqrt{2}i$ for $\sqrt{2i}$. Thus we commonly write $i\sqrt{2}$ instead of $\sqrt{2}i$ to avoid any difficulties with the radical sign.)
3. The number $-4 - 3i$ can be written in the standard form $-4 + (-3i)$ and therefore is a complex number that has a real part of $-4$ and an imaginary part of $-3$. [The form $-4 - 3i$ is often used, but we know that it means $-4 + (-3i)$.]

4. The number $-9i$ can be written as $0 + (-9i)$; thus it is a complex number that has a real part of 0 and an imaginary part of $-9$. (Complex numbers, such as $-9i$, for which $a = 0$ and $b \neq 0$ are called *pure imaginary numbers*.)
5. The real number 4 can be written as $4 + 0i$ and is thus a complex number that has a real part of 4 and an imaginary part of 0.

Look at item 5 in this list. We see that the set of real numbers is a subset of the set of complex numbers. The following diagram indicates the organizational format of the complex numbers.

**Complex numbers** $a + bi$ where $a$ and $b$ are real numbers

**Real numbers** $a + bi$ where $b = 0$

**Imaginary numbers** $a + bi$ where $b \neq 0$

**Pure imaginary numbers** $a + bi$ where $a = 0$ and $b \neq 0$

Two complex numbers $a + bi$ and $c + di$ are said to be **equal** if and only if $a = c$ and $b = d$.

## Adding and Subtracting Complex Numbers

To *add complex numbers*, we simply add their real parts and add their imaginary parts. Thus

$$(a + bi) + (c + di) = (a + c) + (b + d)i$$

The following example shows addition of two complex numbers.

**Classroom Example**
Add the complex numbers:
**(a)** $(2 + 7i) + (4 + i)$
**(b)** $(5 - 2i) + (-3 + 9i)$
**(c)** $\left(\frac{2}{3} + \frac{1}{5}i\right) + \left(\frac{1}{4} + \frac{5}{6}i\right)$

**EXAMPLE 1** Add the complex numbers:

**(a)** $(4 + 3i) + (5 + 9i)$ **(b)** $(-6 + 4i) + (8 - 7i)$

**(c)** $\left(\frac{1}{2} + \frac{3}{4}i\right) + \left(\frac{2}{3} + \frac{1}{5}i\right)$

### Solution

**(a)** $(4 + 3i) + (5 + 9i) = (4 + 5) + (3 + 9)i = 9 + 12i$

**(b)** $(-6 + 4i) + (8 - 7i) = (-6 + 8) + (4 - 7)i$
$= 2 - 3i$

**(c)**
$$\left(\frac{1}{2} + \frac{3}{4}i\right) + \left(\frac{2}{3} + \frac{1}{5}i\right) = \left(\frac{1}{2} + \frac{2}{3}\right) + \left(\frac{3}{4} + \frac{1}{5}\right)i$$
$$= \left(\frac{3}{6} + \frac{4}{6}\right) + \left(\frac{15}{20} + \frac{4}{20}\right)i$$
$$= \frac{7}{6} + \frac{19}{20}i$$

The set of complex numbers is closed with respect to addition; that is, the sum of two complex numbers is a complex number. Furthermore, the commutative and associative properties of addition hold for all complex numbers. The addition identity element is $0 + 0i$ (or simply the real number 0). The additive inverse of $a + bi$ is $-a - bi$, because

$$(a + bi) + (-a - bi) = 0$$

To *subtract complex numbers*, $c + di$ from $a + bi$, add the additive inverse of $c + di$. Thus

$$(a + bi) - (c + di) = (a + bi) + (-c - di) = (a - c) + (b - d)i$$

In other words, we subtract the real parts and subtract the imaginary parts, as in the next examples.

**1.** $(9 + 8i) - (5 + 3i) = (9 - 5) + (8 - 3)i = 4 + 5i$

**2.** $(3 - 2i) - (4 - 10i) = (3 - 4) + (-2 - (-10))i = -1 + 8i$

## Simplifying Radicals Involving Negative Numbers

Because $i^2 = -1$, $i$ is a square root of $-1$, so we let $i = \sqrt{-1}$. It should be evident that $-i$ is also a square root of $-1$, because

$$(-i)^2 = (-i)(-i) = i^2 = -1$$

Thus, in the set of complex numbers, $-1$ has two square roots, $i$ and $-i$. We express these symbolically as

$$\sqrt{-1} = i \quad \text{and} \quad -\sqrt{-1} = -i$$

Let us extend our definition so that in the set of complex numbers every negative real number has two square roots. We simply define $\sqrt{-b}$, where $b$ is a positive real number, to be the number whose square is $-b$. Thus

$$\left(\sqrt{-b}\right)^2 = -b, \quad \text{for } b > 0$$

Furthermore, because $\left(i\sqrt{b}\right)^2 = \left(i\sqrt{b}\right)\left(i\sqrt{b}\right) = i^2(b) = -1(b) = -b$, we see that

$$\sqrt{-b} = i\sqrt{b}$$

We should also observe that $-\sqrt{-b}$ (where $b > 0$) is a square root of $-b$ because

$$\left(-\sqrt{-b}\right)^2 = \left(-i\sqrt{b}\right)^2 = i^2(b) = -1(b) = -b$$

Thus in the set of complex numbers, $-b$ (where $b > 0$) has two square roots, $i\sqrt{b}$ and $-i\sqrt{b}$. We express these symbolically as

$$\sqrt{-b} = i\sqrt{b} \quad \text{and} \quad -\sqrt{-b} = -i\sqrt{b}$$

In other words, a square root of any negative real number can be represented as the product of a real number and the imaginary unit $i$. Consider the following examples.

**Classroom Example**
Simplify each of the following:
**(a)** $\sqrt{-9}$
**(b)** $\sqrt{-19}$
**(c)** $\sqrt{-32}$

**EXAMPLE 2** Simplify each of the following:

**(a)** $\sqrt{-4}$ **(b)** $\sqrt{-17}$ **(c)** $\sqrt{-24}$

### Solution

**(a)** $\sqrt{-4} = i\sqrt{4} = 2i$

**(b)** $\sqrt{-17} = i\sqrt{17}$

**(c)** $\sqrt{-24} = i\sqrt{24} = i\sqrt{4}\sqrt{6} = 2i\sqrt{6}$ — Note that we simplified the radical $\sqrt{24}$ to $2\sqrt{6}$

## Performing Operations on Radicals Involving Negative Numbers

We must be very careful with the use of the symbol $\sqrt{-b}$, where $b > 0$. Some real number properties that involve the square root symbol do not hold if the square root symbol does not represent a real number. For example, $\sqrt{a}\sqrt{b} = \sqrt{ab}$ does not hold if $a$ and $b$ are both negative numbers.

**Correct** $\sqrt{-4}\sqrt{-9} = (2i)(3i) = 6i^2 = 6(-1) = -6$

**Incorrect** $\sqrt{-4}\sqrt{-9} = \sqrt{(-4)(-9)} = \sqrt{36} = 6$

To avoid difficulty with this idea, you should rewrite all expressions of the form $\sqrt{-b}$ (where $b > 0$) in the form $i\sqrt{b}$ before doing any computations. The following example further demonstrates this point.

**Classroom Example**
Simplify each of the following:

(a) $\sqrt{-10}\sqrt{-5}$

(b) $\sqrt{-5}\sqrt{-20}$

(c) $\dfrac{\sqrt{-27}}{\sqrt{-3}}$

(d) $\dfrac{\sqrt{-39}}{\sqrt{13}}$

**EXAMPLE 3** Simplify each of the following:

**(a)** $\sqrt{-6}\sqrt{-8}$ **(b)** $\sqrt{-2}\sqrt{-8}$ **(c)** $\dfrac{\sqrt{-75}}{\sqrt{-3}}$ **(d)** $\dfrac{\sqrt{-48}}{\sqrt{12}}$

### Solution

**(a)** $\sqrt{-6}\sqrt{-8} = (i\sqrt{6})(i\sqrt{8}) = i^2\sqrt{48} = (-1)\sqrt{16}\sqrt{3} = -4\sqrt{3}$

**(b)** $\sqrt{-2}\sqrt{-8} = (i\sqrt{2})(i\sqrt{8}) = i^2\sqrt{16} = (-1)(4) = -4$

**(c)** $\dfrac{\sqrt{-75}}{\sqrt{-3}} = \dfrac{i\sqrt{75}}{i\sqrt{3}} = \dfrac{\sqrt{75}}{\sqrt{3}} = \sqrt{\dfrac{75}{3}} = \sqrt{25} = 5$

**(d)** $\dfrac{\sqrt{-48}}{\sqrt{12}} = \dfrac{i\sqrt{48}}{\sqrt{12}} = i\sqrt{\dfrac{48}{12}} = i\sqrt{4} = 2i$

## Multiplying Complex Numbers

Complex numbers have a binomial form, so we find the product of two complex numbers in the same way that we find the product of two binomials. Then, by replacing $i^2$ with $-1$, we are able to simplify and express the final result in standard form. Consider the following example.

**Classroom Example**
Simplify each of the following:

(a) $(5 + 2i)(1 + 7i)$

(b) $(3 - 7i)(-2 + 3i)$

(c) $(4 - 7i)^2$

(d) $(4 - 6i)(4 + 6i)$

**EXAMPLE 4** Find the product of each of the following:

**(a)** $(2 + 3i)(4 + 5i)$ **(b)** $(-3 + 6i)(2 - 4i)$

**(c)** $(1 - 7i)^2$ **(d)** $(2 + 3i)(2 - 3i)$

### Solution

**(a)**
$$\begin{aligned}(2 + 3i)(4 + 5i) &= 8 + 10i + 12i + 15i^2 && \text{Use distributive property or FOIL}\\ &= 8 + 22i + 15i^2\\ &= 8 + 22i + 15(-1) && \text{Substitute } -1 \text{ for } i^2\\ &= -7 + 22i\end{aligned}$$

**(b)**
$$\begin{aligned}(-3 + 6i)(2 - 4i) &= -6 + 12i + 12i - 24i^2\\ &= -6 + 24i - 24(-1) && \text{Substitute } -1 \text{ for } i^2\\ &= -6 + 24i + 24\\ &= 18 + 24i\end{aligned}$$

**(c)**
$$\begin{aligned}(1 - 7i)^2 &= (1 - 7i)(1 - 7i)\\ &= 1 - 7i - 7i + 49i^2\\ &= 1 - 14i + 49(-1) && \text{Substitute } -1 \text{ for } i^2\\ &= 1 - 14i - 49\\ &= -48 - 14i\end{aligned}$$

**(d)** $(2 + 3i)(2 - 3i) = 4 - 6i + 6i - 9i^2$
$= 4 - 9(-1)$
$= 4 + 9$
$= 13$

Example 4(d) illustrates an important situation: The complex numbers $2 + 3i$ and $2 - 3i$ are conjugates of each other. In general, we say that two complex numbers $a + bi$ and $a - bi$ are called **conjugates** of each other. *The product of a complex number and its conjugate is always a real number*, which can be shown as follows:

$$\begin{aligned}(a + bi)(a - bi) &= a(a - bi) + bi(a - bi)\\ &= a^2 - abi + abi - b^2i^2\\ &= a^2 - b^2(-1)\\ &= a^2 + b^2\end{aligned}$$

## Dividing Complex Numbers

We use conjugates to simplify expressions such as $\frac{3i}{5 + 2i}$ that indicate the quotient of two complex numbers. To eliminate $i$ in the denominator and change the indicated quotient to the standard form of a complex number, we can multiply both the numerator and the denominator by the conjugate of the denominator as follows:

$$\frac{3i}{5 + 2i} = \frac{3i(5 - 2i)}{(5 + 2i)(5 - 2i)}$$ $5 - 2i$ is the conjugate of $5 + 2i$

$$= \frac{15i - 6i^2}{25 - 4i^2}$$

$$= \frac{15i - 6(-1)}{25 - 4(-1)}$$ Substitute $-1$ for $i^2$

$$= \frac{15i + 6}{29}$$

$$= \frac{6}{29} + \frac{15}{29}i$$ Remember $\frac{a + b}{c} = \frac{a}{c} + \frac{b}{c}$

The following example further clarifies the process of dividing complex numbers.

Classroom Example
Find the quotient of each of the following:
(a) $\frac{2 + 4i}{2 - 5i}$
(b) $\frac{9 - 2i}{4i}$

**EXAMPLE 5** Find the quotient of each of the following:

**(a)** $\frac{2 - 3i}{4 - 7i}$ **(b)** $\frac{4 - 5i}{2i}$

### Solution

**(a)** $$\frac{2 - 3i}{4 - 7i} = \frac{(2 - 3i)(4 + 7i)}{(4 - 7i)(4 + 7i)}$$ $4 + 7i$ is the conjugate of $4 - 7i$

$$= \frac{8 + 14i - 12i - 21i^2}{16 - 49i^2}$$

$$= \frac{8 + 2i - 21(-1)}{16 - 49(-1)}$$ Substitute $-1$ for $i^2$

$$= \frac{8 + 2i + 21}{16 + 49}$$

$$= \frac{29 + 2i}{65}$$

$$= \frac{29}{65} + \frac{2}{65}i \qquad \text{Remember } \frac{a+b}{c} = \frac{a}{c} + \frac{b}{c}$$

**(b)** $$\frac{4 - 5i}{2i} = \frac{(4 - 5i)(-2i)}{(2i)(-2i)} \qquad -2i \text{ is the conjugate of } 2i$$

$$= \frac{-8i + 10i^2}{-4i^2}$$

$$= \frac{-8i + 10(-1)}{-4(-1)} \qquad \text{Substitute } -1 \text{ for } i^2$$

$$= \frac{-8i - 10}{4}$$

$$= \frac{-10}{4} - \frac{8i}{4} = -\frac{5}{2} - 2i$$

In Example 5(b), in which the denominator is a pure imaginary number, we can change to standard form by choosing a multiplier other than the conjugate. Consider the following alternative approach for Example 5(b).

$$\frac{4 - 5i}{2i} = \frac{(4 - 5i)(i)}{(2i)(i)}$$

$$= \frac{4i - 5i^2}{2i^2}$$

$$= \frac{4i - 5(-1)}{2(-1)}$$

$$= \frac{4i + 5}{-2}$$

$$= \frac{5}{-2} + \frac{4i}{-2} = -\frac{5}{2} - 2i$$

## Concept Quiz 6.1

For Problems 1–10, answer true or false.

1. The number $i$ is a real number and is called the imaginary unit.
2. The number $4 + 2i$ is a complex number that has a real part of 4.
3. The number $-3 - 5i$ is a complex number that has an imaginary part of 5.
4. Complex numbers that have a real part of 0 are called pure imaginary numbers.
5. The set of real numbers is a subset of the set of complex numbers.
6. Any real number $x$ can be written as the complex number $x + 0i$.
7. By definition, $i^2$ is equal to $-1$.
8. The complex numbers $-2 + 5i$ and $2 - 5i$ are conjugates.
9. The product of two complex numbers is never a real number.
10. In the set of complex numbers, $-16$ has two square roots.

## Problem Set 6.1

For Problems 1–8, label each statement true or false. (Objective 1)

1. Every complex number is a real number.
2. Every real number is a complex number.
3. The real part of the complex number $6i$ is 0.
4. Every complex number is a pure imaginary number.
5. The sum of two complex numbers is always a complex number.
6. The imaginary part of the complex number 7 is 0.
7. The sum of two complex numbers is sometimes a real number.
8. The sum of two pure imaginary numbers is always a pure imaginary number.

For Problems 9–26, add or subtract as indicated. (Objective 2)

9. $(6 + 3i) + (4 + 5i)$
10. $(5 + 2i) + (7 + 10i)$
11. $(-8 + 4i) + (2 + 6i)$
12. $(5 - 8i) + (-7 + 2i)$
13. $(3 + 2i) - (5 + 7i)$
14. $(1 + 3i) - (4 + 9i)$
15. $(-7 + 3i) - (5 - 2i)$
16. $(-8 + 4i) - (9 - 4i)$
17. $(-3 - 10i) + (2 - 13i)$
18. $(-4 - 12i) + (-3 + 16i)$
19. $(4 - 8i) - (8 - 3i)$
20. $(12 - 9i) - (14 - 6i)$
21. $(-1 - i) - (-2 - 4i)$
22. $(-2 - 3i) - (-4 - 14i)$
23. $\left(\frac{3}{2} + \frac{1}{3}i\right) + \left(\frac{1}{6} - \frac{3}{4}i\right)$
24. $\left(\frac{2}{3} - \frac{1}{5}i\right) + \left(\frac{3}{5} - \frac{3}{4}i\right)$
25. $\left(-\frac{5}{9} + \frac{3}{5}i\right) - \left(\frac{4}{3} - \frac{1}{6}i\right)$
26. $\left(\frac{3}{8} - \frac{5}{2}i\right) - \left(\frac{5}{6} + \frac{1}{7}i\right)$

For Problems 27–42, write each of the following in terms of $i$ and simplify. (Objective 3) For example,

$$\sqrt{-20} = i\sqrt{20} = i\sqrt{4}\sqrt{5} = 2i\sqrt{5}$$

27. $\sqrt{-81}$
28. $\sqrt{-49}$
29. $\sqrt{-14}$
30. $\sqrt{-33}$
31. $\sqrt{-\frac{16}{25}}$
32. $\sqrt{-\frac{64}{36}}$
33. $\sqrt{-18}$
34. $\sqrt{-84}$
35. $\sqrt{-75}$
36. $\sqrt{-63}$
37. $3\sqrt{-28}$
38. $5\sqrt{-72}$
39. $-2\sqrt{-80}$
40. $-6\sqrt{-27}$
41. $12\sqrt{-90}$
42. $9\sqrt{-40}$

For Problems 43–60, write each of the following in terms of $i$, perform the indicated operations, and simplify. (Objective 4) For example,

$$\begin{aligned}\sqrt{-3}\sqrt{-8} &= \left(i\sqrt{3}\right)\left(i\sqrt{8}\right)\\ &= i^2\sqrt{24}\\ &= (-1)\sqrt{4}\sqrt{6}\\ &= -2\sqrt{6}\end{aligned}$$

43. $\sqrt{-4}\sqrt{-16}$
44. $\sqrt{-81}\sqrt{-25}$
45. $\sqrt{-3}\sqrt{-5}$
46. $\sqrt{-7}\sqrt{-10}$
47. $\sqrt{-9}\sqrt{-6}$
48. $\sqrt{-8}\sqrt{-16}$
49. $\sqrt{-15}\sqrt{-5}$
50. $\sqrt{-2}\sqrt{-20}$
51. $\sqrt{-2}\sqrt{-27}$
52. $\sqrt{-3}\sqrt{-15}$
53. $\sqrt{6}\sqrt{-8}$
54. $\sqrt{-75}\sqrt{3}$
55. $\dfrac{\sqrt{-25}}{\sqrt{-4}}$
56. $\dfrac{\sqrt{-81}}{\sqrt{-9}}$
57. $\dfrac{\sqrt{-56}}{\sqrt{-7}}$
58. $\dfrac{\sqrt{-72}}{\sqrt{-6}}$
59. $\dfrac{\sqrt{-24}}{\sqrt{6}}$
60. $\dfrac{\sqrt{-96}}{\sqrt{2}}$

For Problems 61–84, find each of the products and express the answers in the standard form of a complex number. (Objective 5)

61. $(5i)(4i)$
62. $(-6i)(9i)$
63. $(7i)(-6i)$
64. $(-5i)(-12i)$
65. $(3i)(2 - 5i)$
66. $(7i)(-9 + 3i)$
67. $(-6i)(-2 - 7i)$
68. $(-9i)(-4 - 5i)$
69. $(3 + 2i)(5 + 4i)$
70. $(4 + 3i)(6 + i)$
71. $(6 - 2i)(7 - i)$
72. $(8 - 4i)(7 - 2i)$
73. $(-3 - 2i)(5 + 6i)$
74. $(-5 - 3i)(2 - 4i)$
75. $(9 + 6i)(-1 - i)$
76. $(10 + 2i)(-2 - i)$
77. $(4 + 5i)^2$
78. $(5 - 3i)^2$
79. $(-2 - 4i)^2$
80. $(-3 - 6i)^2$
81. $(6 + 7i)(6 - 7i)$
82. $(5 - 7i)(5 + 7i)$
83. $(-1 + 2i)(-1 - 2i)$
84. $(-2 - 4i)(-2 + 4i)$

For Problems 85–100, find each of the following quotients, and express the answers in the standard form of a complex number. (Objective 6)

**85.** $\dfrac{3i}{2 + 4i}$

**86.** $\dfrac{4i}{5 + 2i}$

**87.** $\dfrac{-2i}{3 - 5i}$

**88.** $\dfrac{-5i}{2 - 4i}$

**89.** $\dfrac{-2 + 6i}{3i}$

**90.** $\dfrac{-4 - 7i}{6i}$

**91.** $\dfrac{2}{7i}$

**92.** $\dfrac{3}{10i}$

**93.** $\dfrac{2 + 6i}{1 + 7i}$

**94.** $\dfrac{5 + i}{2 + 9i}$

**95.** $\dfrac{3 + 6i}{4 - 5i}$

**96.** $\dfrac{7 - 3i}{4 - 3i}$

**97.** $\dfrac{-2 + 7i}{-1 + i}$

**98.** $\dfrac{-3 + 8i}{-2 + i}$

**99.** $\dfrac{-1 - 3i}{-2 - 10i}$

**100.** $\dfrac{-3 - 4i}{-4 - 11i}$

**101.** Some of the solution sets for quadratic equations in the next sections in this chapter will contain complex numbers such as $\left(-4 + \sqrt{-12}\right)/2$ and $\left(-4 - \sqrt{-12}\right)/2$. We can simplify the first number as follows.

$$\frac{-4 + \sqrt{-12}}{2} = \frac{-4 + i\sqrt{12}}{2}$$

$$= \frac{-4 + 2i\sqrt{3}}{2} = \frac{\not{2}\left(-2 + i\sqrt{3}\right)}{\not{2}}$$

$$= -2 + i\sqrt{3}$$

Simplify each of the following complex numbers. (Objective 3)

**(a)** $\dfrac{-4 - \sqrt{-12}}{2}$

**(b)** $\dfrac{6 + \sqrt{-24}}{4}$

**(c)** $\dfrac{-1 - \sqrt{-18}}{2}$

**(d)** $\dfrac{-6 + \sqrt{-27}}{3}$

**(e)** $\dfrac{10 + \sqrt{-45}}{4}$

**(f)** $\dfrac{4 - \sqrt{-48}}{2}$

## Thoughts Into Words

**102.** Why is the set of real numbers a subset of the set of complex numbers?

**103.** Can the sum of two nonreal complex numbers be a real number? Defend your answer.

**104.** Can the product of two nonreal complex numbers be a real number? Defend your answer.

### Answers to the Concept Quiz

**1.** False **2.** True **3.** False **4.** True **5.** True **6.** True **7.** True **8.** False **9.** False **10.** True

# 6.2 Quadratic Equations

OBJECTIVES

1. Solve quadratic equations by factoring
2. Solve quadratic equations of the form $x^2 = a$
3. Solve problems pertaining to right triangles and 30°–60° triangles

A second-degree equation in one variable contains the variable with an exponent of 2, but no higher power. Such equations are also called *quadratic equations*. The following are examples of quadratic equations.

$$x^2 = 36 \qquad y^2 + 4y = 0 \qquad x^2 + 5x - 2 = 0$$

$$3n^2 + 2n - 1 = 0 \qquad 5x^2 + x + 2 = 3x^2 - 2x - 1$$

A **quadratic equation** in the variable $x$ can also be defined as any equation that can be written in the form

$$ax^2 + bx + c = 0$$

where $a$, $b$, and $c$ are real numbers and $a \neq 0$. The form $ax^2 + bx + c = 0$ is called the *standard form* of a quadratic equation.

In previous chapters you solved quadratic equations (the term *quadratic* was not used at that time) by factoring and applying the property, $ab = 0$ if and only if $a = 0$ or $b = 0$. Let's review a few such examples.

**Classroom Example**
Solve $4x^2 + 11x - 3 = 0$.

**EXAMPLE 1** Solve $3n^2 + 14n - 5 = 0$.

### Solution

$$3n^2 + 14n - 5 = 0$$

$$(3n - 1)(n + 5) = 0 \qquad \text{Factor the left side}$$

$$3n - 1 = 0 \quad \text{or} \quad n + 5 = 0 \qquad \text{Apply: } ab = 0 \text{ if and only if } a = 0 \text{ or } b = 0$$

$$3n = 1 \quad \text{or} \quad n = -5$$

$$n = \frac{1}{3} \quad \text{or} \quad n = -5$$

The solution set is $\left\{-5, \frac{1}{3}\right\}$.

**Classroom Example**
Solve $2\sqrt{y} = y - 3$.

**EXAMPLE 2** Solve $2\sqrt{x} = x - 8$.

### Solution

$$2\sqrt{x} = x - 8$$

$$\left(2\sqrt{x}\right)^2 = (x - 8)^2 \qquad \text{Square both sides}$$

$$4x = x^2 - 16x + 64$$

$$0 = x^2 - 20x + 64$$

$$0 = (x - 16)(x - 4) \qquad \text{Factor the right side}$$

$$x - 16 = 0 \quad \text{or} \quad x - 4 = 0 \qquad \text{Apply: } ab = 0 \text{ if and only if } a = 0 \text{ or } b = 0$$

$$x = 16 \quad \text{or} \quad x = 4$$

### ✔ Check

**If $x = 16$**

$$2\sqrt{x} = x - 8$$

$$2\sqrt{16} \stackrel{?}{=} 16 - 8$$

$$2(4) \stackrel{?}{=} 8$$

$$8 = 8$$

**If $x = 4$**

$$2\sqrt{x} = x - 8$$

$$2\sqrt{4} \stackrel{?}{=} 4 - 8$$

$$2(2) \stackrel{?}{=} -4$$

$$4 \neq -4$$

The solution set is $\{16\}$.

We should make two comments about Example 2. First, remember that applying the property, if $a = b$, then $a^n = b^n$, might produce extraneous solutions. Therefore, we *must* check all potential solutions. Second, the equation $2\sqrt{x} = x - 8$ is said to be of *quadratic form* because it can be written as $2x^{\frac{1}{2}} = \left(x^{\frac{1}{2}}\right)^2 - 8$. More will be said about the phrase *quadratic form* later.

## Solving Quadratic Equations of the Form $x^2 = a$

Let's consider quadratic equations of the form $x^2 = a$, where $x$ is the variable and $a$ is any real number. We can solve $x^2 = a$ as follows:

$$x^2 = a$$
$$x^2 - a = 0$$
$$x^2 - (\sqrt{a})^2 = 0 \qquad a = (\sqrt{a})^2$$
$$(x - \sqrt{a})(x + \sqrt{a}) = 0 \qquad \text{Factor the left side}$$
$$x - \sqrt{a} = 0 \quad \text{or} \quad x + \sqrt{a} = 0 \qquad \text{Apply: } ab = 0 \text{ if and only if } a = 0 \text{ or } b = 0$$
$$x = \sqrt{a} \quad \text{or} \quad x = -\sqrt{a}$$

The solutions are $\sqrt{a}$ and $-\sqrt{a}$. We can state this result as a general property and use it to solve certain types of quadratic equations.

**Property 6.1**

For any real number $a$,

$$x^2 = a \quad \text{if and only if } x = \sqrt{a} \text{ or } x = -\sqrt{a}$$

(The statement $x = \sqrt{a}$ or $x = -\sqrt{a}$ can be written as $x = \pm\sqrt{a}$.)

Property 6.1, along with our knowledge of square roots, makes it very easy to solve quadratic equations of the form $x^2 = a$.

**Classroom Example**
Solve $m^2 = 48$.

**EXAMPLE 3** Solve $x^2 = 45$.

### Solution

$$x^2 = 45$$
$$x = \pm\sqrt{45}$$
$$x = \pm\sqrt{9}\sqrt{5} = \pm 3\sqrt{5} \qquad \sqrt{45} = \sqrt{9}\sqrt{5} = 3\sqrt{5}$$

The solution set is $\{\pm 3\sqrt{5}\}$.

**Classroom Example**
Solve $n^2 = -25$.

**EXAMPLE 4** Solve $x^2 = -9$.

### Solution

$$x^2 = -9$$
$$x = \pm\sqrt{-9}$$
$$x = \pm 3i \qquad \sqrt{-9} = i\sqrt{9} = 3i$$

Thus the solution set is $\{\pm 3i\}$.

**Classroom Example**
Solve $5x^2 = 16$.

**EXAMPLE 5** Solve $7n^2 = 12$.

### Solution

$$7n^2 = 12$$
$$n^2 = \frac{12}{7} \qquad \text{Divide both sides of the equation by 7}$$

$$n = \pm\sqrt{\frac{12}{7}}$$

$$n = \pm\frac{2\sqrt{21}}{7} \qquad \sqrt{\frac{12}{7}} = \frac{\sqrt{12}}{\sqrt{7}} \cdot \frac{\sqrt{7}}{\sqrt{7}} = \frac{\sqrt{84}}{7} = \frac{\sqrt{4}\sqrt{21}}{7} = \frac{2\sqrt{21}}{7}$$

The solution set is $\left\{\pm\frac{2\sqrt{21}}{7}\right\}$.

**Classroom Example**
Solve $(4x - 3)^2 = 49$.

**EXAMPLE 6** Solve $(3n + 1)^2 = 25$.

### Solution

$$(3n + 1)^2 = 25$$
$$(3n + 1) = \pm\sqrt{25} \qquad \text{Apply Property 6.1}$$
$$3n + 1 = \pm 5$$
$$3n + 1 = 5 \quad \text{or} \quad 3n + 1 = -5$$
$$3n = 4 \quad \text{or} \quad 3n = -6$$
$$n = \frac{4}{3} \quad \text{or} \quad n = -2$$

The solution set is $\left\{-2, \frac{4}{3}\right\}$.

**Classroom Example**
Solve $(x + 4)^2 = -18$.

**EXAMPLE 7** Solve $(x - 3)^2 = -10$.

### Solution

$$(x - 3)^2 = -10$$
$$x - 3 = \pm\sqrt{-10}$$
$$x - 3 = \pm i\sqrt{10}$$
$$x = 3 \pm i\sqrt{10}$$

Thus the solution set is $\{3 \pm i\sqrt{10}\}$.

**Remark:** Take another look at the equations in Examples 4 and 7. We should immediately realize that the solution sets will consist only of nonreal complex numbers, because any nonzero real number squared is positive.

Sometimes it may be necessary to change the form before we can apply Property 6.1. Let's consider one example to illustrate this idea.

**Classroom Example**
Solve $2(5x - 1)^2 + 9 = 53$.

**EXAMPLE 8** Solve $3(2x - 3)^2 + 8 = 44$.

### Solution

$$3(2x - 3)^2 + 8 = 44$$
$$3(2x - 3)^2 = 36 \qquad \text{Subtract 8 from both sides}$$
$$(2x - 3)^2 = 12 \qquad \text{Divide both sides by 3}$$
$$2x - 3 = \pm\sqrt{12} \qquad \text{Apply Property 6.1}$$
$$2x - 3 = \pm\sqrt{4}\sqrt{3} = \pm 2\sqrt{3}$$

$$2x = 3 \pm 2\sqrt{3}$$

$$x = \frac{3 \pm 2\sqrt{3}}{2}$$

The solution set is $\left\{\frac{3 \pm 2\sqrt{3}}{2}\right\}$.

## Solving Problems Pertaining to Right Triangles and 30°–60° Triangles

Our work with radicals, Property 6.1, and the Pythagorean theorem form a basis for solving a variety of problems that pertain to right triangles.

**Classroom Example**
A 62-foot guy-wire hangs from the top of a tower. When pulled taut, the guy-wire reaches a point on the ground 25 feet from the base of the tower. Find the height of the tower to the nearest tenth of a foot.

### EXAMPLE 9 Apply Your Skill

A 50-foot rope hangs from the top of a flagpole. When pulled taut to its full length, the rope reaches a point on the ground 18 feet from the base of the pole. Find the height of the pole to the nearest tenth of a foot.

#### Solution

Let's make a sketch (Figure 6.1) and record the given information. Use the Pythagorean theorem to solve for $p$ as follows:

$$p^2 + 18^2 = 50^2$$
$$p^2 + 324 = 2500$$
$$p^2 = 2176$$
$$p = \sqrt{2176} = 46.6 \quad \text{to the nearest tenth}$$

The height of the flagpole is approximately 46.6 feet.

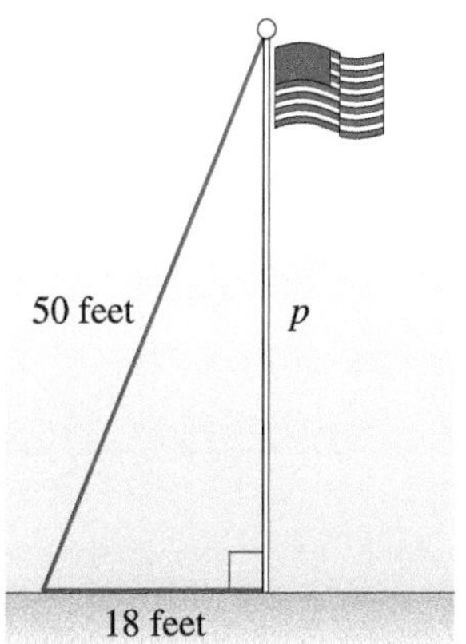


$p$ represents the height of the flagpole.

**Figure 6.1**

There are two special kinds of right triangles that we use extensively in later mathematics courses. The first is the **isosceles right triangle**, which is a right triangle that has both legs of the same length. Let's consider a problem that involves an isosceles right triangle.

**Classroom Example**
Find the length of each leg of an isosceles right triangle that has a hypotenuse of length 16 inches.

### EXAMPLE 10 Apply Your Skill

Find the length of each leg of an isosceles right triangle that has a hypotenuse of length 5 meters.

#### Solution

Let's sketch an isosceles right triangle and let $x$ represent the length of each leg (Figure 6.2). Then we can apply the Pythagorean theorem.

$$x^2 + x^2 = 5^2$$
$$2x^2 = 25$$
$$x^2 = \frac{25}{2}$$
$$x = \pm\sqrt{\frac{25}{2}} = \pm\frac{5}{\sqrt{2}} = \pm\frac{5\sqrt{2}}{2}$$

Each leg is $\frac{5\sqrt{2}}{2}$ meters long.

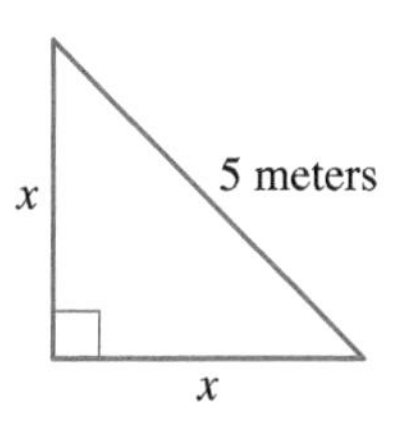


**Figure 6.2**

**Remark:** In Example 9 we made no attempt to express $\sqrt{2176}$ in simplest radical form because the answer was to be given as a rational approximation to the nearest tenth. However, in Example 10 we left the final answer in radical form and therefore expressed it in simplest radical form.

The second special kind of right triangle that we use frequently is one that contains acute angles of 30° and 60°. In such a right triangle, which we refer to as a **30°–60° right triangle**, the side opposite the 30° angle is equal in length to one-half of the length of the hypotenuse. This relationship, along with the Pythagorean theorem, provides us with another problem-solving technique.

**Classroom Example**
Suppose that a 30-foot ladder is leaning against a building and makes an angle of 60° with the ground. How far up the building does the top of the ladder reach? Express your answer to the nearest tenth of a foot.

### EXAMPLE 11 Apply Your Skill

Suppose that a 20-foot ladder is leaning against a building and makes an angle of 60° with the ground. How far up the building does the top of the ladder reach? Express your answer to the nearest tenth of a foot.

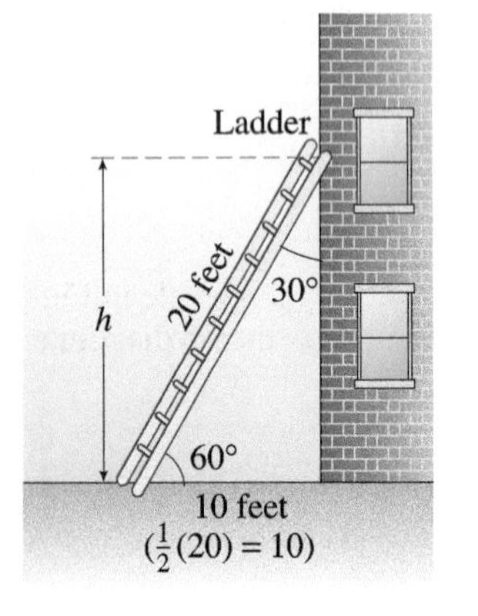


**Figure 6.3**

#### Solution

Figure 6.3 depicts this situation. The side opposite the 30° angle equals one-half of the hypotenuse, so it is of length $\frac{1}{2}(20) = 10$ feet. Now we can apply the Pythagorean theorem.

$$h^2 + 10^2 = 20^2$$
$$h^2 + 100 = 400$$
$$h^2 = 300$$
$$h = \sqrt{300} = 17.3 \quad \text{to the nearest tenth}$$

The ladder touches the building at a point approximately 17.3 feet from the ground.

## Concept Quiz 6.2

For Problems 1–10, answer true or false.

1. The quadratic equation $-3x^2 + 5x - 8 = 0$ is in standard form.
2. The solution set of the equation $(x + 1)^2 = -25$ will consist only of nonreal complex numbers.
3. An isosceles right triangle is a right triangle that has a hypotenuse of the same length as one of the legs.
4. In a 30°–60° right triangle, the hypotenuse is equal in length to twice the length of the side opposite the 30° angle.
5. The equation $2x^2 + x^3 - x + 4 = 0$ is a quadratic equation.
6. The solution set for $4x^2 = 8x$ is $\{2\}$.
7. The solution set for $3x^2 = 8x$ is $\left\{0, \frac{8}{3}\right\}$.
8. The solution set for $x^2 - 8x - 48 = 0$ is $\{-12, 4\}$.
9. If the length of each leg of an isosceles right triangle is 4 inches, then the hypotenuse is of length $4\sqrt{2}$ inches.
10. If the length of the leg opposite the 30° angle in a right triangle is 6 centimeters, then the length of the other leg is 12 centimeters.

## Problem Set 6.2

For Problems 1–20, solve each of the quadratic equations by factoring and applying the property, $ab = 0$ if and only if $a = 0$ or $b = 0$. If necessary, return to Chapter 3 and review the factoring techniques presented there. (Objective 1)

**1.** $x^2 - 9x = 0$

**2.** $x^2 + 5x = 0$

**3.** $x^2 = -3x$

**4.** $x^2 = 15x$

**5.** $3y^2 + 12y = 0$

**6.** $6y^2 - 24y = 0$

**7.** $5n^2 - 9n = 0$

**8.** $4n^2 + 13n = 0$

**9.** $x^2 + x - 30 = 0$

**10.** $x^2 - 8x - 48 = 0$

**11.** $x^2 - 19x + 84 = 0$

**12.** $x^2 - 21x + 104 = 0$

**13.** $2x^2 + 19x + 24 = 0$

**14.** $4x^2 + 29x + 30 = 0$

**15.** $15x^2 + 29x - 14 = 0$

**16.** $24x^2 + x - 10 = 0$

**17.** $25x^2 - 30x + 9 = 0$

**18.** $16x^2 - 8x + 1 = 0$

**19.** $6x^2 - 5x - 21 = 0$

**20.** $12x^2 - 4x - 5 = 0$

For Problems 21–26, solve each radical equation. Don't forget, you *must* check potential solutions. (Objective 1)

**21.** $3\sqrt{x} = x + 2$

**22.** $3\sqrt{2x} = x + 4$

**23.** $\sqrt{2x} = x - 4$

**24.** $\sqrt{x} = x - 2$

**25.** $\sqrt{3x} + 6 = x$

**26.** $\sqrt{5x} + 10 = x$

For Problems 27–62, use Property 6.1 to help solve each quadratic equation. (Objective 2)

**27.** $x^2 = 1$

**28.** $x^2 = 81$

**29.** $x^2 = -36$

**30.** $x^2 = -49$

**31.** $x^2 = 14$

**32.** $x^2 = 22$

**33.** $n^2 - 28 = 0$

**34.** $n^2 - 54 = 0$

**35.** $3t^2 = 54$

**36.** $4t^2 = 108$

**37.** $2t^2 = 7$

**38.** $3t^2 = 8$

**39.** $15y^2 = 20$

**40.** $14y^2 = 80$

**41.** $10x^2 + 48 = 0$

**42.** $12x^2 + 50 = 0$

**43.** $24x^2 = 36$

**44.** $12x^2 = 49$

**45.** $(x - 2)^2 = 9$

**46.** $(x + 1)^2 = 16$

**47.** $(x + 3)^2 = 25$

**48.** $(x - 2)^2 = 49$

**49.** $(x + 6)^2 = -4$

**50.** $(3x + 1)^2 = 9$

**51.** $(2x - 3)^2 = 1$

**52.** $(2x + 5)^2 = -4$

**53.** $(n - 4)^2 = 5$

**54.** $(n - 7)^2 = 6$

**55.** $(t + 5)^2 = 12$

**56.** $(t - 1)^2 = 18$

**57.** $(3y - 2)^2 = -27$

**58.** $(4y + 5)^2 = 80$

**59.** $3(x + 7)^2 + 4 = 79$

**60.** $2(x + 6)^2 - 9 = 63$

**61.** $2(5x - 2)^2 + 5 = 25$

**62.** $3(4x - 1)^2 + 1 = -17$

For Problems 63–68, $a$ and $b$ represent the lengths of the legs of a right triangle, and $c$ represents the length of the hypotenuse. Express answers in simplest radical form. (Objective 3)

**63.** Find $c$ if $a = 4$ centimeters and $b = 6$ centimeters.

**64.** Find $c$ if $a = 3$ meters and $b = 7$ meters.

**65.** Find $a$ if $c = 12$ inches and $b = 8$ inches.

**66.** Find $a$ if $c = 8$ feet and $b = 6$ feet.

**67.** Find $b$ if $c = 17$ yards and $a = 15$ yards.

**68.** Find $b$ if $c = 14$ meters and $a = 12$ meters.

For Problems 69–72, use the isosceles right triangle in Figure 6.4. Express your answers in simplest radical form. (Objective 3)

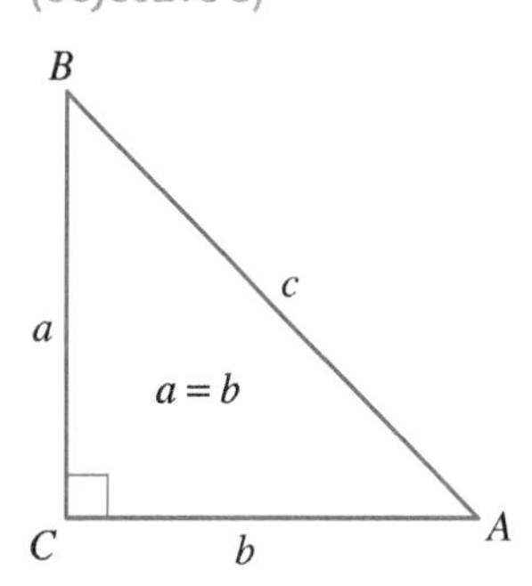


**Figure 6.4**

**69.** If $b = 6$ inches, find $c$.

**70.** If $a = 7$ centimeters, find $c$.

**71.** If $c = 8$ meters, find $a$ and $b$.

**72.** If $c = 9$ feet, find $a$ and $b$.

For Problems 73–78, use the triangle in Figure 6.5. Express your answers in simplest radical form. (Objective 3)

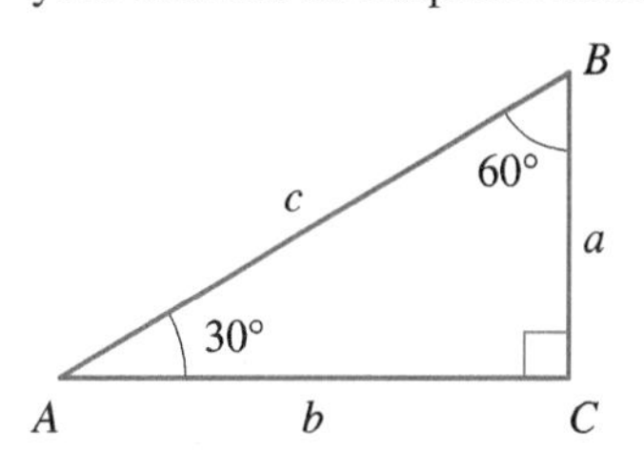


**Figure 6.5**

**73.** If $a = 3$ inches, find $b$ and $c$.

**74.** If $a = 6$ feet, find $b$ and $c$.

**75.** If $c = 14$ centimeters, find $a$ and $b$.

**76.** If $c = 9$ centimeters, find $a$ and $b$.

**77.** If $b = 10$ feet, find $a$ and $c$.

**78.** If $b = 8$ meters, find $a$ and $c$.

**79.** A 24-foot ladder resting against a house reaches a windowsill 16 feet above the ground. How far is the foot of the ladder from the foundation of the house? Express your answer to the nearest tenth of a foot.

**80.** A 62-foot guy-wire makes an angle of 60° with the ground and is attached to a telephone pole (see Figure 6.6). Find the distance from the base of the pole to the point on the pole where the wire is attached. Express your answer to the nearest tenth of a foot.

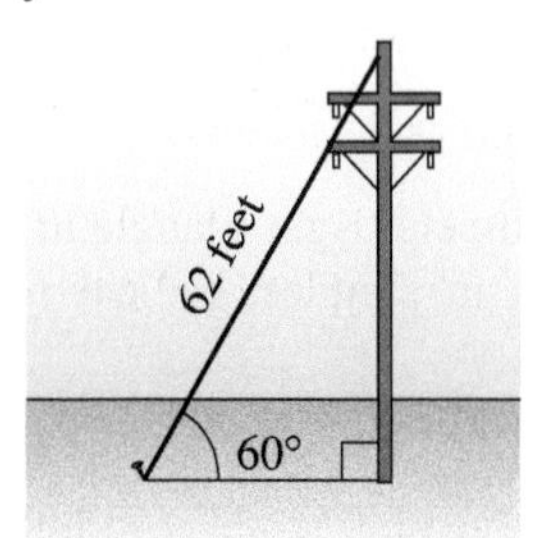


**Figure 6.6**

**81.** A rectangular plot measures 16 meters by 34 meters. Find, to the nearest meter, the distance from one corner of the plot to the corner diagonally opposite.

**82.** Consecutive bases of a square-shaped baseball diamond are 90 feet apart (see Figure 6.7). Find, to the nearest tenth of a foot, the distance from first base diagonally across the diamond to third base.

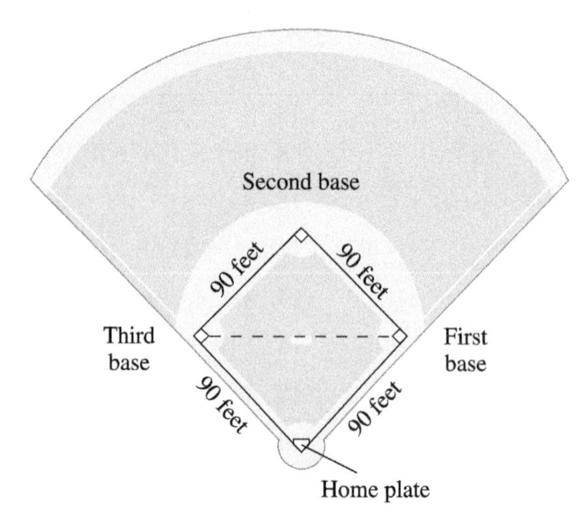


**Figure 6.7**

**83.** A diagonal of a square parking lot is 75 meters. Find, to the nearest meter, the length of a side of the lot.

## Thoughts Into Words

**84.** Explain why the equation $(x + 2)^2 + 5 = 1$ has no real number solutions.

**85.** Suppose that your friend solved the equation $(x + 3)^2 = 25$ as follows:

$$(x + 3)^2 = 25$$
$$x^2 + 6x + 9 = 25$$
$$x^2 + 6x - 16 = 0$$
$$(x + 8)(x - 2) = 0$$
$$x + 8 = 0 \quad \text{or} \quad x - 2 = 0$$
$$x = -8 \quad \text{or} \quad x = 2$$

Is this a correct approach to the problem? Would you offer any suggestion about an easier approach to the problem?

## Further Investigations

**86.** Suppose that we are given a cube with edges 12 centimeters in length. Find the length of a diagonal from a lower corner to the diagonally opposite upper corner. Express your answer to the nearest tenth of a centimeter.

**87.** Suppose that we are given a rectangular box with a length of 8 centimeters, a width of 6 centimeters, and a height of 4 centimeters. Find the length of a diagonal from a lower corner to the upper corner diagonally opposite. Express your answer to the nearest tenth of a centimeter.

**88.** The converse of the Pythagorean theorem is also true. It states, "If the measures $a$, $b$, and $c$ of the sides of a triangle are such that $a^2 + b^2 = c^2$, then the triangle is a right triangle with $a$ and $b$ the measures of the legs and $c$ the measure of the hypotenuse." Use the converse of the Pythagorean theorem to determine which of the triangles with sides of the following measures are right triangles.

**(a)** 9, 40, 41 **(b)** 20, 48, 52

**(c)** 19, 21, 26 **(d)** 32, 37, 49

**(e)** 65, 156, 169 **(f)** 21, 72, 75

**89.** Find the length of the hypotenuse ($h$) of an isosceles right triangle if each leg is $s$ units long. Then use this relationship to redo Problems 69–72.

**90.** Suppose that the side opposite the 30° angle in a 30°–60° right triangle is $s$ units long. Express the length of the hypotenuse and the length of the other leg in terms of $s$. Then use these relationships and redo Problems 73–78.

### Answers to the Concept Quiz

**1.** True **2.** True **3.** False **4.** True **5.** False **6.** False **7.** True **8.** False **9.** True **10.** False

## 6.3 Completing the Square

OBJECTIVE 1 Solve quadratic equations by completing the square

Thus far we have solved quadratic equations by factoring and applying the property, $ab = 0$ if and only if $a = 0$ or $b = 0$, or by applying the property, $x^2 = a$ if and only if $x = \pm\sqrt{a}$. In this section we examine another method called *completing the square*, which will give us the power to solve any quadratic equation.

A factoring technique we studied in Chapter 3 relied on recognizing *perfect-square trinomials*. In each of the following, the perfect-square trinomial on the right side is the result of squaring the binomial on the left side.

$$(x + 4)^2 = x^2 + 8x + 16 \qquad (x - 6)^2 = x^2 - 12x + 36$$
$$(x + 7)^2 = x^2 + 14x + 49 \qquad (x - 9)^2 = x^2 - 18x + 81$$
$$(x + a)^2 = x^2 + 2ax + a^2$$

Note that in each of the square trinomials, the constant term is equal to the square of one-half of the coefficient of the $x$ term. This relationship enables us to form a perfect-square trinomial by adding a proper constant term. To find the constant term, take one-half of the coefficient of the $x$ term and then square the result. For example, suppose that we want to form a perfect-square trinomial from $x^2 + 10x$. The coefficient of the $x$ term is 10. Because $\frac{1}{2}(10) = 5$, and $5^2 = 25$, the constant term should be 25. The perfect-square trinomial that can be formed is $x^2 + 10x + 25$. This perfect-square trinomial can be factored and expressed as $(x + 5)^2$. Let's use the previous ideas to help solve some quadratic equations.

Classroom Example
Solve $x^2 + 8x - 5 = 0$.

**EXAMPLE 1** Solve $x^2 + 10x - 2 = 0$.

### Solution

$x^2 + 10x - 2 = 0$

$x^2 + 10x = 2$ — Isolate the $x^2$ and $x$ terms

$\frac{1}{2}(10) = 5$ and $5^2 = 25$ — Take $\frac{1}{2}$ of the coefficient of the $x$ term and then square the result

$x^2 + 10x + 25 = 2 + 25$ — Add 25 to *both* sides of the equation

$(x + 5)^2 = 27$ — Factor the perfect-square trinomial

$x + 5 = \pm\sqrt{27}$ — Now solve by applying Property 6.1

$x + 5 = \pm 3\sqrt{3}$

$x = -5 \pm 3\sqrt{3}$ — Subtract 5 from both sides

The solution set is $\{-5 \pm 3\sqrt{3}\}$.

The method of completing the square to solve a quadratic equation is merely what the name implies. A perfect-square trinomial is formed, then the equation can be changed to the necessary form for applying the property "$x^2 = a$ if and only if $x = \pm\sqrt{a}$." Let's consider another example.

**Classroom Example**
Solve $x(x + 10) = -33$.

**EXAMPLE 2** Solve $x(x + 8) = -23$.

### Solution

$$x(x + 8) = -23$$

$$x^2 + 8x = -23 \qquad \text{Apply the distributive property}$$

$$\frac{1}{2}(8) = 4 \quad \text{and} \quad 4^2 = 16 \qquad \text{Take } \frac{1}{2} \text{ of the coefficient of the } x \text{ term and then square the result}$$

$$x^2 + 8x + 16 = -23 + 16 \qquad \text{Add 16 to } both \text{ sides of the equation}$$

$$(x + 4)^2 = -7 \qquad \text{Factor the perfect-square trinomial}$$

$$x + 4 = \pm\sqrt{-7} \qquad \text{Now solve by applying Property 6.1}$$

$$x + 4 = \pm i\sqrt{7}$$

$$x = -4 \pm i\sqrt{7} \qquad \text{Subtract 4 from both sides}$$

The solution set is $\{-4 \pm i\sqrt{7}\}$.

**Classroom Example**
Solve $m^2 - 3m - 5 = 0$.

**EXAMPLE 3** Solve $x^2 - 3x + 1 = 0$.

### Solution

$$x^2 - 3x + 1 = 0$$

$$x^2 - 3x = -1$$

$$x^2 - 3x + \frac{9}{4} = -1 + \frac{9}{4} \qquad \frac{1}{2}(-3) = -\frac{3}{2} \text{ and } \left(-\frac{3}{2}\right)^2 = \frac{9}{4}$$

$$\left(x - \frac{3}{2}\right)^2 = \frac{5}{4}$$

$$x - \frac{3}{2} = \pm\sqrt{\frac{5}{4}}$$

$$x - \frac{3}{2} = \pm\frac{\sqrt{5}}{2}$$

$$x = \frac{3}{2} \pm \frac{\sqrt{5}}{2}$$

$$x = \frac{3 \pm \sqrt{5}}{2}$$

The solution set is $\left\{\frac{3 \pm \sqrt{5}}{2}\right\}$.

In Example 3 note that because the coefficient of the $x$ term is odd, we are forced into the realm of fractions. Using common fractions rather than decimals enables us to apply our previous work with radicals.

The relationship for a perfect-square trinomial that states that the constant term is equal to the square of one-half of the coefficient of the $x$ term holds only if the coefficient of $x^2$ is 1. Thus we must make an adjustment when solving quadratic equations that have a coefficient of $x^2$ other than 1. We will need to apply the multiplication property of equality so that the coefficient of the $x^2$ term becomes 1. The next example shows how to make this adjustment.

**Classroom Example**
Solve $3y^2 - 24y + 26 = 0$.

**EXAMPLE 4** Solve $2x^2 + 12x - 5 = 0$.

### Solution

$$2x^2 + 12x - 5 = 0$$
$$2x^2 + 12x = 5$$
$$x^2 + 6x = \frac{5}{2} \quad \text{Multiply both sides by } \frac{1}{2}$$
$$x^2 + 6x + 9 = \frac{5}{2} + 9 \quad \frac{1}{2}(6) = 3, \text{ and } 3^2 = 9$$
$$x^2 + 6x + 9 = \frac{23}{2}$$
$$(x + 3)^2 = \frac{23}{2}$$
$$x + 3 = \pm\sqrt{\frac{23}{2}}$$
$$x + 3 = \pm\frac{\sqrt{46}}{2} \quad \sqrt{\frac{23}{2}} = \frac{\sqrt{23}}{\sqrt{2}} \cdot \frac{\sqrt{2}}{\sqrt{2}} = \frac{\sqrt{46}}{2}$$
$$x = -3 \pm \frac{\sqrt{46}}{2}$$
$$x = \frac{-6}{2} \pm \frac{\sqrt{46}}{2} \quad \text{Common denominator of 2}$$
$$x = \frac{-6 \pm \sqrt{46}}{2}$$

The solution set is $\left\{\frac{-6 \pm \sqrt{46}}{2}\right\}$.

As we mentioned earlier, we can use the method of completing the square to solve *any* quadratic equation. To illustrate, let's use it to solve an equation that could also be solved by factoring.

**Classroom Example**
Solve $t^2 - 10t + 21 = 0$ by completing the square.

**EXAMPLE 5** Solve $x^2 - 2x - 8 = 0$ by completing the square.

### Solution

$$x^2 - 2x - 8 = 0$$
$$x^2 - 2x = 8$$
$$x^2 - 2x + 1 = 8 + 1 \quad \frac{1}{2}(-2) = -1, \text{ and } (-1)^2 = 1$$
$$(x - 1)^2 = 9$$
$$x - 1 = \pm 3$$
$$x - 1 = 3 \quad \text{or} \quad x - 1 = -3$$
$$x = 4 \quad \text{or} \quad x = -2$$

The solution set is $\{-2, 4\}$.

Solving the equation in Example 5 by factoring would be easier than completing the square. Remember, however, that the method of completing the square will work with any quadratic equation.

## Concept Quiz 6.3

For Problems 1–10, answer true or false.

1. In a perfect-square trinomial, the constant term is equal to one-half the coefficient of the $x$ term.
2. The method of completing the square will solve any quadratic equation.
3. Every quadratic equation solved by completing the square will have real number solutions.
4. The completing-the-square method cannot be used if factoring could solve the quadratic equation.
5. To use the completing-the-square method for solving the equation $3x^2 + 2x = 5$, we would first divide both sides of the equation by 3.
6. The equation $x^2 + 2x = 0$ cannot be solved by using the method of completing the square.
7. To solve the equation $x^2 - 5x = 1$ by completing the square, we would start by adding $\frac{25}{4}$ to both sides of the equation.
8. To solve the equation $x^2 - 2x = 14$ by completing the square, we must first change the form of the equation to $x^2 - 2x - 14 = 0$.
9. The solution set of the equation $x^2 - 2x = 14$ is $\{1 \pm \sqrt{15}\}$.
10. The solution set of the equation $x^2 - 5x - 1 = 0$ is $\left\{\frac{5 \pm \sqrt{29}}{2}\right\}$.

## Problem Set 6.3

For Problems 1–14, solve each quadratic equation by using (a) the factoring method and (b) the method of completing the square. (Objective 1)

**1.** $x^2 - 4x - 60 = 0$
**2.** $x^2 + 6x - 16 = 0$
**3.** $x^2 - 14x = -40$
**4.** $x^2 - 18x = -72$
**5.** $x^2 - 5x - 50 = 0$
**6.** $x^2 + 3x - 18 = 0$
**7.** $x(x + 7) = 8$
**8.** $x(x - 1) = 30$
**9.** $2n^2 - n - 15 = 0$
**10.** $3n^2 + n - 14 = 0$
**11.** $3n^2 + 7n - 6 = 0$
**12.** $2n^2 + 7n - 4 = 0$
**13.** $n(n + 6) = 160$
**14.** $n(n - 6) = 216$

For Problems 15–38, use the method of completing the square to solve each quadratic equation. (Objective 1)

**15.** $x^2 + 4x - 2 = 0$
**16.** $x^2 + 2x - 1 = 0$
**17.** $x^2 + 6x - 3 = 0$
**18.** $x^2 + 8x - 4 = 0$
**19.** $y^2 - 10y = 1$
**20.** $y^2 - 6y = -10$
**21.** $n^2 - 8n + 17 = 0$
**22.** $n^2 - 4n + 2 = 0$
**23.** $n(n + 12) = -9$
**24.** $n(n + 14) = -4$
**25.** $n^2 + 2n + 6 = 0$
**26.** $n^2 + n - 1 = 0$
**27.** $x^2 + 3x - 2 = 0$
**28.** $x^2 + 5x - 3 = 0$
**29.** $x^2 + 5x + 1 = 0$
**30.** $x^2 + 7x + 2 = 0$
**31.** $y^2 - 7y + 3 = 0$
**32.** $y^2 - 9y + 30 = 0$
**33.** $2x^2 + 4x - 3 = 0$
**34.** $2t^2 - 4t + 1 = 0$
**35.** $3n^2 - 6n + 5 = 0$
**36.** $3x^2 + 12x - 2 = 0$
**37.** $3x^2 + 5x - 1 = 0$
**38.** $2x^2 + 7x - 3 = 0$

For Problems 39–60, solve each quadratic equation using the method that seems most appropriate.

**39.** $x^2 + 8x - 48 = 0$
**40.** $x^2 + 5x - 14 = 0$
**41.** $2n^2 - 8n = -3$
**42.** $3x^2 + 6x = 1$

**43.** $(3x - 1)(2x + 9) = 0$

**44.** $(5x + 2)(x - 4) = 0$

**45.** $(x + 2)(x - 7) = 10$

**46.** $(x - 3)(x + 5) = -7$

**47.** $(x - 3)^2 = 12$

**48.** $x^2 = 16x$

**49.** $3n^2 - 6n + 4 = 0$

**50.** $2n^2 - 2n - 1 = 0$

**51.** $n(n + 8) = 240$

**52.** $t(t - 26) = -160$

**53.** $3x^2 + 5x = -2$

**54.** $2x^2 - 7x = -5$

**55.** $4x^2 - 8x + 3 = 0$

**56.** $9x^2 + 18x + 5 = 0$

**57.** $x^2 + 12x = 4$

**58.** $x^2 + 6x = -11$

**59.** $4(2x + 1)^2 - 1 = 11$

**60.** $5(x + 2)^2 + 1 = 16$

**61.** Use the method of completing the square to solve $ax^2 + bx + c = 0$ for $x$, where $a$, $b$, and $c$ are real numbers and $a \neq 0$.

## Thoughts Into Words

**62.** Explain the process of completing the square to solve a quadratic equation.

**63.** Give a step-by-step description of how to solve $3x^2 + 9x - 4 = 0$ by completing the square.

## Further Investigations

Solve Problems 64–67 for the indicated variable. Assume that all letters represent positive numbers.

**64.** $\frac{x^2}{a^2} - \frac{y^2}{b^2} = 1$ for $y$

**65.** $\frac{x^2}{a^2} + \frac{y^2}{b^2} = 1$ for $x$

**66.** $s = \frac{1}{2}gt^2$ for $t$

**67.** $A = \pi r^2$ for $r$

Solve each of the following equations for $x$.

**68.** $x^2 + 8ax + 15a^2 = 0$

**69.** $x^2 - 5ax + 6a^2 = 0$

**70.** $10x^2 - 31ax - 14a^2 = 0$

**71.** $6x^2 + ax - 2a^2 = 0$

**72.** $4x^2 + 4bx + b^2 = 0$

**73.** $9x^2 - 12bx + 4b^2 = 0$

### Answers to the Concept Quiz

**1.** False **2.** True **3.** False **4.** False **5.** True **6.** False **7.** True **8.** False **9.** True **10.** True

# 6.4 Quadratic Formula

OBJECTIVES

1. Use the quadratic formula to solve quadratic equations
2. Determine the nature of roots to quadratic equations

As we saw in the last section, the method of completing the square can be used to solve any quadratic equation. Thus if we apply the method of completing the square to the equation $ax^2 + bx + c = 0$, where $a$, $b$, and $c$ are real numbers and $a \neq 0$, we can produce a formula for solving quadratic equations. This formula can then be used to solve any quadratic equation. Let's solve $ax^2 + bx + c = 0$ by completing the square.

$$ax^2 + bx + c = 0$$

$$ax^2 + bx = -c \qquad \text{Isolate the } x^2 \text{ and } x \text{ terms}$$

$$x^2 + \frac{b}{a}x = -\frac{c}{a} \qquad \text{Multiply both sides by } \frac{1}{a}$$

$$x^2 + \frac{b}{a}x + \frac{b^2}{4a^2} = -\frac{c}{a} + \frac{b^2}{4a^2} \qquad \frac{1}{2}\left(\frac{b}{a}\right) = \frac{b}{2a} \text{ and } \left(\frac{b}{2a}\right)^2 = \frac{b^2}{4a^2}$$

Complete the square by adding $\frac{b^2}{4a^2}$ to both sides

$$x^2 + \frac{b}{a}x + \frac{b^2}{4a^2} = -\frac{4ac}{4a^2} + \frac{b^2}{4a^2} \qquad \text{Common denominator of } 4a^2 \text{ on right side}$$

$$x^2 + \frac{b}{a}x + \frac{b^2}{4a^2} = \frac{b^2}{4a^2} - \frac{4ac}{4a^2} \qquad \text{Commutative property on the right side}$$

$$\left(x + \frac{b}{2a}\right)^2 = \frac{b^2 - 4ac}{4a^2} \qquad \text{The right side is combined into a single fraction}$$

$$x + \frac{b}{2a} = \pm\sqrt{\frac{b^2 - 4ac}{4a^2}}$$

$$x + \frac{b}{2a} = \pm\frac{\sqrt{b^2 - 4ac}}{\sqrt{4a^2}}$$

$$x + \frac{b}{2a} = \pm\frac{\sqrt{b^2 - 4ac}}{2a} \qquad \sqrt{4a^2} = |2a| \text{ but } 2a \text{ can be used because of the use of } \pm$$

$$x + \frac{b}{2a} = \frac{\sqrt{b^2 - 4ac}}{2a} \quad \text{or} \quad x + \frac{b}{2a} = -\frac{\sqrt{b^2 - 4ac}}{2a}$$

$$x = -\frac{b}{2a} + \frac{\sqrt{b^2 - 4ac}}{2a} \quad \text{or} \quad x = -\frac{b}{2a} - \frac{\sqrt{b^2 - 4ac}}{2a}$$

$$x = \frac{-b + \sqrt{b^2 - 4ac}}{2a} \quad \text{or} \quad x = \frac{-b - \sqrt{b^2 - 4ac}}{2a}$$

The quadratic formula is usually stated as follows:

### Quadratic Formula

If $a \neq 0$, then the solutions of the equation $ax^2 + bx + c = 0$ are given by

$$x = \frac{-b \pm \sqrt{b^2 - 4ac}}{2a}$$

We can use the quadratic formula to solve *any* quadratic equation by expressing the equation in the standard form $ax^2 + bx + c = 0$ and substituting the values for $a$, $b$, and $c$ into the formula. Let's consider some examples.

**Classroom Example**
Solve $x^2 - 5x - 9 = 0$.

**EXAMPLE 1** Solve $x^2 + 5x + 2 = 0$.

### Solution

$$x^2 + 5x + 2 = 0$$

The given equation is in standard form with $a = 1$, $b = 5$, and $c = 2$. Let's substitute these values into the formula and simplify.

$$x = \frac{-b \pm \sqrt{b^2 - 4ac}}{2a}$$

$$x = \frac{-5 \pm \sqrt{5^2 - 4(1)(2)}}{2(1)}$$

$$x = \frac{-5 \pm \sqrt{25 - 8}}{2}$$

$$x = \frac{-5 \pm \sqrt{17}}{2}$$

The solution set is $\left\{\frac{-5 \pm \sqrt{17}}{2}\right\}$.

**Classroom Example**
Solve $2x^2 + 6x - 5 = 0$.

**EXAMPLE 2** Solve $2x^2 + 4x - 3 = 0$.

### Solution

$$2x^2 + 4x - 3 = 0$$

Here $a = 2$, $b = 4$, and $c = -3$. Solving by using the quadratic formula is unlike solving by completing the square in that there is no need to make the coefficient of $x^2$ equal to 1.

$$x = \frac{-b \pm \sqrt{b^2 - 4ac}}{2a}$$

$$x = \frac{-4 \pm \sqrt{4^2 - 4(2)(-3)}}{2(2)}$$

$$x = \frac{-4 \pm \sqrt{16 + 24}}{4}$$

$$x = \frac{-4 \pm \sqrt{40}}{4}$$

$$x = \frac{-4 \pm 2\sqrt{10}}{4}$$

$$x = \frac{2\left(-2 \pm \sqrt{10}\right)}{4} \quad \text{Factor out a 2 in the numerator}$$

$$x = \frac{\cancel{2}\left(-2 \pm \sqrt{10}\right)}{\underset{2}{\cancel{4}}} = \frac{-2 \pm \sqrt{10}}{2}$$

The solution set is $\left\{\frac{-2 \pm \sqrt{10}}{2}\right\}$.

**Classroom Example**
Solve $x^2 + 8x + 5 = 0$.

**EXAMPLE 3** Solve $x^2 - 2x - 4 = 0$.

### Solution

$$x^2 - 2x - 4 = 0$$

We need to think of $x^2 - 2x - 4 = 0$ as $x^2 + (-2)x + (-4) = 0$ to determine the values $a = 1$, $b = -2$, and $c = -4$. Let's substitute these values into the quadratic formula and simplify.

$$x = \frac{-b \pm \sqrt{b^2 - 4ac}}{2a}$$

$$x = \frac{-(-2) \pm \sqrt{(-2)^2 - 4(1)(-4)}}{2(1)}$$

$$x = \frac{2 \pm \sqrt{4 + 16}}{2}$$

$$x = \frac{2 \pm \sqrt{20}}{2}$$

$$x = \frac{2 \pm 2\sqrt{5}}{2}$$

$$x = \frac{2(1 \pm \sqrt{5})}{2}$$ Factor out a 2 in the numerator

$$x = \frac{\cancel{2}(1 \pm \sqrt{5})}{\cancel{2}} = 1 \pm \sqrt{5}$$

The solution set is $\{1 \pm \sqrt{5}\}$.

**Classroom Example**
Solve $x^2 - 8x + 18 = 0$.

**EXAMPLE 4** Solve $x^2 - 2x + 19 = 0$.

### Solution

$$x^2 - 2x + 19 = 0$$

We can substitute $a = 1$, $b = -2$, and $c = 19$.

$$x = \frac{-b \pm \sqrt{b^2 - 4ac}}{2a}$$

$$x = \frac{-(-2) \pm \sqrt{(-2)^2 - 4(1)(19)}}{2(1)}$$

$$x = \frac{2 \pm \sqrt{4 - 76}}{2}$$

$$x = \frac{2 \pm \sqrt{-72}}{2}$$

$$x = \frac{2 \pm 6i\sqrt{2}}{2}$$ $\sqrt{-72} = i\sqrt{72} = i\sqrt{36}\sqrt{2} = 6i\sqrt{2}$

$$x = \frac{2(1 \pm 3i)}{2}$$ Factor out a 2 in the numerator

$$x = \frac{\cancel{2}(1 \pm 3i\sqrt{2})}{\cancel{2}} = 1 \pm 3i\sqrt{2}$$

The solution set is $\{1 \pm 3i\sqrt{2}\}$.

**Classroom Example**
Solve $x(5x - 7) = 6$.

**EXAMPLE 5** Solve $n(3n - 10) = 25$.

### Solution

$$n(3n - 10) = 25$$

First, we need to change the equation to the standard form $an^2 + bn + c = 0$.

$$n(3n - 10) = 25$$
$$3n^2 - 10n = 25$$
$$3n^2 - 10n - 25 = 0$$

Now we can substitute $a = 3$, $b = -10$, and $c = -25$ into the quadratic formula.

$$n = \frac{-b \pm \sqrt{b^2 - 4ac}}{2a}$$

$$n = \frac{-(-10) \pm \sqrt{(-10)^2 - 4(3)(-25)}}{2(3)}$$

$$n = \frac{10 \pm \sqrt{100 + 300}}{2(3)}$$

$$n = \frac{10 \pm \sqrt{400}}{6}$$

$$n = \frac{10 \pm 20}{6}$$

$$n = \frac{10 + 20}{6} \quad \text{or} \quad n = \frac{10 - 20}{6}$$

$$n = 5 \quad \text{or} \quad n = -\frac{5}{3}$$

The solution set is $\left\{-\frac{5}{3}, 5\right\}$.

In Example 5, note that we used the variable $n$. The quadratic formula is usually stated in terms of $x$, but it certainly can be applied to quadratic equations in other variables. Also note in Example 5 that the polynomial $3n^2 - 10n - 25$ can be factored as $(3n + 5)(n - 5)$. Therefore, we could also solve the equation $3n^2 - 10n - 25 = 0$ by using the factoring approach. Section 6.5 will offer some guidance about which approach to use for a particular equation.

## Determining the Nature of Roots of Quadratic Equations

The quadratic formula makes it easy to determine the nature of the roots of a quadratic equation without completely solving the equation. The number

$$b^2 - 4ac$$

which appears under the radical sign in the quadratic formula, is called the **discriminant** of the quadratic equation. The discriminant is the indicator of the kind of roots the equation has. For example, suppose that you start to solve the equation $x^2 - 4x + 7 = 0$ as follows:

$$x = \frac{-b \pm \sqrt{b^2 - 4ac}}{2a}$$

$$x = \frac{-(-4) \pm \sqrt{(-4)^2 - 4(1)(7)}}{2(1)}$$

$$x = \frac{4 \pm \sqrt{16 - 28}}{2}$$

$$x = \frac{4 \pm \sqrt{-12}}{2}$$

At this stage you should be able to look ahead and realize that you will obtain two nonreal complex solutions for the equation. (Note, by the way, that these solutions are complex conjugates.) In other words, the discriminant $(-12)$ indicates what type of roots you will obtain.

We make the following general statements relative to the roots of a quadratic equation of the form $ax^2 + bx + c = 0$.

1. If $b^2 - 4ac < 0$, then the equation has two nonreal complex solutions.
2. If $b^2 - 4ac = 0$, then the equation has one real solution.
3. If $b^2 - 4ac > 0$, then the equation has two real solutions.

The following examples illustrate each of these situations. (You may want to solve the equations completely to verify the conclusions.)

| Equation | Discriminant | Nature of roots |
|---|---|---|
| $x^2 - 3x + 7 = 0$ | $b^2 - 4ac = (-3)^2 - 4(1)(7)$ $= 9 - 28$ $= -19$ | Two nonreal complex solutions |
| $9x^2 - 12x + 4 = 0$ | $b^2 - 4ac = (-12)^2 - 4(9)(4)$ $= 144 - 144$ $= 0$ | One real solution |
| $2x^2 + 5x - 3 = 0$ | $b^2 - 4ac = (5)^2 - 4(2)(-3)$ $= 25 + 24$ $= 49$ | Two real solutions |

**Remark:** A clarification is called for at this time. Previously, we made the statement that if $b^2 - 4ac = 0$, then the equation has one real solution. Technically, such an equation has two solutions, but they are equal. For example, each factor of $(x - 7)(x - 7) = 0$ produces a solution, but both solutions are the number 7. We sometimes refer to this as one real solution with a *multiplicity of two*. Using the idea of multiplicity of roots, we can say that every quadratic equation has two roots.

**Classroom Example**
Use the discriminant to determine whether the equation $3x^2 - 7x + 2 = 0$ has two nonreal complex solutions, one real solution with a multiplicity of 2, or two real solutions.

### EXAMPLE 6

Use the discriminant to determine if the equation $5x^2 + 2x + 7 = 0$ has two nonreal complex solutions, one real solution with a multiplicity of two, or two real solutions.

### Solution

For the equation $5x^2 + 2x + 7 = 0$, $a = 5$, $b = 2$, and $c = 7$.

$$\begin{aligned} b^2 - 4ac &= (2)^2 - 4(5)(7) \\ &= 4 - 140 \\ &= -136 \end{aligned}$$

Because the discriminant is negative, the solutions will be two nonreal complex numbers.

Most students become very adept at applying the quadratic formula to solve quadratic equations but make errors when reducing the answers. The next example shows two different methods for simplifying the answers.

**Classroom Example**
Solve $7m^2 + 4m - 2 = 0$.

**EXAMPLE 7** Solve $3x^2 - 8x + 2 = 0$.

### Solution

Here $a = 3$, $b = -8$, and $c = 2$. Let's substitute these values into the quadratic formula and simplify.

$$x = \frac{-b \pm \sqrt{b^2 - 4ac}}{2a}$$

$$x = \frac{-(-8) \pm \sqrt{(-8)^2 - 4(3)(2)}}{2(3)}$$

$$x = \frac{8 \pm \sqrt{64 - 24}}{6}$$

$$x = \frac{8 \pm \sqrt{40}}{6} = \frac{8 \pm 2\sqrt{10}}{6} \qquad \sqrt{40} = \sqrt{4}\sqrt{10} = 2\sqrt{10}$$

Now to simplify, one method is to factor 2 out of the numerator and reduce.

$$x = \frac{8 \pm 2\sqrt{10}}{6} = \frac{2\left(4 \pm \sqrt{10}\right)}{6} = \frac{2\left(4 \pm \sqrt{10}\right)}{\cancel{6}_3} = \frac{4 \pm \sqrt{10}}{3}$$

Another method for simplifying the answer is to write the result as two separate fractions and reduce each fraction.

$$x = \frac{8 \pm 2\sqrt{10}}{6} = \frac{8}{6} \pm \frac{2\sqrt{10}}{6} = \frac{4}{3} \pm \frac{\sqrt{10}}{3} = \frac{4 \pm \sqrt{10}}{3}$$

Be very careful when simplifying your result because that is a common source of incorrect answers.

## Concept Quiz 6.4

For Problems 1–10, answer true or false.

1. The quadratic formula can be used to solve any quadratic equation.
2. The number $\sqrt{b^2 - 4ac}$ is called the discriminant of the quadratic equation.
3. Every quadratic equation will have two solutions.
4. The quadratic formula cannot be used if the quadratic equation can be solved by factoring.
5. To use the quadratic formula for solving the equation $3x^2 + 2x - 5 = 0$, you must first divide both sides of the equation by 3.
6. The equation $9x^2 + 30x + 25 = 0$ has one real solution with a multiplicity of 2.
7. The equation $2x^2 + 3x + 4 = 0$ has two nonreal complex solutions.
8. The equation $x^2 + 9 = 0$ has two real solutions.
9. Because the quadratic formula has a denominator, it could be simplified and written as $x = -b \pm \dfrac{\sqrt{b^2 - 4ac}}{2a}$.
10. Rachel reduced the result $x = \dfrac{6 \pm 5\sqrt{7}}{2}$ to obtain $x = 3 \pm \dfrac{5\sqrt{7}}{2}$. Her result is correct.

## Problem Set 6.4

For Problems 1–10, simplify and reduce each expression.

1. $\frac{2 \pm \sqrt{20}}{4}$
2. $\frac{4 \pm \sqrt{20}}{6}$
3. $\frac{-6 \pm \sqrt{27}}{3}$
4. $\frac{-9 \pm \sqrt{54}}{3}$
5. $\frac{6 \pm \sqrt{18}}{9}$
6. $\frac{12 \pm \sqrt{32}}{8}$
7. $\frac{-10 \pm \sqrt{75}}{10}$
8. $\frac{-4 \pm \sqrt{8}}{4}$
9. $\frac{-6 \pm \sqrt{48}}{4}$
10. $\frac{-8 \pm \sqrt{72}}{4}$

For Problems 11–50, use the quadratic formula to solve each of the quadratic equations. (Objective 1)

11. $x^2 + 2x - 1 = 0$
12. $x^2 + 4x - 1 = 0$
13. $n^2 + 5n - 3 = 0$
14. $n^2 + 3n - 2 = 0$
15. $a^2 - 8a = 4$
16. $a^2 - 6a = 2$
17. $n^2 + 5n + 8 = 0$
18. $2n^2 - 3n + 5 = 0$
19. $x^2 - 18x + 80 = 0$
20. $x^2 + 19x + 70 = 0$
21. $-y^2 = -9y + 5$
22. $-y^2 + 7y = 4$
23. $2x^2 + x - 4 = 0$
24. $2x^2 + 5x - 2 = 0$
25. $4x^2 + 2x + 1 = 0$
26. $3x^2 - 2x + 5 = 0$
27. $3a^2 - 8a + 2 = 0$
28. $2a^2 - 6a + 1 = 0$
29. $-2n^2 + 3n + 5 = 0$
30. $-3n^2 - 11n + 4 = 0$
31. $3x^2 + 19x + 20 = 0$
32. $2x^2 - 17x + 30 = 0$
33. $36n^2 - 60n + 25 = 0$
34. $9n^2 + 42n + 49 = 0$
35. $4x^2 - 2x = 3$
36. $6x^2 - 4x = 3$
37. $5x^2 - 13x = 0$
38. $7x^2 + 12x = 0$
39. $3x^2 = 5$
40. $4x^2 = 3$
41. $6t^2 + t - 3 = 0$
42. $2t^2 + 6t - 3 = 0$
43. $n^2 + 32n + 252 = 0$
44. $n^2 - 4n - 192 = 0$
45. $12x^2 - 73x + 110 = 0$
46. $6x^2 + 11x - 255 = 0$
47. $-2x^2 + 4x - 3 = 0$
48. $-2x^2 + 6x - 5 = 0$
49. $-6x^2 + 2x + 1 = 0$
50. $-2x^2 + 4x + 1 = 0$

For each quadratic equation in Problems 51–60, first use the discriminant to determine whether the equation has two nonreal complex solutions, one real solution with a multiplicity of two, or two real solutions. Then solve the equation. (Objective 2)

51. $x^2 + 4x - 21 = 0$
52. $x^2 - 3x - 54 = 0$
53. $9x^2 - 6x + 1 = 0$
54. $4x^2 + 20x + 25 = 0$
55. $x^2 - 7x + 13 = 0$
56. $2x^2 - x + 5 = 0$
57. $15x^2 + 17x - 4 = 0$
58. $8x^2 + 18x - 5 = 0$
59. $3x^2 + 4x = 2$
60. $2x^2 - 6x = -1$

### Thoughts Into Words

61. Your friend states that the equation $-2x^2 + 4x - 1 = 0$ must be changed to $2x^2 - 4x + 1 = 0$ (by multiplying both sides by $-1$) before the quadratic formula can be applied. Is she right about this? If not, how would you convince her she is wrong?

62. Another of your friends claims that the quadratic formula can be used to solve the equation $x^2 - 9 = 0$. How would you react to this claim?

63. Why must we change the equation $3x^2 - 2x = 4$ to $3x^2 - 2x - 4 = 0$ before applying the quadratic formula?

### Further Investigations

The solution set for $x^2 - 4x - 37 = 0$ is $\{2 \pm \sqrt{41}\}$. With a calculator, we found a rational approximation, to the nearest one-thousandth, for each of these solutions.

$$2 - \sqrt{41} = -4.403 \quad \text{and} \quad 2 + \sqrt{41} = 8.403$$

Thus the solution set is $\{-4.403, 8.403\}$, with the answers rounded to the nearest one-thousandth.

Solve each of the equations in Problems 64–73, expressing solutions to the nearest one-thousandth.

64. $x^2 - 6x - 10 = 0$
65. $x^2 - 16x - 24 = 0$
66. $x^2 + 6x - 44 = 0$
67. $x^2 + 10x - 46 = 0$
68. $x^2 + 8x + 2 = 0$
69. $x^2 + 9x + 3 = 0$

**70.** $4x^2 - 6x + 1 = 0$

**71.** $5x^2 - 9x + 1 = 0$

**72.** $2x^2 - 11x - 5 = 0$

**73.** $3x^2 - 12x - 10 = 0$

For Problems 74–76, use the discriminant to help solve each problem.

**74.** Determine $k$ so that the solutions of $x^2 - 2x + k = 0$ are complex but nonreal.

**75.** Determine $k$ so that $4x^2 - kx + 1 = 0$ has two equal real solutions.

**76.** Determine $k$ so that $3x^2 - kx - 2 = 0$ has real solutions.

### Answers to the Concept Quiz

**1.** True **2.** False **3.** True **4.** False **5.** False **6.** True **7.** True **8.** False **9.** False **10.** True

## 6.5 More Quadratic Equations and Applications

OBJECTIVES

1. Solve quadratic equations selecting the most appropriate method
2. Solve word problems involving quadratic equations

Which method should be used to solve a particular quadratic equation? There is no hard and fast answer to that question; it depends on the type of equation and on your personal preference. In the following examples we will state reasons for choosing a specific technique. However, keep in mind that usually this is a decision you must make as the need arises. That's why you need to be familiar with the strengths and weaknesses of each method.

**Classroom Example**
Solve $3x^2 - x - 5 = 0$.

**EXAMPLE 1** Solve $2x^2 - 3x - 1 = 0$.

### Solution

Because of the leading coefficient of 2 and the constant term of $-1$, there are very few factoring possibilities to consider. Therefore, with such problems, first try the factoring approach. Unfortunately, this particular polynomial is not factorable using integers. Let's use the quadratic formula to solve the equation.

$$x = \frac{-b \pm \sqrt{b^2 - 4ac}}{2a}$$

$$x = \frac{-(-3) \pm \sqrt{(-3)^2 - 4(2)(-1)}}{2(2)}$$

$$x = \frac{3 \pm \sqrt{9 + 8}}{4}$$

$$x = \frac{3 \pm \sqrt{17}}{4}$$

The solution set is $\left\{\frac{3 \pm \sqrt{17}}{4}\right\}$.

**Classroom Example**
Solve $\frac{2}{x} + \frac{6}{x+3} = 1$.

**EXAMPLE 2** Solve $\frac{3}{n} + \frac{10}{n+6} = 1$.

## Solution

$$\frac{3}{n} + \frac{10}{n+6} = 1, \quad n \neq 0 \text{ and } n \neq -6$$

$$n(n+6)\left(\frac{3}{n} + \frac{10}{n+6}\right) = 1(n)(n+6) \quad \text{Multiply both sides by } n(n+6)\text{, which is the LCD}$$

$$3(n+6) + 10n = n(n+6)$$

$$3n + 18 + 10n = n^2 + 6n$$

$$13n + 18 = n^2 + 6n$$

$$0 = n^2 - 7n - 18$$

This equation is an easy one to consider for possible factoring, and it factors as follows:

$$0 = (n-9)(n+2)$$

$$n - 9 = 0 \quad \text{or} \quad n + 2 = 0$$

$$n = 9 \quad \text{or} \quad n = -2$$

The solution set is $\{-2, 9\}$.

We should make a comment about Example 2. Note the indication of the initial restrictions $n \neq 0$ and $n \neq -6$. Remember that we need to do this when solving fractional equations.

**Classroom Example**
Solve $m^2 + 20m + 96 = 0$.

**EXAMPLE 3** Solve $x^2 + 22x + 112 = 0$.

## Solution

The size of the constant term makes the factoring approach a little cumbersome for this problem. Furthermore, because the leading coefficient is 1 and the coefficient of the $x$ term is even, the method of completing the square will work effectively.

$$x^2 + 22x + 112 = 0$$

$$x^2 + 22x = -112$$

$$x^2 + 22x + 121 = -112 + 121 \quad \frac{1}{2}(22) = 11 \text{ and } 11^2 = 121$$

$$(x+11)^2 = 9$$

$$x + 11 = \pm\sqrt{9}$$

$$x + 11 = \pm 3$$

$$x + 11 = 3 \quad \text{or} \quad x + 11 = -3$$

$$x = -8 \quad \text{or} \quad x = -14$$

The solution set is $\{-14, -8\}$.

**Classroom Example**
Solve $x^4 + 2x^2 - 360 = 0$.

**EXAMPLE 4** Solve $x^4 - 4x^2 - 96 = 0$.

## Solution

An equation such as $x^4 - 4x^2 - 96 = 0$ is not a quadratic equation, but we can solve it using the techniques that we use on quadratic equations. That is, we can factor the polynomial and apply the property "$ab = 0$ if and only if $a = 0$ or $b = 0$" as follows:

$$x^4 - 4x^2 - 96 = 0$$
$$(x^2 - 12)(x^2 + 8) = 0$$
$$x^2 - 12 = 0 \quad \text{or} \quad x^2 + 8 = 0$$
$$x^2 = 12 \quad \text{or} \quad x^2 = -8$$
$$x = \pm\sqrt{12} \quad \text{or} \quad x = \pm\sqrt{-8}$$
$$x = \pm\sqrt{4}\sqrt{3} \quad \text{or} \quad x = \pm i\sqrt{4}\sqrt{2}$$
$$x = \pm 2\sqrt{3} \quad \text{or} \quad x = \pm 2i\sqrt{2}$$

The solution set is $\{\pm 2\sqrt{3}, \pm 2i\sqrt{2}\}$.

**Remark:** Another approach to Example 4 would be to substitute $y$ for $x^2$ and $y^2$ for $x^4$. The equation $x^4 - 4x^2 - 96 = 0$ becomes the quadratic equation $y^2 - 4y - 96 = 0$. Thus we say that $x^4 - 4x^2 - 96 = 0$ is of *quadratic form*. Then we could solve the quadratic equation $y^2 - 4y - 96 = 0$ and use the equation $y = x^2$ to determine the solutions for $x$.

## Solving Word Problems Involving Quadratic Equations

Before we conclude this section with some word problems that can be solved using quadratic equations, let's restate the suggestions we made in an earlier chapter for solving word problems.

### Suggestions for Solving Word Problems

1. Read the problem carefully, and make certain that you understand the meanings of all the words. Be especially alert for any technical terms used in the statement of the problem.
2. Read the problem a second time (perhaps even a third time) to get an overview of the situation being described and to determine the known facts, as well as what is to be found.
3. Sketch any figure, diagram, or chart that might be helpful in analyzing the problem.
4. Choose a meaningful variable to represent an unknown quantity in the problem (perhaps $l$, if the length of a rectangle is an unknown quantity), and represent any other unknowns in terms of that variable.
5. Look for a guideline that you can use to set up an equation. A guideline might be a formula such as $A = lw$ or a relationship such as "the fractional part of a job done by Bill plus the fractional part of the job done by Mary equals the total job."
6. Form an equation that contains the variable and that translates the conditions of the guideline from English to algebra.
7. Solve the equation and use the solutions to determine all facts requested in the problem.
8. **Check all answers back into the original statement of the problem.**

Keep these suggestions in mind as we now consider some word problems.

Tom Cockrem/Lonely Planet Images/ Getty Images

### EXAMPLE 5 Apply Your Skill

A page for a magazine contains 70 square inches of type. The height of a page is twice the width. If the margin around the type is to be 2 inches uniformly, what are the dimensions of a page?

### Solution

Let $x$ represent the width of a page. Then $2x$ represents the height of a page. Now let's draw and label a model of a page (Figure 6.8).

**Classroom Example**
A margin of 1 inch surrounds the front of a card, which leaves 39 square inches for graphics. If the height of the card is three times the width, what are the dimensions of the card?

Width of typed material — Height of typed material — Area of typed material

$$(x - 4)(2x - 4) = 70$$
$$2x^2 - 12x + 16 = 70$$
$$2x^2 - 12x - 54 = 0$$
$$x^2 - 6x - 27 = 0$$
$$(x - 9)(x + 3) = 0$$
$$x - 9 = 0 \quad \text{or} \quad x + 3 = 0$$
$$x = 9 \quad \text{or} \quad x = -3$$

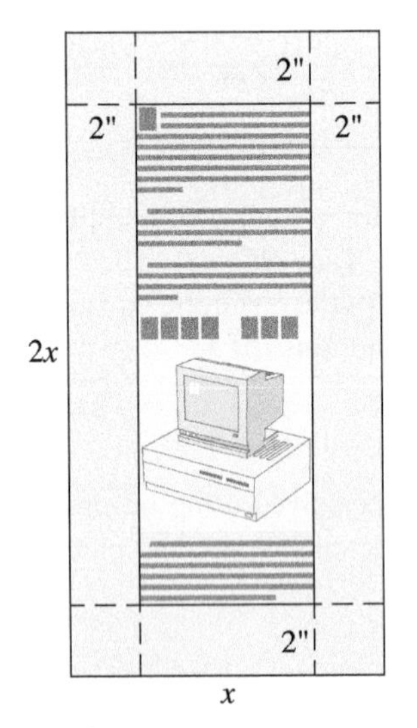

**Figure 6.8**

Disregard the negative solution; the page must be 9 inches wide, and its height is 2(9) = 18 inches.

Let's use our knowledge of quadratic equations to analyze some applications of the business world. For example, if $P$ dollars is invested at $r$ rate of interest compounded annually for $t$ years, then the amount of money, $A$, accumulated at the end of $t$ years is given by the formula

$$A = P(1 + r)^t$$

This compound interest formula serves as a guideline for the next problem.

**Classroom Example**
Suppose that \$2500 is invested at a certain rate of interest compounded annually for 2 years. If the accumulated value at the end of 2 years is \$2704, find the rate of interest.

## EXAMPLE 6 Apply Your Skill

Suppose that \$2000 is invested at a certain rate of interest compounded annually for 2 years. If the accumulated value at the end of 2 years is \$2205, find the rate of interest.

### Solution

Let $r$ represent the rate of interest. Substitute the known values into the compound interest formula to yield

$$A = P(1 + r)^t$$
$$2205 = 2000(1 + r)^2$$

Solving this equation, we obtain

$$\frac{2205}{2000} = (1 + r)^2$$
$$1.1025 = (1 + r)^2$$
$$\pm\sqrt{1.1025} = 1 + r$$
$$\pm 1.05 = 1 + r$$
$$1 + r = 1.05 \quad \text{or} \quad 1 + r = -1.05$$
$$r = -1 + 1.05 \quad \text{or} \quad r = -1 - 1.05$$
$$r = 0.05 \quad \text{or} \quad r = -2.05$$

We must disregard the negative solution, so that $r = 0.05$ is the only solution. Change 0.05 to a percent, and the rate of interest is 5%.

Kenji Arimura/Flickr Select/Getty Images

## EXAMPLE 7 Apply Your Skill

On a 130-mile trip from Orlando to Sarasota, Roberto encountered a heavy thunderstorm for the last 40 miles of the trip. During the thunderstorm he drove an average of 20 miles per hour slower than before the storm. The entire trip took $2\frac{1}{2}$ hours. How fast did he drive before the storm?

**Classroom Example**
After hiking 9 miles of a 10-mile hike, Sam hurt his foot. For the remainder of the hike, his rate was two miles per hour slower than before he hurt his foot. The entire hike took $2\frac{3}{4}$ hours. How fast did he hike before hurting his foot?

### Solution

Let $x$ represent Roberto's rate before the thunderstorm. Then $x - 20$ represents his speed during the thunderstorm. Because $t = \frac{d}{r}$, then $\frac{90}{x}$ represents the time traveling before the storm, and $\frac{40}{x-20}$ represents the time traveling during the storm. The following guideline sums up the situation.

| Time traveling before the storm | + | Time traveling during the storm | = | Total time |
|---|---|---|---|---|
| ↓ | | ↓ | | ↓ |
| $\frac{90}{x}$ | + | $\frac{40}{x-20}$ | = | $\frac{5}{2}$ |

Solving this equation, we obtain

$$2x(x-20)\left(\frac{90}{x} + \frac{40}{x-20}\right) = 2x(x-20)\left(\frac{5}{2}\right)$$

$$2x(x-20)\left(\frac{90}{x}\right) + 2x(x-20)\left(\frac{40}{x-20}\right) = 2x(x-20)\left(\frac{5}{2}\right)$$

$$180(x-20) + 2x(40) = 5x(x-20)$$

$$180x - 3600 + 80x = 5x^2 - 100x$$

$$0 = 5x^2 - 360x + 3600$$

$$0 = 5(x^2 - 72x + 720)$$

$$0 = 5(x-60)(x-12)$$

$$x - 60 = 0 \quad \text{or} \quad x - 12 = 0$$

$$x = 60 \quad \text{or} \quad x = 12$$

We discard the solution of 12 because it would be impossible to drive 20 miles per hour slower than 12 miles per hour; thus Roberto's rate before the thunderstorm was 60 miles per hour.

auremar/Shutterstock.com

## EXAMPLE 8 Apply Your Skill

A computer installer agreed to do an installation for \$150. It took him 2 hours longer than he expected, and therefore he earned \$2.50 per hour less than he anticipated. How long did he expect the installation would take?

**Classroom Example**
James bought a shipment of monitors for \$6000. When he had sold all but 10 monitors at a profit of \$100 per monitor, he had regained the entire cost of the shipment. How many monitors were sold and at what price per monitor?

### Solution

Let $x$ represent the number of hours he expected the installation to take. Then $x + 2$ represents the number of hours the installation actually took. The rate of pay is represented by the pay divided by the number of hours. The following guideline is used to write the equation.

| Anticipated rate of pay | − | \$2.50 | = | Actual rate of pay |
|---|---|---|---|---|
| ↓ | | ↓ | | ↓ |
| $\frac{150}{x}$ | − | $\frac{5}{2}$ | = | $\frac{150}{x+2}$ |

Solving this equation, we obtain

$$2x(x+2)\left(\frac{150}{x}-\frac{5}{2}\right)=2x(x+2)\left(\frac{150}{x+2}\right)$$

$$2(x+2)(150)-x(x+2)(5)=2x(150)$$

$$300(x+2)-5x(x+2)=300x$$

$$300x+600-5x^2-10x=300x$$

$$-5x^2-10x+600=0$$

$$-5(x^2+2x-120)=0$$

$$-5(x+12)(x-10)=0$$

$$x=-12 \quad \text{or} \quad x=10$$

Disregard the negative answer. Therefore he anticipated that the installation would take 10 hours.

This next problem set contains a large variety of word problems. Not only are there some business applications similar to those we discussed in this section, but there are also more problems of the types we discussed in Chapters 3 and 4. Try to give them your best shot without referring to the examples in earlier chapters.

## Concept Quiz 6.5

For Problems 1–5, choose the method that you think is most appropriate for solving the given equation.

1. $2x^2 + 6x - 3 = 0$
2. $(x + 1)^2 = 36$
3. $x^2 - 3x + 2 = 0$
4. $x^2 + 6x = 19$
5. $4x^2 + 2x - 5 = 0$

**A.** Factoring
**B.** Square-root property (Property 6.1)
**C.** Completing the square
**D.** Quadratic formula

## Problem Set 6.5

For Problems 1–20, solve each quadratic equation using the method that seems most appropriate to you. (Objective 1)

1. $x^2 - 4x - 6 = 0$
2. $x^2 - 8x - 4 = 0$
3. $3x^2 + 23x - 36 = 0$
4. $n^2 + 22n + 105 = 0$
5. $x^2 - 18x = 9$
6. $x^2 + 20x = 25$
7. $2x^2 - 3x + 4 = 0$
8. $3y^2 - 2y + 1 = 0$
9. $135 + 24n + n^2 = 0$
10. $28 - x - 2x^2 = 0$
11. $(x - 2)(x + 9) = -10$
12. $(x + 3)(2x + 1) = -3$
13. $2x^2 - 4x + 7 = 0$
14. $3x^2 - 2x + 8 = 0$
15. $x^2 - 18x + 15 = 0$
16. $x^2 - 16x + 14 = 0$
17. $20y^2 + 17y - 10 = 0$
18. $12x^2 + 23x - 9 = 0$
19. $4t^2 + 4t - 1 = 0$
20. $5t^2 + 5t - 1 = 0$

For Problems 21–40, solve each equation. (Objective 1)

21. $n + \frac{3}{n} = \frac{19}{4}$
22. $n - \frac{2}{n} = -\frac{7}{3}$
23. $\frac{3}{x} + \frac{7}{x-1} = 1$
24. $\frac{2}{x} + \frac{5}{x+2} = 1$
25. $\frac{12}{x-3} + \frac{8}{x} = 14$
26. $\frac{16}{x+5} - \frac{12}{x} = -2$

**27.** $\frac{3}{x-1} - \frac{2}{x} = \frac{5}{2}$

**28.** $\frac{4}{x+1} + \frac{2}{x} = \frac{5}{3}$

**29.** $\frac{6}{x} + \frac{40}{x+5} = 7$

**30.** $\frac{12}{t} + \frac{18}{t+8} = \frac{9}{2}$

**31.** $\frac{5}{n-3} - \frac{3}{n+3} = 1$

**32.** $\frac{3}{t+2} + \frac{4}{t-2} = 2$

**33.** $x^4 - 18x^2 + 72 = 0$

**34.** $x^4 - 21x^2 + 54 = 0$

**35.** $3x^4 - 35x^2 + 72 = 0$

**36.** $5x^4 - 32x^2 + 48 = 0$

**37.** $3x^4 + 17x^2 + 20 = 0$

**38.** $4x^4 + 11x^2 - 45 = 0$

**39.** $6x^4 - 29x^2 + 28 = 0$

**40.** $6x^4 - 31x^2 + 18 = 0$

For Problems 41–68, set up an equation and solve each problem. (Objective 2)

**41.** Find two consecutive whole numbers such that the sum of their squares is 145.

**42.** Find two consecutive odd whole numbers such that the sum of their squares is 74.

**43.** Two positive integers differ by 3, and their product is 108. Find the numbers.

**44.** Suppose that the sum of two numbers is 20, and the sum of their squares is 232. Find the numbers.

**45.** Find two numbers such that their sum is 10 and their product is 22.

**46.** Find two numbers such that their sum is 6 and their product is 7.

**47.** Suppose that the sum of two whole numbers is 9, and the sum of their reciprocals is $\frac{1}{2}$. Find the numbers.

**48.** The difference between two whole numbers is 8, and the difference between their reciprocals is $\frac{1}{6}$. Find the two numbers.

**49.** The sum of the lengths of the two legs of a right triangle is 21 inches. If the length of the hypotenuse is 15 inches, find the length of each leg.

**50.** The length of a rectangular floor is 1 meter less than twice its width. If a diagonal of the rectangle is 17 meters, find the length and width of the floor.

**51.** A rectangular plot of ground measuring 12 meters by 20 meters is surrounded by a sidewalk of a uniform width (see Figure 6.9). The area of the sidewalk is 68 square meters. Find the width of the walk.

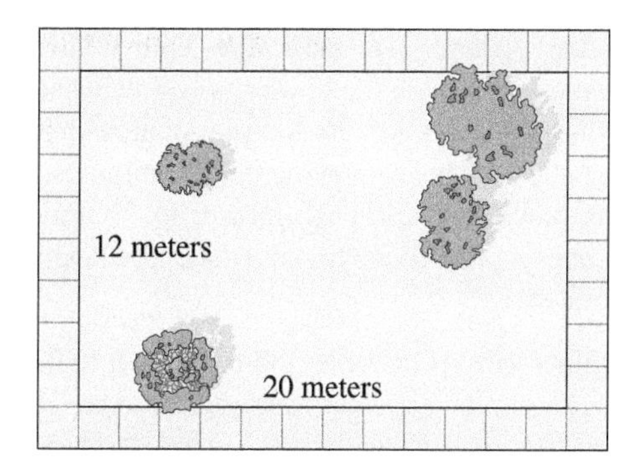


**Figure 6.9**

**52.** A 5-inch by 7-inch picture is surrounded by a frame of uniform width. The area of the picture and frame together is 80 square inches. Find the width of the frame.

**53.** The perimeter of a rectangle is 44 inches, and its area is 112 square inches. Find the length and width of the rectangle.

**54.** A rectangular piece of cardboard is 2 units longer than it is wide. From each of its corners a square piece 2 units on a side is cut out. The flaps are then turned up to form an open box that has a volume of 70 cubic units. Find the length and width of the original piece of cardboard.

**55.** Charlotte's time to travel 250 miles is 1 hour more than Lorraine's time to travel 180 miles. Charlotte drove 5 miles per hour faster than Lorraine. How fast did each one travel?

**56.** Larry's time to travel 156 miles is 1 hour more than Terrell's time to travel 108 miles. Terrell drove 2 miles per hour faster than Larry. How fast did each one travel?

**57.** On a 570-mile trip, Andy averaged 5 miles per hour faster for the last 240 miles than he did for the first 330 miles. The entire trip took 10 hours. How fast did he travel for the first 330 miles?

**58.** On a 135-mile bicycle excursion, Maria averaged 5 miles per hour faster for the first 60 miles than she did for the last 75 miles. The entire trip took 8 hours. Find her rate for the first 60 miles.

**59.** It takes Terry 2 hours longer to do a certain job than it takes Tom. They worked together for 3 hours; then Tom left and Terry finished the job in 1 hour. How long would it take each of them to do the job alone?

**60.** Suppose that Arlene can mow the entire lawn in 40 minutes less time with the power mower than she can with the push mower. One day the power mower broke down after she had been mowing for 30 minutes. She finished the lawn with the push mower in 20 minutes. How long does it take Arlene to mow the entire lawn with the power mower?

**61.** A student did a word processing job for \$24. It took him 1 hour longer than he expected, and therefore he earned \$4 per hour less than he anticipated. How long did he expect that it would take to do the job?

**62.** A group of students agreed that each would chip in the same amount to pay for a party that would cost \$100. Then they found 5 more students interested in the party and in sharing the expenses. This decreased the amount each had to pay by \$1. How many students were involved in the party and how much did each student have to pay?

**63.** A group of students agreed that each would contribute the same amount to buy their favorite teacher an \$80 birthday gift. At the last minute, 2 of the students decided not to chip in. This increased the amount that the remaining students had to pay by \$2 per student. How many students actually contributed to the gift?

**64.** The formula $D = \frac{n(n - 3)}{2}$ yields the number of diagonals, $D$, in a polygon of $n$ sides. Find the number of sides of a polygon that has 54 diagonals.

**65.** The formula $S = \frac{n(n + 1)}{2}$ yields the sum, $S$, of the first $n$ natural numbers 1, 2, 3, 4, . . . . How many consecutive natural numbers starting with 1 will give a sum of 1275?

**66.** At a point 16 yards from the base of a tower, the distance to the top of the tower is 4 yards more than the height of the tower (see Figure 6.10). Find the height of the tower.

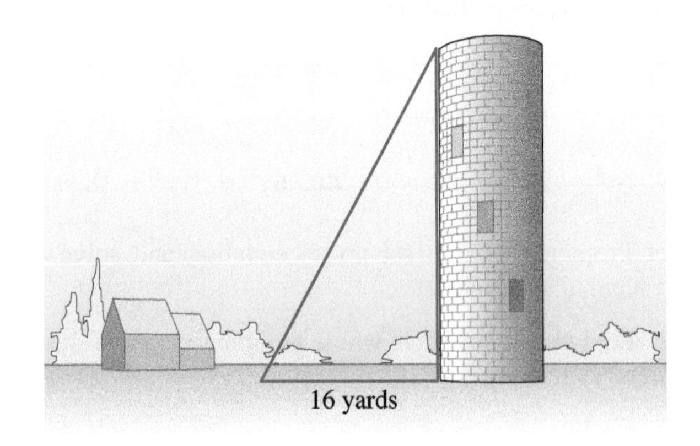


**Figure 6.10**

**67.** Suppose that \$500 is invested at a certain rate of interest compounded annually for 2 years. If the accumulated value at the end of 2 years is \$594.05, find the rate of interest.

**68.** Suppose that \$10,000 is invested at a certain rate of interest compounded annually for 2 years. If the accumulated value at the end of 2 years is \$12,544, find the rate of interest.

## Thoughts Into Words

**69.** How would you solve the equation $x^2 - 4x = 252$? Explain your choice of the method that you would use.

**70.** Explain how you would solve $(x - 2)(x - 7) = 0$ and also how you would solve $(x - 2)(x - 7) = 4$.

**71.** One of our problem-solving suggestions is to look for a guideline that can be used to help determine an equation. What does this suggestion mean to you?

**72.** Can a quadratic equation with integral coefficients have exactly one nonreal complex solution? Explain your answer.

## Further Investigations

For Problems 73–79, solve each equation.

**73.** $x - 9\sqrt{x} + 18 = 0$ [*Hint:* Let $y = \sqrt{x}$.]

**74.** $x - 4\sqrt{x} + 3 = 0$

**75.** $x + \sqrt{x} - 2 = 0$

**76.** $x^{\frac{2}{3}} + x^{\frac{1}{3}} - 6 = 0$ [*Hint:* Let $y = x^{\frac{1}{3}}$.]

**77.** $6x^{\frac{2}{3}} - 5x^{\frac{1}{3}} - 6 = 0$

**78.** $x^{-2} + 4x^{-1} - 12 = 0$

**79.** $12x^{-2} - 17x^{-1} - 5 = 0$

The following equations are also quadratic in form. To solve, begin by raising each side of the equation to the appropriate power so that the exponent will become an integer. Then, to solve the resulting quadratic equation, you may use the square-root property, factoring, or the quadratic formula—whichever is most appropriate. Be aware that raising each side of the equation to a power may introduce extraneous roots; therefore, be sure to check your solutions. Study the following example before you begin the problems.

Solve

$$(x+3)^{\frac{2}{3}} = 1$$
$$\left[(x+3)^{\frac{2}{3}}\right]^3 = 1^3 \quad \text{Raise both sides to the third power}$$
$$(x+3)^2 = 1$$
$$x^2 + 6x + 9 = 1$$
$$x^2 + 6x + 8 = 0$$
$$(x+4)(x+2) = 0$$
$$x + 4 = 0 \quad \text{or} \quad x + 2 = 0$$
$$x = -4 \quad \text{or} \quad x = -2$$

Both solutions do check. The solution set is $\{-4, -2\}$.

For problems 80–88, solve each equation.

**80.** $(5x+6)^{\frac{1}{2}} = x$

**81.** $(3x+4)^{\frac{1}{2}} = x$

**82.** $x^{\frac{2}{3}} = 2$

**83.** $x^{\frac{2}{5}} = 2$

**84.** $(2x+6)^{\frac{1}{2}} = x$

**85.** $(2x-4)^{\frac{2}{3}} = 1$

**86.** $(4x+5)^{\frac{2}{3}} = 2$

**87.** $(6x+7)^{\frac{1}{2}} = x + 2$

**88.** $(5x+21)^{\frac{1}{2}} = x + 3$

### Answers to the Concept Quiz

Answers for Problems 1–5 may vary. **1.** D **2.** B **3.** A **4.** C **5.** D

## 6.6 Quadratic and Other Nonlinear Inequalities

OBJECTIVES
1. Solve quadratic inequalities
2. Solve inequalities of quotients

We refer to the equation $ax^2 + bx + c = 0$ as the standard form of a quadratic equation in one variable. Similarly, the following forms express **quadratic inequalities** in one variable.

$$ax^2 + bx + c > 0 \qquad ax^2 + bx + c < 0$$
$$ax^2 + bx + c \geq 0 \qquad ax^2 + bx + c \leq 0$$

We can use the number line very effectively to help solve quadratic inequalities for which the quadratic polynomial is factorable. Let's consider some examples to illustrate the procedure.

**Classroom Example**
Solve and graph the solutions for $x^2 + 4x - 21 \geq 0$.

**EXAMPLE 1** Solve and graph the solutions for $x^2 + 2x - 8 > 0$.

### Solution

First, let's factor the polynomial:

$$x^2 + 2x - 8 > 0$$
$$(x+4)(x-2) > 0$$

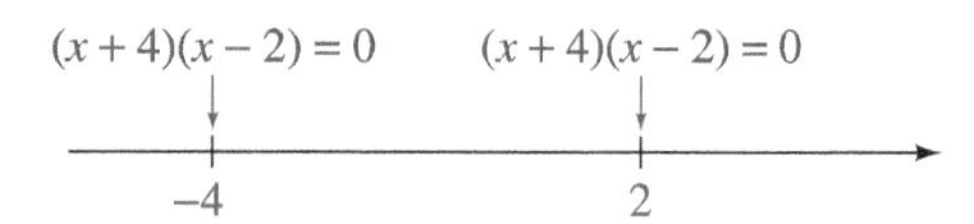


**Figure 6.11**

On a number line (Figure 6.11), we indicate that at $x = 2$ and $x = -4$, the product $(x + 4) \cdot (x - 2)$ equals zero. The numbers $-4$ and 2 divide the number line into three intervals: (1) the numbers less than $-4$, (2) the numbers between $-4$ and 2, and (3) the numbers greater than 2. We can choose a *test number* from each of these intervals and see how it affects the signs of the factors $x + 4$ and $x - 2$ and, consequently, the sign of the product of these factors. For example, if $x < -4$ (try $x = -5$), then $x + 4$ is negative and $x - 2$ is negative, so their product is positive. If $-4 < x < 2$ (try $x = 0$), then $x + 4$ is positive and $x - 2$ is negative, so their product is negative. If $x > 2$ (try $x = 3$), then $x + 4$ is positive and $x - 2$ is positive, so their product is positive. This information can be conveniently arranged using a number line, as shown in Figure 6.12. Note the open circles at $-4$ and 2 to indicate that they are not included in the solution set.

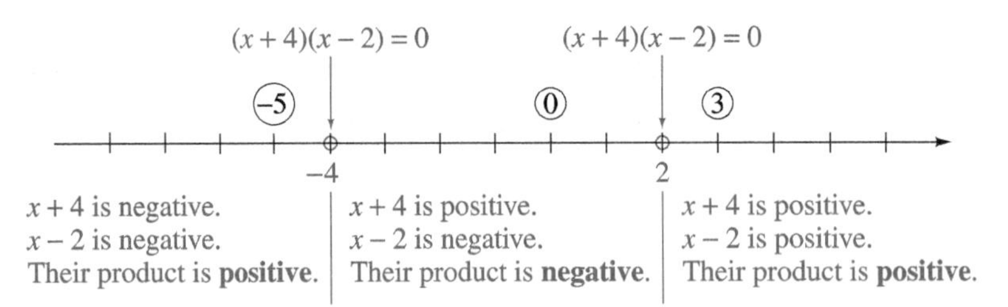


**Figure 6.12**

Thus the given inequality, $x^2 + 2x - 8 > 0$, is satisfied by numbers less than $-4$ along with numbers greater than 2. Using interval notation, the solution set is $(-\infty, -4) \cup (2, \infty)$. These solutions can be shown on a number line (Figure 6.13).

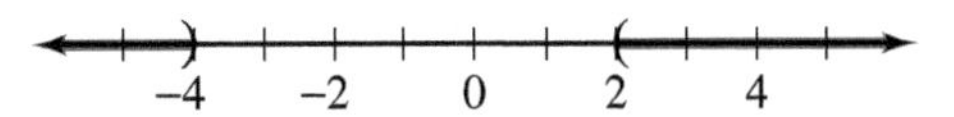


**Figure 6.13**

We refer to numbers such as $-4$ and 2 in the preceding example (where the given polynomial or algebraic expression equals zero or is undefined) as **critical numbers**. Let's consider some additional examples that make use of critical numbers and test numbers.

Classroom Example
Solve and graph the solutions for $x^2 + 3x - 10 < 0$.

**EXAMPLE 2** Solve and graph the solutions for $x^2 + 2x - 3 \leq 0$.

## Solution

First, factor the polynomial:

$$x^2 + 2x - 3 \leq 0$$
$$(x + 3)(x - 1) \leq 0$$

Second, locate the values for which $(x + 3)(x - 1)$ equals zero. We put dots at $-3$ and 1 to remind ourselves that these two numbers are to be included in the solution set because the given statement includes equality. Now let's choose a test number from each of the three intervals, and record the sign behavior of the factors $(x + 3)$ and $(x - 1)$ (Figure 6.14).

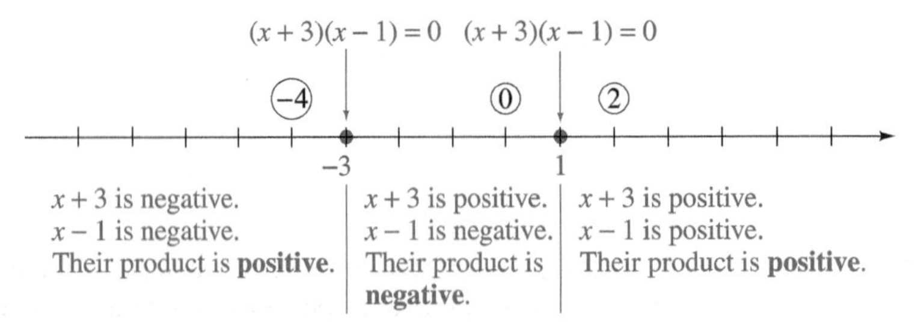


**Figure 6.14**

Therefore, the solution set is $[-3, 1]$, and it can be graphed as in Figure 6.15.

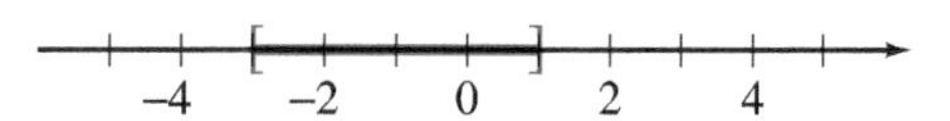


**Figure 6.15**

## Solving Inequalities of Quotients

Examples 1 and 2 have indicated a systematic approach for solving quadratic inequalities when the polynomial is factorable. This same type of number line analysis can also be used to solve indicated quotients such as $\frac{x+1}{x-5} > 0$.

**Classroom Example**
Solve and graph the solutions for $\frac{x-2}{x+6} \geq 0$.

**EXAMPLE 3** Solve and graph the solutions for $\frac{x+1}{x-5} > 0$.

### Solution

First, indicate that at $x = -1$ the given quotient equals zero, and at $x = 5$ the quotient is undefined. Second, choose test numbers from each of the three intervals, and record the sign behavior of $(x + 1)$ and $(x - 5)$ as in Figure 6.16.

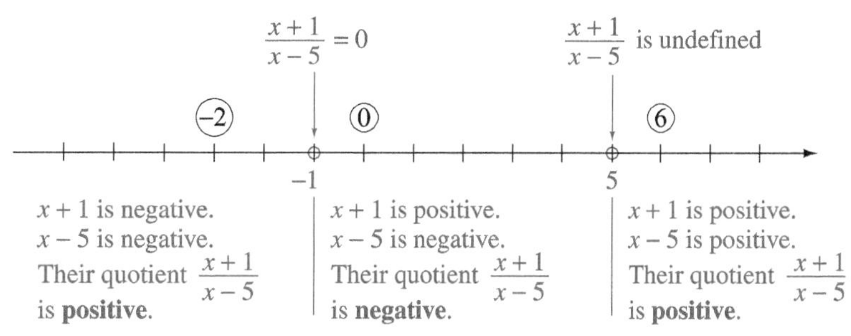


**Figure 6.16**

Therefore, the solution set is $(-\infty, -1) \cup (5, \infty)$, and its graph is shown in Figure 6.17.

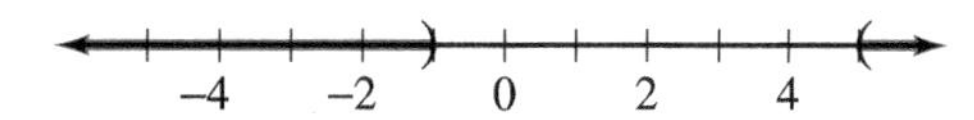


**Figure 6.17**

**Classroom Example**
Solve $\frac{m+1}{m+3} \leq 0$.

**EXAMPLE 4** Solve $\frac{x+2}{x+4} \leq 0$.

### Solution

The indicated quotient equals zero at $x = -2$ and is undefined at $x = -4$. (Note that $-2$ is to be included in the solution set, but $-4$ is not to be included.) Now let's choose some test numbers and record the sign behavior of $(x + 2)$ and $(x + 4)$ as in Figure 6.18.

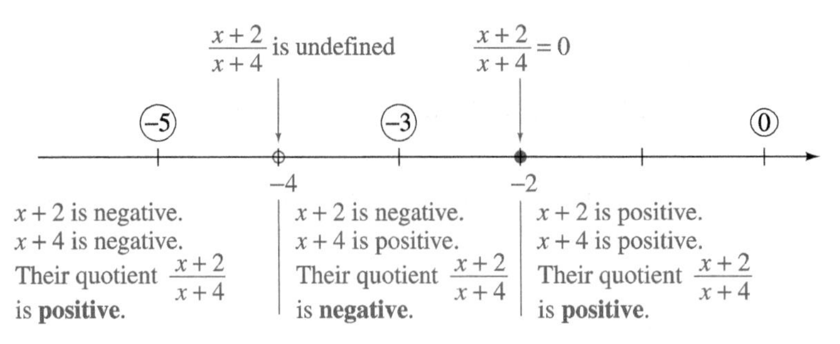


**Figure 6.18**

Therefore, the solution set is $(-4, -2]$.

The final example illustrates that sometimes we need to change the form of the given inequality before we use the number line analysis.

Classroom Example

Solve $\frac{x}{x+4} \le 2$.

**EXAMPLE 5** Solve $\frac{x}{x+2} \ge 3$.

## Solution

First, let's change the form of the given inequality as follows:

$$\frac{x}{x+2} \ge 3$$

$$\frac{x}{x+2} - 3 \ge 0 \qquad \text{Add } -3 \text{ to both sides}$$

$$\frac{x - 3(x+2)}{x+2} \ge 0 \qquad \text{Express the left side over a common denominator}$$

$$\frac{x - 3x - 6}{x+2} \ge 0$$

$$\frac{-2x - 6}{x+2} \ge 0$$

Now we can proceed as we did with the previous examples. If $x = -3$, then $\frac{-2x-6}{x+2}$ equals zero; and if $x = -2$, then $\frac{-2x-6}{x+2}$ is undefined. Then, choosing test numbers, we can record the sign behavior of $(-2x - 6)$ and $(x + 2)$ as in Figure 6.19.

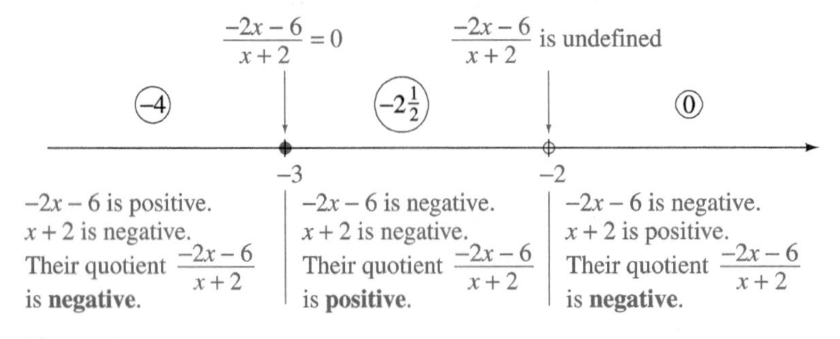


**Figure 6.19**

Therefore, the solution set is $[-3, -2)$. Perhaps you should check a few numbers from this solution set back into the original inequality!

## Concept Quiz 6.6

For Problems 1–10, answer true or false.

1. When solving the inequality $(x + 3)(x - 2) > 0$, we are looking for values of $x$ that make the product of $(x + 3)$ and $(x - 2)$ a positive number.
2. The solution set of the inequality $x^2 + 4 > 0$ is all real numbers.
3. The solution set of the inequality $x^2 \le 0$ is the null set.
4. The critical numbers for the inequality $(x + 4)(x - 1) \le 0$ are $-4$ and $-1$.
5. The number 2 is included in the solution set of the inequality $\frac{x + 4}{x - 2} \ge 0$.
6. The solution set of $(x - 2)^2 \ge 0$ is the set of all real numbers.
7. The solution set of $\frac{x + 2}{x - 3} \le 0$ is $(-2, 3)$.
8. The solution set of $\frac{x - 1}{x} > 2$ is $(-1, 0)$.
9. The solution set of the inequality $(x - 2)^2(x + 1)^2 < 0$ is $\varnothing$.
10. The solution set of the inequality $(x - 4)(x + 3)^2 \le 0$ is $(-\infty, 4]$.

## Problem Set 6.6

For Problems 1–12, solve each inequality and graph its solution set on a number line. (Objective 1)

1. $(x + 2)(x - 1) > 0$
2. $(x - 2)(x + 3) > 0$
3. $(x + 1)(x + 4) < 0$
4. $(x - 3)(x - 1) < 0$
5. $(2x - 1)(3x + 7) \ge 0$
6. $(3x + 2)(2x - 3) \ge 0$
7. $(x + 2)(4x - 3) \le 0$
8. $(x - 1)(2x - 7) \le 0$
9. $(x + 1)(x - 1)(x - 3) > 0$
10. $(x + 2)(x + 1)(x - 2) > 0$
11. $x(x + 2)(x - 4) \le 0$
12. $x(x + 3)(x - 3) \le 0$

For Problems 13–38, solve each inequality. (Objective 1)

13. $x^2 + 2x - 35 < 0$
14. $x^2 + 3x - 54 < 0$
15. $x^2 - 11x + 28 > 0$
16. $x^2 + 11x + 18 > 0$
17. $3x^2 + 13x - 10 \le 0$
18. $4x^2 - x - 14 \le 0$
19. $8x^2 + 22x + 5 \ge 0$
20. $12x^2 - 20x + 3 \ge 0$
21. $x(5x - 36) > 32$
22. $x(7x + 40) < 12$
23. $x^2 - 14x + 49 \ge 0$
24. $(x + 9)^2 \ge 0$
25. $4x^2 + 20x + 25 \le 0$
26. $9x^2 - 6x + 1 \le 0$
27. $(x + 1)(x - 3)^2 > 0$
28. $(x - 4)^2(x - 1) \le 0$
29. $4 - x^2 < 0$
30. $2x^2 - 18 \ge 0$
31. $4(x^2 - 36) < 0$
32. $-4(x^2 - 36) \ge 0$
33. $5x^2 + 20 > 0$
34. $-3x^2 - 27 \ge 0$
35. $x^2 - 2x \ge 0$
36. $2x^2 + 6x < 0$
37. $3x^3 + 12x^2 > 0$
38. $2x^3 + 4x^2 \le 0$

For Problems 39–56, solve each inequality. (Objective 2)

39. $\frac{x + 1}{x - 2} > 0$
40. $\frac{x - 1}{x + 2} > 0$
41. $\frac{x - 3}{x + 2} < 0$
42. $\frac{x + 2}{x - 4} < 0$
43. $\frac{2x - 1}{x} \ge 0$
44. $\frac{x}{3x + 7} \ge 0$
45. $\frac{-x + 2}{x - 1} \le 0$
46. $\frac{3 - x}{x + 4} \le 0$
47. $\frac{2x}{x + 3} > 4$
48. $\frac{x}{x - 1} > 2$
49. $\frac{x - 1}{x - 5} \le 2$
50. $\frac{x + 2}{x + 4} \le 3$
51. $\frac{x + 2}{x - 3} > -2$
52. $\frac{x - 1}{x - 2} < -1$
53. $\frac{3x + 2}{x + 4} \le 2$
54. $\frac{2x - 1}{x + 2} \ge -1$
55. $\frac{x + 1}{x - 2} < 1$
56. $\frac{x + 3}{x - 4} \ge 1$

## Thoughts Into Words

**57.** Explain how to solve the inequality $(x + 1)(x - 2) \cdot (x - 3) > 0$.

**58.** Explain how to solve the inequality $(x - 2)^2 > 0$ by inspection.

**59.** Your friend looks at the inequality $1 + \frac{1}{x} > 2$ and, without any computation, states that the solution set is all real numbers between 0 and 1. How can she do that?

**60.** Why is the solution set for $(x - 2)^2 \geq 0$ the set of all real numbers?

**61.** Why is the solution set for $(x - 2)^2 \leq 0$ the set $\{2\}$?

## Further Investigations

**62.** The product $(x - 2)(x + 3)$ is positive if both factors are negative *or* if both factors are positive. Therefore, we can solve $(x - 2)(x + 3) > 0$ as follows:

$$(x - 2 < 0 \text{ and } x + 3 < 0) \quad \text{or} \quad (x - 2 > 0 \text{ and } x + 3 > 0)$$
$$(x < 2 \text{ and } x < -3) \quad \text{or} \quad (x > 2 \text{ and } x > -3)$$
$$x < -3 \quad \text{or} \quad x > 2$$

The solution set is $(-\infty, -3) \cup (2, \infty)$. Use this type of analysis to solve each of the following.

**(a)** $(x - 2)(x + 7) > 0$ **(b)** $(x - 3)(x + 9) \geq 0$

**(c)** $(x + 1)(x - 6) \leq 0$ **(d)** $(x + 4)(x - 8) < 0$

**(e)** $\frac{x + 4}{x - 7} > 0$ **(f)** $\frac{x - 5}{x + 8} \leq 0$

### Answers to the Concept Quiz

**1.** True **2.** True **3.** False **4.** False **5.** False **6.** True **7.** False **8.** True **9.** True **10.** True

# Chapter 6 Summary

| OBJECTIVE | SUMMARY | EXAMPLE |
|---|---|---|
| Know the set of complex numbers.<br>(Section 6.1/ Objective 1) | A number of the form $a + bi$, where $a$ and $b$ are real numbers and $i$ is the imaginary unit defined by $i = \sqrt{-1}$, is a complex number. Two complex numbers are said to be equal if and only if $a = c$ and $b = d$. | |
| Add and subtract complex numbers.<br>(Section 6.1/ Objective 2) | We describe the addition and subtraction of complex numbers as follows:<br>$(a + bi) + (c + di)$<br>$= (a + c) + (b + d)i$<br>and<br>$(a + bi) - (c + di)$<br>$= (a - c) + (b - d)i$ | Add the complex numbers:<br>$(3 - 6i) + (-7 - 3i)$<br>**Solution**<br>$(3 - 6i) + (-7 - 3i)$<br>$= (3 - 7) + (-6 - 3)i$<br>$= -4 - 9i$<br>**Sample Problem 1**<br>Subtract: $(4 - 3i) - (-8 + 5i)$ |
| Simplify radicals involving negative numbers.<br>(Section 6.1/Objective 3) | We can represent a square root of any negative real number as the product of a real number and the imaginary unit $i$, That is, $\sqrt{-b} = i\sqrt{b}$, where $b$ is a positive real number. | Write $\sqrt{-48}$ in terms of $i$ and simplify.<br>**Solution**<br>$\sqrt{-48} = \sqrt{-1}\sqrt{48}$<br>$= i\sqrt{16}\sqrt{3}$<br>$= 4i\sqrt{3}$<br>**Sample Problem 2**<br>Write $\sqrt{-24}$ in terms of $i$ and simplify. |
| Perform operations on radicals involving negative numbers.<br>(Section 6.1/Objective 4) | Before performing any operations, represent the square root of any negative real number as the product of a real number and the imaginary unit $i$. | Perform the indicated operation and simplify:<br>$\frac{\sqrt{-28}}{\sqrt{-4}}$<br>**Solution**<br>$\frac{\sqrt{-28}}{\sqrt{-4}} = \frac{i\sqrt{28}}{i\sqrt{4}} = \frac{\sqrt{28}}{\sqrt{4}} = \sqrt{\frac{28}{4}} = \sqrt{7}$<br>**Sample Problem 3**<br>Simplify: $\frac{\sqrt{-36}}{\sqrt{-9}}$ |
| Multiply complex numbers.<br>(Section 6.1/Objective 5) | The product of two complex numbers follows the same pattern as the product of two binomials. The conjugate of $a + bi$ is $a - bi$. The product of a complex number and its conjugate is a real number. When simplifying, replace any $i^2$ with $-1$. | Find the product $(2 + 3i)(4 - 5i)$ and express the answer in standard form of a complex number.<br>**Solution**<br>$(2 + 3i)(4 - 5i) = 8 + 2i - 15i^2$<br>$= 8 + 2i - 15(-1)$<br>$= 23 + 2i$<br>**Sample Problem 4**<br>Find the product $(6 + 5i)(3 - i)$. |

Answers to **Sample Problems** are located in the back of the book.

*(continued)*

| OBJECTIVE | SUMMARY | EXAMPLE |
|---|---|---|
| Divide complex numbers. (Section 6.1/Objective 6) | To simplify expressions that indicate the quotient of complex numbers, such as $\frac{4 + 3i}{5 - 2i}$, multiply the numerator and denominator by the conjugate of the denominator. | Find the quotient $\frac{2 + 3i}{4 - i}$ and express the answer in standard form of a complex number.<br>**Solution**<br>Multiply the numerator and denominator by $4 + i$, the conjugate of the denominator.<br>$\frac{2 + 3i}{4 - i} = \frac{(2 + 3i)}{(4 - i)} \cdot \frac{(4 + i)}{(4 + i)}$<br>$= \frac{8 + 14i + 3i^2}{16 - i^2}$<br>$= \frac{8 + 14i + 3(-1)}{16 - (-1)}$<br>$= \frac{5 + 14i}{17} = \frac{5}{17} + \frac{14}{17}i$<br>**Sample Problem 5**<br>Find the quotient $\frac{3 + i}{5 - 2i}$. |
| Solve quadratic equations by factoring. (Section 6.2/Objective 1) | The standard form for a quadratic equation in one variable is $ax^2 + bx + c = 0$, where $a$, $b$, and $c$ are real numbers and $a \neq 0$. Some quadratic equations can be solved by factoring and applying the property, $ab = 0$ if and only if $a = 0$ or $b = 0$. | Solve $2x^2 + x - 3 = 0$.<br>**Solution**<br>$2x^2 + x - 3 = 0$<br>$(2x + 3)(x - 1) = 0$<br>$2x + 3 = 0$ or $x - 1 = 0$<br>$x = -\frac{3}{2}$ or $x = 1$<br>The solution set is $\left\{-\frac{3}{2}, 1\right\}$.<br>**Sample Problem 6**<br>Solve $5x^2 - 14x - 3 = 0$. |
| Solve quadratic equations of the form $x^2 - a$. (Section 6.2/Objective 2) | We can solve some quadratic equations by applying the property, $x^2 = a$ if and only if $x = \pm\sqrt{a}$. | Solve $3(x + 7)^2 = 24$.<br>**Solution**<br>$3(x + 7)^2 = 24$<br>First divide both sides of the equation by 3:<br>$(x + 7)^2 = 8$<br>$x + 7 = \pm\sqrt{8}$<br>$x + 7 = \pm 2\sqrt{2}$<br>$x = -7 \pm 2\sqrt{2}$<br>The solution set is $\{-7 \pm 2\sqrt{2}\}$.<br>**Sample Problem 7**<br>Solve $4(x - 3)^2 = 20$. |

| OBJECTIVE | SUMMARY | EXAMPLE |
|---|---|---|
| Solve quadratic equations by completing the square.<br>(Section 6.3/Objective 1) | To solve a quadratic equation by completing the square, first put the equation in the form $x^2 + bx = k$. Then (1) take one-half of $b$, square that result, and add to each side of the equation; (2) factor the left side; and (3) apply the property, $x^2 = a$ if and only if $x = \pm\sqrt{a}$. | Solve $x^2 + 12x - 2 = 0$.<br>**Solution**<br>$x^2 + 12x - 2 = 0$<br>$x^2 + 12x = 2$<br>$x^2 + 12x + 36 = 2 + 36$<br>$(x + 6)^2 = 38$<br>$x + 6 = \pm\sqrt{38}$<br>$x = -6 \pm \sqrt{38}$<br>The solution set is $\{-6 \pm \sqrt{38}\}$.<br>**Sample Problem 8**<br>Solve $x^2 - 8x - 5 = 0$. |
| Use the quadratic formula to solve quadratic equations.<br>(Section 6.4/Objective 1) | Any quadratic equation of the form $ax^2 + bx + c = 0$, where $a \neq 0$, can be solved by the quadratic formula, which is usually stated as $x = \dfrac{-b \pm \sqrt{b^2 - 4ac}}{2a}$. | Solve $3x^2 - 5x - 6 = 0$.<br>**Solution**<br>$3x^2 - 5x - 6 = 0$<br>$a = 3, b = -5$, and $c = -6$<br>$x = \dfrac{-(-5) \pm \sqrt{(-5)^2 - 4(3)(-6)}}{2(3)}$<br>$x = \dfrac{5 \pm \sqrt{97}}{6}$<br>The solution set is $\left\{\dfrac{5 \pm \sqrt{97}}{6}\right\}$.<br>**Sample Problem 9**<br>Solve $2x^2 + 3x - 7 = 0$. |
| Determine the nature of roots to quadratic equations.<br>(Section 6.4/Objective 2) | The discriminant, $b^2 - 4ac$, can be used to determine the nature of the roots of a quadratic equation.<br>**1.** If $b^2 - 4ac$ is less than zero, then the equation has two nonreal complex solutions.<br>**2.** If $b^2 - 4ac$ is equal to zero, then the equation has two equal real solutions.<br>**3.** If $b^2 - 4ac$ is greater than zero, then the equation has two unequal real solutions. | Use the discriminant to determine the nature of the solutions for the equation $2x^2 + 3x + 5 = 0$.<br>**Solution**<br>$2x^2 + 3x + 5 = 0$<br>For $a = 2$, $b = 3$, and $c = 5$, $b^2 - 4ac = (3)^2 - 4(2)(5) = -31$. Because the discriminant is less than zero, the equation has two nonreal complex solutions.<br>**Sample Problem 10**<br>Use the discriminate to determine the nature of the solutions for the equation $4x^2 + 8x - 5 = 0$. |

(*continued*)

| OBJECTIVE | SUMMARY | EXAMPLE |
|---|---|---|
| Solve quadratic equations by selecting the most appropriate method.<br>(Section 6.5/Objective 1) | There are three major methods for solving a quadratic equation.<br>**1.** Factoring<br>**2.** Completing the square<br>**3.** Quadratic formula<br>Consider which method is most appropriate before you begin solving the equation. | Solve $x^2 - 4x + 9 = 0$.<br>**Solution**<br>This equation does not factor. This equation can easily be solved by completing the square, because $a = 1$ and $b$ is an even number.<br>$x^2 - 4x + 9 = 0$<br>$x^2 - 4x = -9$<br>$x^2 - 4x + 4 = -9 + 4$<br>$(x + 4)^2 = -5$<br>$x + 4 = \pm\sqrt{-5}$<br>$x = -4 \pm i\sqrt{5}$<br>The solution set is $\{-4 \pm i\sqrt{5}\}$.<br>**Sample Problem 11**<br>Solve $x^2 - 6x + 14 = 0$. |
| Solve problems pertaining to right triangles and $30° - 60°$ triangles.<br>(Section 6.2/Objective 3) | There are two special kinds of right triangles that are used in later mathematics courses. The **isosceles right triangle** is a right triangle that has both legs of the same length. In a **30°–60° right triangle**, the side opposite the 30° angle is equal in length to one-half the length of the hypotenuse. | Find the length of each leg of an isosceles right triangle that has a hypotenuse of length 6 inches.<br>**Solution**<br>Let $x$ represent the length of each leg:<br>$x^2 + x^2 = 6^2$<br>$2x^2 = 36$<br>$x^2 = 18$<br>$x = \pm\sqrt{18} = \pm 3\sqrt{2}$<br>Disregard the negative solution. The length of each leg is $3\sqrt{2}$ inches.<br>**Sample Problem 12**<br>Find the length of each leg of an isosceles right triangle that has a hypotenuse of length 8 inches. |
| Solve word problems involving quadratic equations.<br>(Section 6.5/Objective 2) | Keep the following suggestions in mind as you solve word problems.<br>**1.** Read the problem carefully.<br>**2.** Sketch any figure, diagram, or chart that might help you organize and analyze the problem.<br>**3.** Choose a meaningful variable.<br>**4.** Look for a guideline that can be used to set up an equation.<br>**5.** Form an equation that translates the guideline from English into algebra.<br>**6.** Solve the equation and answer the question posed in the problem.<br>**7.** Check all answers back into the original statement of the problem. | Find two consecutive odd whole numbers such that the sum of their squares is 290.<br>**Solution**<br>Let $x$ represent the first whole number. Then $x + 2$ would represent the next consecutive odd whole number.<br>$x^2 + (x + 2)^2 = 290$<br>$x^2 + x^2 + 4x + 4 = 290$<br>$2x^2 + 4x - 286 = 0$<br>$2(x^2 + 2x - 143) = 0$<br>$2(x + 13)(x - 11) = 0$<br>$x = -13$ or $x = 11$<br>Disregard the solution of $-13$ because it is not a whole number. The whole numbers are 11 and 13.<br>**Sample Problem 13**<br>Find two consecutive whole numbers such that the sum of their squares is 41. |

| OBJECTIVE | SUMMARY | EXAMPLE |
|---|---|---|
| Solve quadratic inequalities.<br>(Section 6.6/Objective 1) | To solve quadratic inequalities that are factorable polynomials, the critical numbers are found by factoring the polynomial. The critical numbers partition the number line into regions. A test point from each region is used to determine if the values in that region make the inequality a true statement. The answer is usually expressed in interval notation. | Solve $x^2 + x - 6 \leq 0$.<br>**Solution**<br>Solve the equation $x^2 + x - 6 = 0$ to find the critical numbers.<br>$x^2 + x - 6 = 0$<br>$(x + 3)(x - 2) = 0$<br>$x = -3 \quad \text{or} \quad x = 2$<br>The critical numbers are $-3$ and 2. Choose a test point from each of the intervals $(-\infty, -3)$, $(-3, 2)$, and $(2, \infty)$. Evaluating the inequality $x^2 + x - 6 \leq 0$ for each of the test points shows that $(-3, 2)$ is the only interval of values that makes the inequality a true statement. Because the inequality includes the endpoints of the interval, the solution is $[-3, 2]$.<br>**Sample Problem 14**<br>Solve $x^2 - 2x - 15 \geq 0$. |
| Solve inequalities of quotients.<br>(Section 6.6/Objective 2) | To solve inequalities involving quotients, use the same basic approach as for solving quadratic equations. Be careful to avoid any values that make the denominator zero. | Solve $\frac{x + 1}{2x - 3} \geq 0$.<br>**Solution**<br>Set the numerator equal to zero and then set the denominator equal to zero to find the critical numbers.<br>$x + 1 = 0 \quad \text{and} \quad 2x - 3 = 0$<br>$x = -1 \quad \text{and} \quad x = \frac{3}{2}$<br>The critical numbers are $-1$ and $\frac{3}{2}$.<br>Evaluate the inequality with a test point from each of the intervals $(-\infty, -1)$, $\left(-1, \frac{3}{2}\right)$, and $\left(\frac{3}{2}, \infty\right)$; this shows that the values in the intervals $(-\infty, -1)$ and $\left(\frac{3}{2}, \infty\right)$ make the inequality a true statement. Because the inequality includes the "equal to" statement, the solution should include $-1$ but not $\frac{3}{2}$, because $\frac{3}{2}$ would make the quotient undefined. The solution set is $(-\infty, -1] \cup \left(\frac{3}{2}, \infty\right)$.<br>**Sample Problem 15**<br>Solve $\frac{2x - 1}{x + 7} \leq 0$. |

## Chapter 6 Review Problem Set

For Problems 1–4, perform the indicated operations and express the answers in the standard form of a complex number.

**1.** $(-7 + 3i) + (9 - 5i)$ **2.** $(4 - 10i) - (7 - 9i)$

**3.** $(6 - 3i) - (-2 + 5i)$ **4.** $(-4 + i) + (2 + 3i)$

For Problems 5–8, write each expression in terms of $i$ and simplify.

**5.** $\sqrt{-8}$ **6.** $\sqrt{-25}$

**7.** $3\sqrt{-16}$ **8.** $2\sqrt{-18}$

For Problems 9–18, perform the indicated operation and simplify.

**9.** $\sqrt{-2}\sqrt{-6}$ **10.** $\sqrt{-2}\sqrt{18}$

**11.** $\dfrac{\sqrt{-42}}{\sqrt{-6}}$ **12.** $\dfrac{\sqrt{-6}}{\sqrt{2}}$

**13.** $5i(3 - 6i)$ **14.** $(5 - 7i)(6 + 8i)$

**15.** $(-2 - 3i)(4 - 8i)$ **16.** $(4 - 3i)(4 + 3i)$

**17.** $\dfrac{4 + 3i}{6 - 2i}$ **18.** $\dfrac{-1 - i}{-2 + 5i}$

For Problems 19 and 20, perform the indicated operations and express the answer in the standard form of a complex number.

**19.** $\dfrac{3 + 4i}{2i}$ **20.** $\dfrac{-6 + 5i}{-i}$

For Problems 21–24, solve each of the quadratic equations by factoring.

**21.** $x^2 + 8x = 0$ **22.** $x^2 = 6x$

**23.** $x^2 - 3x - 28 = 0$ **24.** $2x^2 + x - 3 = 0$

For Problems 25–28, use Property 6.1 to help solve each quadratic equation.

**25.** $2x^2 = 90$ **26.** $(y - 3)^2 = -18$

**27.** $(2x + 3)^2 = 24$ **28.** $a^2 - 27 = 0$

For Problems 29–32, use the method of completing the square to solve the quadratic equation.

**29.** $y^2 + 18y - 10 = 0$ **30.** $n^2 + 6n + 20 = 0$

**31.** $x^2 - 10x + 1 = 0$ **32.** $x^2 + 5x - 2 = 0$

For Problems 33–36, use the quadratic formula to solve the equation.

**33.** $x^2 + 6x + 4 = 0$ **34.** $x^2 + 4x + 6 = 0$

**35.** $3x^2 - 2x + 4 = 0$ **36.** $5x^2 - x - 3 = 0$

For Problems 37–40, find the discriminant of each equation and determine whether the equation has (1) two nonreal complex solutions, (2) one real solution with a multiplicity of 2, or (3) two real solutions. Do not solve the equations.

**37.** $4x^2 - 20x + 25 = 0$ **38.** $5x^2 - 7x + 31 = 0$

**39.** $7x^2 - 2x - 14 = 0$ **40.** $5x^2 - 2x = 4$

For Problems 41–59, solve each equation.

**41.** $x^2 - 17x = 0$ **42.** $(x - 2)^2 = 36$

**43.** $(2x - 1)^2 = -64$ **44.** $x^2 - 4x - 21 = 0$

**45.** $x^2 + 2x - 9 = 0$ **46.** $x^2 - 6x = -34$

**47.** $4\sqrt{x} = x - 5$ **48.** $3n^2 + 10n - 8 = 0$

**49.** $n^2 - 10n = 200$ **50.** $3a^2 + a - 5 = 0$

**51.** $x^2 - x + 3 = 0$ **52.** $2x^2 - 5x + 6 = 0$

**53.** $2a^2 + 4a - 5 = 0$ **54.** $t(t + 5) = 36$

**55.** $x^2 + 4x + 9 = 0$ **56.** $(x - 4)(x - 2) = 80$

**57.** $\dfrac{3}{x} + \dfrac{2}{x + 3} = 1$ **58.** $2x^4 - 23x^2 + 56 = 0$

**59.** $\dfrac{3}{n - 2} = \dfrac{n + 5}{4}$

For Problems 60–70, set up an equation and solve each problem.

**60.** The wing of an airplane is in the shape of a 30°–60° right triangle. If the side opposite the 30° angle measures 20 feet, find the measure of the other two sides of the wing. Round the answers to the nearest tenth of a foot.

**61.** An agency is using photo surveillance of a rectangular plot of ground that measures 40 meters by 25 meters. If during the surveillance, someone is observed moving from one corner of the plot to the corner diagonally opposite, how far has the observed person moved? Round the answer to the nearest tenth of a meter.

**62.** One leg of an isosceles right triangle measures 4 inches. Find the length of the hypotenuse of the triangle. Express the answer in radical form.

**63.** Find two numbers whose sum is 6 and whose product is 2.

**64.** A landscaper agreed to design and plant a flower bed for \$40. It took him three hours less than he anticipated, and therefore he earned \$3 per hour more than he anticipated. How long did he expect it would take to design and plant the flower bed?

**65.** Andre traveled 270 miles in 1 hour more than it took Sandy to travel 260 miles. Sandy drove 7 miles per hour faster than Andre. How fast did each one travel?

**66.** The area of a square is numerically equal to twice its perimeter. Find the length of a side of the square.

**67.** Find two consecutive even whole numbers such that the sum of their squares is 164.

**68.** The perimeter of a rectangle is 38 inches, and its area is 84 square inches. Find the length and width of the rectangle.

**69.** It takes Billy 2 hours longer to do a certain job than it takes Reena. They worked together for 2 hours; then Reena left, and Billy finished the job in 1 hour. How long would it take each of them to do the job alone?

**70.** A company has a rectangular parking lot 40 meters wide and 60 meters long. The company plans to increase the area of the lot by 1100 square meters by adding a strip of equal width to one side and one end. Find the width of the strip to be added.

For Problems 71–78, solve each inequality and express the solution set using interval notation.

**71.** $x^2 + 3x - 10 > 0$

**72.** $2x^2 + x - 21 \leq 0$

**73.** $4x^2 - 1 \leq 0$

**74.** $x^2 - 7x + 10 > 0$

**75.** $\dfrac{x - 4}{x + 6} \geq 0$

**76.** $\dfrac{2x - 1}{x + 1} > 4$

**77.** $\dfrac{3x + 1}{x - 4} < 2$

**78.** $\dfrac{3x + 1}{x - 1} \leq 0$

# Chapter 6 Test

1. Find the product $(3 - 4i)(5 + 6i)$, and express the result in the standard form of a complex number.

2. Find the quotient $\frac{2 - 3i}{3 + 4i}$, and express the result in the standard form of a complex number.

For Problems 3–15, solve each equation.

3. $x^2 = 7x$

4. $(x - 3)^2 = 16$

5. $x^2 + 3x - 18 = 0$

6. $x^2 - 2x - 1 = 0$

7. $5x^2 - 2x + 1 = 0$

8. $x^2 + 30x = -224$

9. $(3x - 1)^2 + 36 = 0$

10. $(5x - 6)(4x + 7) = 0$

11. $(2x + 1)(3x - 2) = 55$

12. $n(3n - 2) = 40$

13. $x^4 + 12x^2 - 64 = 0$

14. $\frac{3}{x} + \frac{2}{x + 1} = 4$

15. $3x^2 - 2x - 3 = 0$

16. Does the equation $4x^2 + 20x + 25 = 0$ have (a) two nonreal complex solutions, (b) two equal real solutions, or (c) two unequal real solutions?

17. Does the equation $4x^2 - 3x = -5$ have (a) two nonreal complex solutions, (b) two equal real solutions, or (c) two unequal real solutions?

For Problems 18–20, solve each inequality and express the solution set using interval notation.

18. $x^2 - 3x - 54 \leq 0$

19. $\frac{3x - 1}{x + 2} > 0$

20. $\frac{x - 2}{x + 6} \geq 3$

For Problems 21–25, set up an equation and solve each problem.

21. A 24-foot ladder leans against a building and makes an angle of 60° with the ground. How far up on the building does the top of the ladder reach? Express your answer to the nearest tenth of a foot.

22. A rectangular plot of ground measures 16 meters by 24 meters. Find, to the nearest meter, the distance from one corner of the plot to the diagonally opposite corner.

23. Amy agreed to clean her brother's room for $36. It took her 1 hour longer than she expected, and therefore she earned $3 per hour less than she anticipated. How long did she expect it would take to clean the room?

24. The perimeter of a rectangle is 41 inches, and its area is 91 square inches. Find the length of its shortest side.

25. The sum of two numbers is 6 and their product is 4. Find the larger of the two numbers.

# Chapters 1–6 Cumulative Review Problem Set

For Problems 1–6, evaluate each of the numerical expressions.

**1.** $\left(\frac{3}{2}\right)^{-3}$

**2.** $16^{-\frac{1}{2}}$

**3.** $\frac{2^{-5}}{2^{-6}}$

**4.** $\sqrt[3]{\frac{1}{8}}$

**5.** $\frac{1}{\left(\frac{1}{3}\right)^{-2}}$

**6.** $(4^{-3} \cdot 4)^{-1}$

For Problems 7–14, evaluate each algebraic expression for the given values of the variable.

**7.** $\frac{1}{2x} + \frac{2}{3x} + \frac{3}{x}$ for $x = 5$

**8.** $\frac{3m^2n}{11mn^2}$ for $m = 2$ and $n = -1$

**9.** $(2a + 3b) - 2(6a - 7b)$ for $a = -4$ and $b = -2$

**10.** $4x^2 - y^2$ for $x = \frac{1}{2}$ and $y = -2$

**11.** $(3x + 2y)^2$ for $x = -3$ and $y = \frac{1}{2}$

**12.** $\frac{x^2 + 2y^3}{5x + y}$ for $x = 3$ and $y = -1$

**13.** $\frac{1}{3}n + \frac{1}{2}n - \frac{1}{5}n$ for $n = \frac{2}{3}$

**14.** $3x^2y$ for $x = 1.2$ and $y = 0.2$

For Problems 15–26, perform the indicated operations and express the answers in simplified form.

**15.** $(3ab^2)(-a^3b^3)(4a^2b)$

**16.** $\left(-\frac{4}{3}c^3d^2\right)\left(-\frac{1}{4}cd\right)$

**17.** $(-3mn^3)^4$

**18.** $(a - 2)(3a^2 - a + 7)$

**19.** $-\frac{25t^2k^3}{5tk}$

**20.** $(3x + y)(4x - 5y)$

**21.** $(7m - 6n)^2$

**22.** $\frac{9a^2b}{4ab} \cdot \frac{6a^3b^2}{27ab}$

**23.** $\frac{2x^3 + 2x^2 - 24x}{2x^2 + 19x + 35} \div \frac{4x^3 + 4x^2 - 48x}{2x^2 - x - 15}$

**24.** $\frac{11x - 3}{2} - \frac{6x + 5}{3}$

**25.** $(3x^3 - 7x^2 + x - 6) \div (x - 2)$

**26.** $2 - \frac{x}{4x - 1}$

**27.** Simplify the complex fraction.

$$\frac{\frac{3}{4x} + \frac{7}{9}}{\frac{1}{12} - \frac{5}{3x}}$$

For Problems 28–35, factor completely.

**28.** $2ax - 2ay + 3cx - 3cy$

**29.** $81m^2 - 9n^2$

**30.** $2x^2 - 13x - 7$

**31.** $12y^2 + 28y - 5$

**32.** $6t^2 + 34t - 56$

**33.** $c^2 - y^6$

**34.** $8h^2 - 14h - 15$

**35.** $a^3 + 8b^3$

For Problems 36–58, solve each equation.

**36.** $-7(a + 4) - 3(2a - 9) = 5(3a + 11)$

**37.** $-8(a + 5) = -2(4a - 3)$

**38.** $\frac{t}{5} + \frac{t}{3} - \frac{t}{30} = 1$

**39.** $\frac{4}{5}(a - 2) + \frac{1}{2}(a + 3) = 4$

**40.** $0.035(2000 - x) + 0.04x = 77.50$

**41.** $A = P + Prt$ for $t$

**42.** $\sqrt{3x - 4} = 4$

**43.** $\sqrt[3]{10x - 3} = 3$

**44.** $4c^2 = \frac{4}{3}$

**45.** $\sqrt{x + 3} + 5 = \sqrt{x + 48}$

**46.** $\frac{1}{2}(3x - 5)^2 - 25 = 25$

**47.** $3\sqrt{x} = x - 10$

**48.** $(4x - 3)^2 - 5 = 20$

**49.** $6x^2 + 7x - 20 = 0$

**50.** $(2x + 1)(x - 3) = 9$

**51.** $P = 2l + 2w$ for $w$

**52.** $|9x - 2| = 0$

**53.** $|4x + 7| - 3 = 12$

**54.** $\frac{2x - 5}{12} + \frac{3x + 1}{4} = \frac{x - 4}{6}$

**55.** $\frac{4}{9x + 2} = \frac{7}{3x - 1}$

**56.** $\frac{a + 4}{4} = \frac{3}{a}$

**57.** $\frac{x}{x - 4} - \frac{1}{x + 5} = \frac{1}{x^2 + x - 20}$

**58.** $\frac{n}{n - 5} - \frac{5}{2n + 9} = \frac{25}{2n^2 - n - 45}$

For Problems 59–70, solve each inequality and express the solution set using interval notation.

**59.** $-5x > 10 - x$

**60.** $3(4x - 5) \geq 4(1 - 2x)$

**61.** $9x - 2 < 3(3x + 10)$

**62.** $\frac{2x + 7}{12} - \frac{3x - 8}{8} \leq \frac{1}{3}$

**63.** $0.04x + 0.055(x + 10{,}000) \geq 645$

**64.** $|3x - 4| < 8$

**65.** $\left|5x + \frac{2}{3}\right| < -4$

**66.** $\frac{1}{3}a - \frac{3}{8}a > \frac{1}{6}$

**67.** $2x^2 + x - 15 \geq 0$

**68.** $\frac{3x}{x - 5} < 2$

**69.** $3x + 7 > 10$ or $4x + 1 \leq -19$

**70.** $-3 \leq 5x + 7 \leq 27$

For Problems 71–80, solve by setting up and solving an appropriate equation or inequality.

**71.** Greg leaves Moose Lodge at 1:00 P.M. on snow shoes traveling east at 2.5 miles per hour. His wife, Tricia, leaves the lodge at the same time on cross country skis traveling west at 5 miles per hour. At what time will they be 10 miles apart?

**72.** Sean has \$1000 he wants to invest at 4% interest. How much should he invest at 5% annual interest so that both the investments earn at least \$120 in total annual interest?

**73.** The measure of the smallest angle of a triangle is half the measure of the middle angle. The measure of the largest angle is 5° more than twice the measure of the middle angle. Find the measures of the angles of the triangle.

**74.** Weed-no-More Landscape Company was hired to clean a lot for \$100. The company took 2 hours longer than the estimate indicated, so they earned \$2.50 per hour less than they thought. How many hours did the company estimate it would take to clean the lot?

**75.** A rectangular dog-agility field has a length of 110 feet and a width of 90 feet. The judge stands in one corner and the starting line is in the corner located diagonally across the field. Find the distance between the judge and the starting line to the nearest tenth of a foot.

**76.** Betty Ann invested \$2500 at a certain rate of interest compounded annually. After 2 years, the accumulated value is \$2704. Find the annual rate of interest.

**77.** Together Camden and Aidan can repair a van in 5 hours. If Aidan can complete the job himself in $8\frac{1}{3}$ hours, how long would it take Camden to fix the van by himself?

**78.** The deck of a house is in the shape of a right triangle. The longest side of the deck measures 17 feet. The sum of the measures of the two sides (legs) of the deck is 23 feet. Find the measure of each of the two sides of the deck.

**79.** It takes Samuel 2 hours longer to paddle his canoe 8 miles than it takes him to paddle his kayak 12 miles. His rate when paddling the kayak is $2\frac{2}{5}$ miles per hour greater than his rate when paddling the canoe. Find his rate when paddling each vessel.

**80.** The length of the side of a square is the same as the radius of a circle. The perimeter of the square is numerically equal to 4 times the area of the circle. Find the radius of the circle. Use 3.14 as the value for $\pi$. Round the answer to the nearest hundredth.

# 7 Equations and Inequalities in Two Variables

## Study Skill Tip

Everyone has a preferred learning style when learning math. Students are typically categorized as visual learners, auditory learners, or kinesthetic learners. Which style you are is a personality trait, and no one style is necessarily better than another for learning math. It is helpful to know your own style and to develop learning strategies that embrace that style. The following suggestions can be useful regardless of your preferred learning style.

*"You have all kinds of learners. The sign language is to get to visual learners, the songs get to auditory learners, and hands-on activities, where they are touching things, are for tactile learners."*
DENISE MECHAN

Students who learn math by *seeing* it are visual learners. Visual learners should write down their math problems and study materials. Visual learners may find it helpful to use different colors of ink when writing formulas, notes, or flashcards. Have your tutor show the steps for a problem rather than just talk about the steps.

Students who typically learn math by *hearing* it are auditory learners. Auditory learners should read their formulas and study materials out loud because this will help them remember. You may want to use a recorder yourself to say and hear important math properties and rules. Auditory learners may want to say the numbers in the problems out loud or move their lips while reading problems.

Students who typically learn math best by *touching* and feeling are kinesthetic learners. Kinesthetic learners should use manipulatives whenever possible. One tactile exercise for fractions is cutting up a paper plate or a piece of paper to represent parts of a whole. After graphing equations, trace the various graphs with your finger to tacitly feel the difference in graphs depending on the differences in their equations.

***Based on your learning style, what study techniques have you found helpful to learn math?***

## Chapter Preview

Chapter 7 introduces a new type of equation that has two variables, typically an $x$ and a $y$. An example of this type of equation is $x + y = 10$. The goal of solving any equation is to find values for the variables that make the equation a true statement. Because there are two variables, the solutions are pairs of numbers, a value for $x$ and a value for $y$.

Can you come up with some paired values for $x$ and $y$ that make the equation $x + y = 10$ a true statement? Perhaps you thought of $x = 5$ and $y = 5$, or maybe $x = 1$ and $y = 9$. These are just two of the infinite number of solution pairs.

Because there are an infinite number of solutions, we cannot write down the solution set. Therefore, for equations with two variables, we use a rectangular coordinate system, plot some of the solutions, and then draw a graph to represent the entire set of solutions. This chapter demonstrates techniques for graphing, but you should always remember that a graph is just a picture of the solution set and our goal, as with any equation, is to find the values of the variables that make the equation a true statement.

## 7.1 Rectangular Coordinate System and Linear Equations

OBJECTIVES

1. Find solutions for linear equations in two variables
2. Review the rectangular coordinate system
3. Graph the solutions for linear equations
4. Graph linear equations by finding the x and y intercepts
5. Graph lines passing through the origin, vertical lines, and horizontal lines
6. Apply graphing to linear relationships
7. Introduce graphing utilities

In this chapter we want to solve equations in two variables. Let's begin by considering the solutions for the equation $y = 3x + 2$. A **solution** of an equation in two variables is an ordered pair of real numbers that satisfies the equation. When using the variables $x$ and $y$, we agree that the first number of an ordered pair is a value of $x$ and the second number is a value of $y$. We see that (1, 5) is a solution for $y = 3x + 2$, because if $x$ is replaced by 1 and $y$ by 5, the result is the true numerical statement $5 = 3(1) + 2$. Likewise, (2, 8) is a solution because $8 = 3(2) + 2$ is a true numerical statement. We can find infinitely many pairs of real numbers that satisfy $y = 3x + 2$ by arbitrarily choosing values for $x$, and then, for each chosen value of $x$, determining a corresponding value for $y$. The next example will demonstrate this process.

**Classroom Example**
Determine five ordered-pair solutions for the equation $y = 3x - 4$.

### EXAMPLE 1

Determine some ordered-pair solutions for the equation $y = 2x - 5$ and record the values in a table.

## Solution

We can start by arbitrarily choosing values for $x$ and then determine the corresponding $y$ value. And even though you can arbitrarily choose values for $x$, it is good practice to choose some negative values, zero, and some positive values.

Let $x = -4$; then, according to our equation, $y = 2(-4) - 5 = -13$.
Let $x = -1$; then, according to our equation, $y = 2(-1) - 5 = -7$.
Let $x = 0$; then, according to our equation, $y = 2(0) - 5 = -5$.
Let $x = 2$; then, according to our equation, $y = 2(2) - 5 = -1$.
Let $x = 4$; then, according to our equation, $y = 2(4) - 5 = 3$.

Organizing this information in a chart gives the following table.

| x value | y value determined from y = 2x − 5 | Ordered pair |
|---|---|---|
| −4 | −13 | (−4, −13) |
| −1 | −7 | (−1, −7) |
| 0 | −5 | (0, −5) |
| 2 | −1 | (2, −1) |
| 4 | 3 | (4, 3) |

A table cannot show an infinite number of solutions for a linear equation in two variables, but a graph can display visually the solutions plotted on a coordinate system. Let's review the rectangular coordinate system, and then we can use a graph to display the solutions of an equation in two variables.

## Review of the Rectangular Coordinate System

Consider two number lines, one vertical and one horizontal, perpendicular to each other at the point we associate with zero on both lines (Figure 7.1). We refer to these number lines as the **horizontal and vertical axes** or, together, as the **coordinate axes**. They partition the plane into four regions called **quadrants**. The quadrants are numbered with Roman numerals from I through IV counterclockwise as indicated in Figure 7.1. The point of intersection of the two axes is called the **origin**.

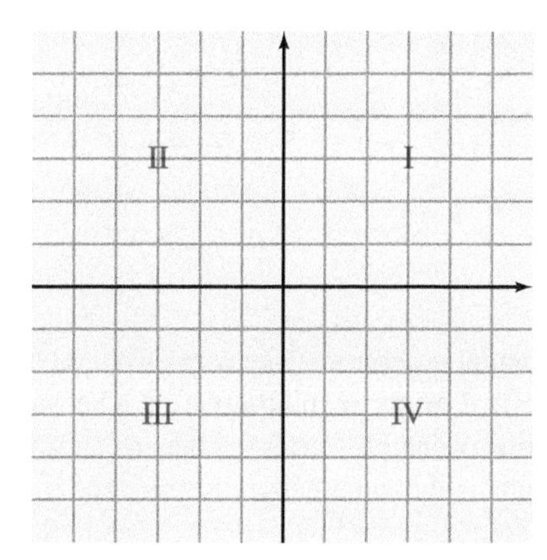


**Figure 7.1**

It is now possible to set up a one-to-one correspondence between **ordered pairs** of real numbers and the points in a plane. To each ordered pair of real numbers there corresponds a unique point in the plane, and to each point in the plane there corresponds a unique ordered pair of real numbers. A part of this correspondence is illustrated in Figure 7.2. The ordered pair (3, 2) denotes that the point $A$ is located three units to the right of, and two units up from, the origin. (The ordered pair (0, 0) is associated with the origin $O$.) The ordered pair $(-3, -5)$ denotes that the point $D$ is located three units to the left and five units down from the origin.

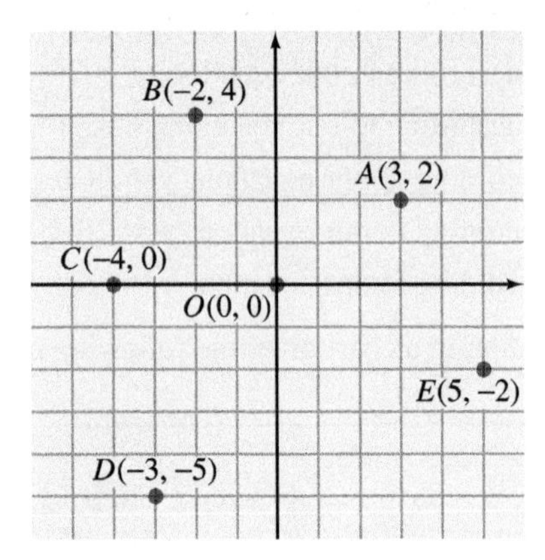


**Figure 7.2**

**Remark:** The notation $(-2, 4)$ was used earlier in this text to indicate an interval of the real number line. Now we are using the same notation to indicate an ordered pair of real numbers. This double meaning should not be confusing because the context of the material will always indicate which meaning of the notation is being used. Throughout this chapter, we will be using the ordered-pair interpretation.

In general we refer to the real numbers $a$ and $b$ in an ordered pair $(a, b)$ associated with a point as the **coordinates of the point**. The first number, $a$, called the **abscissa**, is the directed distance of the point from the vertical axis measured parallel to the horizontal axis. The second number, $b$, called the **ordinate**, is the directed distance of the point from the horizontal axis measured parallel to the vertical axis (Figure 7.3). This system of associating points in a plane with pairs of real numbers is called the **rectangular coordinate system** or the **Cartesian coordinate system**.

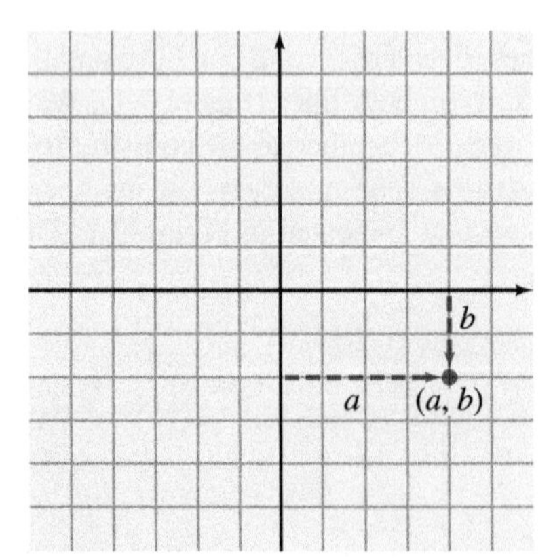


**Figure 7.3**

Historically, the rectangular coordinate system provided the basis for the development of the branch of mathematics called **analytic geometry**, or what we presently refer to as **coordinate geometry**. In this discipline, René Descartes, a French 17th-century mathematician, was able to transform geometric problems into an algebraic setting and then use the tools of algebra to solve the problems. Basically, there are two kinds of problems to solve in coordinate geometry:

1. Given an algebraic equation, find its geometric graph.
2. Given a set of conditions pertaining to a geometric figure, find its algebraic equation.

In this chapter we will discuss problems of both types. Let's begin by plotting the graph of an algebraic equation.

## Graphing the Solutions for Linear Equations

Let's begin by determining some solutions for the equation $y = x + 2$ and then plot the solutions on a rectangular coordinate system to produce a graph of the equation. Let's use a table to record some of the solutions.

| Choose x | Determine y from y = x + 2 | Solutions for y = x + 2 |
|---|---|---|
| 0 | 2 | (0, 2) |
| 1 | 3 | (1, 3) |
| 3 | 5 | (3, 5) |
| 5 | 7 | (5, 7) |
| −2 | 0 | (−2, 0) |
| −4 | −2 | (−4, −2) |
| −6 | −4 | (−6, −4) |

We can plot the ordered pairs as points in a coordinate plane and use the horizontal axis as the $x$ axis and the vertical axis as the $y$ axis, as in Figure 7.4(a). The straight line that contains the points in Figure 7.4(b) is called the **graph of the equation** $y = x + 2$. Every point on the line has coordinates that are solutions of the equation $y = x + 2$. The graph provides a visual display of the infinite solutions for the equation.

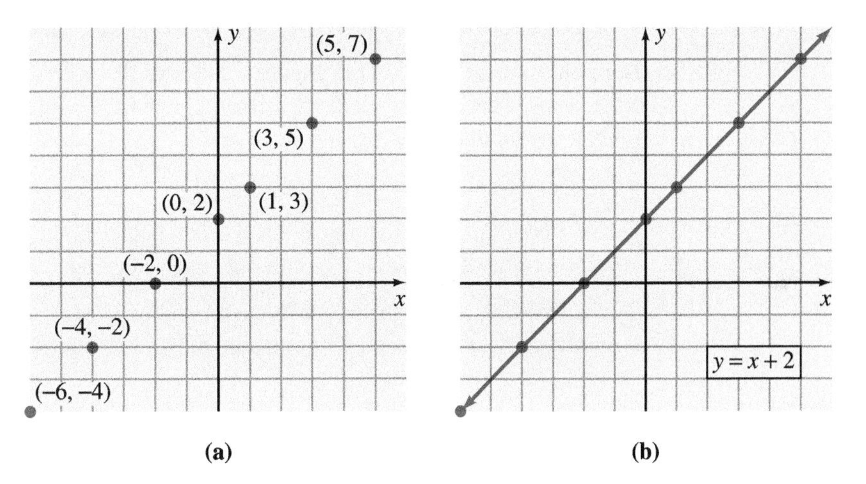


**Figure 7.4**

Classroom Example
Graph the equation $y = -3x + 2$.

**EXAMPLE 2** Graph the equation $y = -x + 4$.

### Solution

Let's begin by determining some solutions for the equation $y = -x + 4$ and then plot the solutions on a rectangular coordinate system to produce a graph of the equation.

Let's use a table to record some of the solutions.

| x value | y value determined from y = −x + 4 | Ordered pair |
|---|---|---|
| −3 | 7 | (−3, 7) |
| −1 | 5 | (−1, 5) |
| 0 | 4 | (0, 4) |
| 2 | 2 | (2, 2) |
| 4 | 0 | (4, 0) |
| 6 | −2 | (6, −2) |

We can plot the ordered pairs on a coordinate system as shown in Figure 7.5(a). The graph of the equation was created by drawing a straight line through the plotted points as in Figure 7.5(b).

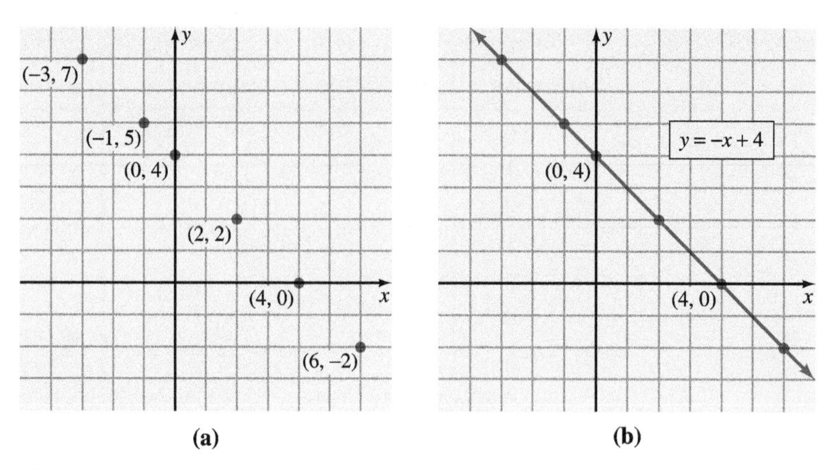


Figure 7.5

## Graphing Linear Equations by Locating the x and y Intercepts

The points (4, 0) and (0, 4) in Figure 7.5(b) are the points of the graph that are on the coordinate axes. That is, they yield the $x$ intercept and the $y$ intercept of the graph. Let's define in general the *intercepts* of a graph.

The $x$ coordinates of the points that a graph has in common with the $x$ axis are called the ***x* intercepts** of the graph. (To compute the $x$ intercepts, let $y = 0$ and solve for $x$.)

The $y$ coordinates of the points that a graph has in common with the $y$ axis are called the ***y* intercepts** of the graph. (To compute the $y$ intercepts, let $x = 0$ and solve for $y$.)

It is advantageous to be able to recognize the kind of graph that a certain type of equation produces. For example, if we recognize that the graph of $3x + 2y = 12$ is a straight line, then it becomes a simple matter to find two points and sketch the line. Let's pursue the graphing of straight lines in a little more detail.

In general, any equation of the form $Ax + By = C$, where $A$, $B$, and $C$ are constants ($A$ and $B$ not both zero) and $x$ and $y$ are variables, is a **linear equation**, and its graph is a straight line. Two points of clarification about this description of a linear equation should be made. First, the choice of $x$ and $y$ for variables is arbitrary. Any two letters could be used to represent the variables. For example, an equation such as $3r + 2s = 9$ can be considered a linear equation in two variables. So that we are not constantly changing the labeling of the coordinate axes when graphing equations, however, it is much easier to use the same two variables in all equations. Thus we will go along with convention and use $x$ and $y$ as variables. Second, the phrase "any equation of the form $Ax + By = C$" technically means "any equation of the form $Ax + By = C$ or equivalent to that form." For example, the equation $y = 2x - 1$ is equivalent to $-2x + y = -1$ and thus is linear and produces a straight-line graph.

The knowledge that any equation of the form $Ax + By = C$ produces a straight-line graph, along with the fact that two points determine a straight line, makes graphing linear equations a simple process. We merely find two solutions (such as the intercepts), plot the corresponding points, and connect the points with a straight line. It is wise to find a third point as a check point. Let's consider an example.

**Classroom Example**
Graph $2x - y = 4$.

**EXAMPLE 3** Graph $3x - 2y = 12$.

### Solution

First, let's find the intercepts. To find the $y$ intercept let $x = 0$.

$$3(0) - 2y = 12$$
$$-2y = 12$$
$$y = -6 \qquad \text{Thus } (0, -6) \text{ is a solution.}$$

To find the $x$ intercept let $y = 0$.

$$3x - 2(0) = 12$$
$$3x = 12$$
$$x = 4 \qquad \text{Thus } (4, 0) \text{ is a solution.}$$

Now let's find a third point to serve as a check point. Let $x = 2$; then

$$3(2) - 2y = 12$$
$$6 - 2y = 12$$
$$-2y = 6$$
$$y = -3 \qquad \text{Thus } (2, -3) \text{ is a solution.}$$

Plot the points associated with these three solutions and connect them with a straight line to produce the graph of $3x - 2y = 12$ in Figure 7.6.

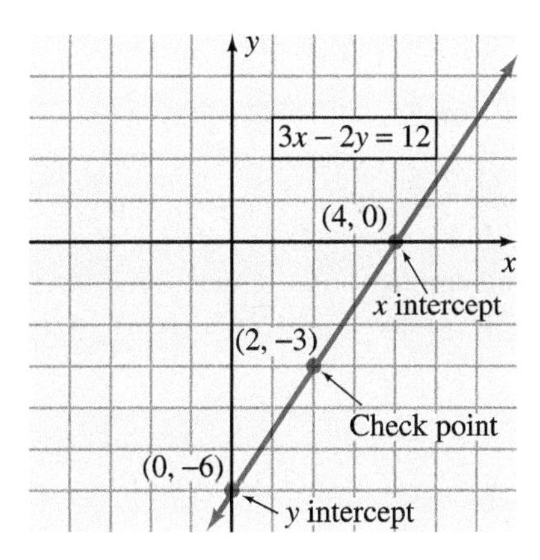


**Figure 7.6**

Let's review our approach to Example 3. Note that we did not solve the equation for $y$ in terms of $x$ or for $x$ in terms of $y$. Because we know the graph is a straight line, there is no need for any extensive table of values; thus there is no need to change the form of the original equation. Furthermore, the solution $(2, -3)$ served as a check point. If it had not been on the line determined by the two intercepts, then we would have known that an error had been made.

**Classroom Example**
Graph $4x + 3y = 6$.

**EXAMPLE 4** Graph $2x + 3y = 7$.

### Solution

Without showing all of our work, the following table indicates the intercepts and a check point. The points from the table are plotted, and the graph of $2x + 3y = 7$ is shown in Figure 7.7.

| $x$ | $y$ | |
|---|---|---|
| 0 | $\frac{7}{3}$ | |
| $\frac{7}{2}$ | 0 | Intercepts |
| 2 | 1 | Check point |

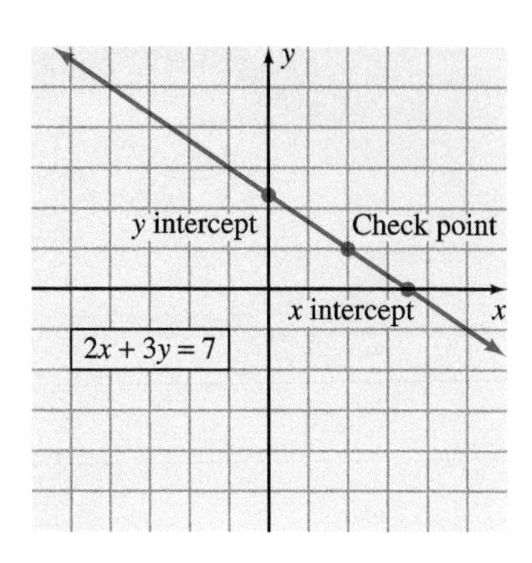


**Figure 7.7**

## Graphing Lines That Pass Through the Origin, Vertical Lines, and Horizontal Lines

It is helpful to recognize some *special* straight lines. For example, the graph of any equation of the form $Ax + By = C$, where $C = 0$ (the constant term is zero), is a straight line that contains the origin. Let's consider an example.

**Classroom Example**
Graph $y = -3x$.

**EXAMPLE 5** Graph $y = 2x$.

### Solution

The equation $y = 2x$ is equivalent to the equation $2x - y = 0$; thus it fits the condition $Ax + By = C$, where $C = 0$. Any equation of this form passes through the origin; hence both the $x$ intercept and $y$ intercept are the point (0, 0). This can be checked by letting $x = 0$ in the equation $y = 2x$. This would give $y = 2(0) = 0$, and it results in the ordered pair (0, 0).

It is necessary to find another point to determine the line. Then a third point should be found as a check point. The graph of $y = 2x$ is shown in Figure 7.8.

| $x$ | $y$ | |
|---|---|---|
| 0 | 0 | Intercepts |
| 2 | 4 | Additional point |
| −1 | −2 | Check point |

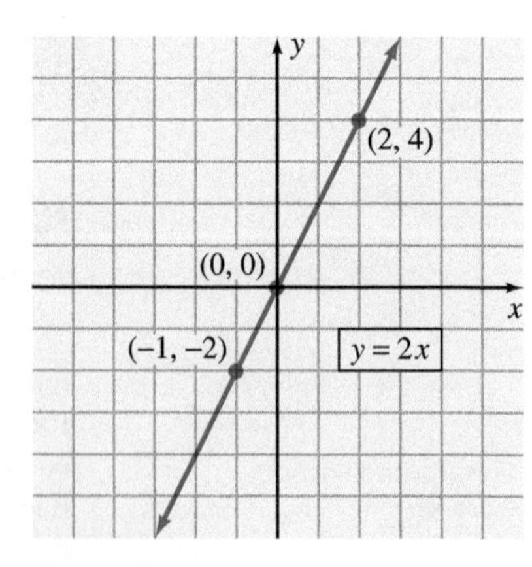


**Figure 7.8**

**Classroom Example**
Graph $x = -3$.

**EXAMPLE 6** Graph $x = 2$.

### Solution

Because we are considering linear equations in *two variables*, the equation $x = 2$ is equivalent to $x + 0(y) = 2$. Any value of $y$ can be used, but the $x$ value will always be 2 regardless of the $y$ value. Therefore, some of the solutions are (2, 0), (2, 1), (2, 2), (2, $-1$), and (2, $-2$). The graph of all solutions of $x = 2$ is the vertical line in Figure 7.9.

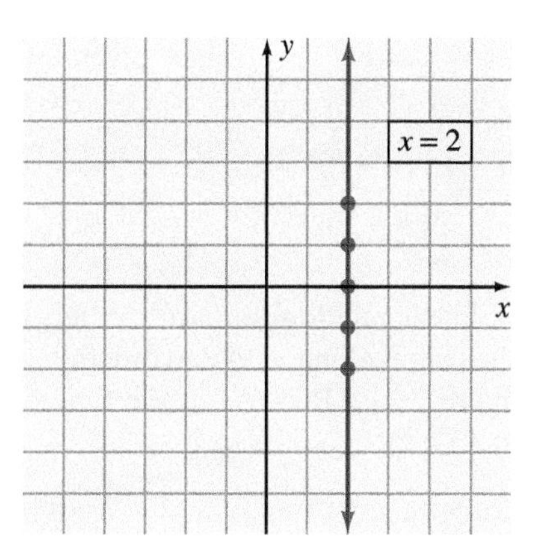


**Figure 7.9**

Classroom Example
Graph $y = 2$.

**EXAMPLE 7** Graph $y = -3$.

## Solution

The equation $y = -3$ is equivalent to $0(x) + y = -3$. Any value of $x$ can be used, but the value of $y$ will always be $-3$ regardless of the $x$ value. Some solutions are (0, $-3$), (1, $-3$), (2, $-3$), ($-1$, $-3$), and ($-2$, $-3$). The graph of $y = -3$ is the horizontal line in Figure 7.10.

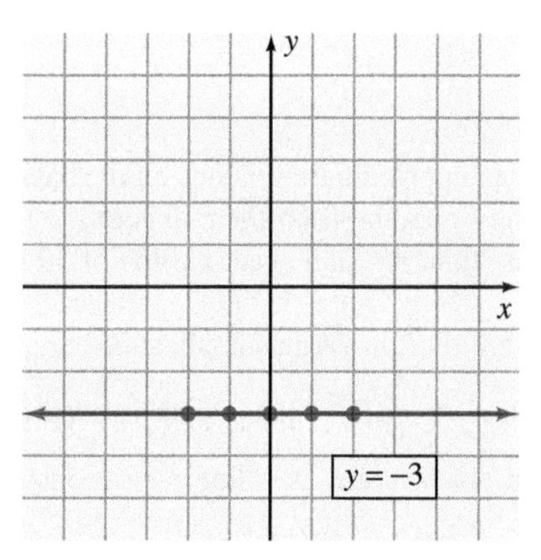


**Figure 7.10**

In general, the graph of any equation of the form $Ax + By = C$, where $A = 0$ or $B = 0$ (not both), is a line parallel to one of the axes. More specifically, any equation of the form $x = a$, where $a$ is a constant, is a vertical line that has an $x$ intercept of $a$. Any equation of the form $y = b$, where $b$ is a constant, is a horizontal line that has a $y$ intercept of $b$.

## Graphing Linear Relationships

There are numerous applications of linear relationships. For example, suppose that a retailer has a number of items that she wants to sell at a profit of 30% of the cost of each item. If we let $s$ represent the selling price and $c$ the cost of each item, then the equation

$$s = c + 0.3c = 1.3c$$

can be used to determine the selling price of each item based on the cost of the item. In other words, if the cost of an item is \$4.50, then it should be sold for $s = (1.3)(4.5) = \$5.85$.

The equation $s = 1.3c$ can be used to determine the following table of values. Reading from the table, we see that if the cost of an item is \$15, then it should be sold for \$19.50 in order to yield a profit of 30% of the cost. Furthermore, because this is a linear relationship, we can obtain exact values between values given in the table.

| c | 1 | 5 | 10 | 15 | 20 |
|---|---|---|---|---|---|
| s | 1.3 | 6.5 | 13 | 19.5 | 26 |

Now let's graph this linear relationship. We can label the horizontal axis $c$, label the vertical axis $s$, and use the origin along with one ordered pair from the table to produce the straight-line graph in Figure 7.11. (Because of the type of application, we use only nonnegative values for $c$ and $s$.)

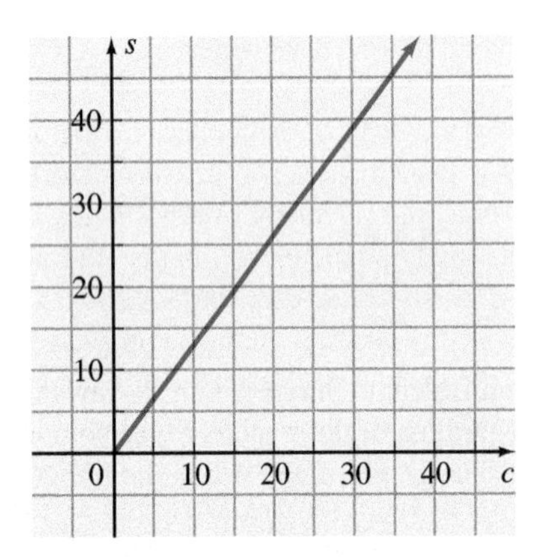


**Figure 7.11**

From the graph we can approximate $s$ values on the basis of given $c$ values. For example, if $c = 30$, then by reading up from 30 on the $c$ axis to the line and then across to the $s$ axis, we see that $s$ is a little less than 40. (An exact $s$ value of 39 is obtained by using the equation $s = 1.3c$.)

Many formulas that are used in various applications are linear equations in two variables. For example, the formula $C = \frac{5}{9}(F - 32)$, which is used to convert temperatures from the Fahrenheit scale to the Celsius scale, is a linear relationship. Using this equation, we can determine that 14°F is equivalent to $C = \frac{5}{9}(14 - 32) = \frac{5}{9}(-18) = -10°C$. Let's use the equation $C = \frac{5}{9}(F - 32)$ to complete the following table.

| F | −22 | −13 | 5 | 32 | 50 | 68 | 86 |
|---|---|---|---|---|---|---|---|
| C | −30 | −25 | −15 | 0 | 10 | 20 | 30 |

Reading from the table, we see, for example, that $-13°F = -25°C$ and $68°F = 20°C$.

To graph the equation $C = \frac{5}{9}(F - 32)$ we can label the horizontal axis F, label the vertical axis C, and plot two ordered pairs (F, C) from the table. Figure 7.12 shows the graph of the equation.

From the graph we can approximate C values on the basis of given F values. For example, if $F = 80°$, then by reading up from 80 on the F axis to the line and then across to the C axis, we see that C is approximately 25°. Likewise, we can obtain approximate F values on the basis of given C values. For example, if $C = -25°$, then by reading across from −25 on the C axis to the line and then up to the F axis, we see that F is approximately −15°.

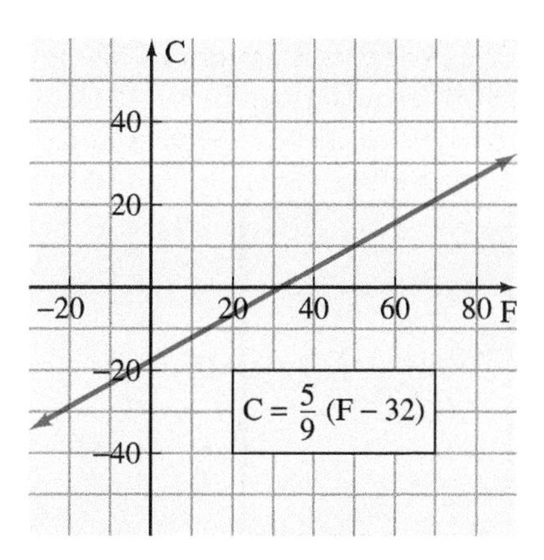


**Figure 7.12**

## Graphing Utilities

The term **graphing utility** is used in current literature to refer to either a graphing calculator (see Figure 7.13) or a computer with a graphing software package. (We use the phrase *use a graphing calculator* to mean "use a graphing calculator or a computer with the appropriate software.")

These devices have a range of capabilities that enable the user not only to obtain a quick sketch of a graph but also to study various characteristics of it, such as the $x$ intercepts, $y$ intercepts, and turning points of a curve. We will introduce some of these features of graphing utilities as we need them in the text. Because there are so many different types of graphing utilities available, we will use generic terminology and let you consult your user's manual for specific key-punching instructions. We urge you to study the graphing utility examples in this text even if you do not have access to a graphing calculator or a computer. The examples were chosen to reinforce concepts under discussion.

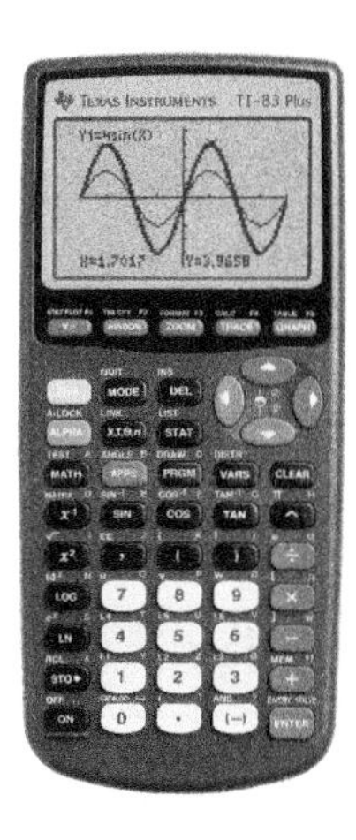
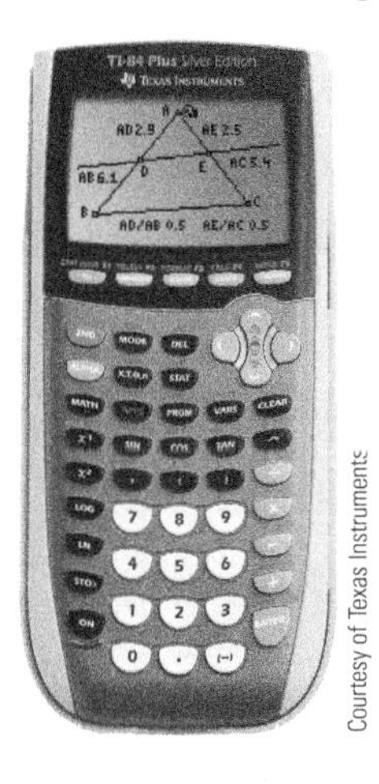

Courtesy of Texas Instruments

**Figure 7.13**

Classroom Example
Use a graphing utility to obtain a graph of the line $1.4x + 2.9y = 10.1$.

### EXAMPLE 8

Use a graphing utility to obtain a graph of the line $2.1x + 5.3y = 7.9$.

### Solution

First, let's solve the equation for $y$ in terms of $x$.

$$2.1x + 5.3y = 7.9$$
$$5.3y = 7.9 - 2.1x$$
$$y = \frac{7.9 - 2.1x}{5.3}$$

Now we can enter the expression $\frac{7.9 - 2.1x}{5.3}$ for $Y_1$ and obtain the graph shown in Figure 7.14.

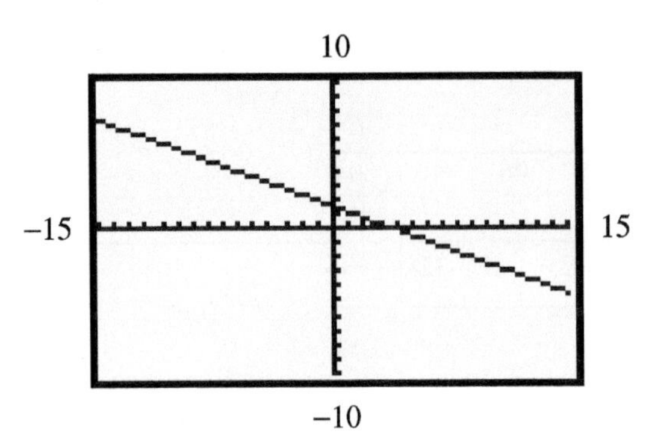


**Figure 7.14**

## Concept Quiz 7.1

For Problems 1–10, answer true or false.

1. In a rectangular coordinate system, the coordinate axes partition the plane into four parts called quadrants.
2. Quadrants are named with Roman numerals and are numbered clockwise.
3. The real numbers in an ordered pair are referred to as the coordinates of the point.
4. If the abscissa of an ordered pair is negative, then the point is in either the third or fourth quadrant.
5. The equation $y = x + 3$ has an infinite number of ordered pairs that satisfy the equation.
6. The graph of $y = x^2$ is a straight line.
7. The $y$ intercept of the graph of $3x + 4y = -4$ is $-4$.
8. The graph of $y = 4$ is a vertical line.
9. The graph of $x = 4$ has an $x$ intercept of 4.
10. The graph of every linear equation has a $y$ intercept.

## Problem Set 7.1

For Problems 1–4, determine which of the ordered pairs are solutions to the given equation. (Objective 1)

1. $y = 3x - 2$ $(2, 4), (-1, -5), (0, 1)$
2. $y = 2x + 3$ $(2, 5), (1, 5), (-1, 1)$
3. $2x + y = 6$ $(-2, 10), (-1, 5), (3, 0)$
4. $-3x + 2y = 2$ $\left(3, \frac{11}{2}\right), (-2, -2)\left(-1, -\frac{1}{2}\right)$

For Problems 5–8, complete the table of values for the equation and graph the equation. (Objective 3)

5. $y = -x + 3$

| x | −2 | −1 | 0 | 4 |
|---|---|---|---|---|
| y | | | | |

6. $y = 2x - 1$

| x | −3 | −1 | 0 | 2 |
|---|---|---|---|---|
| y | | | | |

7. $2x - y = 6$

| x | −2 | 0 | 2 | 4 |
|---|---|---|---|---|
| y | | | | |

8. $2x - 3y = -6$

| x | −3 | 0 | 2 | 3 |
|---|---|---|---|---|
| y | | | | |

For Problems 9–28, graph each of the linear equations by finding the $x$ and $y$ intercepts. (Objective 4)

**9.** $x + 2y = 4$

**10.** $2x + y = 6$

**11.** $2x - y = 2$

**12.** $3x - y = 3$

**13.** $3x + 2y = 6$

**14.** $2x + 3y = 6$

**15.** $5x - 4y = 20$

**16.** $4x - 3y = -12$

**17.** $x + 4y = -6$

**18.** $5x + y = -2$

**19.** $-x - 2y = 3$

**20.** $-3x - 2y = 12$

**21.** $y = x + 3$

**22.** $y = x - 1$

**23.** $y = -2x - 1$

**24.** $y = 4x + 3$

**25.** $y = \frac{1}{2}x + \frac{2}{3}$

**26.** $y = \frac{2}{3}x - \frac{3}{4}$

**27.** $-3y = -x + 3$

**28.** $2y = x - 2$

For Problems 29–40, graph each of the linear equations. (Objective 5)

**29.** $y = -x$

**30.** $y = x$

**31.** $y = 3x$

**32.** $y = -4x$

**33.** $2x - 3y = 0$

**34.** $3x + 4y = 0$

**35.** $x = 0$

**36.** $y = 0$

**37.** $y = 2$

**38.** $x = -3$

**39.** $x = -4$

**40.** $y = -1$

For Problems 41–47, apply graphing to linear relationships. (Objective 6)

**41. (a)** Digital Solutions charges for help-desk services according to the equation $c = 0.25m + 10$, where $c$ represents the cost in dollars and $m$ represents the minutes of service. Complete the following table.

| $m$ | 5 | 10 | 15 | 20 | 30 | 60 |
|---|---|---|---|---|---|---|
| $c$ | | | | | | |

**(b)** Label the horizontal axis $m$ and the vertical axis $c$, and graph the equation $c = 0.25m + 10$ for nonnegative values of $m$.

**(c)** Use the graph from part (b) to approximate values for $c$ when $m = 25$, 40, and 45.

**(d)** Check the accuracy of your readings from the graph in part (c) by using the equation $c = 0.25m + 10$.

**42. (a)** The equation $F = \frac{9}{5}C + 32$ can be used to convert from degrees Celsius to degrees Fahrenheit. Complete the following table.

| C | 0 | 5 | 10 | 15 | 20 | −5 | −10 | −15 | −20 | −25 |
|---|---|---|---|---|---|---|---|---|---|---|
| F | | | | | | | | | | |

**(b)** Graph the equation $F = \frac{9}{5}C + 32$.

**(c)** Use your graph from part (b) to approximate values for F when C = 25°, 30°, −30°, and −40°.

**(d)** Check the accuracy of your readings from the graph in part (c) by using the equation $F = \frac{9}{5}C + 32$.

**43. (a)** A doctor's office wants to chart and graph the linear relationship between the hemoglobin A1c reading and the average blood glucose level. The equation $G = 30h - 60$ describes the relationship, in which $h$ is the hemoglobin A1c reading and $G$ is the average blood glucose reading. Complete this chart of values:

| Hemoglobin A1c, $h$ | 6.0 | 6.5 | 7.0 | 8.0 | 8.5 | 9.0 | 10.0 |
|---|---|---|---|---|---|---|---|
| Blood glucose, $G$ | | | | | | | |

**(b)** Label the horizontal axis $h$ and the vertical axis $G$, then graph the equation $G = 30h - 60$ for $h$ values between 4.0 and 12.0.

**(c)** Use the graph from part (b) to approximate values for $G$ when $h = 5.5$ and 7.5.

**(d)** Check the accuracy of your readings from the graph in part (c) by using the equation $G = 30h - 60$.

**44.** Suppose that the daily profit from an ice cream stand is given by the equation $p = 2n - 4$, where $n$ represents the gallons of ice cream mix used in a day and $p$ represents the dollars of profit. Label the horizontal axis $n$ and the vertical axis $p$, and graph the equation $p = 2n - 4$ for nonnegative values of $n$.

**45.** The cost ($c$) of playing an online computer game for a time ($t$) in hours is given by the equation $c = 3t + 5$. Label the horizontal axis $t$ and the vertical axis $c$, and graph the equation for nonnegative values of $t$.

**46.** The area of a sidewalk whose width is fixed at 3 feet can be given by the equation $A = 3l$, where $A$ represents the area in square feet and $l$ represents the length in feet. Label the horizontal axis $l$ and the vertical axis $A$, and graph the equation $A = 3l$ for nonnegative values of $l$.

**47.** An online grocery store charges for delivery based on the equation $C = 0.30p$, where $C$ represents the cost of delivery in dollars and $p$ represents the weight of the groceries in pounds. Label the horizontal axis $p$ and the vertical axis $C$, and graph the equation $C = 0.30p$ for nonnegative values of $p$.

## Thoughts Into Words

**48.** How do we know that the graph of $y = -3x$ is a straight line that contains the origin?

**49.** How do we know that the graphs of $2x - 3y = 6$ and $-2x + 3y = -6$ are the same line?

**50.** What is the graph of the conjunction $x = 2$ and $y = 4$? What is the graph of the disjunction $x = 2$ or $y = 4$? Explain your answers.

**51.** Your friend claims that the graph of the equation $x = 2$ is the point (2, 0). How do you react to this claim?

## Further Investigations

From our work with absolute value, we know that $|x + y| = 1$ is equivalent to $x + y = 1$ or $x + y = -1$. Therefore, the graph of $|x + y| = 1$ consists of the two lines $x + y = 1$ and $x + y = -1$. Graph each of the following.

**52.** $|x + y| = 1$

**53.** $|x - y| = 4$

**54.** $|2x - y| = 4$

**55.** $|3x + 2y| = 6$

## Graphing Calculator Activities

This is the first of many appearances of a group of problems called graphing calculator activities. These problems are specifically designed for those of you who have access to a graphing calculator or a computer with an appropriate software package. Within the framework of these problems, you will be given the opportunity to reinforce concepts we discussed in the text; lay groundwork for concepts we will introduce later in the text; predict shapes and locations of graphs on the basis of your previous graphing experiences; solve problems that are unreasonable or perhaps impossible to solve without a graphing utility; and in general become familiar with the capabilities and limitations of your graphing utility. (Objective 7)

**56.** **(a)** Graph $y = 3x + 4$, $y = 2x + 4$, $y = -4x + 4$, and $y = -2x + 4$ on the same set of axes.

**(b)** Graph $y = \frac{1}{2}x - 3$, $y = 5x - 3$, $y = 0.1x - 3$, and $y = -7x - 3$ on the same set of axes.

**(c)** What characteristic do all lines of the form $y = ax + 2$ (where $a$ is any real number) share?

**57.** **(a)** Graph $y = 2x - 3$, $y = 2x + 3$, $y = 2x - 6$, and $y = 2x + 5$ on the same set of axes.

**(b)** Graph $y = -3x + 1$, $y = -3x + 4$, $y = -3x - 2$, and $y = -3x - 5$ on the same set of axes.

**(c)** Graph $y = \frac{1}{2}x + 3$, $y = \frac{1}{2}x - 4$, $y = \frac{1}{2}x + 5$, and $y = \frac{1}{2}x - 2$ on the same set of axes.

**(d)** What relationship exists among all lines of the form $y = 3x + b$, where $b$ is any real number?

**58.** **(a)** Graph $2x + 3y = 4$, $2x + 3y = -6$, $4x + 6y = 7$, and $8x + 12y = -1$ on the same set of axes.

**(b)** Graph $5x - 2y = 4$, $5x - 2y = -3$, $10x - 4y = 3$, and $15x - 6y = 30$ on the same set of axes.

**(c)** Graph $x + 4y = 8$, $2x + 8y = 3$, $x + 4y = 6$, and $3x + 12y = 10$ on the same set of axes.

**(d)** Graph $3x - 4y = 6$, $3x - 4y = 10$, $6x - 8y = 20$, and $6x - 8y = 24$ on the same set of axes.

**(e)** For each of the following pairs of lines, (a) predict whether they are parallel lines, and (b) graph each pair of lines to check your prediction.

**(1)** $5x - 2y = 10$ and $5x - 2y = -4$
**(2)** $x + y = 6$ and $x - y = 4$
**(3)** $2x + y = 8$ and $4x + 2y = 2$
**(4)** $y = 0.2x + 1$ and $y = 0.2x - 4$
**(5)** $3x - 2y = 4$ and $3x + 2y = 4$
**(6)** $4x - 3y = 8$ and $8x - 6y = 3$
**(7)** $2x - y = 10$ and $6x - 3y = 6$
**(8)** $x + 2y = 6$ and $3x - 6y = 6$

**59.** Now let's use a graphing calculator to get a graph of $C = \frac{5}{9}(F - 32)$. By letting $F = x$ and $C = y$, we obtain Figure 7.15. Pay special attention to the boundaries on $x$. These values were chosen so that the fraction

$$\frac{(\text{Maximum value of } x) \text{ minus } (\text{Minimum value of } x)}{95}$$

would be equal to 1. The viewing window of the graphing calculator used to produce Figure 7.15 is 95 pixels (dots) wide. Therefore, we use 95 as the denominator of the fraction. We chose the boundaries for $y$ to make sure that the cursor would be visible on the screen when we looked for certain values.

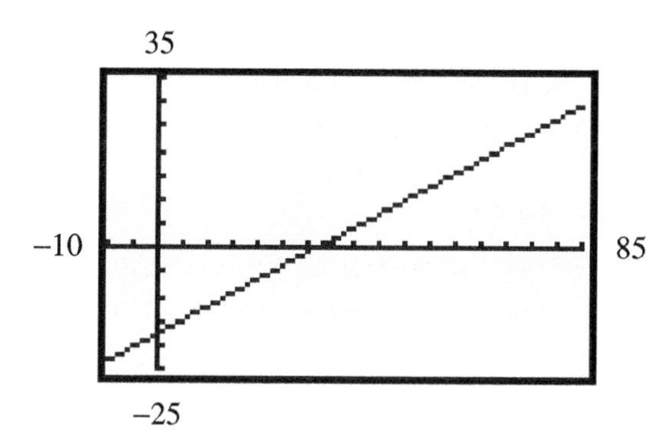

**Figure 7.15**

Now let's use the TRACE feature of the graphing calculator to complete the following table. Note that the cursor moves in increments of 1 as we trace along the graph.

| F | −5 | 5 | 9 | 11 | 12 | 20 | 30 | 45 | 60 |
|---|---|---|---|---|---|---|---|---|---|
| C | | | | | | | | | |

(This was accomplished by setting the aforementioned fraction equal to 1.) By moving the cursor to each of the F values, we can complete the table as follows.

| F | −5 | 5 | 9 | 11 | 12 | 20 | 30 | 45 | 60 |
|---|---|---|---|---|---|---|---|---|---|
| C | −21 | −15 | −13 | −12 | −11 | −7 | −1 | 7 | 16 |

The C values are expressed to the nearest degree. Use your calculator and check the values in the table by using the equation $C = \frac{5}{9}(F - 32)$.

**60.** **(a)** Use your graphing calculator to display the graph of $F = \frac{9}{5}C + 32$. Be sure to set boundaries on the horizontal axis so that when you are using the trace feature, the cursor will move in increments of 1.

**(b)** Use the TRACE feature and check your answers for part (a) of Problem 42.

### Answers to the Concept Quiz

**1.** True **2.** False **3.** True **4.** False **5.** True **6.** False **7.** False **8.** False **9.** True **10.** False

## 7.2 Linear Inequalities in Two Variables

OBJECTIVE **1** Graph linear inequalities in two variables

**Linear inequalities** in two variables are of the form $Ax + By > C$ or $Ax + By < C$, where $A$, $B$, and $C$ are real numbers. (Combined linear equality and inequality statements are of the form $Ax + By \geq C$ or $Ax + By \leq C$.)

Graphing linear inequalities is almost as easy as graphing linear equations. The following discussion leads into a simple, step-by-step process. Let's consider the following equation and related inequalities.

$$x + y = 2 \qquad x + y > 2 \qquad x + y < 2$$

The graph of $x + y = 2$ is shown in Figure 7.16. The line divides the plane into two half-planes, one above the line and one below the line. *All points* in the half-plane above the line satisfy the inequality $x + y > 2$. Therefore, the graph of $x + y > 2$ is the half-plane above the line, as indicated by the shaded portion in Figure 7.17. We use a dashed line to indicate that points on the line do *not* satisfy $x + y > 2$. We would use a solid line if we were graphing $x + y \geq 2$.

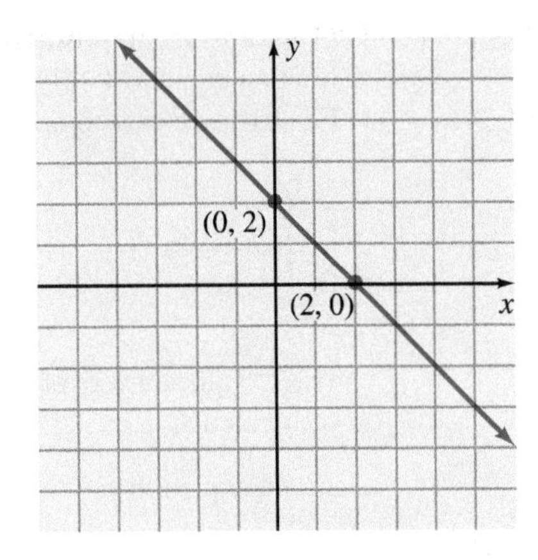


Figure 7.16

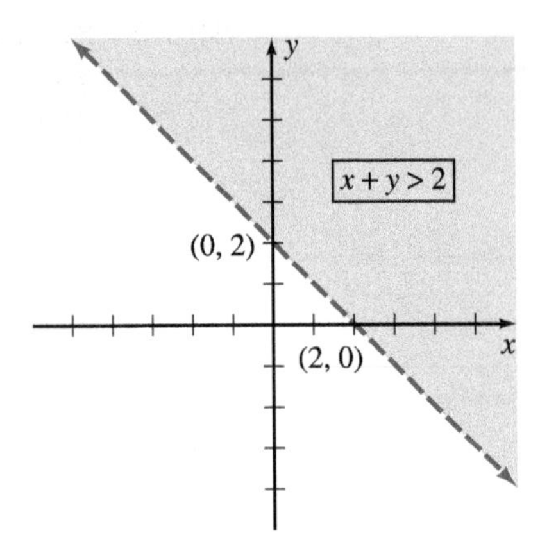


Figure 7.17

*All points* in the half-plane below the line satisfy the inequality $x + y < 2$. Thus the graph of $x + y < 2$ is the half-plane below the line, as indicated in Figure 7.18.

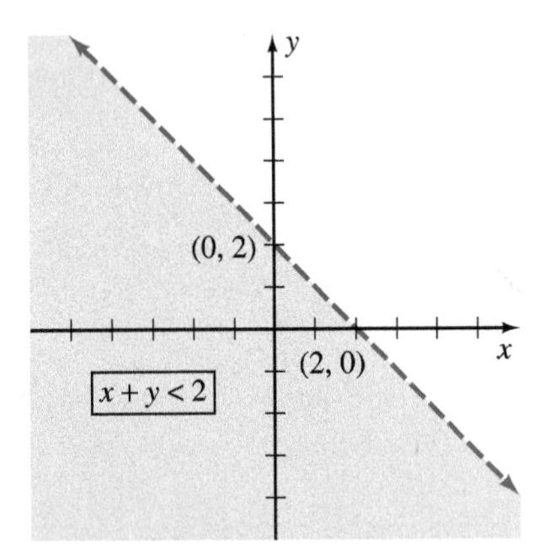


Figure 7.18

To graph a linear inequality, we suggest the following steps.

1. First, graph the corresponding equality. Use a solid line if equality is included in the original statement. Use a dashed line if equality is not included.
2. Choose a "test point" not on the line and substitute its coordinates into the inequality. (The origin is a convenient point to use if it is not on the line.)
3. The graph of the original inequality is
   **(a)** the half-plane that contains the test point if the inequality is satisfied by that point, or
   **(b)** the half-plane that does not contain the test point if the inequality is not satisfied by the point.

Let's apply these steps to some examples.

**Classroom Example**
Graph $2x + 3y < 6$.

**EXAMPLE 1** Graph $x - 2y > 4$.

## Solution

**Step 1** Graph $x - 2y = 4$ as a dashed line because equality is not included in $x - 2y > 4$ (Figure 7.19).

**Step 2** Choose the origin as a test point, and substitute its coordinates into the inequality.

$$x - 2y > 4 \quad \text{becomes } 0 - 2(0) > 4 \text{ equivalent to } 0 > 4 \text{, which is false}$$

**Step 3** Because the test point did not satisfy the given inequality, the graph is the half-plane that does not contain the test point. Thus the graph of $x - 2y > 4$ is the half-plane below the line, as indicated in Figure 7.19.

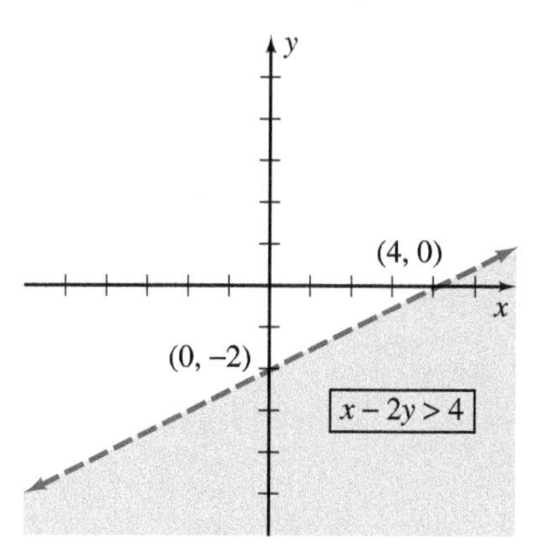


**Figure 7.19**

Classroom Example
Graph $x - 4y \geq 8$.

**EXAMPLE 2** Graph $3x + 2y \leq 6$.

## Solution

**Step 1** Graph $3x + 2y = 6$ as a solid line because equality is included in $3x + 2y \leq 6$ (Figure 7.20).

**Step 2** Choose the origin as a test point and substitute its coordinates into the given statement.

$$3x + 2y \leq 6 \quad \text{becomes } 3(0) + 2(0) \leq 6 \text{ equivalent to } 0 \leq 6\text{, which is true}$$

**Step 3** Because the test point satisfies the given statement, all points in the same half-plane as the test point satisfy the statement. Thus the graph of $3x + 2y \leq 6$ consists of the line and the half-plane below the line (Figure 7.20).

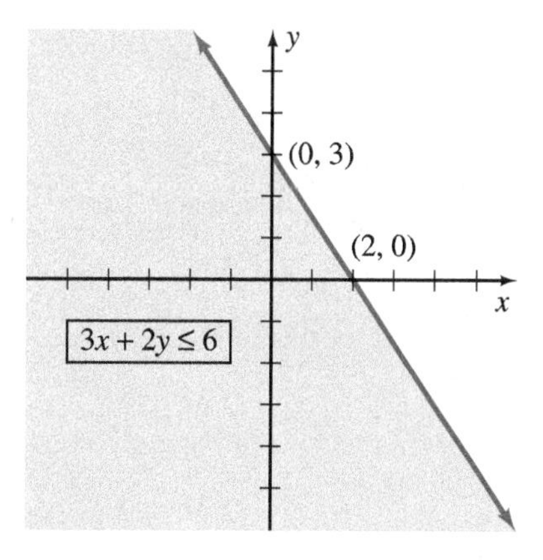


**Figure 7.20**

Classroom Example
Graph $y < -4x$.

**EXAMPLE 3** Graph $y \leq 3x$.

## Solution

**Step 1** Graph $y = 3x$ as a solid line because equality is included in $y \leq 3x$ (Figure 7.21).

**Step 2** The line passes through the origin, so we must choose some other point as a test point. Let's try (2, 1).

$$y \leq 3x \quad \text{becomes } 1 \leq 3(2) \text{ equivalent to } 1 \leq 6\text{, which is a true statement}$$

**Step 3** Because the test point satisfies the given inequality, the graph is the half-plane that contains the test point. Thus the graph of $y \le 3x$ consists of the line and the half-plane below the line, as indicated in Figure 7.21.

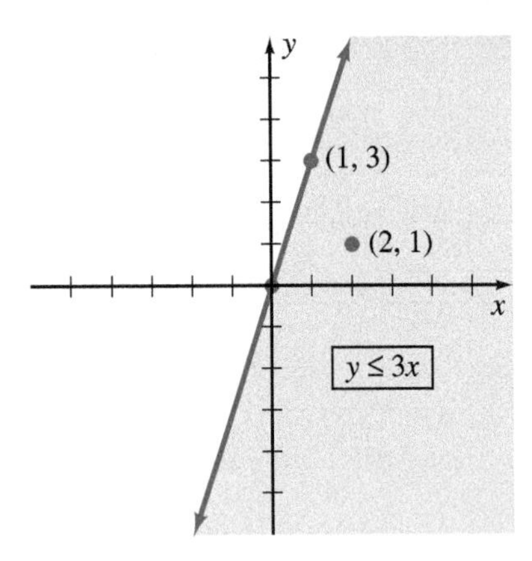


**Figure 7.21**

## Concept Quiz 7.2

For Problems 1–10, answer true or false.

1. The ordered pair $(2, -3)$ satisfies the inequality $2x + y > 1$.
2. A dashed line on the graph indicates that the points on the line do not satisfy the inequality.
3. Any point can be used as a test point to determine the half-plane that is the solution of the inequality.
4. The ordered pair $(3, -2)$ satisfies the inequality $5x - 2y \ge 19$.
5. The ordered pair $(1, -3)$ satisfies the inequality $-2x - 3y < 4$.
6. The graph of $x > 0$ is the half-plane above the $x$ axis.
7. The graph of $y < 0$ is the half-plane below the $x$ axis.
8. The graph of $-x + y > 4$ is the half-plane above the line $-x + y = 4$.
9. The origin can serve as a test point to determine the half-plane that satisfies the inequality $3y > 2x$.
10. The ordered pair $(-2, -1)$ can be used as a test point to determine the half-plane that satisfies the inequality $y < -3x - 7$.

## Problem Set 7.2

For Problems 1–30, graph each of the inequalities. (Objective 1)

1. $x - y > 2$
2. $x + y > 4$
3. $x + 3y < 3$
4. $2x - y > 6$
5. $2x + 5y \ge 10$
6. $3x + 2y \le 4$
7. $y > 2x + 1$
8. $y < -2x + 3$
9. $y \le -3x - 2$
10. $y \ge 3x - 4$
11. $2x - y < -2$
12. $x + 3y \le -6$
13. $y \le -x + 2$
14. $y \ge -2x - 1$
15. $y > -x$
16. $y < x$
17. $2x - y \ge 0$
18. $x + 2y \ge 0$
19. $-x + 4y - 4 \le 0$
20. $-2x + y - 3 \le 0$
21. $y > -\frac{3}{2}x - 3$
22. $2x + 5y > -4$
23. $y < -\frac{1}{2}x + 2$
24. $y < -\frac{1}{3}x + 1$
25. $x \le 3$
26. $y \ge -2$
27. $x > 1$ and $y < 3$
28. $x > -2$ and $y > -1$
29. $x \le -1$ and $y < 1$
30. $x < 2$ and $y \ge -2$

## Thoughts Into Words

**31.** Why is the point $(-4, 1)$ not a good test point to use when graphing $5x - 2y > -22$?

**32.** Explain how you would graph the inequality

$-3 > x - 3y$.

## Further Investigations

**33.** Graph $|x| < 2$. [*Hint*: Remember that $|x| < 2$ is equivalent to $-2 < x < 2$.]

**34.** Graph $|y| > 1$.

**35.** Graph $|x + y| < 1$.

**36.** Graph $|x - y| > 2$.

## Graphing Calculator Activities

**37.** This is a good time for you to become acquainted with the DRAW features of your graphing calculator. Again, you may need to consult your user's manual for specific key-punching instructions. Return to Examples 1, 2, and 3 of this section, and use your graphing calculator to graph the inequalities.

**38.** Use a graphing calculator to check your graphs for Problems 1–30.

**39.** Use the DRAW feature of your graphing calculator to draw each of the following.

**(a)** A line segment between $(-2, -4)$ and $(-2, 5)$

**(b)** A line segment between $(2, 2)$ and $(5, 2)$

**(c)** A line segment between $(2, 3)$ and $(5, 7)$

**(d)** A triangle with vertices at $(1, -2)$, $(3, 4)$, and $(-3, 6)$

### Answers to the Concept Quiz

**1.** False **2.** True **3.** False **4.** True **5.** False **6.** False **7.** True **8.** True **9.** False **10.** False

# 7.3 Distance and Slope

OBJECTIVES

1. Find the distance between two points
2. Find the slope of a line
3. Use slope to graph lines
4. Apply slope to solve problems

As we work with the rectangular coordinate system, it is sometimes necessary to express the length of certain line segments. In other words, we need to be able to find the distance between two points. Let's first consider two specific examples and then develop the general distance formula.

**Classroom Example**
Find the distance between the points $A(-3, 1)$ and $B(-3, 7)$ and also between the points $C(2, 5)$ and $D(-1, 5)$.

### EXAMPLE 1

Find the distance between the points $A(2, 2)$ and $B(5, 2)$ and also between the points $C(-2, 5)$ and $D(-2, -4)$.

### Solution

Let's plot the points and draw $\overline{AB}$ as in Figure 7.22. Because $\overline{AB}$ is parallel to the $x$ axis, its length can be expressed as $|5 - 2|$ or $|2 - 5|$. (The absolute-value symbol is used to ensure a nonnegative value.) Thus the length of $\overline{AB}$ is 3 units. Likewise, the length of $\overline{CD}$ is $|5 - (-4)| = |-4 - 5| = 9$ units.

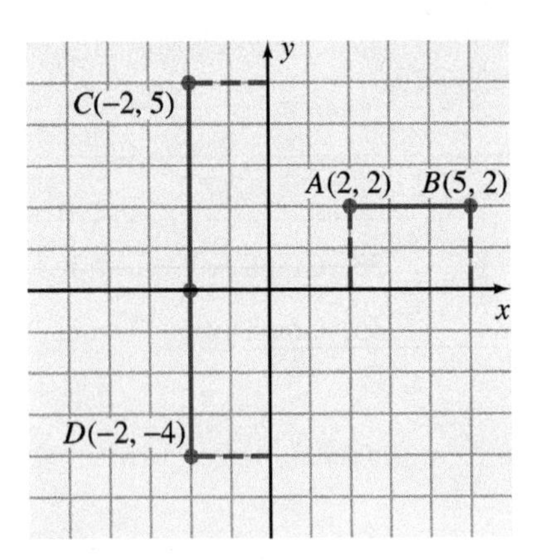


**Figure 7.22**

Classroom Example
Find the distance between the points $A(-2, 2)$ and $B(6, -4)$.

**EXAMPLE 2** Find the distance between the points $A(2, 3)$ and $B(5, 7)$.

### Solution

Let's plot the points and form a right triangle as indicated in Figure 7.23. Note that the coordinates of point $C$ are $(5, 3)$. Because $\overline{AC}$ is parallel to the horizontal axis, its length is easily determined to be 3 units. Likewise, $\overline{CB}$ is parallel to the vertical axis, and its length is 4 units. Let $d$ represent the length of $\overline{AB}$, and apply the Pythagorean theorem to obtain

$$d^2 = 3^2 + 4^2$$
$$d^2 = 9 + 16$$
$$d^2 = 25$$
$$d = \pm\sqrt{25} = \pm 5$$

"Distance between" is a nonnegative value, so the length of $\overline{AB}$ is 5 units.

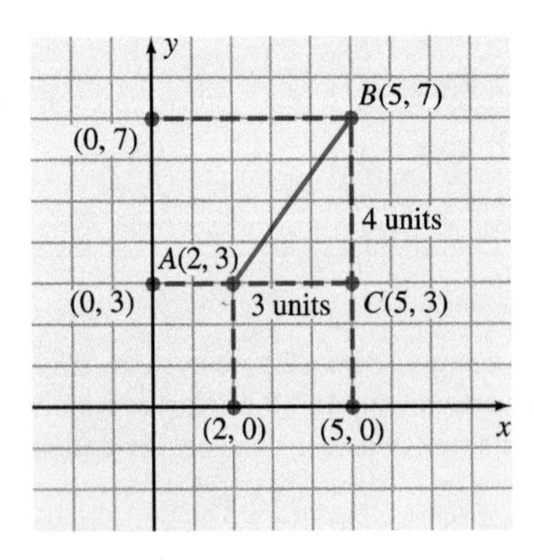


**Figure 7.23**

The approach we used in Example 2 becomes the basis for a general distance formula for finding the distance between any two points in a coordinate plane:

1. Let $P_1(x_1, y_1)$ and $P_2(x_2, y_2)$ represent any two points in a coordinate plane.
2. Form a right triangle as indicated in Figure 7.24. The coordinates of the vertex of the right angle, point $R$, are $(x_2, y_1)$.

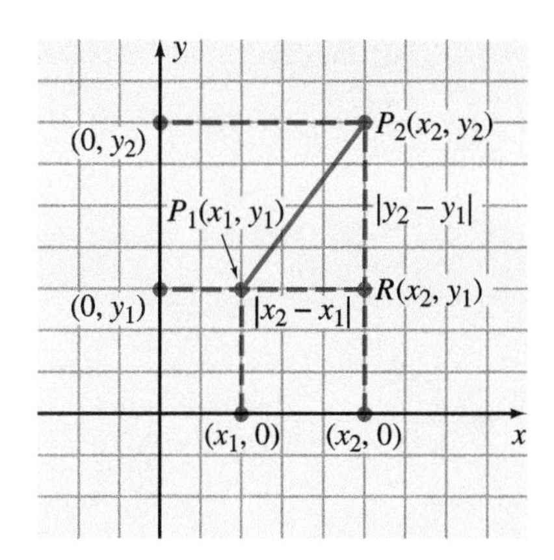


**Figure 7.24**

The length of $\overline{P_1R}$ is $|x_2 - x_1|$, and the length of $\overline{RP_2}$ is $|y_2 - y_1|$. (The absolute-value symbol is used to ensure a nonnegative value.) Let $d$ represent the length of $\overline{P_1P_2}$ and apply the Pythagorean theorem to obtain

$$d^2 = |x_2 - x_1|^2 + |y_2 - y_1|^2$$

Because $|a|^2 = a^2$, the equation can be written as $d^2 = (x_2 - x_1)^2 + (y_2 - y_1)^2$. Applying the square root property would yield $d = \pm\sqrt{(x_2 - x_1)^2 + (y_2 - y_1)^2}$. However, in this application, the distance cannot be negative, so the **distance formula** can be stated as

$$d = \sqrt{(x_2 - x_1)^2 + (y_2 - y_1)^2}$$

It makes no difference which point you call $P_1$ or $P_2$ when using the distance formula. If you forget the formula, don't panic. Just form a right triangle and apply the Pythagorean theorem as we did in Example 2. Let's consider an example that demonstrates the use of the distance formula.

**Classroom Example**
Find the distance between $(-1, 3)$ and $(-6, 8)$.

**EXAMPLE 3** Find the distance between $(-1, 4)$ and $(1, 2)$.

## Solution

Let $(-1, 4)$ be $P_1$ and $(1, 2)$ be $P_2$. Using the distance formula, we obtain

$$\begin{aligned} d &= \sqrt{[1 - (-1)]^2 + (2 - 4)^2} \\ &= \sqrt{2^2 + (-2)^2} \\ &= \sqrt{4 + 4} \\ &= \sqrt{8} = 2\sqrt{2} \end{aligned}$$ Express the answer in simplest radical form

The distance between the two points is $2\sqrt{2}$ units.

In Example 3, we did not sketch a figure because of the simplicity of the problem. However, sometimes it is helpful to use a figure to organize the given information and aid in the analysis of the problem, as we see in the next example.

**Classroom Example**
Verify that the points $(2, -1)$, $(6, 5)$, and $(-4, 3)$ are vertices of an isosceles triangle.

**EXAMPLE 4**

Verify that the points $(-3, 6)$, $(3, 4)$, and $(1, -2)$ are vertices of an isosceles triangle. (An isosceles triangle has two sides of the same length.)

### Solution

Let's plot the points and draw the triangle (Figure 7.25). Use the distance formula to find the lengths $d_1$, $d_2$, and $d_3$, as follows:

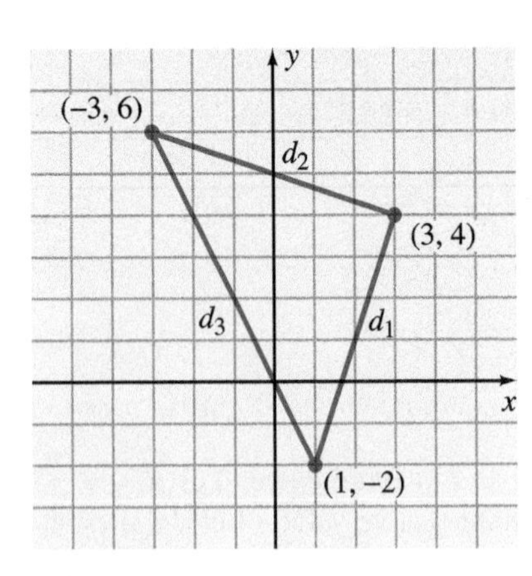


$$d_1 = \sqrt{(3 - 1)^2 + (4 - (-2))^2}$$
$$= \sqrt{2^2 + 6^2} = \sqrt{40} = 2\sqrt{10}$$

$$d_2 = \sqrt{(-3 - 3)^2 + (6 - 4)^2}$$
$$= \sqrt{(-6)^2 + 2^2} = \sqrt{40} = 2\sqrt{10}$$

$$d_3 = \sqrt{(-3 - 1)^2 + (6 - (-2))^2}$$
$$= \sqrt{(-4)^2 + 8^2} = \sqrt{80} = 4\sqrt{5}$$

**Figure 7.25**

Because $d_1 = d_2$, we know that it is an isosceles triangle.

## Finding the Slope of a Line

In coordinate geometry, the concept of *slope* is used to describe the "steepness" of lines. The **slope** of a line is the ratio of the vertical change to the horizontal change as we move from one point on a line to another point. This is illustrated in Figure 7.26 with points $P_1$ and $P_2$.

A precise definition for slope can be given by considering the coordinates of the points $P_1$, $P_2$, and $R$ as indicated in Figure 7.27. The horizontal change as we move from $P_1$ to $P_2$ is $x_2 - x_1$, and the vertical change is $y_2 - y_1$. Thus the following definition for slope is given.

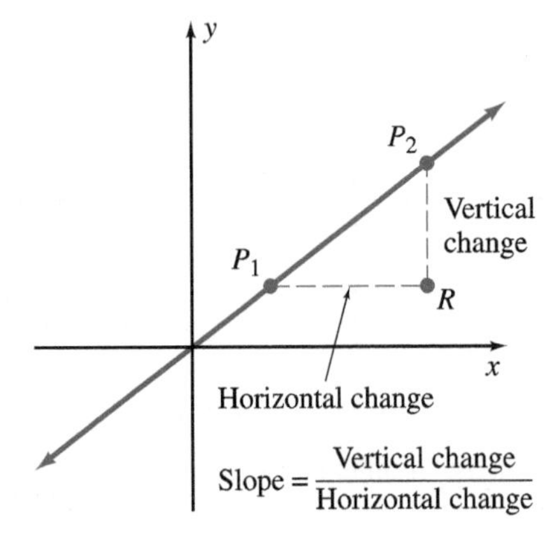


**Figure 7.26**

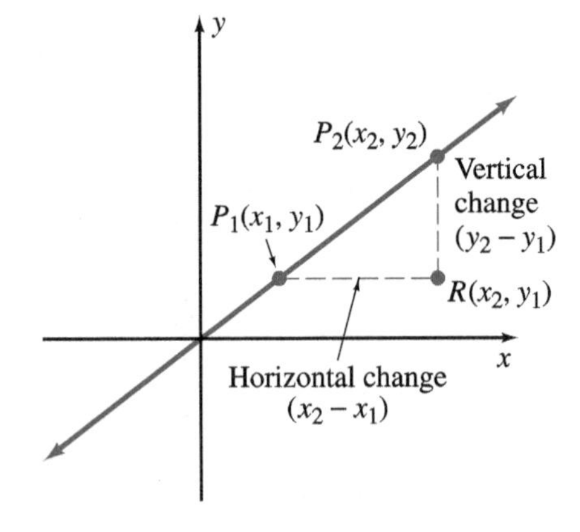


**Figure 7.27**

### Definition 7.1 Slope of a Line

If points $P_1$ and $P_2$ with coordinates $(x_1, y_1)$ and $(x_2, y_2)$, respectively, are any two different points on a line, then the slope of the line (denoted by $m$) is

$$m = \frac{y_2 - y_1}{x_2 - x_1}, \qquad x_2 \neq x_1$$

**Classroom Example**
Find the slope of the line determined by each of the following pairs of points, and graph the lines:
**(a)** $(-3, -4)$ and $(2, 3)$
**(b)** $(-2, 4)$ and $(3, -5)$
**(c)** $(4, 2)$ and $(-3, 2)$

Because $\dfrac{y_2 - y_1}{x_2 - x_1} = \dfrac{y_1 - y_2}{x_1 - x_2}$, how we designate $P_1$ and $P_2$ is not important. Let's use Definition 7.1 to find the slopes of some lines.

## EXAMPLE 5

Find the slope of the line determined by each of the following pairs of points, and graph the lines:

**(a)** $(-1, 1)$ and $(3, 2)$ **(b)** $(4, -2)$ and $(-1, 5)$

**(c)** $(2, -3)$ and $(-3, -3)$

### Solution

**(a)** Let $(-1, 1)$ be $P_1$ and $(3, 2)$ be $P_2$ (Figure 7.28).

$$m = \frac{y_2 - y_1}{x_2 - x_1} = \frac{2 - 1}{3 - (-1)} = \frac{1}{4}$$

**(b)** Let $(4, -2)$ be $P_1$ and $(-1, 5)$ be $P_2$ (Figure 7.29).

$$m = \frac{y_2 - y_1}{x_2 - x_1} = \frac{5 - (-2)}{-1 - 4} = \frac{7}{-5} = -\frac{7}{5}$$

**(c)** Let $(2, -3)$ be $P_1$ and $(-3, -3)$ be $P_2$ (Figure 7.30).

$$m = \frac{y_2 - y_1}{x_2 - x_1} = \frac{-3 - (-3)}{-3 - 2} = \frac{0}{-5} = 0$$

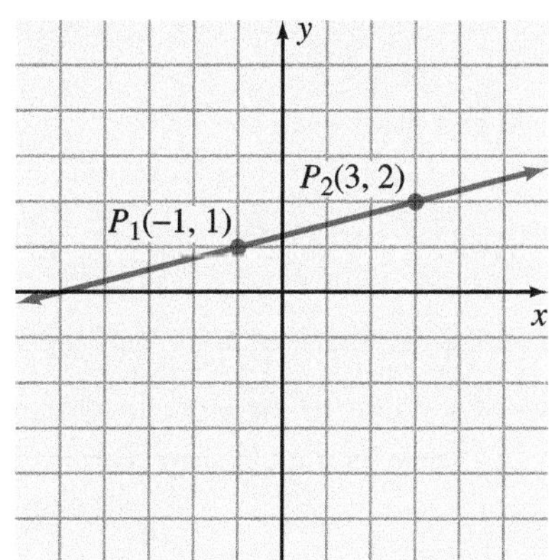


Figure 7.28

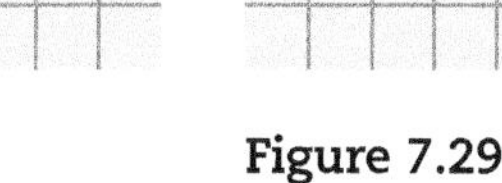

Figure 7.29

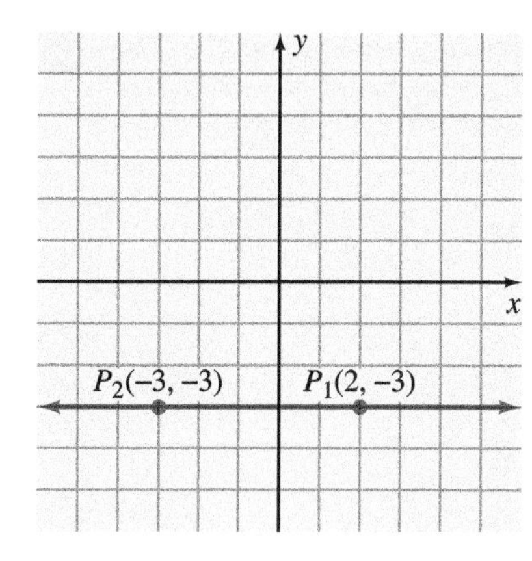


Figure 7.30

The three parts of Example 5 represent the three basic possibilities for slope; that is, the slope of a line can be positive, negative, or zero. A line that has a positive slope rises as we move from left to right, as in Figure 7.28. A line that has a negative slope falls as we move from left to right, as in Figure 7.29. A horizontal line, as in Figure 7.30, has a slope of zero. Finally, we need to realize that *the concept of slope is undefined for vertical lines*. This is because for any vertical line, the horizontal change as we move from one point on the line to another is zero. Thus the ratio $\dfrac{y_2 - y_1}{x_2 - x_1}$ will have a denominator of zero and be undefined. Accordingly, the restriction $x_2 \neq x_1$ is imposed in Definition 7.1.

## Using Slope to Graph Lines

**Classroom Example**
Graph the line that passes through the point (0, 3) and has a slope of $m = -\frac{2}{5}$.

### EXAMPLE 6

Graph the line that passes through the point $(0, -2)$ and has a slope of $\frac{1}{3}$.

### Solution

To graph, plot the point $(0, -2)$. Because the slope $= \frac{\text{vertical change}}{\text{horizontal change}} = \frac{1}{3}$, we can locate another point on the line by starting from the point $(0, -2)$ and moving 1 unit up and 3 units to the right to obtain the point $(3, -1)$. Because two points determine a line, we can draw the line (Figure 7.31).

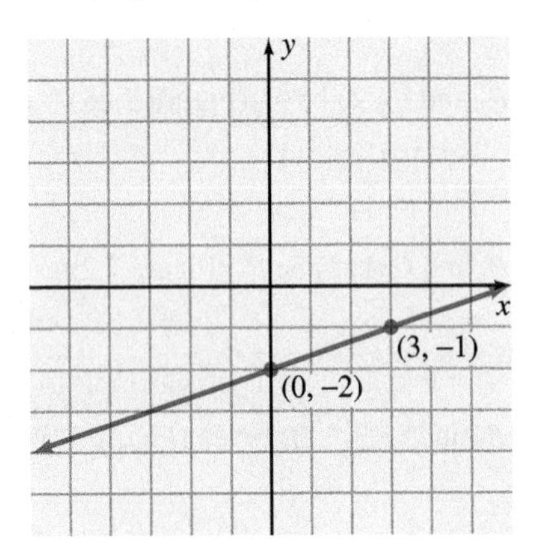


**Figure 7.31**

**Remark:** Because $m = \frac{1}{3} = \frac{-1}{-3}$, we can locate another point by moving 1 unit down and 3 units to the left from the point $(0, -2)$.

**Classroom Example**
Graph the line that passes through the point $(-1, 2)$ and has a slope of $-3$.

### EXAMPLE 7

Graph the line that passes through the point $(1, 3)$ and has a slope of $-2$.

### Solution

To graph the line, plot the point $(1, 3)$. We know that $m = -2 = \frac{-2}{1}$. Furthermore, because the slope $= \frac{\text{vertical change}}{\text{horizontal change}} = \frac{-2}{1}$, we can locate another point on the line by starting from the point $(1, 3)$ and moving 2 units down and 1 unit to the right to obtain the point $(2, 1)$. Because two points determine a line, we can draw the line (Figure 7.32).

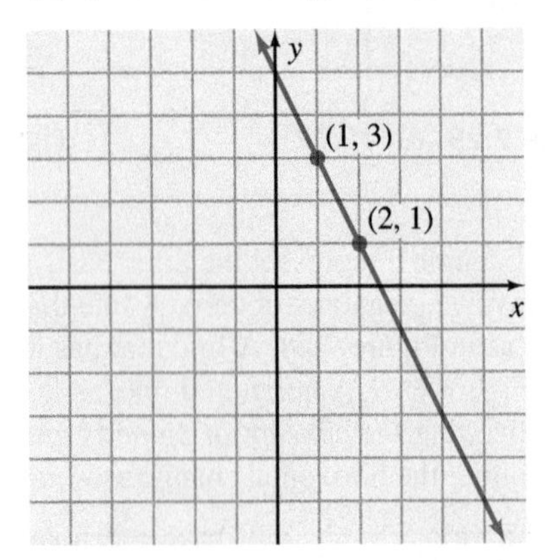


**Figure 7.32**

**Remark:** Because $m = -2 = \frac{-2}{1} = \frac{2}{-1}$, we can locate another point by moving 2 units up and 1 unit to the left from the point (1, 3).

## Applying Slope to Solve Problems

The concept of slope has many real-world applications even though the word *slope* is often not used. The concept of slope is used in most situations that involve an incline. Hospital beds are hinged in the middle so that both the head end and the foot end can be raised or lowered; that is, the slope of either end of the bed can be changed. Likewise, treadmills are designed so that the incline (slope) of the platform can be adjusted. A roofer, when making an estimate to replace a roof, is concerned not only about the total area to be covered but also about the pitch of the roof. (Contractors do not define *pitch* according to the mathematical definition of slope, but both concepts refer to "steepness.") In Figure 7.33, the two roofs might require the same amount of shingles, but the roof on the left will take longer to complete because the pitch is so great that scaffolding will be required.

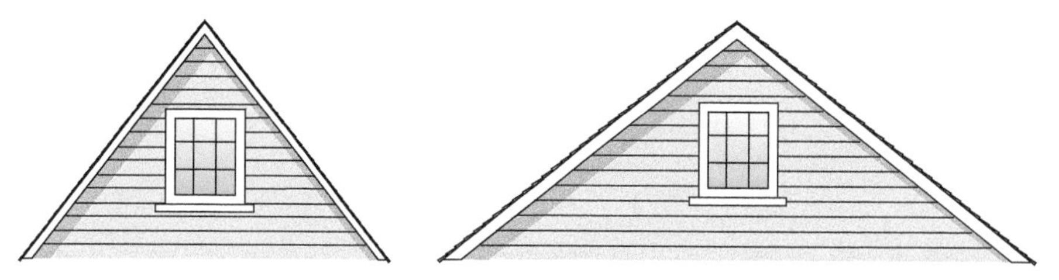

**Figure 7.33**

The concept of slope is also used in the construction of flights of stairs (Figure 7.34). The terms *rise* and *run* are commonly used, and the steepness (slope) of the stairs can be expressed as the ratio of rise to run. In Figure 7.34, the stairs on the left, which have a ratio of rise to run of $\frac{10}{11}$, are steeper than the stairs on the right, which have a ratio of $\frac{7}{11}$.

In highway construction, the word *grade* is used for the concept of slope. For example, in Figure 7.35, the highway is said to have a grade of 17%. This means that for every horizontal distance of 100 feet, the highway rises or drops 17 feet. In other words, the slope of the highway is $\frac{17}{100}$.

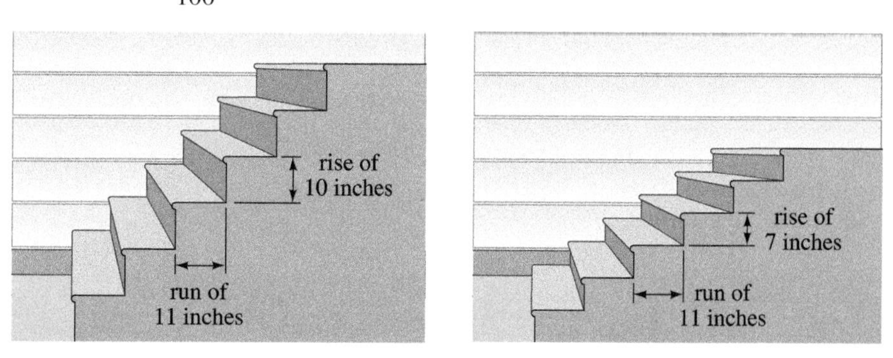


**Figure 7.34**

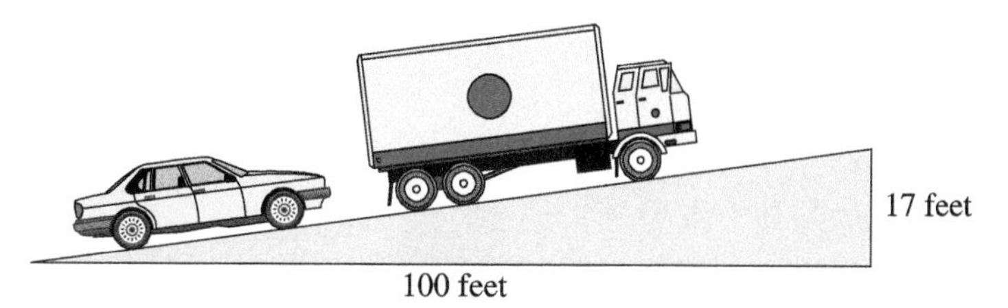


**Figure 7.35**

**Classroom Example**
A certain highway has a 4% grade. How many feet does it rise in a horizontal distance of 2 miles?

**EXAMPLE 8** Apply Your Skill

A certain highway has a 3% grade. How many feet does it rise in a horizontal distance of 1 mile?

### Solution

A 3% grade means a slope of $\frac{3}{100}$. Therefore, if we let $y$ represent the unknown vertical distance, and use the fact that 1 mile = 5280 feet, we can set up and solve the following proportion.

$$\frac{3}{100} = \frac{y}{5280}$$
$$100y = 3(5280) = 15{,}840$$
$$y = 158.4$$

The highway rises 158.4 feet in a horizontal distance of 1 mile.

## Concept Quiz 7.3

For Problems 1–10, answer true or false.

1. When applying the distance formula $d = \sqrt{(x_2 - x_1)^2 + (y_2 - y_1)^2}$ to find the distance between two points, you can designate either of the two points as $P_1$.
2. An isosceles triangle has two sides of the same length.
3. The distance between the points $(-1, 4)$ and $(-1, -2)$ is 2 units.
4. The distance between the points $(3, -4)$ and $(3, 2)$ is undefined.
5. The slope of a line is the ratio of the vertical change to the horizontal change when moving from one point on the line to another point on the line.
6. The slope of a line is always positive.
7. A slope of 0 means that there is no change in the vertical direction when moving from one point on the line to another point on the line.
8. The concept of slope is undefined for horizontal lines.
9. When applying the slope formula $m = \frac{y_2 - y_1}{x_2 - x_1}$ to find the slope of a line between two points, you can designate either of the two points as $P_2$.
10. If the ratio of the rise to the run for some steps is $\frac{3}{4}$ and the rise is 9 inches, then the run is $6\frac{3}{4}$ inches.

## Problem Set 7.3

For Problems 1–12, find the distance between each pair of points. Express answers in simplest radical form. (Objective 1)

1. $(-2, -1), (7, 11)$
2. $(2, 1), (10, 7)$
3. $(1, -1), (3, -4)$
4. $(-1, 3), (2, -2)$
5. $(6, -4), (9, -7)$
6. $(-5, 2), (-1, 6)$
7. $(-3, 3), (0, -3)$
8. $(-2, -4), (4, 0)$
9. $(1, -6), (-5, -6)$
10. $(-2, 3), (-2, -7)$
11. $(1, 7), (4, -2)$
12. $(6, 4), (-4, -8)$
13. Verify that the points $(-3, 1)$, $(5, 7)$, and $(8, 3)$ are vertices of a right triangle. [*Hint*: If $a^2 + b^2 = c^2$, then it is a right triangle with the right angle opposite side $c$.]
14. Verify that the points $(0, 3)$, $(2, -3)$, and $(-4, -5)$ are vertices of an isosceles triangle.
15. Verify that the points $(7, 12)$ and $(11, 18)$ divide the line segment joining $(3, 6)$ and $(15, 24)$ into three segments of equal length.

**16.** Verify that (3, 1) is the midpoint of the line segment joining $(-2, 6)$ and $(8, -4)$.

For Problems 17–28, graph the line determined by the two points, and find the slope of the line. (Objective 2)

**17.** (1, 2), (4, 6)

**18.** $(3, 1), (-2, -2)$

**19.** $(-4, 5), (-1, -2)$

**20.** $(-2, 5), (3, -1)$

**21.** $(2, 6), (6, -2)$

**22.** $(-2, -1), (2, -5)$

**23.** $(-6, 1), (-1, 4)$

**24.** $(-3, 3), (2, 3)$

**25.** $(-2, -4), (2, -4)$

**26.** $(1, -5), (4, -1)$

**27.** $(0, -2), (4, 0)$

**28.** $(-4, 0), (0, -6)$

**29.** Find $x$ if the line through $(-2, 4)$ and $(x, 6)$ has a slope of $\frac{2}{9}$.

**30.** Find $y$ if the line through $(1, y)$ and $(4, 2)$ has a slope of $\frac{5}{3}$.

**31.** Find $x$ if the line through $(x, 4)$ and $(2, -5)$ has a slope of $-\frac{9}{4}$.

**32.** Find $y$ if the line through $(5, 2)$ and $(-3, y)$ has a slope of $-\frac{7}{8}$.

For Problems 33–40, you are given one point on a line and the slope of the line. Find the coordinates of three other points on the line. (Objective 3)

**33.** $(2, 5)$, $m = \frac{1}{2}$

**34.** $(3, 4)$, $m = \frac{5}{6}$

**35.** $(-3, 4)$, $m = 3$

**36.** $(-3, -6)$, $m = 1$

**37.** $(5, -2)$, $m = -\frac{2}{3}$

**38.** $(4, -1)$, $m = -\frac{3}{4}$

**39.** $(-2, -4)$, $m = -2$

**40.** $(-5, 3)$, $m = -3$

For Problems 41–50, find the coordinates of two points on the given line, and then use those coordinates to find the slope of the line. (Objective 2)

**41.** $2x + 3y = 6$

**42.** $4x + 5y = 20$

**43.** $x - 2y = 4$

**44.** $3x - y = 12$

**45.** $4x - 7y = 12$

**46.** $2x + 7y = 11$

**47.** $y = 4$

**48.** $x = 3$

**49.** $y = -5x$

**50.** $y - 6x = 0$

For Problems 51–58, graph the line that passes through the given point and has the given slope. (Objective 3)

**51.** $(3, 1)$ $m = \frac{2}{3}$

**52.** $(-1, 0)$ $m = \frac{3}{4}$

**53.** $(-2, 3)$ $m = -1$

**54.** $(1, -4)$ $m = -3$

**55.** $(0, 5)$ $m = \frac{-1}{4}$

**56.** $(-3, 4)$ $m = \frac{-3}{2}$

**57.** $(2, -2)$ $m = \frac{3}{2}$

**58.** $(3, -4)$ $m = \frac{5}{2}$

For Problems 59–64, use the concept of slope to solve the problems. (Objective 4)

**59.** A certain highway has a 2% grade. How many feet does it rise in a horizontal distance of 1 mile? (1 mile = 5280 feet.)

**60.** The grade of a highway up a hill is 30%. How much change in horizontal distance is there if the vertical height of the hill is 75 feet?

**61.** Suppose that a highway rises a distance of 215 feet in a horizontal distance of 2640 feet. Express the grade of the highway to the nearest tenth of a percent.

**62.** If the ratio of rise to run is to be $\frac{3}{5}$ for some steps and the rise is 19 centimeters, find the run to the nearest centimeter.

**63.** If the ratio of rise to run is to be $\frac{2}{3}$ for some steps, and the run is 28 centimeters, find the rise to the nearest centimeter.

**64.** Suppose that a county ordinance requires a $2\frac{1}{4}\%$ "fall" for a sewage pipe from the house to the main pipe at the street. How much vertical drop must there be for a horizontal distance of 45 feet? Express the answer to the nearest tenth of a foot.

## Thoughts Into Words

**65.** How would you explain the concept of slope to someone who was absent from class the day it was discussed?

**66.** If one line has a slope of $\frac{2}{5}$, and another line has a slope of $\frac{3}{7}$, which line is steeper? Explain your answer.

**67.** Suppose that a line has a slope of $\frac{2}{3}$ and contains the point (4, 7). Are the points (7, 9) and (1, 3) also on the line? Explain your answer.

## Further Investigations

**68.** Sometimes it is necessary to find the coordinate of a point on a number line that is located somewhere between two given points. For example, suppose that we want to find the coordinate ($x$) of the point located two-thirds of the distance from 2 to 8. Because the total distance from 2 to 8 is $8 - 2 = 6$ units, we can start at 2 and move $\frac{2}{3}(6) = 4$ units toward 8. Thus $x = 2 + \frac{2}{3}(6) = 2 + 4 = 6$.

For each of the following, find the coordinate of the indicated point on a number line.

**(a)** Two-thirds of the distance from 1 to 10
**(b)** Three-fourths of the distance from $-2$ to 14
**(c)** One-third of the distance from $-3$ to 7
**(d)** Two-fifths of the distance from $-5$ to 6
**(e)** Three-fifths of the distance from $-1$ to $-11$
**(f)** Five-sixths of the distance from 3 to $-7$

**69.** Now suppose that we want to find the coordinates of point $P$, which is located two-thirds of the distance from $A(1, 2)$ to $B(7, 5)$ in a coordinate plane. We have plotted the given points $A$ and $B$ in Figure 7.36 to help with the analysis of this problem. Point $D$ is two-thirds of the distance from $A$ to $C$ because parallel lines cut off proportional segments on every transversal that intersects the lines.

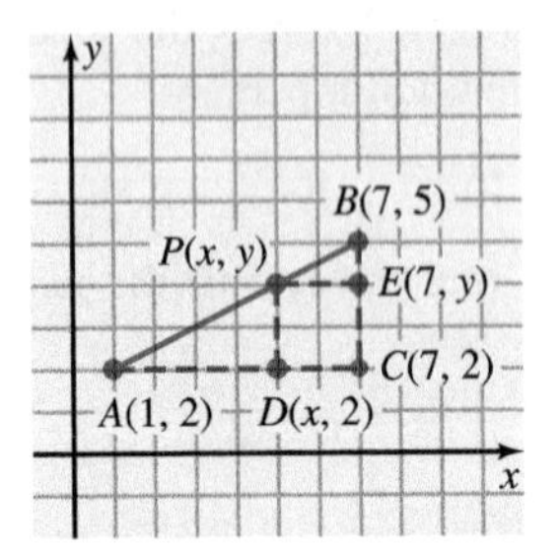


**Figure 7.36**

Thus $\overline{AC}$ can be treated as a segment of a number line, as shown in Figure 7.37. Therefore,

$$x = 1 + \frac{2}{3}(7 - 1) = 1 + \frac{2}{3}(6) = 5$$

**Figure 7.37**

Similarly, $\overline{CB}$ can be treated as a segment of a number line, as shown in Figure 7.38. Therefore,

B 5
E y
C 2

$$y = 2 + \frac{2}{3}(5 - 2) = 2 + \frac{2}{3}(3) = 4$$

The coordinates of point $P$ are (5, 4).

**Figure 7.38**

For each of the following, find the coordinates of the indicated point in the $xy$ plane.

**(a)** One-third of the distance from (2, 3) to (5, 9)
**(b)** Two-thirds of the distance from (1, 4) to (7, 13)
**(c)** Two-fifths of the distance from $(-2, 1)$ to (8, 11)
**(d)** Three-fifths of the distance from $(2, -3)$ to $(-3, 8)$
**(e)** Five-eighths of the distance from $(-1, -2)$ to $(4, -10)$
**(f)** Seven-eighths of the distance from $(-2, 3)$ to $(-1, -9)$

**70.** Suppose we want to find the coordinates of the midpoint of a line segment. Let $P(x, y)$ represent the midpoint of the line segment from $A(x_1, y_1)$ to $B(x_2, y_2)$. Using the method from Problem 68, the formula for the $x$ coordinate of the midpoint is $x = x_1 + \frac{1}{2}(x_2 - x_1)$. This formula can be simplified algebraically to produce a simpler formula.

$$x = x_1 + \frac{1}{2}(x_2 - x_1)$$

$$x = x_1 + \frac{1}{2}x_2 - \frac{1}{2}x_1$$

$$x = \frac{1}{2}x_1 + \frac{1}{2}x_2$$

$$x = \frac{x_1 + x_2}{2}$$

Hence the $x$ coordinate of the midpoint can be interpreted as the average of the $x$ coordinates of the endpoints of the line segment. A similar argument for the $y$ coordinate of the midpoint gives the following formula.

$$y = \frac{y_1 + y_2}{2}$$

For each of the pairs of points, use the formula to find the midpoint of the line segment between the points.

**(a)** (3, 1) and (7, 5)
**(b)** $(-2, 8)$ and (6, 4)
**(c)** $(-3, 2)$ and (5, 8)
**(d)** (4, 10) and (9, 25)
**(e)** $(-4, -1)$ and $(-10, 5)$
**(f)** (5, 8) and $(-1, 7)$

## Graphing Calculator Activities

**71.** Remember that we did some work with parallel lines back in the graphing calculator activities in Problem Set 7.1. Now let's do some work with perpendicular lines. Be sure to set your boundaries so that the distance between tic marks is the same on both axes, or perhaps the calculator has a Zoom Square function.

**(a)** Graph $y = 4x$ and $y = -\frac{1}{4}x$ on the same set of axes. Do they appear to be perpendicular lines?

**(b)** Graph $y = 3x$ and $y = \frac{1}{3}x$ on the same set of axes. Do they appear to be perpendicular lines?

**(c)** Graph $y = \frac{2}{5}x - 1$ and $y = -\frac{5}{2}x + 2$ on the same set of axes. Do they appear to be perpendicular lines?

**(d)** Graph $y = \frac{3}{4}x - 3$, $y = \frac{4}{3}x + 2$, and $y = -\frac{4}{3}x + 2$ on the same set of axes. Does there appear to be a pair of perpendicular lines?

**(e)** On the basis of your results in parts (a) through (d), make a statement about how we can recognize perpendicular lines from their equations.

**72.** For each of the following pairs of equations, (1) predict whether they represent parallel lines, perpendicular lines, or lines that intersect but are not perpendicular, and (2) graph each pair of lines to check your prediction.

**(a)** $5.2x + 3.3y = 9.4$ and $5.2x + 3.3y = 12.6$
**(b)** $1.3x - 4.7y = 3.4$ and $1.3x - 4.7y = 11.6$
**(c)** $2.7x + 3.9y = 1.4$ and $2.7x - 3.9y = 8.2$
**(d)** $5x - 7y = 17$ and $7x + 5y = 19$
**(e)** $9x + 2y = 14$ and $2x + 9y = 17$
**(f)** $2.1x + 3.4y = 11.7$ and $3.4x - 2.1y = 17.3$

### Answers to the Concept Quiz

**1.** True **2.** True **3.** False **4.** False **5.** True **6.** False **7.** True **8.** False **9.** True **10.** False

## 7.4 Determining the Equation of a Line

OBJECTIVES

1. Find the equation of a line given a point and a slope
2. Find the equation of a line given two points
3. Find the equation of a line given the slope and y intercept
4. Use the point-slope form to write equations of lines
5. Apply the slope-intercept form of an equation
6. Find equations for parallel or perpendicular lines

To review, there are basically two types of problems to solve in coordinate geometry:

1. Given an algebraic equation, find its geometric graph.
2. Given a set of conditions pertaining to a geometric figure, find its algebraic equation.

Problems of type 1 have been our primary concern thus far in this chapter. Now let's analyze some problems of type 2 that deal specifically with straight lines. Given certain facts about a line, we need to be able to determine its algebraic equation. Let's consider some examples.

**Classroom Example**
Find the equation of the line that has a slope of $m = \frac{1}{4}$ and contains the point (2, 5).

**EXAMPLE 1**

Find the equation of the line that has a slope of $\frac{2}{3}$ and contains the point (1, 2).

### Solution

First, let's draw the line and record the given information. Then choose a point $(x, y)$ that represents any point on the line other than the given point (1, 2). (See Figure 7.39.)

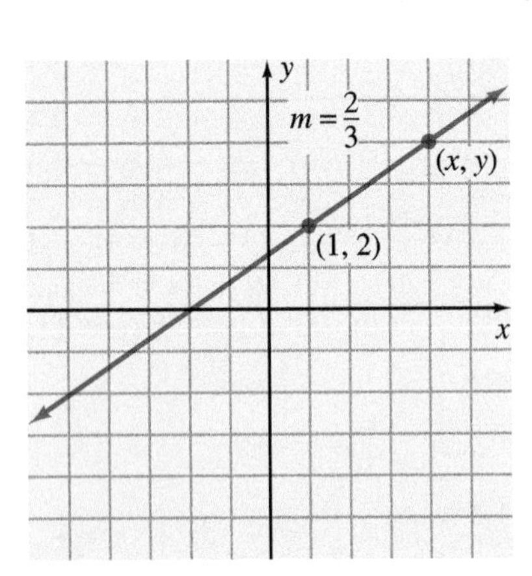


Figure 7.39

The slope determined by (1, 2) and $(x, y)$ is $\frac{2}{3}$. Thus

$$\frac{y-2}{x-1} = \frac{2}{3}$$

$$2(x-1) = 3(y-2) \quad \text{Cross multiply}$$

$$2x - 2 = 3y - 6$$

$$2x - 3y = -4$$

The equation of the line is $2x - 3y = -4$.

## Finding the Equation of a Line, Given Two Points

**Classroom Example**
Find the equation of the line that contains (−4, 3) and (2, −2).

**EXAMPLE 2** Find the equation of the line that contains (3, 2) and (−2, 5).

### Solution

First, let's draw the line determined by the given points (Figure 7.40); if we know two points, we can find the slope.

$$m = \frac{y_2 - y_1}{x_2 - x_1} = \frac{5-2}{-2-3} = \frac{3}{-5} = \frac{-3}{5}$$

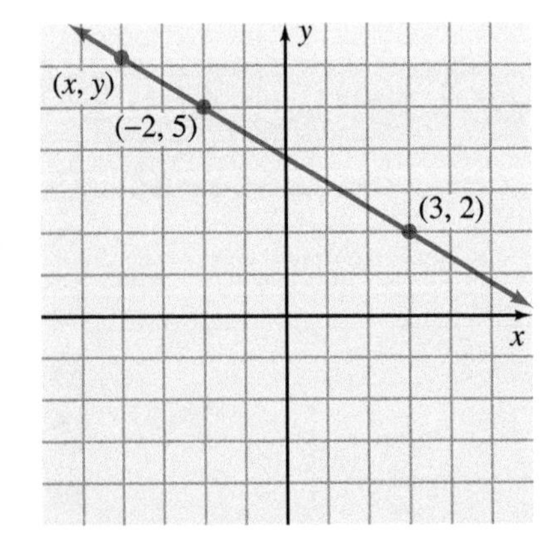


Figure 7.40

Now we can use the same approach as in Example 1. Form an equation using a variable point $(x, y)$, one of the two given points, and the slope of $-\frac{3}{5}$.

$$\frac{y-5}{x+2} = \frac{-3}{5}$$

$$-3(x+2) = 5(y-5) \quad \text{Cross multiply}$$

$$-3x - 6 = 5y - 25$$

$$-3x - 5y = -19$$

Typically we prefer the coefficient of the $x$ term to be positive. Multiply the equation on both sides by −1 to make the coefficient of the $x$ term positive.

$$-1(-3x - 5y) = -1(-19)$$

$$3x + 5y = 19$$

The equation of the line is $3x + 5y = 19$.

## Finding the Equation of a Line, Given the Slope and y Intercept

**Classroom Example**
Find the equation of the line that has a slope of $\frac{4}{5}$ and a $y$ intercept of $-1$.

**EXAMPLE 3**

Find the equation of the line that has a slope of $\frac{1}{4}$ and a $y$ intercept of 2.

### Solution

A $y$ intercept of 2 means that the point (0, 2) is on the line (Figure 7.41).

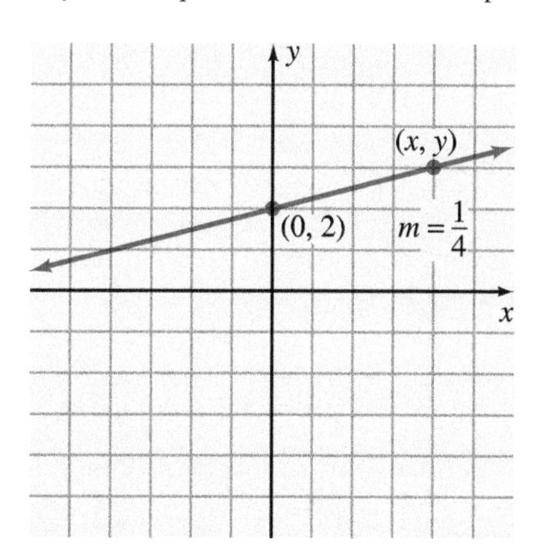


**Figure 7.41**

Choose a variable point $(x, y)$ and proceed as in the previous examples.

$$\frac{y-2}{x-0} = \frac{1}{4}$$

$$1(x-0) = 4(y-2) \quad \text{Cross multiply}$$

$$x = 4y - 8$$

$$x - 4y = -8$$

The equation of the line is $x - 4y = -8$.

Perhaps it would be helpful to pause a moment and look back over Examples 1, 2, and 3. Note that we used the same basic approach in all three situations. We chose a variable point $(x, y)$ and used it to determine the equation that satisfies the conditions given in the problem. The approach we took in the previous examples can be generalized to produce some special forms of equations of straight lines.

## Using the Point-Slope Form to Write Equations of Lines

Generalizing from the previous examples, let's find the equation of a line that has a slope of $m$ and contains the point $(x_1, y_1)$. To use the slope formula we will need two points. Choosing a point $(x, y)$ to represent any other point on the line (Figure 7.42) and using the given point $(x_1, y_1)$, we can determine the slope to be

$$m = \frac{y - y_1}{x - x_1} \quad \text{where } x \neq x_1$$

Simplifying gives us the equation $y - y_1 = m(x - x_1)$.

We refer to the equation

$$y - y_1 = m(x - x_1)$$

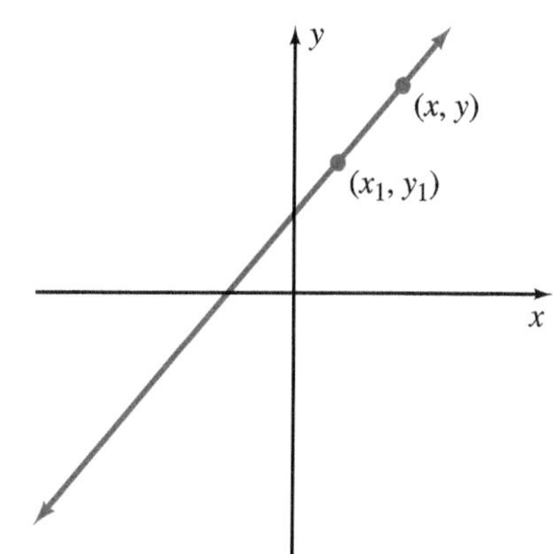


**Figure 7.42**

as the **point-slope form** of the equation of a straight line. Instead of the approach we used in Example 1, we could use the point-slope form to write the equation of a line with a given slope that contains a given point.

**Classroom Example**
Use the point-slope form to find the equation of a line that has a slope of $\frac{2}{3}$ and contains the point $(1, -6)$.

### EXAMPLE 4

Use the point-slope form to find the equation of a line that has a slope of $\frac{3}{5}$ and contains the point (2, 4).

#### Solution

We can determine the equation by substituting $\frac{3}{5}$ for $m$ and (2, 4) for $(x_1, y_1)$ in the point-slope form.

$$y - y_1 = m(x - x_1)$$
$$y - 4 = \frac{3}{5}(x - 2)$$
$$5(y - 4) = 3(x - 2) \quad \text{Multiply both sides of the equation by 5}$$
$$5y - 20 = 3x - 6$$
$$-14 = 3x - 5y$$

Thus the equation of the line is $3x - 5y = -14$.

## Applying the Slope-Intercept Form of an Equation

Another special form of the equation of a line is the slope-intercept form. Let's use the point-slope form to find the equation of a line that has a slope of $m$ and a $y$ intercept of $b$. A $y$ intercept of $b$ means that the line contains the point $(0, b)$, as in Figure 7.43. Therefore, we can use the point-slope form as follows:

$$y - y_1 = m(x - x_1)$$
$$y - b = m(x - 0)$$
$$y - b = mx$$
$$y = mx + b$$

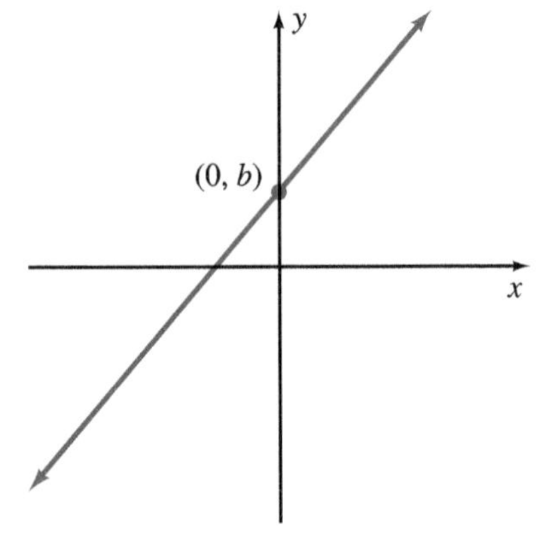

Figure 7.43

We refer to the equation

$$y = mx + b$$

as the **slope-intercept form** of the equation of a straight line. We use it for three primary purposes, as the next three examples illustrate.

**Classroom Example**
Find the equation of the line that has a slope of $m = 2$ and a $y$ intercept of $-3$.

### EXAMPLE 5

Find the equation of the line that has a slope of $\frac{1}{4}$ and a $y$ intercept of 2.

#### Solution

This is a restatement of Example 3, but this time we will use the slope-intercept form $(y = mx + b)$ of a line to write its equation. Because $m = \frac{1}{4}$ and $b = 2$, we can substitute these values into $y = mx + b$.

$$y = mx + b$$

$$y = \frac{1}{4}x + 2$$

$$4y = x + 8 \quad \text{Multiply both sides of the equation by 4}$$

$$x - 4y = -8 \quad \text{Same result as in Example 3}$$

**Classroom Example**
Find the slope of the line when the equation is $2x - 5y = 6$.

**EXAMPLE 6** Find the slope of the line when the equation is $3x + 2y = 6$.

### Solution

We can solve the equation for $y$ in terms of $x$ and then compare it to the slope-intercept form to determine its slope. Thus

$$3x + 2y = 6$$

$$2y = -3x + 6$$

$$y = -\frac{3}{2}x + 3$$

$$y = -\frac{3}{2}x + 3 \qquad y = mx + b$$

The slope of the line is $-\frac{3}{2}$. Furthermore, the $y$ intercept is 3.

**Classroom Example**
Graph the line determined by the equation $y = -\frac{1}{3}x + 2$.

**EXAMPLE 7** Graph the line determined by the equation $y = \frac{2}{3}x - 1$.

### Solution

Comparing the given equation to the general slope-intercept form, we see that the slope of the line is $\frac{2}{3}$, and the $y$ intercept is $-1$. Because the $y$ intercept is $-1$, we can plot the point $(0, -1)$. Because the slope is $\frac{2}{3}$, let's move 3 units to the right and 2 units up from $(0, -1)$ to locate the point $(3, 1)$. The two points $(0, -1)$ and $(3, 1)$ determine the line in Figure 7.44. (Consider picking a third point as a check point.)

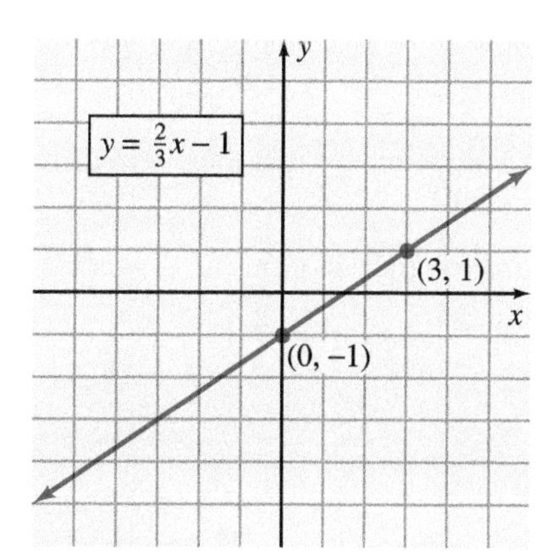


**Figure 7.44**

In general, if the equation of a nonvertical line is written in slope-intercept form ($y = mx + b$), the coefficient of $x$ is the slope of the line, and the constant term is the $y$ intercept. (Remember that the concept of slope is not defined for a vertical line.)

We use two forms of equations of straight lines extensively. They are the **standard form** and the **slope-intercept form**, and we describe them as follows.

**Standard Form** $Ax + By = C$, where $B$ and $C$ are integers, and $A$ is a nonnegative integer ($A$ and $B$ not both zero).

**Slope-Intercept Form** $y = mx + b$, where $m$ is a real number representing the slope, and $b$ is a real number representing the $y$ intercept.

## Finding Equations for Parallel and Perpendicular Lines

We can use two important relationships between lines and their slopes to solve certain kinds of problems. It can be shown that nonvertical parallel lines have the same slope and that two nonvertical lines are perpendicular if the product of their slopes is $-1$. (Details for verifying these facts are left to another course.) In other words, if two lines have slopes $m_1$ and $m_2$, respectively, then

1. The two lines are parallel if and only if $m_1 = m_2$.
2. The two lines are perpendicular if and only if $(m_1)(m_2) = -1$.

The following examples demonstrate the use of these properties.

Classroom Example
**(a)** Verify that the graphs of $4x - 2y = 10$ and $2x - y = 6$ are parallel lines.
**(b)** Verify that the graphs of $5x + 2y = 14$ and $4x - 10y = -3$ are perpendicular lines.

### EXAMPLE 8

**(a)** Verify that the graphs of $2x + 3y = 7$ and $4x + 6y = 11$ are parallel lines.

**(b)** Verify that the graphs of $8x - 12y = 3$ and $3x + 2y = 2$ are perpendicular lines.

### Solution

**(a)** Let's change each equation to slope-intercept form.

$$2x + 3y = 7 \quad \rightarrow \quad 3y = -2x + 7$$
$$y = -\frac{2}{3}x + \frac{7}{3}$$

$$4x + 6y = 11 \quad \rightarrow \quad 6y = -4x + 11$$
$$y = -\frac{4}{6}x + \frac{11}{6}$$
$$y = -\frac{2}{3}x + \frac{11}{6}$$

Both lines have a slope of $-\frac{2}{3}$, but they have different $y$ intercepts. Therefore, the two lines are parallel.

**(b)** Solving each equation for $y$ in terms of $x$, we obtain

$$8x - 12y = 3 \quad \rightarrow \quad -12y = -8x + 3$$
$$y = \frac{8}{12}x - \frac{3}{12}$$
$$y = \frac{2}{3}x - \frac{1}{4}$$

$$3x + 2y = 2 \quad \rightarrow \quad 2y = -3x + 2$$
$$y = -\frac{3}{2}x + 1$$

Because $\left(\frac{2}{3}\right)\left(-\frac{3}{2}\right) = -1$ (the product of the two slopes is $-1$), the lines are therefore perpendicular.

**Remark:** The statement "the product of two slopes is $-1$" has the same meaning as the statement "the two slopes are negative reciprocals of each other"; that is, $m_1 = -\frac{1}{m_2}$.

**Classroom Example**
Find the equation of the line that contains the point $(4, -5)$ and is parallel to the line determined by $8x + 2y = 12$.

## EXAMPLE 9

Find the equation of the line that contains the point $(1, 4)$ and is parallel to the line determined by $x + 2y = 5$.

### Solution

First, let's draw a figure to help in our analysis of the problem (Figure 7.45). Because the line through $(1, 4)$ is to be parallel to the line determined by $x + 2y = 5$, it must have the same slope. Let's find the slope by changing $x + 2y = 5$ to the slope-intercept form.

$$x + 2y = 5$$
$$2y = -x + 5$$
$$y = -\frac{1}{2}x + \frac{5}{2}$$

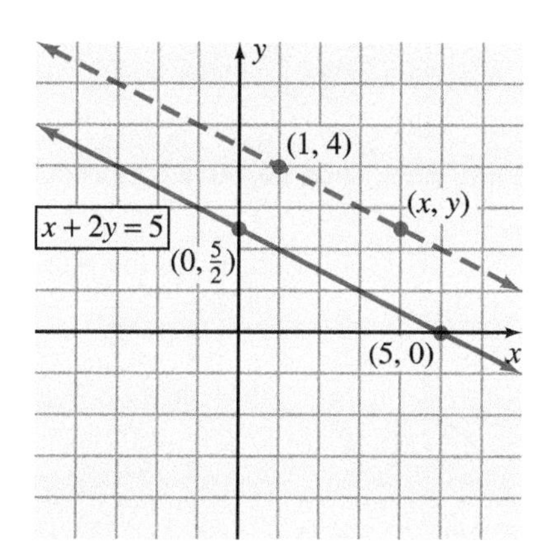


**Figure 7.45**

The slope of both lines is $-\frac{1}{2}$. Now use the point-slope formula substituting $-\frac{1}{2}$ for $m$ and $(1, 4)$ for $(x_1, y_1)$.

$$y - y_1 = m(x - x_1)$$
$$y - 4 = -\frac{1}{2}(x - 1)$$
$$2(y - 4) = -1(x - 1) \quad \text{Multiply both sides of the equation by 2}$$
$$2y - 8 = -x + 1$$
$$x + 2y = 9$$

**Classroom Example**
Find the equation of the line that contains the point $(3, 5)$ and is perpendicular to the line determined by $3x - 4y = 8$.

## EXAMPLE 10

Find the equation of the line that contains the point $(-1, -2)$ and is perpendicular to the line determined by $2x - y = 6$.

### Solution

First, let's draw a figure to help in our analysis of the problem (Figure 7.46). Because the line through $(-1, -2)$ is to be perpendicular to the line determined by $2x - y = 6$, its slope must be the negative reciprocal of the slope of $2x - y = 6$. Let's find the slope of $2x - y = 6$ by changing it to the slope-intercept form.

$$2x - y = 6$$
$$-y = -2x + 6$$
$$y = 2x - 6 \quad \text{The slope is 2}$$

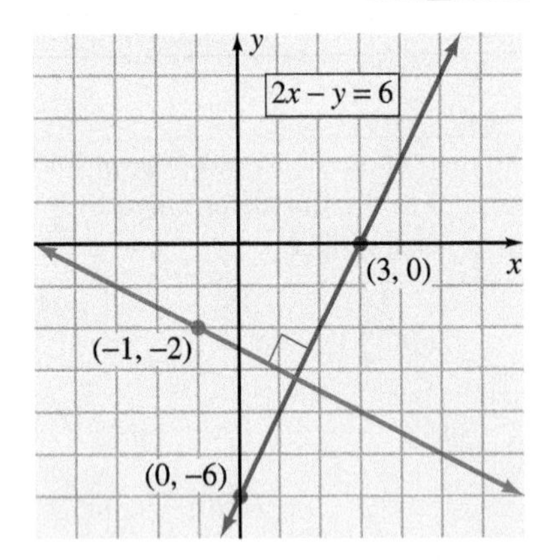


**Figure 7.46**

The slope of the desired line is $-\frac{1}{2}$ (the negative reciprocal of 2), and we can proceed by substituting $-\frac{1}{2}$ for $m$ and $(-1, -2)$ for $(x_1, y_1)$ in the point-slope formula.

$$y - y_1 = m(x - x_1)$$
$$y - (-2) = -\frac{1}{2}[x - (-1)]$$
$$y + 2 = -\frac{1}{2}(x + 1)$$
$$2(y + 2) = -1(x + 1)$$
$$2y + 4 = -x - 1$$
$$x + 2y = -5$$

The equation of the line is $x + 2y = -5$.

## Concept Quiz 7.4

For Problems 1–10, answer true or false.

1. If two distinct lines have the same slope, then the lines are parallel.
2. If the slopes of two lines are reciprocals, then the lines are perpendicular.
3. In the standard form of the equation of a line $Ax + By = C$, $A$ can be a rational number in fractional form.
4. In the slope-intercept form of an equation of a line $y = mx + b$, $m$ is the slope.
5. In the standard form of the equation of a line $Ax + By = C$, $A$ is the slope.
6. The slope of the line determined by the equation $3x - 2y = -4$ is $\frac{3}{2}$.
7. The concept of a slope is not defined for the line $y = 2$.
8. The concept of a slope is not defined for the line $x = 2$.

**9.** The lines determined by the equations $x - 3y = 4$ and $2x - 6y = 11$ are parallel lines.

**10.** The lines determined by the equations $x - 3y = 4$ and $x + 3y = 4$ are perpendicular lines.

## Problem Set 7.4

For Problems 1–14, write the equation of the line that has the indicated slope and contains the indicated point. Express final equations in standard form. (Objective 1)

**1.** $m = \frac{1}{2}$, $(3, 5)$

**2.** $m = \frac{1}{3}$, $(2, 3)$

**3.** $m = 3$, $(-2, 4)$

**4.** $m = -2$, $(-1, 6)$

**5.** $m = -\frac{3}{4}$, $(-1, -3)$

**6.** $m = -\frac{3}{5}$, $(-2, -4)$

**7.** $m = \frac{5}{4}$, $(4, -2)$

**8.** $m = \frac{3}{2}$, $(8, -2)$

**9.** $m = \frac{5}{2}$, $(-3, 4)$

**10.** $m = \frac{2}{3}$, $(1, -4)$

**11.** $m = -2$, $(5, 8)$

**12.** $m = -1$, $(-6, 2)$

**13.** $m = -\frac{1}{3}$, $(5, 0)$

**14.** $m = -\frac{3}{4}$, $(0, 1)$

For Problems 15–24, write the equation of the line that contains the indicated pair of points. Express final equations in standard form. (Objective 2)

**15.** $(2, 1), (6, 5)$

**16.** $(-1, 2), (2, 5)$

**17.** $(-2, -3), (2, 7)$

**18.** $(-3, -4), (1, 2)$

**19.** $(-3, 2), (4, 1)$

**20.** $(-2, 5), (3, -3)$

**21.** $(-1, -4), (3, -6)$

**22.** $(3, 8), (7, 2)$

**23.** $(0, 0), (5, 7)$

**24.** $(0, 0), (-5, 9)$

For Problems 25–32, write the equation of the line that has the indicated slope ($m$) and $y$ intercept ($b$). Express final equations in slope-intercept form. (Objective 3)

**25.** $m = \frac{3}{7}$, $b = 4$

**26.** $m = \frac{2}{9}$, $b = 6$

**27.** $m = 2$, $b = -3$

**28.** $m = -3$, $b = -1$

**29.** $m = -\frac{2}{5}$, $b = 1$

**30.** $m = -\frac{3}{7}$, $b = 4$

**31.** $m = 0$, $b = -4$

**32.** $m = \frac{1}{5}$, $b = 0$

For Problems 33–48, write the equation of the line that satisfies the given conditions. Express final equations in standard form. (Objectives 1, 2, 4, and 6)

**33.** $x$ intercept of 2 and $y$ intercept of $-4$

**34.** $x$ intercept of $-1$ and $y$ intercept of $-3$

**35.** $x$ intercept of $-3$ and slope of $-\frac{5}{8}$

**36.** $x$ intercept of 5 and slope of $-\frac{3}{10}$

**37.** Contains the point $(2, -4)$ and is parallel to the $y$ axis

**38.** Contains the point $(-3, -7)$ and is parallel to the $x$ axis

**39.** Contains the point $(5, 6)$ and is perpendicular to the $y$ axis

**40.** Contains the point $(-4, 7)$ and is perpendicular to the $x$ axis

**41.** Contains the point $(1, 3)$ and is parallel to the line $x + 5y = 9$

**42.** Contains the point $(-1, 4)$ and is parallel to the line $x - 2y = 6$

**43.** Contains the origin and is parallel to the line $4x - 7y = 3$

**44.** Contains the origin and is parallel to the line $-2x - 9y = 4$

**45.** Contains the point $(-1, 3)$ and is perpendicular to the line $2x - y = 4$

**46.** Contains the point $(-2, -3)$ and is perpendicular to the line $x + 4y = 6$

**47.** Is perpendicular to the line $-2x + 3y = 8$ and contains the origin

**48.** Contains the origin and is perpendicular to the line $y = -5x$

For Problems 49–54, change the equation to slope-intercept form and determine the slope and $y$ intercept of the line. (Objective 5)

**49.** $3x + y = 7$

**50.** $5x - y = 9$

**51.** $3x + 2y = 9$

**52.** $x - 4y = 3$

**53.** $x = 5y + 12$

**54.** $-4x - 7y = 14$

For Problems 55–62, use the slope-intercept form to graph the following lines. (Objective 5)

**55.** $y = \frac{2}{3}x - 4$

**56.** $y = \frac{1}{4}x + 2$

**57.** $y = 2x + 1$

**58.** $y = 3x - 1$

**59.** $y = -\frac{3}{2}x + 4$

**60.** $y = -\frac{5}{3}x + 3$

**61.** $y = -x + 2$

**62.** $y = -2x + 4$

For Problems 63–72, graph the following lines using the technique that seems most appropriate.

**63.** $y = -\frac{2}{5}x - 1$

**64.** $y = -\frac{1}{2}x + 3$

**65.** $x + 2y = 5$

**66.** $2x - y = 7$

**67.** $-y = -4x + 7$

**68.** $3x = 2y$

**69.** $7y = -2x$

**70.** $y = -3$

**71.** $x = 2$

**72.** $y = -x$

For Problems 73–78, the situations can be described by the use of linear equations in two variables. If two pairs of values are known, then we can determine the equation by using the approach we used in Example 2 of this section. For each of the following, assume that the relationship can be expressed as a linear equation in two variables, and use the given information to determine the equation. Express the equation in slope-intercept form. (Objectives 2 and 5)

**73.** A diabetic patient was told by her doctor that her hemoglobin A1c reading of 6.5 corresponds to an average blood glucose level of 135. At her next checkup, three months later, the patient was told that her hemoglobin A1c reading of 6.0 corresponds to an average blood glucose level of 120. Let $y$ represent the average blood glucose level, and $x$ represent the hemoglobin A1c reading.

**74.** Hal purchased a 500-minute calling card for \$17.50. After he used all the minutes on that card, he purchased another card from the same company at a price of \$26.25 for 750 minutes. Let $y$ represent the cost of the card in dollars and $x$ represent the number of minutes.

**75.** A company uses 7 pounds of fertilizer for a lawn that measures 5000 square feet and 12 pounds for a lawn that measures 10,000 square feet. Let $y$ represent the pounds of fertilizer and $x$ the square footage of the lawn.

**76.** A new diet guideline claims that a person weighing 140 pounds should consume 1490 daily calories and that a 200-pound person should consume 1700 calories. Let $y$ represent the calories and $x$ the weight of the person in pounds.

**77.** Two banks on opposite corners of a town square had signs that displayed the current temperature. One bank displayed the temperature in degrees Celsius and the other in degrees Fahrenheit. A temperature of 10°C was displayed at the same time as a temperature of 50°F. On another day, a temperature of −5°C was displayed at the same time as a temperature of 23°F. Let $y$ represent the temperature in degrees Fahrenheit and $x$ the temperature in degrees Celsius.

**78.** An accountant has a schedule of depreciation for some business equipment. The schedule shows that after 12 months the equipment is worth \$7600 and that after 20 months it is worth \$6000. Let $y$ represent the worth and $x$ represent the time in months.

## Thoughts Into Words

**79.** What does it mean to say that two points determine a line?

**80.** How would you help a friend determine the equation of the line that is perpendicular to $x - 5y = 7$ and contains the point (5, 4)?

**81.** Explain how you would find the slope of the line $y = 4$.

## Further Investigations

**82.** The equation of a line that contains the two points $(x_1, y_1)$ and $(x_2, y_2)$ is $\frac{y - y_1}{x - x_1} = \frac{y_2 - y_1}{x_2 - x_1}$. We often refer to this as the **two-point form** of the equation of a straight line. Use the two-point form and write the equation of the line that contains each of the indicated pairs of points. Express final equations in standard form.

**(a)** (1, 1) and (5, 2)

**(b)** (2, 4) and (−2, −1)

**(c)** (−3, 5) and (3, 1)

**(d)** (−5, 1) and (2, −7)

**83.** Let $Ax + By = C$ and $A'x + B'y = C'$ represent two lines. Change both of these equations to slope-intercept form, and then verify each of the following properties.

**(a)** If $\frac{A}{A'} = \frac{B}{B'} \neq \frac{C}{C'}$, then the lines are parallel.

**(b)** If $AA' = -BB'$, then the lines are perpendicular.

**84.** The properties in Problem 83 provide us with another way to write the equation of a line parallel or perpendicular to a given line that contains a given point not on the line. For example, suppose that we want the equation of the line perpendicular to $3x + 4y = 6$ that contains the point $(1, 2)$. The form $4x - 3y = k$, where $k$ is a constant, represents a family of lines perpendicular to $3x + 4y = 6$ because we have satisfied the condition $AA' = -BB'$. Therefore, to find what specific line of the family contains $(1, 2)$, we substitute 1 for $x$ and 2 for $y$ to determine $k$.

$$4x - 3y = k$$
$$4(1) - 3(2) = k$$
$$-2 = k$$

Thus the equation of the desired line is $4x - 3y = -2$.

Use the properties from Problem 83 to help write the equation of each of the following lines.

**(a)** Contains $(1, 8)$ and is parallel to $2x + 3y = 6$

**(b)** Contains $(-1, 4)$ and is parallel to $x - 2y = 4$

**(c)** Contains $(2, -7)$ and is perpendicular to $3x - 5y = 10$

**(d)** Contains $(-1, -4)$ and is perpendicular to $2x + 5y = 12$

**85.** The problem of finding the perpendicular bisector of a line segment presents itself often in the study of analytic geometry. As with any problem of writing the equation of a line, you must determine the slope of the line and a point that the line passes through. A perpendicular bisector passes through the midpoint of the line segment and has a slope that is the negative reciprocal of the slope of the line segment. The problem can be solved as follows:

Find the perpendicular bisector of the line segment between the points $(1, -2)$ and $(7, 8)$.

The midpoint of the line segment is $\left(\frac{1+7}{2}, \frac{-2+8}{2}\right) = (4, 3)$.

The slope of the line segment is $m = \frac{8-(-2)}{7-1} = \frac{10}{6} = \frac{5}{3}$.

Hence the perpendicular bisector will pass through the point $(4, 3)$ and have a slope of $m = -\frac{3}{5}$.

$$y - 3 = -\frac{3}{5}(x - 4)$$
$$5(y - 3) = -3(x - 4)$$
$$5y - 15 = -3x + 12$$
$$3x + 5y = 27$$

Thus the equation of the perpendicular bisector of the line segment between the points $(1, -2)$ and $(7, 8)$ is $3x + 5y = 27$.

Find the perpendicular bisector of the line segment between the points for the following. Write the equation in standard form.

**(a)** $(-1, 2)$ and $(3, 0)$
**(b)** $(6, -10)$ and $(-4, 2)$
**(c)** $(-7, -3)$ and $(5, 9)$
**(d)** $(0, 4)$ and $(12, -4)$

## Graphing Calculator Activities

**86.** Predict whether each of the following pairs of equations represents parallel lines, perpendicular lines, or lines that intersect but are not perpendicular. Then graph each pair of lines to check your predictions. (The properties presented in Problem 83 should be very helpful.)

**(a)** $5.2x + 3.3y = 9.4$ and $5.2x + 3.3y = 12.6$
**(b)** $1.3x - 4.7y = 3.4$ and $1.3x - 4.7y = 11.6$
**(c)** $2.7x + 3.9y = 1.4$ and $2.7x - 3.9y = 8.2$
**(d)** $5x - 7y = 17$ and $7x + 5y = 19$
**(e)** $9x + 2y = 14$ and $2x + 9y = 17$
**(f)** $2.1x + 3.4y = 11.7$ and $3.4x - 2.1y = 17.3$
**(g)** $7.1x - 2.3y = 6.2$ and $2.3x + 7.1y = 9.9$
**(h)** $-3x + 9y = 12$ and $9x - 3y = 14$
**(i)** $2.6x - 5.3y = 3.4$ and $5.2x - 10.6y = 19.2$
**(j)** $4.8x - 5.6y = 3.4$ and $6.1x + 7.6y = 12.3$

### Answers to the Concept Quiz

**1.** True **2.** False **3.** False **4.** True **5.** False **6.** True **7.** False **8.** True **9.** True **10.** False

# 7.5 Graphing Nonlinear Equations

OBJECTIVES

1. Graph nonlinear equations
2. Determine if the graph of an equation is symmetric to the x axis, the y axis, or the origin

Equations such as $y = x^2 - 4$, $x = y^2$, $y = \frac{1}{x}$, $x^2y = -2$, and $x = y^3$ are all examples of nonlinear equations. The graphs of these equations are figures other than straight lines, which can be determined by plotting a sufficient number of points. Let's plot the points and observe some characteristics of these graphs that we then can use to supplement the point-plotting process.

**Classroom Example**
Graph $y = x^2 + 3$.

**EXAMPLE 1** Graph $y = x^2 - 4$.

### Solution

Let's begin by finding the intercepts. If $x = 0$, then

$$y = 0^2 - 4 = -4$$

The point $(0, -4)$ is on the graph. If $y = 0$, then

$$0 = x^2 - 4$$
$$0 = (x + 2)(x - 2)$$
$$x + 2 = 0 \quad \text{or} \quad x - 2 = 0$$
$$x = -2 \quad \text{or} \quad x = 2$$

The points $(-2, 0)$ and $(2, 0)$ are on the graph. The given equation is in a convenient form for setting up a table of values.

Plotting these points and connecting them with a smooth curve produces Figure 7.47.

| x | y | |
|---|---|---|
| 0 | −4 | |
| −2 | 0 | Intercepts |
| 2 | 0 | |
| 1 | −3 | |
| −1 | −3 | |
| 3 | 5 | Other points |
| −3 | 5 | |

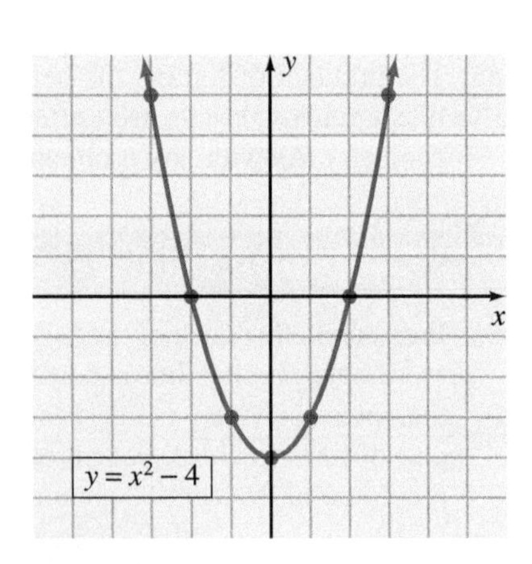


Figure 7.47

The curve in Figure 7.47 is called a parabola; we will study parabolas in more detail in a later chapter. At this time we want to emphasize that the parabola in Figure 7.47 is said to be *symmetric with respect to the y axis*. In other words, the $y$ axis is a line of symmetry. Each half of the curve is a mirror image of the other half through the $y$ axis. Note, in the table of values, that for each ordered pair $(x, y)$, the ordered pair $(-x, y)$ is also a solution. A general test for $y$-axis symmetry can be stated as follows:

### y-Axis Symmetry

The graph of an equation is symmetric with respect to the $y$ axis if replacing $x$ with $-x$ results in an equivalent equation.

The equation $y = x^2 - 4$ exhibits symmetry with respect to the $y$ axis because replacing $x$ with $-x$ produces $y = (-x)^2 - 4 = x^2 - 4$. Let's test some equations for such symmetry. We will replace $x$ with $-x$ and check for an equivalent equation.

| Equation | Test for symmetry with respect to the y axis | Equivalent equation | Symmetric with respect to the y axis |
|---|---|---|---|
| $y = -x^2 + 2$ | $y = -(-x)^2 + 2 = -x^2 + 2$ | Yes | Yes |
| $y = 2x^2 + 5$ | $y = 2(-x)^2 + 5 = 2x^2 + 5$ | Yes | Yes |
| $y = x^4 + x^2$ | $y = (-x)^4 + (-x)^2$ $= x^4 + x^2$ | Yes | Yes |
| $y = x^3 + x^2$ | $y = (-x)^3 + (-x)^2$ $= -x^3 + x^2$ | No | No |
| $y = x^2 + 4x + 2$ | $y = (-x)^2 + 4(-x) + 2$ $= x^2 - 4x + 2$ | No | No |

Some equations yield graphs that have $x$-axis symmetry. In the next example we will see the graph of a parabola that is symmetric with respect to the $x$ axis.

Classroom Example

Graph $x = \frac{1}{2}y^2$.

**EXAMPLE 2** Graph $x = y^2$.

### Solution

First, we see that (0, 0) is on the graph and determines both intercepts. Second, the given equation is in a convenient form for setting up a table of values.

Plotting these points and connecting them with a smooth curve produces Figure 7.48.

| x | y | |
|---|---|---|
| 0 | 0 | Intercepts |
| 1 | 1 | |
| 1 | −1 | Other points |
| 4 | 2 | |
| 4 | −2 | |

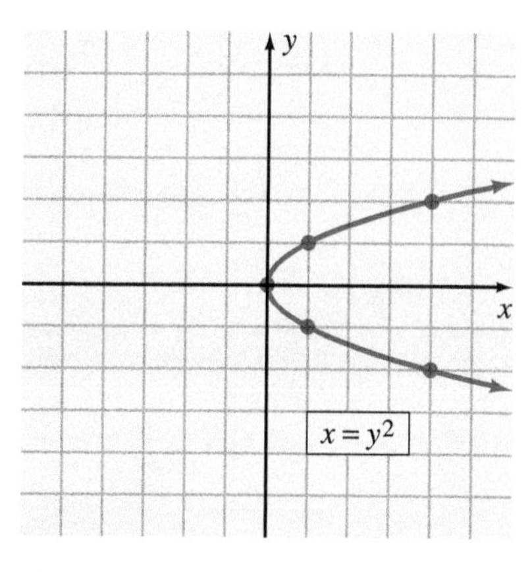


**Figure 7.48**

The parabola in Figure 7.48 is said to be *symmetric with respect to the* x *axis*. Each half of the curve is a mirror image of the other half through the $x$ axis. Also note in the table of values, that for each ordered pair $(x, y)$, the ordered pair $(x, -y)$ is a solution. A general test for $x$-axis symmetry can be stated as follows:

### x-Axis Symmetry

The graph of an equation is symmetric with respect to the $x$ axis if replacing $y$ with $-y$ results in an equivalent equation.

The equation $x = y^2$ exhibits $x$-axis symmetry because replacing $y$ with $-y$ produces $x = (-y)^2 = y^2$. Let's test some equations for $x$-axis symmetry. We will replace $y$ with $-y$ and check for an equivalent equation.

| Equation | Test for symmetry with respect to the x axis | Equivalent equation | Symmetric with respect to the x axis |
|---|---|---|---|
| $x = y^2 + 5$ | $x = (-y)^2 + 5 = y^2 + 5$ | Yes | Yes |
| $x = -3y^2$ | $x = -3(-y)^2 = -3y^2$ | Yes | Yes |
| $x = y^3 + 2$ | $x = (-y)^3 + 2 = -y^3 + 2$ | No | No |
| $x = y^2 - 5y + 6$ | $x = (-y)^2 - 5(-y) + 6$ $= y^2 + 5y + 6$ | No | No |

In addition to $y$-axis and $x$-axis symmetry, some equations yield graphs that have symmetry with respect to the origin. In the next example we will see a graph that is symmetric with respect to the origin.

**Classroom Example**

Graph $y = \frac{2}{x}$.

**EXAMPLE 3** Graph $y = \frac{1}{x}$.

### Solution

First, let's find the intercepts. Let $x = 0$; then $y = \frac{1}{x}$ becomes $y = \frac{1}{0}$, and $\frac{1}{0}$ is undefined. Thus there is no $y$ intercept. Let $y = 0$; then $y = \frac{1}{x}$ becomes $0 = \frac{1}{x}$, and there are no values of $x$ that will satisfy this equation. In other words, this graph has no points on either the $x$ axis or the $y$ axis. Second, let's set up a table of values and keep in mind that neither $x$ nor $y$ can equal zero.

In Figure 7.49(a) we plotted the points associated with the solutions from the table. Because the graph does not intersect either axis, it must consist of two branches. Thus connecting the points in the first quadrant with a smooth curve and then connecting the points in the third quadrant with a smooth curve, we obtain the graph shown in Figure 7.49(b).

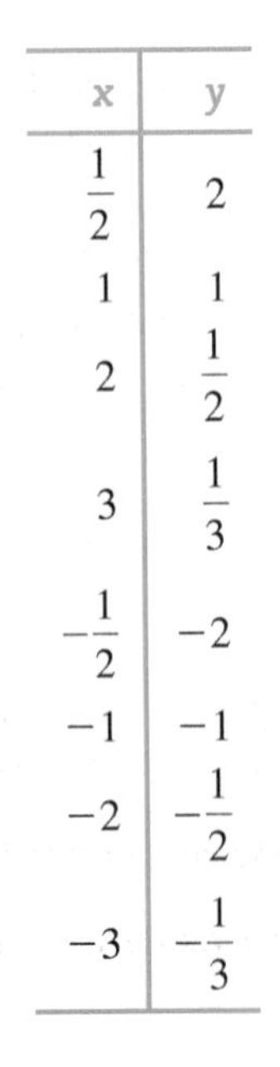

| x | y |
|---|---|
| $\frac{1}{2}$ | 2 |
| 1 | 1 |
| 2 | $\frac{1}{2}$ |
| 3 | $\frac{1}{3}$ |
| $-\frac{1}{2}$ | $-2$ |
| $-1$ | $-1$ |
| $-2$ | $-\frac{1}{2}$ |
| $-3$ | $-\frac{1}{3}$ |

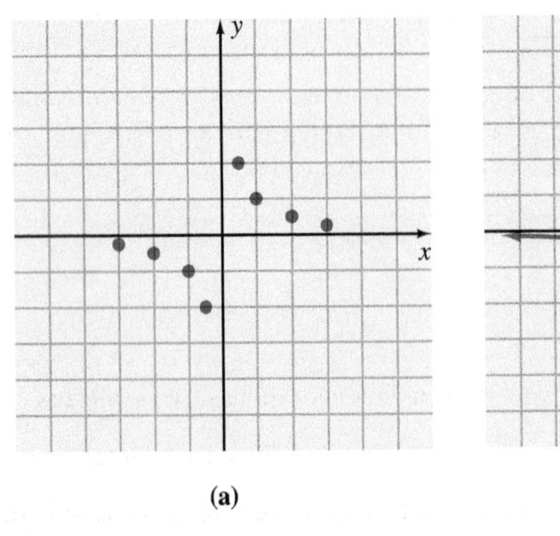


(a)

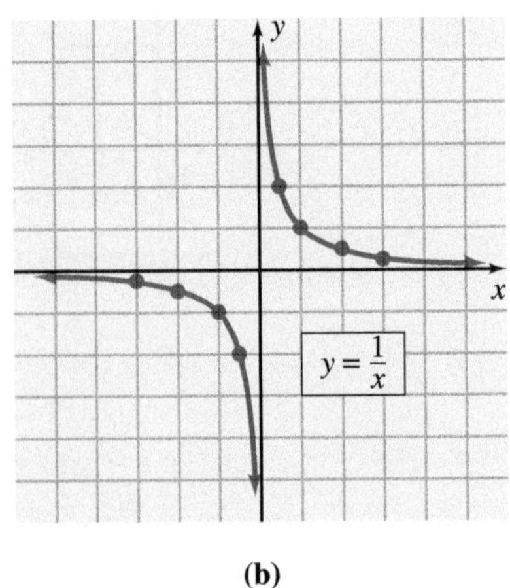


(b)

Figure 7.49

The curve in Figure 7.49 is said to be *symmetric with respect to the origin*. Each half of the curve is a mirror image of the other half through the origin. Note in the table of values that for each ordered pair $(x, y)$, the ordered pair $(-x, -y)$ is also a solution. A general test for origin symmetry can be stated as follows:

### Origin Symmetry

The graph of an equation is symmetric with respect to the origin if replacing $x$ with $-x$ and $y$ with $-y$ results in an equivalent equation.

The equation $y = \frac{1}{x}$ exhibits symmetry with respect to the origin because replacing $y$ with $-y$ and $x$ with $-x$ produces $-y = \frac{1}{-x}$, which is equivalent to $y = \frac{1}{x}$. Let's test some equations for symmetry with respect to the origin. We will replace $y$ with $-y$, replace $x$ with $-x$, and then check for an equivalent equation.

| Equation | Test for symmetry with respect to the origin | Equivalent equation | Symmetric with respect to the origin |
|---|---|---|---|
| $y = x^3$ | $(-y) = (-x)^3$<br>$-y = -x^3$<br>$y = x^3$ | Yes | Yes |
| $x^2 + y^2 = 4$ | $(-x)^2 + (-y)^2 = 4$<br>$x^2 + y^2 = 4$ | Yes | Yes |
| $y = x^2 - 3x + 4$ | $(-y) = (-x)^2 - 3(-x) + 4$<br>$-y = x^2 + 3x + 4$<br>$y = -x^2 - 3x - 4$ | No | No |

Let's pause for a moment and pull together the graphing techniques that we have introduced thus far. The following list is a set of graphing suggestions. The order of the suggestions indicates the order in which we usually attack a new graphing problem.

1. Determine what type of symmetry the equation exhibits.
2. Find the intercepts.
3. Solve the equation for $y$ in terms of $x$ or for $x$ in terms of $y$ if it is not already in such a form.
4. Set up a table of ordered pairs that satisfy the equation. The type of symmetry will affect your choice of values in the table. (We will illustrate this in a moment.)
5. Plot the points associated with the ordered pairs from the table, and connect them with a smooth curve. Then, if appropriate, reflect this part of the curve according to the symmetry shown by the equation.

**Classroom Example**
Graph $x^2y = 3$.

**EXAMPLE 4** Graph $x^2y = -2$.

### Solution

Because replacing $x$ with $-x$ produces $(-x)^2y = -2$ or, equivalently, $x^2y = -2$, the equation exhibits $y$-axis symmetry. It does not exhibit $x$-axis symmetry or symmetry to the origin. Solving the equation for $y$ produces $y = \frac{-2}{x^2}$. There are no intercepts because neither $x$ nor $y$ can equal 0. The equation exhibits $y$-axis symmetry, so let's use only positive values for $x$ and then reflect the curve across the $y$ axis.

Let's plot the points determined by the table, connect them with a smooth curve, and reflect this portion of the curve across the $y$ axis. Figure 7.50 is the result of this process.

| $x$ | $y$ |
|---|---|
| 1 | $-2$ |
| 2 | $-\frac{1}{2}$ |
| 3 | $-\frac{2}{9}$ |
| 4 | $-\frac{1}{8}$ |
| $\frac{1}{2}$ | $-8$ |

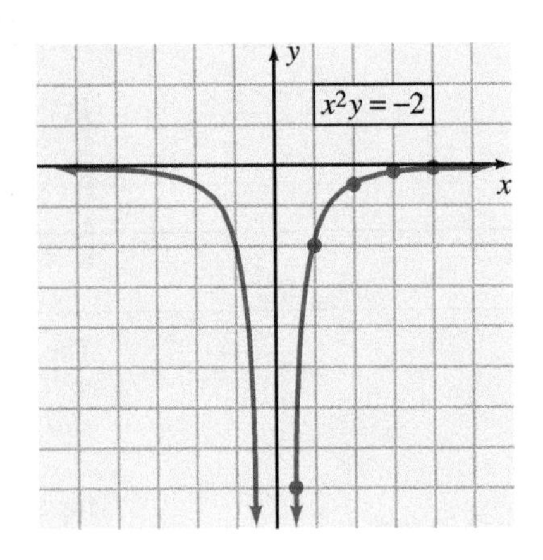


Figure 7.50

**Classroom Example**
Graph $x = \frac{1}{2}y^3$.

**EXAMPLE 5** Graph $x = y^3$.

## Solution

Because replacing $x$ with $-x$ and $y$ with $-y$ produces $-x = (-y)^3 = -y^3$, which is equivalent to $x = y^3$, the given equation exhibits origin symmetry. If $x = 0$, then $y = 0$, so the origin is a point of the graph. The given equation is in an easy form for deriving a table of values.

| $x$ | $y$ |
|---|---|
| 0 | 0 |
| 1 | 1 |
| 8 | 2 |
| $\frac{1}{8}$ | $\frac{1}{2}$ |

Let's plot the points determined by the table, connect them with a smooth curve, and reflect this portion of the curve through the origin to produce Figure 7.51.

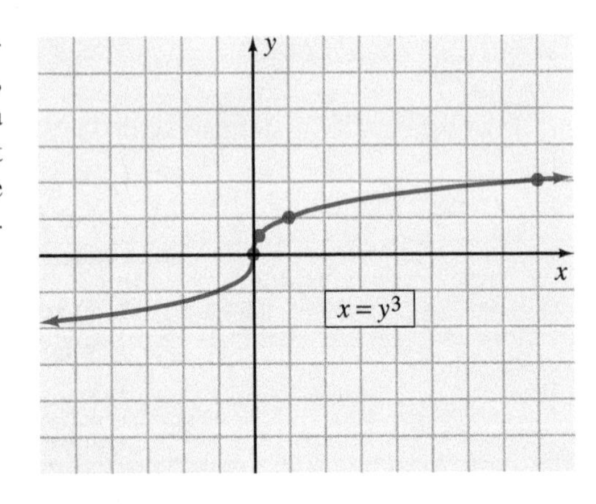


Figure 7.51

**Classroom Example**
Use a graphing utility to obtain a graph of the equation $x = \frac{1}{2}y^3$.

**EXAMPLE 6** Use a graphing utility to obtain a graph of the equation $x = y^3$.

## Solution

First, we may need to solve the equation for $y$ in terms of $x$. (We say we "may need to" because some graphing utilities are capable of graphing two-variable equations without solving for $y$ in terms of $x$.)

$$y = \sqrt[3]{x} = x^{1/3}$$

Now we can enter the expression $x^{1/3}$ for $Y_1$ and obtain the graph shown in Figure 7.52.

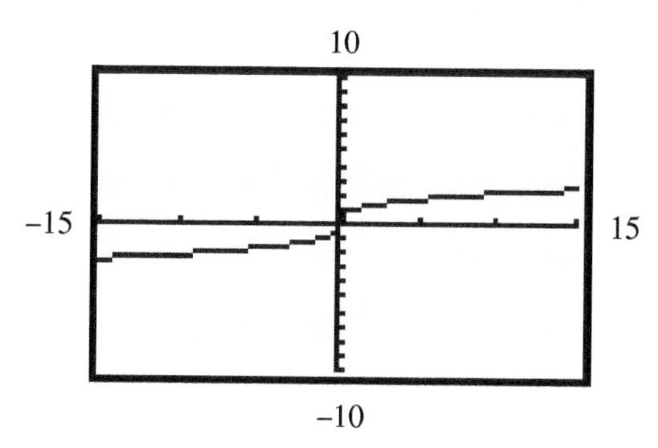


**Figure 7.52**

As indicated in Figure 7.52, the **viewing rectangle** of a graphing utility is a portion of the $xy$ plane shown on the display of the utility. In this display, the boundaries were set so that $-15 \leq x \leq 15$ and $-10 \leq y \leq 10$. These boundaries were set automatically; however, boundaries can be reassigned as necessary, which is an important feature of graphing utilities.

## Concept Quiz 7.5

For Problems 1–10, answer true or false.

1. The equation $y = \sqrt{x}$ is a nonlinear equation.
2. If a graph is symmetric with respect to the $y$ axis, then the $x$ axis is a line of symmetry for the graph.
3. When replacing $y$ with $-y$ in an equation results in an equivalent equation, then the graph of the equation is symmetric with respect to the $x$ axis.
4. If a parabola is symmetric with respect to the $x$ axis, then each half of the curve is a mirror image of the other half through the $x$ axis.
5. If the graph of an equation is symmetric with respect to the $x$ axis, then it cannot be symmetric with respect to the $y$ axis.
6. If the point $(-2, 5)$ is on a graph that is symmetric with respect to the $y$ axis, then the point $(-2, -5)$ is also on the graph.
7. If for each ordered pair $(x, y)$ that is a solution of the equation, the ordered pair $(-x, -y)$ is also a solution, then the graph of the equation is symmetric with respect to the origin.
8. The graph of the line $y = 3x$ is symmetric with respect to the $y$ axis.
9. The graph of a straight line is symmetric with respect to the origin only if the line passes through the origin.
10. Every straight line that passes through the origin is symmetric with respect to the origin.

## Problem Set 7.5

For each of the points in Problems 1–5, determine the points that are symmetric with respect to (a) the $x$ axis, (b) the $y$ axis, and (c) the origin. (Objective 2)

**1.** $(-3, 1)$ **2.** $(-2, -4)$

**3.** $(7, -2)$ **4.** $(0, -4)$

**5.** $(5, 0)$

For Problems 6–25, determine the type(s) of symmetry (symmetry with respect to the $x$ axis, $y$ axis, and/or origin) exhibited by the graph of each of the following equations. Do not sketch the graph. (Objective 2)

**6.** $x^2 + 2y = 4$ **7.** $-3x + 2y^2 = -4$

**8.** $x = -y^2 + 5$ **9.** $y = 4x^2 + 13$

**10.** $xy = -6$

**11.** $2x^2y^2 = 5$

**12.** $2x^2 + 3y^2 = 9$

**13.** $x^2 - 2x - y^2 = 4$

**14.** $y = x^2 - 6x - 4$

**15.** $y = 2x^2 - 7x - 3$

**16.** $y = x$

**17.** $y = 2x$

**18.** $y = x^4 + 4$

**19.** $y = x^4 - x^2 + 2$

**20.** $x^2 + y^2 = 13$

**21.** $x^2 - y^2 = -6$

**22.** $y = -4x^2 - 2$

**23.** $x = -y^2 + 9$

**24.** $x^2 + y^2 - 4x - 12 = 0$

**25.** $2x^2 + 3y^2 + 8y + 2 = 0$

For Problems 26–59, graph each of the equations. (Objective 1)

**26.** $y = x + 1$

**27.** $y = x - 4$

**28.** $y = 3x - 6$

**29.** $y = 2x + 4$

**30.** $y = -2x + 1$

**31.** $y = -3x - 1$

**32.** $y = \frac{2}{3}x - 1$

**33.** $y = -\frac{1}{3}x + 2$

**34.** $y = \frac{1}{3}x$

**35.** $y = \frac{1}{2}x$

**36.** $2x + y = 6$

**37.** $2x - y = 4$

**38.** $x + 3y = -3$

**39.** $x - 2y = 2$

**40.** $y = x^2 - 1$

**41.** $y = x^2 + 2$

**42.** $y = -x^3$

**43.** $y = x^3$

**44.** $y = \frac{2}{x^2}$

**45.** $y = \frac{-1}{x^2}$

**46.** $y = 2x^2$

**47.** $y = -3x^2$

**48.** $xy = -3$

**49.** $xy = 2$

**50.** $x^2y = 4$

**51.** $xy^2 = -4$

**52.** $y^3 = x^2$

**53.** $y^2 = x^3$

**54.** $y = \frac{-2}{x^2 + 1}$

**55.** $y = \frac{4}{x^2 + 1}$

**56.** $x = -y^3$

**57.** $y = x^4$

**58.** $y = -x^4$

**59.** $x = -y^3 + 2$

## Thoughts Into Words

**60.** How would you convince someone that there are infinitely many ordered pairs of real numbers that satisfy $x + y = 7$?

**61.** What is the graph of $x = 0$? What is the graph of $y = 0$? Explain your answers.

**62.** Is a graph symmetric with respect to the origin if it is symmetric with respect to both axes? Defend your answer.

**63.** Is a graph symmetric with respect to both axes if it is symmetric with respect to the origin? Defend your answer.

## Graphing Calculator Activities

This set of activities is designed to help you get started with your graphing utility by setting different boundaries for the viewing rectangle; you will notice the effect on the graphs produced. These boundaries are usually set by using a menu displayed by a key marked either WINDOW or RANGE. You may need to consult the user's manual for specific key-punching instructions.

**64.** Graph the equation $y = \frac{1}{x}$ (Example 3) using the following boundaries.
**(a)** $-15 \le x \le 15$ and $-10 \le y \le 10$
**(b)** $-10 \le x \le 10$ and $-10 \le y \le 10$
**(c)** $-5 \le x \le 5$ and $-5 \le y \le 5$

**65.** Graph the equation $y = \frac{-2}{x^2}$ (Example 4), using the following boundaries.
**(a)** $-15 \le x \le 15$ and $-10 \le y \le 10$
**(b)** $-5 \le x \le 5$ and $-10 \le y \le 10$
**(c)** $-5 \le x \le 5$ and $-10 \le y \le 1$

**66.** Graph the two equations $y = \pm\sqrt{x}$ (Example 2) on the same set of axes, using the following boundaries. (Let $Y_1 = \sqrt{x}$ and $Y_2 = -\sqrt{x}$)
**(a)** $-15 \le x \le 15$ and $-10 \le y \le 10$
**(b)** $-1 \le x \le 15$ and $-10 \le y \le 10$
**(c)** $-1 \le x \le 15$ and $-5 \le y \le 5$

**67.** Graph $y = \frac{1}{x}$, $y = \frac{5}{x}$, $y = \frac{10}{x}$, and $y = \frac{20}{x}$ on the same set of axes. (Choose your own boundaries.) What effect does increasing the constant seem to have on the graph?

**68.** Graph $y = \frac{10}{x}$ and $y = \frac{-10}{x}$ on the same set of axes. What relationship exists between the two graphs?

**69.** Graph $y = \frac{10}{x^2}$ and $y = \frac{-10}{x^2}$ on the same set of axes. What relationship exists between the two graphs?

### Answers to the Concept Quiz

**1.** True **2.** False **3.** True **4.** True **5.** False **6.** False **7.** True **8.** False **9.** True **10.** True

# Chapter 7 Summary

| OBJECTIVE | SUMMARY | EXAMPLE |
|---|---|---|
| Find solutions for linear equations in two variables.<br>(Section 7.1/Objective 1) | A solution of an equation in two variables is an ordered pair of real numbers that satisfies the equation. | Find a solution for the equation $2x - 3y = -6$.<br>**Solution**<br>Choose an arbitrary value for $x$ and determine the corresponding $y$ value. Let $x = 3$; then substitute 3 for $x$ in the equation.<br>$2(3) - 3y = -6$<br>$6 - 3y = -6$<br>$-3y = -12$<br>$y = 4$<br>Therefore, the ordered pair (3, 4) is a solution.<br>**Sample Problem 1**<br>Find the ordered pair for the equation $3x - 2y = 10$ if $x = 2$. |
| Graph the solutions for linear equations.<br>(Section 7.1/Objective 3) | A graph provides a visual display of the infinite solutions of an equation in two variables. The ordered-pair solutions for a linear equation can be plotted as points on a rectangular coordinate system. Connecting the points with a straight line produces a graph of the equation.<br>Any equation of the form $Ax + By = C$, where $A$, $B$, and $C$ are constants ($A$ and $B$ not both zero) and $x$ and $y$ are variables, is a linear equation, and its graph is a straight line. | Graph $y = 2x - 3$.<br>**Solution**<br>Find at least three ordered-pair solutions for the equation. We can determine that $(-1, -5)$, $(0, -3)$, and $(1, -1)$ are solutions. The graph is shown below. 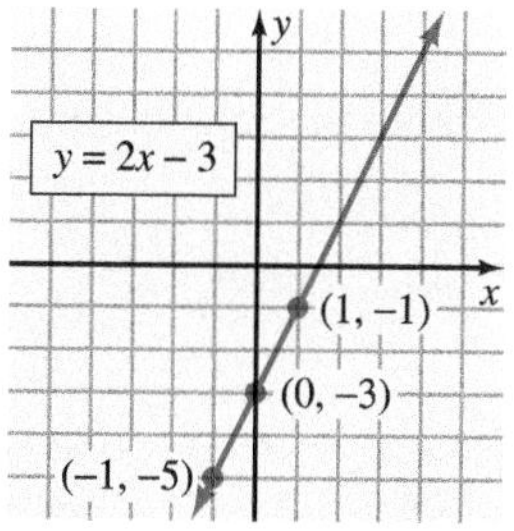<br>$y = 2x - 3$; $(1, -1)$; $(0, -3)$; $(-1, -5)$<br>**Sample Problem 2**<br>Graph $y = -x + 2$. |
| Graph linear equations by finding the $x$ and $y$ intercepts.<br>(Section 7.1/Objective 4) | The $x$ intercept is the $x$ coordinate of the point where the graph intersects the $x$ axis. The $y$ intercept is the $y$ coordinate of the point where the graph intersects the $y$ axis.<br>To find the $x$ intercept, substitute 0 for $y$ in the equation and then solve for $x$. To find the $y$ intercept, substitute 0 for $x$ in the equation and then solve for $y$. Plot the intercepts and connect them with a straight line to produce the graph. | Graph $x - 2y = 4$.<br>**Solution**<br>Let $y = 0$.<br>$x - 2(0) = 4$<br>$x = 4$<br>Let $x = 0$.<br>$0 - 2y = 4$<br>$y = -2$ 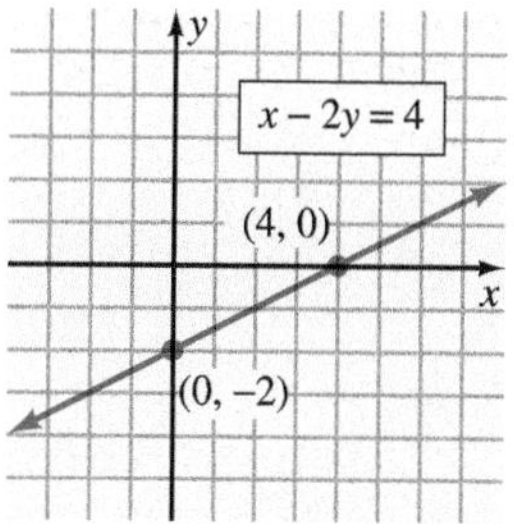<br>$x - 2y = 4$; $(4, 0)$; $(0, -2)$<br>**Sample Problem 3**<br>Graph $2x + y = -4$. |

Answers to **Sample Problems** are located in the back of the book.

*(continued)*

| OBJECTIVE | SUMMARY | EXAMPLE |
|---|---|---|
| Graph lines passing through the origin, vertical lines, and horizontal lines. (Section 7.1/Objective 5) | The graph of any equation of the form $Ax + By = C$, where $C = 0$, is a straight line that passes through the origin.<br><br>Any equation of the form $x = a$, where $a$ is a constant, is a vertical line.<br><br>Any equation of the form $y = b$, where $b$ is a constant, is a horizontal line. | Graph $3x + 2y = 0$.<br><br>**Solution**<br>The equation indicates that the graph will be a line passing through the origin. Solving the equation for $y$ gives us $y = -\frac{3}{2}x$. Find at least three ordered-pair solutions for the equation. We can determine that $(-2, 3)$, $(0, 0)$ and $(2, -3)$ are solutions. The graph is shown below. 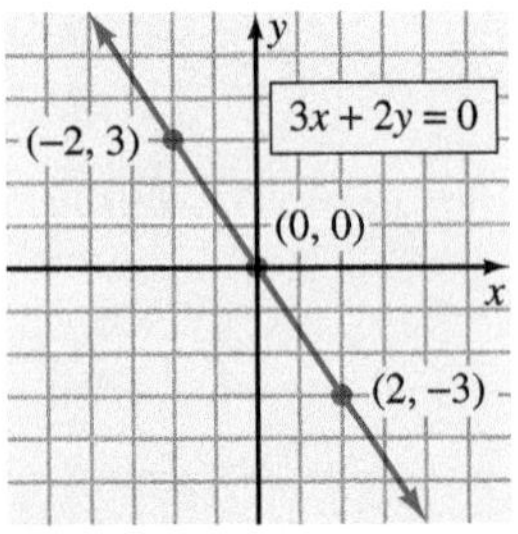<br><br>y; 3x + 2y = 0; (−2, 3); (0, 0); x; (2, −3)<br><br>**Sample Problem 4**<br>Graph $x - 2y = 0$. |
| Apply graphing to linear relationships. (Section 7.1/Objective 6) | Many relationships between two quantities are linear relationships. Graphs of these relationships can be used to present information about the relationship. | Let $c$ represent the cost in dollars, and let $w$ represent the gallons of water used; then the equation $c = 0.004w + 20$ can be used to determine the cost of a water bill for a household. Graph the relationship.<br><br>**Solution**<br>Label the vertical axis $c$ and the horizontal axis $w$. Because of the type of application, we use only nonnegative values for $w$. 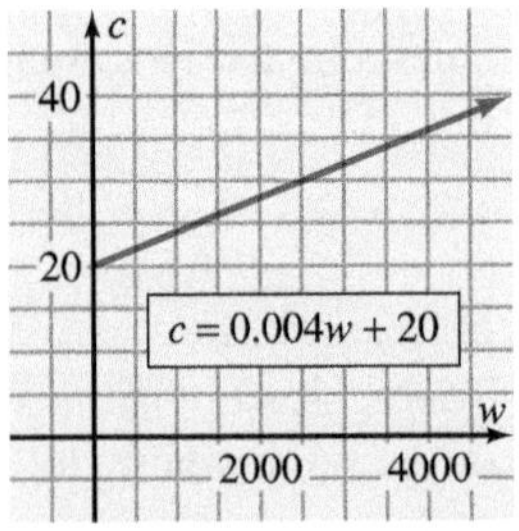<br><br>c; 40; 20; c = 0.004w + 20; w; 2000; 4000<br><br>**Sample Problem 5**<br>A car rental company charges a flat fee of \$30 plus \$0.20 per mile per day. Let $c$ represent the cost of renting a car and let $m$ represent miles driven. Then the equation $c = 0.20m + 30$ can be used to determine the cost of renting the car per day. Graph the relationship. |

| OBJECTIVE | SUMMARY | EXAMPLE |
|---|---|---|
| Graph linear inequalities in two variables. (Section 7.2/Objective 1) | To graph a linear inequality, first graph the line for the corresponding equality. Use a solid line if the equality is included in the given statement or a dashed line if the equality is not included. Then a test point is used to determine which half-plane is included in the solution set. See page 362 for the detailed steps. | Graph $x - 2y \leq -4$.<br>**Solution**<br>First graph $x - 2y = -4$. Choose (0, 0) as a test point. Substituting (0, 0) into the inequality yields $0 \leq -4$. Because the test point (0, 0) makes the inequality a false statement, the half-plane not containing the point (0, 0) is in the solution.<br>$x - 2y \leq -4$; (0, 2); (−4, 0); x; y<br>**Sample Problem 6**<br>Graph $2x - y \geq 4$. |
| Find the distance between two points. (Section 7.3/Objective 1) | The distance between any two points $(x_1, y_1)$ and $(x_2, y_2)$ is given by the distance formula $d = \sqrt{(x_2 - x_1)^2 + (y_2 - y_1)^2}$. | Find the distance between (1, −5) and (4, 2).<br>**Solution**<br>$d = \sqrt{(x_2 - x_1)^2 + (y_2 - y_1)^2}$<br>$d = \sqrt{(4 - 1)^2 + (2 - (-5))^2}$<br>$d = \sqrt{(3)^2 + (7)^2}$<br>$d = \sqrt{9 + 49} = \sqrt{58}$<br>**Sample Problem 7**<br>Find the distance between (−1, 4) and (−2, −3). |
| Find the slope of a line. (Section 7.3/Objective 2) | The slope (denoted by $m$) of a line determined by the points $(x_1, y_1)$ and $(x_2, y_2)$ is given by the slope formula $m = \frac{y_2 - y_1}{x_2 - x_1}$, where $x_2 \neq x_1$. | Find the slope of a line that contains the points (−1, 2) and (7, 8).<br>**Solution**<br>Use the slope formula:<br>$m = \frac{8 - 2}{7 - (-1)} = \frac{6}{8} = \frac{3}{4}$<br>Thus the slope of the line is $\frac{3}{4}$.<br>**Sample Problem 8**<br>Find the slope of a line that contains the points (3, 2) and (−5, 6). |

*(continued)*

| OBJECTIVE | SUMMARY | EXAMPLE |
|---|---|---|
| Use slope to graph lines. (Section 7.3/Objective 3) | A line can be graphed if a point on the line and the slope is known; simply plot the point and from that point use the slope to locate another point on the line. Then those two points can be connected with a straight line to produce the graph. | Graph the line that contains the point $(-3, -2)$ and has a slope of $\frac{5}{2}$.<br>**Solution**<br>From the point $(-3, -2)$, locate another point by moving up 5 units and to the right 2 units to obtain the point $(-1, 3)$. Then draw the line. 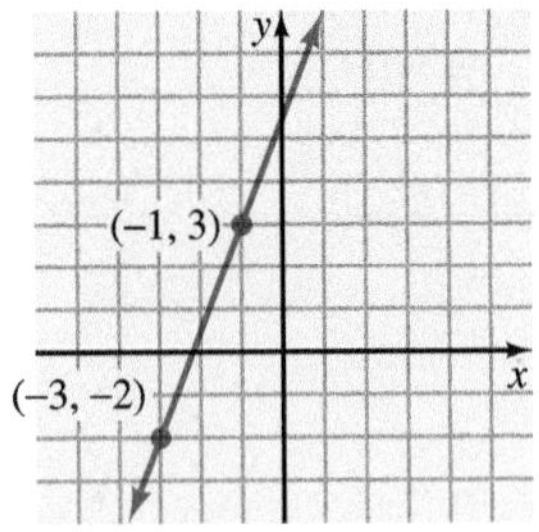<br>y<br>(−1, 3)<br>(−3, −2)<br>x<br>**Sample Problem 9**<br>Graph the line that contains the point $(-1, 5)$ and has a slope of $-3$. |
| Apply slope to solve problems. (Section 7.3/Objective 4) | The concept of slope is used in most situations in which an incline is involved. In highway construction the word "grade" is often used instead of "slope." | A certain highway has a grade of 2%. How many feet does it rise in a horizontal distance of one-third of a mile (1760 feet)?<br>**Solution**<br>A 2% grade is equivalent to a slope of $\frac{2}{100}$. We can set up the proportion $\frac{2}{100} = \frac{y}{1760}$; then solving for $y$ gives us $y = 35.2$. So the highway rises 35.2 feet in one-third of a mile.<br>**Sample Problem 10**<br>A wheelchair ramp will be installed that needs to connect to a porch that is elevated 3 feet above the ground. How far from the porch must the ramp be started so that the slope of the ramp will be at a 10% grade? |

| OBJECTIVE | SUMMARY | EXAMPLE |
|---|---|---|
| Apply the slope-intercept form of an equation of a line.<br>(Section 7.4/Objective 5) | The equation $y = mx + b$ is referred to as the slope-intercept form of the equation of a line. If the equation of a nonvertical line is written in this form, then the coefficient of $x$ is the slope and the constant term is the $y$ intercept. | Change the equation $2x + 7y = -21$ to slope-intercept form and determine the slope and $y$ intercept.<br>**Solution**<br>Solve the equation $2x + 7y = -21$ for $y$:<br>$2x + 7y = -21$<br>$7y = -2x - 21$<br>$y = \frac{-2}{7}x - 3$<br>The slope is $-\frac{2}{7}$, and the $y$ intercept is $-3$.<br>**Sample Problem 11**<br>Change the equation $3x - 2y = 14$ to slope-intercept form and determine the slope and $y$ intercept. |
| Find the equation of a line given the slope and a point contained in the line.<br>(Section 7.4/Objective 1) | To determine the equation of a straight line given a set of conditions, we can use the point-slope form $y - y_1 = m(x - x_1)$, or $m = \frac{y - y_1}{x - x_1}$. The result can be expressed in standard form or slope-intercept form.<br>**Standard Form** $Ax + By = C$, where $B$ and $C$ are integers, and $A$ is a nonnegative integer ($A$ and $B$ not both zero).<br>**Slope-Intercept Form**<br>$y = mx + b$, where $m$ is a real number representing the slope, and $b$ is a real number representing the $y$ intercept. | Find the equation of a line that contains the point $(1,-4)$ and has a slope of $\frac{3}{2}$.<br>**Solution**<br>Substitute $\frac{3}{2}$ for $m$ and $(1, -4)$ for $(x_1, y_1)$ into the formula $m = \frac{y - y_1}{x - x_1}$:<br>$\frac{3}{2} = \frac{y - (-4)}{x - 1}$<br>Simplifying the equation yields $3x - 2y = 11$.<br>**Sample Problem 12**<br>Find the equation of the line that contains the point $(-3, 2)$ and has a slope of $-\frac{1}{2}$. |
| Find the equation of a line given two points contained in the line.<br>(Section 7.4/Objective 2) | First calculate the slope of the line. Substitute the slope and the coordinates of one of the points into either of the following equations.<br>$y - y_1 = m(x - x_1)$ or<br>$m = \frac{y - y_1}{x - x_1}$. | Find the equation of a line that contains the points $(-3, 4)$ and $(-6, 10)$.<br>**Solution**<br>First calculate the slope:<br>$m = \frac{10 - 4}{-6 - (-3)} = \frac{6}{-3} = -2$<br>Now substitute $-2$ for $m$ and $(-3, 4)$ for $(x_1, y_1)$ in the formula $y - y_1 = m(x - x_1)$:<br>$y - 4 = -2(x -(-3))$<br>Simplifying this equation yields $2x + y = -2$.<br>**Sample Problem 13**<br>Find the equation of the line that contains the points $(2, -1)$ and $(-3,-5)$. |

*(continued)*

| OBJECTIVE | SUMMARY | EXAMPLE |
|---|---|---|
| Find the equations for parallel and perpendicular lines.<br>(Section 7.4/Objective 6) | If two distinct lines have slopes $m_1$ and $m_2$, respectively, then:<br>**1.** The two lines are parallel if and only if $m_1 = m_2$.<br>**2.** The two lines are perpendicular if and only if $(m_1)(m_2) = -1$. | Find the equation of a line that contains the point (2, 1) and is parallel to the line $y = 3x + 4$.<br>**Solution**<br>The slope of the parallel line is 3. Therefore, use this slope and the point (2, 1) to determine the equation:<br>$y - 1 = 3(x - 2)$<br>Simplifying this equation yields $y = 3x - 5$.<br>**Sample Problem 14**<br>Find the equation of the line that contains the point $(2, -3)$ and is perpendicular to the line $y = \frac{1}{2}x + 3$. |
| Determine if the graph of an equation is symmetric to the $x$ axis, the $y$ axis, or the origin.<br>(Section 7.5/Objective 2) | The graph of an equation is symmetric with respect to the $y$ axis if replacing $x$ with $-x$ results in an equivalent equation.<br>The graph of an equation is symmetric with respect to the $x$ axis if replacing $y$ with $-y$ results in an equivalent equation.<br>The graph of an equation is symmetric with respect to origin if replacing $x$ with $-x$ and $y$ with $-y$ results in an equivalent equation. | Determine the type of symmetry exhibited by the graph of the equation $x = y^2 + 4$.<br>**Solution**<br>Replacing $x$ with $-x$ gives $-x = y^2 + 4$. This is not an equivalent equation, so the graph will not exhibit $y$-axis symmetry.<br>Replacing $y$ with $-y$ gives $x = (-y)^2 + 4 = y^2 + 4$. This is an equivalent equation, so the graph will exhibit $x$-axis symmetry.<br>Replacing $x$ with $-x$ and $y$ with $-y$ gives:<br>$(-x) = (-y)^2 + 4$<br>$-x = y^2 + 4$<br>This is not an equivalent equation, so the graph will not exhibit symmetry with respect to the origin.<br>**Sample Problem 15**<br>Determine the type of symmetry exhibited by the graph of the equation $y = x^2 - 2$. |

| OBJECTIVE | SUMMARY | EXAMPLE |
|---|---|---|
| Graph nonlinear equations.<br>(Section 7.5/Objective 1) | The following suggestions are offered for graphing an equation in two variables.<br>**1.** Determine what type of symmetry the equation exhibits.<br>**2.** Find the intercepts.<br>**3.** Solve the equation for $y$ or $x$ if it is not already in such a form.<br>**4.** Set up a table of ordered pairs that satisfies the equation. The type of symmetry will affect your choice of values in the table.<br>**5.** Plot the points associated with the ordered pairs and connect them with a smooth curve. Then, if appropriate, reflect this part of the curve according to the symmetry shown by the equation. | Graph $x - y^2 + 4 = 0$.<br>**Solution**<br>Replacing $y$ with $-y$ gives an equivalent equation, so the graph will be symmetric with respect to the $x$ axis.<br>To find the $x$ intercept, let $y = 0$ and solve for $x$. This gives an $x$ intercept of $-4$.<br>To find the $y$ intercept, let $x = 0$ and solve for $y$. This gives $y$ intercepts of 2 and $-2$.<br>Solving the equation for $x$ gives the equation $x = y^2 - 4$.<br>Choose values for $y$ to obtain the table of points.<br>$x$: $-3$, 0, 5; $y$: 1, 2, 3<br>(5, 3) (−3, 1) (0, 2) (0, −4)<br>**Sample Problem 16**<br>Graph $x^2 + y - 3 = 0$. |

## Chapter 7 Review Problem Set

For Problems 1–4, determine which of the ordered pairs are solutions of the given equation.

1. $4x + y = 6$; (1, 2), (6, 0), (−1, 10)
2. $-x + 2y = 4$; (−4, 1), (−4, −1), (0, 2)
3. $3x + 2y = 12$; (2, 3), (−2, 9), (3, 2)
4. $2x - 3y = -6$; (0, −2), (−3, 0), (1, 2)

For Problems 5–8, complete the table of values for the equation and graph the equation.

5. $y = 2x - 5$

| $x$ | −1 | 0 | 1 | 4 |
|---|---|---|---|---|
| $y$ | | | | |

6. $y = -2x - 1$

| $x$ | −3 | −1 | 0 | 2 |
|---|---|---|---|---|
| $y$ | | | | |

7. $y = \dfrac{3x - 4}{2}$

| $x$ | −2 | 0 | 2 | 4 |
|---|---|---|---|---|
| $y$ | | | | |

8. $2x - 3y = 3$

| $x$ | −3 | 0 | 3 |
|---|---|---|---|
| $y$ | | | |

For Problems 9–16, graph each equation by finding the $x$ and $y$ intercepts.

9. $2x - y = 6$
10. $-3x - 2y = 6$
11. $x - 2y = 4$
12. $5x - y = -5$
13. $3x + 2y = -6$
14. $x + 4y = -4$
15. $y = -\frac{1}{2}x + 3$
16. $y = \frac{1}{4}x - 1$

For Problems 17–28, graph each equation.

17. $y = 3x - 1$
18. $y = -x + 4$
19. $x = -3$
20. $y = 5$
21. $y = -4x$
22. $2x + 3y = 0$
23. $x = 1$
24. $y = -2$
25. $y = -\frac{1}{3}x$
26. $y = \frac{2}{3}x$
27. $y = 4$
28. $x = -3$

**29.** **(a)** An apartment moving company charges according to the equation $c = 75h + 150$, where $c$ represents the charge in dollars and $h$ represents the number of hours for the move. Complete the following table.

| $h$ | 1 | 2 | 3 | 4 |
|---|---|---|---|---|
| $c$ | | | | |

**(b)** Labeling the horizontal axis $h$ and the vertical axis $c$, graph the equation $c = 75h + 150$ for nonnegative values of $h$.

**(c)** Use the graph from part (b) to approximate values of $c$ when $h = 1.5$ and 3.5.

**(d)** Check the accuracy of your reading from the graph in part (c) by using the equation $c = 75h + 150$.

**30.** **(a)** The value-added tax is computed by the equation $t = 0.15v$ where $t$ represents the tax and $v$ represents the value of the goods. Complete the following table.

| $v$ | 100 | 200 | 350 | 400 |
|---|---|---|---|---|
| $t$ | | | | |

**(b)** Labeling the horizontal axis $v$ and the vertical axis $t$, graph the equation $t = 0.15v$ for nonnegative values of $v$.

**(c)** Use the graph from part (b) to approximate values of $t$ when $v = 250$ and 300.

**(d)** Check the accuracy of your reading from the graph in part (c) by using the equation $t = 0.15v$.

For Problems 31–38, graph each inequality.

**31.** $-x + 3y < -6$

**32.** $x + 2y \geq 4$

**33.** $2x - 3y \leq 6$

**34.** $y > -\frac{1}{2}x + 3$

**35.** $y < 2x - 5$

**36.** $y \geq \frac{2}{3}x$

**37.** $y > 2x$

**38.** $y < -3x$

**39.** Find the distance between each of the pairs of points.

**(a)** $(-1, 5)$ and $(1, -2)$

**(b)** $(5, 0)$ and $(2, 7)$

**40.** Find the lengths of the sides of a triangle whose vertices are at $(2, 3)$, $(5, -1)$, and $(-4, -5)$.

**41.** Verify that $(1, 2)$ is the midpoint of the line segment joining $(-3, -1)$ and $(5, 5)$.

**42.** Find the slope of the line determined by each pair of points.

**(a)** $(3, 4)$, $(-2, -2)$ **(b)** $(-2, 3)$, $(4, -1)$

**43.** Find $y$ if the line through $(-4, 3)$ and $(12, y)$ has a slope of $\frac{1}{8}$.

**44.** Find $x$ if the line through $(x, 5)$ and $(3, -1)$ has a slope of $-\frac{3}{2}$.

For Problems 45–48, graph the line that has the indicated slope and contains the indicated point.

**45.** $m = -\frac{1}{2}$, $(0, 3)$

**46.** $m = \frac{3}{5}$, $(0, -4)$

**47.** $m = 3$, $(-1, 2)$

**48.** $m = -2$, $(1, 4)$

**49.** A certain highway has a 6% grade. How many feet does it rise in a horizontal distance of 1 mile (5280 feet)?

**50.** If the ratio of rise to run is to be $\frac{2}{3}$ for the steps of a staircase, and the run is 12 inches, find the rise.

**51.** Find the slope of each of the following lines.

**(a)** $4x + y = 7$ **(b)** $2x - 7y = 3$

**52.** Find the slope of any line that is perpendicular to the line $-3x + 5y = 7$.

**53.** Find the slope of any line that is parallel to the line $4x + 5y = 10$.

For Problems 54–61, write the equation of the line that satisfies the stated conditions. Express final equations in standard form.

**54.** Having a slope of $-\frac{3}{7}$ and a $y$ intercept of 4

**55.** Containing the point $(-1, -6)$ and having a slope of $\frac{2}{3}$

**56.** Containing the point $(3, -5)$ and having a slope of $-1$

**57.** Containing the points $(-1, 2)$ and $(3, -5)$

**58.** Containing the points $(0, 4)$ and $(2, 6)$

**59.** Containing the point $(2, 5)$ and parallel to the line $x - 2y = 4$

**60.** Containing the point $(-2, -6)$ and perpendicular to the line $3x + 2y = 12$

**61.** Containing the point $(-8, 3)$ and parallel to the line $4x + y = 7$

**62.** The taxes for a primary residence can be described by a linear relationship. Find the equation for the relationship if the taxes for a home valued at $200,000 are $2400, and the taxes are $3150 when the home is valued at $250,000. Let $y$ be the taxes and $x$ the value of the home. Write the equation in slope-intercept form.

**63.** The freight charged by a trucking firm for a parcel under 200 pounds depends on the distance it is being shipped. To ship a 150-pound parcel 300 miles, it costs $40. If the same parcel is shipped 1000 miles, the cost is $180. Assume the relationship between the cost and distance is linear. Find the equation for the relationship. Let $y$ be the cost and $x$ be the miles. Write the equation in slope-intercept form.

**64.** On a final exam in math class, the number of points earned has a linear relationship with the number of correct answers. John got 96 points when he answered 12 questions correctly. Kimberly got 144 points when she answered 18 questions correctly. Find the equation for the relationship. Let $y$ be the number of points and $x$ be the number of correct answers. Write the equation in slope-intercept form.

**65.** The time needed to install computer cables has a linear relationship with the number of feet of cable being installed. It takes $1\frac{1}{2}$ hours to install 300 feet, and 1050 feet can be installed in 4 hours. Find the equation for the relationship. Let $y$ be the feet of cable installed and $x$ be the time in hours. Write the equation in slope-intercept form.

**66.** Determine the type(s) of symmetry (symmetry with respect to the $x$ axis, $y$ axis, and/or origin) exhibited by the graph of each of the following equations. Do not sketch the graph.

**(a)** $y = x^2 + 4$ **(b)** $xy = -4$
**(c)** $y = -x^3$ **(d)** $x = y^4 + 2y^2$

For Problems 67–72, graph each equation.

**67.** $y = x^3 + 2$ **68.** $y = -x^3$

**69.** $y = x^2 + 3$ **70.** $y = -2x^2 - 1$

**71.** $y = -x^2 + 6$ **72.** $y = x^2 - 4$

# Chapter 7 Test

1. Find the slope of the line determined by the points $(-2, 4)$ and $(3, -2)$.

2. Find the slope of the line determined by the equation $3x - 7y = 12$.

3. Find the length of a line segment with endpoints of $(4, 2)$ and $(-3, -1)$. Express the answer in simplest radical form.

4. Find the equation of the line that has a slope of $-\frac{3}{2}$ and contains the point $(4, -5)$. Express the equation in standard form.

5. Find the equation of the line that contains the points $(-4, 2)$ and $(2, 1)$. Express the equation in slope-intercept form.

6. Find the equation of the line that is parallel to the line $5x + 2y = 7$ and contains the point $(-2, -4)$. Express the equation in standard form.

7. Find the equation of the line that is perpendicular to the line $x - 6y = 9$ and contains the point $(4, 7)$. Express the equation in standard form.

8. What kind(s) of symmetry does the graph of $y = 9x$ exhibit?

9. What kind(s) of symmetry does the graph of $y^2 = x^2 + 6$ exhibit?

10. What kind(s) of symmetry does the graph of $x^2 + 6x + 2y^2 - 8 = 0$ exhibit?

11. What is the slope of all lines that are parallel to the line $7x - 2y = 9$?

12. What is the slope of all lines that are perpendicular to the line $4x + 9y = -6$?

13. Find the $x$ intercept of the line $y = \frac{3}{5}x - \frac{2}{3}$.

14. Find the $y$ intercept of the line $\frac{3}{4}x - \frac{2}{5}y = \frac{1}{4}$.

15. The grade of a highway up a hill is 25%. How much change in horizontal distance is there if the vertical height of the hill is 120 feet?

16. Suppose that a highway rises 200 feet in a horizontal distance of 3000 feet. Express the grade of the highway to the nearest tenth of a percent.

17. If the ratio of rise to run is to be $\frac{3}{4}$ for the steps of a staircase, and the rise is 32 centimeters, find the run to the nearest centimeter.

For Problems 18–23, graph each equation.

18. $y = -x^2 - 3$

19. $y = -x - 3$

20. $-3x + y = 5$

21. $y = \frac{2}{3}x$

22. $2x + 3y = 12$

23. $y = -\frac{1}{4}x - \frac{1}{4}$

For Problems 24 and 25, graph each inequality.

24. $2x - y < 4$

25. $3x + 2y \geq 6$

# 10 Systems of Equations

## Study Skill Tip

Hopefully at the start of your math course you determined if your course has a final exam. Most math courses have a cumulative final exam. Cumulative means that the exam covers every topic taught in the course. The final exam may be written by your instructor, or it may be a departmental final exam made up by the math department. In either case, you should check with your instructor to get practice tests for the final exam. The percent of your grade determined by the final exam typically varies from 10% to 30%. Your final course grade may depend on your grade on the final exam.

Preparing for the final exam is similar to preparing for any test, except you need even more time to get ready. Start preparing for the final exam about two weeks before the test. Write an outline listing the topics to be covered on the final exam. Review those topics in your notes and also do practice problems on those topics. If you have access to your previous tests, go over them and correct any errors that you made. Arrive early for the final exam to avoid being late. Final exams are often timed tests, and you don't want to lose any time by being late.

***"Be prepared, work hard, and hope for a little luck. Recognize that the harder you work and the better prepared you are, the more luck you might have."***
ED BRADLEY

***Is your success due to hard work and preparation or luck, or a combination of both?***

## Chapter Preview

Equations in two variables were previously introduced in Chapter 7. This chapter considers systems of two or more equations. That is, we are looking for solutions that satisfy more than one equation at the same time. Systems of equations are also referred to as simultaneous equations.

Three methods for solving systems covered in this chapter are

- Graphing
- Substitution method
- Elimination-by-addition method

# 10.1 Systems of Two Linear Equations and Linear Inequalities in Two Variables

OBJECTIVES

1. Solve systems of two linear equations by graphing
2. Solve systems of linear inequalities

In Chapter 7, we stated that any equation of the form $Ax + By = C$, where $A$, $B$, and $C$ are real numbers ($A$ and $B$ not both zero) is a *linear equation* in the two variables $x$ and $y$, and its graph is a straight line. Two linear equations in two variables considered together form a **system of two linear equations in two variables**. Here are a few examples:

$$\begin{pmatrix} x + y = 6 \\ x - y = 2 \end{pmatrix} \quad \begin{pmatrix} 3x + 2y = 1 \\ 5x - 2y = 23 \end{pmatrix} \quad \begin{pmatrix} 4x - 5y = 21 \\ 3x + 7y = -38 \end{pmatrix}$$

To *solve a system*, such as one of the above, means to find all of the ordered pairs that satisfy both equations in the system. For example, if we graph the two equations $x + y = 6$ and $x - y = 2$ on the same set of axes, as in Figure 10.1, then the ordered pair associated with the point of intersection of the two lines is the solution of the system. Thus we say that $\{(4, 2)\}$ is the solution set of the system

$$\begin{pmatrix} x + y = 6 \\ x - y = 2 \end{pmatrix}$$

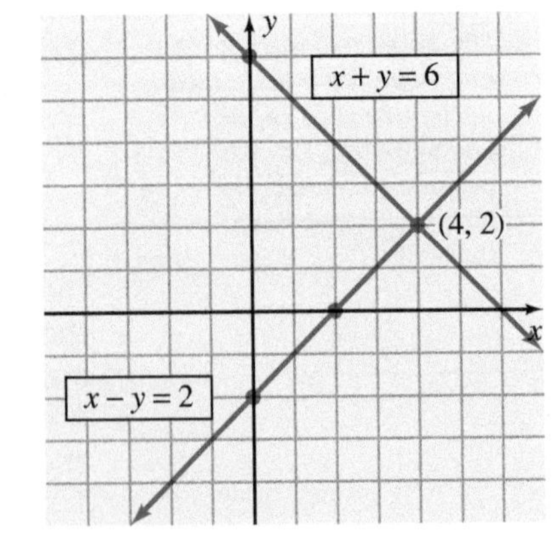

**Figure 10.1**

To check, substitute 4 for $x$ and 2 for $y$ in the two equations, which yields

$x + y = 6$ becomes $4 + 2 = 6$ A true statement

$x - y = 2$ becomes $4 - 2 = 2$ A true statement

Because the graph of a linear equation in two variables is a straight line, there are three possible situations that can occur when we solve a system of two linear equations in two variables. We illustrate these cases in Figure 10.2.

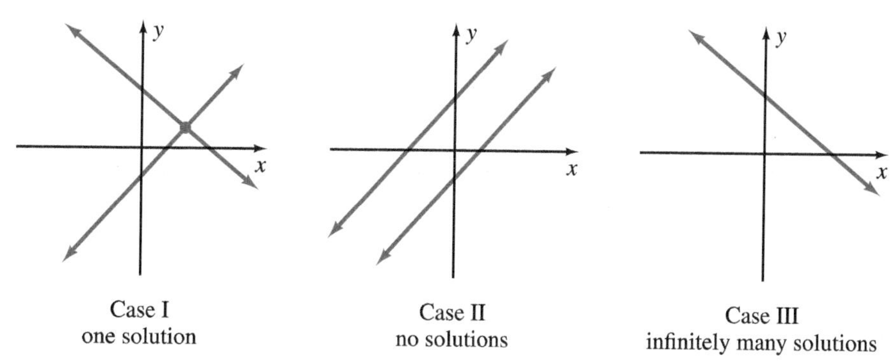


**Figure 10.2**

**Case I** The graphs of the two equations are two lines intersecting in *one* point, and there is *one solution*. This system is called a **consistent system**.

**Case II** The graphs of the two equations are parallel lines, and there is *no solution*. This system is called an **inconsistent system**.

**Case III** The graphs of the two equations are the same line, and there are *infinitely many solutions* to the system. Any pair of real numbers that satisfies one of the equations will also satisfy the other equation. We say that the equations are **dependent**.

Thus as we solve a system of two linear equations in two variables, we know what to expect. The system will have no solutions, one ordered pair as a solution, or infinitely many ordered pairs as solutions.

**Classroom Example**
Solve the system:

$$\begin{pmatrix} x + 2y = 5 \\ 2x - y = -5 \end{pmatrix}$$

**EXAMPLE 1** Solve the system $\begin{pmatrix} 2x - y = -2 \\ 4x + y = 8 \end{pmatrix}$.

### Solution

Graph both lines on the same coordinate system. Let's graph the lines by determining intercepts and a check point for each of the lines.

**$2x - y = -2$**

| $x$ | $y$ |
|---|---|
| 0 | 2 |
| −1 | 0 |
| 2 | 6 |

**$4x + y = 8$**

| $x$ | $y$ |
|---|---|
| 0 | 8 |
| 2 | 0 |
| 1 | 4 |

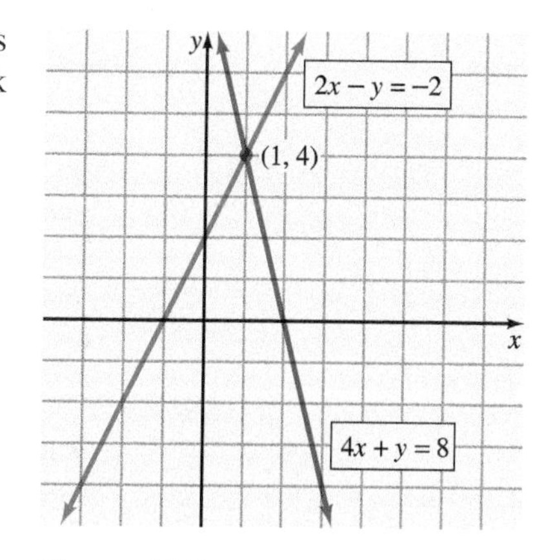


**Figure 10.3**

Figure 10.3 shows the graphs of the two equations. It appears that (1, 4) is the solution of the system.

To check it, we can substitute 1 for $x$ and 4 for $y$ in both equations.

$2x - y = -2$ becomes $2(1) - 4 = -2$ A true statement

$4x + y = 8$ becomes $4(1) + 4 = 8$ A true statement

Therefore, $\{(1, 4)\}$ is the solution set.

**Classroom Example**
Solve the system:

$$\begin{pmatrix} 4x + 6y = 8 \\ 2x + 3y = 4 \end{pmatrix}$$

**EXAMPLE 2** Solve the system $\begin{pmatrix} x - 3y = 3 \\ 2x - 6y = 6 \end{pmatrix}$.

### Solution

Graph both lines on the same coordinate system. Let's graph the lines by determining intercepts and a check point for each of the lines.

**$x - 3y = 3$**

| $x$ | $y$ |
|---|---|
| 0 | $-1$ |
| 3 | 0 |
| $-3$ | $-2$ |

**$2x - 6y = 6$**

| $x$ | $y$ |
|---|---|
| 0 | $-1$ |
| 3 | 0 |
| 1 | $-\frac{2}{3}$ |

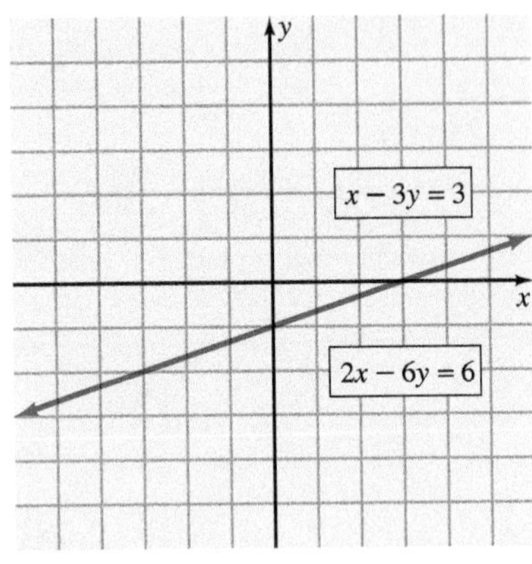


Figure 10.4

Figure 10.4 shows the graph of this system. Since the graphs of both equations are the same line, the coordinates of any point on the line satisfy both equations. Hence the system has infinitely many solutions. Informally, the solution is stated as "infinitely many solutions." In set notation the solution would be written as $\{(x, y)|x - 3y = 3\}$. This is read as "the set of ordered pairs $(x, y)$ such that $x - 3y = 3$," which means that the coordinates of every point on the line $x - 3y = 3$ are solutions to the system.

**Classroom Example**
Solve the system:

$$\begin{pmatrix} 3x - y = 5 \\ y = 3x - 2 \end{pmatrix}$$

**EXAMPLE 3** Solve the system $\begin{pmatrix} y = 2x + 3 \\ 2x - y = 8 \end{pmatrix}$.

### Solution

Graph both lines on the same coordinate system. Let's graph the lines by determining intercepts and a check point for each of the lines.

**$y = 2x + 3$**

| $x$ | $y$ |
|---|---|
| 0 | 3 |
| $-\frac{3}{2}$ | 0 |
| 1 | 5 |

**$2x - y = 8$**

| $x$ | $y$ |
|---|---|
| 0 | $-8$ |
| 4 | 0 |
| 2 | $-4$ |

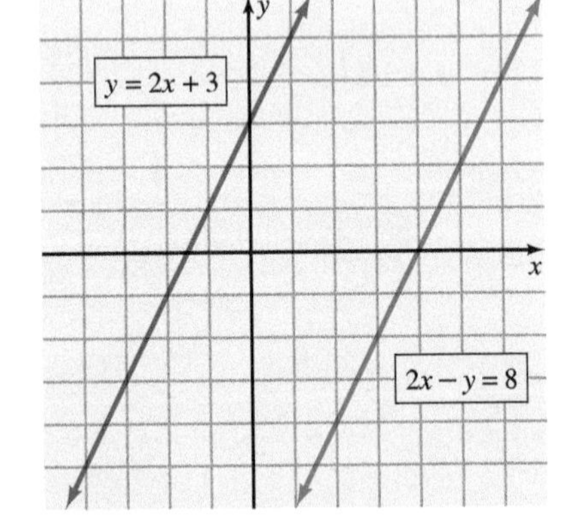


Figure 10.5

Figure 10.5 shows the graph of this system. Since the lines are parallel, there is no solution to the system. The solution set is $\varnothing$.

## Solving Systems of Linear Inequalities

Finding solution sets for systems of linear inequalities relies heavily on the graphing approach. The solution set of a system of linear inequalities, such as

$$\begin{pmatrix} x + y > 2 \\ x - y < 2 \end{pmatrix}$$

is the intersection of the solution sets of the individual inequalities. In Figure 10.6(a) we indicated the solution set for $x + y > 2$, and in Figure 10.6(b) we indicated the solution set for $x - y < 2$. Then, in Figure 10.6(c), we shaded the region that represents the intersection of the

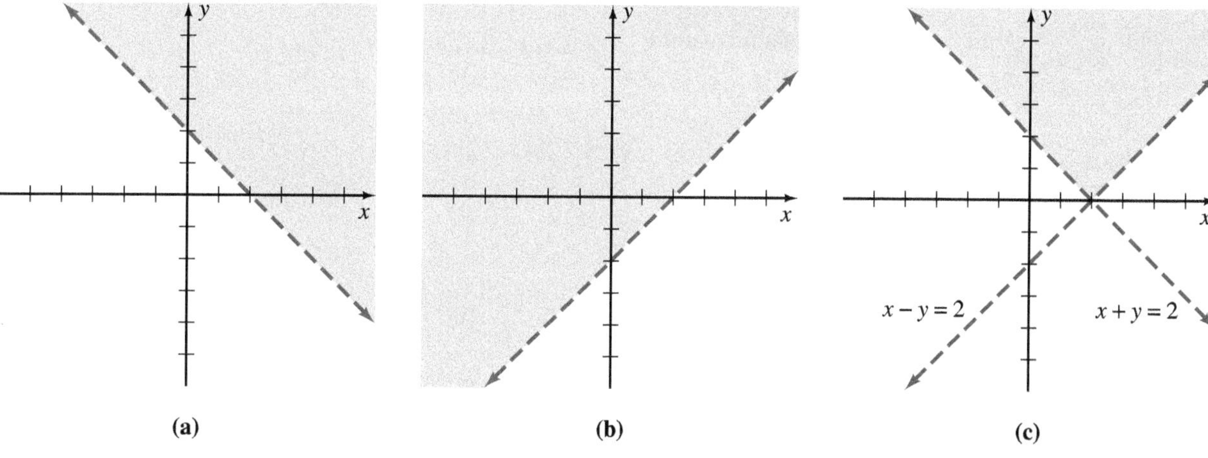


Figure 10.6

two solution sets from parts (a) and (b); thus it is the graph of the system. Remember that dashed lines are used to indicate that the points on the lines are not included in the solution set.

In the following examples, we indicated only the final solution set for the system.

Classroom Example
Solve this system by graphing:
$\begin{pmatrix} x - 2y < 3 \\ 3x + 4y \geq -8 \end{pmatrix}$

**EXAMPLE 4** Solve this system by graphing $\begin{pmatrix} 2x - \ \ y \geq 4 \\ x + 2y < 2 \end{pmatrix}$.

## Solution

The graph of $2x - y \geq 4$ consists of all points *on or below* the line $2x - y = 4$. The graph of $x + 2y < 2$ consists of all points *below* the line $x + 2y = 2$. The graph of the system is indicated by the shaded region in Figure 10.7. Note that all points in the shaded region are on or below the line $2x - y = 4$ *and* below the line $x + 2y = 2$.

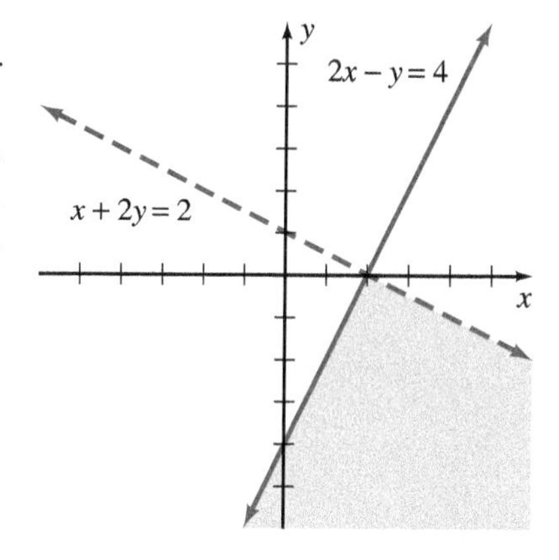


Figure 10.7

Classroom Example
Solve this system by graphing:
$\begin{pmatrix} y < -\frac{1}{3}x + 1 \\ y > \frac{4}{5}x - 3 \end{pmatrix}$

**EXAMPLE 5** Solve this system by graphing $\begin{pmatrix} y > \frac{2}{5}x - 4 \\ y < -\frac{1}{3}x - 1 \end{pmatrix}$.

## Solution

The graph of $y > \frac{2}{5}x - 4$ consists of all points *above* the line $y = \frac{2}{5}x - 4$. The graph of $y < -\frac{1}{3}x - 1$ consists of all points *below* the line $y = -\frac{1}{3}x - 1$. The graph of the system is indicated by the shaded region in Figure 10.8. Note that all points in the shaded region are *above* the line $y = \frac{2}{5}x - 4$ and *below* the line $y = -\frac{1}{3}x - 1$.

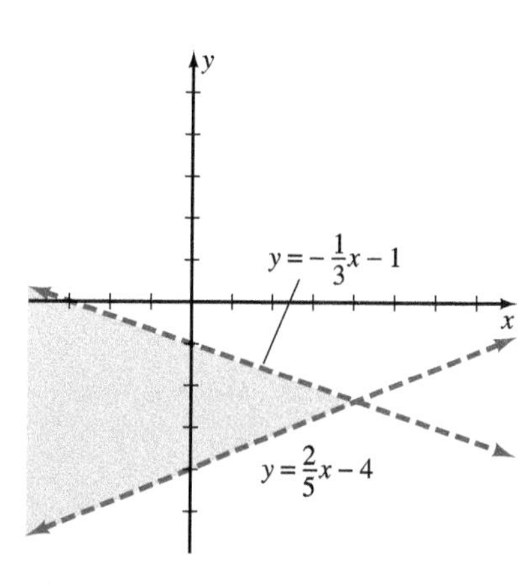


Figure 10.8

**Classroom Example**
Solve this system by graphing:

$$\begin{pmatrix} x > -3 \\ y \leq 4 \end{pmatrix}$$

**EXAMPLE 6** Solve this system by graphing $\begin{pmatrix} x \leq \ \ 2 \\ y \geq -1 \end{pmatrix}$.

### Solution

Remember that even though each inequality contains only one variable, we are working in a rectangular coordinate system that involves ordered pairs. That is, the system could be written as

$$\begin{pmatrix} x + 0(y) \leq 2 \\ 0(x) + y \geq -1 \end{pmatrix}$$

The graph of the system is the shaded region in Figure 10.9. Note that all points in the shaded region are on or to the left of the line $x = 2$ and on or above the line $y = -1$.

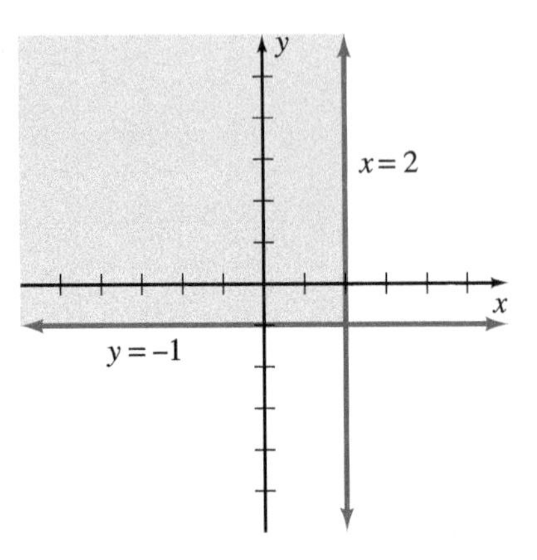


**Figure 10.9**

In our final example of this section, we will use a graphing utility to help solve a system of equations.

**Classroom Example**
Solve the system:

$$\begin{pmatrix} 3.25x + 5.01y = \ \ 31.55 \\ 1.73x - 2.45y = -8.79 \end{pmatrix}$$

**EXAMPLE 7** Solve the system $\begin{pmatrix} 1.14x + 2.35y = -7.12 \\ 3.26x - 5.05y = \ \ 26.72 \end{pmatrix}$.

### Solution

First, we need to solve each equation for $y$ in terms of $x$. Thus the system becomes

$$\begin{pmatrix} y = \dfrac{-7.12 - 1.14x}{2.35} \\ y = \dfrac{3.26x - 26.72}{5.05} \end{pmatrix}$$

Now we can enter both of these equations into a graphing utility and obtain Figure 10.10. In this figure it appears that the point of intersection is at approximately $x = 2$ and $y = -4$. By direct substitution into the given equations, we can verify that the point of intersection is exactly $(2, -4)$.

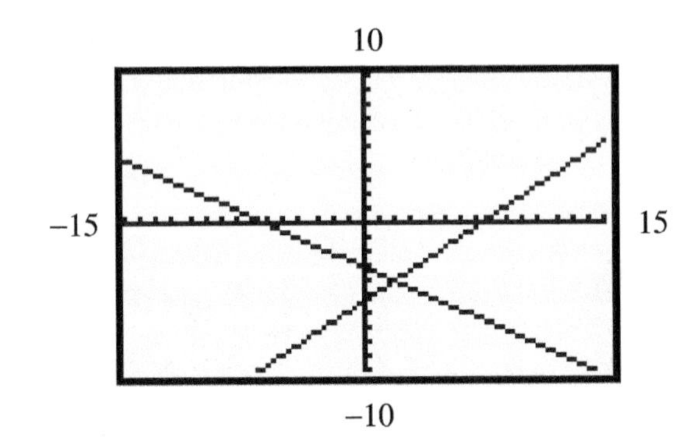


**Figure 10.10**

## Concept Quiz 10.1

For Problems 1–10, answer true or false.

1. To solve a system of equations means to find all the ordered pairs that satisfy all the equations in the system.
2. A consistent system of linear equations will have more than one solution.
3. If the graph of a system of two linear equations results in two distinct parallel lines, then the system has no solutions.
4. Every system of equations has a solution.
5. If the graphs of the two equations in a system are the same line, then the equations in the system are dependent.

**6.** The solution set of a system of linear inequalities is the intersection of the solution sets of the individual inequalities.

**7.** For the system of inequalities $\begin{pmatrix} 2x + y > 4 \\ x - 3y < 6 \end{pmatrix}$, the points on the line $2x + y = 4$ are included in the solution.

**8.** The solution set of the system of inequalities $\begin{pmatrix} y > 2x + 5 \\ y < 2x + 1 \end{pmatrix}$ is the null set.

**9.** The ordered pair (1, 4) satisfies the system of inequalities $\begin{pmatrix} x + y > 2 \\ -2x + y < 3 \end{pmatrix}$.

**10.** The ordered pair (−4, −1) satisfies the system of inequalities $\begin{pmatrix} x + y < 5 \\ 2x - 3y > 6 \end{pmatrix}$.

## Problem Set 10.1

For Problems 1–16, use the graphing approach to determine whether the system is consistent, the system is inconsistent, or the equations are dependent. If the system is consistent, find the solution set from the graph and check it. (Objective 1)

**1.** $\begin{pmatrix} x - y = 1 \\ 2x + y = 8 \end{pmatrix}$

**2.** $\begin{pmatrix} 3x + y = 0 \\ x - 2y = -7 \end{pmatrix}$

**3.** $\begin{pmatrix} 4x + 3y = -5 \\ 2x - 3y = -7 \end{pmatrix}$

**4.** $\begin{pmatrix} 2x - y = 9 \\ 4x - 2y = 11 \end{pmatrix}$

**5.** $\begin{pmatrix} \frac{1}{2}x + \frac{1}{4}y = 9 \\ 4x + 2y = 72 \end{pmatrix}$

**6.** $\begin{pmatrix} 5x + 2y = -9 \\ 4x - 3y = 2 \end{pmatrix}$

**7.** $\begin{pmatrix} \frac{1}{2}x - \frac{1}{3}y = 3 \\ x + 4y = -8 \end{pmatrix}$

**8.** $\begin{pmatrix} 4x - 9y = -60 \\ \frac{1}{3}x - \frac{3}{4}y = -5 \end{pmatrix}$

**9.** $\begin{pmatrix} x - \frac{y}{2} = -4 \\ 8x - 4y = -1 \end{pmatrix}$

**10.** $\begin{pmatrix} 3x - 2y = 7 \\ 6x + 5y = -4 \end{pmatrix}$

**11.** $\begin{pmatrix} x + 2y = 4 \\ 2x - y = 3 \end{pmatrix}$

**12.** $\begin{pmatrix} 2x - y = -8 \\ x + y = 2 \end{pmatrix}$

**13.** $\begin{pmatrix} y = 2x + 5 \\ x + 3y = -6 \end{pmatrix}$

**14.** $\begin{pmatrix} y = 4 - 2x \\ y = 7 - 3x \end{pmatrix}$

**15.** $\begin{pmatrix} y = 2x \\ 3x - 2y = -2 \end{pmatrix}$

**16.** $\begin{pmatrix} y = -2x \\ 3x - y = 0 \end{pmatrix}$

For Problems 17–32, indicate the solution set for each system of inequalities by shading the appropriate region. (Objective 2)

**17.** $\begin{pmatrix} 3x - 4y \geq 0 \\ 2x + 3y \leq 0 \end{pmatrix}$

**18.** $\begin{pmatrix} 3x + 2y \leq 6 \\ 2x - 3y \geq 6 \end{pmatrix}$

**19.** $\begin{pmatrix} x - 3y < 6 \\ x + 2y \geq 4 \end{pmatrix}$

**20.** $\begin{pmatrix} 2x - y \leq 4 \\ 2x + y > 4 \end{pmatrix}$

**21.** $\begin{pmatrix} x + y < 4 \\ x - y > 2 \end{pmatrix}$

**22.** $\begin{pmatrix} x + y > 1 \\ x - y < 1 \end{pmatrix}$

**23.** $\begin{pmatrix} y < x + 1 \\ y \geq x \end{pmatrix}$

**24.** $\begin{pmatrix} y > x - 3 \\ y < x \end{pmatrix}$

**25.** $\begin{pmatrix} y > x \\ y > 2 \end{pmatrix}$

**26.** $\begin{pmatrix} 2x + y > 6 \\ 2x + y < 2 \end{pmatrix}$

**27.** $\begin{pmatrix} x \geq -1 \\ y < 4 \end{pmatrix}$

**28.** $\begin{pmatrix} x < 3 \\ y > 2 \end{pmatrix}$

**29.** $\begin{pmatrix} 2x - y > 4 \\ 2x - y > 0 \end{pmatrix}$

**30.** $\begin{pmatrix} x + y > 4 \\ x + y > 6 \end{pmatrix}$

**31.** $\begin{pmatrix} 3x - 2y < 6 \\ 2x - 3y < 6 \end{pmatrix}$

**32.** $\begin{pmatrix} 2x + 5y > 10 \\ 5x + 2y > 10 \end{pmatrix}$

## Thoughts Into Words

**33.** How do you know by inspection, without graphing, that the solution set of the system $\begin{pmatrix} 3x - 2y > 5 \\ 3x - 2y < 2 \end{pmatrix}$ is the null set?

**34.** Is it possible for a system of two linear equations in two variables to have exactly two solutions? Defend your answer.

## Graphing Calculator Activities

**35.** Use your graphing calculator to help determine whether, in Problems 1–16, the system is consistent, the system is inconsistent, or the equations are dependent.

**36.** Use your graphing calculator to help determine the solution set for each of the following systems. Be sure to check your answers.

**(a)** $\begin{pmatrix} 3x - y = 30 \\ 5x - y = 46 \end{pmatrix}$ **(b)** $\begin{pmatrix} 1.2x + 3.4y = 25.4 \\ 3.7x - 2.3y = 14.4 \end{pmatrix}$

**(c)** $\begin{pmatrix} 1.98x + 2.49y = 13.92 \\ 1.19x + 3.45y = 16.18 \end{pmatrix}$

**(d)** $\begin{pmatrix} 2x - 3y = 10 \\ 3x + 5y = 53 \end{pmatrix}$ **(e)** $\begin{pmatrix} 4x - 7y = -49 \\ 6x + 9y = 219 \end{pmatrix}$

**(f)** $\begin{pmatrix} 3.7x - 2.9y = -14.3 \\ 1.6x + 4.7y = -30 \end{pmatrix}$

### Answers to the Concept Quiz

**1.** True **2.** False **3.** True **4.** False **5.** True **6.** True **7.** False **8.** True **9.** True **10.** False

# 10.2 Substitution Method

OBJECTIVES

1. Solve systems of linear equations by substitution
2. Use systems of equations to solve problems

It should be evident that solving systems of equations by graphing requires accurate graphs. In fact, unless the solutions are integers, it is quite difficult to obtain exact solutions from a graph. Thus we will consider some other methods for solving systems of equations.

We describe the **substitution method**, which works quite well with systems of two linear equations in two unknowns, as follows:

**Step 1** Solve one of the equations for one variable in terms of the other variable if neither equation is in such a form. (If possible, make a choice that will avoid fractions.)

**Step 2** Substitute the expression obtained in Step 1 into the other equation to produce an equation with one variable.

**Step 3** Solve the equation obtained in step 2.

**Step 4** Use the solution obtained in step 3, along with the expression obtained in step 1, to determine the solution of the system.

Now let's look at some examples that illustrate the substitution method.

Classroom Example
Solve the system:

$\begin{pmatrix} 2x + y = 2 \\ y = x - 7 \end{pmatrix}$

**EXAMPLE 1** Solve the system $\begin{pmatrix} x + y = 16 \\ y = x + 2 \end{pmatrix}$.

### Solution

Because the second equation states that $y$ equals $x + 2$, we can substitute $x + 2$ for $y$ in the first equation.

$$x + y = 16 \xrightarrow{\text{Substitute } x + 2 \text{ for } y} x + (x + 2) = 16$$

Now we have an equation with one variable that we can solve in the usual way.

$$\begin{aligned} x + (x + 2) &= 16 \\ 2x + 2 &= 16 \\ 2x &= 14 \\ x &= 7 \end{aligned}$$

Substituting 7 for $x$ in one of the two original equations (let's use the second one) yields

$$y = 7 + 2 = 9$$

To check, we can substitute 7 for $x$ and 9 for $y$ in both of the original equations.

$$7 + 9 = 16 \quad \text{A true statement}$$
$$9 = 7 + 2 \quad \text{A true statement}$$

The solution set is $\{(7, 9)\}$.

**Classroom Example**
Solve the system:
$\begin{pmatrix} 5x + 2y = 4 \\ x = 4y - 30 \end{pmatrix}$

**EXAMPLE 2** Solve the system $\begin{pmatrix} x = 3y - 25 \\ 4x + 5y = 19 \end{pmatrix}$.

### Solution

In this case the first equation states that $x$ equals $3y - 25$. Therefore, we can substitute $3y - 25$ for $x$ in the second equation.

$$4x + 5y = 19 \xrightarrow{\text{Substitute } 3y - 25 \text{ for } x} 4(3y - 25) + 5y = 19$$

Solving this equation yields

$$4(3y - 25) + 5y = 19$$
$$12y - 100 + 5y = 19$$
$$17y = 119$$
$$y = 7$$

Substituting 7 for $y$ in the first equation produces

$$x = 3(7) - 25$$
$$= 21 - 25 = -4$$

The solution set is $\{(-4, 7)\}$; check it.

**Classroom Example**
Solve the system:
$\begin{pmatrix} 6x + 5y = 4 \\ 3x + \phantom{5}y = 9 \end{pmatrix}$

**EXAMPLE 3** Solve the system $\begin{pmatrix} 3x - 7y = 2 \\ x + 4y = 1 \end{pmatrix}$.

### Solution

Let's solve the second equation for $x$ in terms of $y$.

$$x + 4y = 1$$
$$x = 1 - 4y$$

Now we can substitute $1 - 4y$ for $x$ in the first equation.

$$3x - 7y = 2 \xrightarrow{\text{Substitute } 1 - 4y \text{ for } x} 3(1 - 4y) - 7y = 2$$

Let's solve this equation for $y$.

$$3(1 - 4y) - 7y = 2$$
$$3 - 12y - 7y = 2$$
$$-19y = -1$$
$$y = \frac{1}{19}$$

Finally, we can substitute $\frac{1}{19}$ for $y$ in the equation $x = 1 - 4y$.

$$x = 1 - 4\left(\frac{1}{19}\right)$$
$$= 1 - \frac{4}{19}$$
$$= \frac{15}{19}$$

The solution set is $\left\{\left(\frac{15}{19}, \frac{1}{19}\right)\right\}$.

**Classroom Example**
Solve the system:
$\begin{pmatrix} 9x + 4y = 44 \\ 3x - 5y = 2 \end{pmatrix}$

**EXAMPLE 4** Solve the system $\begin{pmatrix} 5x - 6y = -4 \\ 3x + 2y = -8 \end{pmatrix}$.

### Solution

Note that solving either equation for either variable will produce a fractional form. Let's solve the second equation for $y$ in terms of $x$.

$$3x + 2y = -8$$
$$2y = -8 - 3x$$
$$y = \frac{-8 - 3x}{2}$$

Now we can substitute $\frac{-8 - 3x}{2}$ for $y$ in the first equation.

$$5x - 6y = -4 \xrightarrow{\text{Substitute } \frac{-8-3x}{2} \text{ for } y} 5x - 6\left(\frac{-8 - 3x}{2}\right) = -4$$

Solving this equation yields

$$5x - 6\left(\frac{-8 - 3x}{2}\right) = -4$$
$$5x - 3(-8 - 3x) = -4$$
$$5x + 24 + 9x = -4$$
$$14x = -28$$
$$x = -2$$

Substituting $-2$ for $x$ in $y = \frac{-8 - 3x}{2}$ yields

$$y = \frac{-8 - 3(-2)}{2}$$
$$= \frac{-8 + 6}{2}$$
$$= \frac{-2}{2}$$
$$= -1$$

The solution set is $\{(-2, -1)\}$.

## Using Systems of Equations to Solve Problems

Many word problems that we solved earlier in this text using one variable and one equation can also be solved using a system of two linear equations in two variables. In fact, in many of these problems you may find it more natural to use two variables. Let's consider some examples.

**Classroom Example**
Jan invested some money at 3.5% and $1100 more than that amount at 4.8%. The yearly interest from the two investments was $252. How much did Jan invest at each rate?

### EXAMPLE 5 Apply Your Skill

Anita invested some money at 8% and $400 more than that amount at 9%. The yearly interest from the two investments was $87. How much did Anita invest at each rate?

### Solution

Let $x$ represent the amount invested at 8%, and let $y$ represent the amount invested at 9%. The problem translates into the following system.

Amount invested at 9% was $400 more than at 8% ⟶ $y = x + 400$
Yearly interest from the two investments was $87 ⟶ $0.08x + 0.09y = 87$

$$\begin{pmatrix} y = x + 400 \\ 0.08x + 0.09y = 87 \end{pmatrix}$$

From the first equation we can substitute $x + 400$ for $y$ in the second equation and solve for $x$.

$$\begin{aligned} 0.08x + 0.09(x + 400) &= 87 \\ 0.08x + 0.09x + 36 &= 87 \\ 0.17x &= 51 \\ x &= 300 \end{aligned}$$

Therefore, $300 is invested at 8% and $300 + $400 = $700 is invested at 9%.

**Classroom Example**
If the perimeter of a rectangle is 98 centimeters, and the length is 15 centimeters more than the width, find the dimensions of the rectangle.

### EXAMPLE 6 Apply Your Skill

The perimeter of a rectangle is 66 inches. The width of the rectangle is 7 inches less than the length of the rectangle. Find the dimensions of the rectangle.

### Solution

Let $l$ represent the length of the rectangle and $w$ represent the width of the rectangle. The problem translates into the following system.

$$\begin{pmatrix} 2w + 2l = 66 \\ w = l - 7 \end{pmatrix}$$

From the second equation, we can substitute $l - 7$ for $w$ in the first equation and solve.

$$\begin{aligned} 2w + 2l &= 66 \\ 2(l - 7) + 2l &= 66 \\ 2l - 14 + 2l &= 66 \\ 4l &= 80 \\ l &= 20 \end{aligned}$$

Substitute 20 for $l$ in $w = l - 7$ to obtain

$$w = 20 - 7 = 13.$$

Therefore, the dimensions of the rectangle are 13 inches by 20 inches.

## Concept Quiz 10.2

For Problems 1–5, answer true or false.

1. Graphing a system of equations is the most accurate method to find the solution of a system.
2. To begin solving a system of equations by substitution, one of the equations is solved for one variable in terms of the other variable.
3. When solving a system of equations by substitution, deciding which variable to solve for may allow you to avoid working with fractions.
4. When finding the solution of the system $\begin{pmatrix} x = 2y + 4 \\ x = -y + 5 \end{pmatrix}$, you only have to find a value for $x$.
5. The ordered pairs (1, 3) and (5, 11) are both solutions of the system $\begin{pmatrix} y = 2x + 1 \\ 4x - 2y = -2 \end{pmatrix}$.

## Problem Set 10.2

For Problems 1–26, solve each system by using the substitution method. (Objective 1)

1. $\begin{pmatrix} x + y = 20 \\ x = y - 4 \end{pmatrix}$

2. $\begin{pmatrix} x + y = 23 \\ y = x - 5 \end{pmatrix}$

3. $\begin{pmatrix} y = -3x - 18 \\ 5x - 2y = -8 \end{pmatrix}$

4. $\begin{pmatrix} 4x - 3y = 33 \\ x = -4y - 25 \end{pmatrix}$

5. $\begin{pmatrix} x = -3y \\ 7x - 2y = -69 \end{pmatrix}$

6. $\begin{pmatrix} 9x - 2y = -38 \\ y = -5x \end{pmatrix}$

7. $\begin{pmatrix} 2x + 3y = 11 \\ 3x - 2y = -3 \end{pmatrix}$

8. $\begin{pmatrix} 3x - 4y = -14 \\ 4x + 3y = 23 \end{pmatrix}$

9. $\begin{pmatrix} 3x - 4y = 9 \\ x = 4y - 1 \end{pmatrix}$

10. $\begin{pmatrix} y = 3x - 5 \\ 2x + 3y = 6 \end{pmatrix}$

11. $\begin{pmatrix} y = \frac{2}{5}x - 1 \\ 3x + 5y = 4 \end{pmatrix}$

12. $\begin{pmatrix} y = \frac{3}{4}x - 5 \\ 5x - 4y = 9 \end{pmatrix}$

13. $\begin{pmatrix} 7x - 3y = -2 \\ x = \frac{3}{4}y + 1 \end{pmatrix}$

14. $\begin{pmatrix} 5x - y = 9 \\ x = \frac{1}{2}y - 3 \end{pmatrix}$

15. $\begin{pmatrix} 2x + y = 12 \\ 3x - y = 13 \end{pmatrix}$

16. $\begin{pmatrix} -x + 4y = -22 \\ x - 7y = 34 \end{pmatrix}$

17. $\begin{pmatrix} 4x + 3y = -40 \\ 5x - y = -12 \end{pmatrix}$

18. $\begin{pmatrix} x - 5y = 33 \\ -4x + 7y = -41 \end{pmatrix}$

19. $\begin{pmatrix} 3x + y = 2 \\ 11x - 3y = 5 \end{pmatrix}$

20. $\begin{pmatrix} 2x - y = 9 \\ 7x + 4y = 1 \end{pmatrix}$

21. $\begin{pmatrix} 3x + 5y = 22 \\ 4x - 7y = -39 \end{pmatrix}$

22. $\begin{pmatrix} 2x - 3y = -16 \\ 6x + 7y = 16 \end{pmatrix}$

23. $\begin{pmatrix} 4x - 5y = 3 \\ 8x + 15y = -24 \end{pmatrix}$

24. $\begin{pmatrix} 2x + 3y = 3 \\ 4x - 9y = -4 \end{pmatrix}$

25. $\begin{pmatrix} 6x - 3y = 4 \\ 5x + 2y = -1 \end{pmatrix}$

26. $\begin{pmatrix} 7x - 2y = 1 \\ 4x + 5y = 2 \end{pmatrix}$

For Problems 27–40, solve each problem by setting up and solving an appropriate system of equations. (Objective 2)

27. Doris invested some money at 7% and some money at 8%. She invested $6000 more at 8% than she did at 7%. Her total yearly interest from the two investments was $780. How much did Doris invest at each rate?

28. Suppose that Gus invested a total of $8000, part of it at 8% and the remainder at 9%. His yearly income from the two investments was $690. How much did he invest at each rate?

29. Two numbers are added together, and the sum is 131. One number is five less than three times the other. Find the two numbers.

30. The length of a rectangle is twice the width of the rectangle. Given that the perimeter of the rectangle is 72 centimeters, find the dimensions.

31. Two angles are complementary, and the measure of one of the angles is 10° less than four times the measure of the other angle. Find the measure of each angle.

32. The difference of two numbers is 75. The larger number is three less than four times the smaller number. Find the numbers.

33. In a class of 50 students, the number of females is two more than five times the number of males. How many females are there in the class?

**34.** In a recent survey, one thousand registered voters were asked about their political preferences. The number of males in the survey was five less than one-half of the number of females. Find the number of males in the survey.

**35.** The perimeter of a rectangle is 94 inches. The length of the rectangle is 7 inches more than the width. Find the dimensions of the rectangle.

**36.** Two angles are supplementary, and the measure of one of them is 20° less than three times the measure of the other angle. Find the measure of each angle.

**37.** A deposit slip listed $700 in cash to be deposited. There were 100 bills, some of them five-dollar bills and the remainder ten-dollar bills. How many bills of each denomination were deposited?

**38.** Cindy has 30 coins, consisting of dimes and quarters, that total $5.10. How many coins of each kind does she have?

**39.** The income from a student production was $47,500. The price of a student ticket was $15, and nonstudent tickets were sold at $20 each. Three thousand tickets were sold. How many tickets of each kind were sold?

**40.** Sue bought 3 packages of cookies and 2 sacks of potato chips for $13.50. Later she bought 2 more packages of cookies and 5 additional sacks of potato chips for $20.00. Find the price of a package of cookies.

## Thoughts Into Words

**41.** Give a general description of how to use the substitution method to solve a system of two linear equations in two variables.

**42.** Is it possible for a system of two linear equations in two variables to have exactly two solutions? Defend your answer.

**43.** Explain how you would solve the system $\begin{pmatrix} 2x + 5y = 5 \\ 5x - y = 9 \end{pmatrix}$ using the substitution method.

## Graphing Calculator Activities

**44.** Use your graphing calculator to help determine whether, in Problems 1–10, the system is consistent, the system is inconsistent, or the equations are dependent.

**45.** Use your graphing calculator to help determine the solution set for each of the following systems. Be sure to check your answers.

**(a)** $\begin{pmatrix} 3x - y = 30 \\ 5x - y = 46 \end{pmatrix}$ **(b)** $\begin{pmatrix} 1.2x + 3.4y = 25.4 \\ 3.7x - 2.3y = 14.4 \end{pmatrix}$

**(c)** $\begin{pmatrix} 1.98x + 2.49y = 13.92 \\ 1.19x + 3.45y = 16.18 \end{pmatrix}$ **(d)** $\begin{pmatrix} 2x - 3y = 10 \\ 3x + 5y = 53 \end{pmatrix}$

**(e)** $\begin{pmatrix} 4x - 7y = -49 \\ 6x + 9y = 219 \end{pmatrix}$ **(f)** $\begin{pmatrix} 3.7x - 2.9y = -14.3 \\ 1.6x + 4.7y = -30 \end{pmatrix}$

### Answers to the Concept Quiz

**1.** False **2.** True **3.** True **4.** False **5.** True

# 10.3 Elimination-by-Addition Method

OBJECTIVES
1. Solve systems of equations by the elimination-by-addition method
2. Determine which method to use to solve a system of equations
3. Use systems of equations to solve problems

We found in the previous section that the substitution method for solving a system of two equations and two unknowns works rather well. However, as the number of equations and unknowns increases, the substitution method becomes quite unwieldy. In this section we are going to introduce another method, called the *elimination-by-addition* method. We shall introduce it here using systems of two linear equations in two unknowns and then, in the next section, extend its use to three linear equations in three unknowns.

The **elimination-by-addition** method involves applying an addition property of equations. This process leads to eliminating one variable, and then the resulting equation can be solved for the other variable. Property 10.1 is as follows.

> **Property 10.1**
>
> If $a = b$ and $c = d$, then $a + c = b + d$.

Now let's see how to apply this property to solve a linear system of equations in two variables.

**Classroom Example**
Solve the system:
$\begin{pmatrix} 4x + 5y = 23 \\ 3x - 5y = -9 \end{pmatrix}$

**EXAMPLE 1** Solve the system $\begin{pmatrix} 3x + 2y = 1 \\ 5x - 2y = 23 \end{pmatrix}$.

## Solution

Because $3x + 2y$ and 1 are equals and also $5x - 2y$ and 23 are equals, we can add the two equations together by adding the left sides of each equation and adding the right sides of each equation to give a valid equation.

$$\begin{aligned} 3x + 2y &= 1 && \text{First equation} \\ 5x - 2y &= 23 && \text{Second equation} \\ \hline 8x \qquad &= 24 && \text{Add the equations} \end{aligned}$$

Because the $y$ terms were opposites, when the equations were added the $y$ variable was eliminated. The result is an equation that can be solved for $x$.

$$\begin{aligned} 8x &= 24 \\ x &= 3 \end{aligned}$$

The $x$ coordinate of the solution is 3. To find the $y$ coordinate of the solution, we can substitute 3 for $x$ in either one of the equations of the system and solve for $y$. Let's use the first equation.

$$\begin{aligned} 3x + 2y &= 1 && \text{First equation} \\ 3(3) + 2y &= 1 \\ 2y &= -8 \\ y &= -4 \end{aligned}$$

The solution set is $\{(3, -4)\}$. Check it!

**Classroom Example**
Solve the system:
$\begin{pmatrix} 2x + \phantom{3}y = -3 \\ 4x - 3y = -11 \end{pmatrix}$

**EXAMPLE 2** Solve the system $\begin{pmatrix} x + 5y = -2 \\ 3x - 4y = -25 \end{pmatrix}$.

## Solution

By inspection, we notice that adding the equations together will not eliminate a variable. However, if we multiply both sides of the first equation by $-3$, then adding the equations will eliminate the $x$ variable terms.

$$\begin{aligned} x + 5y &= -2 && \text{First equation multiplied by } -3 && -3x - 15y = 6 \\ 3x - 4y &= -25 && \text{Second equation} && \underline{3x - 4y = -25} \\ &&&& -19y = -19 \quad \text{Add the equations} \end{aligned}$$

Because the $x$ terms were opposites, when the equations were added the $x$ variable was eliminated. The result is an equation that can be solved for $y$.

$$-19y = -19$$
$$y = 1$$

The $y$ coordinate of the solution is 1. To find the $x$ coordinate of the solution, we can substitute 1 for $y$ in either one of the equations of the system and solve for $x$. Let's use the first equation.

$$x + 5y = -2 \quad \text{First equation}$$
$$x + 5(1) = -2$$
$$x = -7$$

The solution set is $\{(-7, 1)\}$.

Note that our objective has been to produce an equivalent system of equations from which one of the variables can be *eliminated* from one equation. We accomplish this by multiplying one equation of the system by an appropriate number and then *adding* that result to the other equation. Thus the method is called *elimination by addition*. Let's look at another example.

**Classroom Example**
Solve the system:
$\begin{pmatrix} 2x - 7y = 22 \\ 3x + 5y = 2 \end{pmatrix}$

**EXAMPLE 3** Solve the system $\begin{pmatrix} 2x + 5y = 4 \\ 5x - 7y = -29 \end{pmatrix}$.

## Solution

We notice that if we multiply both sides of the first equation by $-5$ and also multiply both sides of the second equation by 2, then adding the equations will eliminate the $x$ variable terms.

$$2x + 5y = 4 \quad \text{First equation multiplied by } -5 \quad -10x - 25y = -20$$
$$5x - 7y = -29 \quad \text{Second equation multiplied by 2} \quad 10x - 14y = -58$$
$$-39y = -78 \quad \text{Add the equations}$$

Solving the resulting equation we can find the value of $y$.

$$-39y = -78$$
$$y = 2$$

Now we can substitute 2 for $y$ in the first equation.

$$2x + 5y = 4$$
$$2x + 5(2) = 4$$
$$2x = -6$$
$$x = -3$$

The solution set is $\{(-3, 2)\}$.

**Classroom Example**
Solve the system:
$\begin{pmatrix} 7x - 3y = 6 \\ 2x + 5y = 8 \end{pmatrix}$

**EXAMPLE 4** Solve the system $\begin{pmatrix} 3x - 2y = 5 \\ 2x + 7y = 9 \end{pmatrix}$.

## Solution

We can start by multiplying both sides of the first equation by 2 and also multiplying both sides of the second equation by $-3$. This will begin the process of eliminating the $x$ variable terms.

$$3x - 2y = 5 \quad \text{First equation multiplied by 2} \quad 6x - 4y = 10$$
$$2x + 7y = 9 \quad \text{Second equation multiplied by } -3 \quad -6x - 21y = -27$$
$$-25y = -17 \quad \text{Add the equations}$$

From the equation we can find the value of $y$.

$$-25y = -17$$
$$y = \frac{17}{25}$$

Now we can substitute $\frac{17}{25}$ for $y$ in the first equation. However, rather than substituting a fraction, it may be easier to solve the system again and eliminate the $y$ variable. Let's do that by multiplying both sides of the first equation by 7 and also multiplying both sides of the second equation by 2.

| | | |
|---|---|---|
| $3x - 2y = 5$ | First equation multiplied by 7 | $21x - 14y = 35$ |
| $2x + 7y = 9$ | Second equation multiplied by 2 | $4x + 14y = 18$ |
| | | $25x = 53$ Add the equations |

Therefore, $\frac{53}{25}$ is the $x$ coordinate of the solution.

The solution set is $\left\{\left(\frac{53}{25}, \frac{17}{25}\right)\right\}$. (Perhaps you should check this result!)

## Determining Which Method to Use to Solve a System of Equations

Both the elimination-by-addition and the substitution methods can be used to obtain exact solutions for any system of two linear equations in two unknowns. Sometimes the issue is that of deciding which method to use on a particular system. As we have seen with examples thus far in this section and those of the previous section, many systems lend themselves to one or the other method by the original format of the equations. Let's emphasize that point with some more examples.

**Classroom Example**
Solve the system:
$\begin{pmatrix} 14x - 4y = -10 \\ 4x + 8y = 4 \end{pmatrix}$.

**EXAMPLE 5** Solve the system $\begin{pmatrix} 4x - 3y = 4 \\ 10x + 9y = -1 \end{pmatrix}$.

### Solution

Because changing the form of either equation in preparation for the substitution method would produce a fractional form, we are probably better off using the elimination-by-addition method. Let's multiply both sides of the first equation by 3, then add the equations to eliminate the $y$ variable terms.

| | | |
|---|---|---|
| $4x - 3y = 4$ | First equation multiplied by 3 | $12x - 9y = 12$ |
| $10x + 9y = -1$ | Second equation | $10x + 9y = -1$ |
| | | $22x = 11$ Add the equations |
| | | $x = \frac{11}{22} = \frac{1}{2}$ |

The $x$ coordinate of the solution is $\frac{1}{2}$. To find the $y$ coordinate of the solution, we can substitute $\frac{1}{2}$ for $x$ in either one of the equations of the system and solve for $x$. Let's use the first equation.

$$4x - 3y = 4 \quad \text{First equation}$$
$$4\left(\frac{1}{2}\right) - 3y = 4$$
$$2 - 3y = 4$$
$$-3y = 2$$
$$y = -\frac{2}{3}$$

The solution set is $\left\{\left(\frac{1}{2}, -\frac{2}{3}\right)\right\}$.

Classroom Example
Solve the system:
$\begin{pmatrix} 9x + 2y = -3 \\ x = 3y + 19 \end{pmatrix}$.

**EXAMPLE 6** Solve the system $\begin{pmatrix} 6x + 5y = -3 \\ y = -2x - 7 \end{pmatrix}$.

### Solution

Because the second equation is of the form "$y$ equals," let's use the substitution method. From the second equation we can substitute $-2x - 7$ for $y$ in the first equation.

$$6x + 5y = -3 \xrightarrow{\text{Substitute } -2x - 7 \text{ for } y} 6x + 5(-2x - 7) = -3$$

Solving this equation yields

$$\begin{aligned} 6x + 5(-2x - 7) &= -3 \\ 6x - 10x - 35 &= -3 \\ -4x - 35 &= -3 \\ -4x &= 32 \\ x &= -8 \end{aligned}$$

Substitute $-8$ for $x$ in the second equation to obtain

$$y = -2(-8) - 7 = 16 - 7 = 9$$

The solution set is $\{(-8, 9)\}$.

Sometimes we need to simplify the equations of a system before we can decide which method to use for solving the system. Let's consider an example of that type.

Classroom Example
Solve the system:
$\begin{pmatrix} \frac{x+1}{2} + \frac{y-3}{6} = 4 \\ \frac{x+3}{6} + \frac{y-1}{3} = 4 \end{pmatrix}$

**EXAMPLE 7** Solve the system $\begin{pmatrix} \frac{x-2}{4} + \frac{y+1}{3} = 2 \\ \frac{x+1}{7} + \frac{y-3}{2} = \frac{1}{2} \end{pmatrix}$.

### Solution

First, we need to simplify the two equations. Let's multiply both sides of the first equation by 12 and simplify.

$$\begin{aligned} 12\left(\frac{x-2}{4} + \frac{y+1}{3}\right) &= 12(2) \\ 3(x - 2) + 4(y + 1) &= 24 \\ 3x - 6 + 4y + 4 &= 24 \\ 3x + 4y - 2 &= 24 \\ 3x + 4y &= 26 \end{aligned}$$

Let's multiply both sides of the second equation by 14.

$$\begin{aligned} 14\left(\frac{x+1}{7} + \frac{y-3}{2}\right) &= 14\left(\frac{1}{2}\right) \\ 2(x + 1) + 7(y - 3) &= 7 \\ 2x + 2 + 7y - 21 &= 7 \\ 2x + 7y - 19 &= 7 \\ 2x + 7y &= 26 \end{aligned}$$

Now we have the following system to solve.

$$\begin{pmatrix} 3x + 4y = 26 \\ 2x + 7y = 26 \end{pmatrix}$$

Probably the easiest approach is to use the elimination-by-addition method. We can start by multiplying both sides of the first equation by 2 and also multiplying both sides of the second equation by $-3$. This will begin the process of eliminating the $x$ variable terms.

$$\begin{array}{lll} 3x + 4y = 26 & \text{First equation multiplied by 2} & 6x + 8y = 52 \\ 2x + 7y = 26 & \text{Second equation multiplied by } -3 & -6x - 21y = -78 \\ & & \overline{\quad -13y = -26} \quad \text{Add the equations} \\ & & \quad y = 2 \end{array}$$

The $y$ coordinate of the solution is 2. We can substitute 2 for $y$ in either one of the equations of the system and solve for $x$. Let's use the first equation.

$$\begin{aligned} 3x + 4y &= 26 \\ 3x + 4(2) &= 26 \\ 3x &= 18 \\ x &= 6 \end{aligned}$$

The solution set is $\{(6, 2)\}$.

**Remark:** Don't forget that to check a problem like Example 7 you must check the potential solutions back in the original equations.

In Section 10.1, we discussed the fact that you can tell whether a system of two linear equations in two unknowns has no solution, one solution, or infinitely many solutions by graphing the equations of the system. That is, the two lines may be parallel (no solution), or they may intersect in one point (one solution), or they may coincide (infinitely many solutions).

From a practical viewpoint, the systems that have one solution deserve most of our attention. However, we need to be able to deal with the other situations; they do occur occasionally. Let's use two examples to illustrate the type of thing that happens when we encounter *no solution* or *infinitely many solutions* when using either the elimination-by-addition method or the substitution method.

Classroom Example
Solve the system:
$$\begin{pmatrix} y = -5x + 3 \\ 10x + 2y = 7 \end{pmatrix}$$

**EXAMPLE 8** Solve the system $\begin{pmatrix} y = 3x - 1 \\ -9x + 3y = 4 \end{pmatrix}$.

## Solution

Using the substitution method, we can proceed as follows:

$$-9x + 3y = 4 \xrightarrow{\text{Substitute } 3x - 1 \text{ for } y} -9x + 3(3x - 1) = 4$$

Solving this equation yields

$$\begin{aligned} -9x + 3(3x - 1) &= 4 \\ -9x + 9x - 3 &= 4 \\ -3 &= 4 \end{aligned}$$

The *false numerical statement*, $-3 = 4$, implies that the system has *no solution*. The system of equations is inconsistent, hence the solution set is $\varnothing$. (You may want to graph the two lines to verify this conclusion!)

Classroom Example
Solve the system:
$$\begin{pmatrix} 4x - 12y = 16 \\ x - 3y = 4 \end{pmatrix}$$

**EXAMPLE 9** Solve the system $\begin{pmatrix} 5x + y = 2 \\ 10x + 2y = 4 \end{pmatrix}$.

## Solution

Use the elimination-by-addition method and proceed as follows: Let's multiply both sides of the first equation by $-2$, then add the equations to eliminate the $x$ variable terms.

| | | | |
|---|---|---|---|
| $5x + y = 2$ | First equation multiplied by $-2$ | $-10x - 2y = -4$ | |
| $10x + 2y = 4$ | Second equation | $10x + 2y = 4$ | |
| | | $0 = 0$ | Add the equations |

The *true numerical statement*, $0 = 0$, implies that the system has *infinitely many solutions.* Any ordered pair that satisfies one of the equations will also satisfy the other equation. Thus the solution set can be expressed as

$$\{(x, y) | 5x + y = 2\}$$

## Concept Quiz 10.3

For Problems 1–8, answer true or false.

1. Equivalent systems of equations are systems that have exactly the same solution set.
2. Any equation of a system can be multiplied on both sides by zero to obtain an equivalent system.
3. Any equation of the system can be replaced by the difference of that equation and a nonzero multiple of another equation.
4. The objective of the elimination-by-addition method is to produce an equivalent system with an equation in which one of the variables has been eliminated.
5. The elimination-by-addition method is only used for solving a system of equations if the substitution method cannot be used.
6. If an equivalent system for an original system is $\begin{pmatrix} 3x - 5y = 7 \\ 0 + 0 = 0 \end{pmatrix}$, then the original system is inconsistent and has no solution.
7. The system $\begin{pmatrix} 5x - 2y = 3 \\ 5x - 2y = 9 \end{pmatrix}$ has infinitely many solutions.
8. The solution set of the system $\begin{pmatrix} x = 3y + 7 \\ 2x - 6y = 9 \end{pmatrix}$ is the null set.

## Problem Set 10.3

For Problems 1–18, use the elimination-by-addition method to solve each system. (Objective 1)

1. $\begin{pmatrix} 2x + 3y = -1 \\ 5x - 3y = 29 \end{pmatrix}$
2. $\begin{pmatrix} 3x - 4y = -30 \\ 7x + 4y = 10 \end{pmatrix}$
3. $\begin{pmatrix} 6x - 7y = 15 \\ 6x + 5y = -21 \end{pmatrix}$
4. $\begin{pmatrix} 5x + 2y = -4 \\ 5x - 3y = 6 \end{pmatrix}$
5. $\begin{pmatrix} x - 2y = -12 \\ 2x + 9y = 2 \end{pmatrix}$
6. $\begin{pmatrix} x - 4y = 29 \\ 3x + 2y = -11 \end{pmatrix}$
7. $\begin{pmatrix} 4x + 7y = -16 \\ 6x - y = -24 \end{pmatrix}$
8. $\begin{pmatrix} 6x + 7y = 17 \\ 3x + y = -4 \end{pmatrix}$
9. $\begin{pmatrix} 3x - 2y = 5 \\ 2x + 5y = -3 \end{pmatrix}$
10. $\begin{pmatrix} 4x + 3y = -4 \\ 3x - 7y = 34 \end{pmatrix}$
11. $\begin{pmatrix} 7x - 2y = 4 \\ 7x - 2y = 9 \end{pmatrix}$
12. $\begin{pmatrix} 2x + 3y = -5 \\ 4x + 6y = 10 \end{pmatrix}$
13. $\begin{pmatrix} 5x + 4y = 2 \\ 3x - 2y = -12 \end{pmatrix}$
14. $\begin{pmatrix} 2x - 7y = -1 \\ 3x + y = 10 \end{pmatrix}$
15. $\begin{pmatrix} 8x - 3y = 13 \\ 4x + 9y = 3 \end{pmatrix}$
16. $\begin{pmatrix} 10x - 8y = -11 \\ 8x + 4y = -1 \end{pmatrix}$
17. $\begin{pmatrix} 3x - y = 4 \\ 6x - 2y = 8 \end{pmatrix}$
18. $\begin{pmatrix} 5x - y = 6 \\ 10x - 2y = 12 \end{pmatrix}$

For Problems 19–48, solve each system by using either the substitution or the elimination-by-addition method, whichever seems more appropriate. (Objective 2)

19. $\begin{pmatrix} 5x + 3y = -7 \\ 7x - 3y = 55 \end{pmatrix}$
20. $\begin{pmatrix} 4x - 7y = 21 \\ -4x + 3y = -9 \end{pmatrix}$

**21.** $\begin{pmatrix} x = 5y + 7 \\ 4x + 9y = 28 \end{pmatrix}$

**22.** $\begin{pmatrix} 11x - 3y = -60 \\ y = -38 - 6x \end{pmatrix}$

**23.** $\begin{pmatrix} x = -6y + 79 \\ x = 4y - 41 \end{pmatrix}$

**24.** $\begin{pmatrix} y = 3x + 34 \\ y = -8x - 54 \end{pmatrix}$

**25.** $\begin{pmatrix} 4x - 3y = 2 \\ 5x - y = 3 \end{pmatrix}$

**26.** $\begin{pmatrix} 3x - y = 9 \\ 5x + 7y = 1 \end{pmatrix}$

**27.** $\begin{pmatrix} 5x - 2y = 1 \\ 10x - 4y = 7 \end{pmatrix}$

**28.** $\begin{pmatrix} y = \frac{2}{3}x - 4 \\ 2x - 3y = 1 \end{pmatrix}$

**29.** $\begin{pmatrix} 3x - 2y = 14 \\ 5x + 7y = -18 \end{pmatrix}$

**30.** $\begin{pmatrix} 4x + 7y = -7 \\ 9x - 2y = 2 \end{pmatrix}$

**31.** $\begin{pmatrix} y = -2x + 1 \\ 6x + 3y = 3 \end{pmatrix}$

**32.** $\begin{pmatrix} 2x - 3y = 4 \\ y = \frac{2}{3}x - \frac{4}{3} \end{pmatrix}$

**33.** $\begin{pmatrix} -2x + 5y = -16 \\ x = \frac{3}{4}y + 1 \end{pmatrix}$

**34.** $\begin{pmatrix} y = \frac{2}{3}x - \frac{3}{4} \\ 2x + 4y = 11 \end{pmatrix}$

**35.** $\begin{pmatrix} y = \frac{2}{3}x - 4 \\ 5x - 3y = 9 \end{pmatrix}$

**36.** $\begin{pmatrix} 5x - 3y = 9 \\ x = \frac{y}{2} + 1 \end{pmatrix}$

**37.** $\begin{pmatrix} \frac{x}{6} + \frac{y}{3} = 3 \\ \frac{5x}{2} - \frac{y}{6} = -17 \end{pmatrix}$

**38.** $\begin{pmatrix} \frac{3x}{4} - \frac{2y}{3} = 31 \\ \frac{7x}{5} + \frac{y}{4} = 22 \end{pmatrix}$

**39.** $\begin{pmatrix} -(x - 6) + 6(y + 1) = 58 \\ 3(x + 1) - 4(y - 2) = -15 \end{pmatrix}$

**40.** $\begin{pmatrix} -2(x + 2) + 4(y - 3) = -34 \\ 3(x + 4) - 5(y + 2) = 23 \end{pmatrix}$

**41.** $\begin{pmatrix} 5(x + 1) - (y + 3) = -6 \\ 2(x - 2) + 3(y - 1) = 0 \end{pmatrix}$

**42.** $\begin{pmatrix} 2(x - 1) - 3(y + 2) = 30 \\ 3(x + 2) + 2(y - 1) = -4 \end{pmatrix}$

**43.** $\begin{pmatrix} \frac{1}{2}x - \frac{1}{3}y = 12 \\ \frac{3}{4}x + \frac{2}{3}y = 4 \end{pmatrix}$

**44.** $\begin{pmatrix} \frac{2}{3}x + \frac{1}{5}y = 0 \\ \frac{3}{2}x - \frac{3}{10}y = -15 \end{pmatrix}$

**45.** $\begin{pmatrix} \frac{2x}{3} - \frac{y}{2} = -\frac{5}{4} \\ \frac{x}{4} + \frac{5y}{6} = \frac{17}{16} \end{pmatrix}$

**46.** $\begin{pmatrix} \frac{x}{2} + \frac{y}{3} = \frac{5}{72} \\ \frac{x}{4} + \frac{5y}{2} = -\frac{17}{48} \end{pmatrix}$

**47.** $\begin{pmatrix} \frac{3x + y}{2} + \frac{x - 2y}{5} = 8 \\ \frac{x - y}{3} - \frac{x + y}{6} = \frac{10}{3} \end{pmatrix}$

**48.** $\begin{pmatrix} \frac{x - y}{4} - \frac{2x - y}{3} = -\frac{1}{4} \\ \frac{2x + y}{3} + \frac{x + y}{2} = \frac{17}{6} \end{pmatrix}$

For Problems 49–60, solve each problem by setting up and solving an appropriate system of equations. (Objective 3)

**49.** A 10%-salt solution is to be mixed with a 20%-salt solution to produce 20 gallons of a 17.5%-salt solution. How many gallons of the 10% solution and how many gallons of the 20% solution will be needed?

**50.** A small-town library buys a total of 35 books that cost \$644. Some of the books cost \$16 each, and the remainder cost \$20 each. How many books of each price did the library buy?

**51.** Suppose that on a particular day the cost of 3 tennis balls and 2 golf balls is \$7. The cost of 6 tennis balls and 3 golf balls is \$12. Find the cost of 1 tennis ball and the cost of 1 golf ball.

**52.** For moving purposes, the Hendersons bought 25 cardboard boxes for \$97.50. There were two kinds of boxes; the large ones cost \$7.50 per box, and the small ones were \$3 per box. How many boxes of each kind did they buy?

**53.** A motel in a suburb of Chicago rents double rooms for \$120 per day and single rooms for \$90 per day. If a total of 55 rooms were rented for \$6150, how many of each kind were rented?

**54.** Suppose that one solution contains 50% alcohol and another solution contains 80% alcohol. How many liters of each solution should be mixed to make 10.5 liters of a 70%-alcohol solution?

**55.** A college fraternity house spent \$670 for an order of 85 pizzas. The order consisted of cheese pizzas, which cost \$5 each, and Supreme pizzas, which cost \$12 each. Find the number of each kind of pizza ordered.

**56.** Part of \$8400 is invested at 5%, and the remainder is invested at 8%. The total yearly interest from the two investments is \$576. Determine how much is invested at each rate.

**57.** If the numerator of a certain fraction is increased by 5, and the denominator is decreased by 1, the resulting fraction is $\frac{8}{3}$. However, if the numerator of the original fraction is doubled, and the denominator is increased by 7, then the resulting fraction is $\frac{6}{11}$. Find the original fraction.

58. A man bought 2 pounds of coffee and 1 pound of butter for a total of \$9.25. A month later, the prices had not changed (this makes it a fictitious problem), and he bought 3 pounds of coffee and 2 pounds of butter for \$15.50. Find the price per pound of both the coffee and the butter.

59. Suppose that we have a rectangular-shaped book cover. If the width is increased by 2 centimeters, and the length is decreased by 1 centimeter, the area is increased by 28 square centimeters. However, if the width is decreased by 1 centimeter, and the length is increased by 2 centimeters, then the area is increased by 10 square centimeters. Find the dimensions of the book cover.

60. A blueprint indicates a master bedroom in the shape of a rectangle. If the width is increased by 2 feet and the length remains the same, then the area is increased by 36 square feet. However, if the width is increased by 1 foot and the length is increased by 2 feet, then the area is increased by 48 square feet. Find the dimensions of the room as indicated on the blueprint.

## Thoughts Into Words

61. Give a general description of how to use the elimination-by-addition method to solve a system of two linear equations in two variables.

62. Explain how you would solve the system

$$\begin{pmatrix} 3x - 4y = -1 \\ 2x - 5y = 9 \end{pmatrix}$$

using the elimination-by-addition method.

63. How do you decide whether to solve a system of linear equations in two variables by using the substitution method or by using the elimination-by-addition method?

## Further Investigations

64. There is another way of telling whether a system of two linear equations in two unknowns is consistent or inconsistent, or whether the equations are dependent, without taking the time to graph each equation. It can be shown that any system of the form

$$a_1x + b_1y = c_1$$
$$a_2x + b_2y = c_2$$

has one and only one solution if

$$\frac{a_1}{a_2} \neq \frac{b_1}{b_2}$$

that it has no solution if

$$\frac{a_1}{a_2} = \frac{b_1}{b_2} \neq \frac{c_1}{c_2}$$

and that it has infinitely many solutions if

$$\frac{a_1}{a_2} = \frac{b_1}{b_2} = \frac{c_1}{c_2}$$

For each of the following systems, determine whether the system is consistent, the system is inconsistent, or the equations are dependent.

**(a)** $\begin{pmatrix} 4x - 3y = 7 \\ 9x + 2y = 5 \end{pmatrix}$ **(b)** $\begin{pmatrix} 5x - y = 6 \\ 10x - 2y = 19 \end{pmatrix}$

**(c)** $\begin{pmatrix} 5x - 4y = 11 \\ 4x + 5y = 12 \end{pmatrix}$ **(d)** $\begin{pmatrix} x + 2y = 5 \\ x - 2y = 9 \end{pmatrix}$

**(e)** $\begin{pmatrix} x - 3y = 5 \\ 3x - 9y = 15 \end{pmatrix}$ **(f)** $\begin{pmatrix} 4x + 3y = 7 \\ 2x - y = 10 \end{pmatrix}$

**(g)** $\begin{pmatrix} 3x + 2y = 4 \\ y = -\frac{3}{2}x - 1 \end{pmatrix}$ **(h)** $\begin{pmatrix} y = \frac{4}{3}x - 2 \\ 4x - 3y = 6 \end{pmatrix}$

65. A system such as

$$\begin{pmatrix} \frac{3}{x} + \frac{2}{y} = 2 \\ \frac{2}{x} - \frac{3}{y} = \frac{1}{4} \end{pmatrix}$$

is not a system of linear equations but can be transformed into a linear system by changing variables. For example, when we substitute $u$ for $\frac{1}{x}$ and $v$ for $\frac{1}{y}$, the system cited becomes

$$\begin{pmatrix} 3u + 2v = 2 \\ 2u - 3v = \frac{1}{4} \end{pmatrix}$$

We can solve this "new" system either by elimination by addition or by substitution (we will leave the details for you) to produce $u = \frac{1}{2}$ and $v = \frac{1}{4}$. Therefore, because $u = \frac{1}{x}$ and $v = \frac{1}{y}$, we have

$$\frac{1}{x} = \frac{1}{2} \quad \text{and} \quad \frac{1}{y} = \frac{1}{4}$$

Solving these equations yields

$$x = 2 \quad \text{and} \quad y = 4$$

The solution set of the original system is $\{(2, 4)\}$. Solve each of the following systems.

**(a)** $\begin{pmatrix} \frac{1}{x} + \frac{2}{y} = \frac{7}{12} \\ \frac{3}{x} - \frac{2}{y} = \frac{5}{12} \end{pmatrix}$ **(b)** $\begin{pmatrix} \frac{2}{x} + \frac{3}{y} = \frac{19}{15} \\ -\frac{2}{x} + \frac{1}{y} = -\frac{7}{15} \end{pmatrix}$

**(c)** $\begin{pmatrix} \frac{3}{x} - \frac{2}{y} = \frac{13}{6} \\ \frac{2}{x} + \frac{3}{y} = 0 \end{pmatrix}$ **(d)** $\begin{pmatrix} \frac{4}{x} + \frac{1}{y} = 11 \\ \frac{3}{x} - \frac{5}{y} = -9 \end{pmatrix}$

**(e)** $\begin{pmatrix} \frac{5}{x} - \frac{2}{y} = 23 \\ \frac{4}{x} + \frac{3}{y} = \frac{23}{2} \end{pmatrix}$ **(f)** $\begin{pmatrix} \frac{2}{x} - \frac{7}{y} = \frac{9}{10} \\ \frac{5}{x} + \frac{4}{y} = -\frac{41}{20} \end{pmatrix}$

## Graphing Calculator Activities

**66.** Use a graphing calculator to check your answers for Problem 64.

**67.** Use a graphing calculator to check your answers for Problem 65.

### Answers to the Concept Quiz

**1.** True **2.** False **3.** True **4.** True **5.** False **6.** False **7.** False **8.** True

# 10.4 Systems of Three Linear Equations in Three Variables

OBJECTIVES

1. Solve systems of three linear equations
2. Use systems of three linear equations to solve problems

Consider a linear equation in three variables $x$, $y$, and $z$, such as $3x - 2y + z = 7$. Any **ordered triple** $(x, y, z)$ that makes the equation a true numerical statement is said to be a solution of the equation. For example, the ordered triple $(2, 1, 3)$ is a solution because $3(2) - 2(1) + 3 = 7$. However, the ordered triple $(5, 2, 4)$ is not a solution because $3(5) - 2(2) + 4 \neq 7$. There are infinitely many solutions in the solution set.

**Remark:** The concept of a *linear* equation is generalized to include equations of more than two variables. Thus an equation such as $5x - 2y + 9z = 8$ is called a linear equation in three variables; the equation $5x - 7y + 2z - 11w = 1$ is called a linear equation in four variables; and so on.

To *solve* a system of three linear equations in three variables, such as

$$\begin{pmatrix} 3x - y + 2z = 13 \\ 4x + 2y + 5z = 30 \\ 5x - 3y - z = 3 \end{pmatrix}$$

means to find all of the ordered triples that satisfy all three equations. In other words, the solution set of the system is the intersection of the solution sets of all three equations in the system.

The graph of a linear equation in three variables is a **plane**, not a line. In fact, graphing equations in three variables requires the use of a three-dimensional coordinate system. Thus using a graphing approach to solve systems of three linear equations in three variables is not at all practical. However, a simple graphical analysis does give us some idea of what we can expect as we begin solving such systems.

In general, because each linear equation in three variables produces a plane, a system of three such equations produces three planes. There are various ways in which three planes can be related. For example, they may be mutually parallel, or two of the planes may be parallel and the third one intersect each of the two. (You may want to analyze all of the other possibilities for the three planes!) However, for our purposes at this time, we need to realize that from a solution set viewpoint, a system of three linear equations in three variables produces one of the following possibilities.

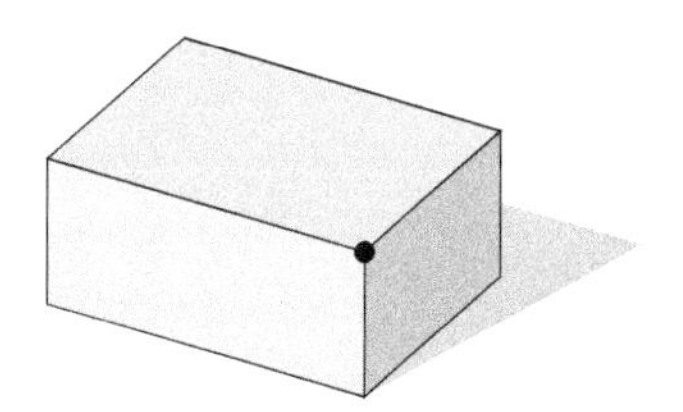

Figure 10.11

1. There is *one ordered triple* that satisfies all three equations. The three planes have a common point of intersection, as indicated in Figure 10.11.
2. There are *infinitely many* ordered triples in the solution set, all of which are coordinates of points on a line common to the planes. This can happen if three planes have a common line of intersection (Figure 10.12a), or if two of the planes coincide and the third plane intersects them (Figure 10.12b).

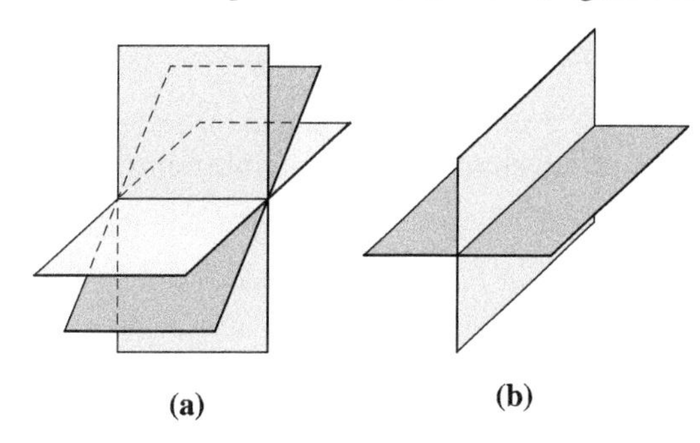

(a) (b)

Figure 10.12

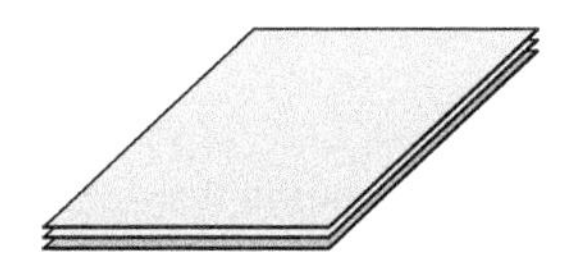

Figure 10.13

3. There are *infinitely many* ordered triples in the solution set, all of which are coordinates of points on a plane. This happens if the three planes coincide, as illustrated in Figure 10.13.
4. The solution set is *empty*; it is ∅. This can happen in various ways, as we see in Figure 10.14. Note that in each situation there are no points common to all three planes.

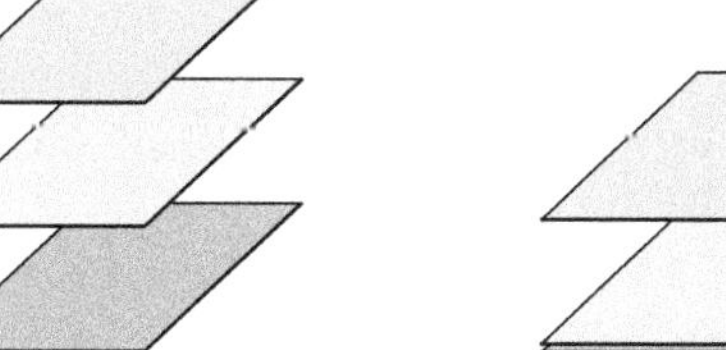

(a) Three parallel planes

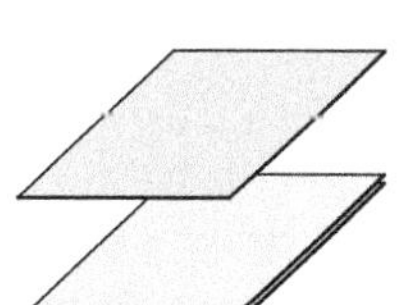

(b) Two planes coincide and the third one is parallel to the coinciding planes.

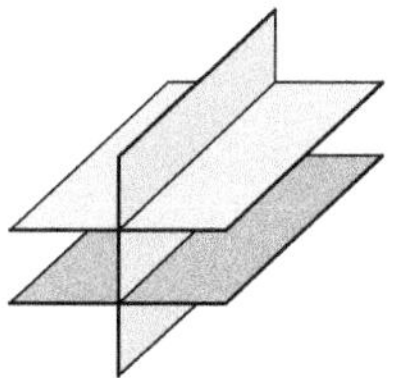

(c) Two planes are parallel and the third intersects them in parallel lines.

(d) No two planes are parallel, but two of them intersect in a line that is parallel to the third plane.

Figure 10.14

Now that we know what possibilities exist, let's consider finding the solution sets for some systems. Our approach will be the elimination-by-addition method, whereby we replace systems with equivalent systems until we obtain a system for which we can easily determine the solution set. Let's start with an example that allows us to determine the solution set without changing to another, equivalent system.

**Classroom Example**
Solve the system:

$$\begin{pmatrix} 4x + 3y + 2z = 1 \\ 2x + y + z = 2 \\ x - 2y + 3z = 11 \end{pmatrix}$$

**EXAMPLE 1**

Solve the system $\begin{pmatrix} 3x + 2y + z = 8 \\ x + 3y - z = -5 \\ 2x + y + 2z = 11 \end{pmatrix}$.   Equation 1, Equation 2, Equation 3

### Solution

Let's add the first two equations to eliminate the $z$ variable terms.

$$\begin{aligned} 3x + 2y + z &= 8 \\ x + 3y - z &= -5 \\ \hline 4x + 5y &= 3 \end{aligned}$$ Equation 4

Now we choose another pair of equations and eliminate the same variable. Let's choose equations 2 and 3. To eliminate the $z$ variable terms, multiply equation 2 by two and add that result to equation 3.

$$\begin{aligned} 2x + 6y - 2z &= -10 \\ 2x + y + 2z &= 11 \\ \hline 4x + 7y &= 1 \end{aligned}$$ Equation 5

Now we can form a system of two equations in two variables using equations 4 and 5.

$$\begin{pmatrix} 4x + 5y = 3 \\ 4x + 7y = 1 \end{pmatrix}$$

To solve this system, let's eliminate the $x$ variable terms by multiplying the equation 4 by $-1$ and then adding the result to equation 5.

$4x + 5y = 3$   Multiply by $-1$

$$\begin{aligned} -4x - 5y &= -3 \\ 4x + 7y &= 1 \\ \hline 2y &= -2 \quad \text{Add the equations} \\ y &= -1 \end{aligned}$$

Now that we know $y = -1$, substitute $-1$ for $y$ in equation 4.

$$\begin{aligned} 4x + 5y &= 3 \\ 4x + 5(-1) &= 3 \\ 4x - 5 &= 3 \\ 4x &= 8 \\ x &= 2 \end{aligned}$$

To find the value of $z$, substitute $-1$ for $y$ and 2 for $x$ in any of the original equations in the system. Let's use equation 2.

$$\begin{aligned} x + 3y - z &= -5 \\ 2 + 3(-1) - z &= -5 \\ -1 - z &= -5 \\ -z &= -4 \\ z &= 4 \end{aligned}$$

The solution set is $\{(2, -1, 4)\}$.

Let's summarize the process used in Example 1 with a list of steps.

**Step 1** Choose a pair of equations and eliminate one of the variables.

**Step 2** Choose a **different** pair of equations and eliminate the **same** variable.

**Step 3** Use the equations from Step 1 and Step 2 to form a system of two equations in two variables.

**Step 4** Solve the system of two equations by either elimination-by-addition or substitution.

**Step 5** Substitute the two solutions found in Step 4 into any of the original equations to find the value of the remaining variable.

**Step 6** Check your solution in all three of the original equations.

**Classroom Example**
Solve the system:

$$\begin{pmatrix} 2x + 2y - 3z = -1 \\ 4x - 3y + 2z = 20 \\ x + 5y - z = -8 \end{pmatrix}$$

**EXAMPLE 2**

Solve the system $\begin{pmatrix} x - y + 4z = -29 \\ 3x - 2y - z = -6 \\ 2x - 5y + 6z = -55 \end{pmatrix}$. (1) (2) (3)

### Solution

Choosing equations 1 and 2, we can eliminate the $x$ variable terms by multiplying equation 1 by $-3$ and adding the result to equation 2.

$$x - y + 4z = -29 \quad \text{Multiply by } -3 \quad \begin{aligned} -3x + 3y - 12z &= 87 \\ 3x - 2y - z &= -6 \\ \hline y - 13z &= 81 \end{aligned} \quad \text{Equation 4}$$

Now choosing a different pair of equations, 1 and 3, we can eliminate the $x$ variable terms by multiplying equation 1 by $-2$ and adding the result to equation 2.

$$x - y + 4z = -29 \quad \text{Multiply by } -2 \quad \begin{aligned} -2x + 2y - 8z &= 58 \\ 2x - 5y + 6z &= -55 \\ \hline -3y - 2z &= 3 \end{aligned} \quad \text{Equation 5}$$

Now we can form a system of two equations in two variables using equations 4 and 5.

$$\begin{pmatrix} y - 13z = 81 \\ -3y - 2z = 3 \end{pmatrix}$$

To solve this system, let's eliminate the $y$ variable terms by multiplying equation 4 by 3 and then add the result to equation 5.

$$y - 13z = 81 \quad \text{Multiply by } 3 \quad \begin{aligned} 3y - 39z &= 243 \\ -3y - 2z &= 3 \\ \hline -41z &= 246 \quad \text{Add the equations} \\ z &= -6 \end{aligned}$$

Now that we know $z = -6$, substitute $-6$ for $z$ in equation 4.

$$\begin{aligned} y - 13z &= 81 \\ y - 13(-6) &= 81 \\ y + 78 &= 81 \\ y &= 3 \end{aligned}$$

To find the value of $x$, substitute 3 for $y$ and $-6$ for $z$ in any of the original equations in the system. Let's use equation 1.

$$\begin{aligned} x - y + 4z &= -29 \\ x - (3) + 4(-6) &= -29 \\ x - 3 - 24 &= -29 \end{aligned}$$

$$x - 27 = -29$$
$$x = -2$$

The solution set is $\{(-2, 3, -6)\}$.

**Classroom Example**
Solve the system:
$$\begin{pmatrix} 3x + 6y - 4z = -12 \\ 2y + z = 1 \\ 6y - 4z = -18 \end{pmatrix}$$

**EXAMPLE 3**

Solve the system $\begin{pmatrix} 3x + 2y - 7z = -34 \\ y + 5z = 21 \\ 3y - 2z = -22 \end{pmatrix}$. (1) (2) (3)

## Solution

In this problem, we already have two equations with $x$ variable terms eliminated. So let's form a system of equations with equations 2 and 3.

$$\begin{pmatrix} y + 5z = 21 \\ 3y - 2z = -22 \end{pmatrix}$$

Let's eliminate the $y$ variable by multiplying the equation 2 by $-3$ and adding the equations.

$y + 5z = 21$ Multiply by $-3$ $\quad -3y - 15z = -63$

$$3y - 2z = -22$$
$$-17z = -85 \quad \text{Add the equations}$$
$$z = 5$$

Now that we know $z = 5$, substitute 5 for $z$ in equation 2 to find the value of $y$.

$$y + 5z = 21$$
$$y + 5(5) = 21$$
$$y + 25 = 21$$
$$y = -4$$

Finally, we can substitute 5 for $z$ and $-4$ for $y$ in equation 1.

$$3x + 2y - 7z = -34$$
$$3x + 2(-4) - 7(5) = -34$$
$$3x - 8 - 35 = -34$$
$$3x - 43 = -34$$
$$3x = 9$$
$$x = 3$$

The solution set is $\{(3, -4, 5)\}$.

Now let's consider some examples in which we make more than one replacement of equivalent systems.

**Classroom Example**
Solve the system:
$$\begin{pmatrix} 4x - 10y + 6z = 10 \\ 7x + 3y - 2z = 12 \\ 2x - 5y + 3z = 6 \end{pmatrix}$$

**EXAMPLE 4**

Solve the system $\begin{pmatrix} x - 2y + 3z = 1 \\ 2x - 4y + 6z = 7 \\ 3x - 5y - 2z = 4 \end{pmatrix}$. (1) (2) (3)

## Solution

Choosing equations 1 and 2, we can eliminate the $x$ variable terms by multiplying equation 1 by $-2$ and adding the result to equation 2.

$x - 2y + 3z = 1$ Multiply by $-2$ $\quad -2x + 4y - 6z = -2$

$$2x - 4y + 6z = 7$$
$$0 = 5 \quad \text{Equation 3}$$

The false statement, $0 = 5$, indicates that the system is inconsistent and that the solution set is therefore $\varnothing$. (If you were to graph this system, equations 1 and 3 would produce parallel planes, which is the situation depicted back in Figure 10.14c.)

**Classroom Example**
Solve the system:

$$\begin{pmatrix} 3x - 4y - z = -2 \\ 2x + 10y + 2z = -2 \\ x - 14y - 3z = 0 \end{pmatrix}$$

### EXAMPLE 5

Solve the system $\begin{pmatrix} 2x - y + 4z = 1 \\ 3x + 2y - z = 5 \\ 5x - 6y + 17z = -1 \end{pmatrix}$. (1) (2) (3)

### Solution

Choosing equations 1 and 2, we can eliminate the $y$ variable terms by multiplying equation 1 by 2 and adding the result to equation 2.

$$2x - y + 4z = 1 \quad \text{Multiply by 2} \quad 4x - 2y + 8z = 2$$
$$3x + 2y - z = 5$$
$$7x + 7z = 7 \quad \text{Equation 4}$$

Choosing equations 1 and 3, we can eliminate the $y$ variable terms by multiplying equation 1 by $-6$ and adding the result to equation 3.

$$2x - y + 4z = 1 \quad \text{Multiply by } -6 \quad -12x + 6y - 24z = -6$$
$$5x - 6y + 17z = -1$$
$$-7x - 7z = -7 \quad \text{Equation 5}$$

Now we can form a system of equations using equations 4 and 5.

$$\begin{pmatrix} 7x + 7z = 7 \\ -7x - 7z = -7 \end{pmatrix}$$

Let's add the equations in the system.

$$7x + 7z = 7$$
$$-7x - 7z = -7$$
$$0 + 0 = 0$$

The true numerical statement, $0 + 0 = 0$, indicates that the system has infinitely many solutions. (The graph of this system is shown in Figure 10.12a).

**Remark:** It can be shown that the solutions for the system in Example 5 are of the form $(t, 3 - 2t, 1 - t)$, where $t$ is any real number. For example, if we let $t = 2$, then we get the ordered triple $(2, -1, -1)$, and this triple will satisfy all three of the original equations. For our purposes in this text, we shall simply indicate that such a system has infinitely many solutions.

## Using Systems of Three Linear Equations to Solve Problems

When using a system of equations to solve a problem that involves three variables, it is necessary to have three equations in the system of equations. In the next example, a system of three equations will be set up; and we will omit the problem-solving details.

**Classroom Example**
Part of \$60,000 is invested at 2%, another part at 1%, and the remainder at 1.6%. The total yearly income from the three investments is \$1150. Four times the amount invested at 1% less \$500 equals the amount invested at 1.6%. Determine how much is invested at each rate.

### EXAMPLE 6 Apply Your Skill

Part of \$50,000 is invested at 4%, another part at 6%, and the remainder at 7%. The total yearly income from the three investments is \$3050. The sum of the amounts invested at 4% and 6% equals the amount invested at 7%. Determine how much is invested at each rate.

### Solution

Let $x$ represent the amount invested at 4%, let $y$ represent the amount invested at 6%, and let $z$ represent the amount invested at 7%. Knowing that all three parts equal the total amount invested, \$50,000, we can form the equation $x + y + z = 50{,}000$. We can determine the yearly interest from each part by multiplying the amount invested times the interest rate. Hence, the next equation is $0.04x + 0.06y + 0.07z = 3050$. We obtain the third equation from the information that the sum of the amounts invested at 4% and 6% equals the amount invested at 7%. So the third equation is $x + y = z$. These equations form a system of equations as follows.

$$\begin{pmatrix} x + y + z = 50{,}000 \\ 0.04x + 0.06y + 0.07z = 3050 \\ x + y - z = 0 \end{pmatrix}$$

Solving this system, it can be determined that \$10,000 is invested at 4%, \$15,000 is invested at 6%, and \$25,000 is invested at 7%.

## Concept Quiz 10.4

For Problems 1–10, answer true or false.

1. The graph of a linear equation in three variables is a line.
2. A system of three linear equations in three variables produces three planes when graphed.
3. Three planes can be related by intersecting in exactly two points.
4. One way three planes can be related is if two of the planes are parallel and the third plane intersects them in parallel lines.
5. A system of three linear equations in three variables always has an infinite number of solutions.
6. A system of three linear equations in three variables can have one ordered triple as a solution.
7. The solution set of the system $\begin{pmatrix} x - y + z = 4 \\ x - y + z = 6 \\ 3y - 2z = 9 \end{pmatrix}$ is $\{(5, 15, 3)\}$.
8. The solution set of the system is $\{(3, 1, 2)\}$.
9. It is possible for a system of three linear equations in three variables to have a solution set consisting of $\{(0, 0, 0)\}$.
10. A system of three equations with three unknowns named $x$, $y$, and $z$ is always solved by first finding the value of $z$.

## Problem Set 10.4

Solve each of the following systems. If the solution set is $\varnothing$ or if it contains infinitely many solutions, then so indicate. (Objective 1)

1. $\begin{pmatrix} x + 2y - 3z = 2 \\ 3y - z = 13 \\ 3y + 5z = 25 \end{pmatrix}$

2. $\begin{pmatrix} 2x + 3y - 4z = -10 \\ 2y + 3z = 16 \\ 2y - 5z = -16 \end{pmatrix}$

3. $\begin{pmatrix} 3x + 2y - 2z = 14 \\ x - 6z = 16 \\ 2x + 5z = -2 \end{pmatrix}$

4. $\begin{pmatrix} 3x + 2y - z = -11 \\ 2x - 3y = -1 \\ 4x + 5y = -13 \end{pmatrix}$

5. $\begin{pmatrix} 2x - y + z = 0 \\ 3x - 2y + 4z = 11 \\ 5x + y - 6z = -32 \end{pmatrix}$

6. $\begin{pmatrix} x - 2y + 3z = 7 \\ 2x + y + 5z = 17 \\ 3x - 4y - 2z = 1 \end{pmatrix}$

7. $\begin{pmatrix} 4x - y + z = 5 \\ 3x + y + 2z = 4 \\ x - 2y - z = 1 \end{pmatrix}$

8. $\begin{pmatrix} 2x - y + 3z = -14 \\ 4x + 2y - z = 12 \\ 6x - 3y + 4z = -22 \end{pmatrix}$

9. $\begin{pmatrix} x - y + 2z = 4 \\ 2x - 2y + 4z = 7 \\ 3x - 3y + 6z = 1 \end{pmatrix}$

10. $\begin{pmatrix} x + y - z = 2 \\ 3x - 4y + 2z = 5 \\ 2x + 2y - 2z = 7 \end{pmatrix}$

**11.** $\begin{pmatrix} x - 2y + z = -4 \\ 2x + 4y - 3z = -1 \\ -3x - 6y + 7z = 4 \end{pmatrix}$

**12.** $\begin{pmatrix} 2x - y + 3z = 1 \\ 4x + 7y - z = 7 \\ x + 4y - 2z = 3 \end{pmatrix}$

**13.** $\begin{pmatrix} 3x - 2y + 4z = 6 \\ 9x + 4y - z = 0 \\ 6x - 8y - 3z = 3 \end{pmatrix}$

**14.** $\begin{pmatrix} 2x - y + 3z = 0 \\ 3x + 2y - 4z = 0 \\ 5x - 3y + 2z = 0 \end{pmatrix}$

**15.** $\begin{pmatrix} 3x - y + 4z = 9 \\ 3x + 2y - 8z = -12 \\ 9x + 5y - 12z = -23 \end{pmatrix}$

**16.** $\begin{pmatrix} 5x - 3y + z = 1 \\ 2x - 5y = -2 \\ 3x - 2y - 4z = -27 \end{pmatrix}$

**17.** $\begin{pmatrix} 4x - y + 3z = -12 \\ 2x + 3y - z = 8 \\ 6x + y + 2z = -8 \end{pmatrix}$

**18.** $\begin{pmatrix} x + 3y - 2z = 19 \\ 3x - y - z = 7 \\ -2x + 5y + z = 2 \end{pmatrix}$

**19.** $\begin{pmatrix} x + y + z = 1 \\ 2x - 3y + 6z = 1 \\ -x + y + z = 0 \end{pmatrix}$

**20.** $\begin{pmatrix} 3x + 2y - 2z = -2 \\ x - 3y + 4z = -13 \\ -2x + 5y + 6z = 29 \end{pmatrix}$

Solve each of the following problems by setting up and solving a system of three linear equations in three variables. (Objective 2)

**21.** Two bottles of catsup, 2 jars of peanut butter, and 1 jar of pickles cost \$7.78. Three bottles of catsup, 4 jars of peanut butter, and 2 jars of pickles cost \$14.34. Four bottles of catsup, 3 jars of peanut butter, and 5 jars of pickles cost \$19.19. Find the cost per bottle of catsup, the cost per jar of peanut butter, and the cost per jar of pickles.

**22.** Five pounds of potatoes, 1 pound of onions, and 2 pounds of apples cost \$5.12. Two pounds of potatoes, 3 pounds of onions, and 4 pounds of apples cost \$7.14. Three pounds of potatoes, 4 pounds of onions, and 1 pound of apples cost \$5.96. Find the price per pound for each item.

**23.** The sum of three numbers is 20. The sum of the first and third numbers is 2 more than twice the second number. The third number minus the first yields three times the second number. Find the numbers.

**24.** The sum of three numbers is 40. The third number is 10 less than the sum of the first two numbers. The second number is 1 larger than the first. Find the numbers.

**25.** The sum of the measures of the angles of a triangle is 180°. The largest angle is twice the smallest angle. The sum of the smallest and the largest angle is twice the other angle. Find the measure of each angle.

**26.** A box contains \$2 in nickels, dimes, and quarters. There are 19 coins in all, and there are twice as many nickels as dimes. How many coins of each kind are there?

**27.** Part of \$3000 is invested at 4%, another part at 5%, and the remainder at 6%. The total yearly income from the three investments is \$160. The sum of the amounts invested at 4% and 5% equals the amount invested at 6%. Determine how much is invested at each rate.

**28.** The perimeter of a triangle is 45 centimeters. The longest side is 4 centimeters less than twice the shortest side. The sum of the lengths of the shortest and longest sides is 7 centimeters less than three times the length of the remaining side. Find the lengths of all three sides of the triangle.

## Thoughts Into Words

**29.** Give a step-by-step description of how to solve the system

$$\begin{pmatrix} x - 2y + 3z = -23 \\ 5y - 2z = 32 \\ 4z = -24 \end{pmatrix}$$

**30.** Describe how you would solve the system

$$\begin{pmatrix} x - 3z = 4 \\ 3x - 2y + 7z = -1 \\ 2x + z = 9 \end{pmatrix}$$

### Answers to the Concept Quiz

**1.** False **2.** True **3.** False **4.** True **5.** False **6.** True **7.** True **8.** False **9.** True **10.** False

# 10.5 Systems Involving Nonlinear Equations

OBJECTIVES
1. Graph systems of nonlinear equations
2. Solve systems of nonlinear equations

Thus far in this chapter, we have solved systems of linear equations. In this section, we shall consider some systems where at least one of the equations is *nonlinear*. Let's begin by considering a system of one linear equation and one quadratic equation.

**Classroom Example**
Solve the system:
$\begin{pmatrix} x^2 + y^2 = 40 \\ x + y = 4 \end{pmatrix}$

**EXAMPLE 1** Solve the system $\begin{pmatrix} x^2 + y^2 = 17 \\ x + y = 5 \end{pmatrix}$.

## Solution

First, let's graph the system so that we can predict approximate solutions. From our previous graphing experiences in Chapters 8 and 9, we should recognize $x^2 + y^2 = 17$ as a circle, and $x + y = 5$ as a straight line (Figure 10.15).

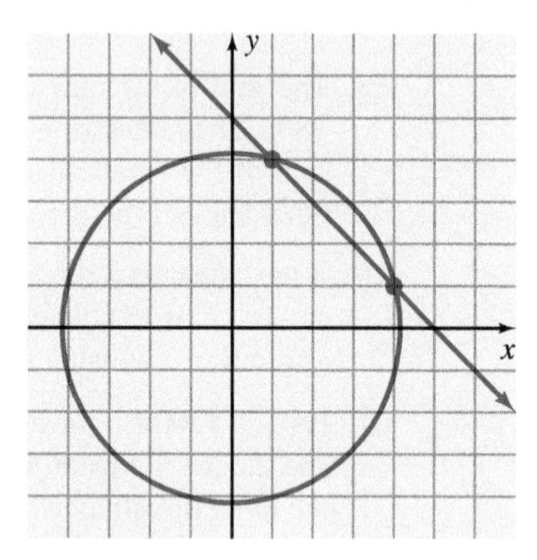


**Figure 10.15**

The graph indicates that there should be two ordered pairs with positive components (the points of intersection occur in the first quadrant) as solutions for this system. In fact, we could guess that these solutions are (1, 4) and (4, 1), and verify our guess by checking them in the given equations.

Let's also solve the system analytically using the substitution method as follows: Change the form of $x + y = 5$ to $y = 5 - x$, and substitute $5 - x$ for $y$ in the first equation.

$$x^2 + y^2 = 17$$
$$x^2 + (5 - x)^2 = 17$$
$$x^2 + 25 - 10x + x^2 = 17$$
$$2x^2 - 10x + 8 = 0$$
$$x^2 - 5x + 4 = 0$$
$$(x - 4)(x - 1) = 0$$
$$x - 4 = 0 \quad \text{or} \quad x - 1 = 0$$
$$x = 4 \quad \text{or} \quad x = 1$$

Substitute 4 for $x$ and then 1 for $x$ in the second equation of the system to produce

$$x + y = 5 \qquad x + y = 5$$
$$4 + y = 5 \qquad 1 + y = 5$$
$$y = 1 \qquad y = 4$$

Therefore, the solution set is $\{(1, 4), (4, 1)\}$.

**Classroom Example**
Solve the system:
$\begin{pmatrix} y = x^2 - 3 \\ y = -x^2 + 2 \end{pmatrix}$

**EXAMPLE 2** Solve the system $\begin{pmatrix} y = -x^2 + 1 \\ y = x^2 - 2 \end{pmatrix}$.

### Solution

Again, let's get an idea of approximate solutions by graphing the system. Both equations produce parabolas, as indicated in Figure 10.16. From the graph, we can predict two nonintegral ordered-pair solutions, one in the third quadrant and the other in the fourth quadrant.

Substitute $-x^2 + 1$ for $y$ in the second equation to obtain

$$\begin{aligned} y &= x^2 - 2 \\ -x^2 + 1 &= x^2 - 2 \\ 3 &= 2x^2 \\ \frac{3}{2} &= x^2 \\ \pm\sqrt{\frac{3}{2}} &= x \\ \pm\frac{\sqrt{6}}{2} &= x \end{aligned}$$

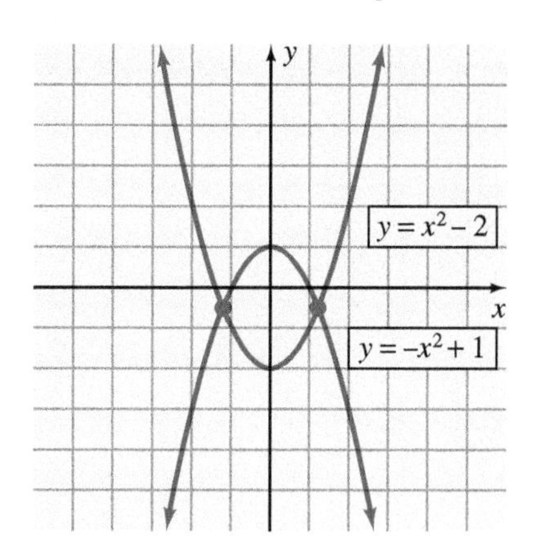


**Figure 10.16**

Substitute $\frac{\sqrt{6}}{2}$ for $x$ in the second equation to yield

$$\begin{aligned} y &= x^2 - 2 \\ y &= \left(\frac{\sqrt{6}}{2}\right)^2 - 2 \\ &= \frac{6}{4} - 2 \\ &= -\frac{1}{2} \end{aligned}$$

Substitute $-\frac{\sqrt{6}}{2}$ for $x$ in the second equation to yield

$$\begin{aligned} y &= x^2 - 2 \\ y &= \left(-\frac{\sqrt{6}}{2}\right)^2 - 2 \\ &= \frac{6}{4} - 2 = -\frac{1}{2} \end{aligned}$$

The solution set is $\left\{\left(-\frac{\sqrt{6}}{2}, -\frac{1}{2}\right), \left(\frac{\sqrt{6}}{2}, -\frac{1}{2}\right)\right\}$. Check it!

**Classroom Example**
Solve the system:
$\begin{pmatrix} y = x^2 + 1 \\ 10x - 5y = 1 \end{pmatrix}$

**EXAMPLE 3** Solve the system $\begin{pmatrix} y = x^2 + 2 \\ 6x - 4y = -5 \end{pmatrix}$.

### Solution

From previous graphing experiences, we recognize that $y = x^2 + 2$ is the basic parabola shifted upward 2 units and that $6x - 4y = -5$ is a straight line (see Figure 10.17). Because of the close proximity of the curves, it is difficult to tell whether they intersect. In other words, the graph does not definitely indicate any real number solutions for the system.

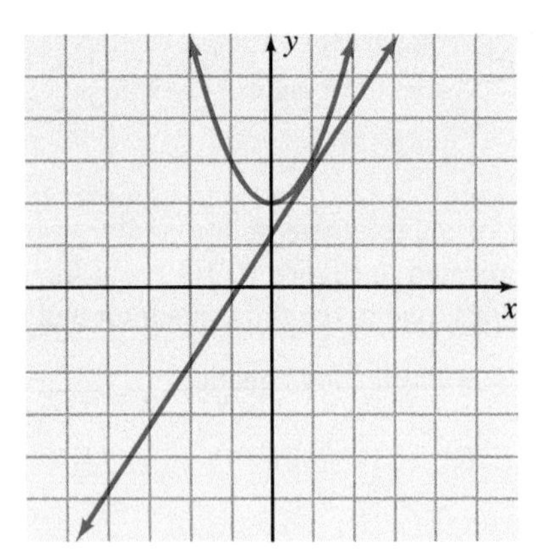


**Figure 10.17**

Let's solve the system using the substitution method. We can substitute $x^2 + 2$ for $y$ in the second equation, which produces two values for $x$.

$$6x - 4(x^2 + 2) = -5$$
$$6x - 4x^2 - 8 = -5$$
$$-4x^2 + 6x - 3 = 0$$
$$4x^2 - 6x + 3 = 0$$
$$x = \frac{6 \pm \sqrt{36 - 48}}{8}$$
$$x = \frac{6 \pm \sqrt{-12}}{8}$$
$$x = \frac{6 \pm 2i\sqrt{3}}{8}$$
$$x = \frac{3 \pm i\sqrt{3}}{4}$$

It is now obvious that the system has no real number solutions. That is, the line and the parabola do not intersect in the real number plane. However, there will be two pairs of complex numbers in the solution set. We can substitute $\frac{3 + i\sqrt{3}}{4}$ for $x$ in the first equation.

$$y = \left(\frac{3 + i\sqrt{3}}{4}\right)^2 + 2$$
$$= \frac{6 + 6i\sqrt{3}}{16} + 2$$
$$= \frac{6 + 6i\sqrt{3} + 32}{16}$$
$$= \frac{38 + 6i\sqrt{3}}{16} = \frac{19 + 3i\sqrt{3}}{8}$$

Likewise, we can substitute $\frac{3 - i\sqrt{3}}{4}$ for $x$ in the first equation.

$$y = \left(\frac{3 - i\sqrt{3}}{4}\right)^2 + 2$$
$$= \frac{6 - 6i\sqrt{3}}{16} + 2$$

$$= \frac{6 - 6i\sqrt{3} + 32}{16}$$

$$= \frac{38 - 6i\sqrt{3}}{16}$$

$$= \frac{19 - 3i\sqrt{3}}{8}$$

The solution set is $\left\{\left(\frac{3 + i\sqrt{3}}{4}, \frac{19 + 3i\sqrt{3}}{8}\right), \left(\frac{3 - i\sqrt{3}}{4}, \frac{19 - 3i\sqrt{3}}{8}\right)\right\}$.

In Example 3, the use of a graphing utility may not, at first, indicate whether or not the system has any real number solutions. Suppose that we graph the system using a viewing rectangle such that $-15 \le x \le 15$ and $-10 \le y \le 10$. In Figure 10.18, we cannot tell whether or not the line and parabola intersect.

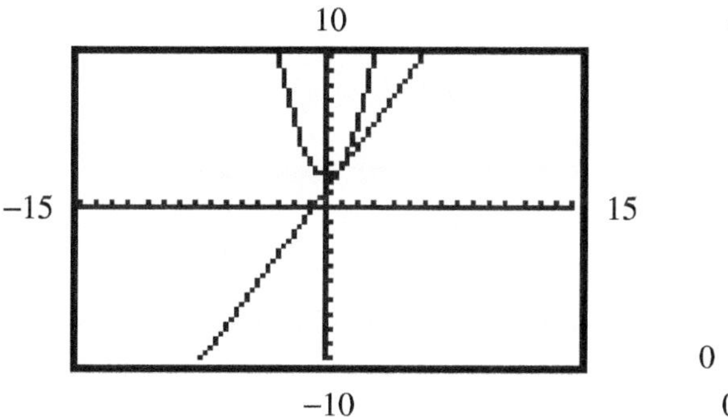


**Figure 10.18**

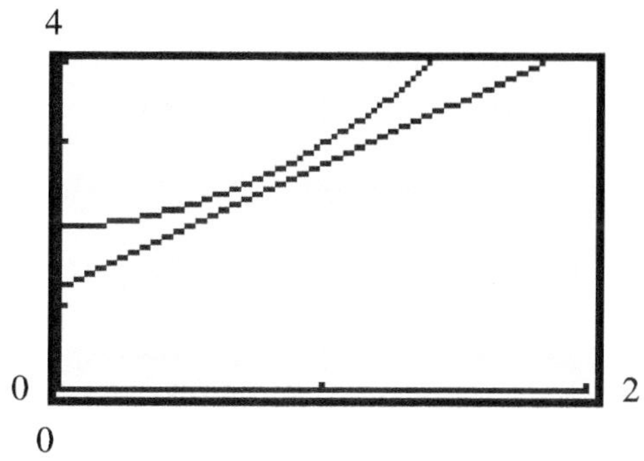

**Figure 10.19**

However, if we change the viewing rectangle so that $0 \le x \le 2$ and $0 \le y \le 4$, as shown in Figure 10.19, then it becomes apparent that the two graphs do not intersect.

## Concept Quiz 10.5

For Problems 1–8, answer true or false.

1. Graphing a system of equations is a method of approximating the solutions.
2. Every system of nonlinear equations has a real number solution.
3. Every nonlinear system of equations can be solved by substitution.
4. Every nonlinear system of equations can be solved by the elimination method.
5. Graphs of a circle and a line will have one, two, or no points of intersection.
6. The solution set for the system $\begin{pmatrix} y = x^2 + 1 \\ y = -x^2 + 1 \end{pmatrix}$ is $\{(0, 1)\}$.
7. The solution set for the system $\begin{pmatrix} 3x^2 + 4y^2 = 12 \\ x^2 - y^2 = 9 \end{pmatrix}$ is $\{(4, \sqrt{7})\}$.
8. The solution set for the system $\begin{pmatrix} 2x^2 + y^2 = 8 \\ x^2 + y^2 = 4 \end{pmatrix}$ is the null set.

## Problem Set 10.5

For Problems 1–28, (a) graph each system so that approximate real number solutions (if there are any) can be predicted, and (b) solve each system using the substitution method or the elimination-by-addition method. (Objectives 1 and 2)

**1.** $\begin{pmatrix} y = (x+2)^2 \\ y = -2x - 4 \end{pmatrix}$

**2.** $\begin{pmatrix} y = x - 1 \\ x = (y-1)^2 \end{pmatrix}$

**3.** $\begin{pmatrix} y = -x^2 \\ y = -x - 2 \end{pmatrix}$

**4.** $\begin{pmatrix} y = x^2 \\ y = x + 2 \end{pmatrix}$

**5.** $\begin{pmatrix} x^2 + y^2 = 13 \\ 3x + 2y = 0 \end{pmatrix}$

**6.** $\begin{pmatrix} x^2 + y^2 = 26 \\ x + y = 6 \end{pmatrix}$

**7.** $\begin{pmatrix} y = \frac{5}{2}x \\ x^2 + y^2 = 29 \end{pmatrix}$

**8.** $\begin{pmatrix} x + y = 4 \\ x^2 + y^2 = 10 \end{pmatrix}$

**9.** $\begin{pmatrix} y = x^2 + 6x + 7 \\ 2x + y = -5 \end{pmatrix}$

**10.** $\begin{pmatrix} y = x^2 - 4x + 5 \\ -x + y = 1 \end{pmatrix}$

**11.** $\begin{pmatrix} y = x^2 \\ y = x^2 - 4x + 4 \end{pmatrix}$

**12.** $\begin{pmatrix} y = -x^2 + 3 \\ y = x^2 + 1 \end{pmatrix}$

**13.** $\begin{pmatrix} x + y = -8 \\ x^2 - y^2 = 16 \end{pmatrix}$

**14.** $\begin{pmatrix} x - y = 2 \\ x^2 - y^2 = 16 \end{pmatrix}$

**15.** $\begin{pmatrix} x + y = 2 \\ x^2 - y^2 = 36 \end{pmatrix}$

**16.** $\begin{pmatrix} x - y = 2 \\ x^2 - y^2 = 36 \end{pmatrix}$

**17.** $\begin{pmatrix} y = x^2 + 2x - 1 \\ y = x^2 + 4x + 5 \end{pmatrix}$

**18.** $\begin{pmatrix} y = x^2 + 4x - 9 \\ y = x^2 - 2x + 3 \end{pmatrix}$

**19.** $\begin{pmatrix} x^2 + y^2 = 9 \\ 4x^2 + y^2 = 36 \end{pmatrix}$

**20.** $\begin{pmatrix} 2x^2 + y^2 = 8 \\ x^2 + y^2 = 4 \end{pmatrix}$

**21.** $\begin{pmatrix} xy = 4 \\ y = x \end{pmatrix}$

**22.** $\begin{pmatrix} y = -x \\ xy = -9 \end{pmatrix}$

**23.** $\begin{pmatrix} y = -x^2 - 3 \\ y = -2x^2 + 1 \end{pmatrix}$

**24.** $\begin{pmatrix} y = x^2 + 2 \\ y = 2x^2 + 1 \end{pmatrix}$

**25.** $\begin{pmatrix} x^2 + y^2 = 2 \\ x - y = 4 \end{pmatrix}$

**26.** $\begin{pmatrix} y = -x^2 + 1 \\ x + y = 2 \end{pmatrix}$

**27.** $\begin{pmatrix} 2x + y = 6 \\ x^2 + y^2 = 4 \end{pmatrix}$

**28.** $\begin{pmatrix} y = x^2 - 1 \\ x - y = 3 \end{pmatrix}$

## Thoughts Into Words

**29.** What happens if you try to graph the system

$$\begin{pmatrix} x^2 + 4y^2 = 16 \\ 2x^2 + 5y^2 = -12 \end{pmatrix}$$

**30.** Explain how you would solve the system

$$\begin{pmatrix} x^2 + y^2 = 9 \\ y^2 = x^2 + 4 \end{pmatrix}$$

## Graphing Calculator Activities

**31.** Use a graphing calculator to graph the systems in Problems 1–16, and check the reasonableness of your answers.

**32.** For each of the following systems, (a) use your graphing calculator to show that there are no real number solutions, and (b) solve the system by the substitution method or the elimination-by-addition method to find the complex solutions.

**(a)** $\begin{pmatrix} y = x^2 + 1 \\ y = -3 \end{pmatrix}$

**(b)** $\begin{pmatrix} y = -x^2 + 1 \\ y = 3 \end{pmatrix}$

**(c)** $\begin{pmatrix} y = x^2 \\ x - y = 4 \end{pmatrix}$

**(d)** $\begin{pmatrix} y = x^2 + 1 \\ y = -x^2 \end{pmatrix}$

**(e)** $\begin{pmatrix} x^2 + y^2 = 1 \\ x + y = 2 \end{pmatrix}$

**(f)** $\begin{pmatrix} x^2 + y^2 = 2 \\ x^2 - y^2 = 6 \end{pmatrix}$

**33.** Graph the system $\begin{pmatrix} y = x^2 + 2 \\ 6x - 4y = -5 \end{pmatrix}$ and use the TRACE and ZOOM features of your calculator to demonstrate clearly that this system has no real number solutions.

### Answers to the Concept Quiz

**1.** True **2.** False **3.** True **4.** False **5.** True **6.** True **7.** False **8.** False

# Chapter 10 Summary

| OBJECTIVE | SUMMARY | EXAMPLE |
|---|---|---|
| Solve systems of two linear equations by graphing. (Section 10.1/Objective 1) | Graphing a system of two linear equations in two variables produces one of the following results. **1.** The graphs of the two equations are two intersecting lines, which indicates that there is one unique solution of the system. Such a system is called a consistent system. **2.** The graphs of the two equations are two parallel lines, which indicates that there is no solution for the system. It is called an inconsistent system. **3.** The graphs of the two equations are the same line, which indicates infinitely many solutions for the system. The equations are called dependent equations. | Solve $\begin{pmatrix} x - 3y = 6 \\ 2x + 3y = 3 \end{pmatrix}$ by graphing. |

**Solution**

Graph the lines by determining the $x$ and $y$ intercepts and a check point.

$x - 3y = 6$

| x | y |
|---|---|
| 0 | −2 |
| 6 | 0 |
| −3 | −3 |

$2x + 3y = 3$

| x | y |
|---|---|
| 0 | 1 |
| $\frac{3}{2}$ | 0 |
| −1 | $\frac{5}{3}$ |

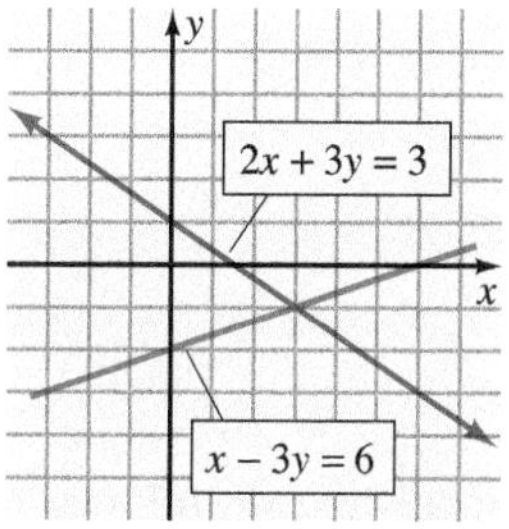


It appears that $(3, -1)$ is the solution. Checking these values in the equations, we can determine that the solution set is $\{(3, -1)\}$.

**Sample Problem 1**

Solve $\begin{pmatrix} 2x + y = 4 \\ x + 2y = -1 \end{pmatrix}$ by graphing.

Answers to **Sample Problems** are located in the back of the book.

(*continued*)

| OBJECTIVE | SUMMARY | EXAMPLE |
|---|---|---|
| Solve systems of linear inequalities. (Section 10.1/Objective 2) | The solution set of a system of linear inequalities is the intersection of the solution sets of the individual inequalities. | Solve the system $\begin{pmatrix} y < -x + 2 \\ y > \frac{1}{2}x + 1 \end{pmatrix}$.<br>**Solution**<br>The graph of $y < -x + 2$ consists of all the points below the line $y = -x + 2$.<br>The graph of $y > \frac{1}{2}x + 1$ consists of all the points above the line $y = \frac{1}{2}x + 1$.<br>y; $y = \frac{1}{2}x + 1$; x; $y = -x + 2$<br>The graph of the system is indicated by the shaded region.<br>**Sample Problem 2**<br>Solve the system $\begin{pmatrix} y \le -\frac{1}{2}x + 2 \\ y > x - 3 \end{pmatrix}$. |
| Solve systems of linear equations using substitution. (Section 10.2/Objective 1) | We can describe the substitution method of solving a system of equations as follows.<br>**Step 1** Solve one of the equations for one variable in terms of the other variable if neither equation is in such a form. (If possible, make a choice that will avoid fractions.)<br>**Step 2** Substitute the expression obtained in step 1 into the other equation to produce an equation with one variable.<br>**Step 3** Solve the equation obtained in step 2.<br>**Step 4** Use the solution obtained in step 3, along with the expression obtained in step 1, to determine the solution of the system. | Solve the system $\begin{pmatrix} 3x + y = -9 \\ 2x + 3y = 8 \end{pmatrix}$.<br>**Solution**<br>Solving the first equation for $y$ gives the equation $y = -3x - 9$. In the second equation, substitute $-3x - 9$ for $y$ and solve.<br>$2x + 3(-3x - 9) = 8$<br>$2x - 9x - 27 = 8$<br>$-7x = 35$<br>$x = -5$<br>Now, to find the value of $y$, substitute $-5$ for $x$ in the equation<br>$y = -3x - 9$<br>$y = -3(-5) - 9$<br>$y = 6$<br>The solution set of the system is $\{(-5, 6)\}$.<br>**Sample Problem 3**<br>Solve the system $\begin{pmatrix} 12x - 2y = 0 \\ -7x + y = -1 \end{pmatrix}$. |

| OBJECTIVE | SUMMARY | EXAMPLE |
| --- | --- | --- |
| Solve systems of equations by the elimination-by-addition method. (Section 10.3/Objective 1) | The elimination-by-addition method has its basis in the addition property of equations where *If* $a = b$ *and* $c = d$, *then* $a + c = b + d$. We apply the property to eliminate one of the variable terms so that we can solve for the other variable. The equations may need to be multiplied by a nonzero number so that the coefficients of the variable to be eliminated are opposites. | Solve the system $\begin{pmatrix} 2x - 5y = 31 \\ 4x + 3y = 23 \end{pmatrix}$.<br><br>**Solution**<br>Let's multiply both sides of the first equation by $-2$, then adding the equations will eliminate the $x$ variable terms.<br>$-4x + 10y = -62$<br>$4x + 3y = 23$<br>$13y = -39$<br>$y = -3$<br>The $y$ coordinate of the solution is $-3$.<br><br>Substitute $-3$ for $y$ in either of the original equations and solve for $x$:<br>$2x - 5(-3) = 31$<br>$2x + 15 = 31$<br>$2x = 16$<br>$x = 8$<br>The solution set of the system is $\{(8, -3)\}$.<br><br>**Sample Problem 4**<br>Solve the system $\begin{pmatrix} 2x - 5y = 7 \\ 3x - 10y = 13 \end{pmatrix}$. |
| Determine which method to use to solve a system of equations. (Section 10.3/Objective 2) | Graphing a system provides visual support for the solution, but it may be impossible to get exact solutions from a graph. Both the substitution method and the elimination-by-addition method provide exact solutions. Many systems lend themselves to one or the other method. Substitution is usually the preferred method if one of the equations in the system is already solved for a variable. | Solve $\begin{pmatrix} x - 3y = -13 \\ 2x + 5y = 18 \end{pmatrix}$.<br><br>**Solution**<br>Because the first equation can be solved for $x$ without involving any fractions, the system is a good candidate for solving by the substitution method. The system could also be solved very easily using the elimination-by-addition method by multiplying the first equation by $-2$ and adding the result to the second equation. Either method will produce the solution set of $\{(-1, 4)\}$.<br><br>**Sample Problem 5**<br>Solve $\begin{pmatrix} 3x + 12y = 24 \\ x - 5y = 17 \end{pmatrix}$. |

(*continued*)

| OBJECTIVE | SUMMARY | EXAMPLE |
|---|---|---|
| Use systems of equations to solve problems. (Section 10.2/Objective 2; Section 10.3/Objective 3) | Many problems that were solved earlier using only one variable may seem easier to solve by using two variables and a system of equations. | A car dealership has 220 vehicles on the lot. The number of cars on the lot is 5 less than twice the number of trucks. Find the number of cars and the number of trucks on the lot.<br>**Solution**<br>Letting $x$ represent the number of cars and $y$ the number of trucks, we obtain the following system:<br>$\begin{pmatrix} x + y = 220 \\ x = 2y - 5 \end{pmatrix}$<br>Solving the system, we can determine that the dealership has 145 cars and 75 trucks on the lot.<br>**Sample Problem 6**<br>A car dealership has 510 cars and trucks on the lot. There are 190 more cars than trucks on the lot. Find the number of cars and the number of trucks on the lot. |
| Solve systems of three linear equations. (Section 10.4/Objective 1) | Solving a system of three linear equations in three variables produces one of the following results:<br>**1.** There is one ordered triple that satisfies all three equations.<br>**2.** There are infinitely many ordered triples in the solution set, all of which are coordinates of points on a line common to the planes.<br>**3.** There are infinitely many ordered triples in the solution set, all of which are coordinates of points on a plane.<br>**4.** The solution set is empty; it is $\varnothing$.<br>The steps for the elimination-by-addition method are listed on page 527. | Solve $\begin{pmatrix} 4x + 3y - 2z = -5 \\ 2y + 3z = -7 \\ y - 3z = -8 \end{pmatrix}$.<br>**Solution**<br>Adding equations 2 and 3 yields $3y = -15$. Therefore we can determine that $y = -5$. Substituting $-5$ for $y$ in the third equation gives $-5 - 3z = -8$. Solving this equation yields $z = 1$. Substituting $-5$ for $y$ and 1 for $z$ in the first equation gives $4x + 3(-5) - 2(1) = -5$. Solving this equation gives $x = 3$. The solution set for the system is $\{(3, -5, 1)\}$.<br>**Sample Problem 7**<br>Solve $\begin{pmatrix} x - 2y = 2 \\ 2x + 3y = 11 \\ y - 4z = -7 \end{pmatrix}$. |

| OBJECTIVE | SUMMARY | EXAMPLE |
| --- | --- | --- |
| Use systems of three linear equations to solve problems.<br>(Section 10.4/Objective 2) | Many word problems involving three variables can be solved using a system of three linear equations. | The sum of the measures of the angles in a triangle is 180°. The largest angle is eight times the smallest angle. The sum of the smallest and the largest angle is three times the other angle. Find the measure of each angle.<br>**Solution**<br>Let $x$ represent the measure of the largest angle, let $y$ represent the measure of the middle angle, and let $z$ represent the measure of the smallest angle. From the information in the problem, we can write the following system of equations:<br>$\begin{pmatrix} x + y + z = 180 \\ x = 8z \\ x + z = 3y \end{pmatrix}$<br>By solving this system, we can determine that the measures of the angles of the triangle are 15°, 45°, and 120°.<br>**Sample Problem 8**<br>A triangle has a perimeter of 55 inches. The measure of the smallest side is 8 less than the middle side. The measure of the longest side is equal to 1 less than the sum of the other two sides. Find the length of the sides. |
| Solve systems of nonlinear equations.<br>(Section 10.5/Objective 2) | Graphing a system of equations involving nonlinear equations is useful for predicting the number of solutions and approximating the solutions. The substitution and elimination methods can be used to find exact solutions for systems of equations involving nonlinear equations. | Solve $\begin{pmatrix} y = x^2 - 2 \\ 2x + y = 1 \end{pmatrix}$.<br>**Solution**<br>Substituting $x^2 - 2$ for $y$ in the second equation gives $2x + x^2 - 2 = 1$. Rewrite the equation in standard form and solve for $x$:<br>$2x + x^2 - 2 = 1$<br>$x^2 + 2x - 3 = 0$<br>$(x + 3)(x - 1) = 0$<br>$x = -3 \quad \text{or} \quad x = 1$<br>To find $y$ when $x = -3$, substitute $-3$ for $x$ in the first equation:<br>$y = (-3)^2 - 2 = 7$<br>To find $y$ when $x = 1$, substitute 1 for $x$ in the first equation:<br>$y = 1^2 - 2 = -1$<br>The solution set is $\{(-3, 7), (1, -1)\}$.<br>**Sample Problem 9**<br>Solve $\begin{pmatrix} y = x^2 + 1 \\ x - y = -1 \end{pmatrix}$. |

# Chapter 10 Review Problem Set

For Problems 1–14, solve each system using the method that seems most appropriate to you.

1. $\begin{pmatrix} 3x - 2y = -6 \\ 2x + 5y = 34 \end{pmatrix}$

2. $\begin{pmatrix} x + 4y = 25 \\ y = -3x - 2 \end{pmatrix}$

3. $\begin{pmatrix} x = 5y - 49 \\ 4x + 3y = -12 \end{pmatrix}$

4. $\begin{pmatrix} x - 6y = 7 \\ 3x + 5y = 9 \end{pmatrix}$

5. $\begin{pmatrix} x - 3y = 25 \\ -3x + 2y = -26 \end{pmatrix}$

6. $\begin{pmatrix} 5x - 7y = -66 \\ x + 4y = 30 \end{pmatrix}$

7. $\begin{pmatrix} 4x + 3y = -9 \\ 3x - 5y = 15 \end{pmatrix}$

8. $\begin{pmatrix} 2x + 5y = 47 \\ 4x - 7y = -25 \end{pmatrix}$

9. $\begin{pmatrix} 7x - 3y = 25 \\ y = 3x - 9 \end{pmatrix}$

10. $\begin{pmatrix} x = -4 - 5y \\ y = 4x + 16 \end{pmatrix}$

11. $\begin{pmatrix} \frac{1}{2}x + \frac{2}{3}y = 6 \\ \frac{3}{4}x - \frac{5}{6}y = -24 \end{pmatrix}$

12. $\begin{pmatrix} \frac{3}{4}x - \frac{1}{2}y = 14 \\ \frac{5}{12}x + \frac{3}{4}y = 16 \end{pmatrix}$

13. $\begin{pmatrix} 6x - 4y = 7 \\ 9x + 8y = 0 \end{pmatrix}$

14. $\begin{pmatrix} 4x - 5y = -5 \\ 6x - 10y = -9 \end{pmatrix}$

For Problems 15–20, solve each system.

15. $\begin{pmatrix} x - 2y + 4z = -14 \\ 3x - 5y + z = 20 \\ -2x + y - 5z = 22 \end{pmatrix}$

16. $\begin{pmatrix} x + 3y - 2z = 28 \\ 2x - 8y + 3z = -63 \\ 3x + 8y - 5z = 72 \end{pmatrix}$

17. $\begin{pmatrix} x + y - z = -2 \\ 2x - 3y + 4z = 17 \\ -3x + 2y + 5z = -7 \end{pmatrix}$

18. $\begin{pmatrix} -x - y + z = -3 \\ 3x + 2y - 4z = 12 \\ 5x + y + 2z = 5 \end{pmatrix}$

19. $\begin{pmatrix} 3x + y - z = -6 \\ 3x + 2y + 3z = 9 \\ 6x - 2y + 2z = 8 \end{pmatrix}$

20. $\begin{pmatrix} x - 3y + z = 2 \\ 2x - 5y - 3z = 22 \\ -4x + 3y + 5z = -26 \end{pmatrix}$

For Problems 21–22, graph the following system, and then find the solution set by using either the substitution or the elimination method.

21. $\begin{pmatrix} y = 2x^2 - 1 \\ 2x + y = 3 \end{pmatrix}$

22. $\begin{pmatrix} x^2 + y^2 = 25 \\ y = x - 3 \end{pmatrix}$

For Problems 23–26, indicate the solution set of the following system of inequalities by graphing the system and shading the appropriate region.

23. $\begin{pmatrix} 3x + y > 6 \\ x - 2y \leq 4 \end{pmatrix}$

24. $\begin{pmatrix} 4x - y > -4 \\ x + 3y > 3 \end{pmatrix}$

25. $\begin{pmatrix} 2x + y > 4 \\ y > -2x \end{pmatrix}$

26. $\begin{pmatrix} x - 3y > 3 \\ y \leq -x \end{pmatrix}$

For Problems 27–42, solve each problem by setting up and solving a system of two equations and two unknowns or a system of three equations and three unknowns.

27. The sum of the squares of two numbers is 13. If one number is 1 larger than the other number, find the numbers.

28. The sum of the squares of two numbers is 34. The difference of the squares of the same two numbers is 16. Find the numbers.

29. A number is 1 larger than the square of another number. The sum of the two numbers is 7. Find the numbers.

30. The area of a rectangular region is 54 square meters and its perimeter is 30 meters. Find the length and width of the rectangle.

31. At a local confectionery, 7 pounds of cashews and 5 pounds of Spanish peanuts cost $88, and 3 pounds of cashews and 2 pounds of Spanish peanuts cost $37. Find the price per pound for cashews and for Spanish peanuts.

32. We bought 2 liters of pop and 4 pounds of candy for $12. The next day we bought 3 liters of pop and 2 pounds of candy for $9. Find the price of a liter of pop and also the price of a pound of candy.

33. Suppose that a mail-order company charges a fixed fee for shipping merchandise that weighs 1 pound or less,

plus an additional fee for each pound over 1 pound. If the shipping charge for 5 pounds is \$2.40 and for 12 pounds is \$3.10, find the fixed fee and the additional fee.

**34.** How many quarts of 1% milk must be mixed with 4% milk to obtain 10 quarts of 2% milk?

**35.** The perimeter of a rectangle is 56 centimeters. The length of the rectangle is three times the width. Find the dimensions of the rectangle.

**36.** Antonio had a total of \$4200 debt on two credit cards. One of the cards charged 1% interest per month, and the other card charged 1.5% interest per month. Find the amount of debt on each card if he paid \$57 in interest charges for the month.

**37.** After working her shift as a waitress, Kelly had collected tips of 30 bills consisting of one-dollar bills and five-dollar bills. If her tips amounted to \$50, how many of each type did she have?

**38.** In an ideal textbook, every problem set has a fixed number of review problems and a fixed number of problems on the new material. Professor Kelly always assigned 80% of the review problems and 40% of the problems on the new material, which amounted to 56 problems. Professor Edward always assigned 100% of the review problems and 60% of the problems on the new material, which amounted to 78 problems. How many problems of each type are in the problem sets?

**39.** After an evening of selling flowers, a vendor had collected 64 bills consisting of five-dollar bills, ten-dollar bills, and twenty-dollar bills, which amounted to \$620. The number of ten-dollar bills was three times the number of twenty-dollar bills. Find the number of each kind of bill.

**40.** The measure of the largest angle of a triangle is twice the measure of the smallest angle of the triangle. The sum of the measures of the largest angle and the smallest angle of a triangle is twice the measure of the remaining angle of the triangle. Find the measure of all three angles of the triangle.

**41.** Kenisha has a Bank of U.S. credit card that charges 1% interest per month, a Community Bank credit card that charges 1.5% interest per month, and a First National credit card that charges 2% interest per month. In total she has \$6400 charged among the three credit cards. The total interest for the month for all three cards is \$99. The amount charged on the Community Bank card is \$500 less than the amount charged on the Bank of U.S. card. Find the amount charged on each card.

**42.** The perimeter of a triangle is 33 inches. The longest side is 3 inches more than twice the shortest side. The sum of the lengths of the shortest side and the longest side is 9 more than the remaining side. Find the length of all three sides of the triangle.

# Chapter 10 Test

For Problems 1–4, refer to the following systems of equations:

I. $\begin{pmatrix} 5x - 2y = 12 \\ 2x + 5y = 7 \end{pmatrix}$ II. $\begin{pmatrix} x - 4y = 1 \\ 2x - 8y = 2 \end{pmatrix}$

III. $\begin{pmatrix} 4x - 5y = 6 \\ 4x - 5y = 1 \end{pmatrix}$ IV. $\begin{pmatrix} 2x + 3y = 9 \\ 7x - 4y = 9 \end{pmatrix}$

1. For which of these systems are the equations said to be dependent?

2. For which of these systems does the solution set consist of a single ordered pair?

3. For which of these systems are the graphs parallel lines?

4. For which of these systems are the graphs perpendicular lines?

For Problems 5–12, solve each of the systems.

5. $\begin{pmatrix} 2x - 3y = -17 \\ 5x + y = 17 \end{pmatrix}$

6. $\begin{pmatrix} -5x + 4y = 35 \\ x - 3y = -18 \end{pmatrix}$

7. $\begin{pmatrix} 6x + 7y = -40 \\ 4x - 3y = 4 \end{pmatrix}$

8. $\begin{pmatrix} x - 4y = 15 \\ 3x + 8y = -15 \end{pmatrix}$

9. $\begin{pmatrix} 2x - 7y = -8 \\ 4x + 5y = 3 \end{pmatrix}$

10. $\begin{pmatrix} \frac{2}{3}x - \frac{1}{2}y = 7 \\ \frac{1}{4}x + \frac{1}{3}y = 12 \end{pmatrix}$

11. $\begin{pmatrix} x - 2y - z = 2 \\ 2x + y + z = 3 \\ 3x - 4y - 2z = 5 \end{pmatrix}$

12. $\begin{pmatrix} x - 2y + z = -8 \\ 2x + y - z = 3 \\ -3x - 4y + 2z = -11 \end{pmatrix}$

13. Find the value of $x$ in the solution for the system:

$$\begin{pmatrix} x = -2y + 5 \\ 7x + 3y = 46 \end{pmatrix}$$

14. Find the value of $y$ in the solution for the system:

$$\begin{pmatrix} x - y + 4z = 25 \\ 3x + 2y - z = 5 \\ 5y + 2z = 5 \end{pmatrix}$$

15. Find the value of $z$ in the solution for the system:

$$\begin{pmatrix} x - y - z = -6 \\ 4x + y + 3z = -4 \\ 5x - 2y - 4z = -18 \end{pmatrix}$$

For Problems 16–20, determine how many ordered pairs of real numbers are in the solution set for each of the systems.

16. $\begin{pmatrix} y = 2x^2 + 1 \\ 3x + 4y = 12 \end{pmatrix}$ 17. $\begin{pmatrix} x^2 + 2y^2 = 8 \\ 2x^2 + y^2 = 8 \end{pmatrix}$

18. $\begin{pmatrix} y = x^3 \\ y = (x - 1)^2 - 1 \end{pmatrix}$

19. $\begin{pmatrix} y = -x^2 - 2 \\ y = x^2 + 1 \end{pmatrix}$

20. Solve the system $\begin{pmatrix} y = x^2 - 4x + 7 \\ y = -2x^2 + 8x - 5 \end{pmatrix}$.

21. Graph the solution set for the system:

$$\begin{pmatrix} x + 3y < 3 \\ 2x - y > 2 \end{pmatrix}$$

For Problems 22–25, set up and solve a system of equations to help solve each problem.

22. Stefan went to the convenience store and bought four sodas and three bottles of water for \$8.60. The next day he returned and bought three sodas and two bottles of water for \$6.20. Find the price of a soda and of a bottle of water.

23. A box contains \$7.80 in nickels, dimes, and quarters. There are 6 more dimes than nickels and three times as many quarters as nickels. Find the number of quarters.

24. One solution contains 30% alcohol and another solution contains 80% alcohol. Some of each of the two solutions is mixed to produce 5 liters of a 60%-alcohol solution. How many liters of the 80%-alcohol solution are used?

25. The perimeter of a rectangle is 82 inches. The length of the rectangle is 4 inches less than twice the width of the rectangle. Find the dimensions of the rectangle.

# Chapters 1–10 Cumulative Review Problem Set

For Problems 1–5, evaluate each algebraic expression for the given values of the variables.

1. $-5(x - 1) - 3(2x + 4) + 3(3x - 1)$ for $x = -2$

2. $\dfrac{14a^3b^2}{7a^2b}$ for $a = -1$ and $b = 4$

3. $\dfrac{2}{n} - \dfrac{3}{2n} + \dfrac{5}{3n}$ for $n = 4$

4. $-4\sqrt{2x - y} + 5\sqrt{3x + y}$ for $x = 16$ and $y = 16$

5. $\dfrac{3}{x - 2} - \dfrac{5}{x + 3}$ for $x = 3$

For Problems 6–15, perform the indicated operations and express the answers in simplified form.

6. $(-5\sqrt{6})(3\sqrt{12})$

7. $(2\sqrt{x} - 3)(\sqrt{x} + 4)$

8. $(3\sqrt{2} - \sqrt{6})(\sqrt{2} + 4\sqrt{6})$

9. $(2x - 1)(x^2 + 6x - 4)$

10. $\dfrac{x^2 - x}{x + 5} \cdot \dfrac{x^2 + 5x + 4}{x^4 - x^2}$

11. $\dfrac{16x^2y}{24xy^3} \div \dfrac{9xy}{8x^2y^2}$

12. $\dfrac{x + 3}{10} + \dfrac{2x + 1}{15} - \dfrac{x - 2}{18}$

13. $\dfrac{7}{12ab} - \dfrac{11}{15a^2}$

14. $\dfrac{8}{x^2 - 4x} + \dfrac{2}{x}$

15. $(8x^3 - 6x^2 - 15x + 4) \div (4x - 1)$

For Problems 16–19, simplify each of the complex fractions.

16. $\dfrac{\dfrac{5}{x^2} - \dfrac{3}{x}}{\dfrac{1}{y} + \dfrac{2}{y^2}}$

17. $\dfrac{\dfrac{2}{x} - 3}{\dfrac{3}{y} + 4}$

18. $\dfrac{2 - \dfrac{1}{n + 2}}{3 + \dfrac{4}{n + 3}}$

19. $\dfrac{3a}{2 - \dfrac{1}{a}} - 1$

For Problems 20–25, factor each of the algebraic expressions completely.

20. $20x^2 + 7x - 6$

21. $16x^3 + 54$

22. $4x^4 - 25x^2 + 36$

23. $12x^3 - 52x^2 - 40x$

24. $xy - 6x + 3y - 18$

25. $10 + 9x - 9x^2$

For Problems 26–33, evaluate each of the numerical expressions.

26. $\left(\dfrac{2}{3}\right)^{-4}$

27. $\dfrac{3}{\left(\dfrac{4}{3}\right)^{-1}}$

28. $\sqrt[3]{-\dfrac{27}{64}}$

29. $-\sqrt{0.09}$

30. $(27)^{-\frac{4}{3}}$

31. $4^0 + 4^{-1} + 4^{-2}$

32. $\left(\dfrac{3^{-1}}{2^{-3}}\right)^{-2}$

33. $(2^{-3} - 3^{-2})^{-1}$

For Problems 34–36, find the indicated products and quotients, and express the final answers with positive integral exponents only.

34. $(-3x^{-1}y^2)(4x^{-2}y^{-3})$

35. $\dfrac{48x^{-4}y^2}{6xy}$

36. $\left(\dfrac{27a^{-4}b^{-3}}{-3a^{-1}b^{-4}}\right)^{-1}$

For Problems 37–44, express each radical expression in simplest radical form.

37. $\sqrt{80}$

38. $-2\sqrt{54}$

39. $\sqrt{\dfrac{75}{81}}$

40. $\dfrac{4\sqrt{6}}{3\sqrt{8}}$

41. $\sqrt[3]{56}$

42. $\dfrac{\sqrt[3]{3}}{\sqrt[3]{4}}$

43. $4\sqrt{52x^3y^2}$

44. $\sqrt{\dfrac{2x}{3y}}$

For Problems 45–47, use the distributive property to help simplify each of the following.

45. $-3\sqrt{24} + 6\sqrt{54} - \sqrt{6}$

46. $\dfrac{\sqrt{8}}{3} - \dfrac{3\sqrt{18}}{4} - \dfrac{5\sqrt{50}}{2}$

47. $8\sqrt[3]{3} - 6\sqrt[3]{24} - 4\sqrt[3]{81}$

For Problems 48 and 49, rationalize the denominator and simplify.

48. $\dfrac{\sqrt{3}}{\sqrt{6} - 2\sqrt{2}}$

49. $\dfrac{3\sqrt{5} - \sqrt{3}}{2\sqrt{3} + \sqrt{7}}$

For Problems 50–52, use scientific notation to help perform the indicated operations.

50. $\dfrac{(0.00016)(300)(0.028)}{0.064}$

**51.** $\dfrac{0.00072}{0.0000024}$

**52.** $\sqrt{0.00000009}$

For Problems 53–56, find each of the indicated products or quotients, and express the answers in standard form.

**53.** $(5 - 2i)(4 + 6i)$

**54.** $(-3 - i)(5 - 2i)$

**55.** $\dfrac{5}{4i}$

**56.** $\dfrac{-1 + 6i}{7 - 2i}$

**57.** Find the slope of the line determined by the points $(2, -3)$ and $(-1, 7)$.

**58.** Find the slope of the line determined by the equation $4x - 7y = 9$.

**59.** Find the length of the line segment whose endpoints are $(4, 5)$ and $(-2, 1)$.

**60.** Write the equation of the line that contains the points $(3, -1)$ and $(7, 4)$.

**61.** Write the equation of the line that is perpendicular to the line $3x - 4y = 6$ and contains the point $(-3, -2)$.

**62.** Find the center and the length of a radius of the circle $x^2 + 4x + y^2 - 12y + 31 = 0$.

**63.** For the parabola $y = x^2 + 10x + 21$, find the coordinates of the vertex.

**64.** For the ellipse $x^2 + 4y^2 = 16$, find the length of the major axis.

For Problems 65–70, graph each of the equations.

**65.** $-x + 2y = -4$

**66.** $x^2 + y^2 = 9$

**67.** $x^2 - y^2 = 9$

**68.** $x^2 + 2y^2 = 8$

**69.** $y = -3x$

**70.** $x^2y = 4$

For Problems 71–76, graph each of the functions.

**71.** $f(x) = -2x - 4$

**72.** $f(x) = -2x^2 - 2$

**73.** $f(x) = x^2 - 2x - 2$

**74.** $f(x) = \sqrt{x + 1} + 2$

**75.** $f(x) = 2x^2 + 8x + 9$

**76.** $f(x) = -|x - 2| + 1$

**77.** If $f(x) = x - 3$ and $g(x) = 2x^2 - x - 1$, find $(g \circ f)(x)$ and $(f \circ g)(x)$.

**78.** Find the inverse $(f^{-1})$ of $f(x) = 3x - 7$.

**79.** Find the inverse of $f(x) = -\frac{1}{2}x + \frac{2}{3}$.

**80.** Find the constant of variation if $y$ varies directly as $x$, and $y = 2$ when $x = -\frac{2}{3}$.

**81.** If $y$ is inversely proportional to the square of $x$, and $y = 4$ when $x = 3$, find $y$ when $x = 6$.

**82.** The volume of a gas at a constant temperature varies inversely as the pressure. What is the volume of a gas under a pressure of 25 pounds if the gas occupies 15 cubic centimeters under a pressure of 20 pounds?

For Problems 83–103, solve each of the equations.

**83.** $3(2x - 1) - 2(5x + 1) = 4(3x + 4)$

**84.** $n + \dfrac{3n - 1}{9} - 4 = \dfrac{3n + 1}{3}$

**85.** $0.92 + 0.9(x - 0.3) = 2x - 5.95$

**86.** $|4x - 1| = 11$

**87.** $3x^2 = 7x$

**88.** $x^3 - 36x = 0$

**89.** $30x^2 + 13x - 10 = 0$

**90.** $8x^3 + 12x^2 - 36x = 0$

**91.** $x^4 + 8x^2 - 9 = 0$

**92.** $(n + 4)(n - 6) = 11$

**93.** $2 - \dfrac{3x}{x - 4} = \dfrac{14}{x + 7}$

**94.** $\dfrac{2n}{6n^2 + 7n - 3} - \dfrac{n - 3}{3n^2 + 11n - 4} = \dfrac{5}{2n^2 + 11n + 12}$

**95.** $\sqrt{3y} - y = -6$

**96.** $\sqrt{x + 19} - \sqrt{x + 28} = -1$

**97.** $(3x - 1)^2 = 45$

**98.** $(2x + 5)^2 = -32$

**99.** $2x^2 - 3x + 4 = 0$

**100.** $3n^2 - 6n + 2 = 0$

**101.** $\dfrac{5}{n - 3} - \dfrac{3}{n + 3} = 1$

**102.** $12x^4 - 19x^2 + 5 = 0$

**103.** $2x^2 + 5x + 5 = 0$

For Problems 104–113, solve each of the inequalities.

**104.** $-5(y - 1) + 3 > 3y - 4 - 4y$

**105.** $0.06x + 0.08(250 - x) \geq 19$

**106.** $|5x - 2| > 13$

**107.** $|6x + 2| < 8$

**108.** $\dfrac{x - 2}{5} - \dfrac{3x - 1}{4} \leq \dfrac{3}{10}$

**109.** $(x - 2)(x + 4) \leq 0$

**110.** $(3x - 1)(x - 4) > 0$

**111.** $x(x + 5) < 24$

**112.** $\dfrac{x - 3}{x - 7} \geq 0$

**113.** $\dfrac{2x}{x + 3} > 4$

For Problems 114–118, solve each of the systems of equations.

**114.** $\begin{pmatrix} 4x - 3y = 18 \\ 3x - 2y = 15 \end{pmatrix}$

**115.** $\begin{pmatrix} y = \frac{2}{5}x - 1 \\ 3x + 5y = 4 \end{pmatrix}$

**116.** $\begin{pmatrix} \frac{x}{2} - \frac{y}{3} = 1 \\ \frac{2x}{5} + \frac{y}{2} = 2 \end{pmatrix}$

**117.** $\begin{pmatrix} 4x - y + 3z = -12 \\ 2x + 3y - z = 8 \\ 6x + y + 2z = -8 \end{pmatrix}$

**118.** $\begin{pmatrix} x - y + 5z = -10 \\ 5x + 2y - 3z = 6 \\ -3x + 2y - z = 12 \end{pmatrix}$

For Problems 119–130, set up an equation, an inequality, or a system of equations to help solve each problem.

**119.** Find three consecutive odd integers whose sum is 57.

**120.** Suppose that Eric has a collection of 63 coins consisting of nickels, dimes, and quarters. The number of dimes is 6 more than the number of nickels, and the number of quarters is 1 more than twice the number of nickels. How many coins of each kind are in the collection?

**121.** One of two supplementary angles is 4° more than one-third of the other angle. Find the measure of each of the angles.

**122.** If a ring costs a jeweler \$300, at what price should it be sold for the jeweler to make a profit of 50% on the selling price?

**123.** Last year Beth invested a certain amount of money at 8% and \$300 more than that amount at 9%. Her total yearly interest was \$316. How much did she invest at each rate?

**124.** Two trains leave the same depot at the same time, one traveling east and the other traveling west. At the end of $4\frac{1}{2}$ hours, they are 639 miles apart. If the rate of the train traveling east is 10 miles per hour greater than that of the other train, find their rates.

**125.** Sam shot rounds of 70, 73, and 76 on the first three days of a golf tournament. What must he shoot on the fourth day of the tournament to average 72 or less for the four days?

**126.** The cube of a number equals nine times the same number. Find the number.

**127.** A strip of uniform width is to be cut off of both sides and both ends of a sheet of paper that is 8 inches by 14 inches to reduce the size of the paper to an area of 72 square inches. (See Figure 10.20.) Find the width of the strip.

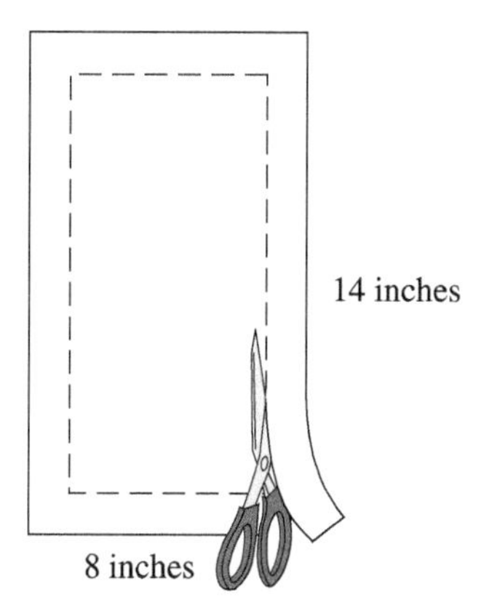


**Figure 10.20**

**128.** A sum of \$2450 is to be divided between two people in the ratio of 3 to 4. How much does each person receive?

**129.** Working together, Crystal and Dean can complete a task in $1\frac{1}{5}$ hours. Dean can do the task by himself in 2 hours. How long would it take Crystal to complete the task by herself?

**130.** The sum of the two smallest angles of a triangle is 40° less than the other angle. The sum of the smallest and largest angles is twice the other angle. Find the measures of the three angles of the triangle.

# Answer Section

## Chapter 1

### Mastery Set 1 (page 2)

**1.** Composite **2.** Composite **3.** Prime **4.** Composite
**5.** $2 \cdot 2 \cdot 2 \cdot 5$ **6.** $2 \cdot 2 \cdot 3 \cdot 7$ **7.** $2 \cdot 3 \cdot 5 \cdot 7$ **8.** 36
**9.** 126 **10.** 60

### Mastery Set 2 (page 5)

**1.** $\frac{2}{5}$ **2.** $\frac{6}{17}$ **3.** $\frac{1}{3}$ **4.** $\frac{2}{35}$ **5.** $\frac{1}{2}$ **6.** $\frac{4}{15}$ **7.** 6 **8.** $\frac{9}{5}$
**9.** $\frac{9}{2}$ **10.** 20

### Mastery Set 3 (page 8)

**1.** $\frac{3}{4}$ **2.** $\frac{7}{8}$ **3.** $\frac{19}{15}$ **4.** $\frac{23}{24}$ **5.** $\frac{29}{40}$ **6.** $\frac{4}{7}$ **7.** $\frac{2}{3}$ **8.** $\frac{3}{40}$
**9.** $\frac{13}{72}$ **10.** $\frac{5}{8}$

### Mastery Set 4 (page 11)

**1.** 2 **2.** $\frac{9}{5}$ **3.** $\frac{9}{28}$ **4.** $\frac{43}{30}$ **5.** $\frac{5}{3}$ **6.** $\frac{29}{10}$ **7.** 9 **8.** $\frac{13}{12}$
**9.** $\frac{25}{32}$ **10.** 17

### Problem Set 1.0 (page 12)

**1.** $2 \cdot 13$ **2.** $2 \cdot 2 \cdot 2 \cdot 2$ **3.** $2 \cdot 2 \cdot 3 \cdot 3$
**4.** $2 \cdot 2 \cdot 2 \cdot 2 \cdot 5$ **5.** $7 \cdot 7$ **6.** $2 \cdot 2 \cdot 23$
**7.** $2 \cdot 2 \cdot 2 \cdot 7$ **8.** $2 \cdot 2 \cdot 2 \cdot 2 \cdot 3 \cdot 3$
**9.** $2 \cdot 2 \cdot 2 \cdot 3 \cdot 5$ **10.** $2 \cdot 2 \cdot 3 \cdot 7$ **11.** $3 \cdot 3 \cdot 3 \cdot 5$
**12.** $2 \cdot 7 \cdot 7$ **13.** 24 **14.** 24 **15.** 48 **16.** 36 **17.** 140
**18.** 462 **19.** 392 **20.** 72 **21.** 168 **22.** 60 **23.** 90
**24.** 168 **25.** $\frac{2}{3}$ **26.** $\frac{3}{4}$ **27.** $\frac{2}{3}$ **28.** $\frac{9}{16}$ **29.** $\frac{5}{3}$ **30.** $\frac{4}{3}$
**31.** $\frac{15}{28}$ **32.** $\frac{12}{55}$ **33.** $\frac{10}{21}$ **34.** $\frac{65}{66}$ **35.** $\frac{3}{10}$ **36.** $\frac{2}{3}$
**37.** $\frac{3}{8}$ cup **38.** $\frac{1}{6}$ of the bottle **39.** $\frac{2}{9}$ of the disk space
**40.** $\frac{1}{3}$ **41.** $\frac{5}{7}$ **42.** $\frac{8}{11}$ **43.** $\frac{5}{9}$ **44.** $\frac{5}{13}$ **45.** 3 **46.** 2
**47.** $\frac{2}{3}$ **48.** $\frac{3}{8}$ **49.** $\frac{2}{3}$ **50.** $\frac{5}{9}$ **51.** $\frac{8}{15}$ **52.** $\frac{7}{24}$ **53.** $\frac{9}{16}$
**54.** $\frac{11}{12}$ **55.** $\frac{37}{30}$ **56.** $\frac{29}{24}$ **57.** $\frac{59}{96}$ **58.** $\frac{19}{24}$ **59.** $\frac{13}{12}$ pounds
**60.** $\frac{5}{16}$ ounce **61.** $\frac{1}{4}$ **62.** $\frac{5}{3}$ **63.** $\frac{37}{30}$ **64.** $\frac{4}{5}$ **65.** $\frac{1}{3}$
**66.** $\frac{27}{35}$ **67.** $\frac{7}{26}$ **68.** 30 **69.** $\frac{7}{20}$ **70.** $\frac{11}{32}$ ounce

### Problem Set 1.1 (page 20)

**1.** True **3.** False **5.** True **7.** False **9.** True
**11.** 0 and 14 **13.** 0, 14, $\frac{2}{3}$, $-\frac{11}{14}$, 2.34, $3.2\overline{1}$, $\frac{55}{8}$, $-19$, and $-2.6$ **15.** 0 and 14 **17.** All of them **19.** $\not\subseteq$ **21.** $\subseteq$
**23.** $\not\subseteq$ **25.** $\subseteq$ **27.** $\not\subseteq$ **29.** Real, rational, an integer, and negative **31.** Real, irrational, and negative **33.** $\{1, 2\}$
**35.** $\{0, 1, 2, 3, 4, 5\}$ **37.** $\{\ldots, -1, 0, 1, 2\}$ **39.** $\varnothing$
**41.** $\{0, 1, 2, 3, 4\}$ **43.** $-6$ **45.** 2 **47.** $3x + 1$ **49.** $5x$
**51.** 26 **53.** 84 **55.** 23 **57.** 65 **59.** 60 **61.** 33 **63.** 4
**65.** 1320 **67.** 20 **69.** 119 **71.** 9 **73.** 18 **75.** 4 **77.** 31

### Problem Set 1.2 (page 29)

**1.** −5 −4 −3 −2 −1 0 1 2 3 4 5

**3.** **(a)** 7 **(b)** 0 **(c)** 15 **5.** $-7$ **7.** $-19$ **9.** $-22$ **11.** $-7$
**13.** 108 **15.** $-70$ **17.** 14 **19.** $-7$ **21.** $3\frac{1}{2}$ **23.** $5\frac{1}{2}$
**25.** $-\frac{2}{15}$ **27.** $-4$ **29.** 0 **31.** Undefined **33.** $-60$
**35.** $-4.8$ **37.** 14.13 **39.** $-6.5$ **41.** $-38.88$ **43.** 0.2
**45.** $-\frac{13}{12}$ **47.** $-\frac{3}{4}$ **49.** $-\frac{13}{9}$ **51.** $-\frac{3}{5}$ **53.** $-\frac{3}{2}$ **55.** $-12$
**57.** $-24$ **59.** $\frac{35}{4}$ **61.** 15 **63.** $-17$ **65.** $\frac{47}{12}$ **67.** 5
**69.** 0 **71.** 26 **73.** 6 **75.** 25 **77.** 78 **79.** $-10$ **81.** 5
**83.** $-5$ **85.** 10.5 **87.** $-3.3$ **89.** 19.5 **91.** $\frac{3}{4}$ **93.** $\frac{5}{2}$
**97.** 10 over par **99.** Lost \$16.50 **101.** A gain of 0.88 dollar
**103.** No; they made it 49.1 pounds lighter

### Problem Set 1.3 (page 37)

**1.** Associative property of addition
**3.** Commutative property of addition
**5.** Additive inverse property
**7.** Multiplication property of negative one
**9.** Commutative property of multiplication
**11.** Distributive property
**13.** Associative property of multiplication
**15.** 18 **17.** 2 **19.** $-1300$ **21.** 1700 **23.** $-47$ **25.** 3200
**27.** $-19$ **29.** $-41$ **31.** $-17$ **33.** $-39$ **35.** 24 **37.** 20
**39.** 55 **41.** 16 **43.** 49 **45.** $-216$ **47.** $-14$ **49.** $-8$
**51.** $\frac{3}{16}$ **53.** $-\frac{10}{9}$ **57.** 2187 **59.** $-2048$ **61.** $-15{,}625$
**63.** 3.9525416

### Problem Set 1.4 (page 45)

**1.** $4x$ **3.** $-a^2$ **5.** $-6n$ **7.** $-5x + 2y$ **9.** $6a^2 + 5b^2$
**11.** $21x - 13$ **13.** $-2a^2b - ab^2$ **15.** $8x + 21$

**17.** $-5a + 2$ **19.** $-5n^2 + 11$ **21.** $-7x^2 + 32$
**23.** $22x - 3$ **25.** $-14x - 7$ **27.** $-10n^2 + 4$
**29.** $4x - 30y$ **31.** $-13x - 31$ **33.** $-21x - 9$
**35.** $-17$ **37.** 12 **39.** 4 **41.** 3 **43.** $-38$ **45.** $-14$
**47.** 64 **49.** 104 **51.** 5 **53.** 4 **55.** $-\frac{22}{3}$ **57.** $\frac{29}{4}$
**59.** 221.6 **61.** 1092.4 **63.** 1420.5 **65.** $n + 12$
**67.** $n - 5$ **69.** $50n$ **71.** $\frac{1}{2}n - 4$ **73.** $\frac{n}{8}$ **75.** $2n - 9$
**77.** $10(n - 6)$ **79.** $n + 20$ **81.** $2t - 3$ **83.** $n + 47$
**85.** $8y$ **87.** 25 cm **89.** $\frac{c}{25}$ **91.** $n + 2$ **93.** $\frac{c}{5}$
**95.** $12d$ **97.** $3y + f$ **99.** $5280m$

### Chapter 1 Sample Problems (page 48)

**1.** $-0.03$, 8, and 0
**2.** Symmetric property of equality
**3.** **(a)** $\frac{1}{2}$ **(b)** 0 **(c)** 4 **4.** **(a)** 5 **(b)** $-12$ **(c)** $-6$
**5.** $-56$ **6.** 38 **7.** $x^2 - 5x + 9$ **8.** 21 **9.** $3x - 4$
**10.** \$64 a share

### Chapter 1 Review Problem Set (page 51)

**1.** **(a)** 67 **(b)** 0, $-8$, and 67 **(c)** 0 and 67
**(d)** $0, \frac{3}{4}, -\frac{5}{6}, \frac{25}{3}, -8, 0.34, 0.2\overline{3}, 67$, and $\frac{9}{7}$
**(e)** $\sqrt{2}$ and $-\sqrt{3}$
**2.** Associative property of addition
**3.** Substitution property of equality
**4.** Multiplication property of negative one
**5.** Distributive property
**6.** Associative property of multiplication
**7.** Commutative property of addition
**8.** Distributive property
**9.** Multiplicative inverse property
**10.** Symmetric property of equality
**11.** 6.2 **12.** $\frac{7}{3}$ **13.** $\sqrt{15}$ **14.** 8 **15.** $-6\frac{1}{2}$ **16.** $-6\frac{1}{6}$
**17.** $-8$ **18.** $-15$ **19.** 20 **20.** 49 **21.** $-56$ **22.** 8
**23.** $-24$ **24.** 6 **25.** 4 **26.** 100 **27.** $-4a^2 - 5b^2$
**28.** $3x - 2$ **29.** $ab^2$ **30.** $-\frac{7}{3}x^2y$ **31.** $10n^2 - 17$
**32.** $-13a + 4$ **33.** $-2n + 2$ **34.** $-7x - 29y$
**35.** $-7a - 9$ **36.** $-9x^2 + 7$ **37.** $-6\frac{1}{2}$ **38.** $-\frac{5}{16}$
**39.** $-55$ **40.** 144 **41.** $-16$ **42.** $-44$ **43.** 19.4
**44.** 59.6 **45.** $-\frac{59}{3}$ **46.** $\frac{9}{2}$ **47.** $4 + 2n$ **48.** $3n - 50$
**49.** $\frac{2}{3}n - 6$ **50.** $10(n - 14)$ **51.** $5n - 8$ **52.** $\frac{n}{n - 3}$
**53.** $5(n + 2) - 3$ **54.** $\frac{3}{4}(n + 12)$ **55.** $37 - n$ **56.** $\frac{w}{60}$
**57.** $2y - 7$ **58.** $n + 3$ **59.** $p + 5n + 25q$ **60.** $\frac{i}{48}$ feet
**61.** $24f + 72y$ **62.** $10d$ **63.** $12f + i$ **64.** $25 - c$
**65.** 1 minute **66.** Loss of \$0.03 **67.** 0.2 ounce
**68.** 32 pounds

### Chapter 1 Test (page 53)

**1.** Symmetric property **2.** Distributive property **3.** $-3$
**4.** $-23$ **5.** $-\frac{23}{6}$ **6.** 11 **7.** 8 **8.** $-94$ **9.** $-4$ **10.** 960
**11.** $-32$ **12.** $-x^2 - 8x - 2$ **13.** $-19n - 20$ **14.** 27
**15.** $\frac{11}{16}$ **16.** $\frac{2}{3}$ **17.** 77 **18.** $-22.5$ **19.** 93 **20.** $-5$
**21.** $6n - 30$ **22.** $3n + 28$ or $3(n + 8) + 4$ **23.** $\frac{72}{n}$
**24.** $5n + 10d + 25q$ **25.** $6x + 2y$

## Chapter 2

### Problem Set 2.1 (page 62)

**1.** $\{4\}$ **3.** $\{-3\}$ **5.** $\{-14\}$ **7.** $\{6\}$ **9.** $\{0\}$ **11.** $\{1\}$
**13.** $\left\{-\frac{10}{3}\right\}$ **15.** $\varnothing$ **17.** $\left\{-\frac{13}{3}\right\}$ **19.** $\{3\}$ **21.** $\{8\}$
**23.** $\{-9\}$ **25.** $\{-3\}$ **27.** $\{0\}$ **29.** {All reals} **31.** $\{-2\}$
**33.** $\left\{-\frac{5}{3}\right\}$ **35.** $\left\{\frac{33}{2}\right\}$ **37.** $\{-35\}$ **39.** $\left\{\frac{1}{2}\right\}$ **41.** $\left\{\frac{1}{6}\right\}$
**43.** $\{0\}$ **45.** $\{-1\}$ **47.** $\left\{-\frac{21}{16}\right\}$ **49.** $\left\{\frac{12}{7}\right\}$ **51.** 14
**53.** 13, 14, and 15 **55.** 9, 11, and 13 **57.** 14 and 81
**59.** \$11 per hour **61.** 30 pennies, 50 nickels, and 70 dimes
**63.** \$300 **65.** 20 three-bedroom, 70 two-bedroom, and 140 one-bedroom **73.** **(a)** $\varnothing$ **(c)** $\{0\}$ **(e)** $\varnothing$

### Problem Set 2.2 (page 69)

**1.** $\{12\}$ **3.** $\left\{-\frac{3}{5}\right\}$ **5.** $\{3\}$ **7.** $\frac{3}{4}x$ **9.** $\{-2\}$ **11.** $\{-36\}$
**13.** $\left\{\frac{20}{9}\right\}$ **15.** $\frac{7y}{20}$ **17.** $\{3\}$ **19.** $\{0\}$ **21.** $\{-2\}$
**23.** $\left\{\frac{8}{5}\right\}$ **25.** $\{-3\}$ **27.** $\frac{20x + 10}{9}$ **29.** $\left\{\frac{48}{17}\right\}$
**31.** $\left\{\frac{103}{6}\right\}$ **33.** $\{3\}$ **35.** $\left\{\frac{40}{3}\right\}$ **37.** $\left\{-\frac{20}{7}\right\}$
**39.** $\left\{\frac{24}{5}\right\}$ **41.** $\{-10\}$ **43.** $-\frac{5}{12}x + \frac{17}{12}$ or $\frac{-5x + 17}{12}$
**45.** $\left\{-\frac{25}{4}\right\}$ **47.** $\{0\}$ **49.** 18 **51.** 16 inches long and 5 inches wide **53.** 14, 15, and 16 **55.** 8 feet
**57.** 80, 90, and 94 **59.** 48° and 132° **61.** 78°

### Problem Set 2.3 (page 77)

**1.** $\{20\}$ **3.** $\{50\}$ **5.** $\{40\}$ **7.** $\{12\}$ **9.** $\{6\}$ **11.** $\{400\}$
**13.** $\{400\}$ **15.** $\{38\}$ **17.** $\{6\}$ **19.** $\{3000\}$ **21.** $\{3000\}$
**23.** $\{400\}$ **25.** $\{14\}$ **27.** $\{15\}$ **29.** \$90 **31.** \$54.40
**33.** \$48 **35.** \$2400 **37.** 65% **39.** 62.5% **41.** \$42,000

**43.** \$3000 at 4% and \$4500 at 6% **45.** \$53,000
**47.** 8 pennies, 15 nickels, and 18 dimes
**49.** 15 dimes, 45 quarters, and 10 half-dollars **55.** {7.5}
**57.** {−4775} **59.** {8.7} **61.** {17.1} **63.** {13.5}
**65.** Yes, based on the selling price

## Problem Set 2.4 (page 86)

**1.** \$600 **3.** 3 years **5.** 6% **7.** \$800 **9.** \$1350 **11.** 8%
**13.** \$200 **15.** $b_2 = \frac{2A - hb_1}{h}$; 6 feet; 14 feet; 10 feet; 20 feet; 7 feet; 2 feet **17.** $h = \frac{V}{B}$ **19.** $h = \frac{V}{\pi r^2}$ **21.** $r = \frac{C}{2\pi}$
**23.** $C = \frac{100M}{I}$ **25.** $C = \frac{5}{9}(F - 32)$ or $C = \frac{5F - 160}{9}$
**27.** $x = \frac{y - b}{m}$ **29.** $x = \frac{y - y_1 + mx_1}{m}$ **31.** $x = \frac{ab + bc}{b - a}$
**33.** $x = a + bc$ **35.** $x = \frac{3b - 6a}{2}$ **37.** $x = \frac{5y + 7}{2}$
**39.** $y = -7x - 4$ **41.** $x = \frac{6y + 4}{3}$ **43.** $x = \frac{cy - ac - b^2}{b}$
**45.** $y = \frac{x - a + 1}{a - 3}$ **47.** 22 meters long and 6 meters wide
**49.** $16\frac{2}{3}$ years **51.** $16\frac{2}{3}$ years **53.** 4 hours **55.** 3 hours
**57.** 40 miles **59.** 8 ounces of 6% solution and 8 ounces of 14% solution **61.** 25 milliliters **67.** \$281.25
**69.** 1.5 years **71.** 8% **73.** \$1850

## Problem Set 2.5 (page 95)

**1.** $(1, \infty)$

**3.** $[-1, \infty)$

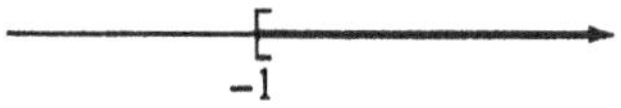

**5.** $(-\infty, -2)$

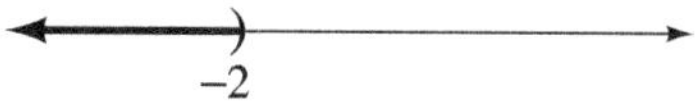

**7.** $(-\infty, 2]$

**9.** $x < 4$ **11.** $x \leq -7$ **13.** $x > 8$ **15.** $x \geq -7$
**17.** $(1, \infty)$

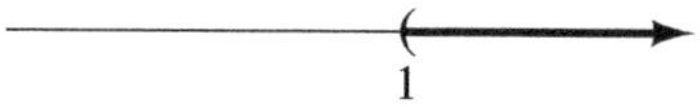

**19.** $(-\infty, -4]$

**21.** $(-\infty, -2]$

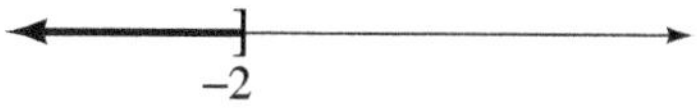

**23.** $(-\infty, 2)$

**25.** $(-1, \infty)$

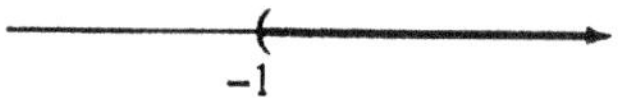

**27.** $[-1, \infty)$

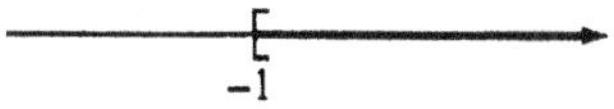

**29.** $(-2, \infty)$

**31.** $(-2, \infty)$

**33.** $(-\infty, -2)$

**35.** $[-3, \infty)$

**37.** $(0, \infty)$

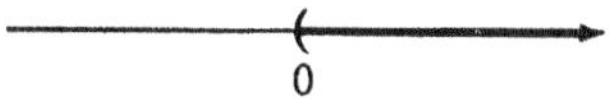

**39.** $[4, \infty)$

**41.** $\left(\frac{7}{2}, \infty\right)$ **43.** $\left(\frac{12}{5}, \infty\right)$ **45.** $\left(-\infty, -\frac{5}{2}\right]$ **47.** $[0, \infty)$
**49.** $(-6, \infty)$ **51.** $(-5, \infty)$ **53.** $\left(-\infty, \frac{5}{3}\right]$ **55.** $(-36, \infty)$
**57.** $\left(-\infty, -\frac{8}{17}\right]$ **59.** $\left(-\frac{11}{2}, \infty\right)$ **61.** $(23, \infty)$ **63.** $(-\infty, 3)$
**65.** $\left(-\infty, -\frac{1}{7}\right]$ **67.** $(-22, \infty)$ **69.** $\left(-\infty, \frac{6}{5}\right)$

## Problem Set 2.6 (page 103)

**1.** $(4, \infty)$ **3.** $\left(-\infty, \frac{23}{3}\right)$ **5.** $[5, \infty)$ **7.** $[-9, \infty)$
**9.** $\left(-\infty, -\frac{37}{3}\right]$ **11.** $\left(-\infty, -\frac{19}{6}\right)$ **13.** $(-\infty, 50]$
**15.** $(300, \infty)$ **17.** $[4, \infty)$
**19.** $(-1, 2)$

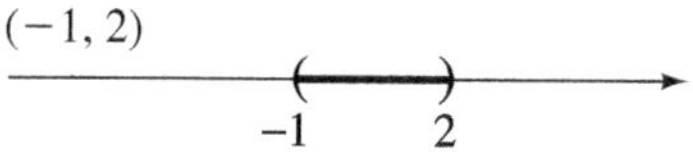

**21.** $(-1, 2]$

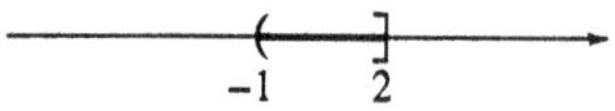

**23.** $(-\infty, -1) \cup (2, \infty)$

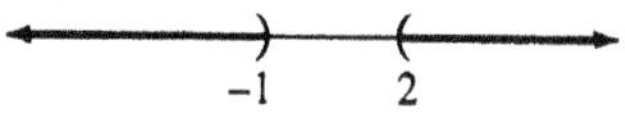

**25.** $(-\infty, 1] \cup (3, \infty)$

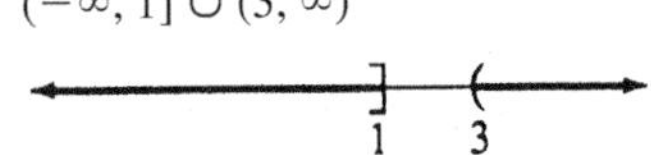


**27.** $(0, \infty)$

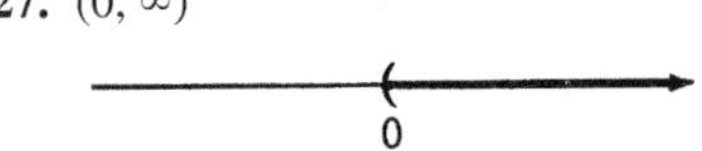


**29.** $\varnothing$

**31.** $(-\infty, \infty)$

**33.** $(-1, \infty)$

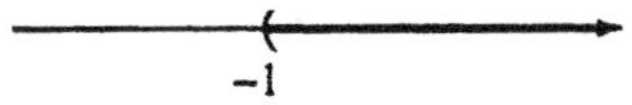

**35.** $(1, 3)$

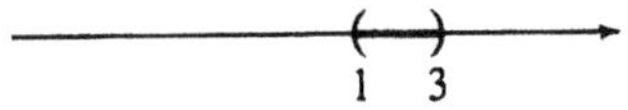

**37.** $(-\infty, -5) \cup (1, \infty)$

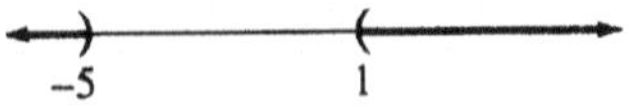

**39.** $[3, \infty)$

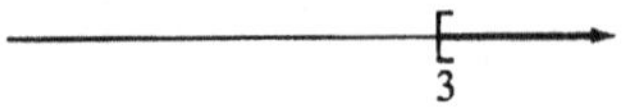

**41.** $\left(\frac{1}{3}, \frac{2}{5}\right)$

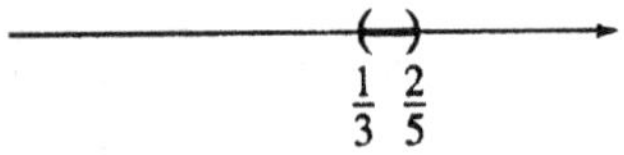

**43.** $(-\infty, -1) \cup \left(-\frac{1}{3}, \infty\right)$

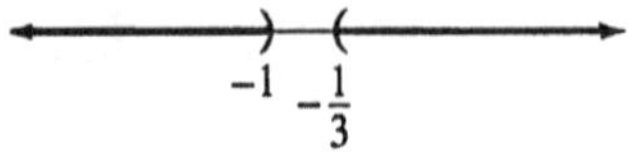

**45.** $(-2, 2)$ **47.** $[-5, 4]$ **49.** $\left(-\frac{1}{2}, \frac{3}{2}\right)$

**51.** $\left(-\frac{1}{4}, \frac{11}{4}\right)$ **53.** $[-11, 13]$ **55.** $(-1, 5)$

**57.** More than 7.5% **59.** 5 feet and 10 inches or better **61.** 168 or better **63.** 77 or less **65.** $163° \le C \le 218°$ **67.** $6.3 \le M \le 11.25$

## Problem Set 2.7 (page 110)

**1.** $\{-7, 9\}$ **3.** $\{-1, 5\}$ **5.** $\left\{-5, \frac{7}{3}\right\}$ **7.** $\{-1, 5\}$

**9.** $\left\{\frac{1}{12}, \frac{7}{12}\right\}$ **11.** $\{0, 3\}$ **13.** $\{-6, 2\}$ **15.** $\left\{\frac{3}{4}\right\}$

**17.** $(-5, 5)$

**19.** $[-2, 2]$

**21.** $(-\infty, -2) \cup (2, \infty)$

**23.** $(-1, 3)$

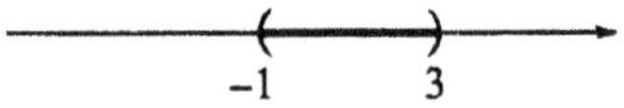

**25.** $[-6, 2]$

**27.** $(-\infty, -3) \cup (-1, \infty)$

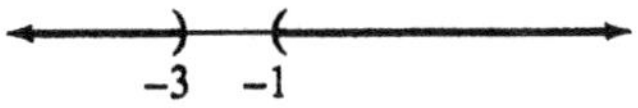

**29.** $(-\infty, 1] \cup [5, \infty)$

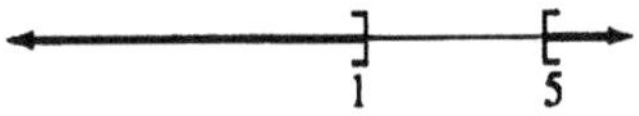

**31.** $(-\infty, -4) \cup (8, \infty)$ **33.** $(-8, 2)$ **35.** $[-4, 5]$

**37.** $\left(-\infty, -\frac{7}{2}\right] \cup \left[\frac{5}{2}, \infty\right)$ **39.** $(-\infty, -2) \cup (6, \infty)$

**41.** $\left(-\frac{1}{2}, \frac{3}{2}\right)$ **43.** $\left[-5, \frac{7}{5}\right]$ **45.** $[-3, 10]$ **47.** $(-5, 11)$

**49.** $\left(-\infty, -\frac{3}{2}\right) \cup \left(\frac{1}{2}, \infty\right)$ **51.** $(-\infty, -14] \cup [0, \infty)$

**53.** $[-2, 3]$ **55.** $\varnothing$ **57.** $(-\infty, \infty)$ **59.** $\left\{\frac{2}{5}\right\}$ **61.** $\varnothing$

**63.** $\varnothing$ **69.** $\left\{-2, -\frac{4}{3}\right\}$ **71.** $\{-2\}$ **73.** $\{0\}$

## Chapter 2 Sample Problems (page 112)

**1.** $\left\{\frac{2}{3}\right\}$ **2.** $\left\{\frac{9}{4}\right\}$ **3.** $\{2\}$ **4.** The sides are 7 feet, 7 feet, and 21 feet **5.** \$62.30 **6.** 3 years

**7.** $l = \frac{P - 2w}{2}$ or $l = \frac{P}{2} - w$ **8.** 25% **9.** $(5, \infty)$

**10.** $[-25, \infty)$ **11.** $\left[\frac{14}{3}, \infty\right)$ **12.** $[3, 5)$

**13.** 13 points or greater **14.** $\left\{-\frac{23}{3}, 3\right\}$ **15.** $[-2, 18]$

## Chapter 2 Review Problem Set (page 118)

**1.** $\{18\}$ **2.** $\{-14\}$ **3.** $\{0\}$ **4.** $\left\{\frac{1}{2}\right\}$ **5.** $\{10\}$ **6.** $\left\{\frac{7}{3}\right\}$

**7.** $\left\{\frac{28}{17}\right\}$ **8.** $\frac{5x - 2}{6}$ **9.** $\left\{\frac{27}{17}\right\}$ **10.** $\left\{-\frac{21}{13}\right\}$

**11.** $\varnothing$ **12.** {All reals} **13.** $\{-38\}$ **14.** $\{-11\}$

**15.** $\frac{1}{3}x - \frac{7}{3}$ or $\frac{x - 7}{3}$ **16.** $\left\{-\frac{1}{38}\right\}$ **17.** $\{43\}$ **18.** $\left\{-\frac{58}{9}\right\}$

**19.** {All reals} **20.** $\varnothing$ **21.** $\{0\}$ **22.** $\{-1\}$ **23.** $0.07x + 60$

**24.** $0.21x + 340$ **25.** $\{50\}$ **26.** $\left\{-\frac{39}{2}\right\}$ **27.** $\{200\}$

**28.** $\{-8\}$ **29.** The length is 14 m and the width is 9 m **30.** 4, 5, and 6 **31.** \$10.50 per hour

**32.** 20 nickels, 50 dimes, 75 quarters **33.** 80° **34.** \$200 invested at 7%, \$300 invested at 8% **35.** \$45.60 **36.** \$300.00 **37.** 60% **38.** \$104.00 **39.** \$8000 **40.** 4.5% **41.** 11 m **42.** −20° **43.** $x = \frac{2b+2}{a}$ **44.** $x = \frac{c}{a-b}$ **45.** $x = \frac{pb - ma}{m - p}$ **46.** $x = \frac{11 + 7y}{5}$ **47.** $x = \frac{by + b + ac}{c}$ **48.** $s = \frac{A - \pi r^2}{\pi r}$ **49.** $b_2 = \frac{2A - hb_1}{h}$ **50.** $n = \frac{2S_n}{a_1 + a_2}$ **51.** $R = \frac{R_1 R_2}{R_1 + R_2}$ **52.** $y = \frac{c - ax}{b}$ **53.** $6\frac{2}{3}$ pints **54.** 55 mph **55.** Sonya for $3\frac{1}{4}$ hr, Rita for $4\frac{1}{2}$ hr **56.** $6\frac{1}{4}$ cups of orange juice **57.** $[-2, \infty)$ **58.** $(6, \infty)$ **59.** $(-\infty, -1)$ **60.** $(-\infty, 0]$ **61.** $[-5, \infty)$ **62.** $(4, \infty)$ **63.** $\left(-\frac{7}{3}, \infty\right)$ **64.** $\left[\frac{17}{2}, \infty\right)$ **65.** $(-\infty, -17)$ **66.** $\left(-\infty, \frac{1}{3}\right)$ **67.** $\left(\frac{53}{11}, \infty\right)$ **68.** $\left(-\infty, -\frac{15}{4}\right)$ **69.** $[6, \infty)$ **70.** $(-\infty, 100]$

**71.** 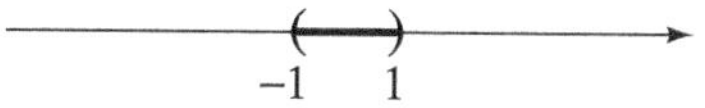 −1 1

**72.**  −3 2

**73.**  3

**74.**

**75.** 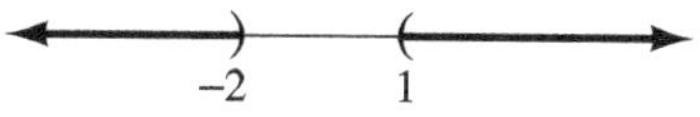 −2 1

**76.** 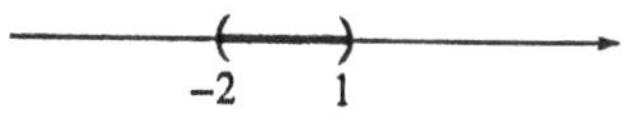 −2 1

**77.** 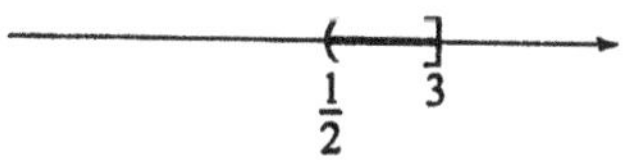 $\frac{1}{2}$ 3

**78.** ∅ **79.** 88 or better **80.** More than \$4000 **81.** $\left\{-\frac{10}{3}, 4\right\}$ **82.** $\left\{-\frac{7}{2}, \frac{1}{2}\right\}$ **83.** $\left\{-\frac{11}{3}, 3\right\}$ **84.** $\{-18, 6\}$ **85.** $(-5, 6)$ **86.** $\left(-\infty, -\frac{11}{3}\right) \cup (3, \infty)$ **87.** $\left(-\infty, -\frac{4}{5}\right] \cup \left[\frac{12}{5}, \infty\right)$ **88.** $[-28, 20]$

### Chapter 2 Test (page 120)

**1.** $\{-3\}$ **2.** $\{5\}$ **3.** $\left\{\frac{1}{2}\right\}$ **4.** $\left\{\frac{16}{5}\right\}$ **5.** $\left\{-\frac{14}{5}\right\}$ **6.** $\{-1\}$ **7.** $\left\{-\frac{3}{2}, 3\right\}$ **8.** $\{3\}$ **9.** $\left\{\frac{31}{3}\right\}$ **10.** $\{650\}$ **11.** $y = \frac{8x - 24}{9}$ **12.** $h = \frac{S - 2\pi r^2}{2\pi r}$ **13.** $(-2, \infty)$ **14.** $[-4, \infty)$ **15.** $(-\infty, -35]$ **16.** $(-\infty, 10)$ **17.** $(3, \infty)$ **18.** $(-\infty, 200]$ **19.** $\left(-1, \frac{7}{3}\right)$ **20.** $\left(-\infty, -\frac{11}{4}\right] \cup \left[\frac{1}{4}, \infty\right)$ **21.** \$72 **22.** 19 centimeters **23.** $\frac{2}{3}$ of a cup **24.** 97 or better **25.** 70°

### Chapters 1–2 Cumulative Test (page 121)

**1.**

| | Natural numbers | Whole numbers | Integers | Rational numbers | Irrational numbers | Real numbers |
|---|---|---|---|---|---|---|
| $9$ | X | X | X | X | | X |
| $-\frac{1}{2}$ | | | | X | | X |
| $-\sqrt{7}$ | | | | | X | X |
| $0.\overline{3}$ | | | | X | | X |
| $\frac{8}{3}$ | | | | X | | X |
| $-2$ | | | X | X | | X |
| $0$ | | X | X | X | | X |

**2. (a)** Commutative Property of Multiplication **(b)** Distributive Property **(c)** Symmetric Property of Equality **3.** 3 **4.** 24 **5.** 84 **6.** 1 **7.** $-\frac{67}{15}$ **8.** −20 **9.** −2 **10.** $-6c^2 + 1$ **11.** $25a + 9$ **12.** $\frac{5}{6}cd^2$ **13.** −1 **14.** 7.4 **15.** $\frac{19}{18}$ **16.** $\frac{2x}{3x - 2}$ **17.** $\left\{\frac{50}{11}\right\}$ **18.** $\left\{-\frac{9}{2}\right\}$ **19.** $\{1200\}$ **20.** $x = \frac{3y + 6z}{4}$ **21.** $\left\{-\frac{25}{3}, 11\right\}$ **22.** ∅ **23.** $\left\{\frac{2}{3}\right\}$ **24.** $[-2, \infty)$ **25.** $(-\infty, 4)$ **26.** $\left[-\frac{1}{3}, 3\right]$ **27.** $(-\infty, 0) \cup (5, \infty)$ **28.** ∅ **29.** \$18 **30.** 25 leashes, 9 collars **31.** 4 tens, 3 fifties, 12 twenties; \$430 **32.** \$2625 **33.** 84° **34.** 6.5 hours **35.** \$6000 **36.** \$26,518.50

## Chapter 3

### Problem Set 3.1 (page 128)

**1.** 2 **3.** 3 **5.** 2 **7.** 6 **9.** 0 **11.** $10x - 3$ **13.** $-11t + 5$ **15.** $-x^2 + 2x - 2$ **17.** $17a^2b^2 - 5ab$ **19.** $-9x + 7$ **21.** $-2x + 6$ **23.** $10a + 7$ **25.** $4x^2 + 10x + 6$ **27.** $-6a^2 + 12a + 14$ **29.** $3x^3 + x^2 + 13x - 11$ **31.** $7x + 8$ **33.** $-3x - 16$ **35.** $2x^2 - 2x - 8$ **37.** $-3x^3 + 5x^2 - 2x + 9$ **39.** $5x^2 - 4x + 11$ **41.** $-6x^2 + 9x + 7$ **43.** $-2x^2 + 9x + 4$ **45.** $-10n^2 + n + 9$ **47.** $8x - 2$ **49.** $8x - 14$ **51.** $-9x^2 - 12x + 4$ **53.** $10x^2 + 13x - 18$ **55.** $-n^2 - 4n - 4$ **57.** $-x + 6$ **59.** $6x^2 - 4$ **61.** $-7n^2 + n + 6$ **63.** $t^2 - 4t + 8$ **65.** $4n^2 - n - 12$ **67.** $-4x - 2y$ **69.** $-x^3 - x^2 + 3x$ **71. (a)** $8x + 4$ **(c)** $12x + 6$ **73.** $8\pi h + 32\pi$ **(a)** 226.1 **(c)** 452.2

## Problem Set 3.2 (page 134)

**1.** $36x^4$ **3.** $-12x^5$ **5.** $4a^3b^4$ **7.** $-3x^3y^2z^6$ **9.** $-30xy^4$
**11.** $27a^4b^5$ **13.** $-m^3n^3$ **15.** $\frac{3}{10}x^3y^6$ **17.** $-\frac{3}{20}a^3b^4$
**19.** $-\frac{1}{6}x^3y^4$ **21.** $30x^6$ **23.** $-18x^9$ **25.** $-3x^6y^6$ **27.** $-24y^9$
**29.** $-56a^4b^2$ **31.** $-18a^3b^3$ **33.** $-10x^7y^7$ **35.** $50x^5y^2$
**37.** $27x^3y^6$ **39.** $-32x^{10}y^5$ **41.** $x^{16}y^{20}$ **43.** $a^6b^{12}c^{18}$
**45.** $64a^{12}b^{18}$ **47.** $81x^2y^8$ **49.** $81a^4b^{12}$ **51.** $-16a^4b^4$
**53.** $-x^6y^{12}z^{18}$ **55.** $-125a^6b^6c^3$ **57.** $-x^7y^{28}z^{14}$ **59.** $3x^3y^3$
**61.** $-5x^3y^2$ **63.** $9bc^2$ **65.** $-18xyz^4$ **67.** $-a^2b^3c^2$ **69.** 9
**71.** $-b^2$ **73.** $-18x^3$ **75.** $6x^{3n}$ **77.** $a^{5n+3}$ **79.** $x^{4n}$
**81.** $a^{5n+1}$ **83.** $-10x^{2n}$ **85.** $12a^{n+4}$ **87.** $6x^{3n+2}$
**89.** $12x^{n+2}$ **91.** $22x^2$; $6x^3$ **93.** $\pi r^2 - 36\pi$

## Problem Set 3.3 (page 142)

**1.** $10x^2y^3 + 6x^3y^4$ **3.** $-12a^3b^3 + 15a^5b$
**5.** $24a^4b^5 - 16a^4b^6 + 32a^5b^6$ **7.** $-6x^3y^3 - 3x^4y^4 + x^5y^2$
**9.** $ax + ay + 2bx + 2by$ **11.** $ac + 4ad - 3bc - 12bd$
**13.** $x^2 + 16x + 60$ **15.** $y^2 + 6y - 55$ **17.** $n^2 - 5n - 14$
**19.** $x^2 - 36$ **21.** $x^2 - 12x + 36$ **23.** $x^2 - 14x + 48$
**25.** $x^3 - 4x^2 + x + 6$ **27.** $x^3 - x^2 - 9x + 9$
**29.** $t^2 + 18t + 81$ **31.** $y^2 - 14y + 49$ **33.** $4x^2 + 33x + 35$
**35.** $9y^2 - 1$ **37.** $14x^2 + 3x - 2$ **39.** $5 + 3t - 2t^2$
**41.** $9t^2 + 42t + 49$ **43.** $4 - 25x^2$ **45.** $49x^2 - 56x + 16$
**47.** $18x^2 - 39x - 70$ **49.** $2x^2 + xy - 15y^2$ **51.** $25x^2 - 4a^2$
**53.** $t^3 - 14t - 15$ **55.** $x^3 + x^2 - 24x + 16$
**57.** $2x^3 + 9x^2 + 2x - 30$ **59.** $12x^3 - 7x^2 + 25x - 6$
**61.** $x^4 + 5x^3 + 11x^2 + 11x + 4$
**63.** $2x^4 - x^3 - 12x^2 + 5x + 4$ **65.** $x^3 + 6x^2 + 12x + 8$
**67.** $x^3 - 12x^2 + 48x - 64$ **69.** $8x^3 + 36x^2 + 54x + 27$
**71.** $64x^3 - 48x^2 + 12x - 1$
**73.** $125x^3 + 150x^2 + 60x + 8$ **75.** $x^{2n} - 16$
**77.** $x^{2a} + 4x^a - 12$ **79.** $6x^{2n} + x^n - 35$
**81.** $x^{4a} - 10x^{2a} + 21$ **83.** $4x^{2n} + 20x^n + 25$
**87.** $2x^2 + 6$ **89.** $4x^3 - 64x^2 + 256x$; $256 - 4x^2$
**93.** **(a)** $a^6 + 6a^5b + 15a^4b^2 + 20a^3b^3 + 15a^2b^4 + 6ab^5 + b^6$
**(c)** $a^8 + 8a^7b + 28a^6b^2 + 56a^5b^3 + 70a^4b^4 + 56a^3b^5 + 28a^2b^6 + 8ab^7 + b^8$

## Problem Set 3.4 (page 150)

**1.** No **3.** No **5.** $3(2x + y)$ **7.** $2x(3x + 7)$ **9.** $4y(7y - 1)$
**11.** $5x(4y - 3)$ **13.** $3y(5x - 2z)$ **15.** $mn(7m^4 + 16n^2)$
**17.** $x^2(7x + 10)$ **19.** $9ab(2a + 3b)$ **21.** $3x^3y^3(4y - 13x)$
**23.** $3m(4m^2 - 2m + 1)$ **25.** $4x^2(2x^2 + 3x - 6)$
**27.** $x(5 + 7x + 9x^3)$ **29.** $5ab^3(1 + 2ab - 3a^2b^2)$
**31.** $5xy^2(3xy + 4 + 7x^2y^2)$ **33.** $(y + 2)(x + 3)$
**35.** $(x + 8)(4a - 5b)$ **37.** $(b^3 - c^3)(6a^2 - x^2y)$
**39.** $(2a + b)(3x - 2y)$ **41.** $(x + 2)(x + 5)$
**43.** $(a + 4)(x + y)$ **45.** $(a - 2b)(x + y)$
**47.** $(a - b)(3x - y)$ **49.** $(a + 1)(2x + y)$
**51.** $(a - 1)(x^2 + 2)$ **53.** $(a + b)(2c + 3d)$
**55.** $(a + b)(x - y)$ **57.** $(x + 9)(x + 6)$ **59.** $(x + 4)(2x + 1)$
**61.** $(a^2 + 3)(b + 6x)$ **63.** $(4z + 1)(x + 2y^2)$ **65.** $\{-7, 0\}$
**67.** $\{0, 1\}$ **69.** $\{0, 5\}$ **71.** $\left\{-\frac{1}{2}, 0\right\}$ **73.** $\left\{-\frac{7}{3}, 0\right\}$
**75.** $\left\{0, \frac{5}{4}\right\}$ **77.** $\left\{0, \frac{1}{4}\right\}$ **79.** $\{-12, 0\}$ **81.** $\left\{0, \frac{3a}{5b}\right\}$
**83.** $\left\{-\frac{3a}{2b}, 0\right\}$ **85.** $\{a, -2b\}$ **87.** 0 or 7 **89.** 6 units
**91.** $\frac{4}{\pi}$ units **93.** The square is 100 feet by 100 feet, and the rectangle is 50 feet by 100 feet. **95.** 6 units
**103.** $A = 836$ cm$^2$ **105.** $A = 132$ ft$^2$ **107.** $x^a(2x^a - 3)$
**109.** $y^{2m}(y^m + 5)$ **111.** $x^{4a}(2x^{2a} - 3x^a + 7)$

## Problem Set 3.5 (page 157)

**1.** $(x + 1)(x - 1)$ **3.** $(4x + 5)(4x - 5)$
**5.** $(3x + 5y)(3x - 5y)$ **7.** $(5a + b)(5a - b)$
**9.** $(2m + 3n)(2m - 3n)$ **11.** $(1 + 4y)(1 - 4y)$
**13.** $(5xy + 6)(5xy - 6)$ **15.** $(2x + y^2)(2x - y^2)$
**17.** $(1 + 12n)(1 - 12n)$ **19.** $(x + 2 + y)(x + 2 - y)$
**21.** $(2x + y + 1)(2x - y - 1)$
**23.** $(3a + 2b + 3)(3a - 2b - 3)$ **25.** $-5(2x + 9)$
**27.** $9(x + 2)(x - 2)$ **29.** $5(x^2 + 1)$ **31.** $8(y + 2)(y - 2)$
**33.** $ab(a + 3)(a - 3)$ **35.** Not factorable
**37.** $(n + 3)(n - 3)(n^2 + 9)$ **39.** $3x(x^2 + 9)$
**41.** $4xy(x + 4y)(x - 4y)$ **43.** $6x(1 + x)(1 - x)$
**45.** $(1 + xy)(1 - xy)(1 + x^2y^2)$ **47.** $4(x + 4y)(x - 4y)$
**49.** $3(x + 2)(x - 2)(x^2 + 4)$ **51.** $(a - 4)(a^2 + 4a + 16)$
**53.** $(x + 1)(x^2 - x + 1)$ **55.** $(2a - 1)(4a^2 + 2a + 1)$
**57.** $(4m + 5)(16m^2 - 20m + 25)$
**59.** $(2a - 5b)(4a^2 + 10ab + 25b^2)$
**61.** $(3x + 4y)(9x^2 - 12xy + 16y^2)$
**63.** $(1 - 3a)(1 + 3a + 9a^2)$ **65.** $(xy - 1)(x^2y^2 + xy + 1)$
**67.** $(x + y)(x - y)(x^2 - xy + y^2)(x^2 + xy + y^2)$
**69.** $\{-5, 5\}$ **71.** $\left\{-\frac{7}{3}, \frac{7}{3}\right\}$ **73.** $\{-2, 2\}$ **75.** $\{-1, 0, 1\}$
**77.** $\{-2, 2\}$ **79.** $\{-3, 3\}$ **81.** $\{0\}$ **83.** $-3$, 0, or 3
**85.** 4 centimeters and 8 centimeters **87.** 10 meters long and 5 meters wide **89.** 6 inches **91.** 8 yards

## Problem Set 3.6 (page 165)

**1.** $(x + 5)(x + 4)$ **3.** $(x - 4)(x - 7)$ **5.** $(a + 9)(a - 4)$
**7.** $(y + 6)(y + 14)$ **9.** $(x - 7)(x + 2)$ **11.** Not factorable
**13.** $(6 - x)(1 + x)$ **15.** $(x + 3y)(x + 12y)$
**17.** $(a - 8b)(a + 7b)$ **19.** $(x + 10)(x + 15)$
**21.** $(n - 16)(n - 20)$ **23.** $(t + 15)(t - 12)$
**25.** $(t^2 - 3)(t^2 - 2)$ **27.** $(x + 1)(x - 1)(x^2 - 8)$
**29.** $(x + 1)(x - 1)(x + 4)(x - 4)$ **31.** $(3x + 1)(5x + 6)$
**33.** $(4x - 3)(3x + 2)$ **35.** $(a + 3)(4a - 9)$
**37.** $(n - 4)(3n + 5)$ **39.** Not factorable **41.** $(2n - 7)(5n + 3)$
**43.** $(4x - 5)(2x + 9)$ **45.** $(1 - 6x)(6 + x)$
**47.** $(5y + 9)(4y - 1)$ **49.** $(12n + 5)(2n - 1)$
**51.** $(5n + 3)(n + 6)$ **53.** $(2x^2 - 1)(5x^2 + 4)$
**55.** $(3n + 1)(3n - 1)(2n^2 + 3)$ **57.** $(y - 8)^2$ **59.** $(2x + 3y)^2$
**61.** $2(2y - 1)^2$ **63.** $(x - 12)(x - 2)$ **65.** $(a^2 + 6)(b^2 + 3)$
**67.** $(3y - 2)(9y^2 + 6y + 4)$ **69.** $2(t + 2)(t - 2)$
**71.** $(4x + 5y)(3x - 2y)$ **73.** $3n(2n + 5)(3n - 1)$
**75.** $(n - 12)(n - 5)$ **77.** $(6a - 1)^2$ **79.** $6(x^2 + 9)$
**81.** Not factorable **83.** $(x - 9)(x - 2)$ **85.** $(5n + 7)(4m + 1)$
**87.** $(5a - 4b)(25a^2 + 20ab + 16b^2)$
**89.** $(x + y - 7)(x - y + 7)$ **91.** $(1 + 4x^2)(1 + 2x)(1 - 2x)$
**93.** $(4n + 9)(n + 4)$ **95.** $n(n + 7)(n - 7)$

**97.** $(x - 8)(x + 1)$ **99.** $3x(x - 3)(x^2 + 3x + 9)$
**101.** $(x^2 + 3)^2$ **103.** $(x + 3)(x - 3)(x^2 + 4)$
**105.** $(2w - 7)(3w + 5)$ **107.** Not factorable
**109.** $2n(n^2 + 7n - 10)$ **111.** $(2x + 1)(y + 3)$
**117.** $(x^a + 3)(x^a + 7)$ **119.** $(2x^a + 5)^2$
**121.** $(5x^n - 1)(4x^n + 5)$ **123.** $(x - 2)(x - 4)$
**125.** $(3x - 11)(3x + 2)$ **127.** $(5x + 9)(3x + 4)$

## Problem Set 3.7 (page 170)

**1.** $\{-3, -1\}$ **3.** $\{-12, -6\}$ **5.** $\{4, 9\}$ **7.** $\{-6, 2\}$
**9.** $\{-1, 5\}$ **11.** $\{-13, -12\}$ **13.** $\left\{-5, \frac{1}{3}\right\}$
**15.** $\left\{-\frac{7}{2}, -\frac{2}{3}\right\}$ **17.** $\{0, 4\}$ **19.** $\left\{\frac{1}{6}, 2\right\}$ **21.** $\{-6, 0, 6\}$
**23.** $\{-4, 6\}$ **25.** $\{-4, 4\}$ **27.** $\{-11, 4\}$ **29.** $\{-5, 5\}$
**31.** $\left\{-\frac{5}{3}, -\frac{3}{5}\right\}$ **33.** $\left\{-\frac{1}{8}, 6\right\}$ **35.** $\left\{\frac{3}{7}, \frac{5}{4}\right\}$
**37.** $\left\{-\frac{2}{7}, \frac{4}{5}\right\}$ **39.** $\left\{-7, \frac{2}{3}\right\}$ **41.** $\{-20, 18\}$
**43.** $\left\{-2, -\frac{1}{3}, \frac{1}{3}, 2\right\}$ **45.** $\left\{-\frac{2}{3}, 16\right\}$ **47.** $\left\{-\frac{3}{2}, 1\right\}$
**49.** $\left\{-\frac{5}{2}, -\frac{4}{3}, 0\right\}$ **51.** $\left\{-1, \frac{5}{3}\right\}$ **53.** $\left\{-\frac{3}{2}, \frac{1}{2}\right\}$
**55.** 8 and 9 or $-9$ and $-8$ **57.** 7 and 15
**59.** 10 inches by 6 inches **61.** $-7$ and $-6$ or 6 and 7
**63.** 4 centimeters by 4 centimeters and 6 centimeters by 8 centimeters **65.** 3, 4, and 5 units **67.** 9 inches and 12 inches **69.** An altitude of 4 inches and a side 14 inches long

## Chapter 3 Sample Problems (page 173)

**1.** The degree of the polynomial is 5. **2.** $-6y^2 + 26y + 18$
**3.** **(a)** $-6x^7y^2$ **(b)** $-8x^3y^9$ **4.** $-8x^5z^2$
**5.** $3y^3 + 10y^2 - 29y + 14$ **6.** $21x^2 - 17x - 30$
**7.** $9x^2 + 12x + 4$ **8.** $16x^2 - y^2$ **9.** $64x^3 - 48x^2 + 12x - 1$
**10.** $V = (12 - 2x)^2(x)$ or $V = 4x^3 - 48x^2 + 144x$
**11.** (b) **12.** $-3x^2y(2x - 1 + 3y)$ **13.** $(x - 7)(2y + 3)$
**14.** $(2x - 3y)(3a - 4b)$ **15.** $(7x + 9y)(7x - 9y)$
**16.** $(x - 4y)(x^2 + 4xy + 16y^2)$ **17.** $(x - 3)(x + 9)$
**18.** $(3x - 5)(2x + 1)$ **19.** $(5x - 2)^2$ **20.** $5(2x - 3)(2x + 3)$
**21.** $\{-4, 3\}$ **22.** 2 units

## Chapter 3 Review Problem Set (page 179)

**1.** Third degree **2.** Fourth degree **3.** Sixth degree
**4.** Fifth degree **5.** $5x - 3$ **6.** $3x^2 + 12x - 2$
**7.** $12x^2 - x + 5$ **8.** $11x^2 + 2x + 4$ **9.** $2x + y - 2$
**10.** $5x + 5y - 2$ **11.** $-20x^5y^7$ **12.** $-6a^5b^5$
**13.** $-8a^7b^3$ **14.** $27x^5y^8$ **15.** $256x^8y^{12}$ **16.** $-8x^6y^9z^3$
**17.** $-12a^5b^7$ **18.** $6x^{4n}$ **19.** $-13x^2y$ **20.** $2x^3y^3$ **21.** $-4b^2$
**22.** $4a^3b^3$ **23.** $15a^4 - 10a^3 - 5a^2$ **24.** $-8x^5 + 6x^4 + 10x^3$
**25.** $3x^3 + 7x^2 - 21x - 4$ **26.** $6x^3 - 11x^2 - 7x + 2$
**27.** $x^4 + x^3 - 18x^2 - x + 35$
**28.** $3x^4 + 5x^3 - 21x^2 - 3x + 20$
**29.** $24x^2 + 2xy - 15y^2$ **30.** $7x^2 + 19x - 36$
**31.** $21 + 26x - 15x^2$ **32.** $x^4 + 5x^2 - 24$
**33.** $4x^2 - 12x + 9$ **34.** $25x^2 - 10x + 1$
**35.** $16x^2 + 24xy + 9y^2$ **36.** $4x^2 + 20xy + 25y^2$
**37.** $4x^2 - 49$ **38.** $9x^2 - 1$ **39.** $x^3 - 6x^2 + 12x - 8$
**40.** $8x^3 + 60x^2 + 150x + 125$ **41.** $2x^2 + 7x - 2$
**42.** $2x^3 + 2x^2$ **43.** $5ab(2a - b^2 - 3a^2b)$
**44.** $xy(3 - 5xy - 15x^2y^2)$ **45.** $(x + 4)(a + b)$
**46.** $(3x - 1)(y + 7)$ **47.** $(2x + y)(3x^2 + z^2)$
**48.** $(m + 5n)(n - 4)$ **49.** $(7a - 5b)(7a + 5b)$
**50.** $(6x - y)(6x + y)$ **51.** $(5a - 2)(25a^2 + 10a + 4)$
**52.** $(3x + 4y)(9x^2 - 12xy + 16y^2)$ **53.** $(x - 3)(x - 6)$
**54.** $(x + 4)(x + 7)$ **55.** $(x - 7)(x + 3)$ **56.** $(x + 8)(x - 2)$
**57.** $(2x + 1)(x + 4)$ **58.** $(3x - 4)(2x - 1)$
**59.** $(3x - 2)(4x + 1)$ **60.** $(4x + 1)(2x - 3)$ **61.** $(2x - 3y)^2$
**62.** $(x + 8y)^2$ **63.** $(x + 7)(x - 4)$ **64.** $2(t + 3)(t - 3)$
**65.** Not factorable **66.** $(4n - 1)(3n - 1)$
**67.** $x^2(x^2 + 1)(x + 1)(x - 1)$ **68.** $x(x - 12)(x + 6)$
**69.** $2a^2b(3a + 2b - c)$ **70.** $(x - y + 1)(x + y - 1)$
**71.** $4(2x^2 + 3)$ **72.** $(4x + 7)(3x - 5)$ **73.** $(4n - 5)^2$
**74.** $4n(n - 2)$ **75.** $3w(w^2 + 6w - 8)$ **76.** $(5x + 2y)(4x - y)$
**77.** $16a(a - 4)$ **78.** $3x(x + 1)(x - 6)$ **79.** $(n + 8)(n - 16)$
**80.** $(t + 5)(t - 5)(t^2 + 3)$ **81.** $(5x - 3)(7x + 2)$
**82.** $(3 - x)(5 - 3x)$ **83.** $(4n - 3)(16n^2 + 12n + 9)$
**84.** $2(2x + 5)(4x^2 - 10x + 25)$ **85.** $\{-3, 3\}$ **86.** $\{-6, 1\}$
**87.** $\left\{\frac{2}{7}\right\}$ **88.** $\left\{-\frac{2}{5}, \frac{1}{3}\right\}$ **89.** $\left\{-\frac{1}{3}, 3\right\}$ **90.** $\{-3, 0, 3\}$
**91.** $\{-1, 0, 1\}$ **92.** $\{-7, 9\}$ **93.** $\left\{-\frac{4}{7}, \frac{2}{7}\right\}$ **94.** $\left\{-\frac{4}{5}, \frac{5}{6}\right\}$
**95.** $\{-2, 2\}$ **96.** $\left\{\frac{5}{3}\right\}$ **97.** $\{-8, 6\}$ **98.** $\left\{-5, \frac{2}{7}\right\}$
**99.** $\{-8, 5\}$ **100.** $\{-12, 1\}$ **101.** $\varnothing$ **102.** $\left\{-5, \frac{6}{5}\right\}$
**103.** $\{0, 1, 8\}$ **104.** $\left\{-10, \frac{1}{4}\right\}$
**105.** 8, 9, and 10 or $-1$, 0, and 1 **106.** $-6$ and 8
**107.** 13 and 15 **108.** 12 miles and 16 miles **109.** 4 m by 12 m **110.** 9 rows and 16 chairs per row
**111.** The side is 13 ft long and the altitude is 6 ft.
**112.** 3 ft **113.** 5 cm by 5 cm and 8 cm by 8 cm **114.** 6 in.

## Chapter 3 Test (page 181)

**1.** $2x - 11$ **2.** $-48x^4y^4$ **3.** $-27x^6y^{12}$
**4.** $20x^2 + 17x - 63$ **5.** $6n^2 - 13n + 6$
**6.** $x^3 - 12x^2y + 48xy^2 - 64y^3$ **7.** $2x^3 + 11x^2 - 11x - 30$
**8.** $-14x^3y$ **9.** $(6x - 5)(x + 4)$ **10.** $3(2x + 1)(2x - 1)$
**11.** $(4 + t)(16 - 4t + t^2)$ **12.** $2x(3 - 2x)(5 + 4x)$
**13.** $(x - y)(x + 4)$ **14.** $(3n + 8)(8n - 3)$ **15.** $\{-12, 4\}$
**16.** $\left\{0, \frac{1}{4}\right\}$ **17.** $\left\{\frac{3}{2}\right\}$ **18.** $\{-4, -1\}$ **19.** $\{-9, 0, 2\}$
**20.** $\left\{-\frac{3}{7}, \frac{4}{5}\right\}$ **21.** $\left\{-\frac{1}{3}, 2\right\}$ **22.** $\{-2, 2\}$ **23.** 9 inches
**24.** 15 rows **25.** 8 feet

# Chapter 4

## Problem Set 4.1 (page 188)

**1.** $\frac{3}{4}$ **3.** $\frac{5}{6}$ **5.** $-\frac{2}{5}$ **7.** $\frac{2}{7}$ **9.** $\frac{2x}{7}$ **11.** $\frac{2a}{5b}$ **13.** $-\frac{y}{4x}$

**15.** $-\frac{9c}{13d}$ **17.** $\frac{5x^2}{3y^3}$ **19.** $\frac{x-2}{x}$ **21.** $\frac{3x+2}{2x-1}$ **23.** $\frac{a+5}{a-9}$

**25.** $\frac{n-3}{5n-1}$ **27.** $\frac{5x^2+7}{10x}$ **29.** $\frac{3x+5}{4x+1}$

**31.** $\frac{3x}{x^2+4x+16}$ **33.** $\frac{x+6}{3x-1}$ **35.** $\frac{x(2x+7)}{y(x+9)}$

**37.** $\frac{y+4}{5y-2}$ **39.** $\frac{3x(x-1)}{x^2+1}$ **41.** $\frac{2(x+3y)}{3x(3x+y)}$

**43.** $\frac{3n-2}{7n+2}$ **45.** $\frac{4-x}{5+3x}$ **47.** $\frac{9x^2+3x+1}{2(x+2)}$

**49.** $\frac{-2(x-1)}{x+1}$ **51.** $\frac{y+b}{y+c}$ **53.** $\frac{x+2y}{2x+y}$ **55.** $\frac{x+1}{x-6}$

**57.** $\frac{2s+5}{3s+1}$ **59.** $-1$ **61.** $-n-7$ **63.** $-\frac{2}{x+1}$

**65.** $-2$ **67.** $-\frac{n+3}{n+5}$

## Problem Set 4.2 (page 194)

**1.** $\frac{1}{10}$ **3.** $-\frac{4}{15}$ **5.** $\frac{3}{16}$ **7.** $-\frac{5}{6}$ **9.** $-\frac{2}{3}$

**11.** $\frac{10}{11}$ **13.** $-\frac{5x^3}{12y^2}$ **15.** $\frac{2a^3}{3b}$ **17.** $\frac{3x^3}{4}$

**19.** $\frac{25x^3}{108y^2}$ **21.** $\frac{ac^2}{2b^2}$ **23.** $\frac{3x}{4y}$ **25.** $\frac{3(x^2+4)}{5y(x+8)}$

**27.** $\frac{5(a+3)}{a(a-2)}$ **29.** $\frac{3}{2}$ **31.** $\frac{3xy}{4(x+6)}$

**33.** $\frac{5(x-2y)}{7y}$ **35.** $\frac{5+n}{3-n}$ **37.** $\frac{x^2+1}{x^2-10}$

**39.** $\frac{6x+5}{3x+4}$ **41.** $\frac{2t^2+5}{2(t^2+1)(t+1)}$ **43.** $\frac{t(t+6)}{4t+5}$

**45.** $\frac{n+3}{n(n-2)}$ **47.** $\frac{25x^3y^3}{4(x+1)}$ **49.** $\frac{2(a-2b)}{a(3a-2b)}$

## Problem Set 4.3 (page 201)

**1.** $\frac{13}{12}$ **3.** $\frac{11}{40}$ **5.** $\frac{19}{20}$ **7.** $\frac{49}{75}$ **9.** $\frac{17}{30}$ **11.** $-\frac{11}{84}$

**13.** $\frac{2x+4}{x-1}$ **15.** 4 **17.** $\frac{7y-10}{7y}$ **19.** $\frac{5x+3}{6}$

**21.** $\frac{12a+1}{12}$ **23.** $\frac{n+14}{18}$ **25.** $-\frac{11}{15}$ **27.** $\frac{3x-25}{30}$

**29.** $\frac{43}{40x}$ **31.** $\frac{20y-77x}{28xy}$ **33.** $\frac{16y+15x-12xy}{12xy}$

**35.** $\frac{21+22x}{30x^2}$ **37.** $\frac{10n-21}{7n^2}$ **39.** $\frac{45-6n+20n^2}{15n^2}$

**41.** $\frac{11x-10}{6x^2}$ **43.** $\frac{42t+43}{35t^3}$ **45.** $\frac{20b^2-33a^3}{96a^2b}$

**47.** $\frac{14-24y^3+45xy}{18xy^3}$ **49.** $\frac{2x^2+3x-3}{x(x-1)}$

**51.** $\frac{a^2-a-8}{a(a+4)}$ **53.** $\frac{-41n-55}{(4n+5)(3n+5)}$

**55.** $\frac{-3x+17}{(x+4)(7x-1)}$ **57.** $\frac{-x+74}{(3x-5)(2x+7)}$

**59.** $\frac{38x+13}{(3x-2)(4x+5)}$ **61.** $\frac{5x+5}{2x+5}$ **63.** $\frac{x+15}{x-5}$

**65.** $\frac{-2x-4}{2x+1}$ **67. (a)** $-1$ **(c)** 0

## Problem Set 4.4 (page 209)

**1.** $\frac{7x+20}{x(x+4)}$ **3.** $\frac{-x-3}{x(x+7)}$ **5.** $\frac{6x-5}{(x+1)(x-1)}$

**7.** $\frac{1}{a+1}$ **9.** $\frac{5n+15}{4(n+5)(n-5)}$ **11.** $\frac{x^2+60}{x(x+6)}$

**13.** $\frac{11x+13}{(x+2)(x+7)(2x+1)}$ **15.** $\frac{-3a+1}{(a-5)(a+2)(a+9)}$

**17.** $\frac{3a^2+14a+1}{(4a-3)(2a+1)(a+4)}$ **19.** $\frac{3x^2+20x-111}{(x^2+3)(x+7)(x-3)}$

**21.** $\frac{x+6}{(x-3)^2}$ **23.** $\frac{14x-4}{(x-1)(x+1)^2}$

**25.** $\frac{-7y-14}{(y+8)(y-2)}$ **27.** $\frac{-2x^2-4x+3}{(x+2)(x-2)}$

**29.** $\frac{2x^2+14x-19}{(x+10)(x-2)}$ **31.** $\frac{2n+1}{n-6}$

**33.** $\frac{2x^2-32x+16}{(x+1)(2x-1)(3x-2)}$ **35.** $\frac{1}{(n^2+1)(n+1)}$

**37.** $\frac{-16x}{(5x-2)(x-1)}$ **39.** $\frac{t+1}{t-2}$ **41.** $\frac{2}{11}$ **43.** $-\frac{7}{27}$ **45.** $\frac{x}{4}$

**47.** $\frac{3y-2x}{4x-7}$ **49.** $\frac{6ab^2-5a^2}{12b^2+2a^2b}$ **51.** $\frac{2y-3xy}{3x+4xy}$ **53.** $\frac{3n+14}{5n+19}$

**55.** $\frac{5n-17}{4n-13}$ **57.** $\frac{-x+5y-10}{3y-10}$ **59.** $\frac{-x+15}{-2x-1}$

**61.** $\frac{3a^2-2a+1}{2a-1}$ **63.** $\frac{-x^2+6x-4}{3x-2}$

## Problem Set 4.5 (page 216)

**1.** $3x^3+6x^2$ **3.** $-6x^4+9x^6$ **5.** $3a^2-5a-8$
**7.** $-13x^2+17x-28$ **9.** $-3xy+4x^2y-8xy^2$

**11.** $x-13$ **13.** $x+20$ **15.** $2x+1-\frac{3}{x-1}$

**17.** $5x-1$ **19.** $3x^2-2x-7$ **21.** $x^2+5x-6$

**23.** $4x^2+7x+12+\frac{30}{x-2}$ **25.** $x^3-4x^2-5x+3$

**27.** $x^2+5x+25$ **29.** $x^2-x+1+\frac{63}{x+1}$

**31.** $2x^2-4x+7-\frac{20}{x+2}$ **33.** $4a-4b$

**35.** $4x+7+\frac{23x-6}{x^2-3x}$ **37.** $8y-9+\frac{8y+5}{y^2+y}$

**39.** $2x-1$ **41.** $x-3$ **43.** $5a-8+\frac{42a-41}{a^2+3a-4}$

**45.** $2n^2 + 3n - 4$ **47.** $x^4 + x^3 + x^2 + x + 1$

**49.** $x^3 - x^2 + x - 1$ **51.** $3x^2 + x + 1 + \frac{7}{x^2 - 1}$

**53.** $x - 6$ **55.** $x + 6$, R = 14 **57.** $x^2 - 1$

**59.** $x^2 - 2x - 3$ **61.** $2x^2 - x - 6$, R = −6

**63.** $x^3 + 7x^2 + 21x + 56$, R = 167

### Problem Set 4.6 (page 222)

**1.** $\{2\}$ **3.** $\{-3\}$ **5.** $\{6\}$ **7.** $\left\{-\frac{85}{18}\right\}$ **9.** $\left\{\frac{7}{10}\right\}$

**11.** $\{5\}$ **13.** $\{58\}$ **15.** $\left\{\frac{1}{4}, 4\right\}$ **17.** $\left\{-\frac{2}{5}, 5\right\}$ **19.** $\{-16\}$

**21.** $\left\{-\frac{13}{3}\right\}$ **23.** $\{-3, 1\}$ **25.** $\left\{-\frac{5}{2}\right\}$ **27.** $\{-51\}$

**29.** $\left\{-\frac{5}{3}, 4\right\}$ **31.** $\varnothing$ **33.** $\left\{-\frac{11}{8}, 2\right\}$ **35.** $\{-29, 0\}$

**37.** $\{-9, 3\}$ **39.** $\left\{-2, \frac{23}{8}\right\}$ **41.** $\left\{\frac{11}{23}\right\}$ **43.** $\left\{3, \frac{7}{2}\right\}$

**45.** \$750 and \$1000 **47.** 48° and 72°

**49.** \$3500 **51.** 300 overweight employees

**53.** 14 feet and 6 feet **55.** 690 females and 460 males

### Problem Set 4.7 (page 230)

**1.** $\{-21\}$ **3.** $\{-1, 2\}$ **5.** $\{2\}$ **7.** $\left\{\frac{37}{15}\right\}$ **9.** $\{-1\}$

**11.** $\{-1\}$ **13.** $\left\{0, \frac{13}{2}\right\}$ **15.** $\left\{-2, \frac{19}{2}\right\}$ **17.** $\{-2\}$

**19.** $\left\{-\frac{1}{5}\right\}$ **21.** $\varnothing$ **23.** $\left\{\frac{7}{2}\right\}$ **25.** $\{-3\}$ **27.** $\left\{-\frac{7}{9}\right\}$

**29.** $\left\{-\frac{7}{6}\right\}$ **31.** $x = \frac{18y - 4}{15}$ **33.** $y = \frac{-5x + 22}{2}$

**35.** $M = \frac{IC}{100}$ **37.** $R = \frac{ST}{S + T}$ **39.** $y = \frac{bx - x - 3b + a}{a - 3}$

**41.** $y = \frac{ab - bx}{a}$ **43.** $y = \frac{-2x - 9}{3}$

**45.** 50 miles per hour for Dave and 54 miles per hour for Kent

**47.** 60 minutes

**49.** 60 words per minute for Connie and 40 words per minute for Katie

**51.** Plane B could travel at 400 miles per hour for 5 hours and plane A at 350 miles per hour for 4 hours, or plane B could travel at 250 miles per hour for 8 hours and plane A at 200 miles per hour for 7 hours.

**53.** 60 minutes for Nancy and 120 minutes for Amy

**55.** 16 miles per hour on the way out and 12 miles per hour on the way back, or 12 miles per hour out and 8 miles per hour back

### Chapter 4 Sample Problems (page 232)

**1.** $\frac{x - 5}{x + 7}$ **2.** $\frac{x - 7}{2(x + 7)}$ **3.** $\frac{2x - 1}{4x(2x + 1)}$ **4.** $\frac{5x}{3y^2}$

**5.** $\frac{4x + 14}{(x - 1)(x + 5)(x + 2)}$ **6.** $\frac{y^2 + 6x}{5y^2 - 3xy}$ **7.** $3x - 5 + \frac{8}{x + 2}$

**8.** $x^2 - x - 2$ **9.** $\{15\}$ **10.** $\{-15\}$ **11.** $\{60\}$

**12.** $a = \frac{3bx + 24xy}{2y}$ **13.** 6 female firefighters **14.** $2\frac{2}{5}$ hr

### Chapter 4 Review Problem Set (page 238)

**1.** $\frac{2y}{3x^2}$ **2.** $\frac{a - 3}{a}$ **3.** $\frac{n - 5}{n - 1}$ **4.** $\frac{x^2 + 1}{x}$ **5.** $\frac{2x + 1}{3}$

**6.** $\frac{x^2 - 10}{2x^2 + 1}$ **7.** $\frac{3}{22}$ **8.** $\frac{18y + 20x}{48y - 9x}$ **9.** $\frac{3x + 2}{3x - 2}$ **10.** $\frac{x - 1}{2x - 1}$

**11.** $\frac{2x}{7y^2}$ **12.** $3b$ **13.** $\frac{n(n + 5)}{n - 1}$ **14.** $\frac{x(x - 3y)}{x^2 + 9y^2}$

**15.** $\frac{23x - 6}{20}$ **16.** $\frac{57 - 2n}{18n}$ **17.** $\frac{3x^2 - 2x - 14}{x(x + 7)}$ **18.** $\frac{2}{x - 5}$

**19.** $\frac{5n - 21}{(n - 9)(n + 4)(n - 1)}$ **20.** $\frac{6y - 23}{(2y + 3)(y - 6)}$

**21.** $xy^4$ **22.** $\frac{3x^2y^5}{4}$ **23.** $\frac{4x(x + 1)}{x + 4}$ **24.** $-\frac{5}{3(x + 2)(x + 1)}$

**25.** $6x - 1$ **26.** $3x^2 - 7x + 22 - \frac{90}{x + 4}$

**27.** $3x^3 - 2x^2 - x + 2$ **28.** $2x^3 - 2x^2 + 3x - 4 + \frac{7}{x + 1}$

**29.** $5x^2 + 4x + 1 - \frac{8}{x + 2}$ **30.** $2x^2 - x - 1$ **31.** $\left\{\frac{4}{13}\right\}$

**32.** $\left\{\frac{3}{16}\right\}$ **33.** $\varnothing$ **34.** $\{-17\}$ **35.** $\left\{\frac{2}{7}, \frac{7}{2}\right\}$ **36.** $\{22\}$

**37.** $\left\{-\frac{6}{7}, 3\right\}$ **38.** $\left\{\frac{3}{4}, \frac{5}{2}\right\}$ **39.** $\left\{\frac{9}{7}\right\}$ **40.** $\left\{-\frac{5}{4}\right\}$

**41.** $y = \frac{3x + 27}{4}$ **42.** $y = \frac{bx - ab}{a}$ **43.** \$525 and \$875

**44.** Busboy \$36; waiter \$126 **45.** 1200 patients on Medicare

**46.** 15 true/false; 25 multiple-choice problems

**47.** 20 minutes for Julio and 30 minutes for Dan

**48.** $2\frac{11}{12}$ hours **49.** It takes George 3 hours, and it takes Maria $1\frac{1}{2}$ hours **50.** 9 hours **51.** Speedboat 16 mph; jon boat 8 mph

**52.** Duck 50 mph; heron 15 mph **53.** 50 miles per hour and 55 miles per hour or $8\frac{1}{3}$ miles per hour and $13\frac{1}{3}$ miles per hour

**54.** 13 miles per hour

### Chapter 4 Test (page 240)

**1.** $\frac{13y^2}{24x}$ **2.** $\frac{3x - 1}{x(x - 6)}$ **3.** $\frac{2n - 3}{n + 4}$ **4.** $-\frac{2x}{x + 1}$ **5.** $\frac{3y^2}{8}$

**6.** $\frac{a - b}{4(2a + b)}$ **7.** $\frac{x + 4}{5x - 1}$ **8.** $\frac{13x + 7}{12}$ **9.** $\frac{3x}{2}$

**10.** $\frac{10n - 26}{15n}$ **11.** $\frac{3x^2 + 2x - 12}{x(x - 6)}$ **12.** $\frac{11 - 2x}{x(x - 1)}$

**13.** $\frac{13n + 46}{(2n + 5)(n - 2)(n + 7)}$ **14.** $3x^2 - 2x - 1$

**15.** $\frac{18 - 2x}{8 + 9x}$ **16.** $y = \frac{4x + 20}{3}$ **17.** $\{1\}$ **18.** $\left\{\frac{1}{10}\right\}$

**19.** $\{-35\}$ **20.** $\{-1, 5\}$ **21.** $\left\{\frac{5}{3}\right\}$ **22.** $\left\{-\frac{9}{13}\right\}$ **23.** $\frac{27}{72}$

**24.** 60 minutes **25.** 15 miles per hour

### Chapters 1–4 Cumulative Review Problem Set (page 241)

**1.** 16 **2.** −13 **3.** 104 **4.** $\frac{5}{2}$ **5.** $7a^2 + 11a + 1$

**6.** $-2x^2 + 9x - 4$ **7.** $-2x^3y^5$ **8.** $16x^2y^6$ **9.** $36a^7b^2$ **10.** $-24a^6b^4$ **11.** $-18x^4 + 3x^3 - 12x^2$ **12.** $10x^2 + xy - 3y^2$ **13.** $x^2 + 8xy + 16y^2$ **14.** $a^3 - a^2b - 11ab^2 + 3b^3$ **15.** $(x - 2)(x - 3)$ **16.** $(2x + 1)(3x - 4)$ **17.** $2(x - 1)(x - 3)$ **18.** $3(x + 8)(x - 2)$ **19.** $(3m - 4n)(3m + 4n)$ **20.** $(3a + 2)(9a^2 - 6a + 4)$ **21.** $-\frac{7y^4}{x^2}$ **22.** $-x$ **23.** $\frac{x(x-5)}{x-1}$ **24.** $\frac{x+7}{x+3}$ **25.** $\frac{9n-26}{10}$ **26.** $\frac{8x-19}{(x-2)(x+3)(x-3)}$ **27.** $\frac{2y+3x}{6xy}$ **28.** $\frac{m-n}{mn}$ **29.** $3x^2 - x + 4$ **30.** $2x^2 + 5x - 3 + \frac{2}{x-4}$ **31.** $\{6\}$ **32.** $\{24\}$ **33.** $\left\{\frac{15}{7}\right\}$ **34.** $\{6\}$ **35.** $\left\{-2, \frac{10}{3}\right\}$ **36.** $\{-28, 12\}$ **37.** $\{-8, 1\}$ **38.** $\left\{-5, -\frac{3}{2}\right\}$ **39.** $\left\{-\frac{1}{3}, 9\right\}$ **40.** $\varnothing$ **41.** $P = \frac{A}{1+rt}$ **42.** $B = \frac{3V}{H}$ **43.** $(-\infty, 2]$ **44.** $(-4, 2)$ **45.** $\left(-\frac{9}{3}, 3\right)$ **46.** $(-\infty, -13] \cup [7, \infty)$ **47.** $(-\infty, -10] \cup (2, \infty)$ **48.** \$9.00 **49.** \$690 **50.** Width is 23 in., length is 38 in. **51.** 4 hr **52.** \$13,600 and \$54,400 **53.** 12 min **54.** 5 in., 12 in., and 13 in. **55.** $16\frac{2}{3}$ yr **56.** 27 dimes and 13 quarters

## Chapter 5

### Problem Set 5.1 (page 248)

**1.** $\frac{1}{27}$ **3.** $-\frac{1}{100}$ **5.** 81 **7.** $-27$ **9.** $-8$ **11.** 1 **13.** $\frac{9}{49}$ **15.** 16 **17.** $\frac{1}{1000}$ **19.** $\frac{1}{1000}$ **21.** 27 **23.** $\frac{1}{125}$ **25.** $\frac{9}{8}$ **27.** $\frac{256}{25}$ **29.** $\frac{2}{25}$ **31.** $\frac{81}{4}$ **33.** 81 **35.** $\frac{1}{10{,}000}$ **37.** $\frac{13}{36}$ **39.** $\frac{1}{2}$ **41.** $\frac{72}{17}$ **43.** $\frac{1}{x^6}$ **45.** $\frac{1}{a^3}$ **47.** $\frac{1}{a^8}$ **49.** $\frac{y^6}{x^2}$ **51.** $\frac{c^8}{a^4b^{12}}$ **53.** $\frac{y^{12}}{8x^9}$ **55.** $\frac{x^3}{y^{12}}$ **57.** $\frac{4a^4}{9b^2}$ **59.** $\frac{1}{x^2}$ **61.** $a^5b^2$ **63.** $\frac{6y^3}{x}$ **65.** $7b^2$ **67.** $\frac{7x}{y^2}$ **69.** $-\frac{12b^3}{a}$ **71.** $\frac{x^5y^5}{5}$ **73.** $\frac{b^{20}}{81}$ **75.** $\frac{x+1}{x^3}$ **77.** $\frac{y-x^3}{x^3y}$ **79.** $\frac{3b+4a^2}{a^2b}$ **81.** $\frac{1-x^2y}{xy^2}$ **83.** $\frac{2x-3}{x^2}$

### Problem Set 5.2 (page 258)

**1.** 8 **3.** $-10$ **5.** 3 **7.** $-4$ **9.** 3 **11.** $\frac{4}{5}$ **13.** $-\frac{6}{7}$ **15.** $\frac{1}{2}$ **17.** $\frac{3}{4}$ **19.** 8 **21.** $3\sqrt{3}$ **23.** $4\sqrt{2}$ **25.** $4\sqrt{5}$ **27.** $4\sqrt{10}$ **29.** $12\sqrt{2}$ **31.** $-12\sqrt{5}$ **33.** $2\sqrt{3}$ **35.** $3\sqrt{6}$ **37.** $-\frac{5}{3}\sqrt{7}$ **39.** $\frac{\sqrt{19}}{2}$ **41.** $\frac{3\sqrt{3}}{4}$ **43.** $\frac{5\sqrt{3}}{9}$ **45.** $\frac{\sqrt{14}}{7}$ **47.** $\frac{\sqrt{6}}{3}$ **49.** $\frac{\sqrt{15}}{6}$ **51.** $\frac{\sqrt{66}}{12}$ **53.** $\frac{\sqrt{6}}{3}$ **55.** $\sqrt{5}$ **57.** $\frac{2\sqrt{21}}{7}$ **59.** $-\frac{8\sqrt{15}}{5}$ **61.** $\frac{\sqrt{6}}{4}$ **63.** $-\frac{12}{25}$ **65.** $2\sqrt[3]{2}$ **67.** $6\sqrt[3]{3}$ **69.** $\frac{2\sqrt[3]{3}}{3}$ **71.** $\frac{3\sqrt[3]{2}}{2}$ **73.** $\frac{\sqrt[3]{12}}{2}$ **75.** 42 miles per hour; 49 miles per hour; 65 miles per hour **77.** 107 square centimeters **79.** 140 square inches **85.** (a) 1.414 (c) 12.490 (e) 57.000 (g) 0.374 (i) 0.930

### Problem Set 5.3 (page 263)

**1.** $13\sqrt{2}$ **3.** $54\sqrt{3}$ **5.** $-30\sqrt{2}$ **7.** $-\sqrt{5}$ **9.** $-21\sqrt{6}$ **11.** $-\frac{7\sqrt{7}}{12}$ **13.** $\frac{37\sqrt{10}}{10}$ **15.** $\frac{41\sqrt{2}}{20}$ **17.** $-9\sqrt[3]{3}$ **19.** $10\sqrt[3]{2}$ **21.** $4\sqrt{2x}$ **23.** $5x\sqrt{3}$ **25.** $2x\sqrt{5y}$ **27.** $8xy^3\sqrt{xy}$ **29.** $3a^2b\sqrt{6b}$ **31.** $3x^3y^4\sqrt{7}$ **33.** $4a\sqrt{10a}$ **35.** $\frac{8y}{3}\sqrt{6xy}$ **37.** $\frac{\sqrt{10xy}}{5y}$ **39.** $\frac{\sqrt{15}}{6x^2}$ **41.** $\frac{5\sqrt{2y}}{6y}$ **43.** $\frac{\sqrt{14xy}}{4y^3}$ **45.** $\frac{3y\sqrt{2xy}}{4x}$ **47.** $\frac{2\sqrt{42ab}}{7b^2}$ **49.** $2\sqrt[3]{3y}$ **51.** $2x\sqrt[3]{2x}$ **53.** $2x^2y^2\sqrt[3]{7y^2}$ **55.** $\frac{\sqrt[3]{21x}}{3x}$ **57.** $\frac{\sqrt[3]{12x^2y}}{4x^2}$ **59.** $\frac{\sqrt[3]{4x^2y^2}}{xy^2}$ **61.** $2\sqrt{2x+3y}$ **63.** $4\sqrt{x+3y}$ **65.** $33\sqrt{x}$ **67.** $-30\sqrt{2x}$ **69.** $7\sqrt{3n}$ **71.** $-40\sqrt{ab}$ **73.** $-7x\sqrt{2x}$ **79.** (a) $5|x|\sqrt{5}$ (b) $4x^2$ (c) $2b\sqrt{2b}$ (d) $y^2\sqrt{3y}$ (e) $12|x^3|\sqrt{2}$ (f) $2m^4\sqrt{7}$ (g) $8|c^5|\sqrt{2}$ (h) $3d^3\sqrt{2d}$ (i) $7|x|$ (j) $4n^{10}\sqrt{5}$ (k) $9h\sqrt{h}$

### Problem Set 5.4 (page 269)

**1.** $6\sqrt{2}$ **3.** $18\sqrt{2}$ **5.** $-24\sqrt{10}$ **7.** $24\sqrt{6}$ **9.** 120 **11.** 24 **13.** $56\sqrt[3]{3}$ **15.** $\sqrt{6} + \sqrt{10}$ **17.** $6\sqrt{10} - 3\sqrt{35}$ **19.** $24\sqrt{3} - 60\sqrt{2}$ **21.** $-40 - 32\sqrt{15}$ **23.** $15\sqrt{2x} + 3\sqrt{xy}$ **25.** $5xy - 6x\sqrt{y}$ **27.** $2\sqrt{10xy} + 2y\sqrt{15y}$ **29.** $-25\sqrt{6}$ **31.** $-25 - 3\sqrt{3}$ **33.** $23 - 9\sqrt{5}$ **35.** $6\sqrt{35} + 3\sqrt{10} - 4\sqrt{21} - 2\sqrt{6}$ **37.** $8\sqrt{3} - 36\sqrt{2} + 6\sqrt{10} - 18\sqrt{15}$ **39.** $11 + 13\sqrt{30}$ **41.** $141 - 51\sqrt{6}$ **43.** $-10$ **45.** $-8$ **47.** $2x - 3y$ **49.** $10\sqrt[3]{12} + 2\sqrt[3]{18}$ **51.** $12 - 36\sqrt[3]{2}$ **53.** $\frac{\sqrt{7}-1}{3}$ **55.** $\frac{-3\sqrt{2}-15}{23}$ **57.** $\frac{\sqrt{7}-\sqrt{2}}{5}$ **59.** $\frac{2\sqrt{5}+\sqrt{6}}{7}$ **61.** $\frac{\sqrt{15}-2\sqrt{3}}{2}$ **63.** $\frac{6\sqrt{7}+4\sqrt{6}}{13}$ **65.** $\sqrt{3} - \sqrt{2}$ **67.** $\frac{2\sqrt{x}-8}{x-16}$ **69.** $\frac{x+5\sqrt{x}}{x-25}$

**71.** $\frac{x - 8\sqrt{x} + 12}{x - 36}$ **73.** $\frac{x - 2\sqrt{xy}}{x - 4y}$ **75.** $\frac{6\sqrt{xy} + 9y}{4x - 9y}$

### Problem Set 5.5 (page 275)

**1.** $\{20\}$ **3.** $\varnothing$ **5.** $\left\{\frac{25}{4}\right\}$ **7.** $\left\{\frac{4}{9}\right\}$ **9.** $\{5\}$ **11.** $\left\{\frac{39}{4}\right\}$
**13.** $\left\{\frac{10}{3}\right\}$ **15.** $\{-1\}$ **17.** $\varnothing$ **19.** $\{1\}$ **21.** $\left\{\frac{3}{2}\right\}$ **23.** $\{3\}$
**25.** $\left\{\frac{61}{25}\right\}$ **27.** $\{-3, 3\}$ **29.** $\{-9, -4\}$ **31.** $\{0\}$ **33.** $\{3\}$
**35.** $\{4\}$ **37.** $\{-4, -3\}$ **39.** $\{12\}$ **41.** $\{25\}$ **43.** $\{29\}$
**45.** $\{-15\}$ **47.** $\left\{-\frac{1}{3}\right\}$ **49.** $\{-3\}$ **51.** $\{0\}$ **53.** $\{5\}$
**55.** $\{2, 6\}$ **57.** 56 feet; 106 feet; 148 feet
**59.** 3.2 feet; 5.1 feet; 7.3 feet

### Problem Set 5.6 (page 279)

**1.** 9 **3.** 3 **5.** $-2$ **7.** $-5$ **9.** $\frac{1}{6}$ **11.** 3 **13.** 8 **15.** 81
**17.** $-1$ **19.** $-32$ **21.** $\frac{81}{16}$ **23.** 4 **25.** $\frac{1}{128}$ **27.** $-125$
**29.** 625 **31.** $\sqrt[3]{x^4}$ **33.** $3\sqrt{x}$ **35.** $\sqrt[3]{2y}$ **37.** $\sqrt{2x - 3y}$
**39.** $\sqrt[3]{(2a - 3b)^2}$ **41.** $\sqrt[3]{x^2y}$ **43.** $-3\sqrt[5]{xy^2}$ **45.** $5^{\frac{1}{2}}y^{\frac{1}{2}}$
**47.** $3y^{\frac{1}{2}}$ **49.** $x^{\frac{1}{3}}y^{\frac{2}{3}}$ **51.** $a^{\frac{1}{2}}b^{\frac{3}{4}}$ **53.** $(2x - y)^{\frac{3}{5}}$ **55.** $5xy^{\frac{1}{2}}$
**57.** $-(x + y)^{\frac{1}{3}}$ **59.** $12x^{\frac{13}{20}}$ **61.** $y^{\frac{5}{12}}$ **63.** $\frac{4}{x^{\frac{1}{10}}}$ **65.** $16xy^2$
**67.** $2x^2y$ **69.** $4x^{\frac{4}{15}}$ **71.** $\frac{4}{b^{\frac{5}{12}}}$ **73.** $\frac{36x^{\frac{4}{5}}}{49y^{\frac{4}{3}}}$ **75.** $\frac{y^{\frac{3}{2}}}{x}$ **77.** $4x^{\frac{1}{6}}$
**79.** $\frac{16}{a^{\frac{11}{10}}}$ **81.** $\sqrt[6]{243}$ **83.** $\sqrt[4]{216}$ **85.** $\sqrt[12]{3}$ **87.** $\sqrt{2}$
**89.** $\sqrt[4]{3}$ **93.** **(a)** 12 **(c)** 7 **(e)** 11 **95.** **(a)** 1024
**(c)** 512 **(e)** 49

### Problem Set 5.7 (page 284)

**1.** $(8.9)(10^1)$ **3.** $(4.29)(10^3)$ **5.** $(6.12)(10^6)$ **7.** $(4)(10^7)$
**9.** $(3.764)(10^2)$ **11.** $(3.47)(10^{-1})$ **13.** $(2.14)(10^{-2})$
**15.** $(5)(10^{-5})$ **17.** $(1.94)(10^{-9})$ **19.** 23 **21.** 4190
**23.** 500,000,000 **25.** 31,400,000,000 **27.** 0.43
**29.** 0.000914 **31.** 0.00000005123 **33.** 0.000000074
**35.** 0.77 **37.** 300,000,000,000 **39.** 0.000000004
**41.** 1000 **43.** 1000 **45.** 3000 **47.** 20 **49.** 27,000,000
**51.** $(6.02)(10^{23})$ **53.** 831 **55.** \$37,060
**57.** 0.000137 in$^2$ **59.** 1833 **63.** **(a)** 7000 **(c)** 120 **(e)** 30
**65.** **(a)** $(4.385)(10^{14})$ **(c)** $(2.322)(10^{17})$ **(e)** $(3.052)(10^{12})$

### Chapter 5 Sample Problems (page 287)

**1.** 27 **2.** $\frac{y}{16x^2}$ **3.** $-\frac{6x}{y^2}$ **4.** $\frac{3y^2 - 4x}{xy^2}$ **5.** $2x^4y^2\sqrt{5y}$
**6.** $\frac{6\sqrt{21}}{7}$ **7.** $15\sqrt{2}$ **8.** $2xy\sqrt[3]{x^2}$ **9.** $6x - 3x\sqrt{2y}$
**10.** $\frac{5(\sqrt{6} + \sqrt{3})}{3}$ **11.** $\{5\}$ **12.** 0.27 **13.** 32 **14.** $\sqrt[3]{y^2}$
**15.** $x^{\frac{1}{5}}y^{\frac{3}{5}}$ **16.** $-10x^{\frac{1}{4}}$ **17.** $x\sqrt[6]{x^5}$ **18.** **(a)** $(7.02)(10^{-4})$
**(b)** $(3.981)(10^8)$ **19.** **(a)** 0.324 **(b)** 923,000 **20.** 0.0016

### Chapter 5 Review Problem Set (page 292)

**1.** $\frac{1}{64}$ **2.** $\frac{9}{4}$ **3.** 3 **4.** 1 **5.** 27 **6.** $\frac{1}{125}$ **7.** $\frac{x^6}{y^8}$
**8.** $\frac{27a^3b^{12}}{8}$ **9.** $\frac{9a^4}{16b^4}$ **10.** $\frac{y^6}{125x^9}$ **11.** $\frac{x^{12}}{9}$ **12.** $\frac{1}{4y^3}$
**13.** $-10x^3$ **14.** $\frac{3b^5}{a^3}$ **15.** $\frac{b^3}{a^5}$ **16.** $\frac{x^4}{y}$ **17.** $-\frac{2}{x^2}$
**18.** $-\frac{2a}{b}$ **19.** $\frac{y + x^2}{x^2y}$ **20.** $\frac{b - 2a}{a^2b}$ **21.** $\frac{2y^2 + 3x}{xy^2}$
**22.** $\frac{y^2 + 6x}{2xy^2}$ **23.** $3\sqrt{6}$ **24.** $4x\sqrt{3xy}$ **25.** $2\sqrt[3]{7}$
**26.** $3xy^2\sqrt[3]{4xy^2}$ **27.** $\frac{15\sqrt{6}}{4}$ **28.** $2y\sqrt{5xy}$ **29.** $2\sqrt{2}$
**30.** $\frac{\sqrt{15x}}{6x^2}$ **31.** $\frac{\sqrt[3]{6}}{3}$ **32.** $\frac{3\sqrt{5}}{5}$ **33.** $\frac{x\sqrt{21x}}{7}$ **34.** $2\sqrt{x}$
**35.** $\sqrt{5}$ **36.** $5\sqrt[3]{3}$ **37.** $\frac{29\sqrt{6}}{5}$ **38.** $-15\sqrt{3x}$
**39.** $24\sqrt{10}$ **40.** 60 **41.** $2x\sqrt{15y}$ **42.** $-18y^2\sqrt{x}$
**43.** $24\sqrt{3} - 6\sqrt{14}$ **44.** $x - 2\sqrt{x} - 15$ **45.** 17
**46.** $12 - 8\sqrt{3}$ **47.** $6a - 5\sqrt{ab} - 4b$ **48.** 70
**49.** $\frac{2(\sqrt{7} + 1)}{3}$ **50.** $\frac{2\sqrt{6} - \sqrt{15}}{3}$ **51.** $\frac{3\sqrt{5} - 2\sqrt{3}}{11}$
**52.** $\frac{6\sqrt{3} + 3\sqrt{5}}{7}$ **53.** $\left\{\frac{19}{7}\right\}$ **54.** $\{4\}$ **55.** $\{8\}$ **56.** $\varnothing$
**57.** $\{14\}$ **58.** $\{-10, 1\}$ **59.** $\{2\}$ **60.** $\{8\}$ **61.** 493 ft
**62.** 4.7 ft **63.** 32 **64.** 1 **65.** $\frac{4}{9}$ **66.** $-64$ **67.** $\frac{1}{9}$ **68.** $\frac{1}{4}$
**69.** 27 **70.** 8 **71.** $\sqrt[3]{xy^2}$ **72.** $\sqrt[4]{a^3}$ **73.** $4\sqrt{y}$
**74.** $\sqrt[3]{(x + 5y)^2}$ **75.** $x^{\frac{3}{5}}y^{\frac{1}{5}}$ **76.** $4^{\frac{1}{3}}a^{\frac{2}{3}}$ **77.** $6y^{\frac{1}{2}}$
**78.** $(3a + b)^{\frac{5}{3}}$ **79.** $20x^{\frac{7}{10}}$ **80.** $7a^{\frac{5}{12}}$ **81.** $\frac{y^{\frac{4}{3}}}{x}$ **82.** $-\frac{6}{a^{\frac{1}{4}}}$
**83.** $\frac{1}{x^{\frac{2}{5}}}$ **84.** $-6y^{\frac{5}{12}}$ **85.** $\sqrt[4]{3^3}$ **86.** $3\sqrt[6]{3}$ **87.** $\sqrt[12]{5}$
**88.** $\sqrt[6]{2^5}$ **89.** $(5.4)(10^8)$ **90.** $(8.4)(10^4)$ **91.** $(3.2)(10^{-8})$
**92.** $(7.68)(10^{-4})$ **93.** 0.0000014 **94.** 0.000638
**95.** 41,200,000 **96.** 125,000 **97.** 0.000000006
**98.** 36,000,000,000 **99.** 6 **100.** 0.15 **101.** 0.000028
**102.** 0.002 **103.** 0.002 **104.** 8,000,000,000

### Chapter 5 Test (page 294)

**1.** $\frac{1}{32}$ **2.** $-32$ **3.** $\frac{81}{16}$ **4.** $\frac{1}{4}$ **5.** $3\sqrt{7}$ **6.** $3\sqrt[3]{4}$
**7.** $2x^2y\sqrt{13y}$ **8.** $\frac{5\sqrt{6}}{6}$ **9.** $\frac{\sqrt{42x}}{12x^2}$ **10.** $72\sqrt{2}$ **11.** $-5\sqrt{6}$
**12.** $-38\sqrt{2}$ **13.** $\frac{3\sqrt{6} + 3}{10}$ **14.** $\frac{9x^2y^2}{4}$ **15.** $-\frac{12}{a^{\frac{3}{10}}}$
**16.** $\frac{y^3 + x}{xy^3}$ **17.** $\frac{-12}{x^{\frac{5}{4}}}$ **18.** 33 **19.** 600 **20.** 0.003
**21.** $\left\{\frac{8}{3}\right\}$ **22.** $\{2\}$ **23.** $\{4\}$ **24.** $\{5\}$ **25.** $\{4, 6\}$

# Chapter 6

## Problem Set 6.1 (page 302)

**1.** False **3.** True **5.** True **7.** True **9.** $10 + 8i$ **11.** $-6 + 10i$ **13.** $-2 - 5i$ **15.** $-12 + 5i$ **17.** $-1 - 23i$ **19.** $-4 - 5i$ **21.** $1 + 3i$ **23.** $\frac{5}{3} - \frac{5}{12}i$ **25.** $-\frac{17}{9} + \frac{23}{30}i$ **27.** $9i$ **29.** $i\sqrt{14}$ **31.** $\frac{4}{5}i$ **33.** $3i\sqrt{2}$ **35.** $5i\sqrt{3}$ **37.** $6i\sqrt{7}$ **39.** $-8i\sqrt{5}$ **41.** $36i\sqrt{10}$ **43.** $-8$ **45.** $-\sqrt{15}$ **47.** $-3\sqrt{6}$ **49.** $-5\sqrt{3}$ **51.** $-3\sqrt{6}$ **53.** $4i\sqrt{3}$ **55.** $\frac{5}{2}$ **57.** $2\sqrt{2}$ **59.** $2i$ **61.** $-20 + 0i$ **63.** $42 + 0i$ **65.** $15 + 6i$ **67.** $-42 + 12i$ **69.** $7 + 22i$ **71.** $40 - 20i$ **73.** $-3 - 28i$ **75.** $-3 - 15i$ **77.** $-9 + 40i$ **79.** $-12 + 16i$ **81.** $85 + 0i$ **83.** $5 + 0i$ **85.** $\frac{3}{5} + \frac{3}{10}i$ **87.** $\frac{5}{17} - \frac{3}{17}i$ **89.** $2 + \frac{2}{3}i$ **91.** $0 - \frac{2}{7}i$ **93.** $\frac{22}{25} - \frac{4}{25}i$ **95.** $-\frac{18}{41} + \frac{39}{41}i$ **97.** $\frac{9}{2} - \frac{5}{2}i$ **99.** $\frac{4}{13} - \frac{1}{26}i$ **101. (a)** $-2 - i\sqrt{3}$ **(c)** $\frac{-1 - 3i\sqrt{2}}{2}$ **(e)** $\frac{10 + 3i\sqrt{5}}{4}$

## Problem Set 6.2 (page 309)

**1.** $\{0, 9\}$ **3.** $\{-3, 0\}$ **5.** $\{-4, 0\}$ **7.** $\left\{0, \frac{9}{5}\right\}$ **9.** $\{-6, 5\}$ **11.** $\{7, 12\}$ **13.** $\left\{-8, -\frac{3}{2}\right\}$ **15.** $\left\{-\frac{7}{3}, \frac{2}{5}\right\}$ **17.** $\left\{\frac{3}{5}\right\}$ **19.** $\left\{-\frac{3}{2}, \frac{7}{3}\right\}$ **21.** $\{1, 4\}$ **23.** $\{8\}$ **25.** $\{12\}$ **27.** $\{\pm 1\}$ **29.** $\{\pm 6i\}$ **31.** $\{\pm\sqrt{14}\}$ **33.** $\{\pm 2\sqrt{7}\}$ **35.** $\{\pm 3\sqrt{2}\}$ **37.** $\left\{\pm\frac{\sqrt{14}}{2}\right\}$ **39.** $\left\{\pm\frac{2\sqrt{3}}{3}\right\}$ **41.** $\left\{\pm\frac{2i\sqrt{30}}{5}\right\}$ **43.** $\left\{\pm\frac{\sqrt{6}}{2}\right\}$ **45.** $\{-1, 5\}$ **47.** $\{-8, 2\}$ **49.** $\{-6 \pm 2i\}$ **51.** $\{1, 2\}$ **53.** $\{4 \pm \sqrt{5}\}$ **55.** $\{-5 \pm 2\sqrt{3}\}$ **57.** $\left\{\frac{2 \pm 3i\sqrt{3}}{3}\right\}$ **59.** $\{-12, -2\}$ **61.** $\left\{\frac{2 \pm \sqrt{10}}{5}\right\}$ **63.** $2\sqrt{13}$ centimeters **65.** $4\sqrt{5}$ inches **67.** 8 yards **69.** $6\sqrt{2}$ inches **71.** $a = b = 4\sqrt{2}$ meters **73.** $b = 3\sqrt{3}$ inches and $c = 6$ inches **75.** $a = 7$ centimeters and $b = 7\sqrt{3}$ centimeters **77.** $a = \frac{10\sqrt{3}}{3}$ feet and $c = \frac{20\sqrt{3}}{3}$ feet **79.** 17.9 feet **81.** 38 meters **83.** 53 meters **87.** 10.8 centimeters **89.** $h = s\sqrt{2}$

## Problem Set 6.3 (page 314)

**1.** $\{-6, 10\}$ **3.** $\{4, 10\}$ **5.** $\{-5, 10\}$ **7.** $\{-8, 1\}$ **9.** $\left\{-\frac{5}{2}, 3\right\}$ **11.** $\left\{-3, \frac{2}{3}\right\}$ **13.** $\{-16, 10\}$ **15.** $\{-2 \pm \sqrt{6}\}$ **17.** $\{-3 \pm 2\sqrt{3}\}$ **19.** $\{5 \pm \sqrt{26}\}$ **21.** $\{4 \pm i\}$ **23.** $\{-6 \pm 3\sqrt{3}\}$ **25.** $\{-1 \pm i\sqrt{5}\}$ **27.** $\left\{\frac{-3 \pm \sqrt{17}}{2}\right\}$ **29.** $\left\{\frac{-5 \pm \sqrt{21}}{2}\right\}$ **31.** $\left\{\frac{7 \pm \sqrt{37}}{2}\right\}$ **33.** $\left\{\frac{-2 \pm \sqrt{10}}{2}\right\}$ **35.** $\left\{\frac{3 \pm i\sqrt{6}}{3}\right\}$ **37.** $\left\{\frac{-5 \pm \sqrt{37}}{6}\right\}$ **39.** $\{-12, 4\}$ **41.** $\left\{\frac{4 \pm \sqrt{10}}{2}\right\}$ **43.** $\left\{-\frac{9}{2}, \frac{1}{3}\right\}$ **45.** $\{-3, 8\}$ **47.** $\{3 \pm 2\sqrt{3}\}$ **49.** $\left\{\frac{3 \pm i\sqrt{3}}{3}\right\}$ **51.** $\{-20, 12\}$ **53.** $\left\{-1, -\frac{2}{3}\right\}$ **55.** $\left\{\frac{1}{2}, \frac{3}{2}\right\}$ **57.** $\{-6 \pm 2\sqrt{10}\}$ **59.** $\left\{\frac{-1 \pm \sqrt{3}}{2}\right\}$ **61.** $\left\{\frac{-b \pm \sqrt{b^2 - 4ac}}{2a}\right\}$ **65.** $x = \frac{a\sqrt{b^2 - y^2}}{b}$ **67.** $r = \frac{\sqrt{A\pi}}{\pi}$ **69.** $\{2a, 3a\}$ **71.** $\left\{\frac{a}{2}, -\frac{2a}{3}\right\}$ **73.** $\left\{\frac{2b}{3}\right\}$

## Problem Set 6.4 (page 322)

**1.** $\frac{1 \pm \sqrt{5}}{2}$ **3.** $-2 \pm \sqrt{3}$ **5.** $\frac{2 \pm \sqrt{2}}{3}$ **7.** $\frac{-2 \pm \sqrt{3}}{2}$ **9.** $\frac{-3 \pm 2\sqrt{3}}{2}$ **11.** $\{-1 \pm \sqrt{2}\}$ **13.** $\left\{\frac{-5 \pm \sqrt{37}}{2}\right\}$ **15.** $\{4 \pm 2\sqrt{5}\}$ **17.** $\left\{\frac{-5 \pm i\sqrt{7}}{2}\right\}$ **19.** $\{8, 10\}$ **21.** $\left\{\frac{9 \pm \sqrt{61}}{2}\right\}$ **23.** $\left\{\frac{-1 \pm \sqrt{33}}{4}\right\}$ **25.** $\left\{\frac{-1 \pm i\sqrt{3}}{4}\right\}$ **27.** $\left\{\frac{4 \pm \sqrt{10}}{3}\right\}$ **29.** $\left\{-1, \frac{5}{2}\right\}$ **31.** $\left\{-5, -\frac{4}{3}\right\}$ **33.** $\left\{\frac{5}{6}\right\}$ **35.** $\left\{\frac{1 \pm \sqrt{13}}{4}\right\}$ **37.** $\left\{0, \frac{13}{5}\right\}$ **39.** $\left\{\pm\frac{\sqrt{15}}{3}\right\}$ **41.** $\left\{\frac{-1 \pm \sqrt{73}}{12}\right\}$ **43.** $\{-18, -14\}$ **45.** $\left\{\frac{11}{4}, \frac{10}{3}\right\}$ **47.** $\left\{\frac{2 \pm i\sqrt{2}}{2}\right\}$ **49.** $\left\{\frac{1 \pm \sqrt{7}}{6}\right\}$

**51.** Two real solutions; $\{-7, 3\}$ **53.** One real solution; $\left\{\frac{1}{3}\right\}$

**55.** Two complex solutions; $\left\{\frac{7 \pm i\sqrt{3}}{2}\right\}$

**57.** Two real solutions; $\left\{-\frac{4}{3}, \frac{1}{5}\right\}$

**59.** Two real solutions; $\left\{\frac{-2 \pm \sqrt{10}}{3}\right\}$

**65.** $\{-1.381, 17.381\}$ **67.** $\{-13.426, 3.426\}$
**69.** $\{-8.653, -0.347\}$ **71.** $\{0.119, 1.681\}$
**73.** $\{-0.708, 4.708\}$ **75.** $k = 4$ or $k = -4$

## Problem Set 6.5 (page 328)

**1.** $\{2 \pm \sqrt{10}\}$ **3.** $\left\{-9, \frac{4}{3}\right\}$ **5.** $\{9 \pm 3\sqrt{10}\}$

**7.** $\left\{\frac{3 \pm i\sqrt{23}}{4}\right\}$ **9.** $\{-15, -9\}$ **11.** $\{-8, 1\}$

**13.** $\left\{\frac{2 \pm i\sqrt{10}}{2}\right\}$ **15.** $\{9 \pm \sqrt{66}\}$ **17.** $\left\{-\frac{5}{4}, \frac{2}{5}\right\}$

**19.** $\left\{\frac{-1 \pm \sqrt{2}}{2}\right\}$ **21.** $\left\{\frac{3}{4}, 4\right\}$ **23.** $\left\{\frac{11 \pm \sqrt{109}}{2}\right\}$

**25.** $\left\{\frac{3}{7}, 4\right\}$ **27.** $\left\{\frac{7 \pm \sqrt{129}}{10}\right\}$ **29.** $\left\{-\frac{10}{7}, 3\right\}$

**31.** $\{1 \pm \sqrt{34}\}$ **33.** $\{\pm\sqrt{6}, \pm 2\sqrt{3}\}$

**35.** $\left\{\pm 3, \pm\frac{2\sqrt{6}}{3}\right\}$ **37.** $\left\{\pm\frac{i\sqrt{15}}{3}, \pm 2i\right\}$

**39.** $\left\{\pm\frac{\sqrt{14}}{2}, \pm\frac{2\sqrt{3}}{3}\right\}$ **41.** 8 and 9 **43.** 9 and 12

**45.** $5 + \sqrt{3}$ and $5 - \sqrt{3}$ **47.** 3 and 6
**49.** 9 inches and 12 inches **51.** 1 meter
**53.** 8 inches by 14 inches **55.** 20 miles per hour for Lorraine and 25 miles per hour for Charlotte, or 45 miles per hour for Lorraine and 50 miles per hour for Charlotte
**57.** 55 miles per hour **59.** 6 hours for Tom and 8 hours for Terry **61.** 2 hours **63.** 8 students **65.** 50 numbers **67.** 9%

**73.** $\{9, 36\}$ **75.** $\{1\}$ **77.** $\left\{-\frac{8}{27}, \frac{27}{8}\right\}$ **79.** $\left\{-4, \frac{3}{5}\right\}$

**81.** $\{4\}$ **83.** $\{\pm 4\sqrt{2}\}$ **85.** $\left\{\frac{3}{2}, \frac{5}{2}\right\}$ **87.** $\{-1, 3\}$

## Problem Set 6.6 (page 335)

**1.** $(-\infty, -2) \cup (1, \infty)$

−2 1

**3.** $(-4, -1)$

−4 −1

**5.** $\left(-\infty, -\frac{7}{3}\right] \cup \left[\frac{1}{2}, \infty\right)$

$-\frac{7}{3}$ $\frac{1}{2}$

**7.** $\left[-2, \frac{3}{4}\right]$

−2 $\frac{3}{4}$

**9.** $(-1, 1) \cup (3, \infty)$

−1 1 3

**11.** $(-\infty, -2] \cup [0, 4]$

−2 0 4

**13.** $(-7, 5)$ **15.** $(-\infty, 4) \cup (7, \infty)$ **17.** $\left[-5, \frac{2}{3}\right]$

**19.** $\left(-\infty, -\frac{5}{2}\right] \cup \left[-\frac{1}{4}, \infty\right)$ **21.** $\left(-\infty, -\frac{4}{5}\right) \cup (8, \infty)$

**23.** $(-\infty, \infty)$ **25.** $\left\{-\frac{5}{2}\right\}$ **27.** $(-1, 3) \cup (3, \infty)$

**29.** $(-\infty, -2) \cup (2, \infty)$ **31.** $(-6, 6)$ **33.** $(-\infty, \infty)$
**35.** $(-\infty, 0] \cup [2, \infty)$ **37.** $(-4, 0) \cup (0, \infty)$

**39.** $(-\infty, -1) \cup (2, \infty)$ **41.** $(-2, 3)$ **43.** $(-\infty, 0) \cup \left[\frac{1}{2}, \infty\right)$

**45.** $(-\infty, 1) \cup [2, \infty)$ **47.** $(-6, -3)$

**49.** $(-\infty, 5) \cup [9, \infty)$ **51.** $\left(-\infty, \frac{4}{3}\right) \cup (3, \infty)$

**53.** $(-4, 6]$ **55.** $(-\infty, 2)$

## Chapter 6 Sample Problems (page 337)

**1.** $12 - 8i$ **2.** $2i\sqrt{6}$ **3.** 2 **4.** $23 + 9i$

**5.** $\frac{13}{29} + \frac{11}{29}i$ **6.** $\left\{-\frac{1}{5}, 3\right\}$ **7.** $\{3 \pm \sqrt{5}\}$

**8.** $\{4 \pm \sqrt{21}\}$ **9.** $\left\{\frac{-3 \pm \sqrt{65}}{4}\right\}$

**10.** Two unequal real solutions.
**11.** $\{3 \pm i\sqrt{5}\}$ **12.** The length of each leg is $4\sqrt{2}$ inches
**13.** Disregard the solution of $-5$ because it is not a whole number. The two consecutive whole numbers are 4 and 5.

**14.** $(-\infty, -3] \cup [5, \infty)$ **15.** $\left(-7, \frac{1}{2}\right]$

## Chapter 6 Review Problem Set (page 342)

**1.** $2 - 2i$ **2.** $-3 - i$ **3.** $8 - 8i$ **4.** $-2 + 4i$ **5.** $2i\sqrt{2}$
**6.** $5i$ **7.** $12i$ **8.** $6i\sqrt{2}$ **9.** $-2\sqrt{3}$ **10.** $6i$ **11.** $\sqrt{7}$
**12.** $i\sqrt{3}$ **13.** $30 + 15i$ **14.** $86 - 2i$ **15.** $-32 + 4i$
**16.** $25 + 0i$ **17.** $\frac{9}{20} + \frac{13}{20}i$ **18.** $-\frac{3}{29} + \frac{7}{29}i$ **19.** $2 - \frac{3}{2}i$
**20.** $-5 - 6i$ **21.** $\{-8, 0\}$ **22.** $\{0, 6\}$ **23.** $\{-4, 7\}$

**24.** $\left\{-\frac{3}{2}, 1\right\}$ **25.** $\{\pm 3\sqrt{5}\}$ **26.** $3 \pm 3i\sqrt{2}$

**27.** $\left\{\frac{-3 \pm 2\sqrt{6}}{2}\right\}$ **28.** $\{\pm 3\sqrt{3}\}$ **29.** $\{-9 \pm \sqrt{91}\}$

**30.** $\{-3 \pm i\sqrt{11}\}$ **31.** $\{5 \pm 2\sqrt{6}\}$ **32.** $\left\{\frac{-5 \pm \sqrt{33}}{2}\right\}$

**33.** $\{-3 \pm \sqrt{5}\}$ **34.** $\{-2 \pm i\sqrt{2}\}$

**35.** $\left\{\frac{1 \pm i\sqrt{11}}{3}\right\}$ **36.** $\left\{\frac{1 \pm \sqrt{61}}{10}\right\}$

**37.** One real solution with a multiplicity of 2
**38.** Two nonreal complex solutions
**39.** Two unequal real solutions
**40.** Two unequal real solutions **41.** $\{0, 17\}$ **42.** $\{-4, 8\}$

**43.** $\left\{\frac{1 \pm 8i}{2}\right\}$ **44.** $\{-3, 7\}$ **45.** $\{-1 \pm \sqrt{10}\}$

**46.** $\{3 \pm 5i\}$ **47.** $\{25\}$ **48.** $\left\{-4, \frac{2}{3}\right\}$ **49.** $\{-10, 20\}$

**50.** $\left\{\frac{-1 \pm \sqrt{61}}{6}\right\}$ **51.** $\left\{\frac{1 \pm i\sqrt{11}}{2}\right\}$

**52.** $\left\{\frac{5 \pm i\sqrt{23}}{4}\right\}$ **53.** $\left\{\frac{-2 \pm \sqrt{14}}{2}\right\}$ **54.** $\{-9, 4\}$

**55.** $\{-2 \pm i\sqrt{5}\}$ **56.** $\{-6, 12\}$ **57.** $\{1 \pm \sqrt{10}\}$

**58.** $\left\{\pm\frac{\sqrt{14}}{2}, \pm 2\sqrt{2}\right\}$ **59.** $\left\{\frac{-3 \pm \sqrt{97}}{2}\right\}$

**60.** 34.6 ft and 40.0 ft **61.** 47.2 m **62.** $4\sqrt{2}$ in.
**63.** $3 + \sqrt{7}$ and $3 - \sqrt{7}$ **64.** 8 hr
**65.** Andre: 45 mph and Sandy: 52 mph
**66.** 8 units **67.** 8 and 10 **68.** 7 in. by 12 in.
**69.** 4 hr for Reena and 6 hr for Billy **70.** 10 m

**71.** $(-\infty, -5) \cup (2, \infty)$ **72.** $\left[-\frac{7}{2}, 3\right]$ **73.** $\left[-\frac{1}{2}, \frac{1}{2}\right]$

**74.** $(-\infty, 2) \cup (5, \infty)$ **75.** $(-\infty, -6) \cup [4, \infty)$

**76.** $\left(-\frac{5}{2}, -1\right)$ **77.** $(-9, 4)$ **78.** $\left[-\frac{1}{3}, 1\right)$

## Chapter 6 Test (page 344)

**1.** $39 - 2i$ **2.** $-\frac{6}{25} - \frac{17}{25}i$ **3.** $\{0, 7\}$ **4.** $\{-1, 7\}$

**5.** $\{-6, 3\}$ **6.** $\{1 - \sqrt{2}, 1 + \sqrt{2}\}$ **7.** $\left\{\frac{1 - 2i}{5}, \frac{1 + 2i}{5}\right\}$

**8.** $\{-16, -14\}$ **9.** $\left\{\frac{1 - 6i}{3}, \frac{1 + 6i}{3}\right\}$ **10.** $\left\{-\frac{7}{4}, \frac{6}{5}\right\}$

**11.** $\left\{-3, \frac{19}{6}\right\}$ **12.** $\left\{-\frac{10}{3}, 4\right\}$ **13.** $\{-2, 2, -4i, 4i\}$

**14.** $\left\{-\frac{3}{4}, 1\right\}$ **15.** $\left\{\frac{1 - \sqrt{10}}{3}, \frac{1 + \sqrt{10}}{3}\right\}$

**16.** Two equal real solutions
**17.** Two nonreal complex solutions

**18.** $[-6, 9]$ **19.** $(-\infty, -2) \cup \left(\frac{1}{3}, \infty\right)$

**20.** $[-10, -6)$ **21.** 20.8 ft **22.** 29 m **23.** 3 hr

**24.** $6\frac{1}{2}$ in. **25.** $3 + \sqrt{5}$

## Chapters 1–6 Cumulative Review Problem Set (page 345)

**1.** $\frac{8}{27}$ **2.** $\frac{1}{4}$ **3.** 2 **4.** $\frac{1}{2}$ **5.** $\frac{1}{9}$ **6.** 16 **7.** $\frac{5}{6}$ **8.** $-\frac{6}{11}$

**9.** 6 **10.** $-3$ **11.** 64 **12.** $\frac{1}{2}$ **13.** $\frac{19}{45}$ **14.** 0.864

**15.** $-12a^6b^6$ **16.** $\frac{1}{3}c^4d^3$ **17.** $81m^4n^{12}$

**18.** $3a^3 - 7a^2 + 9a - 14$ **19.** $-5tk^2$
**20.** $12x^2 - 11xy - 5y^2$ **21.** $49m^2 - 84mn + 36n^2$

**22.** $\frac{a^3b}{2}$ **23.** $\frac{x - 3}{2(x + 7)}$ **24.** $\frac{21x - 19}{6}$

**25.** $3x^2 - x - 1 + \frac{-8}{x - 2}$ **26.** $\frac{7x - 2}{4x - 1}$ **27.** $\frac{27 + 28x}{3x - 60}$

**28.** $(x - y)(2a + 3c)$ **29.** $9(3m + n)(3m - n)$
**30.** $(2x + 1)(x - 7)$ **31.** $(2y + 5)(6y - 1)$
**32.** $2(3t - 4)(t + 7)$ **33.** $(c + y^3)(c - y^3)$
**34.** $(4h + 3)(2h - 5)$ **35.** $(a + 2b)(a^2 - 2ab + 4b^2)$

**36.** $\{-2\}$ **37.** $\varnothing$ **38.** $\{2\}$ **39.** $\left\{\frac{41}{13}\right\}$ **40.** $\{1500\}$

**41.** $t = \frac{A - P}{Pr}$ **42.** $\left\{\frac{20}{3}\right\}$ **43.** $\{3\}$ **44.** $\left\{-\frac{\sqrt{3}}{3}, \frac{\sqrt{3}}{3}\right\}$

**45.** $\{1\}$ **46.** $\left\{-\frac{5}{3}, 5\right\}$ **47.** $\{25\}$ **48.** $\left\{-\frac{1}{2}, 2\right\}$

**49.** $\left\{-\frac{5}{2}, \frac{4}{3}\right\}$ **50.** $\left\{-\frac{3}{2}, 4\right\}$ **51.** $w = \frac{P - 2l}{2}$ **52.** $\left\{\frac{2}{9}\right\}$

**53.** $\left\{-\frac{11}{2}, 2\right\}$ **54.** $\left\{-\frac{2}{3}\right\}$ **55.** $\left\{-\frac{6}{17}\right\}$ **56.** $\{-6, 2\}$

**57.** $\{-3, -1\}$ **58.** $\{-2, 0\}$ **59.** $\left(-\infty, -\frac{5}{2}\right)$ **60.** $\left[\frac{19}{20}, \infty\right)$

**61.** $(-\infty, \infty)$ **62.** $[6, \infty)$ **63.** $x \geq 1000$ or $[1000, \infty)$

**64.** $\left(-\frac{4}{3}, 4\right)$ **65.** $\varnothing$ **66.** $(-\infty, -4)$

**67.** $(-\infty, -3] \cup \left[\frac{5}{2}, \infty\right)$ **68.** $(-10, 5)$

**69.** $(-\infty, -5] \cup (1, \infty)$ **70.** $[-2, 4]$ **71.** At 2:20 p.m.
**72.** $1,600 or more **73.** 25°, 50°, 105° **74.** 8 hours

**75.** 142.1 feet **76.** 4% **77.** $12\frac{1}{2}$ hours

**78.** 8 feet and 15 feet

**79.** Canoe: $1\frac{3}{5}$ miles per hour; Kayak: 4 miles per hour

**80.** 0.32 units

## Chapter 7

### Problem Set 7.1 (page 358)

**1.** (2, 4), (−1, −5)
**3.** (−2, 10), (3, 0)
**5.** 5, 4, 3, −1
**7.** −10, −6, −2, 2

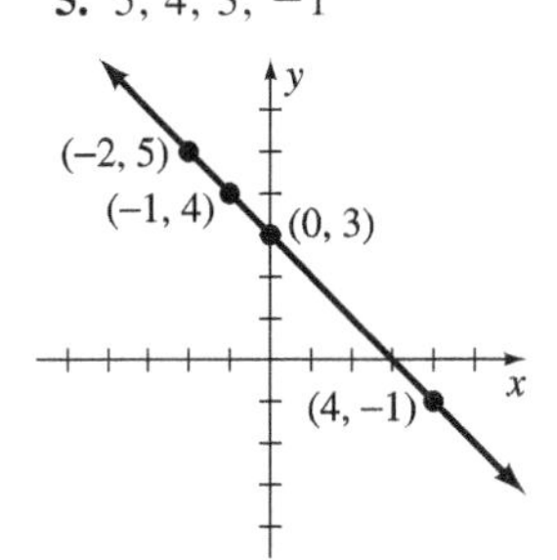


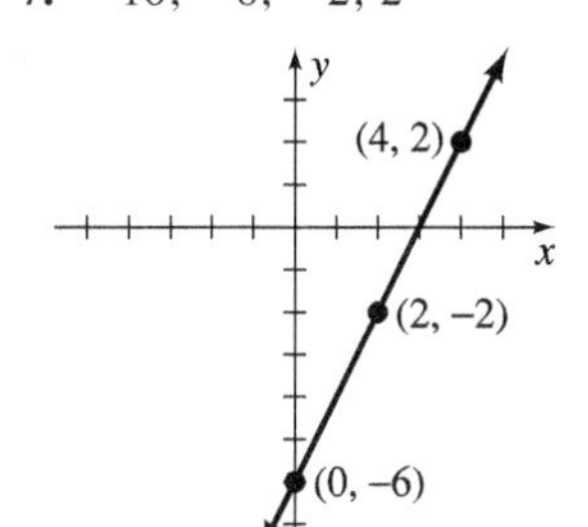


**9.**

**11.**

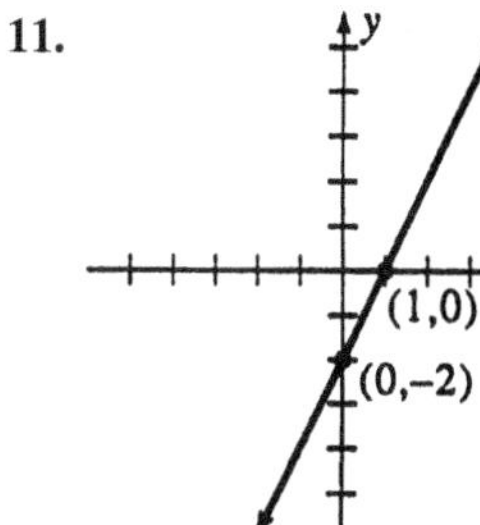


**13.**

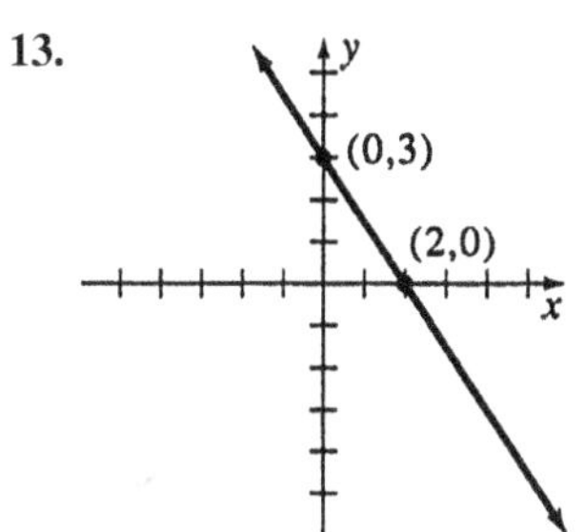


**15.**

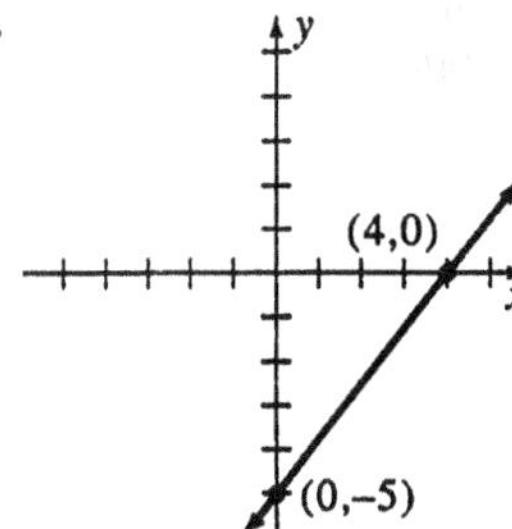


**17.**

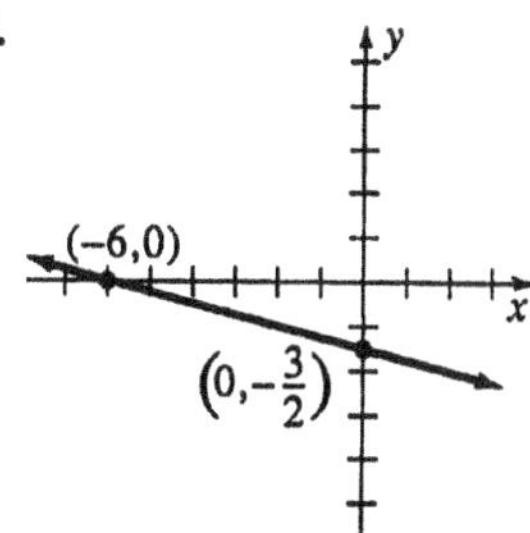


**19.**

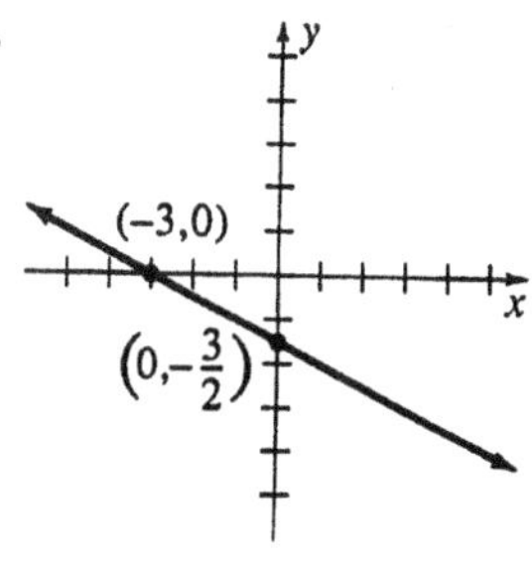


**21.**

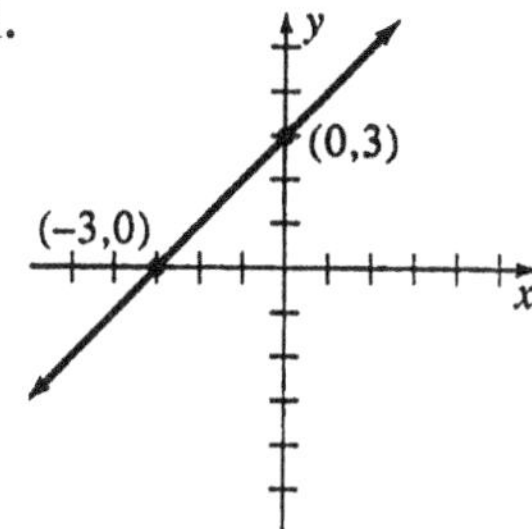


**23.**

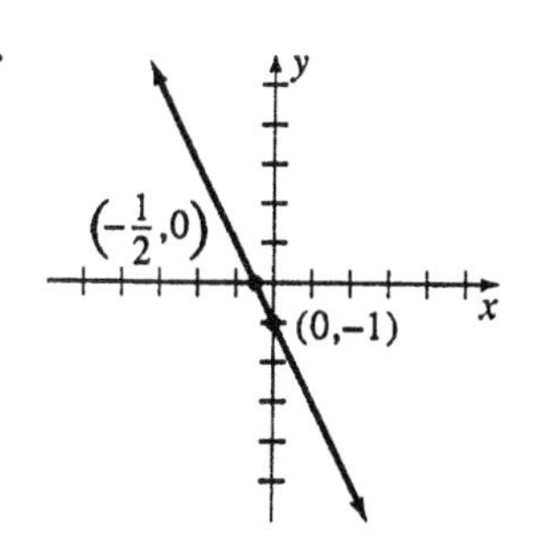


**25.**

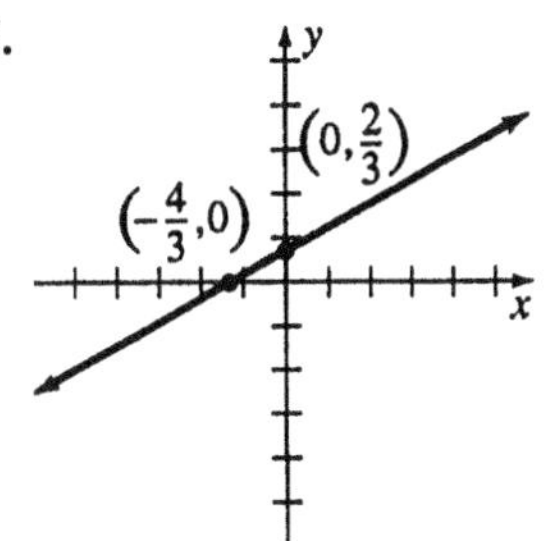


**27.**

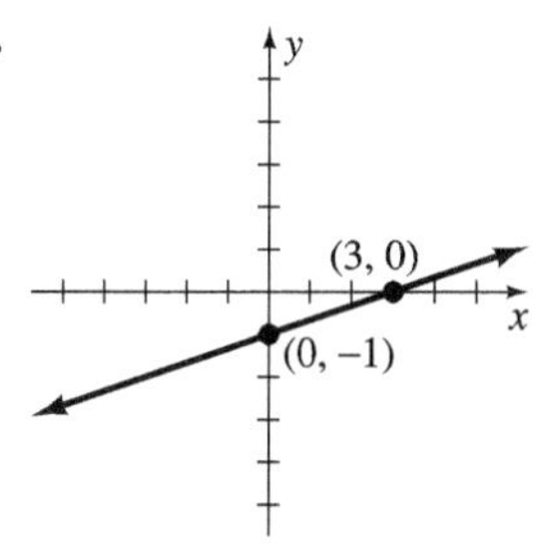


**29.**

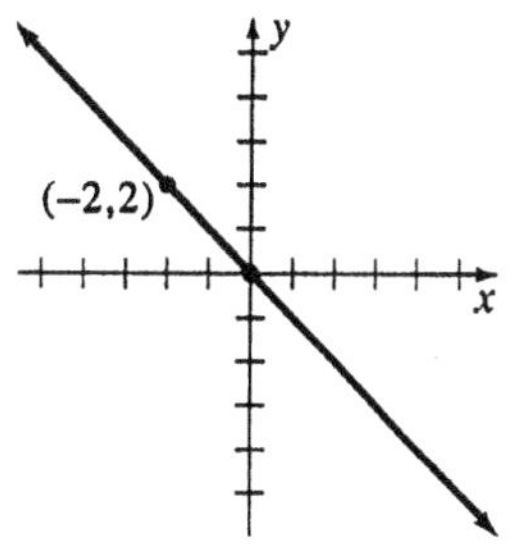


**31.**

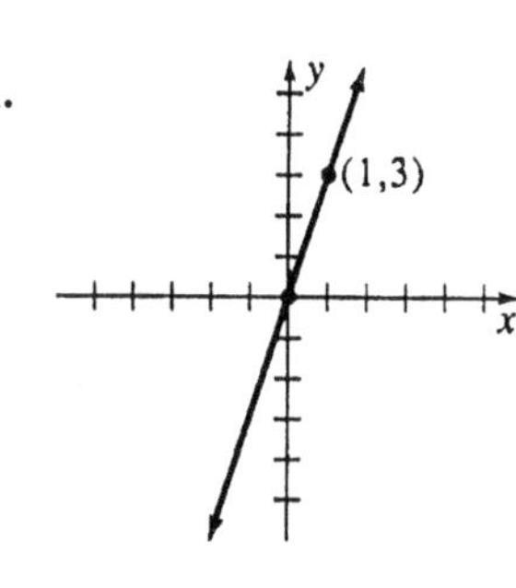


**33.**

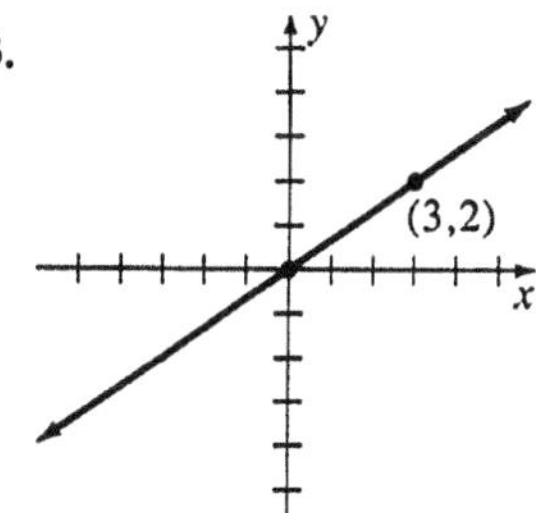


**35.**

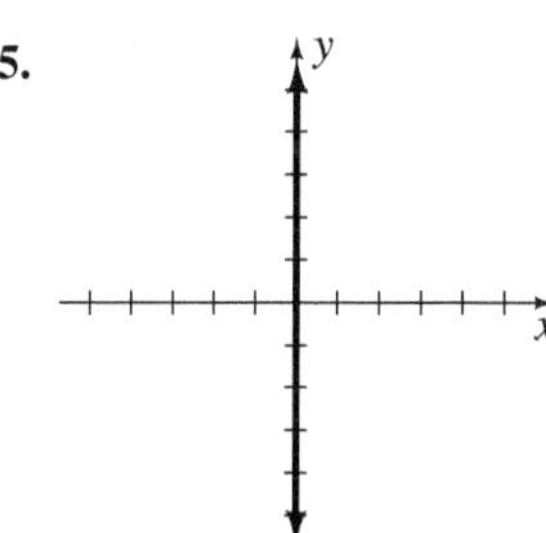


**37.**

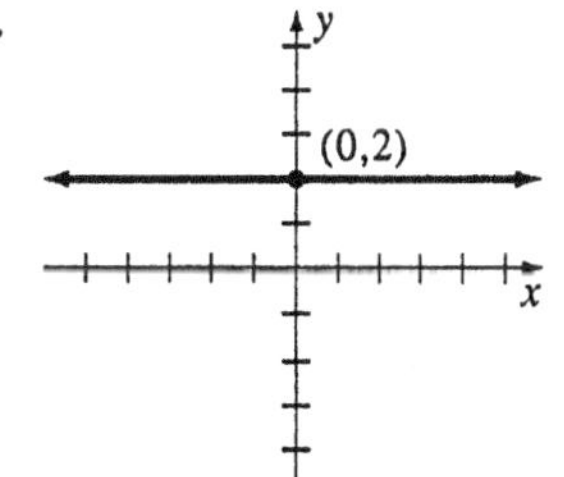


**39.**

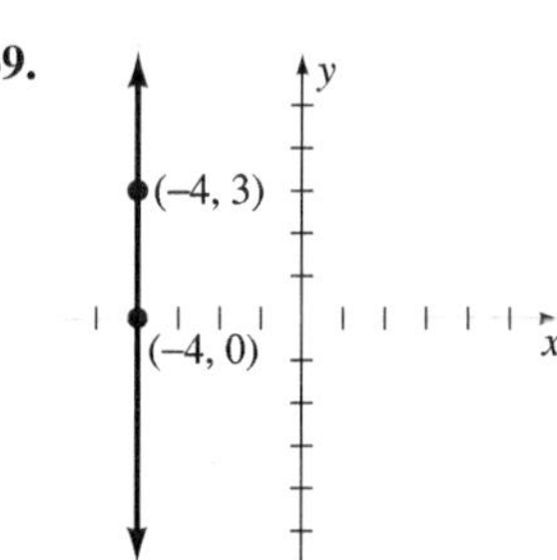


**41. (a)**

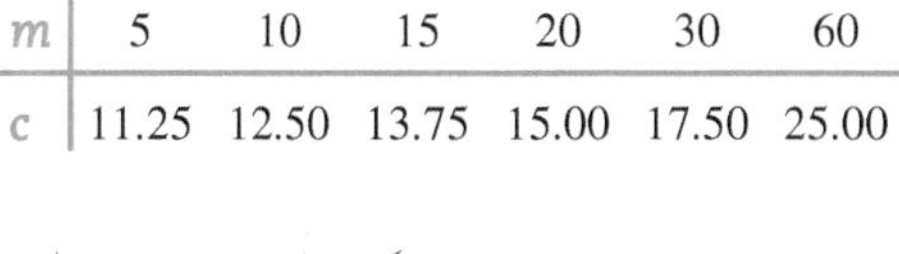

| *m* | 5 | 10 | 15 | 20 | 30 | 60 |
|---|---|---|---|---|---|---|
| *c* | 11.25 | 12.50 | 13.75 | 15.00 | 17.50 | 25.00 |

**(b)**

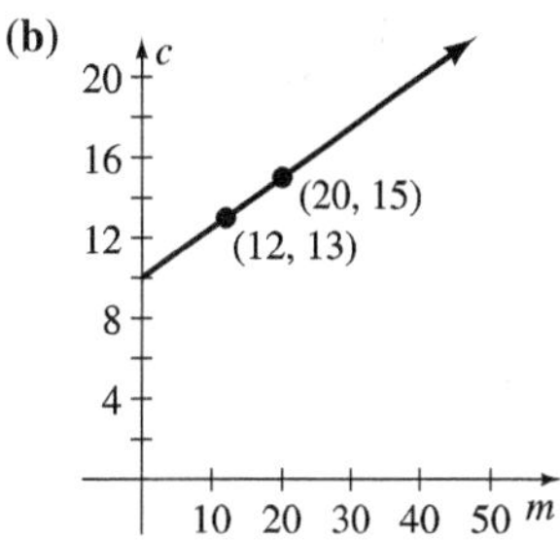


**(d)** 16.25, 20.00, 21.25

**43. (a)**

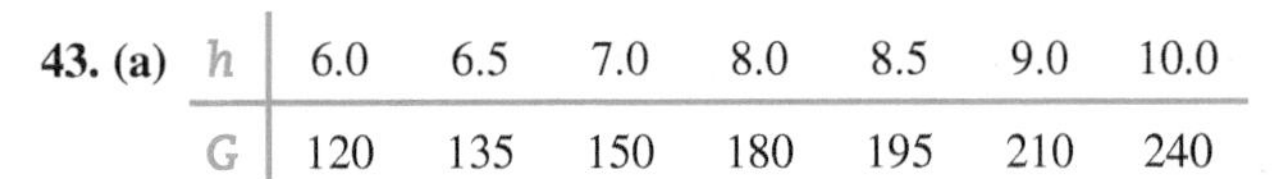

| h | 6.0 | 6.5 | 7.0 | 8.0 | 8.5 | 9.0 | 10.0 |
|---|---|---|---|---|---|---|---|
| G | 120 | 135 | 150 | 180 | 195 | 210 | 240 |

**(b)**

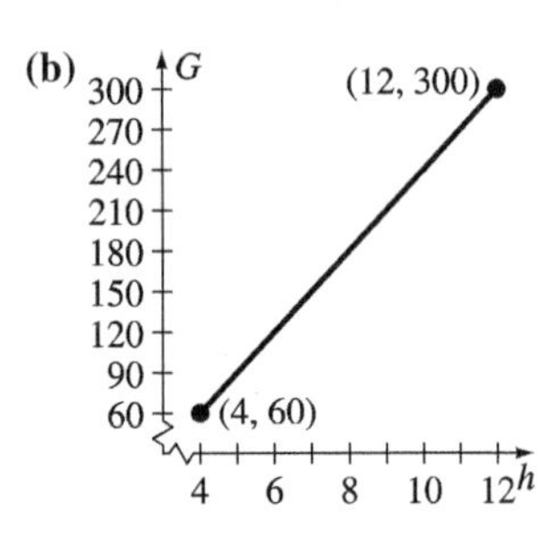


**(d)** 105, 165

**45.**

c
(1, 8)
(0, 5)
t

**47.**

C
(30, 9)
(10, 3)
p

**53.**

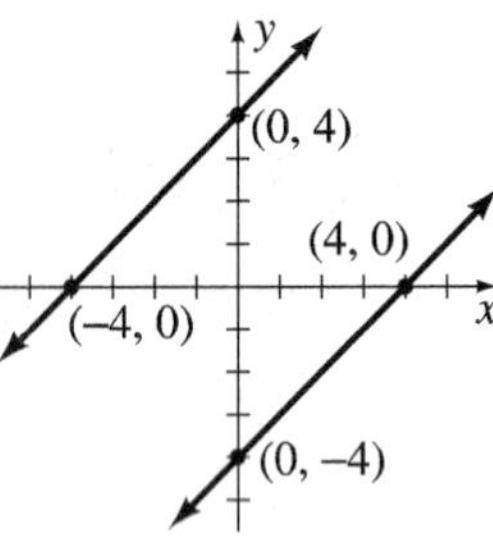


**55.**

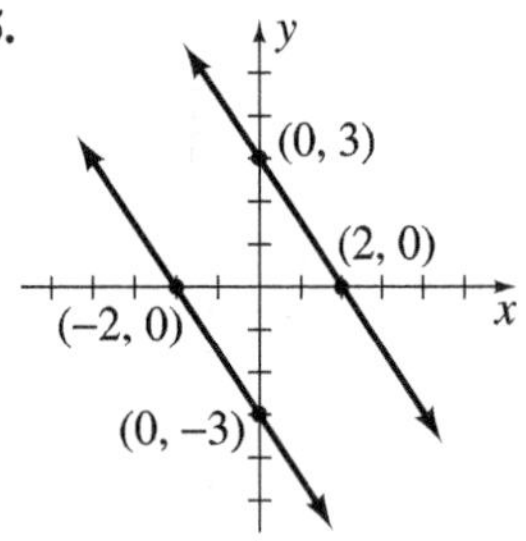

## Problem Set 7.2 (page 364)

**1.**

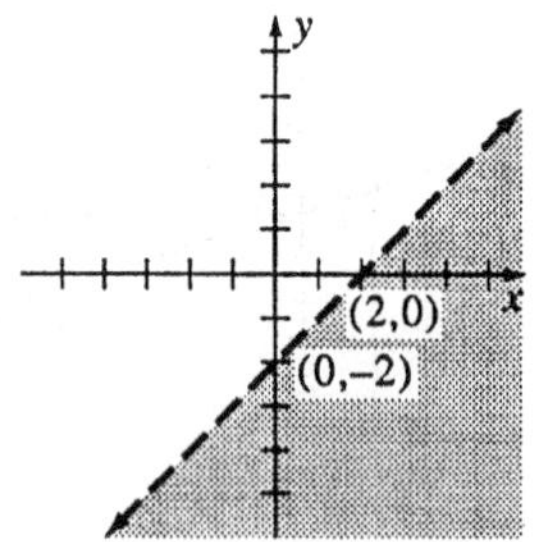


**3.**

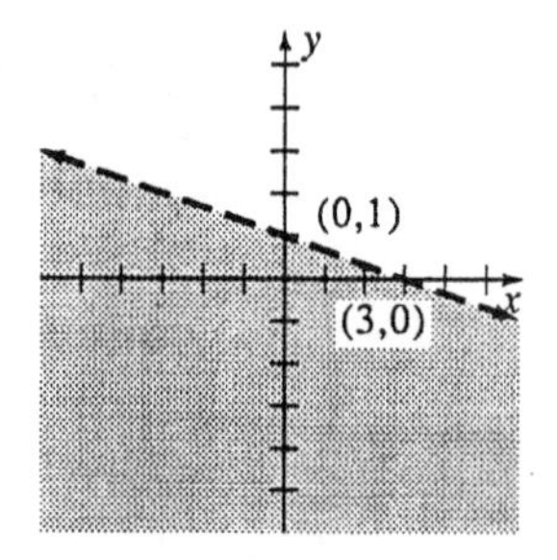


**5.**

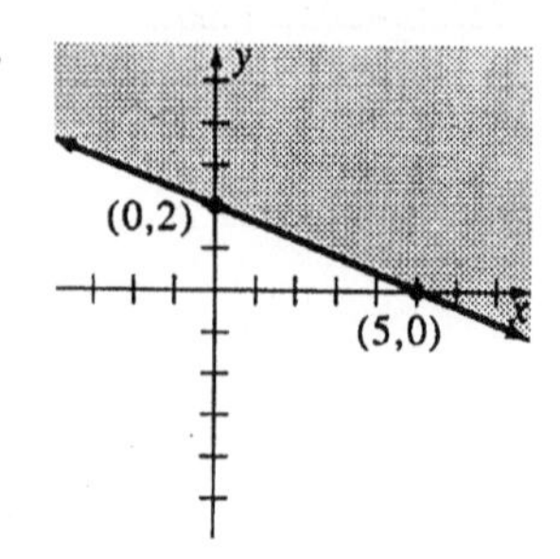


**7.**

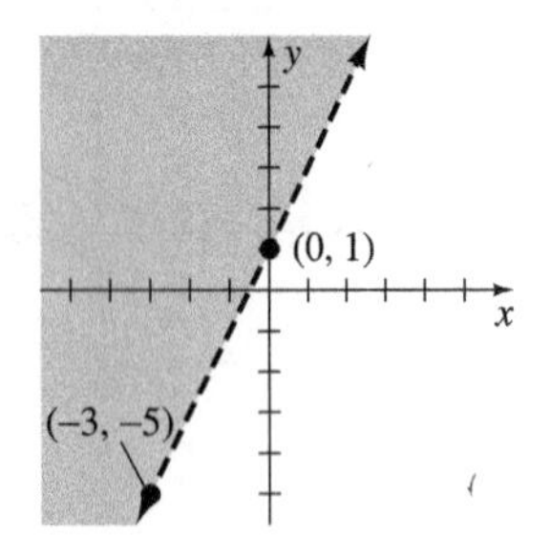


**9.**

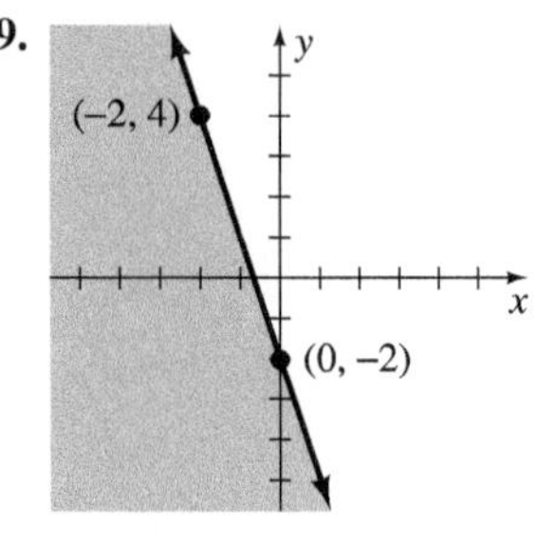


**11.**

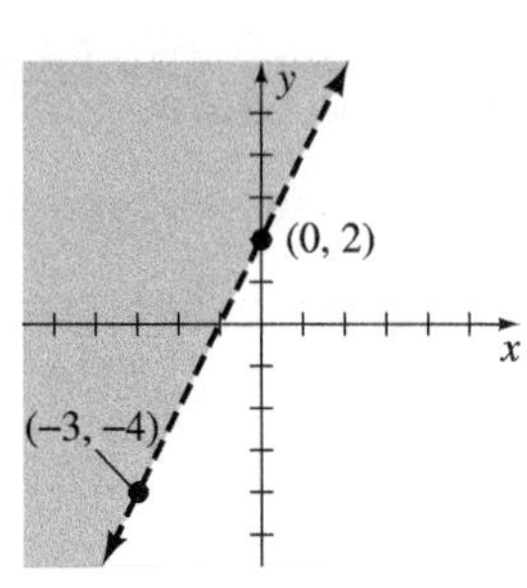


**13.**

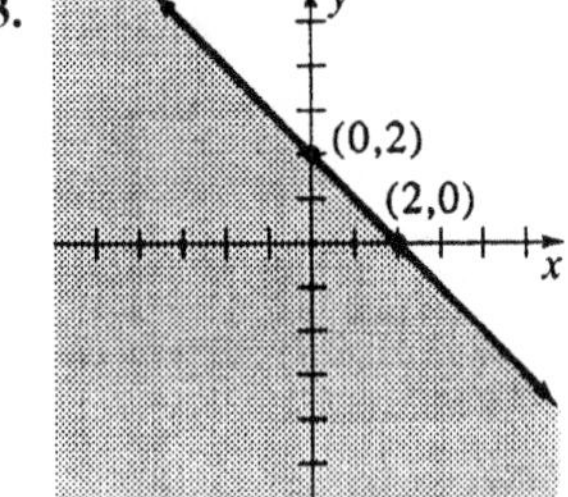


**15.**

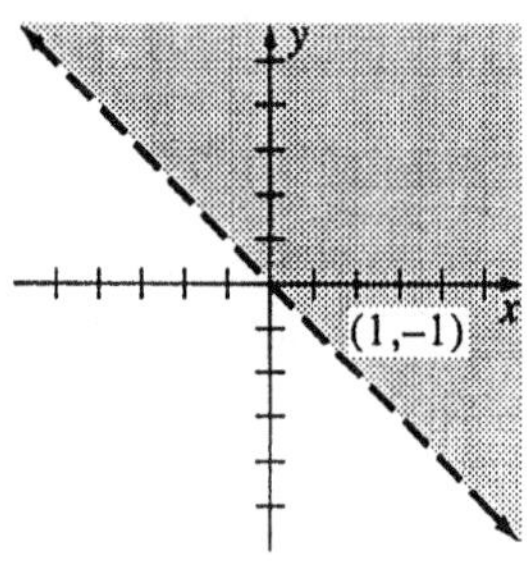


**17.**

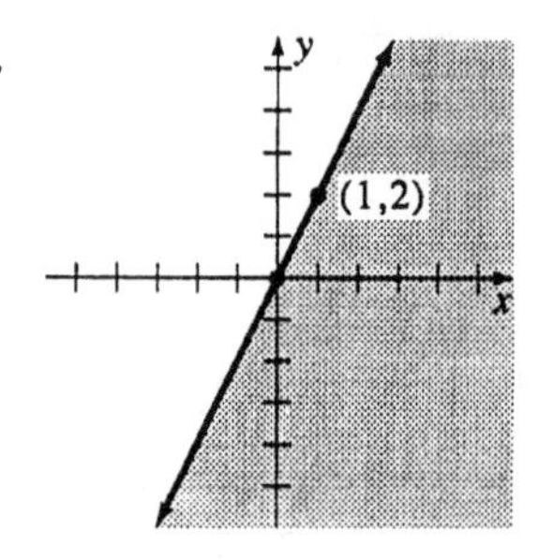


**19.**

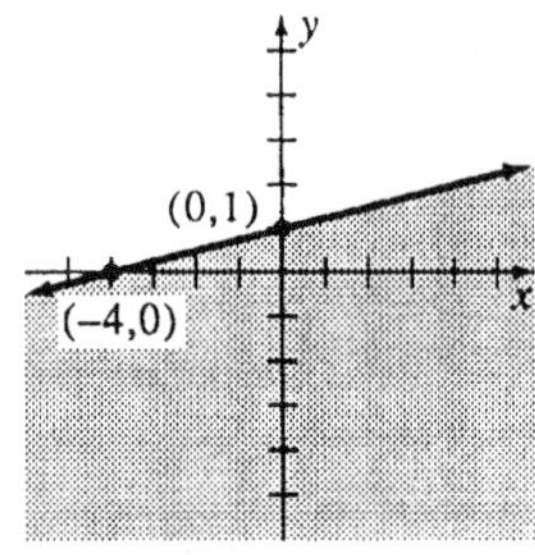


**21.**

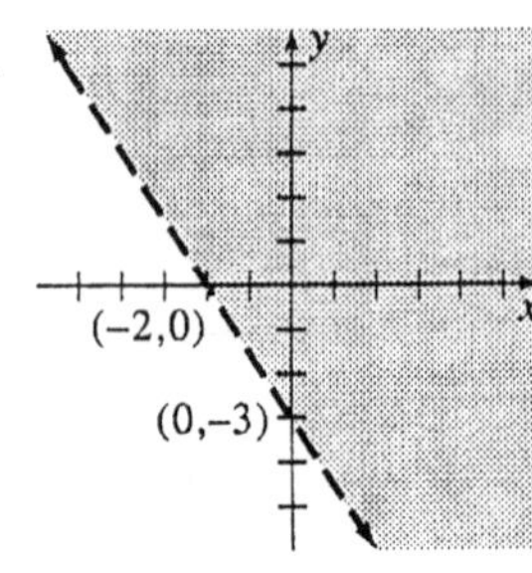


**23.**

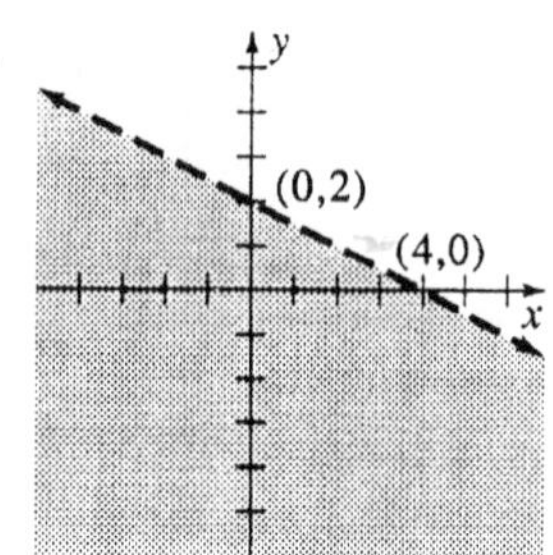


**25.**

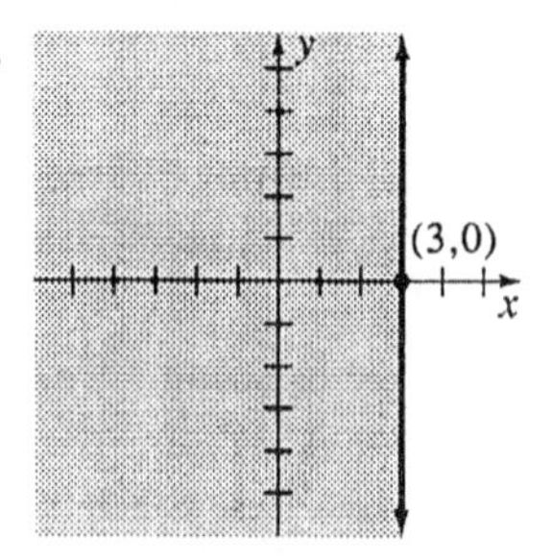


**27.**

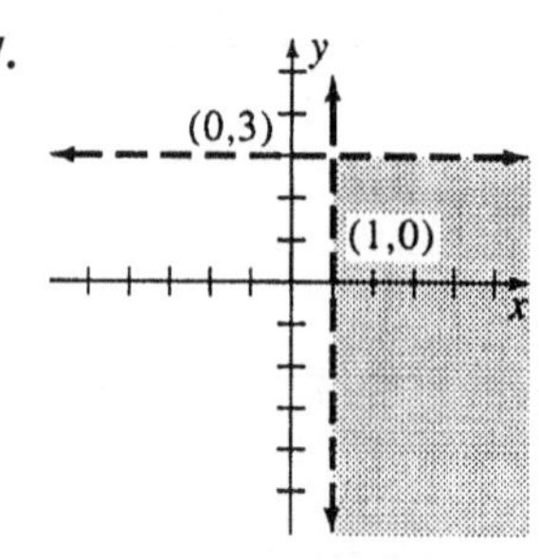


**29.**

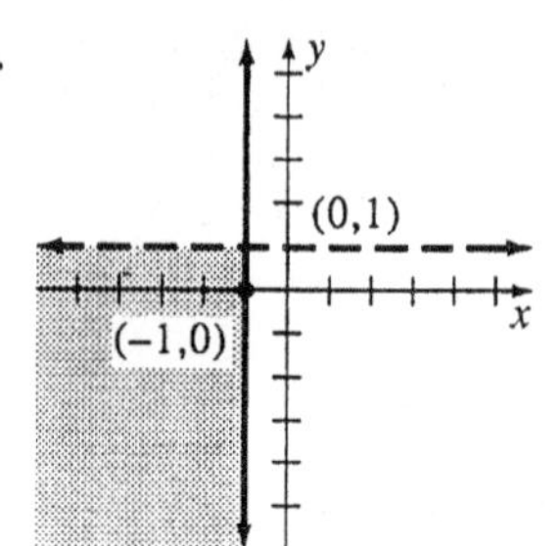


**33.**

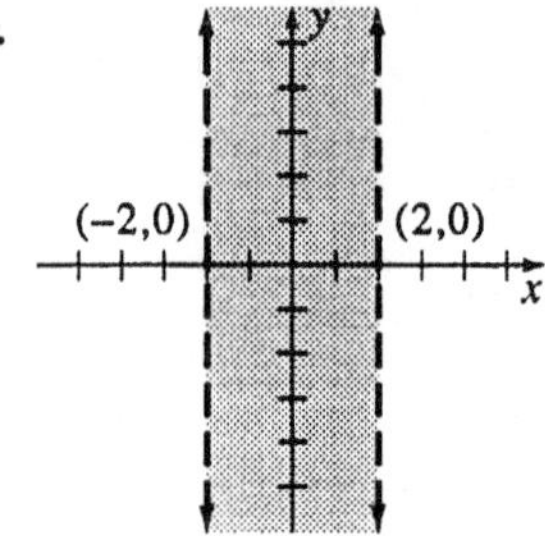

35. 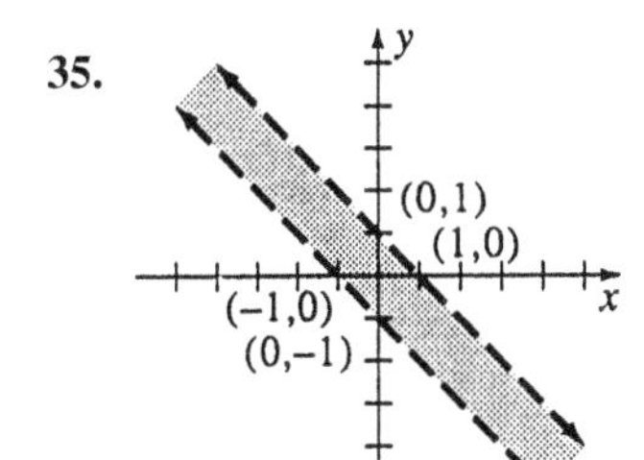


## Problem Set 7.3 (page 372)

**1.** 15 **3.** $\sqrt{13}$ **5.** $3\sqrt{2}$ **7.** $3\sqrt{5}$ **9.** 6 **11.** $3\sqrt{10}$
**13.** The lengths of the sides are 10, $5\sqrt{5}$, and 5. Because $10^2 + 5^2 = (5\sqrt{5})^2$, it is a right triangle.
**15.** The distances between (3, 6), and (7, 12), between (7, 12) and (11, 18), and between (11, 18) and (15, 24) are all $2\sqrt{13}$ units.

**17.** $\frac{4}{3}$ **19.** $-\frac{7}{3}$ **21.** $-2$ **23.** $\frac{3}{5}$ **25.** 0 **27.** $\frac{1}{2}$

**29.** 7 **31.** $-2$ **33–39.** Answers will vary.

**41.** $-\frac{2}{3}$ **43.** $\frac{1}{2}$ **45.** $\frac{4}{7}$ **47.** 0 **49.** $-5$

**51.** 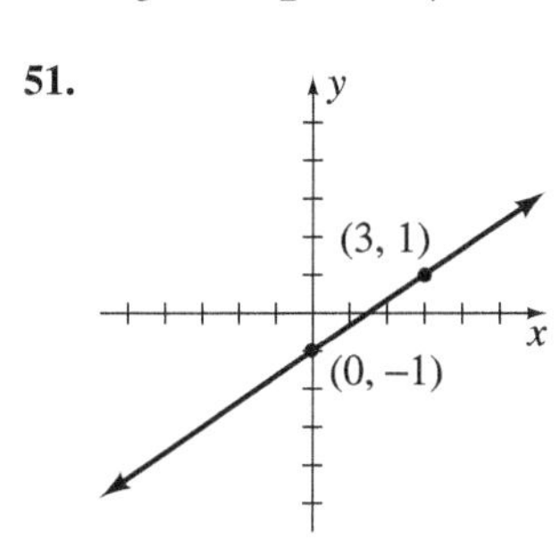


**53.** 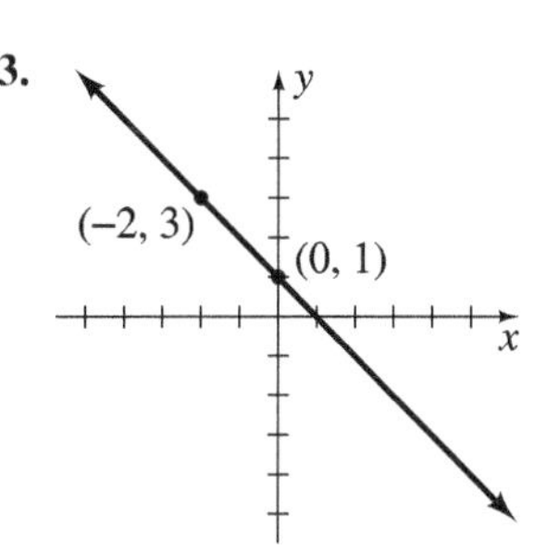


**55.** 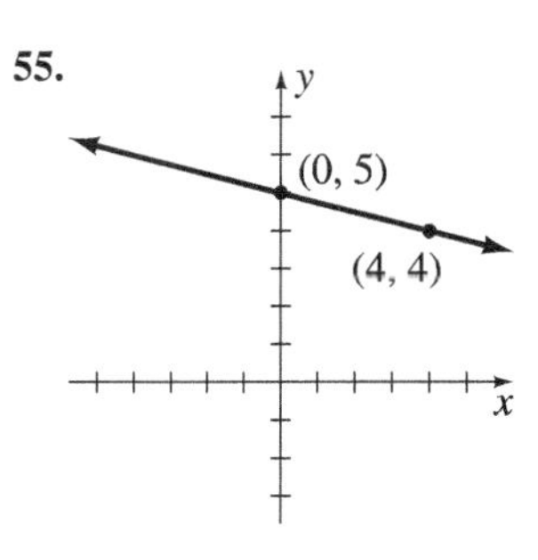


**57.** 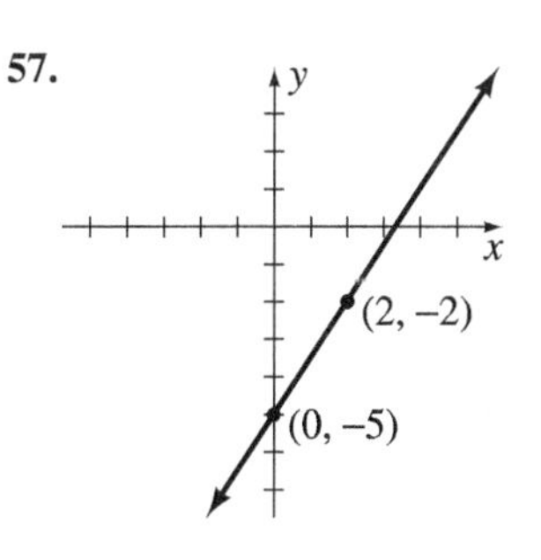


**59.** 105.6 feet **61.** 8.1% **63.** 19 centimeters

**69.** **(a)** (3, 5) **(c)** (2, 5) **(e)** $\left(\frac{17}{8}, -7\right)$

## Problem Set 7.4 (page 383)

**1.** $x - 2y = -7$ **3.** $3x - y = -10$ **5.** $3x + 4y = -15$
**7.** $5x - 4y = 28$ **9.** $5x - 2y = -23$ **11.** $2x + y = 18$
**13.** $x + 3y = 5$ **15.** $x - y = 1$ **17.** $5x - 2y = -4$
**19.** $x + 7y = 11$ **21.** $x + 2y = -9$ **23.** $7x - 5y = 0$
**25.** $y = \frac{3}{7}x + 4$ **27.** $y = 2x - 3$ **29.** $y = -\frac{2}{5}x + 1$
**31.** $y = 0(x) - 4$ **33.** $2x - y = 4$ **35.** $5x + 8y = -15$
**37.** $x + 0(y) = 2$ **39.** $0(x) + y = 6$ **41.** $x + 5y = 16$
**43.** $4x - 7y = 0$ **45.** $x + 2y = 5$ **47.** $3x + 2y = 0$

**49.** $m = -3$ and $b = 7$ **51.** $m = -\frac{3}{2}$ and $b = \frac{9}{2}$

**53.** $m = \frac{1}{5}$ and $b = -\frac{12}{5}$

**55.** 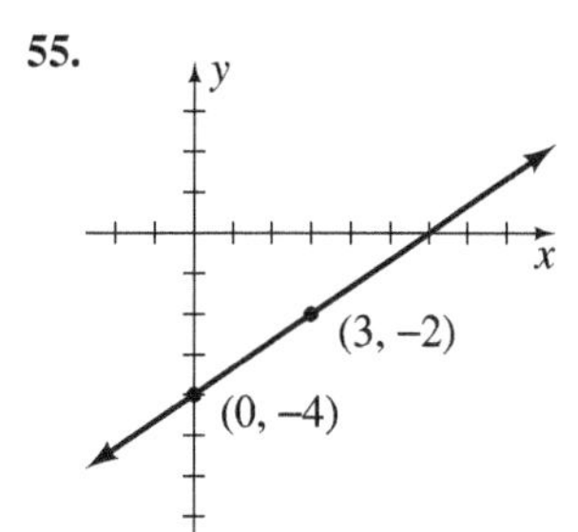


**57.** 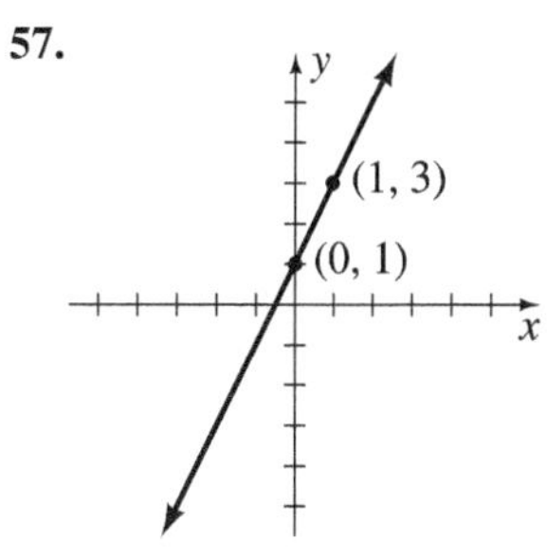


**59.** 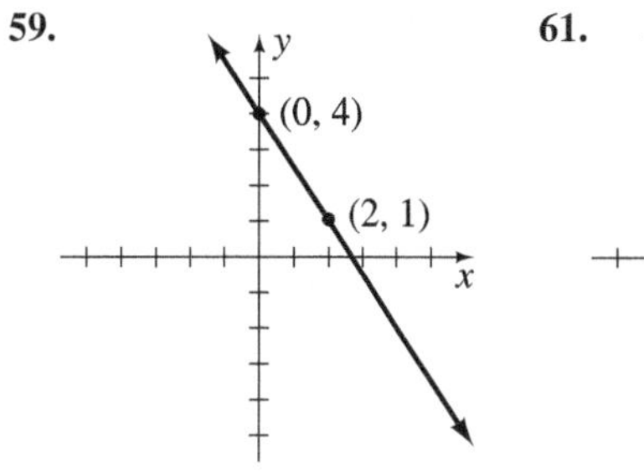


**61.** 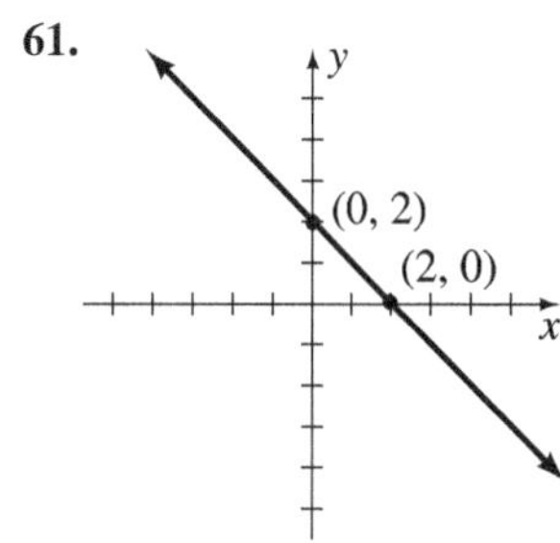


**63.** 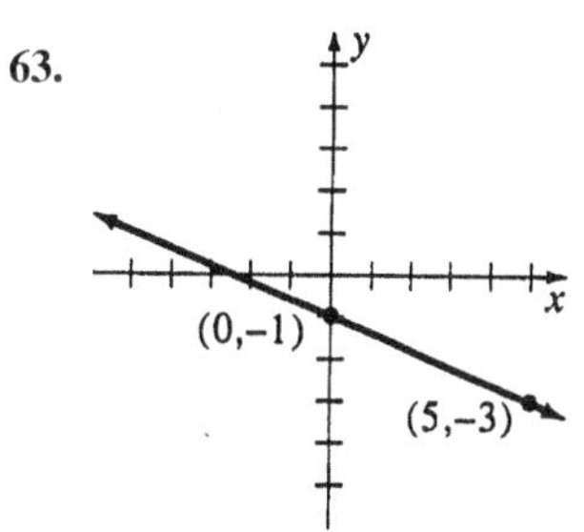


**65.** 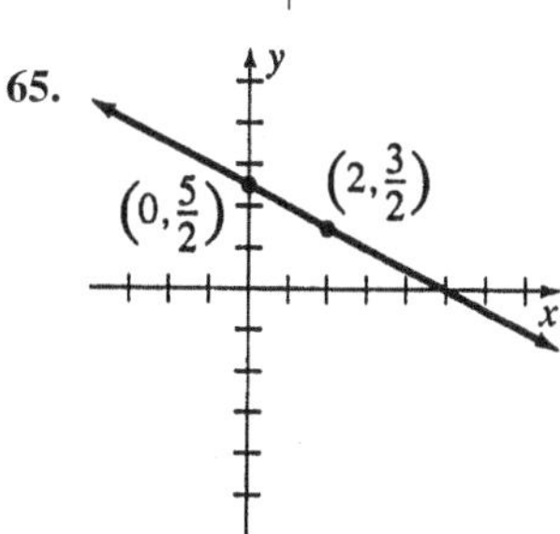


**67.** 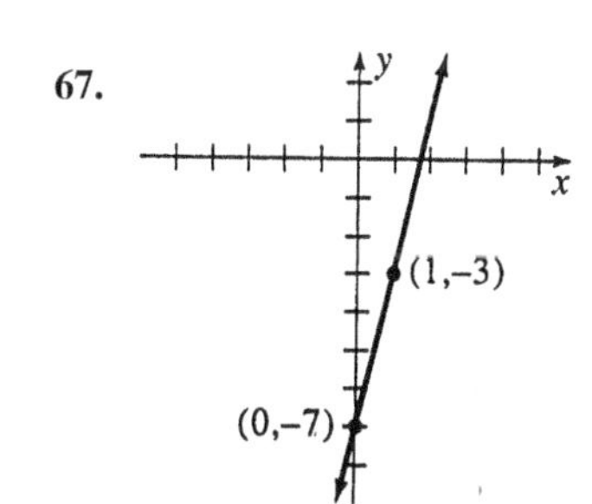


**69.** 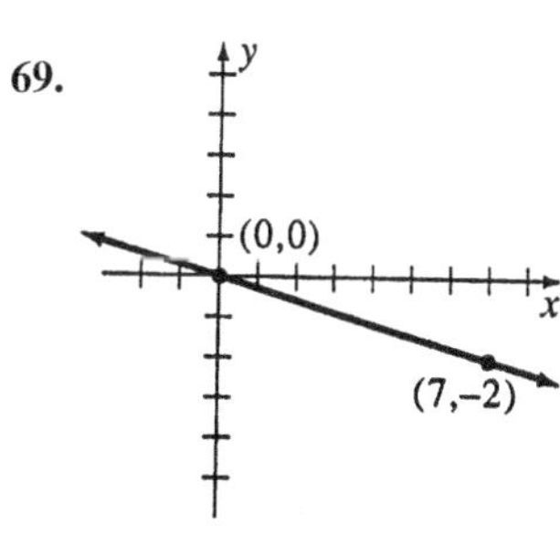


**71.** 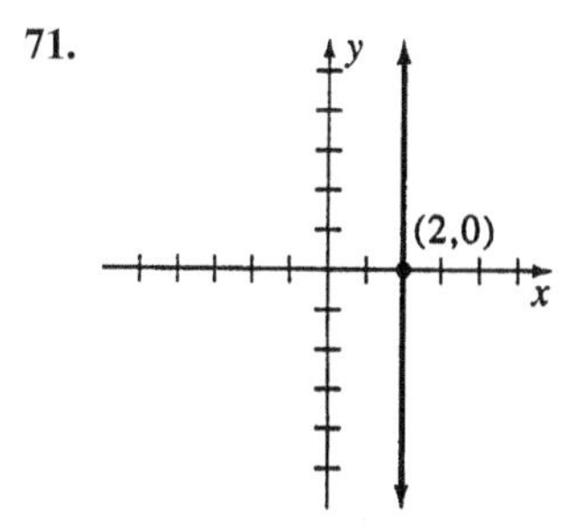


**73.** $y = 30x - 60$

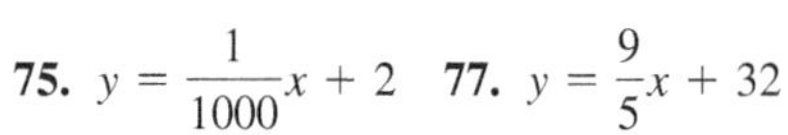
**75.** $y = \frac{1}{1000}x + 2$ **77.** $y = \frac{9}{5}x + 32$

**85.** **(a)** $2x - y = 1$ **(b)** $5x - 6y = 29$ **(c)** $x + y = 2$ **(d)** $3x - 2y = 18$

## Problem Set 7.5 (page 391)

**1.** $(-3, -1)$; $(3, 1)$; $(3, -1)$ **3.** $(7, 2)$; $(-7, -2)$; $(-7, 2)$
**5.** $(5, 0)$; $(-5, 0)$; $(-5, 0)$ **7.** $x$ axis **9.** $y$ axis
**11.** $x$ axis, $y$ axis, and origin **13.** $x$ axis **15.** None
**17.** Origin **19.** $y$ axis **21.** All three **23.** $x$ axis **25.** $y$ axis

**27.**

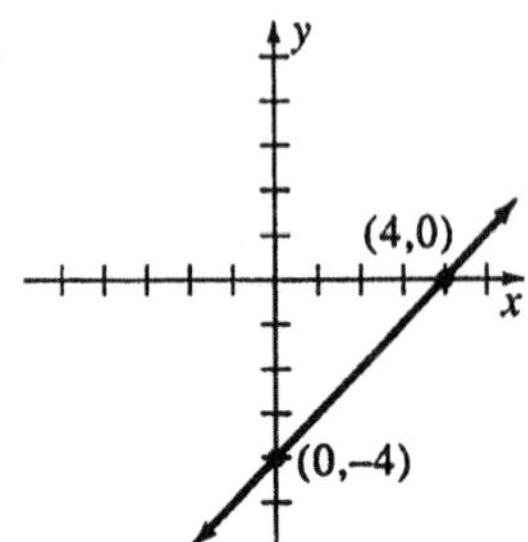


**29.**

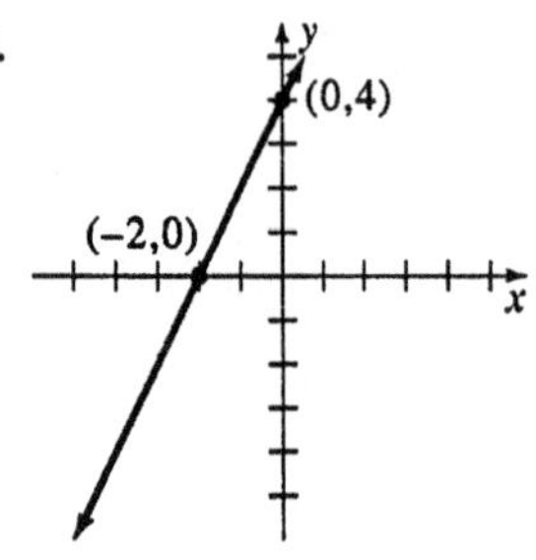


**31.**

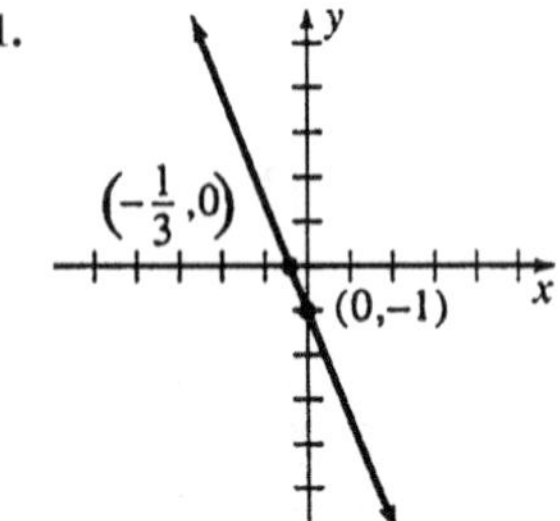


**33.**

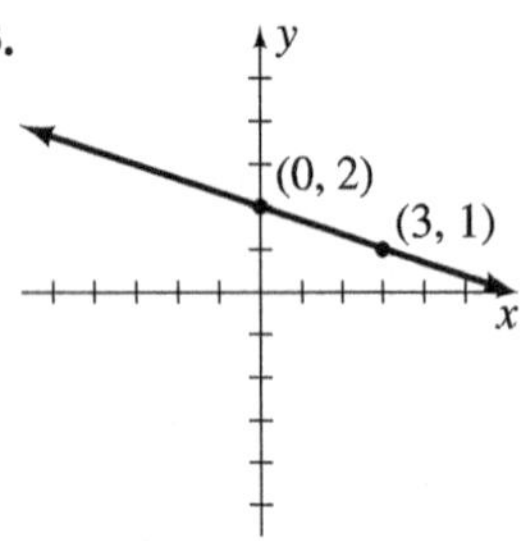


**35.**

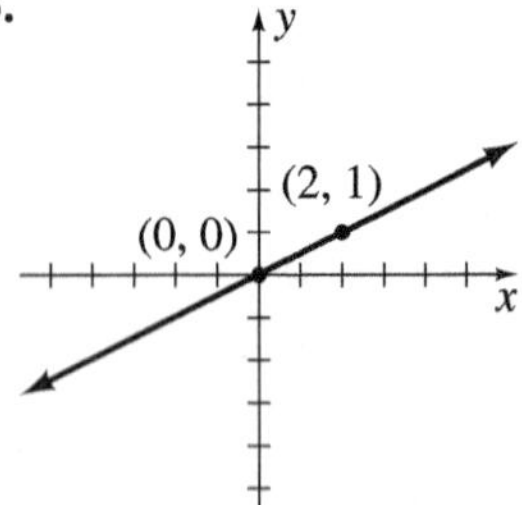


**37.**

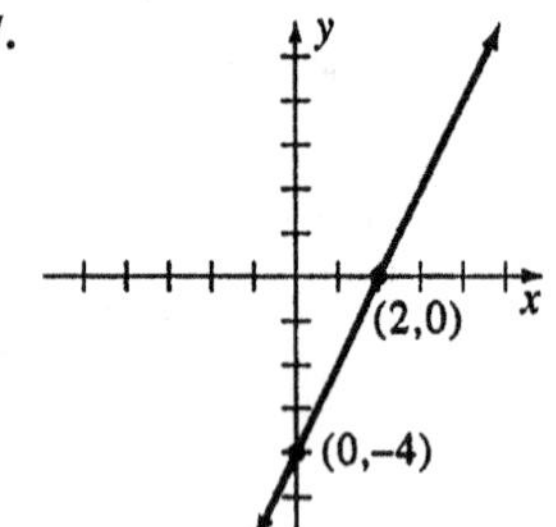


**39.**

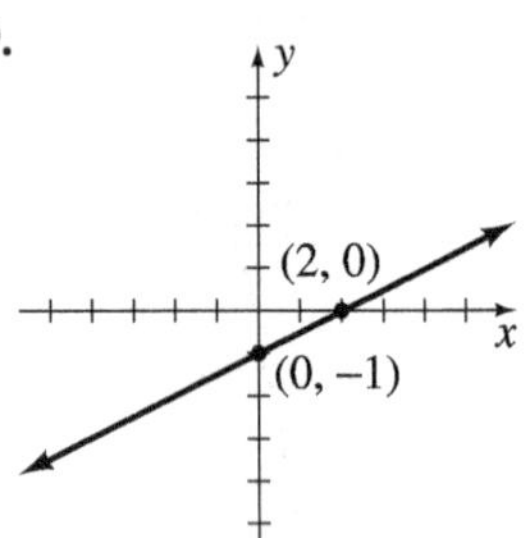


**41.**

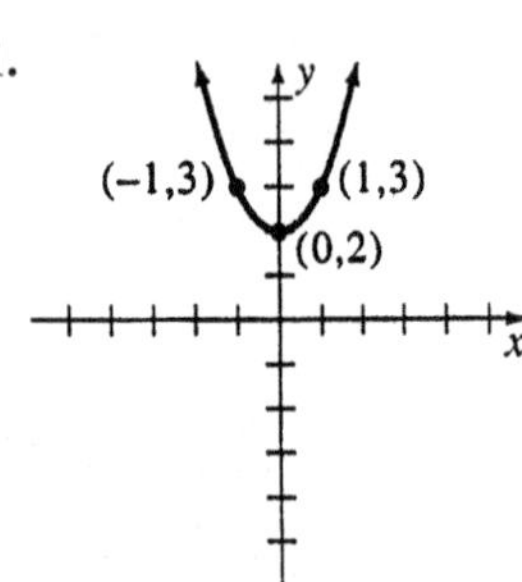


**43.**

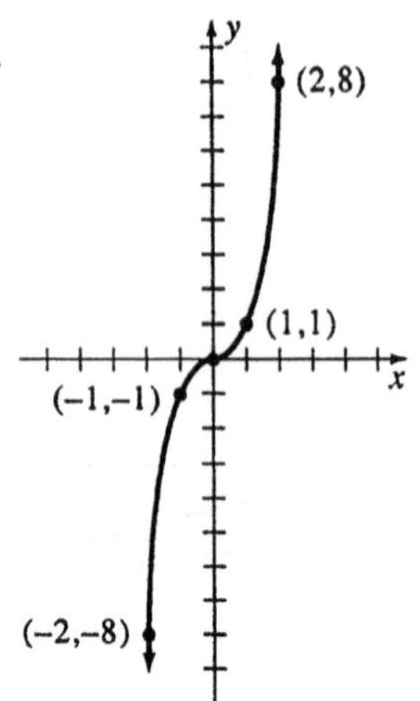


**45.**

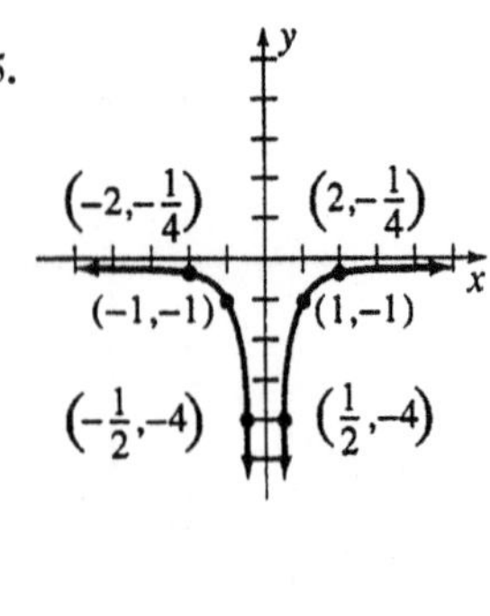


**47.**

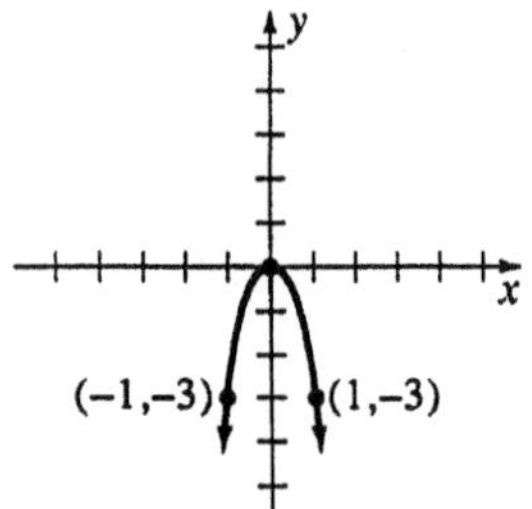


**49.**

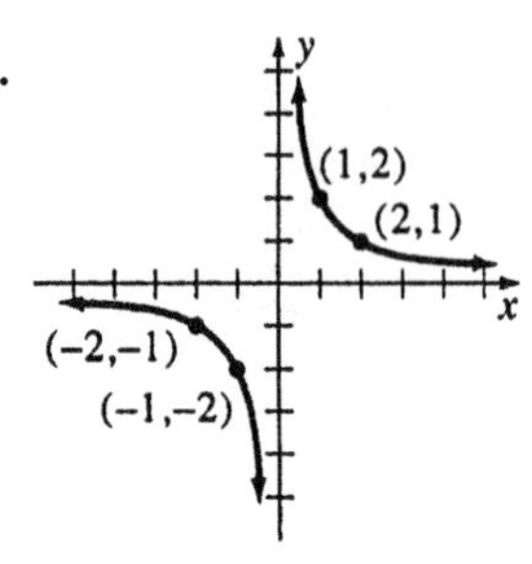


**51.**

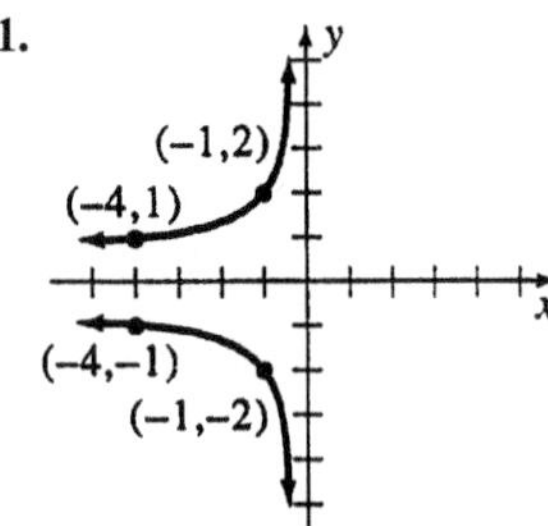


**53.**

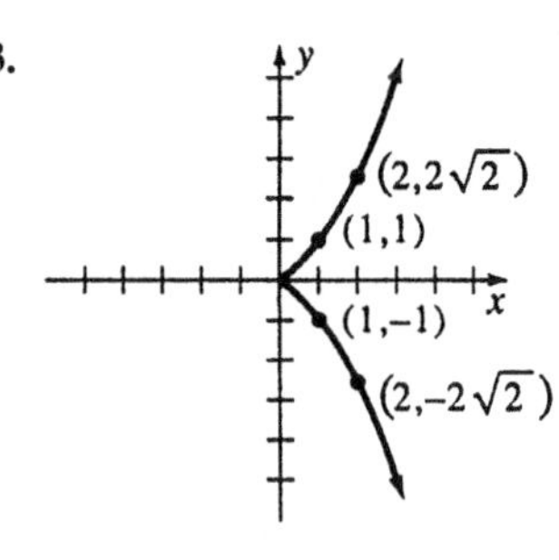


**55.**

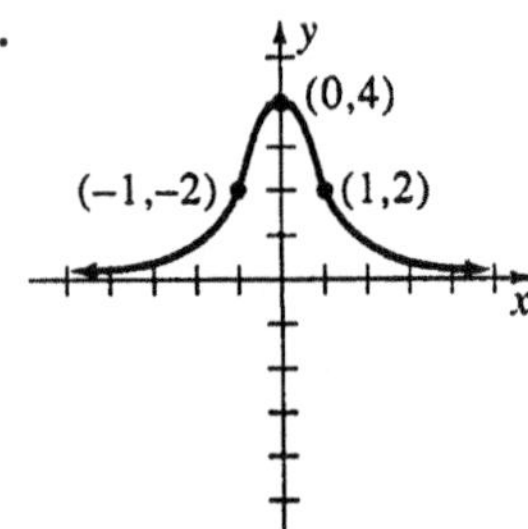


**57.**

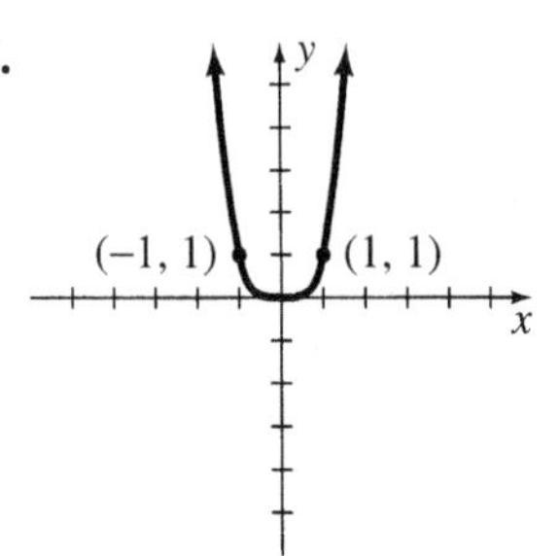


**59.**

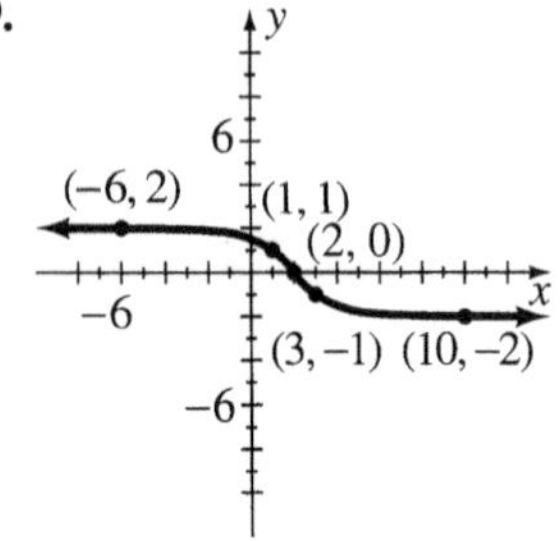


## Chapter 7 Sample Problems (page 393)

**1.** $(2, -2)$

**2.**

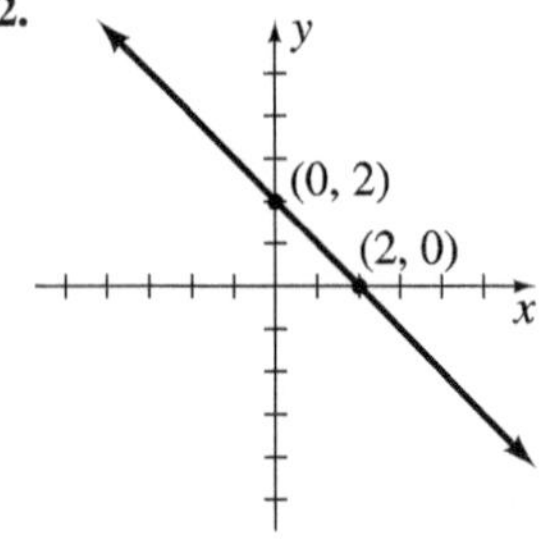


**3.**

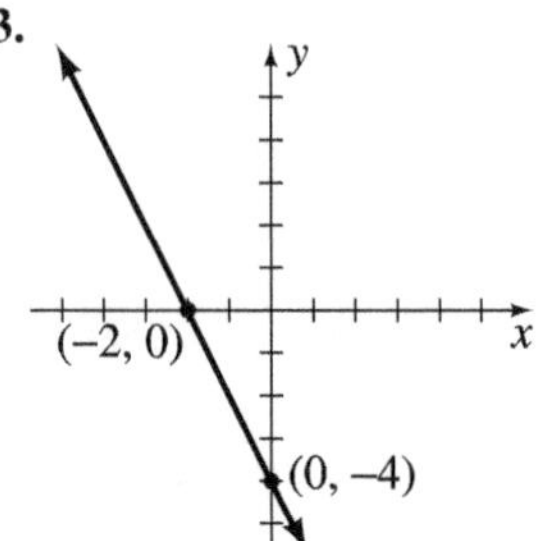

**4.**

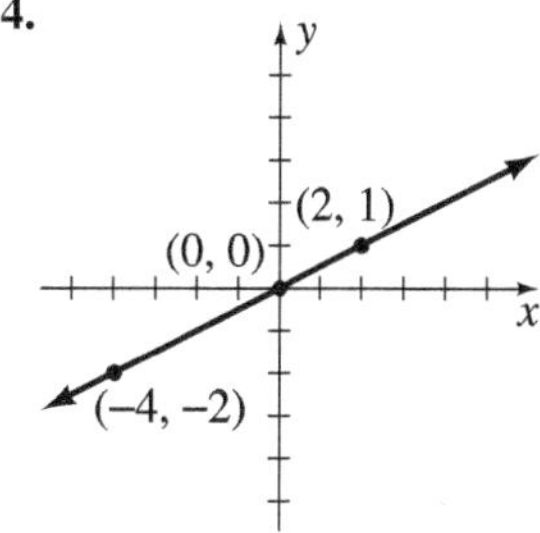


**5.**

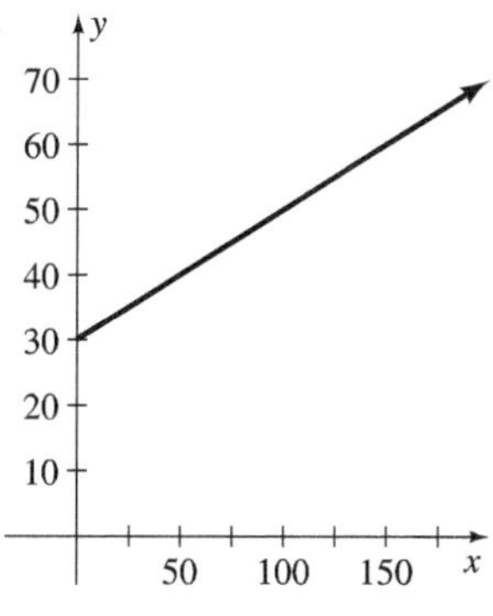

**6.**

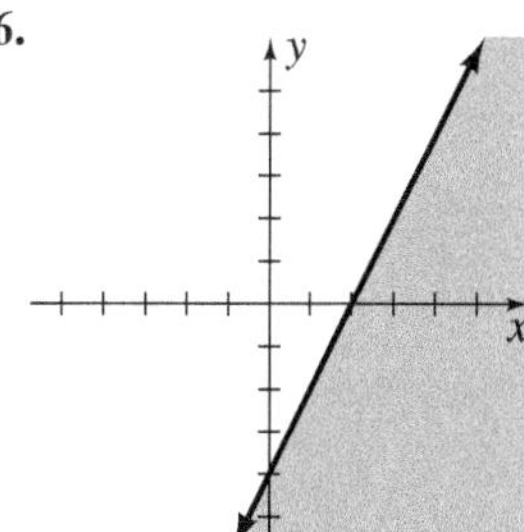


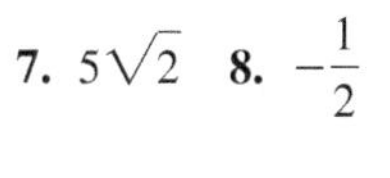

**7.** $5\sqrt{2}$ **8.** $-\frac{1}{2}$

**9.**

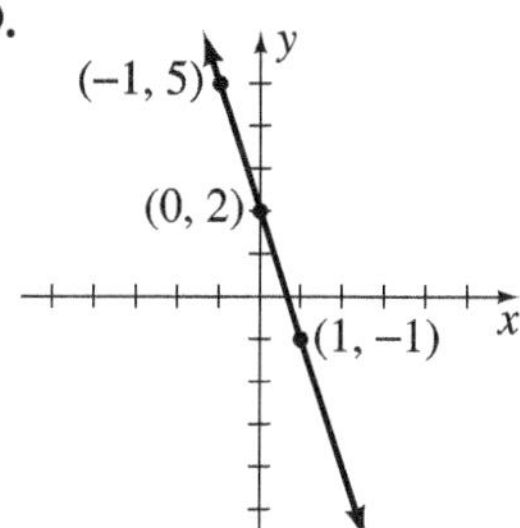


**10.** 30 feet

**11.** $m = \frac{3}{2}$; $y$ intercept is $-7$. **12.** $x + 2y = 1$

**13.** $4x - 5y = 13$ **14.** $2x + y = 1$ **15.** $y$-axis symmetry

**16.**

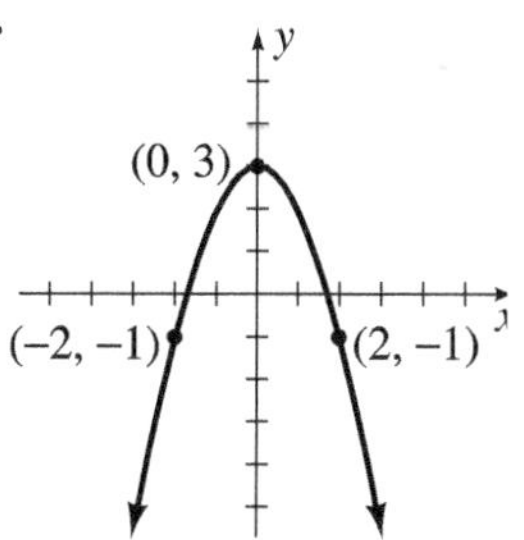


## Chapter 7 Review Problem Set (page 399)

**1.** (1, 2), (−1, 10) **2.** (0, 2)
**3.** (2, 3), (−2, 9) **4.** (−3, 0)

**5.**

| x | −1 | 0 | 1 | 4 |
|---|---|---|---|---|
| y | −7 | −5 | −3 | 3 |

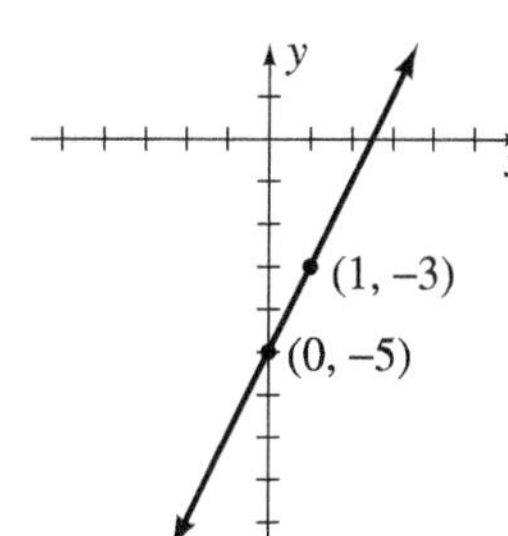


**6.**

| x | −3 | −1 | 0 | 2 |
|---|---|---|---|---|
| y | 5 | 1 | −1 | −5 |

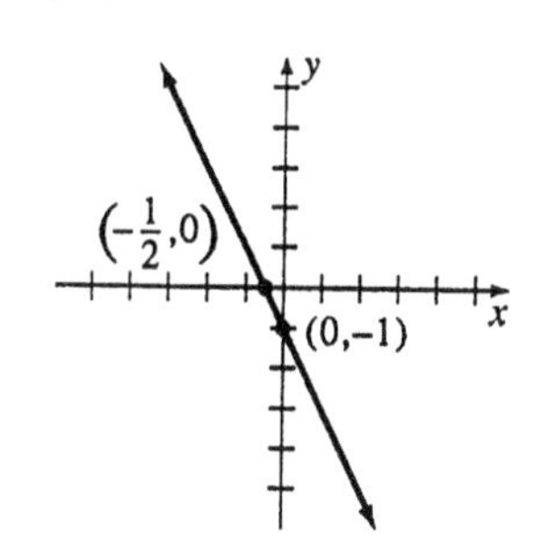


**7.**

| x | −2 | 0 | 2 | 4 |
|---|---|---|---|---|
| y | −5 | −2 | 1 | 4 |

**8.**

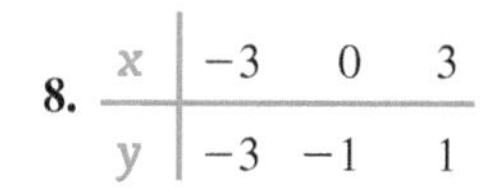

| x | −3 | 0 | 3 |
|---|---|---|---|
| y | −3 | −1 | 1 |

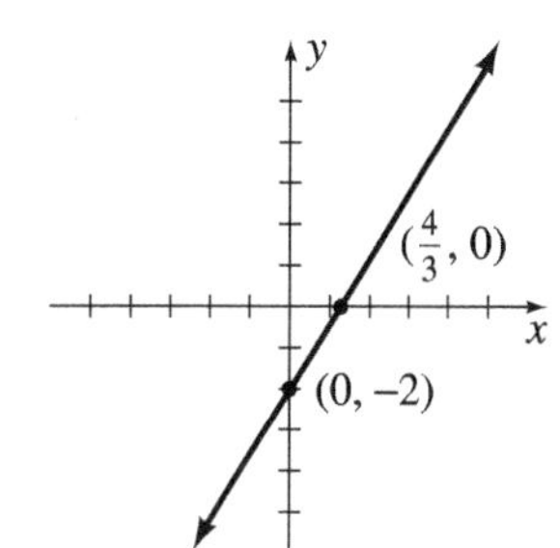


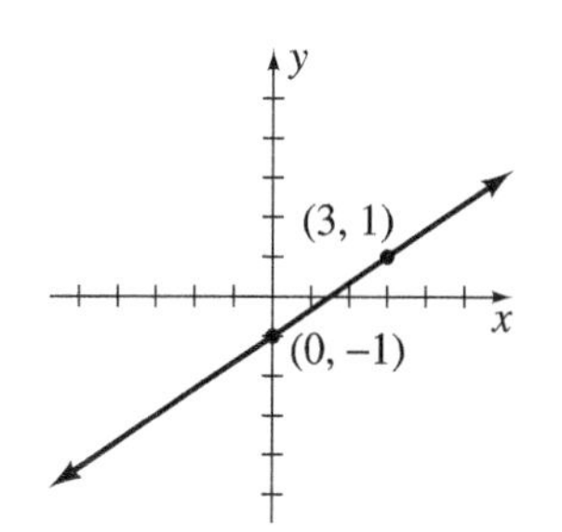


**9.**

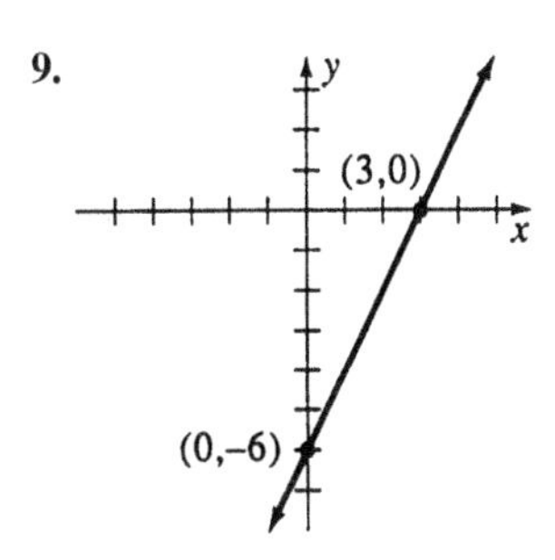


**10.**

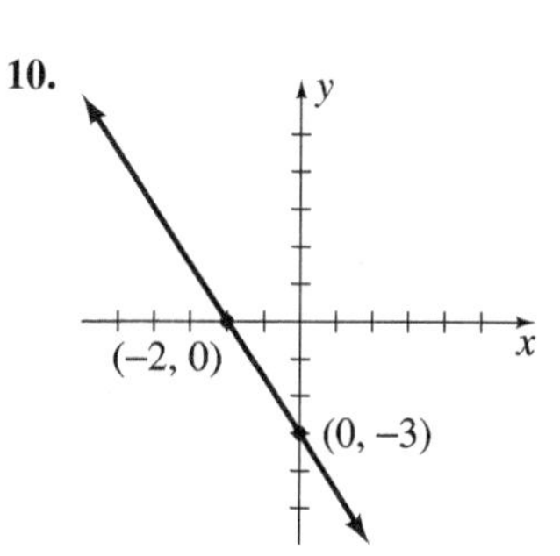


**11.**

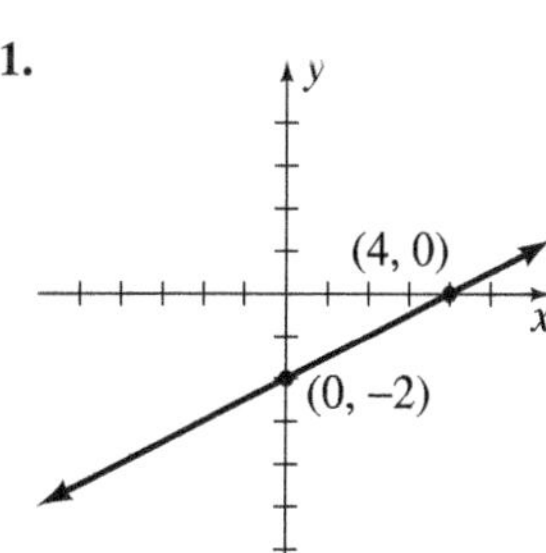


**12.**

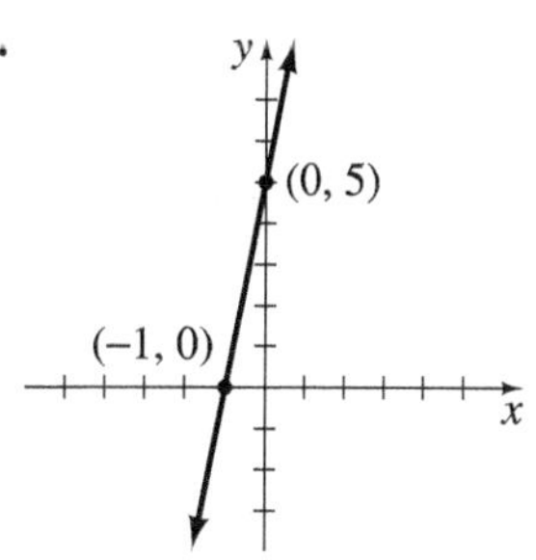


**13.** $3x + 2y = 6$

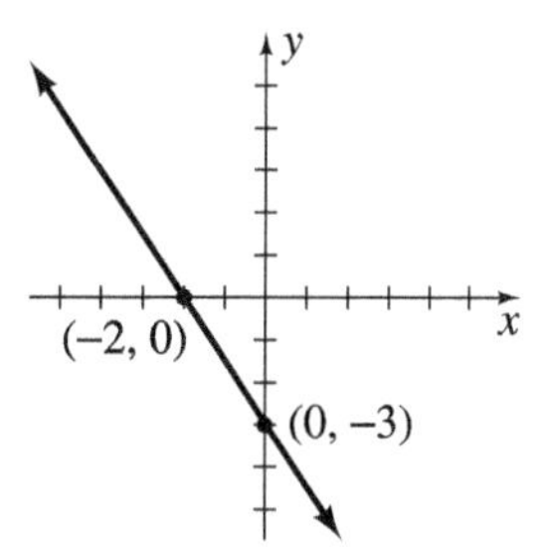


**14.** $x + 4y = -4$

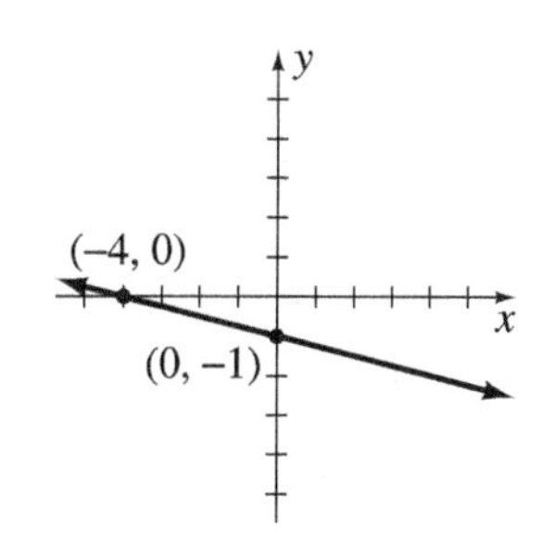

**15.** $y = -\frac{1}{2}x + 3$

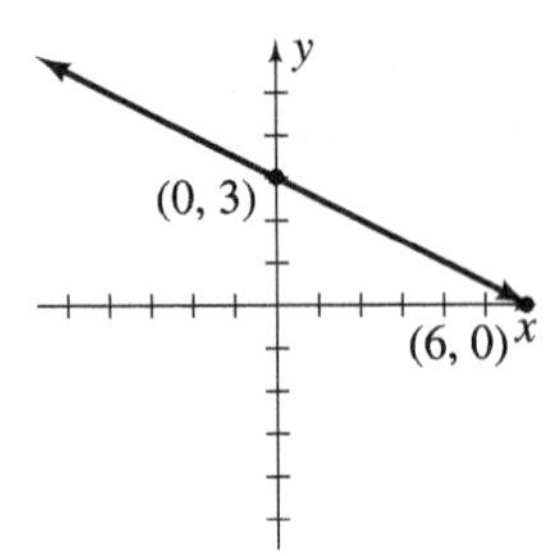


**16.** $y = \frac{1}{4}x - 1$

**17.** $y = 3x - 1$

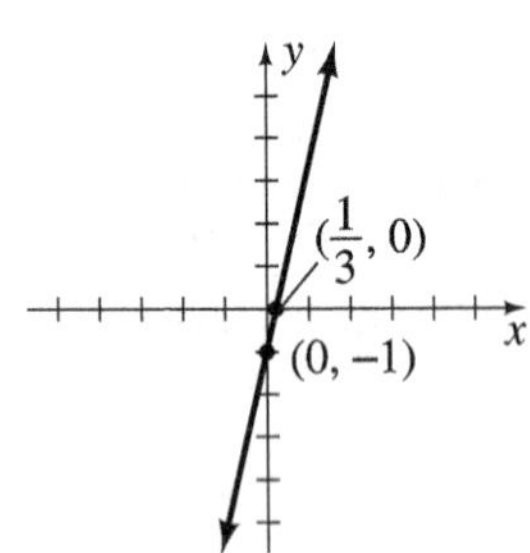


**18.** $y = -x + 4$

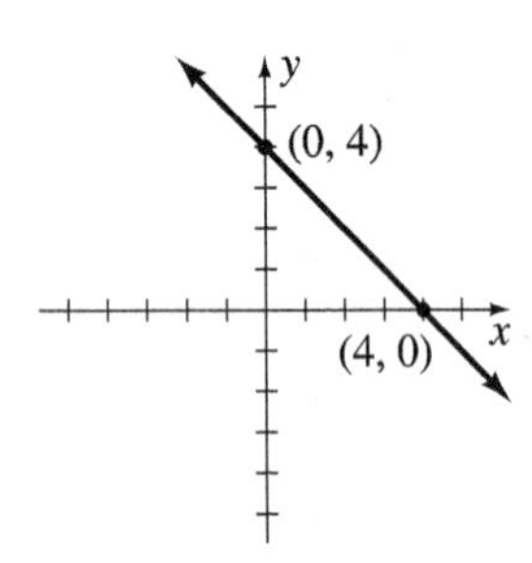


**19.** $x = -3$

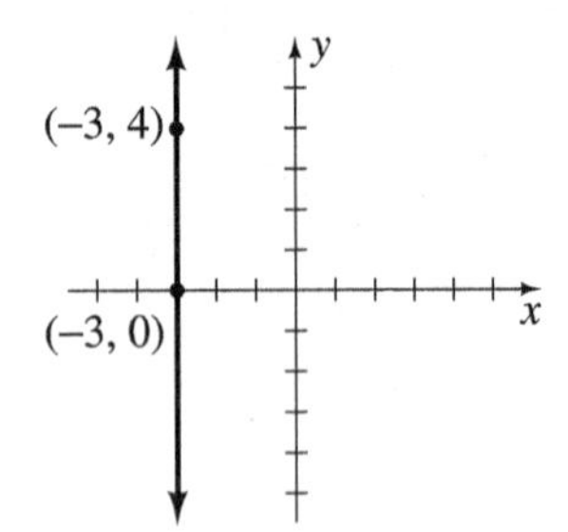


**20.** $y = 5$

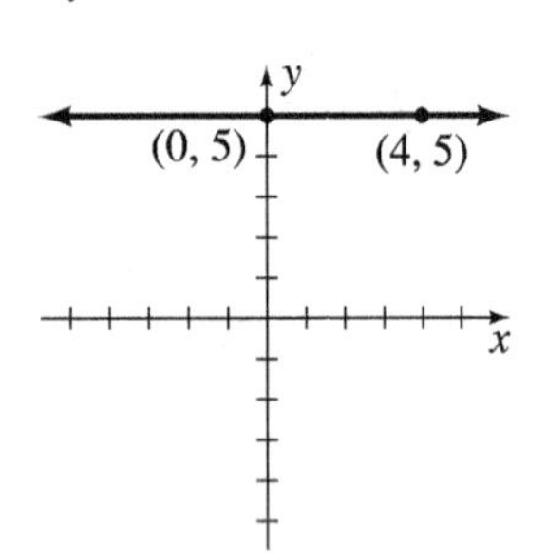


**21.**

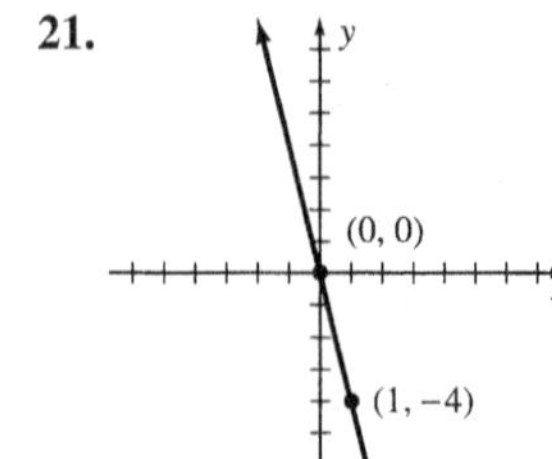


**22.**

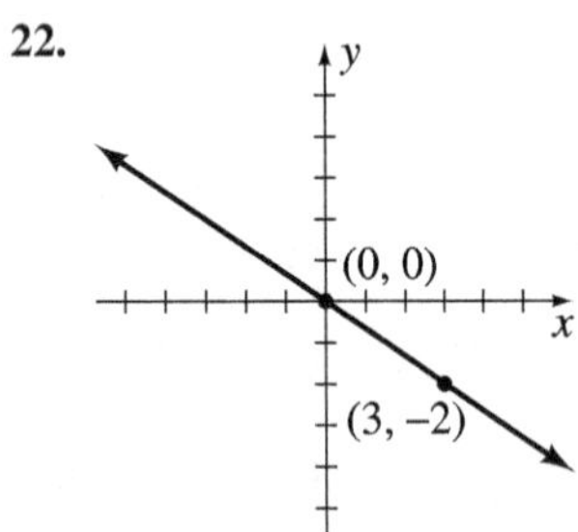


**23.**

**24.**

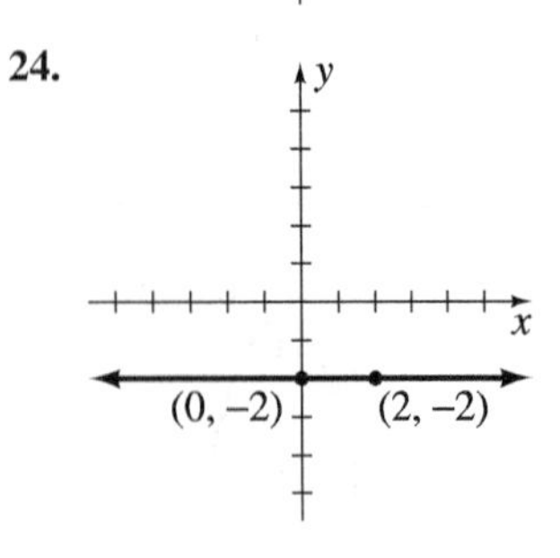


**25.** $y = -\frac{1}{3}x$

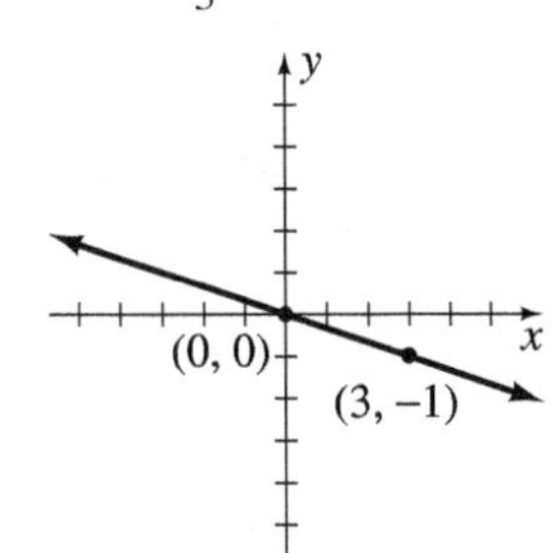


**26.** $y = \frac{2}{3}x$

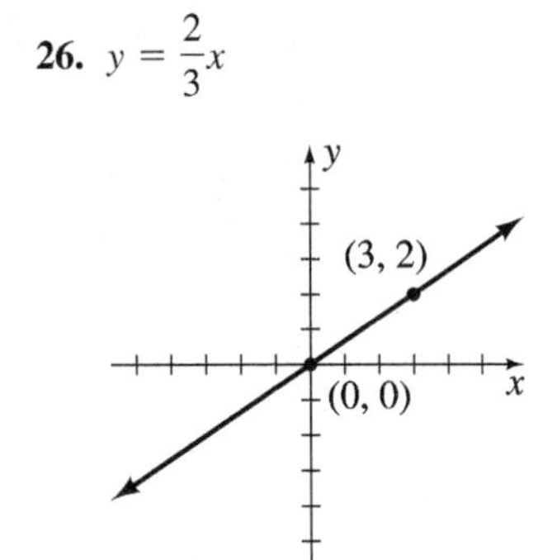


**27.**

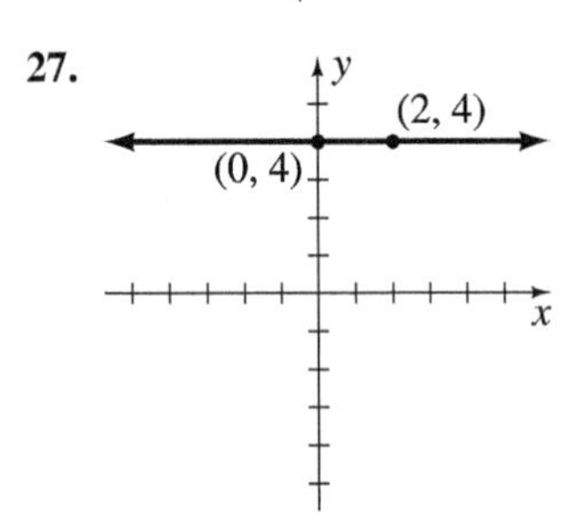


**28.**

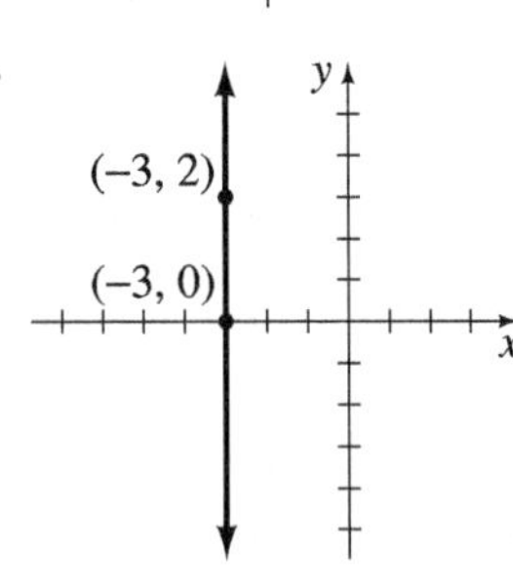


**29. (a)**

| h | 1 | 2 | 3 | 4 |
|---|---|---|---|---|
| C | 225 | 300 | 375 | 450 |

**30. (a)**

| v | 100 | 200 | 350 | 400 |
|---|---|---|---|---|
| t | 15 | 30 | 52.50 | 60 |

**29. (b)**

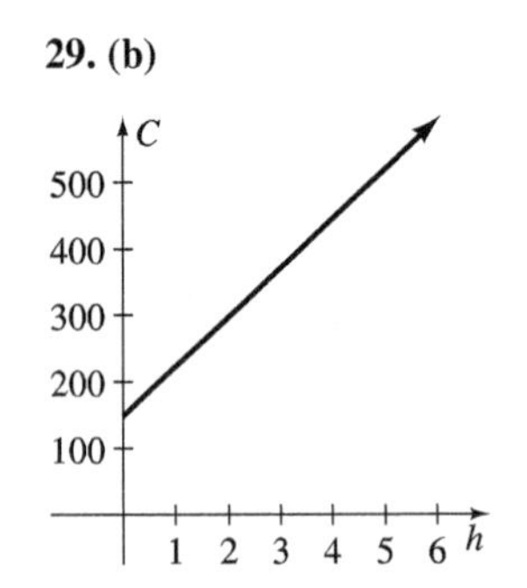


**30. (b)**

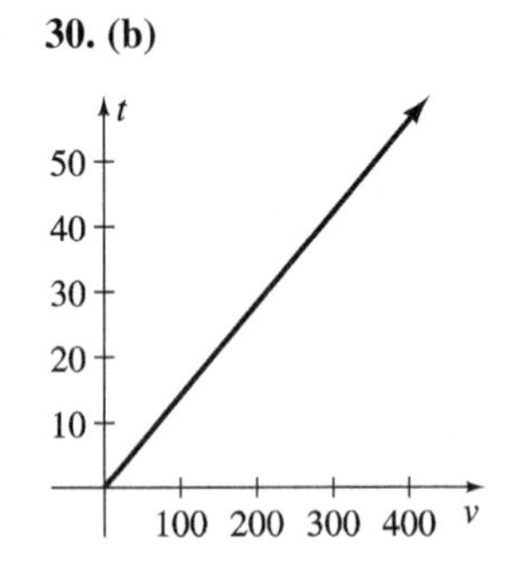


**29. (d)** \$262.50, \$412.50

**30. (d)** \$37.50, 45

**31.**

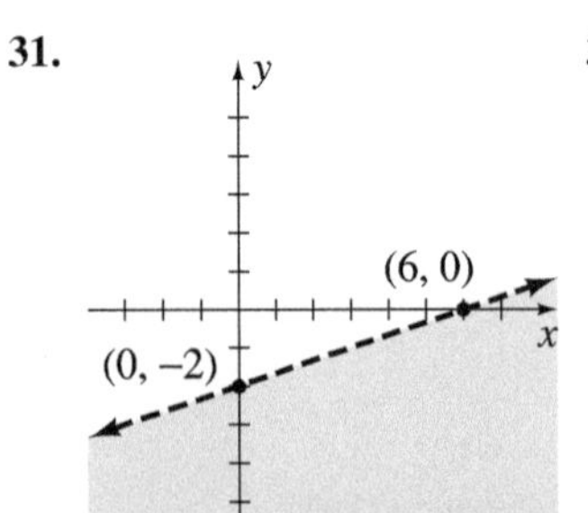


**32.**

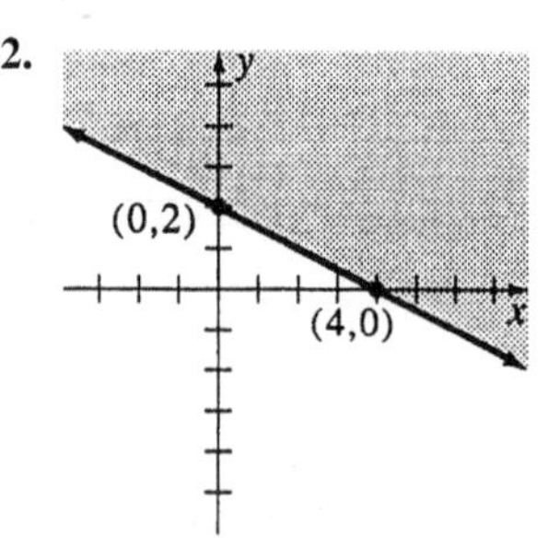


**33.**

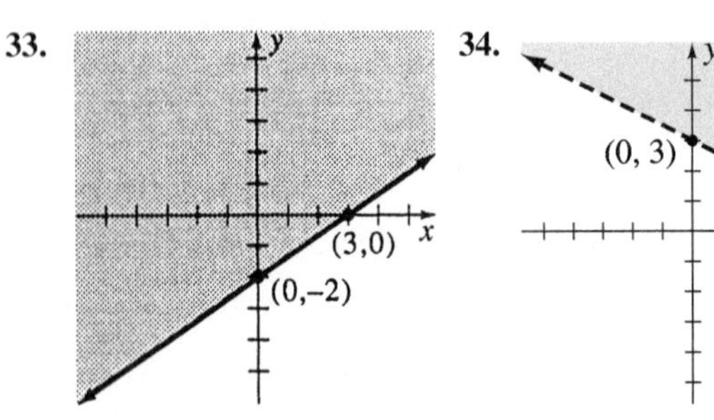


**34.**

**35.**

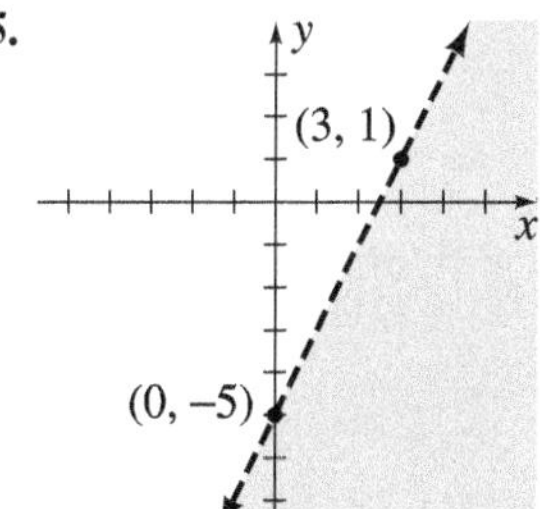


**36.**

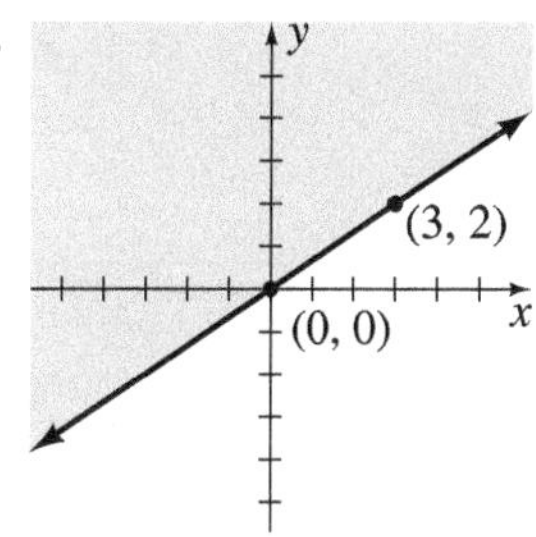


**37.**

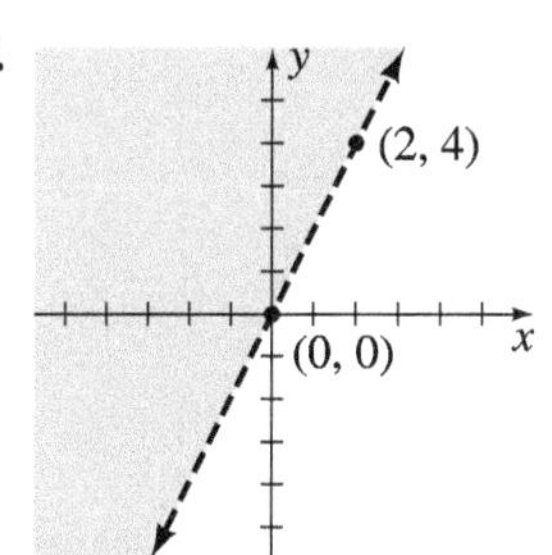


**38.**

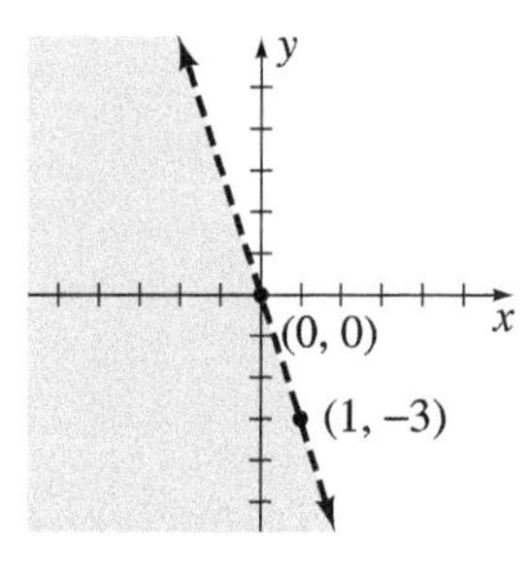


**39.** **(a)** $\sqrt{53}$ **(b)** $\sqrt{58}$ **40.** 5, 10, and $\sqrt{97}$

**41.** The distances between $(-3, -1)$ and $(1, 2)$ and between $(1, 2)$ and $(5, 5)$ are 5.

**42.** **(a)** $\frac{6}{5}$ **(b)** $-\frac{2}{3}$ **43.** 5 **44.** $-1$

**45.**

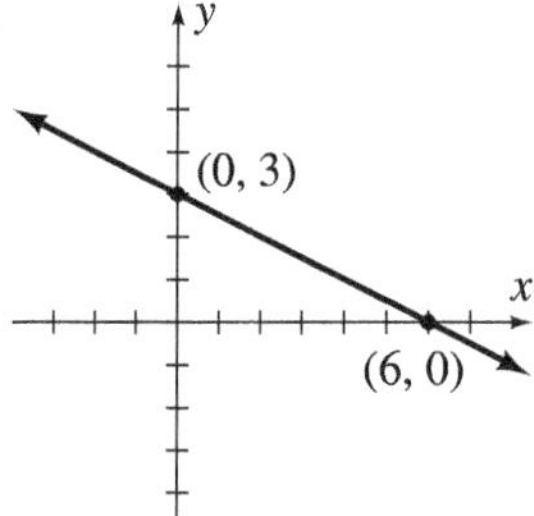


**46.**

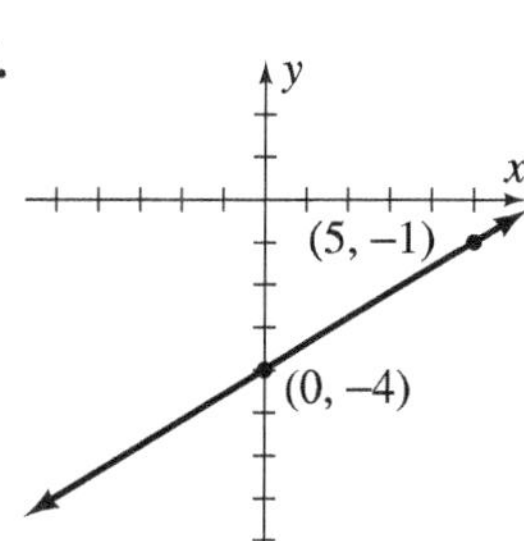


**47.**

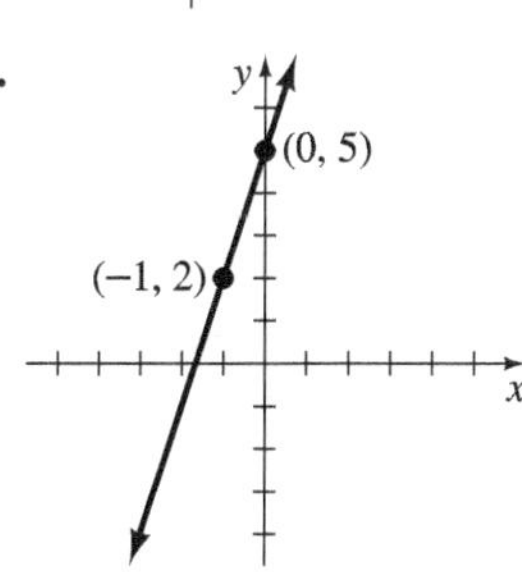


**48.**

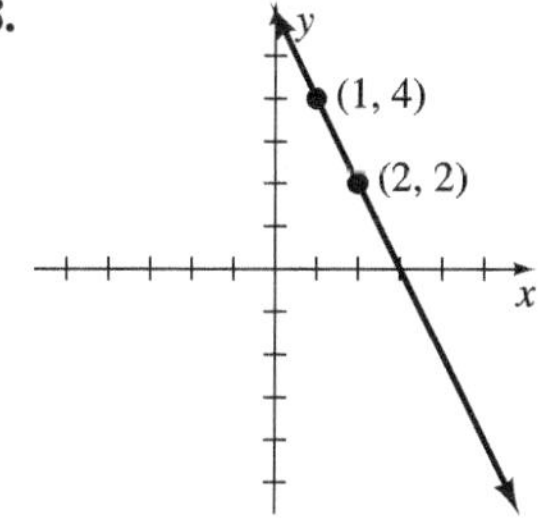


**49.** 316.8 ft **50.** 8 in. **51.** **(a)** $m = -4$ **(b)** $m = \frac{2}{7}$

**52.** $m = -\frac{5}{3}$ **53.** $m = -\frac{4}{5}$ **54.** $3x + 7y = 28$

**55.** $2x - 3y = 16$ **56.** $x + y = -2$ **57.** $7x + 4y = 1$

**58.** $x - y = -4$ **59.** $x - 2y = -8$ **60.** $2x - 3y = 14$

**61.** $4x + y = -29$ **62.** $y = \frac{3}{200}x - 600$ **63.** $y = \frac{1}{5}x - 20$

**64.** $y = 8x$ **65.** $y = 300x - 150$

**66.** **(a)** $y$ axis **(b)** origin **(c)** origin **(d)** $x$ axis

**67.**

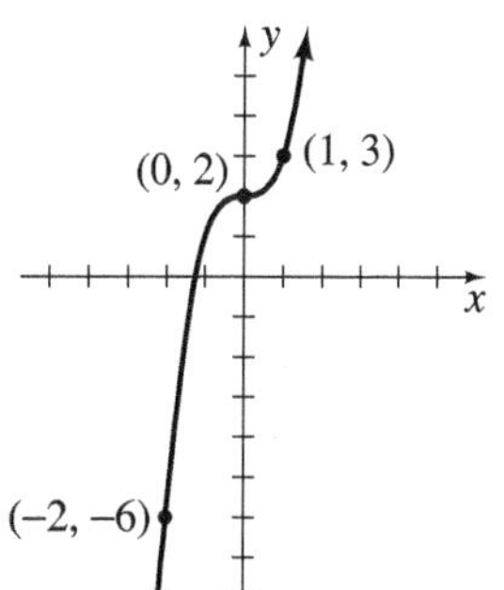


**68.**

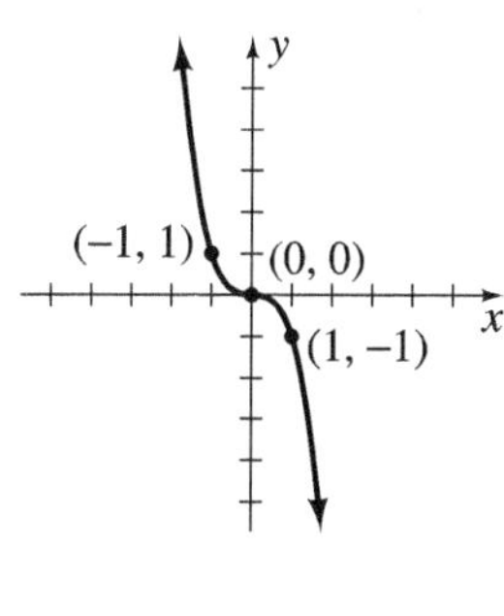


**69.**

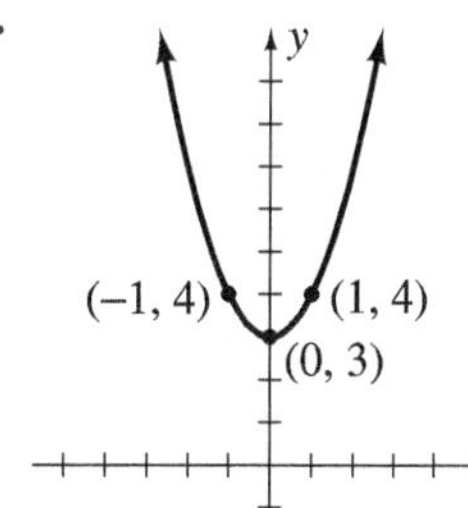


**70.**

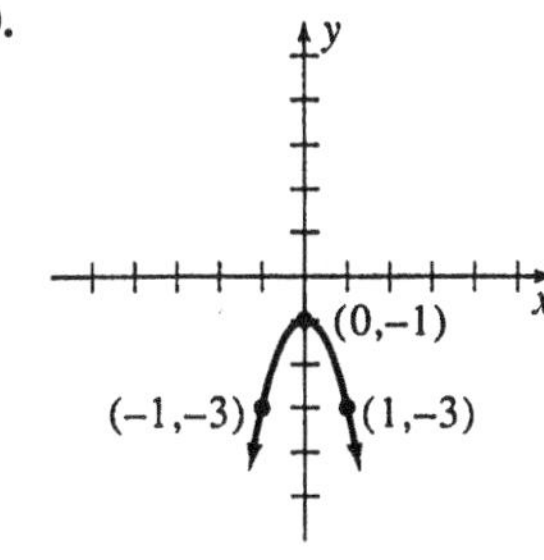


**71.** $y = -x^2 + 6$

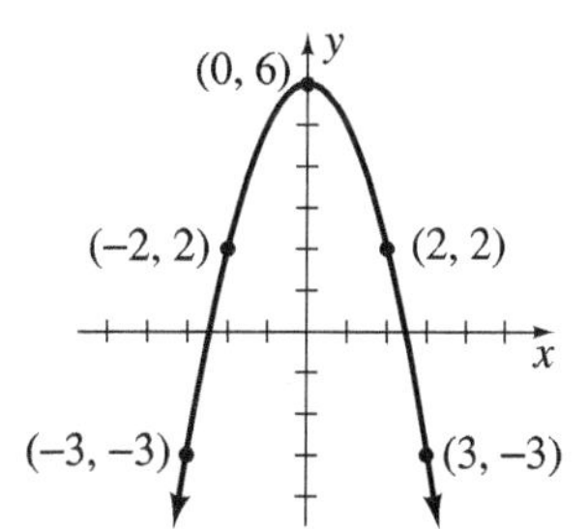


**72.** $y = x^2 - 4$

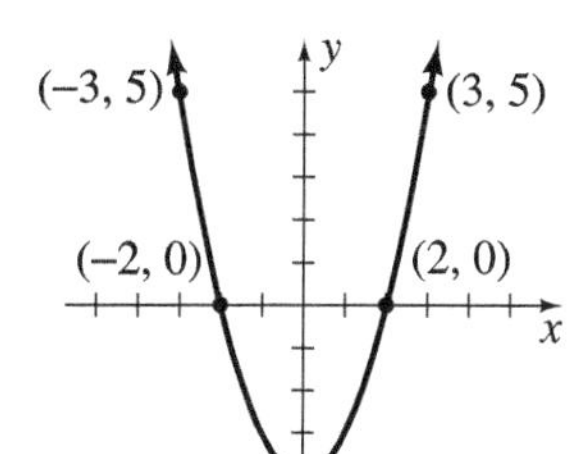


## Chapter 7 Test (page 402)

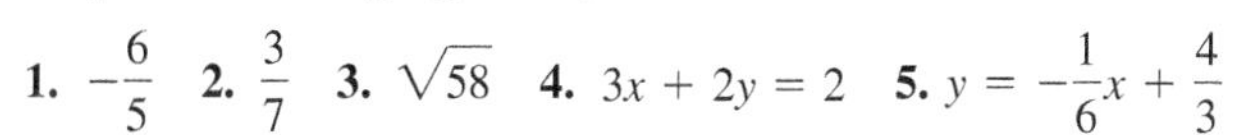

**1.** $-\frac{6}{5}$ **2.** $\frac{3}{7}$ **3.** $\sqrt{58}$ **4.** $3x + 2y = 2$ **5.** $y = -\frac{1}{6}x + \frac{4}{3}$

**6.** $5x + 2y = -18$ **7.** $6x + y = 31$ **8.** Origin symmetry

**9.** $x$-axis, $y$-axis, and origin symmetry

**10.** $x$-axis symmetry **11.** $\frac{7}{2}$ **12.** $\frac{9}{4}$ **13.** $\frac{10}{9}$

**14.** $-\frac{5}{8}$ **15.** 480 feet **16.** 6.7% **17.** 43 centimeters

**18.**

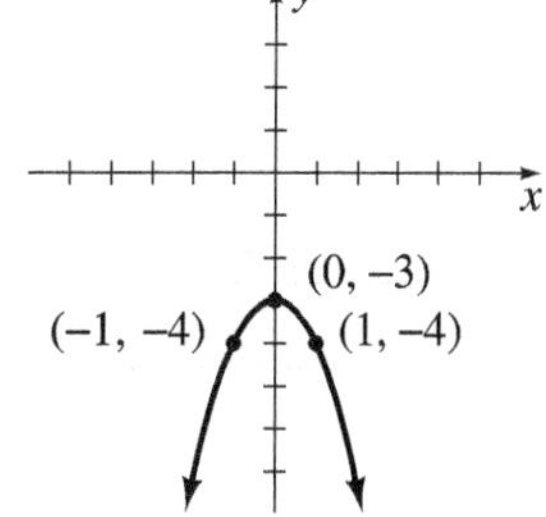


**19.**

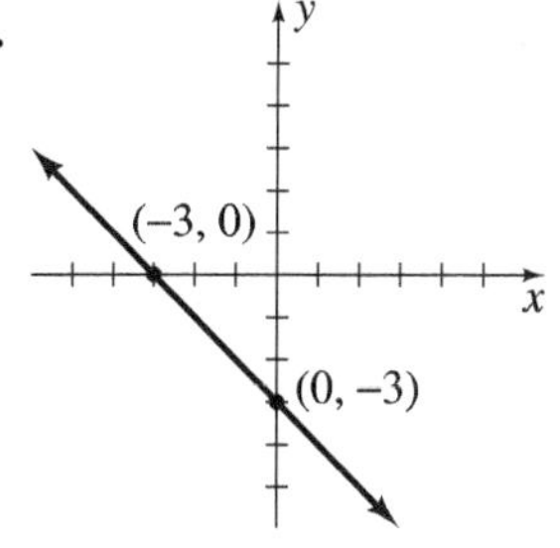

20. 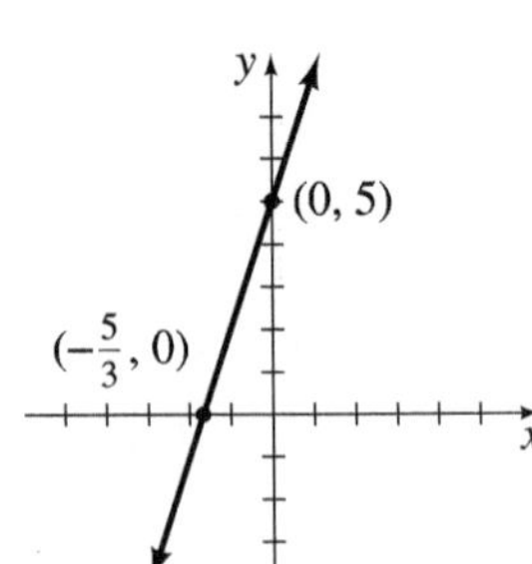


21. 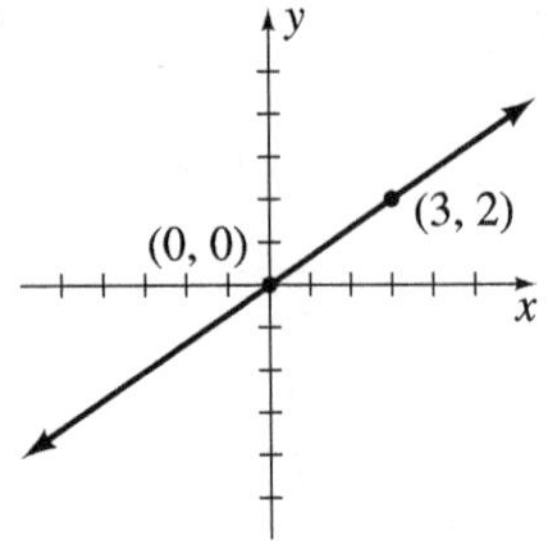


22. 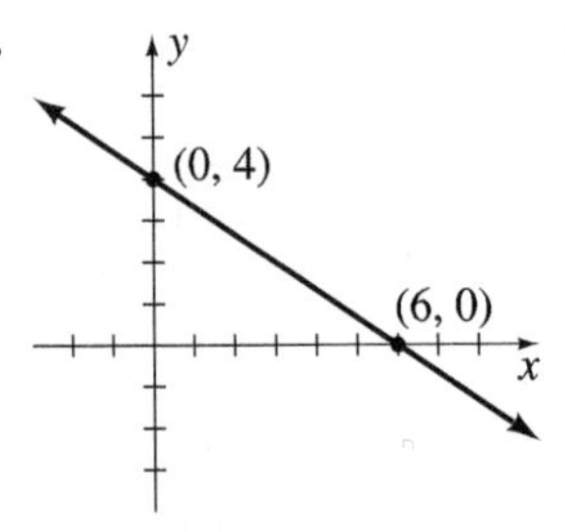


23. 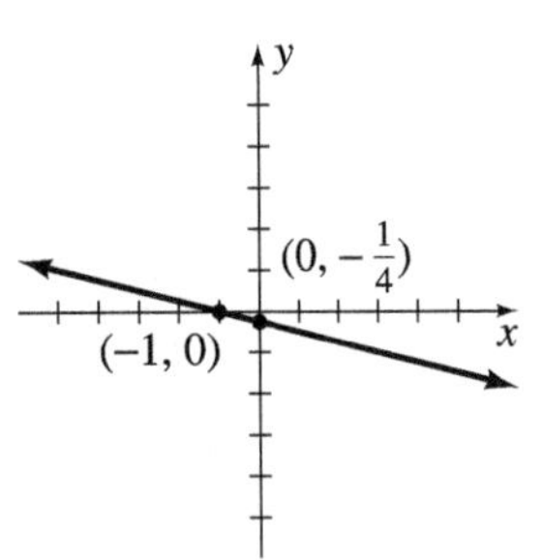


24. 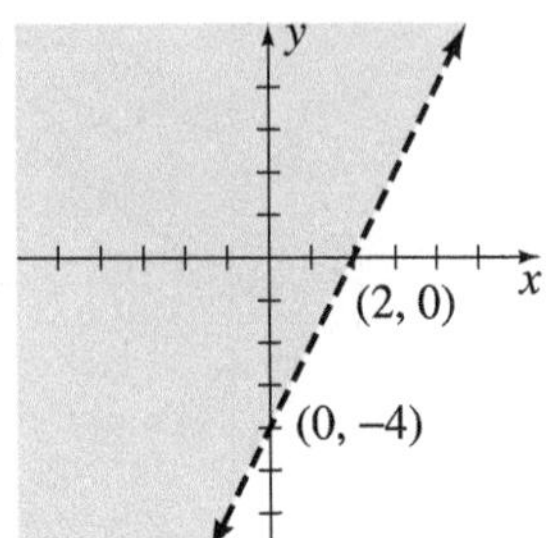


25. 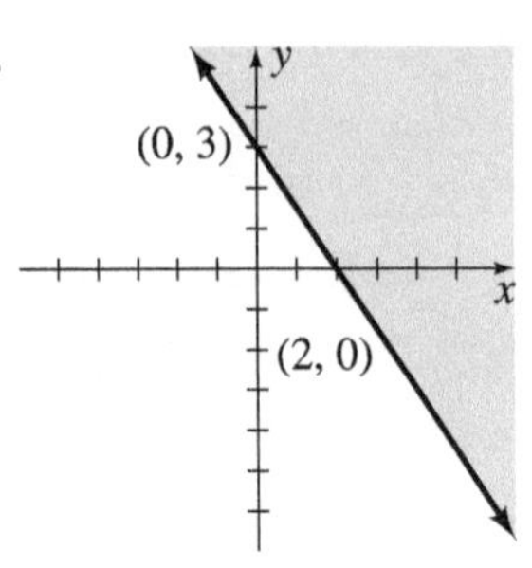


## Chapter 8

### Problem Set 8.1 (page 411)

1. 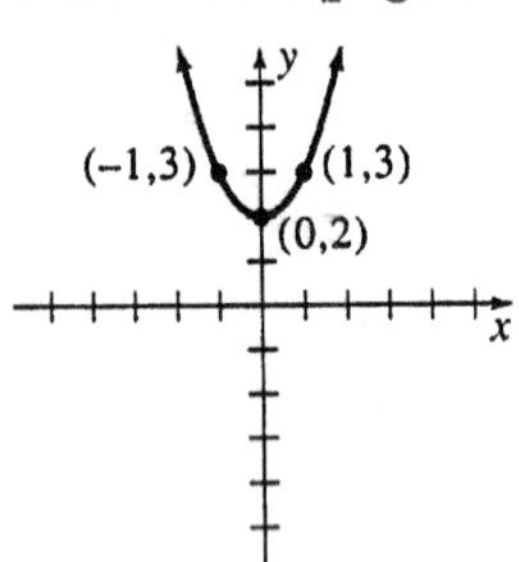


3. 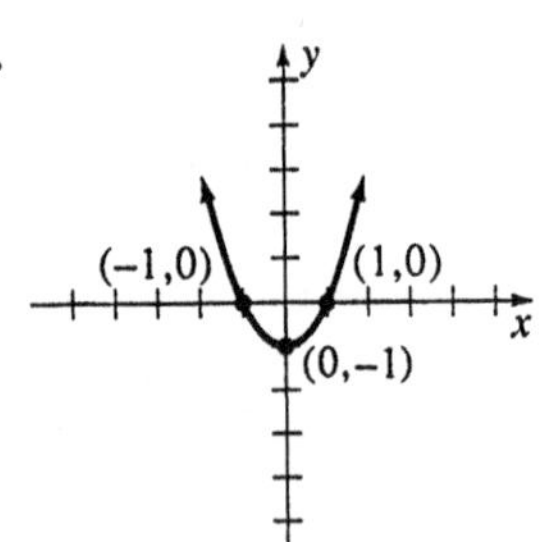


5. 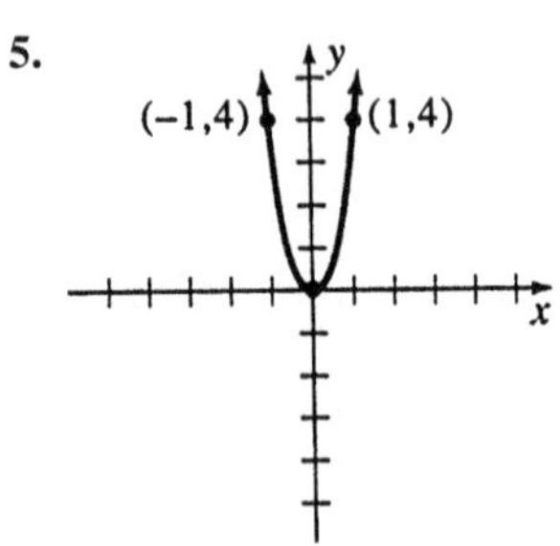


7. 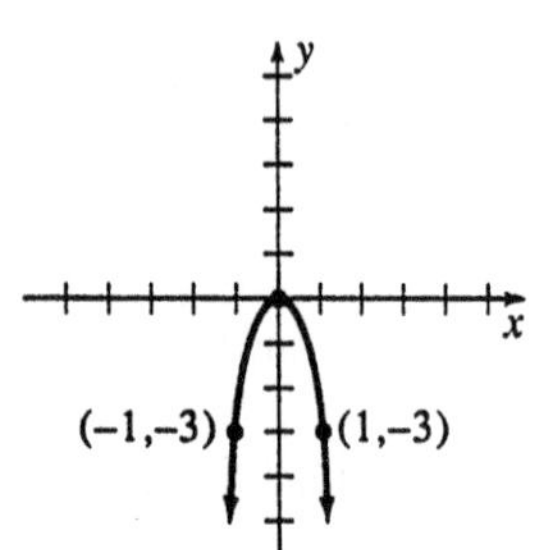


9. 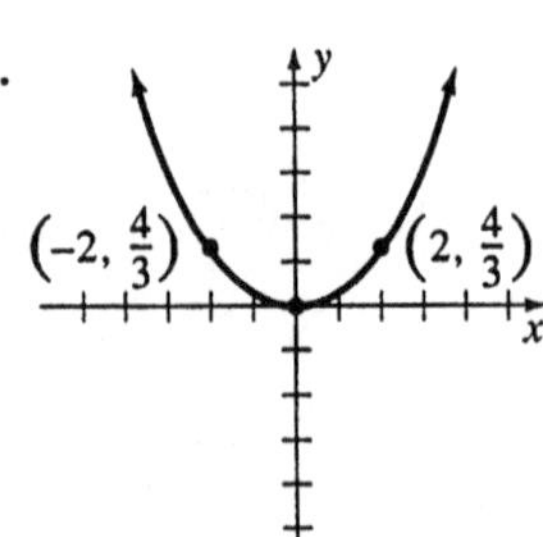


11. 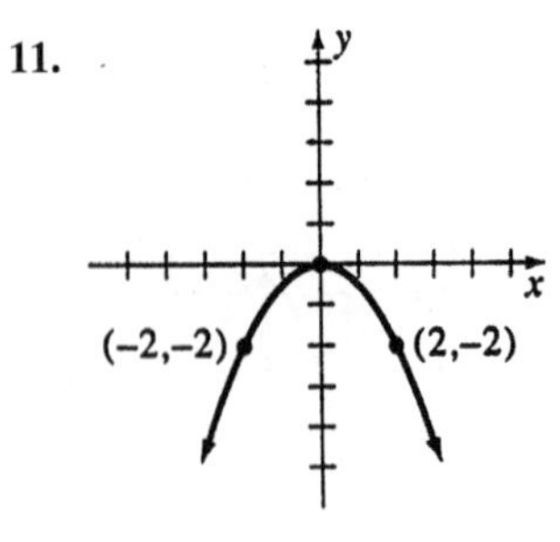


13. 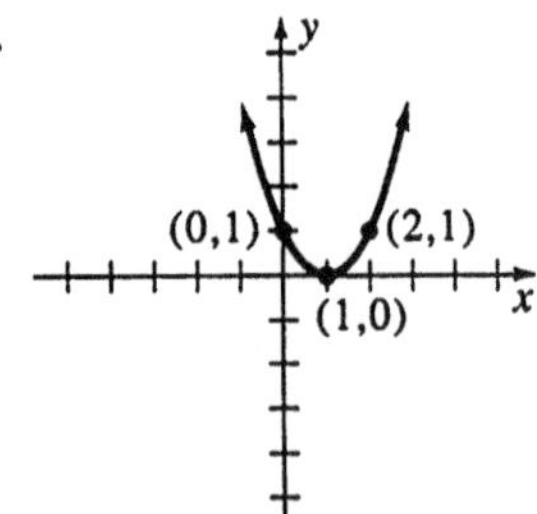


15. 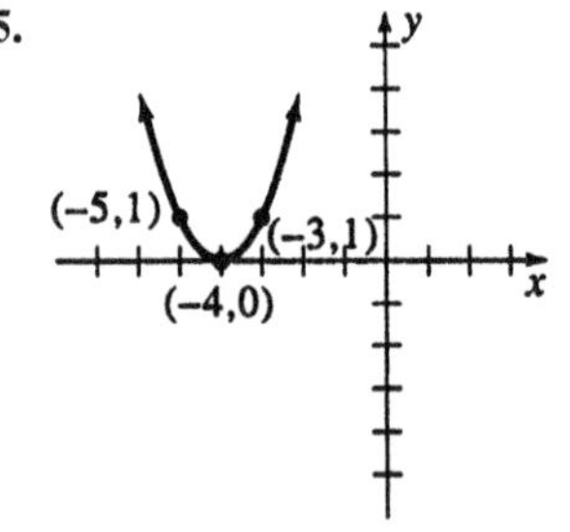


17. 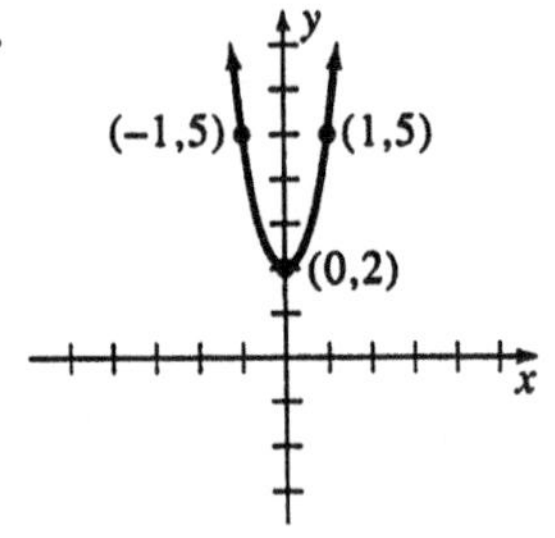


19. 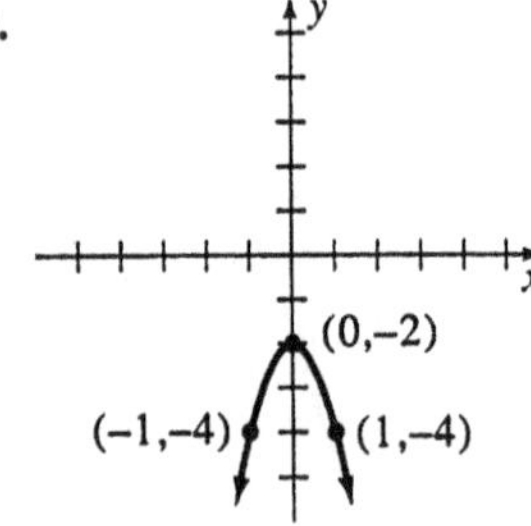


21. 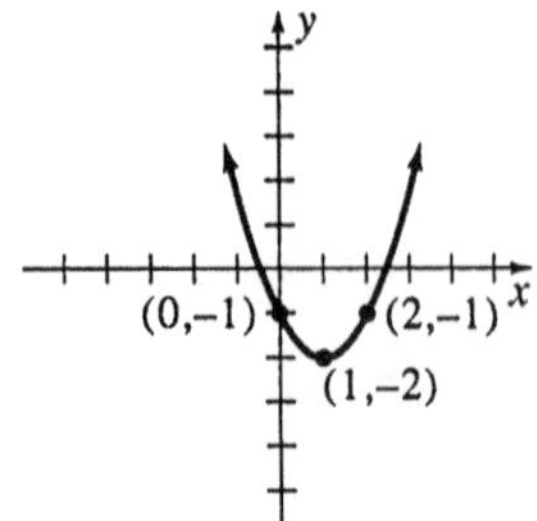


23. 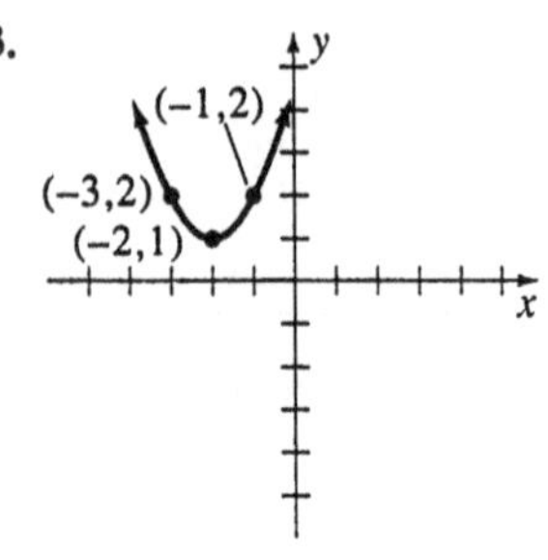

**25.**

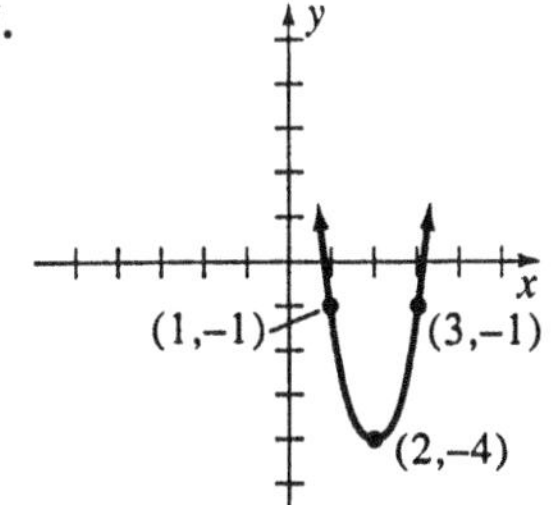


**27.**

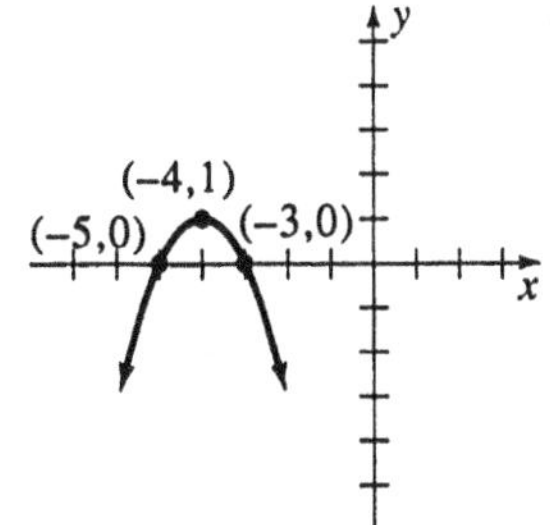


**29.**

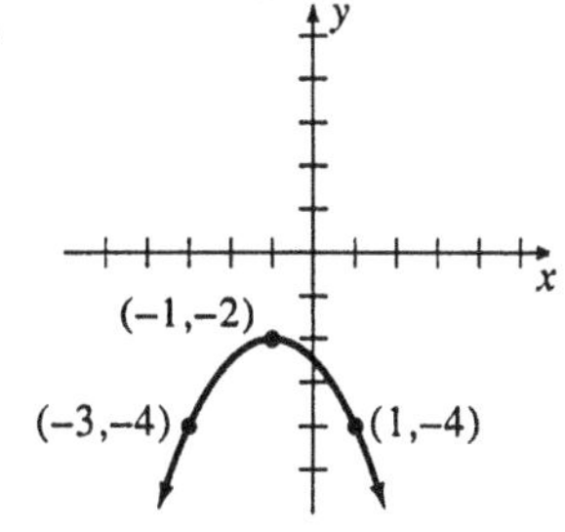


## Problem Set 8.2 (page 417)

**1.**

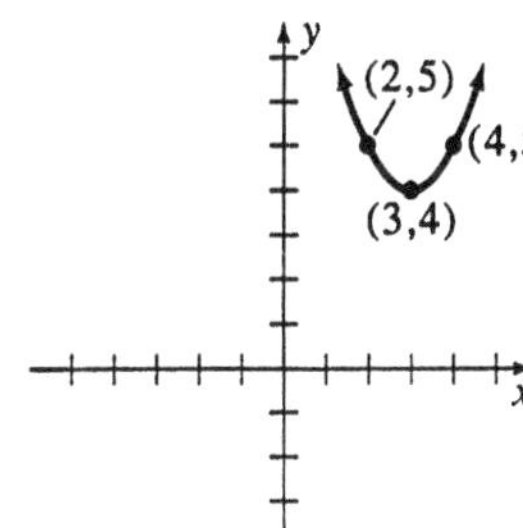


**3.**

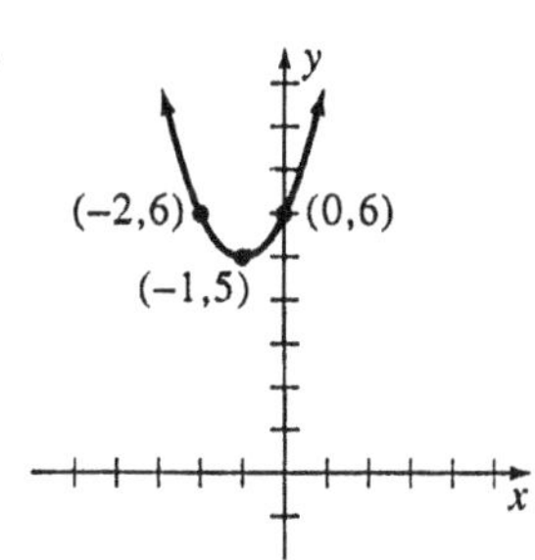


**5.**

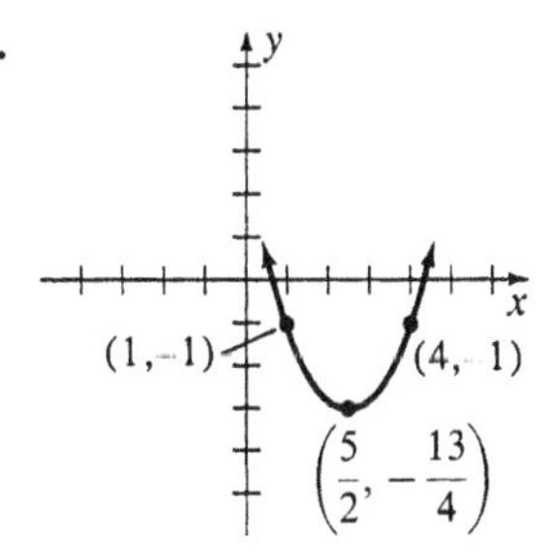


**7.**

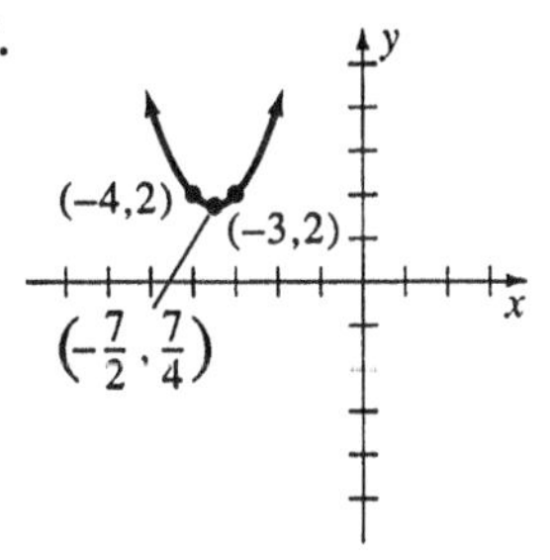


**9.**

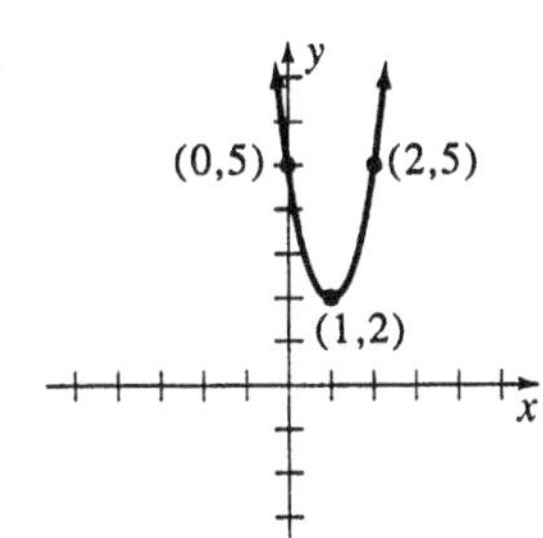


**11.**

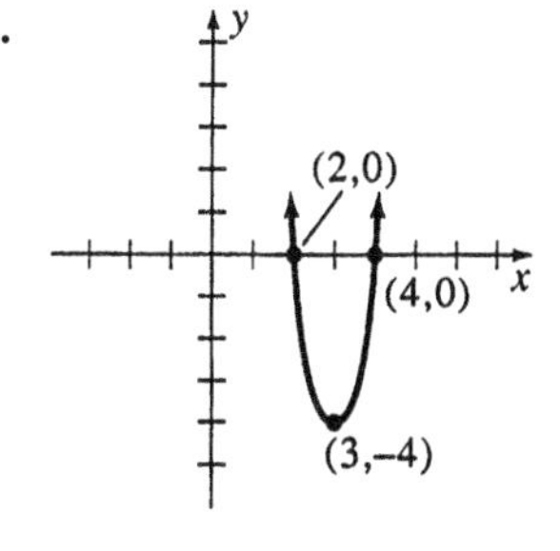


**13.**

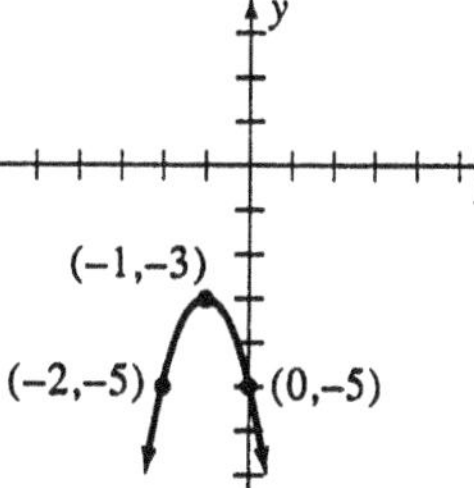


**15.**

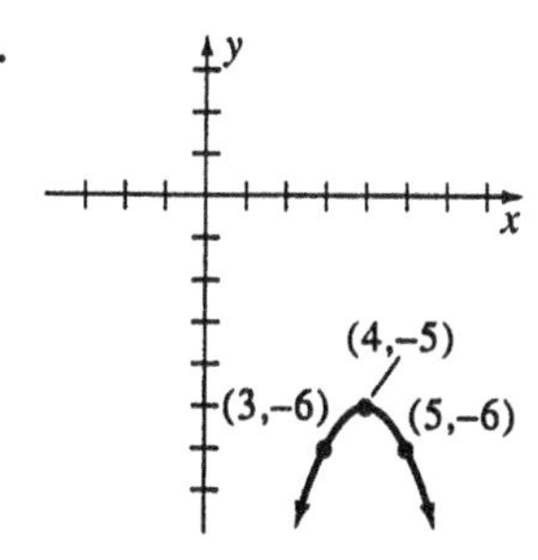


**17.**

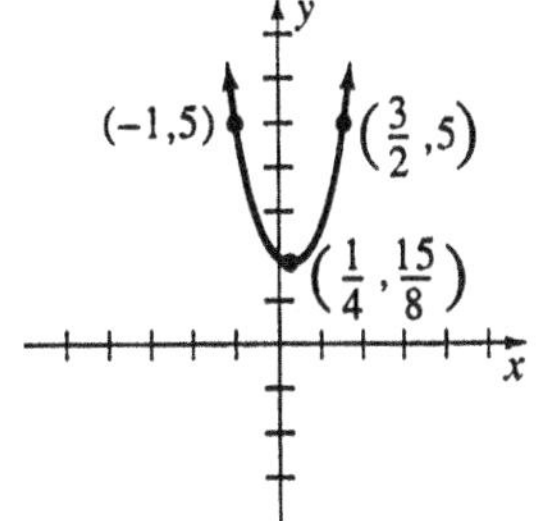


**19.**

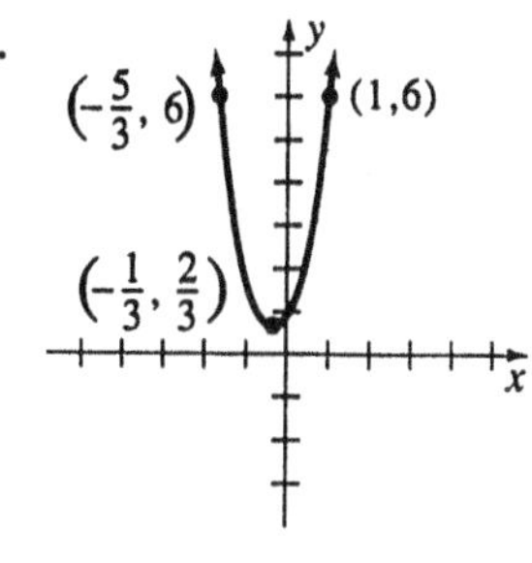


**21.**

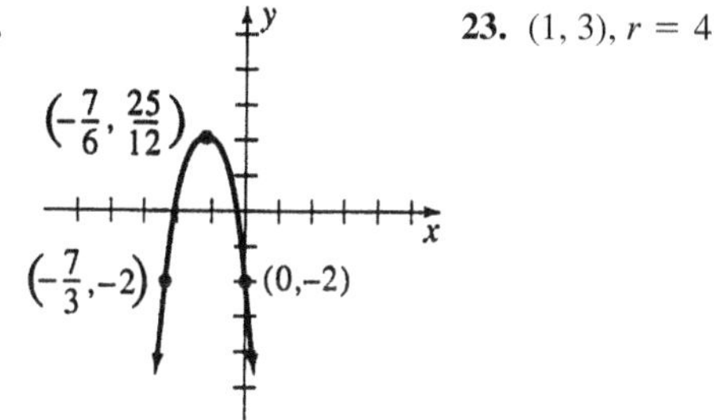


**23.** $(1, 3), r = 4$

**25.** $(-3, -5), r = 4$ **27.** $(0, 0), r = \sqrt{10}$

**29.** $(8, -3), r = \sqrt{2}$ **31.** $(-3, 4), r = 5$

**33.** $\left(-\frac{1}{2}, 4\right), r = 2\sqrt{2}$

**35.** **37.**

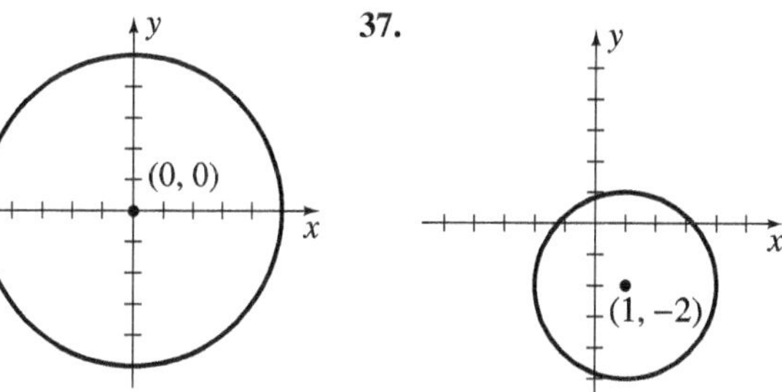


**39.**

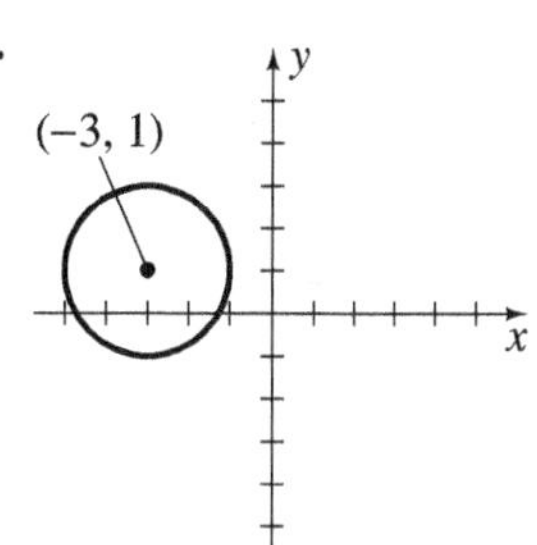


**41.**

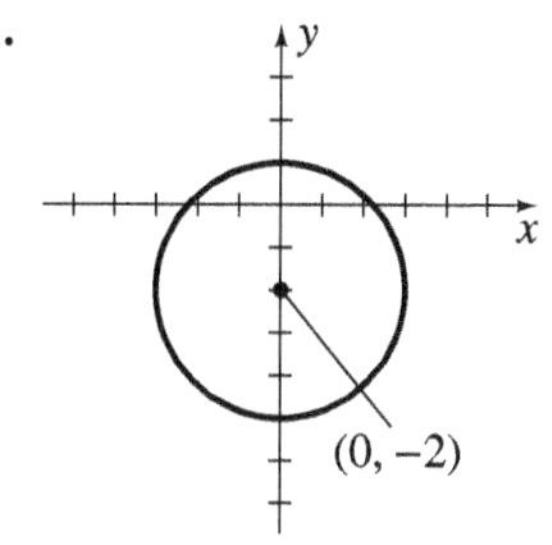

**43.**

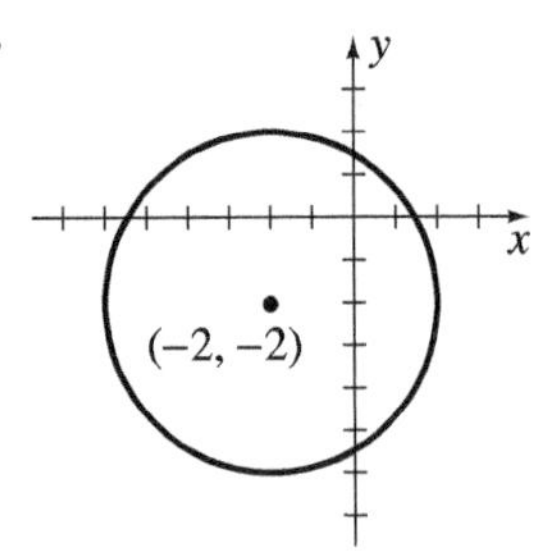


**45.** $x^2 + y^2 - 6x - 10y + 9 = 0$
**47.** $x^2 + y^2 + 8x - 2y - 47 = 0$
**49.** $x^2 + y^2 + 4x + 12y + 22 = 0$ **51.** $x^2 + y^2 - 20 = 0$
**53.** $x^2 + y^2 - 10x + 16y - 7 = 0$
**55.** $x^2 + y^2 - 8y = 0$ **57.** $x^2 + y^2 + 8x - 6y = 0$
**63.** **(a)** $(1, 4), r = 3$ **(c)** $(-6, -4), r = 8$
**(e)** $(0, 6), r = 9$

## Problem Set 8.3 (page 422)

**1.**

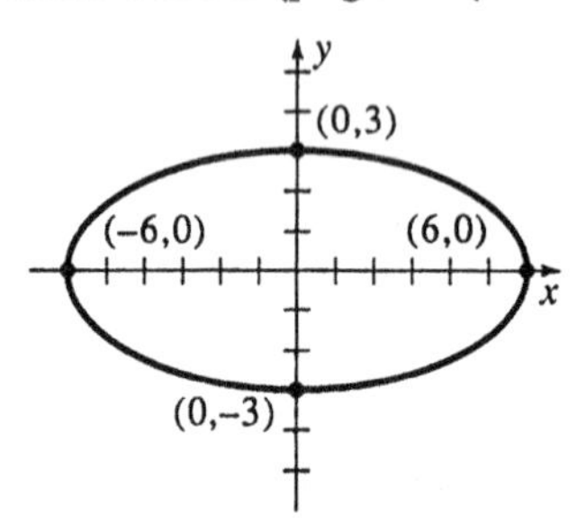


**3.**

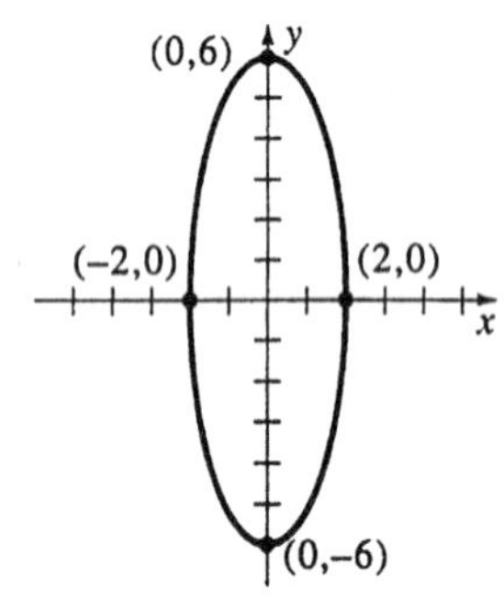


**5.**

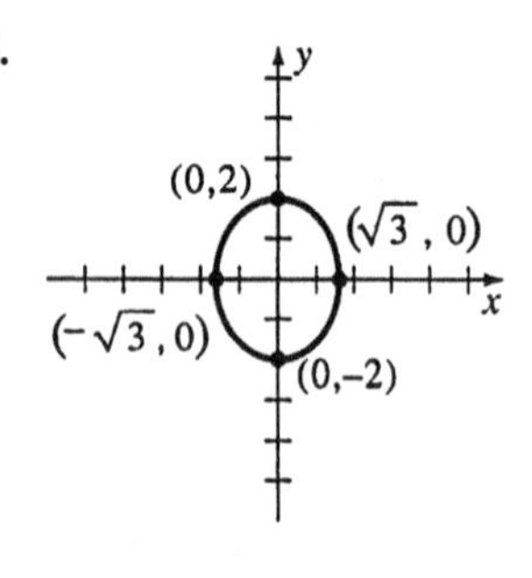


**7.**

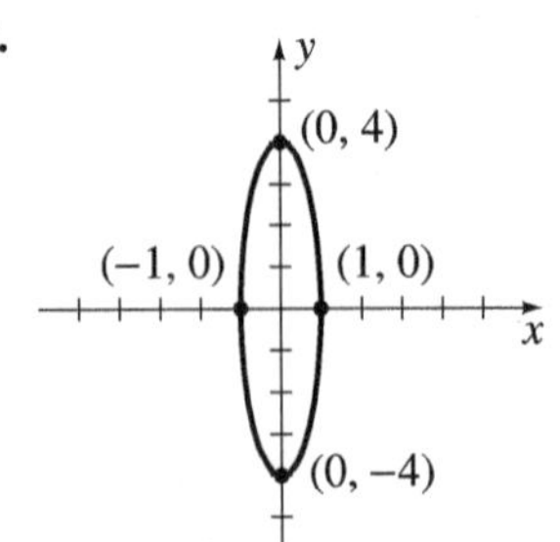


**9.**

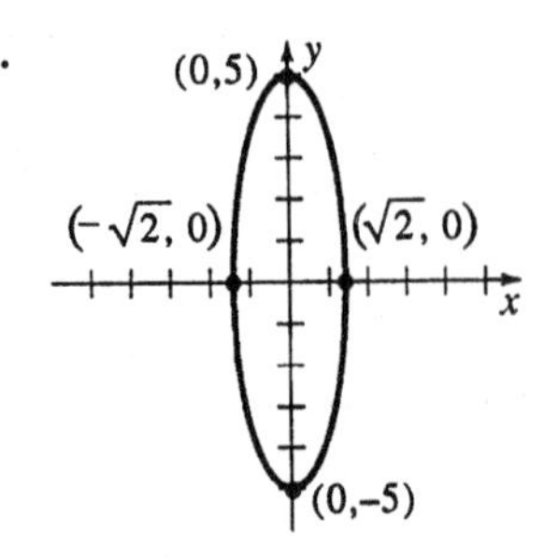


**11.**

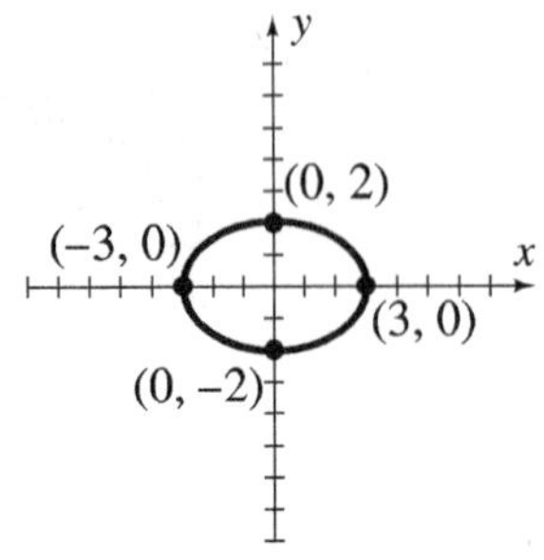


**13.**

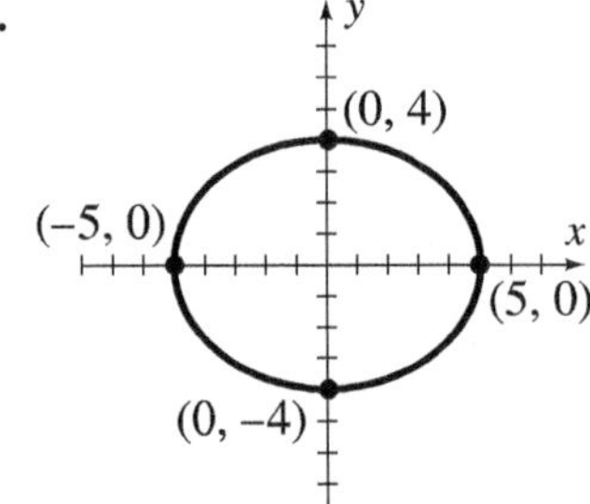


**15.**

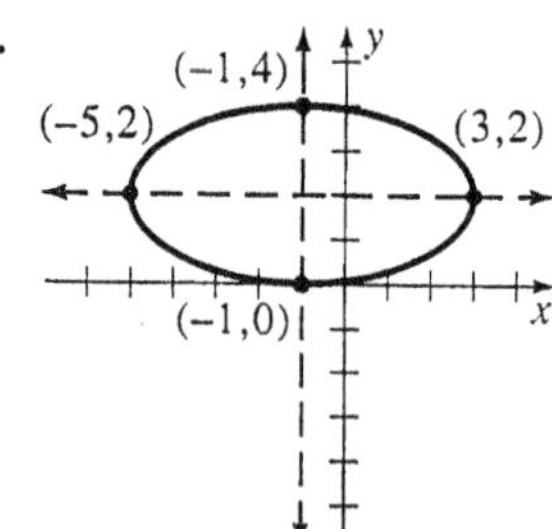


**17.**

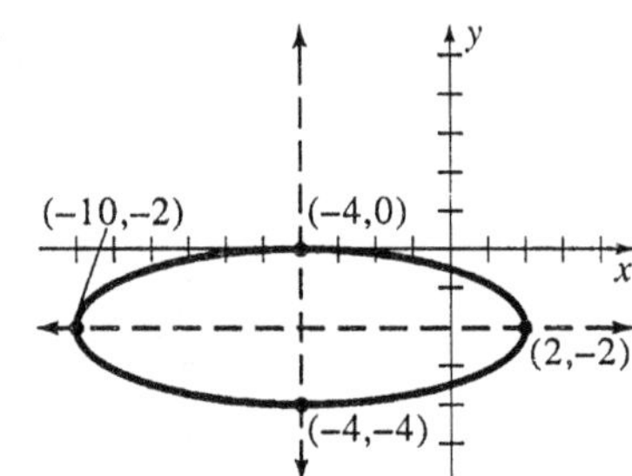


**19.**

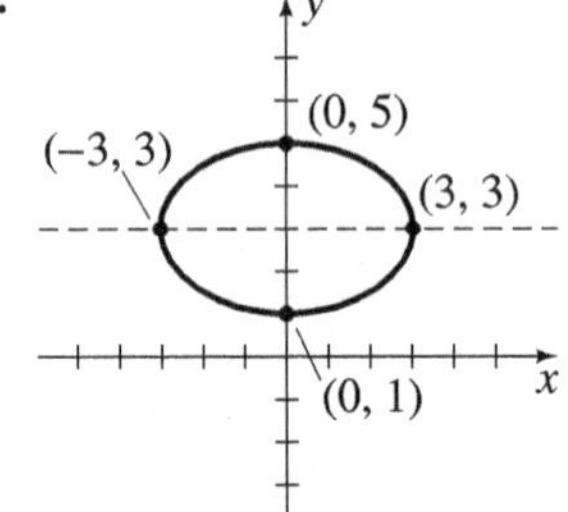


**21.**

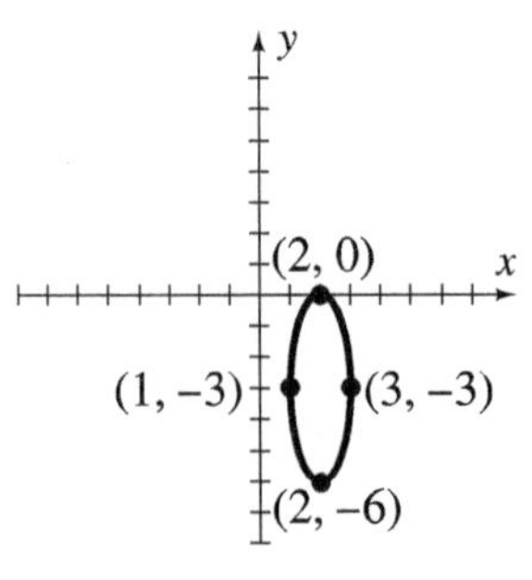


**23.**

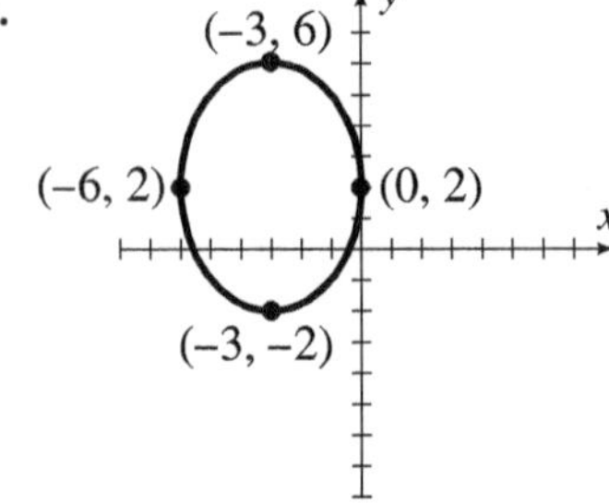

## Problem Set 8.4 (page 429)

**1.** $(4, 0), (-4, 0)$, $y = \frac{1}{3}x$ and $y = -\frac{1}{3}x$

**3.** $(0, 6), (0, -6)$ $y = 3x$ and $y = -3x$

**5.** $\left(\frac{2}{5}, 0\right), \left(-\frac{2}{5}, 0\right)$, $y = \frac{5}{3}x$ and $y = -\frac{5}{3}x$

**7.**

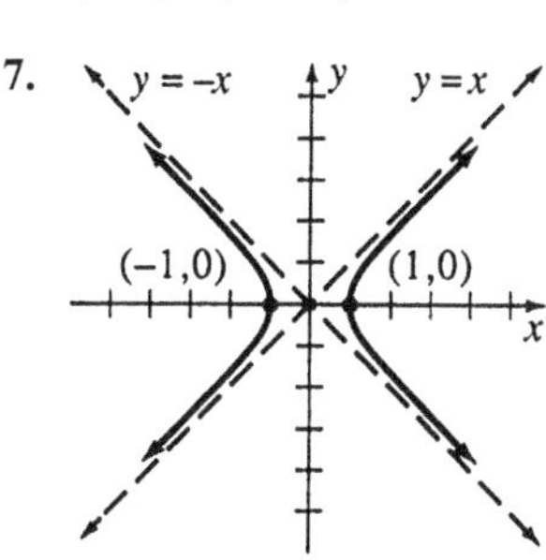


**9.**

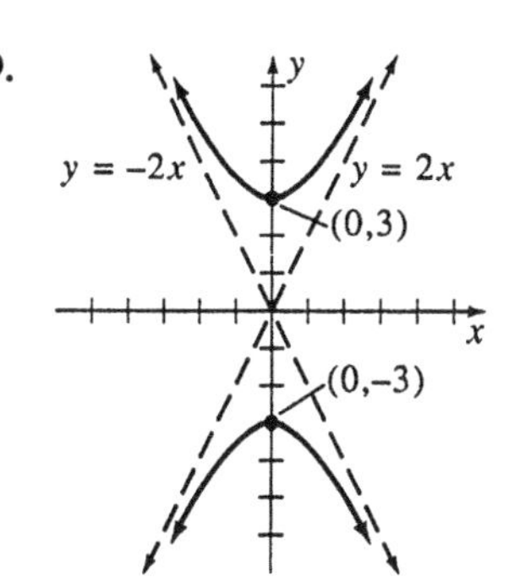


**11.**

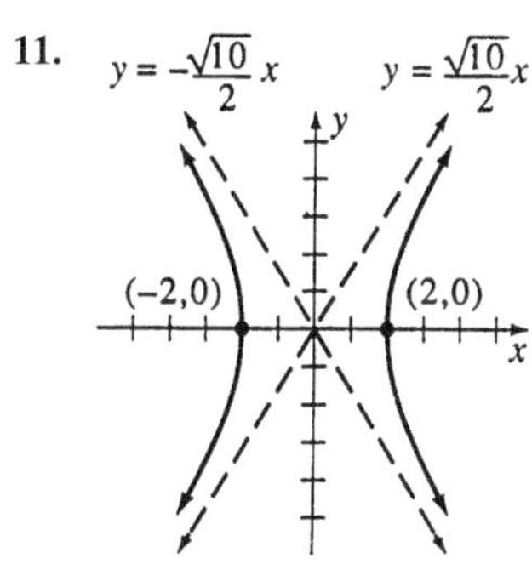


**13.**

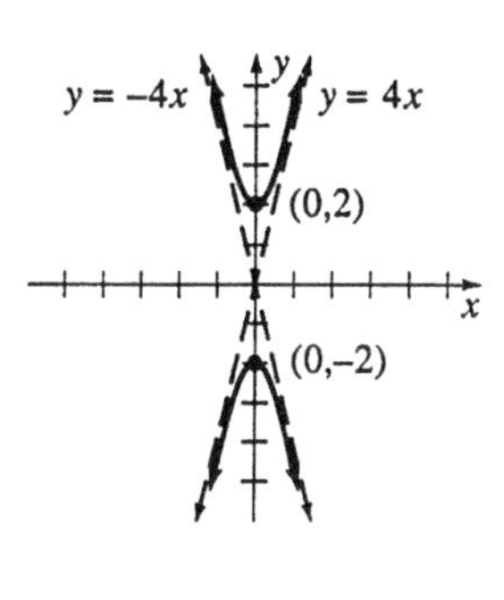


**15.**

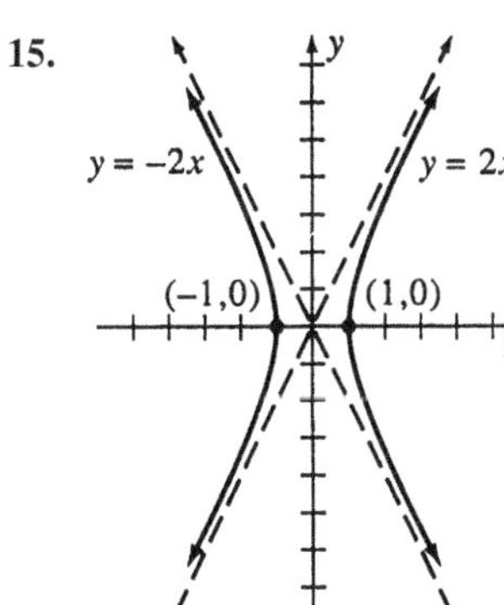


**17.**

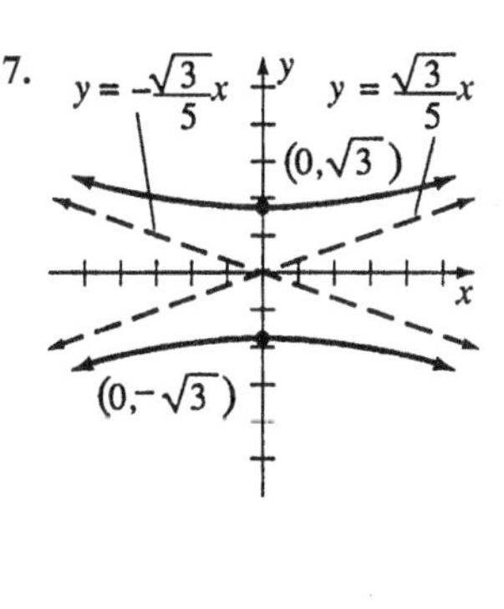


**19.** $y = x - 1$ and $y = -x - 5$

**21.** $y = -\frac{3}{2}x - \frac{3}{2}$ and $y = \frac{3}{2}x - \frac{9}{2}$

**23.**

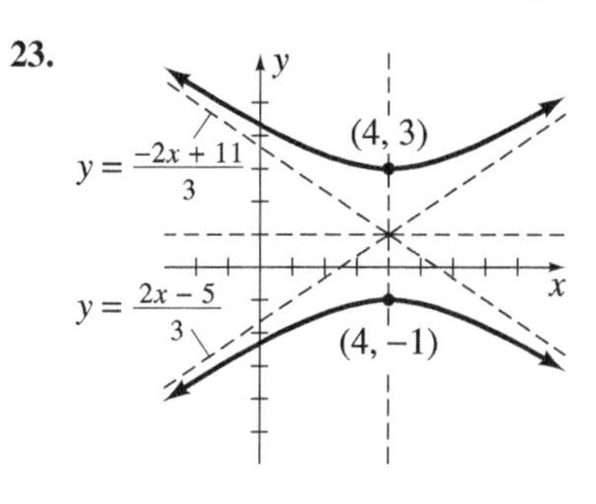


**25.**

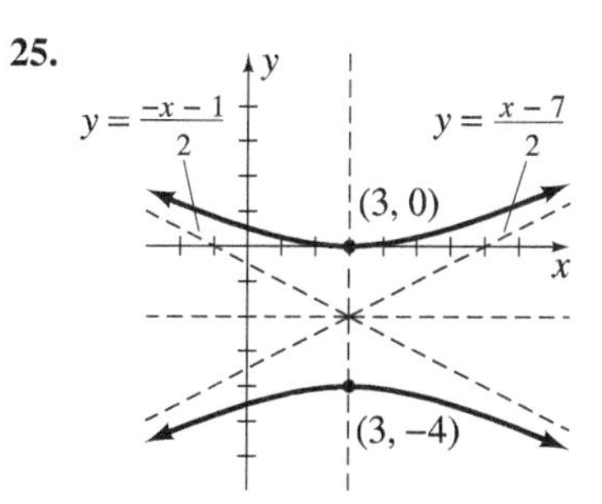


**27.**

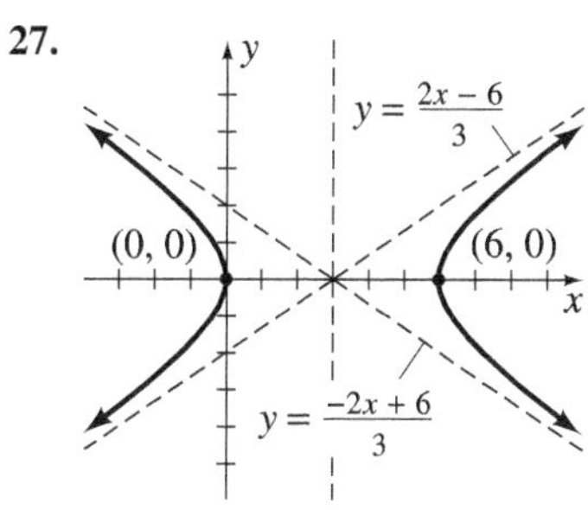


**29.**

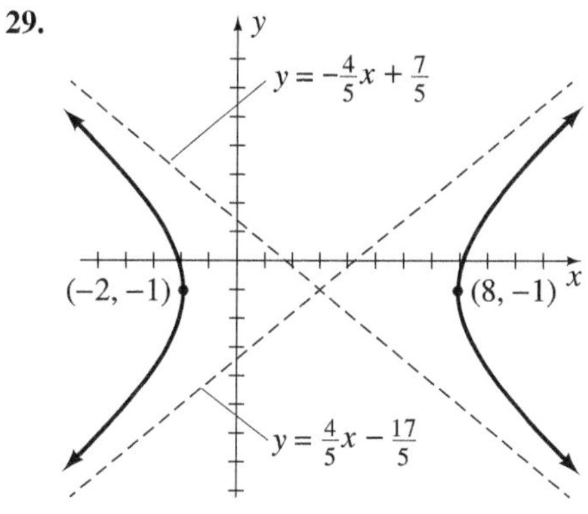


**31.**

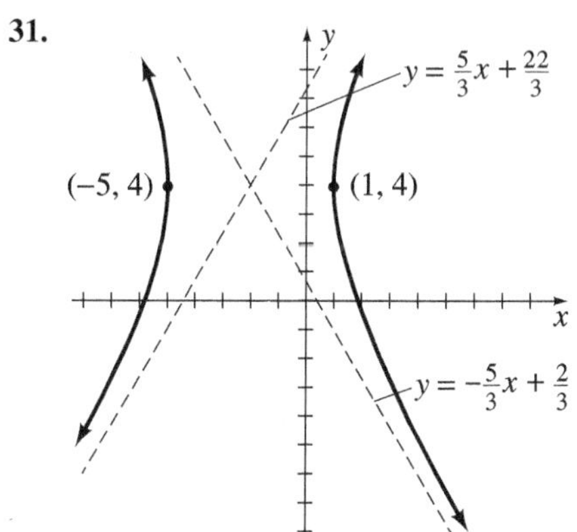


**33.**

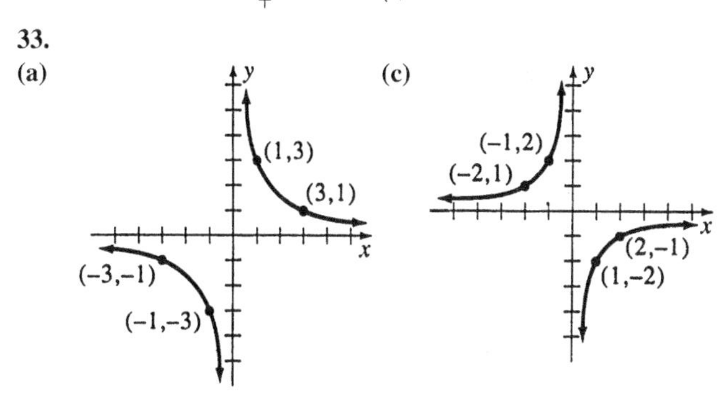

**35.** (a) Origin (c)

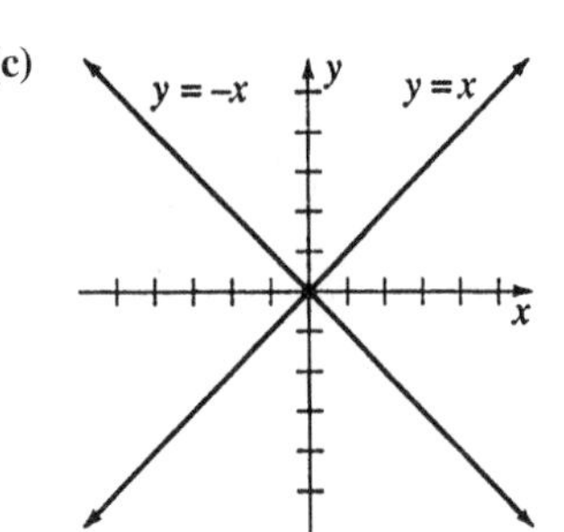


## Chapter 8 Sample Problems (page 430)

**1.**

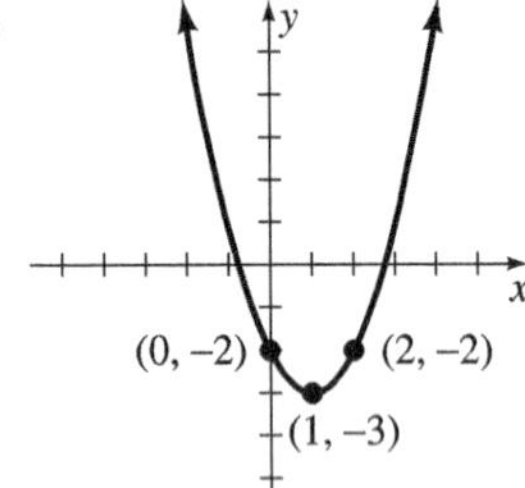


**2.**

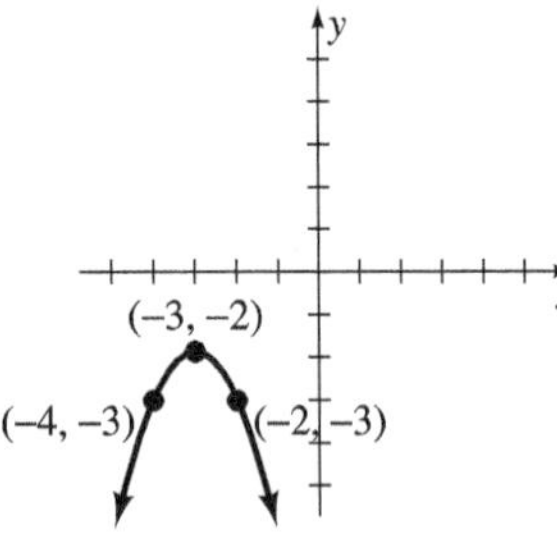


**3.** $(x - 5)^2 + (y + 2)^2 = 9$ or $x^2 + y^2 - 10x + 4y + 20 = 0$

**4.**

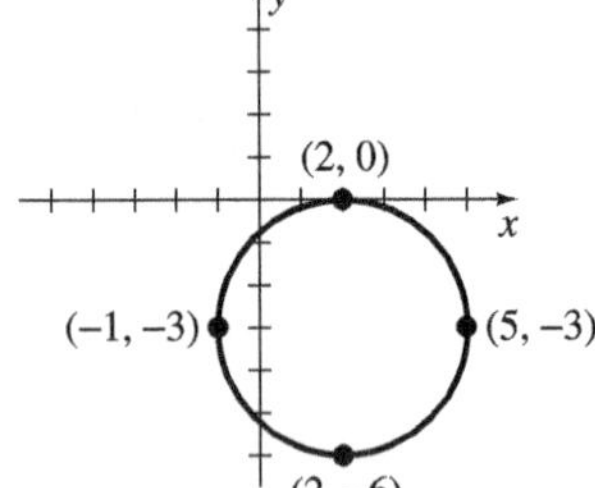


**5.**

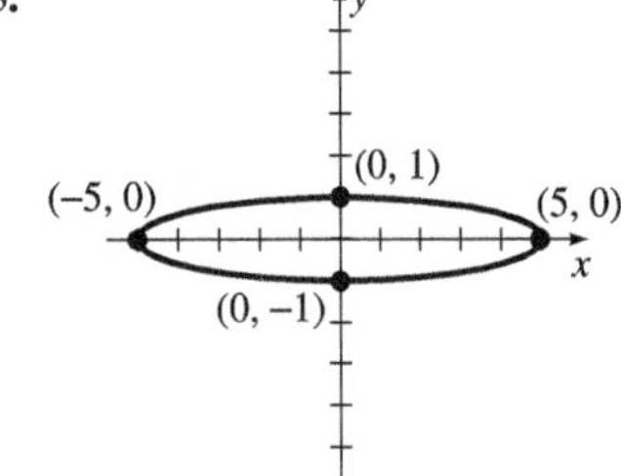


**6.**

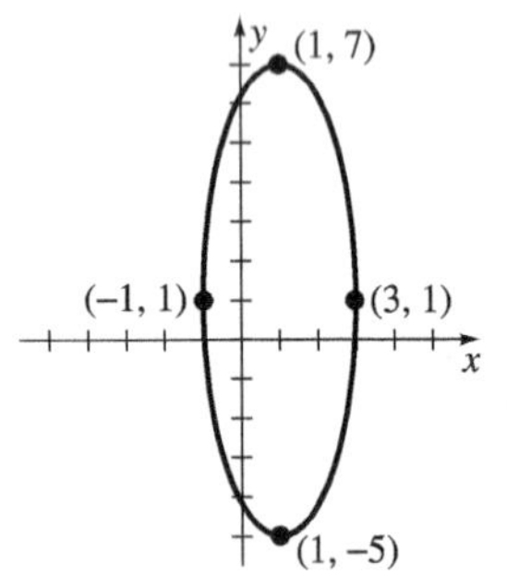


**7.**

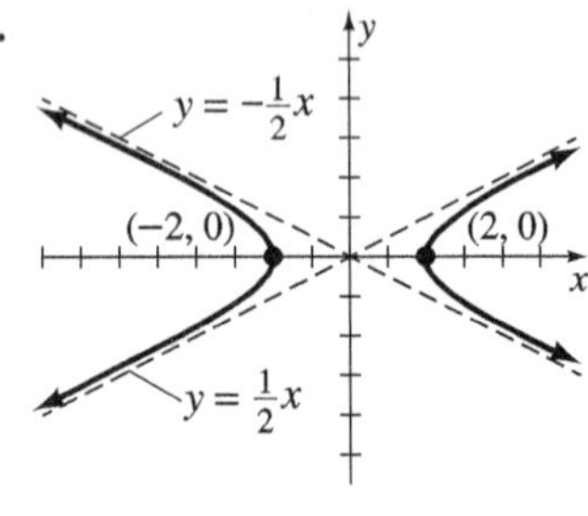


**8.**

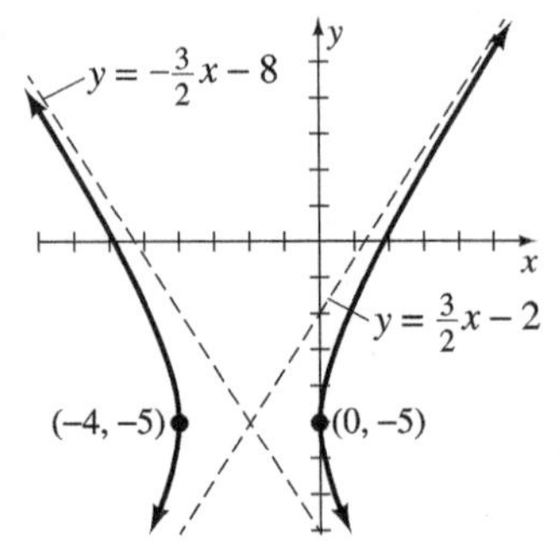


## Chapter 8 Review Problem Set (page 433)

**1.** (0, 6) **2.** (0, −8) **3.** (−3, −1) **4.** (−1, 0) **5.** (7, 5) **6.** (6, −8) **7.** (−4, −9) **8.** (−2, −3) **9.** $x^2 + y^2 - 36 = 0$ **10.** $x^2 + y^2 - 4x + 12y + 15 = 0$

**11.** $x^2 + y^2 + 8x + 16y + 68 = 0$ **12.** $x^2 + y^2 - 10y = 0$
**13.** (−7, 4) and $r = 7$ **14.** (−8, 0) and $r = 5$
**15.** (6, −8) and $r = 10$ **16.** (0, 0) and $r = 2\sqrt{6}$
**17.** Major = 16 and minor = 4 **18.** Major = 8 and minor = 6
**19.** Major = 10 and minor = 4
**20.** Major = $2\sqrt{14}$ and minor = 4
**21.** Major = 6 and minor = 2 **22.** Major = 8, minor = 4
**23.** Major = 6 and minor = 4
**24.** Major = 8, minor = 6 **25.** $y = \pm\frac{1}{3}x$ **26.** $y = \pm 2x$
**27.** $y = \pm\frac{5}{3}x$ **28.** $y = \pm\frac{1}{2}x$
**29.** $y = -\frac{5}{2}x - 2$ and $y = \frac{5}{2}x + 8$
**30.** $y = -\frac{1}{6}x + \frac{25}{6}$ and $y = \frac{1}{6}x + \frac{23}{6}$

**31.**

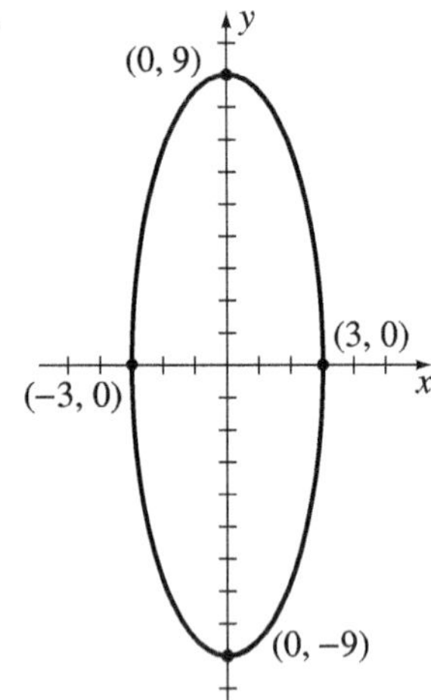


**32.**

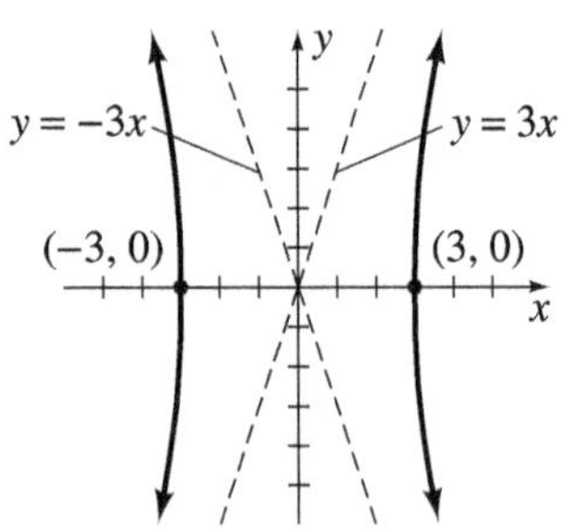


**33.**

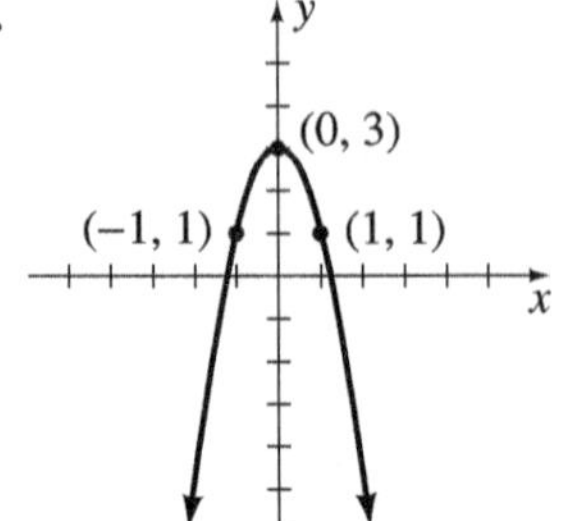


**34.**

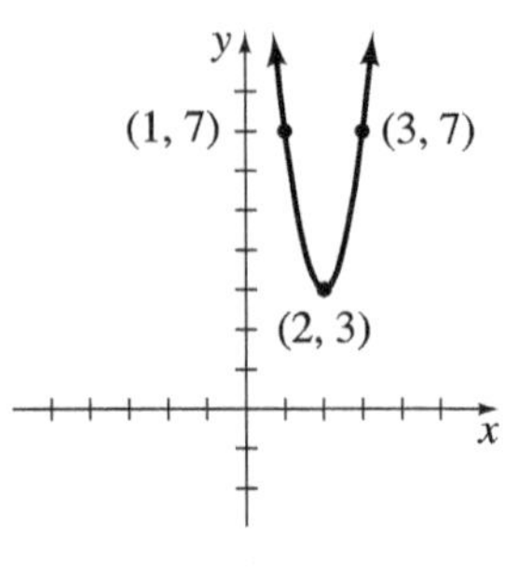

**35.**

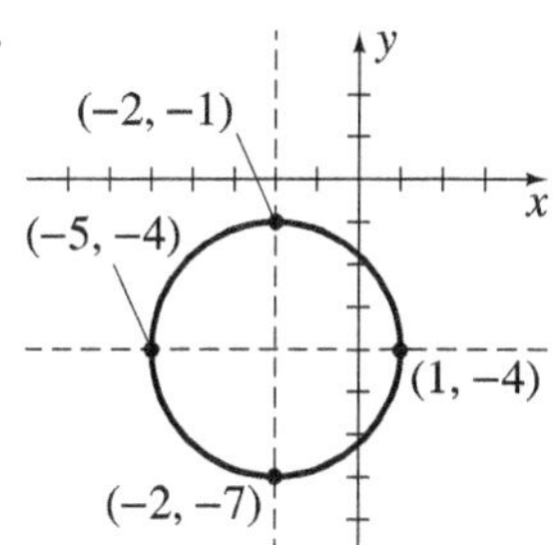


**36.**

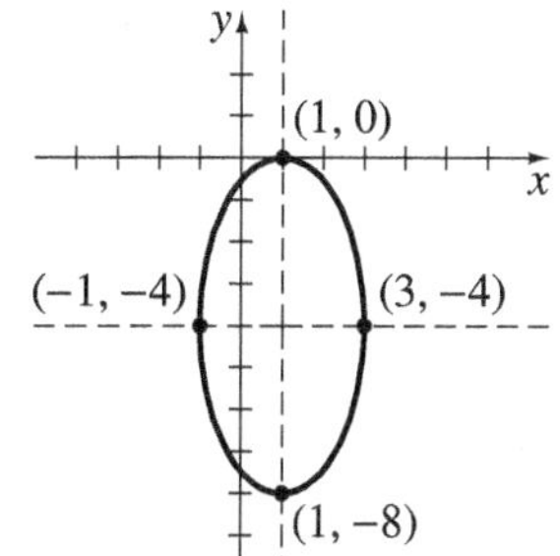


**37.**

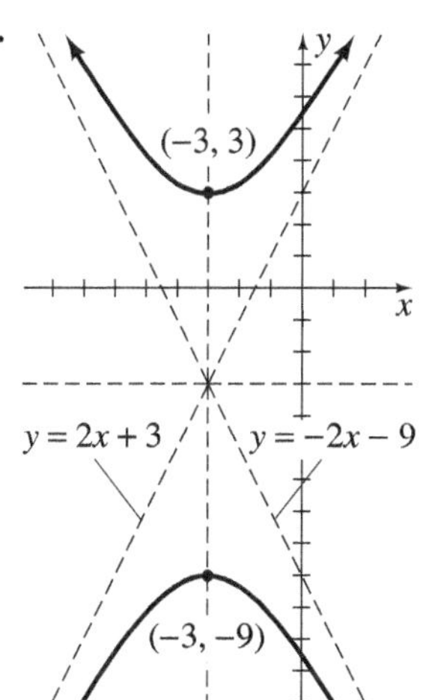


**38.**

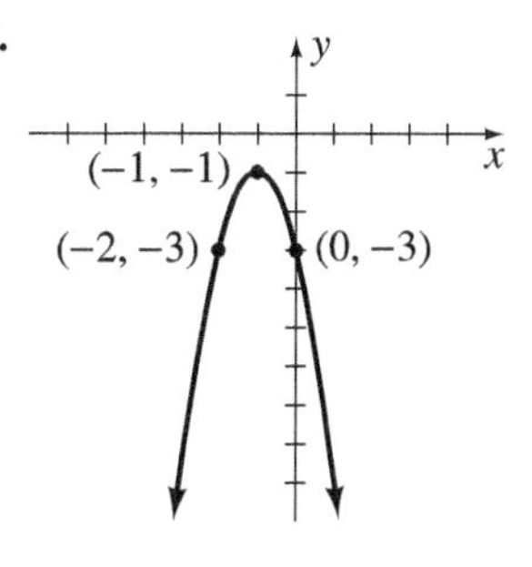


**39.**

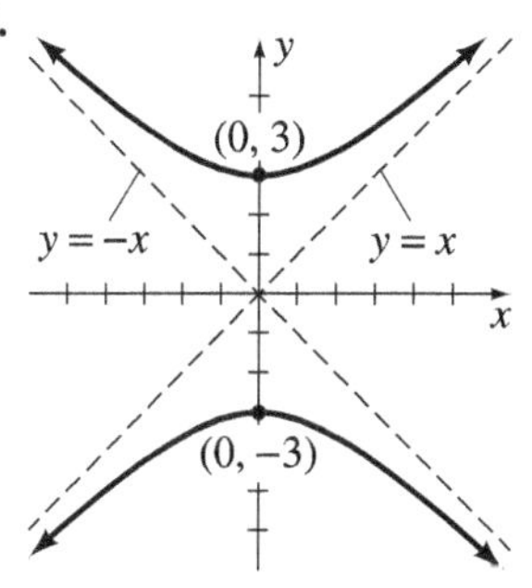


**40.**

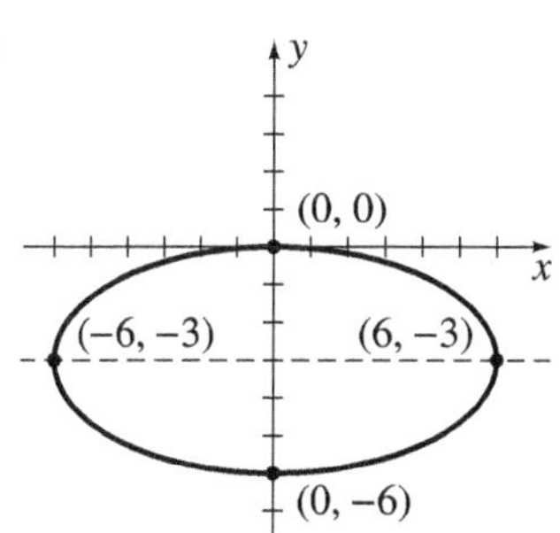


**41.**

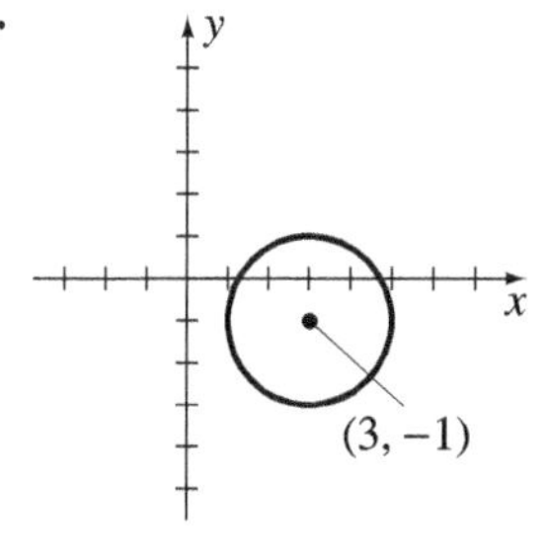


**42.**

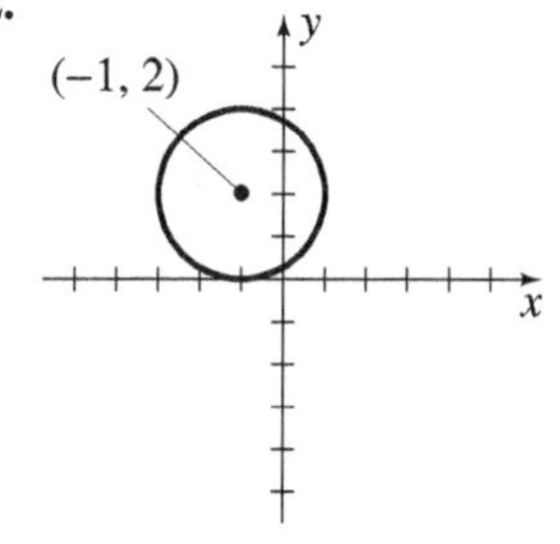


**43.**

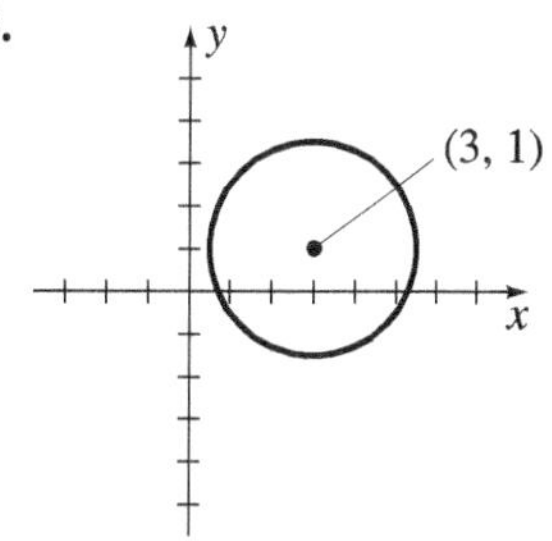


**44.**

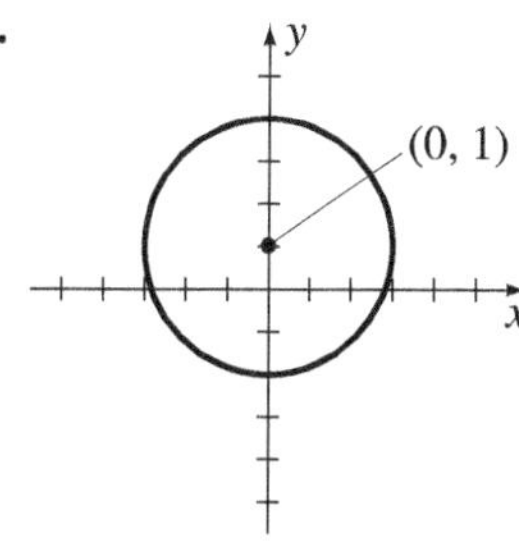


**45.**

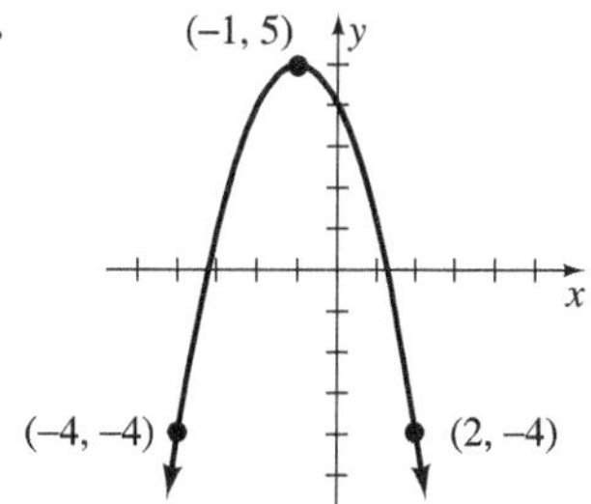


**46.**

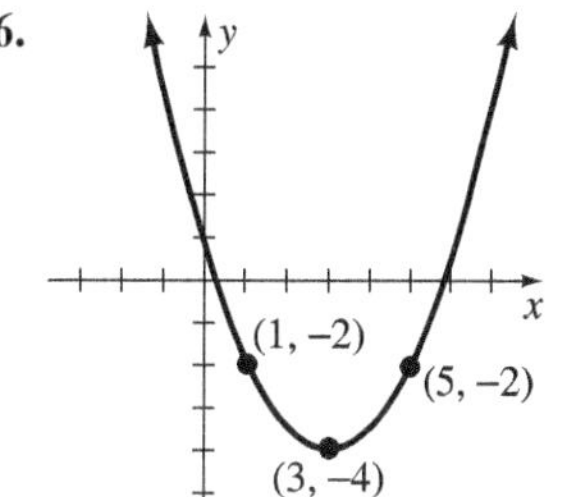


**47.**

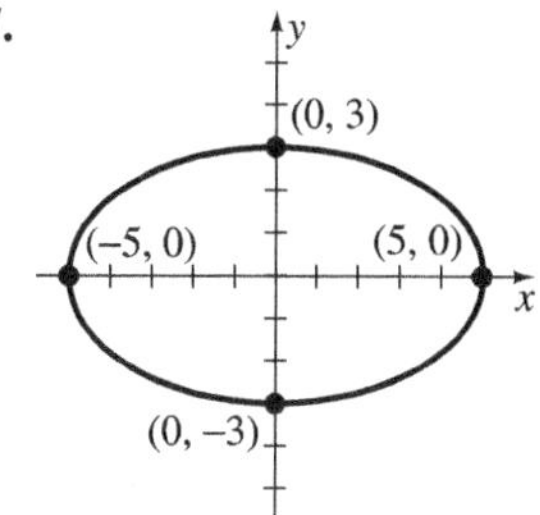


**48.**

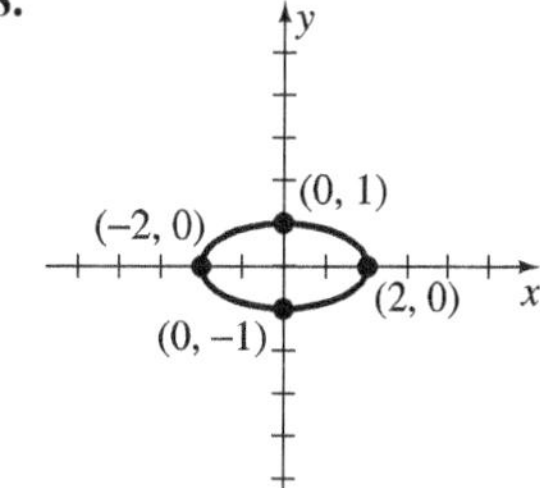


**49.**

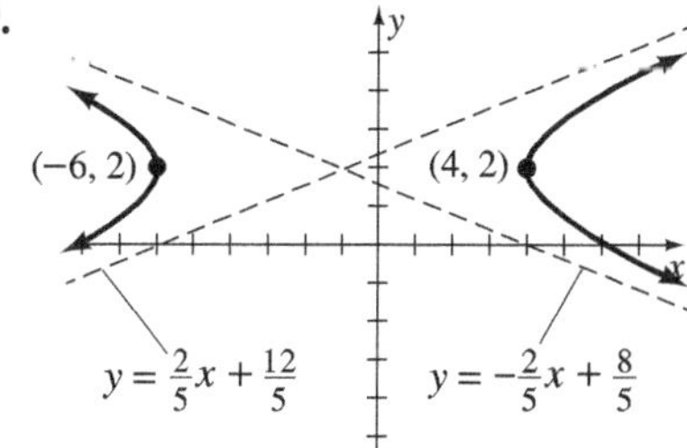


**50.**

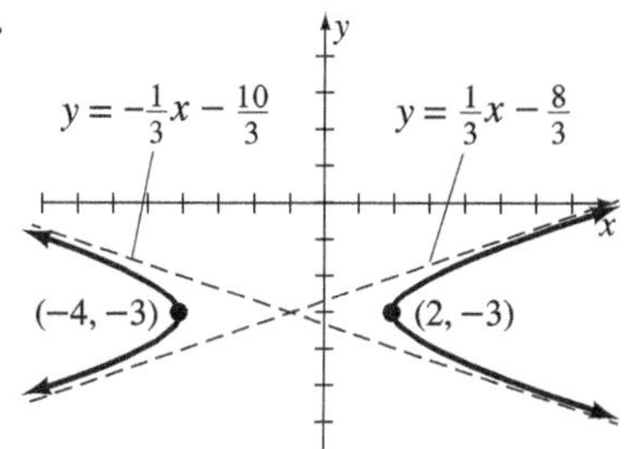


## Chapter 8 Test (page 435)

**1.** (0, 9) **2.** (1, 7) **3.** (−4, −2)

**4.** (3, 0) **5.** $x^2 + y^2 + 8x - 29 = 0$
**6.** $x^2 + y^2 - 4x - 16y + 59 = 0$
**7.** $x^2 + y^2 + 6x + 8y = 0$
**8.** (0, 0) and $r = 4\sqrt{2}$ **9.** (6, −4) and $r = 7$
**10.** (−5, −1) and $r = 8$ **11.** 8 units **12.** 6 units

**13.** 6 units **14.** 4 units **15.** $y = \pm 4x$ **16.** $y = \pm\frac{5}{4}x$

**17.** $y = \frac{1}{5}x - \frac{6}{5}$ and $y = -\frac{1}{5}x - \frac{4}{5}$

**18.**

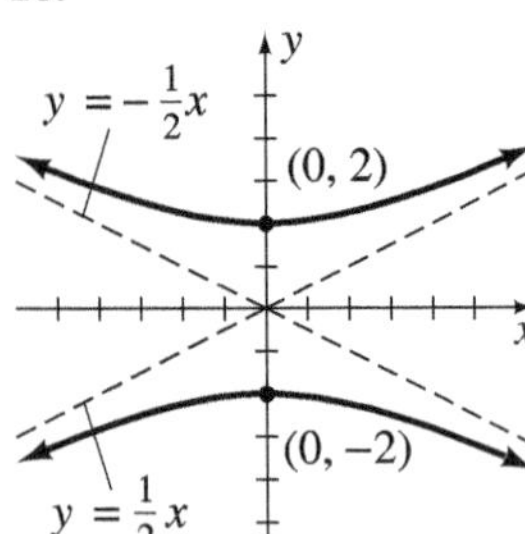


**19.**

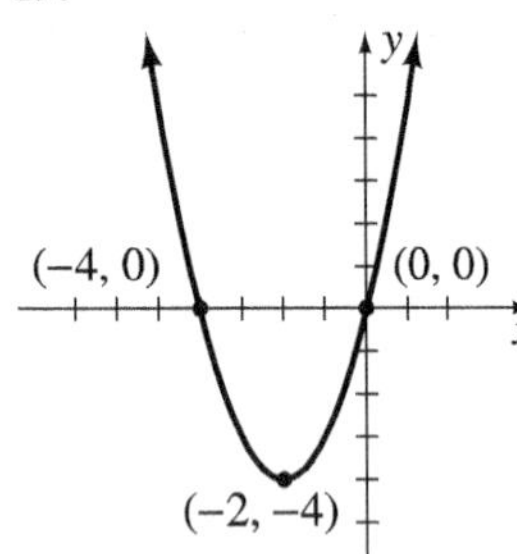


**20.**

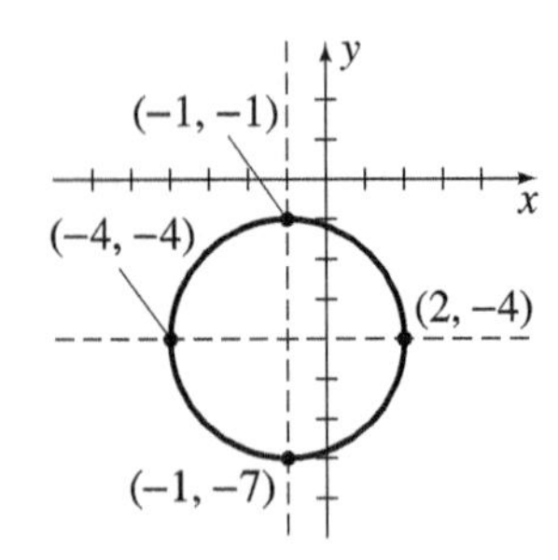


**21.**

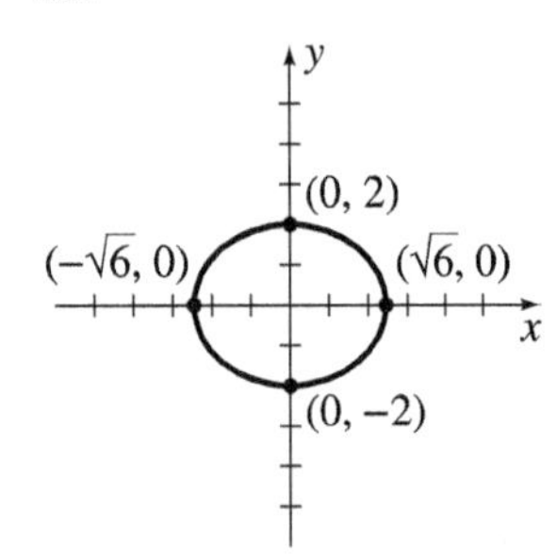


**22.**

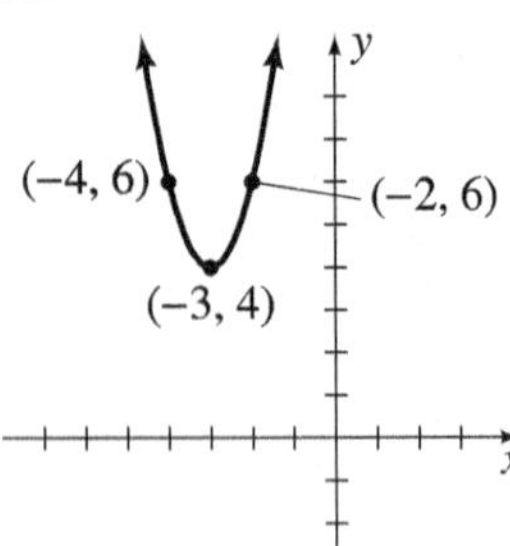


**23.**

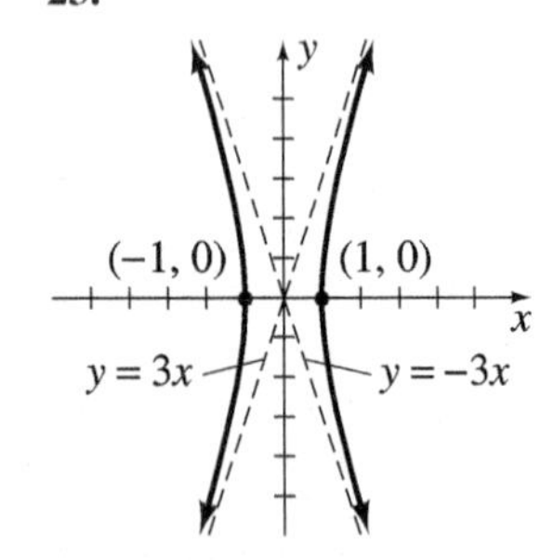


**24.**

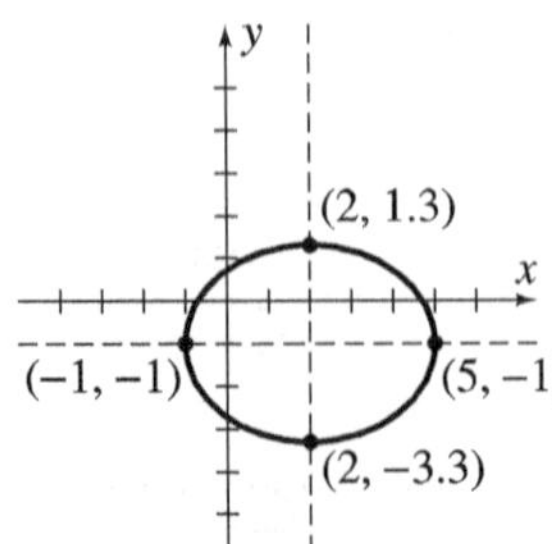


**25.**

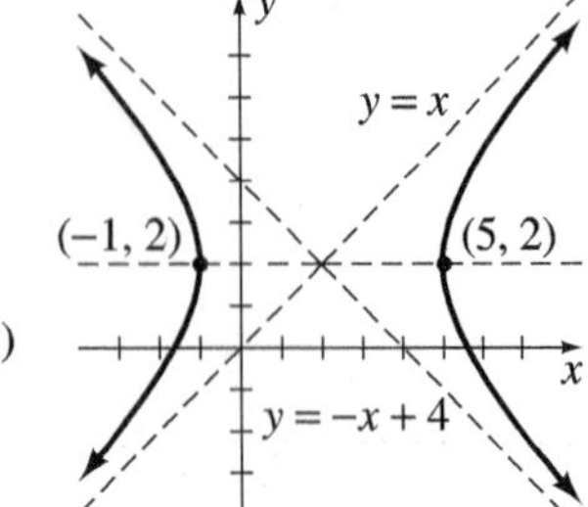


## Chapters 1–8 Cumulative Review Problem Set (page 436)

**1.** $\frac{64}{15}$ **2.** $\frac{11}{3}$ **3.** $\frac{1}{6}$ **4.** −7 **5.** $-24a^4b^5$

**6.** $2x^3 + 5x^2 - 7x - 12$ **7.** $\frac{3x^2y^2}{8}$ **8.** $\frac{a(a+1)}{2a-1}$

**9.** $\frac{-x+14}{18}$ **10.** $\frac{5x+19}{x(x+3)}$ **11.** $\frac{2}{n+8}$

**12.** $\frac{x-14}{(5x-2)(x+1)(x-4)}$ **13.** $y^2 - 5y + 6$

**14.** $x^2 - 3x - 2$ **15.** $20 + 7\sqrt{10}$

**16.** $2x - 2\sqrt{xy} - 12y$ **17.** $-\frac{3}{8}$ **18.** $-\frac{2}{3}$ **19.** 0.2

**20.** $\frac{1}{2}$ **21.** $\frac{13}{9}$ **22.** −27 **23.** $\frac{16}{9}$ **24.** $\frac{8}{27}$

**25.** $3x(x+3)(x^2 - 3x + 9)$ **26.** $(6x - 5)(x + 4)$
**27.** $(4 + 7x)(3 - 2x)$ **28.** $(3x + 2)(3x - 2)(x^2 + 8)$
**29.** $(2x - y)(a - b)$ **30.** $(3x - 2y)(9x^2 + 6xy + 4y^2)$

**31.** $\left\{-\frac{12}{7}\right\}$ **32.** {150} **33.** {25} **34.** {0} **35.** {−2, 2}

**36.** {−7} **37.** $\left\{-6, \frac{4}{3}\right\}$ **38.** $\left\{\frac{5}{4}\right\}$ **39.** {3} **40.** $\left\{\frac{4}{5}, 1\right\}$

**41.** $\left\{-\frac{10}{3}, 4\right\}$ **42.** $\left\{\frac{1}{4}, \frac{2}{3}\right\}$ **43.** $\left\{-\frac{3}{2}, 3\right\}$ **44.** $\left\{\frac{1}{5}\right\}$

**45.** $\left\{\frac{5}{7}\right\}$ **46.** {−2, 2} **47.** {0} **48.** {−6, 19}

**49.** $\left\{-\frac{3}{4}, \frac{2}{3}\right\}$ **50.** $\{1 \pm 5i\}$ **51.** {1, 3} **52.** $\left\{-4, \frac{1}{3}\right\}$

**53.** $\{-2 \pm 4i\}$ **54.** $\left\{\frac{1 \pm \sqrt{33}}{4}\right\}$ **55.** $(-\infty, -2]$

**56.** $\left(-\infty, \frac{19}{5}\right)$ **57.** $\left(\frac{1}{4}, \infty\right)$ **58.** (−2, 3)

**59.** $\left(-\infty, -\frac{13}{3}\right) \cup (3, \infty)$ **60.** $(-\infty, 29]$ **61.** [−2, 4]

**62.** $(-\infty, -5) \cup \left(\frac{1}{3}, \infty\right)$ **63.** $(-\infty, -2] \cup (7, \infty)$

**64.** (−3, 4)

**65.**

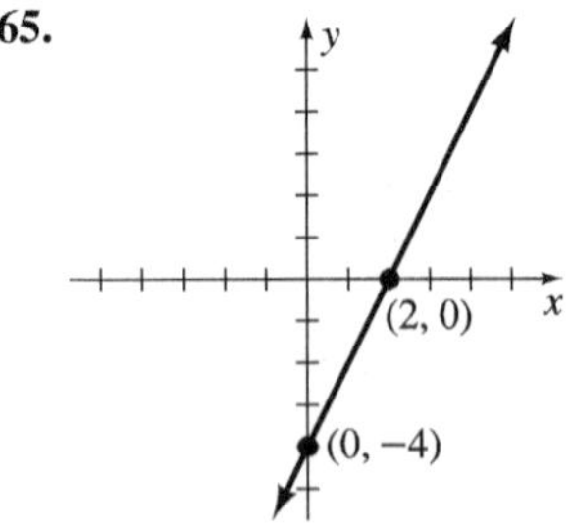


**66.**

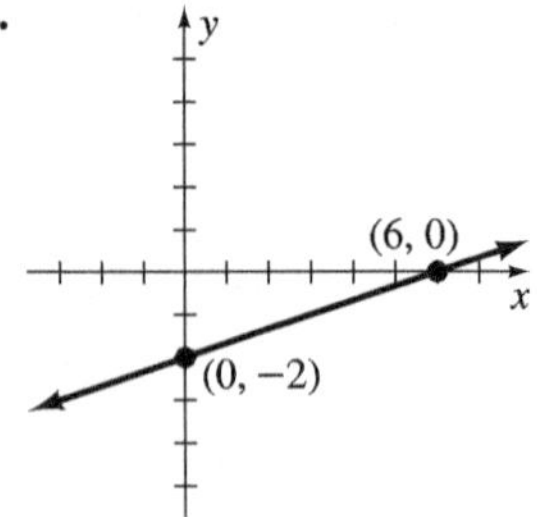


**67.**

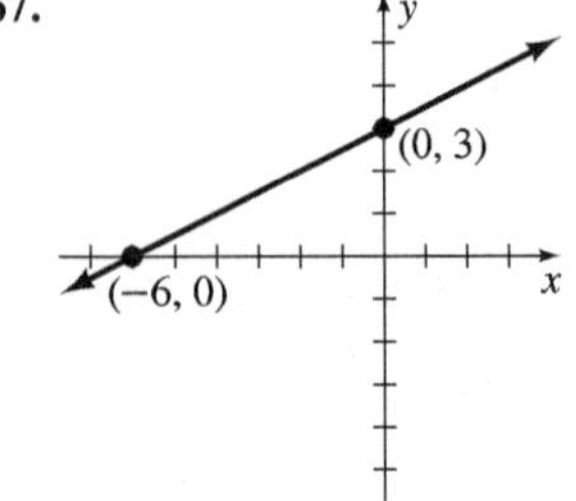


**68.**

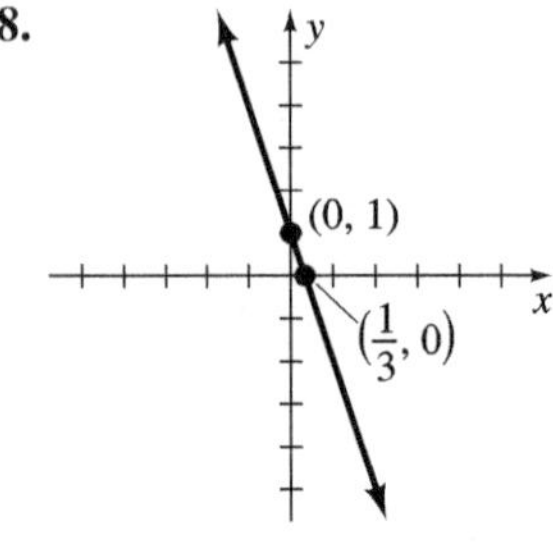

69.
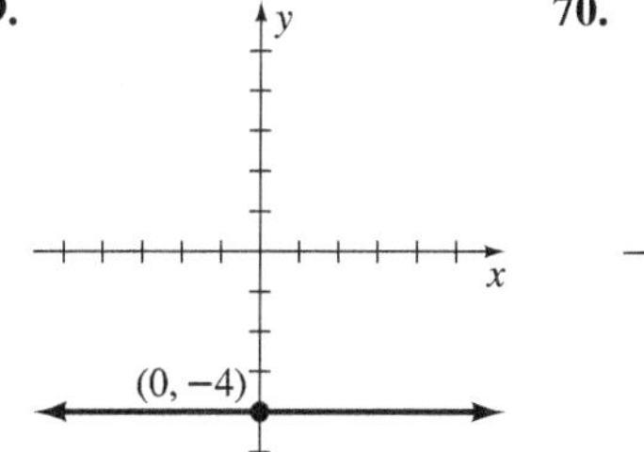


70.
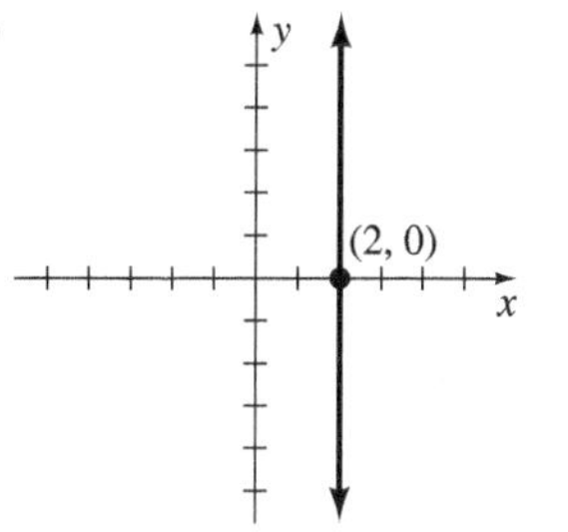


**71.** $7\sqrt{2}$ **72.** $\sqrt{41}$ **73.** $3x - 5y = 6$ **74.** $x + y = 3$

**75.** $2x + y = -1$ **76.** $3x + y = 1$

**77.** $C(-4, -2), r = 5$ units **78.** $(1, 5)$ **79.** $y = \pm\frac{4}{3}x$

80.
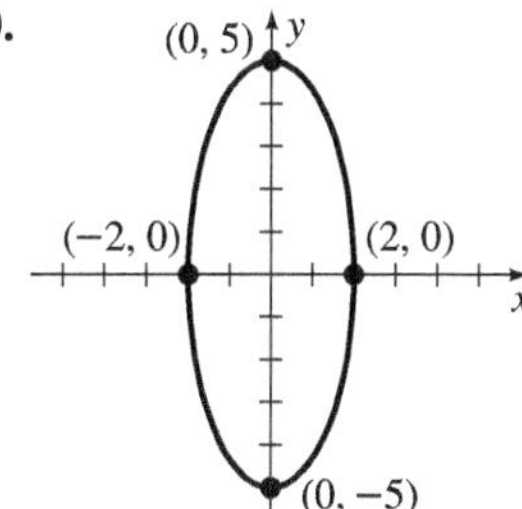


81.
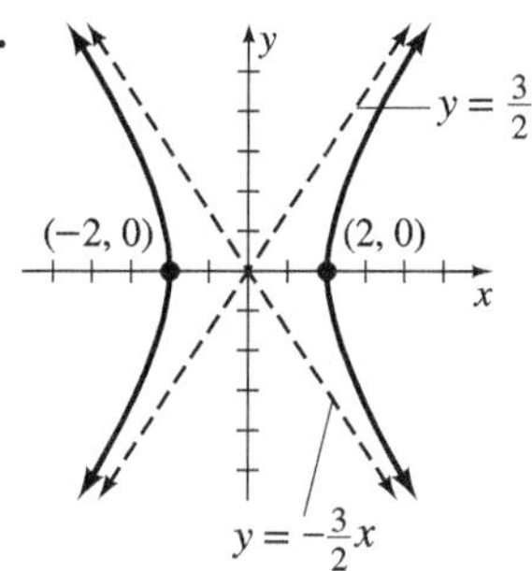


82.
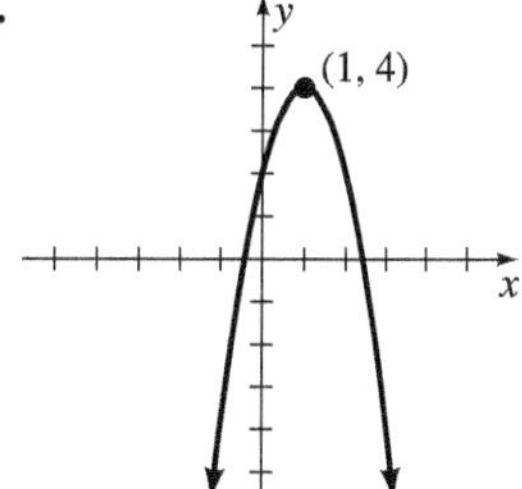


**83.** \$900, \$1350 **84.** 12 in. by 17 in. **85.** 5 hr **86.** 12 min

**87.** 7% **88.** 15 chairs per row

# Chapter 9

## Problem Set 9.1 (page 445)

**1.** $D = \{1, 2, 3, 4\}$, $R = \{5, 8, 11, 14\}$ It is a function.
**3.** $D = \{0, 1\}$, $R = \{-2\sqrt{6}, -5, 5, 2\sqrt{6}\}$ It is not a function.
**5.** $D = \{1, 2, 3, 4, 5\}$, $R = \{2, 5, 10, 17, 26\}$ It is a function.
**7.** $D = \{\text{all reals}\}$ or $D: (-\infty, \infty)$, $R = \{\text{all reals}\}$ or $R: (-\infty, \infty)$. It is a function.
**9.** $D = \{\text{all reals}\}$ or $D: (-\infty, \infty)$, $R = \{y|y \geq 0\}$ or $R: [0, \infty)$. It is a function. **11.** $\{\text{all reals}\}$ or $(-\infty, \infty)$
**13.** $\{x|x \neq 1\}$ or $(-\infty, 1) \cup (1, \infty)$
**15.** $\left\{x|x \neq \frac{3}{4}\right\}$ or $\left(-\infty, \frac{3}{4}\right) \cup \left(\frac{3}{4}, \infty\right)$
**17.** $\{x|x \neq -1, x \neq 4\}$ or $(-\infty, -1) \cup (-1, 4) \cup (4, \infty)$
**19.** $\{x|x \neq -8, x \neq 5\}$ or $(-\infty, -8) \cup (-8, 5) \cup (5, \infty)$
**21.** $\{x|x \neq -6, x \neq 0\}$ or $(-\infty, -6) \cup (-6, 0) \cup (0, \infty)$
**23.** $\{\text{all reals}\}$ or $(-\infty, \infty)$
**25.** $\{t|t \neq -2, t \neq 2\}$ or $(-\infty, -2) \cup (-2, 2) \cup (2, \infty)$
**27.** $\{x|x \geq -4\}$ or $[-4, \infty)$ **29.** $\left\{s|s \geq \frac{5}{4}\right\}$ or $\left[\frac{5}{4}, \infty\right)$
**31.** $\{x|x \leq -4 \text{ or } x \geq 4\}$ or $(-\infty, -4] \cup [4, \infty)$
**33.** $\{x|x \leq -3 \text{ or } x \geq 6\}$ or $(-\infty, -3] \cup [6, \infty)$
**35.** $\{x|-1 \leq x \leq 1\}$ or $[-1, 1]$
**37.** $f(0) = -2$, $f(2) = 8$, $f(-1) = -7$, $f(-4) = -22$
**39.** $f(-2) = -\frac{7}{4}, f(0) = -\frac{3}{4}, f\left(\frac{1}{2}\right) = -\frac{1}{2}, f\left(\frac{2}{3}\right) = -\frac{5}{12}$
**41.** $g(-1) = 0$; $g(2) = -9$; $g(-3) = 26$; $g(4) = 5$
**43.** $h(-2) = -2$; $h(-3) = -11$; $h(4) = -32$; $h(5) = -51$
**45.** $f(3) = \sqrt{7}; f(4) = 3; f(10) = \sqrt{21}; f(12) = 5$
**47.** $f(1) = -1$; $f(-1) = -2$; $f(3) = -\frac{2}{3}; f(-6) = \frac{4}{3}$
**49.** $f(a) = a^2 - 7a, f(a - 3) = a^2 - 13a + 30$, $f(a + h) = a^2 + 2ah + h^2 - 7a - 7h$
**51.** $f(-a) = 2a^2 + a - 1, f(a + 1) = 2a^2 + 3a$, $f(a + h) = 2a^2 + 4ah + 2h^2 - a - h - 1$
**53.** $f(-a) = -a^2 + 2a - 7$, $f(-a - 2) = -a^2 - 2a - 7$, $f(a + 7) = -a^2 - 16a - 70$
**55.** $f(-2) = 27$; $f(3) = 42$; $g(-4) = -37$; $g(6) = -17$
**57.** $f(-2) = 5$; $f(3) = 8$; $g(-4) = -3$; $g(5) = -4$
**59.** $-3$ **61.** $-2a - h$ **63.** $4a - 1 + 2h$
**65.** $-8a - 7 - 4h$
**67.** $h(1) = 48$; $h(2) = 64$; $h(3) = 48$; $h(4) = 0$
**69.** $C(75) = \$74$; $C(150) = \$98$; $C(225) = \$122$; $C(650) = \$258$
**71.** $I(0.04) = \$20$; $I(0.06) = \$30$; $I(0.075) = \$37.50$; $I(0.09) = \$45$
**73.** $M(2) = 4.4$; $M(6) = 29.2$; $M(10) = 54$

## Problem Set 9.2 (page 457)

1.
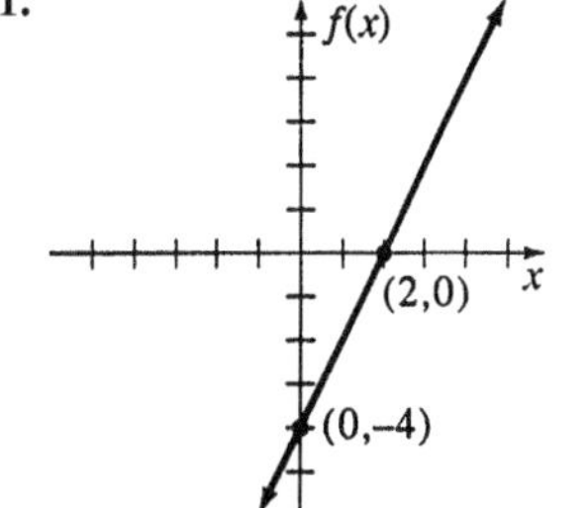


3.
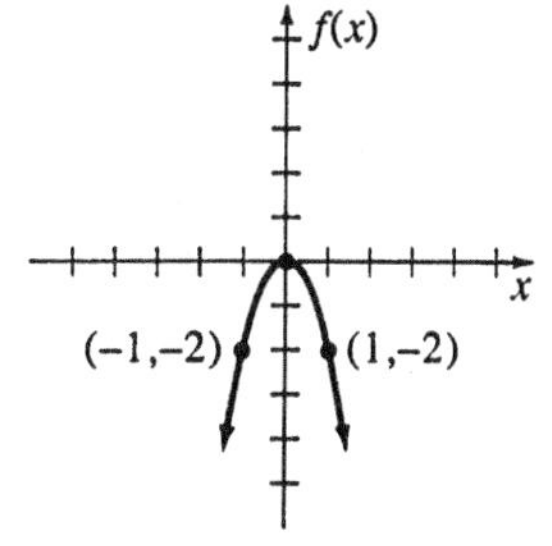

**5.**

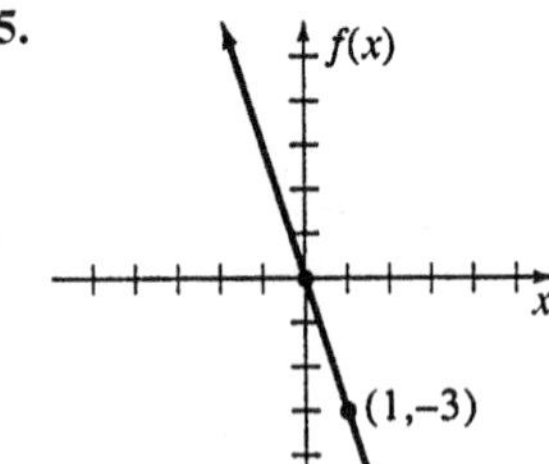


**7.**

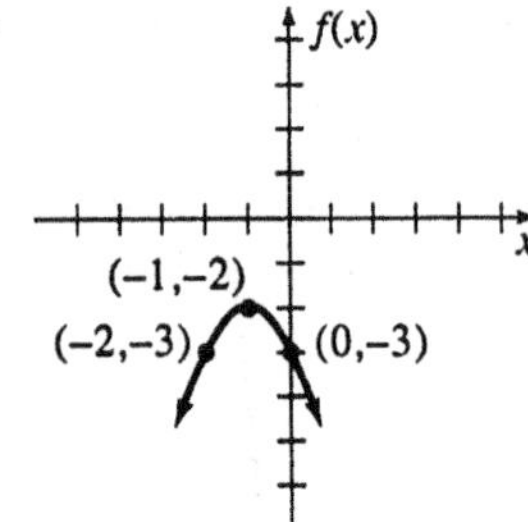


**9.**

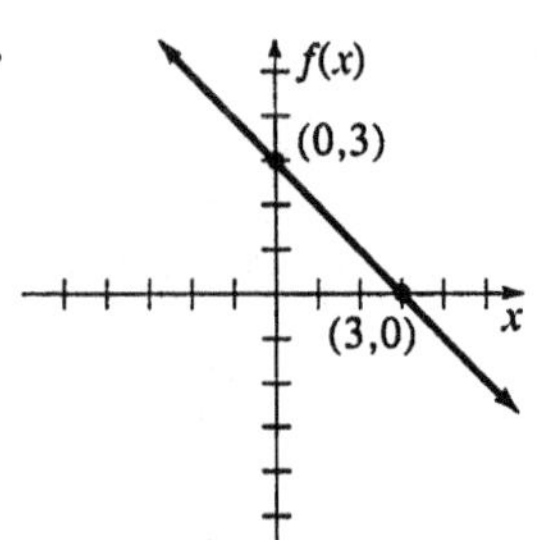


**11.**

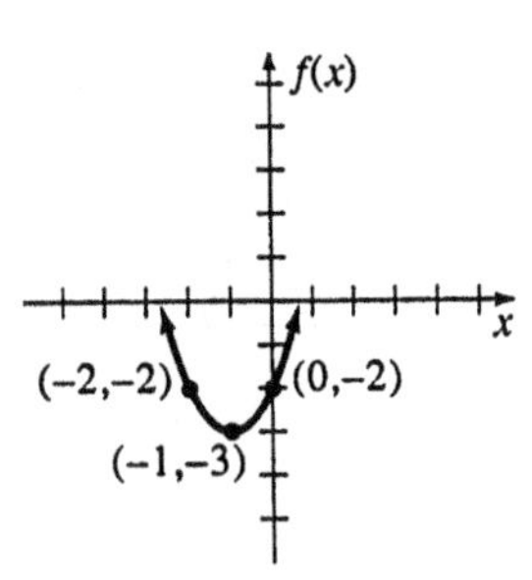


**13.**

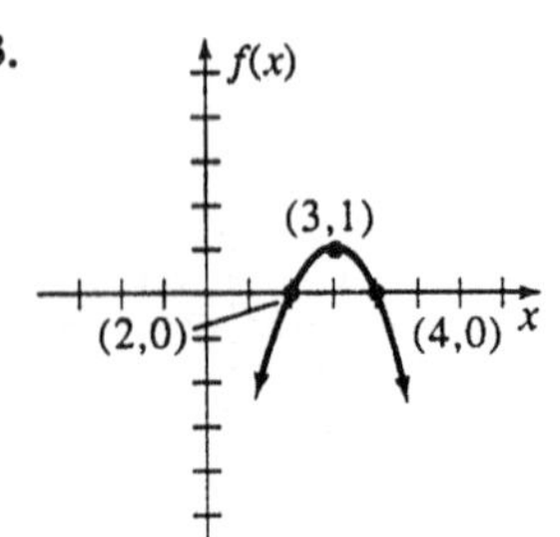


**15.**

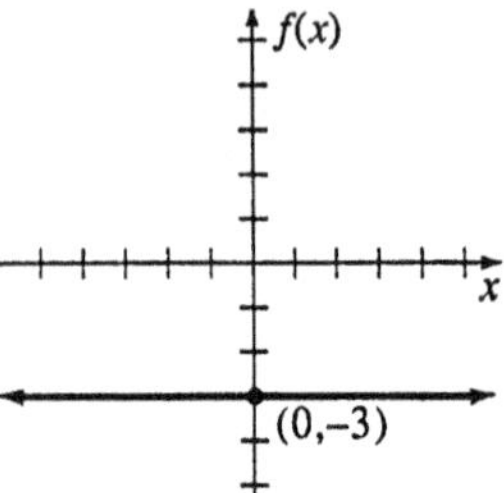
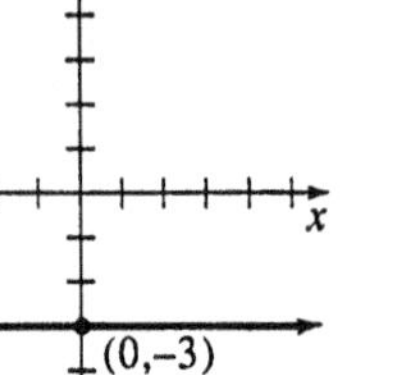


**17.**



**19.**



**21.**

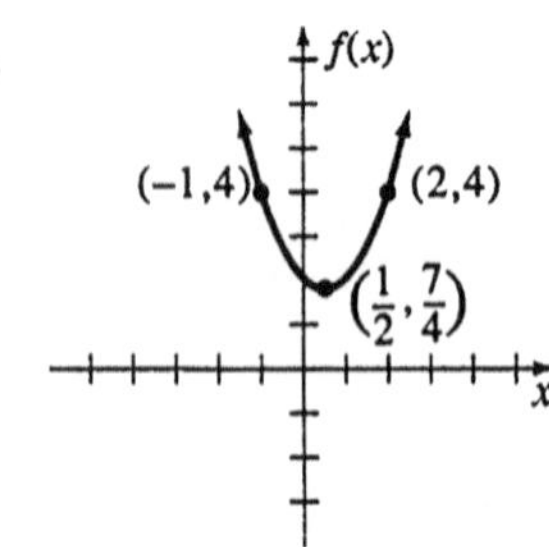


**23.**

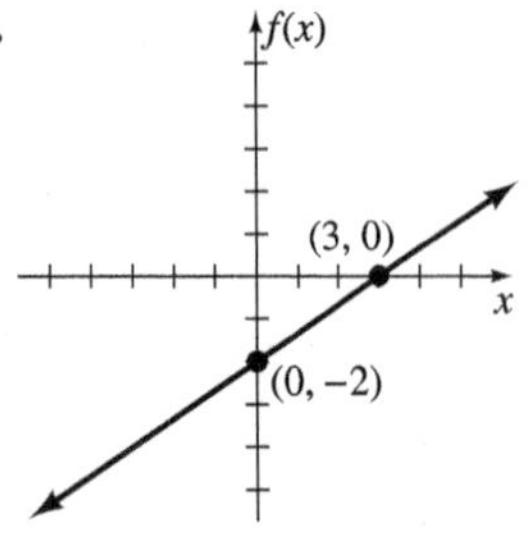


**25.**

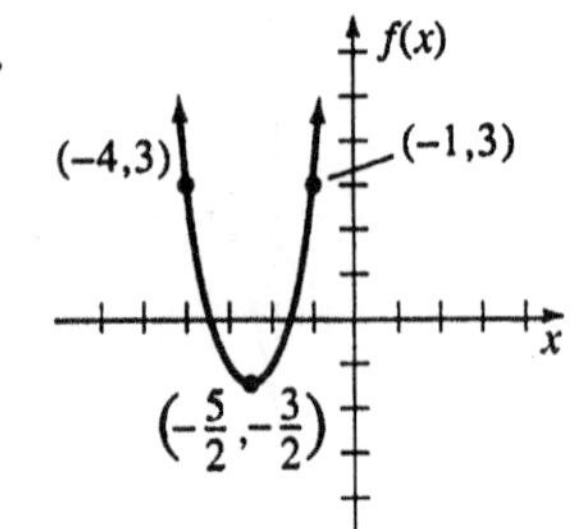


**27.**

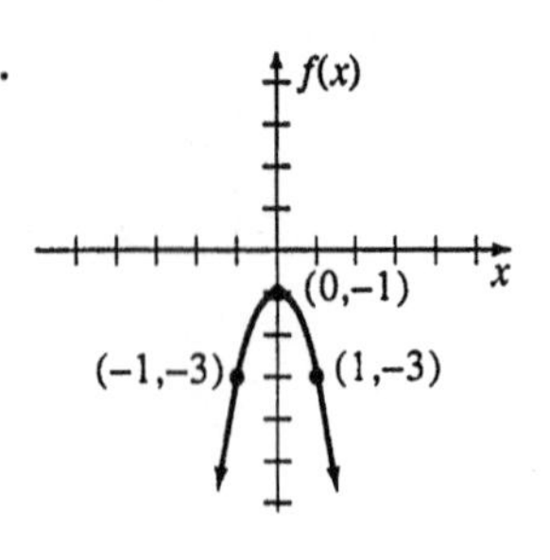


**29.**

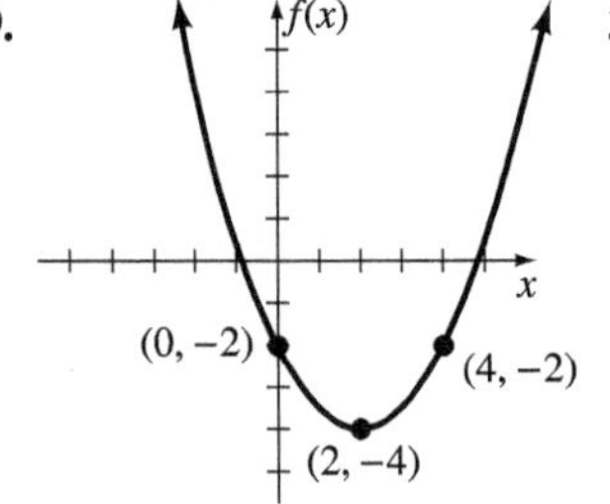


**31.**

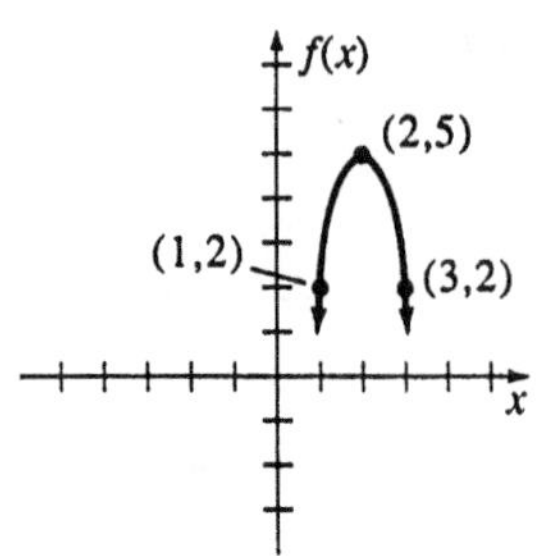


**33.**

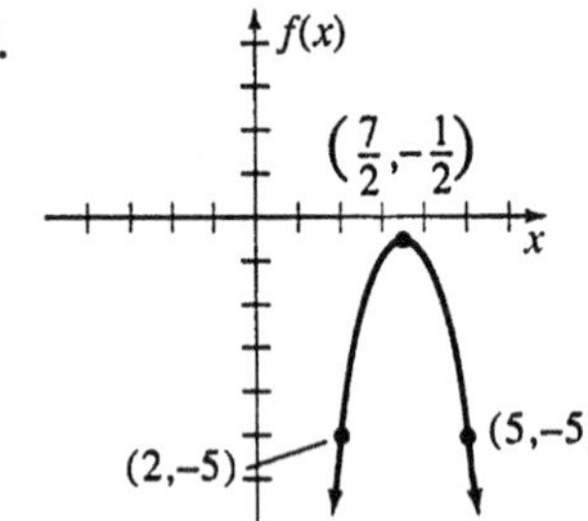


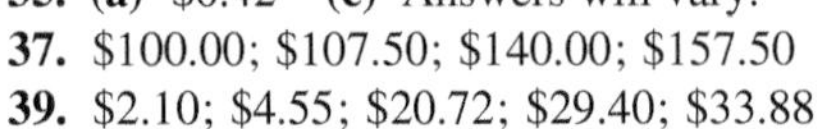

**35.** **(a)** \$0.42 **(c)** Answers will vary.
**37.** \$100.00; \$107.50; \$140.00; \$157.50
**39.** \$2.10; \$4.55; \$20.72; \$29.40; \$33.88
**41.** 21 days; 29 days; 45 days
**43.** $f(p) = 0.8p$; \$7.60; \$12; \$60; \$10; \$600
**45.** 80 items **47.** 5 and 25 **49.** 60 meters by 60 meters
**51.** 40 yards by 80 yards
**53.** 1100 subscribers at \$13.75 per month

## Problem Set 9.3 (page 468)

**1.** B **3.** A **5.** F

**7.**

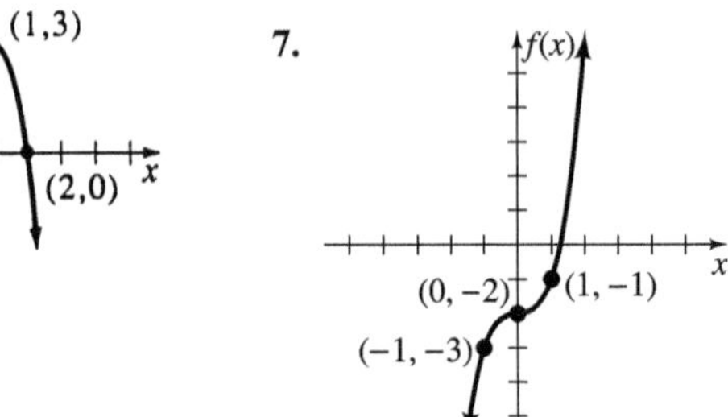


**9.**

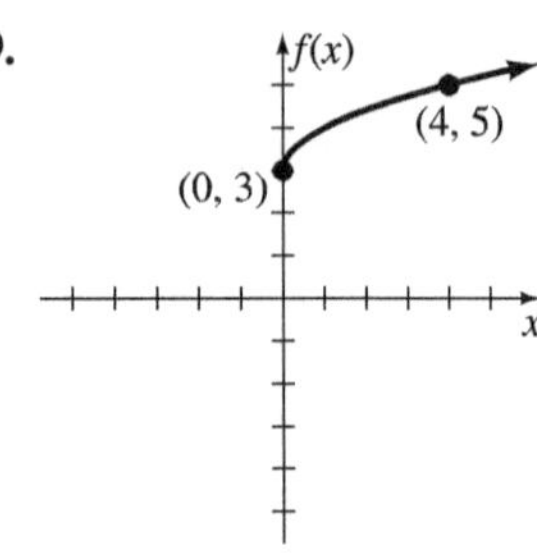


**11.**

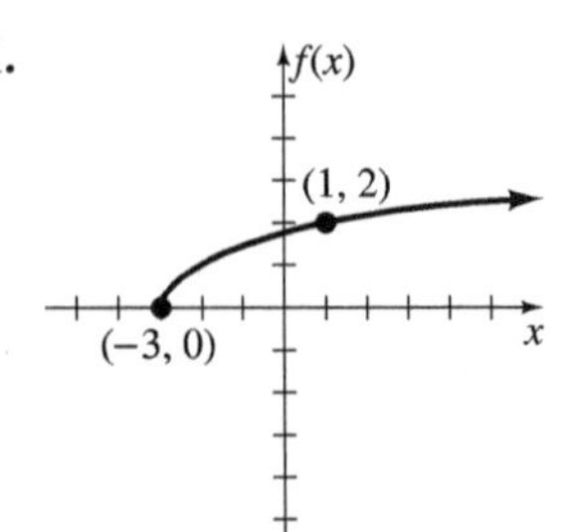


**13.**

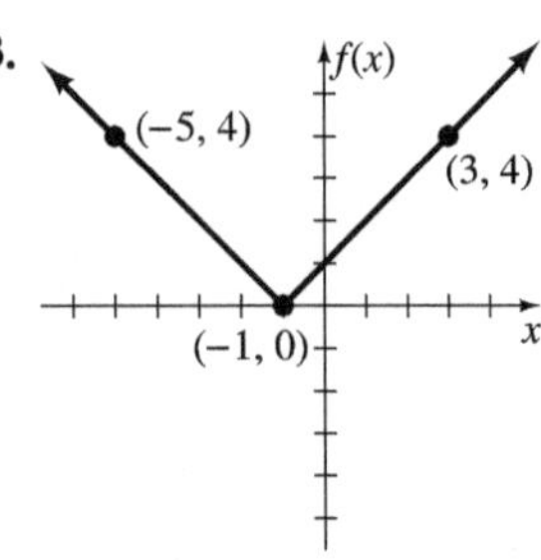


**15.**

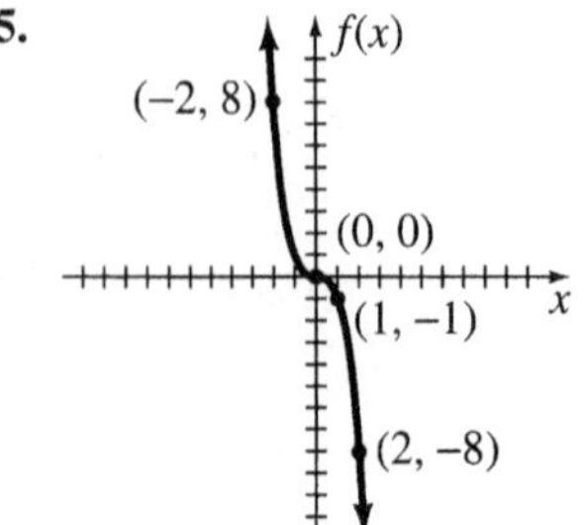


**17.**

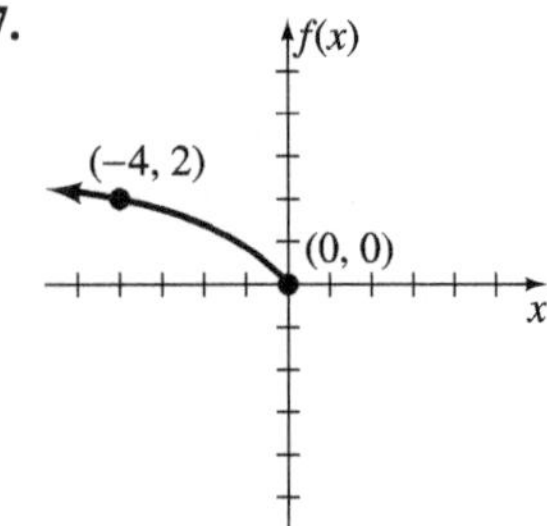

**19.**
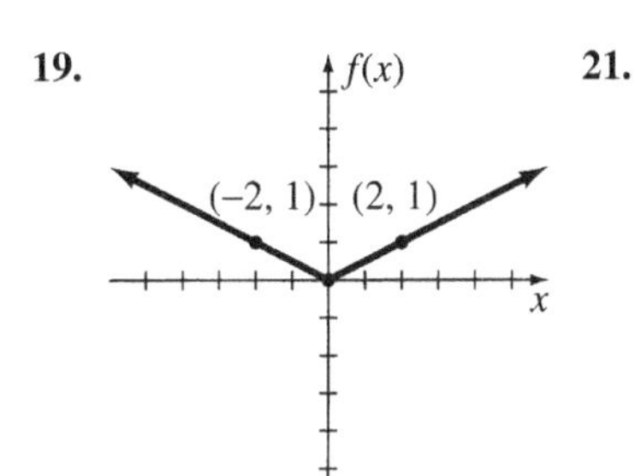


**21.**
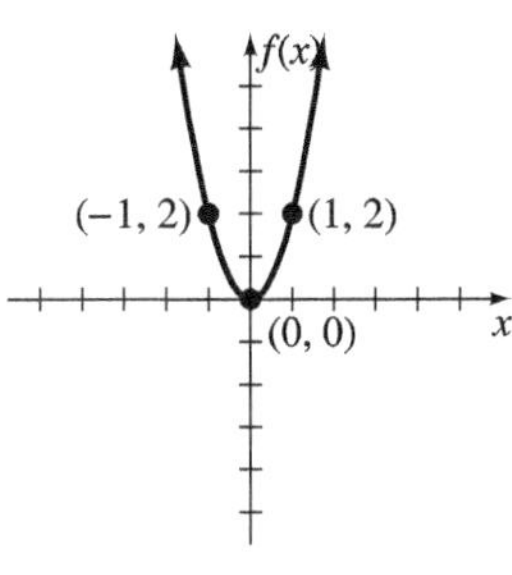


**23.**
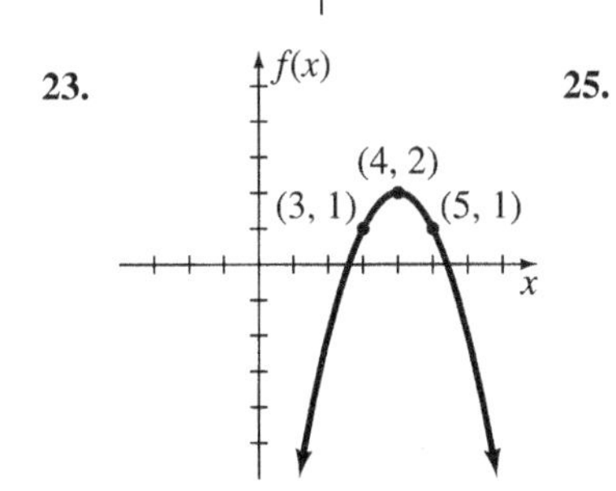


**25.**
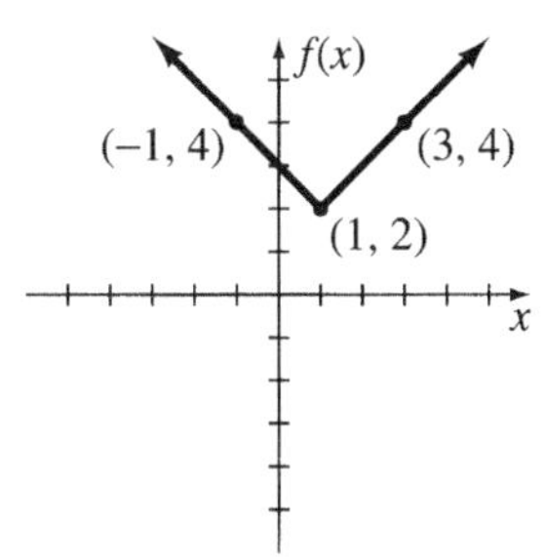


**27.**
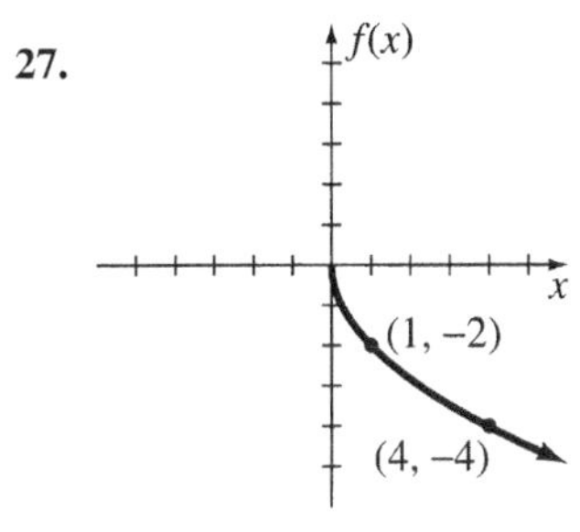


**29.**
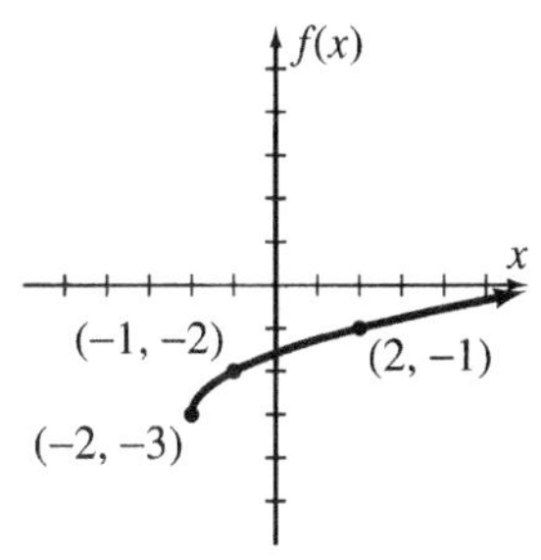


**31.**
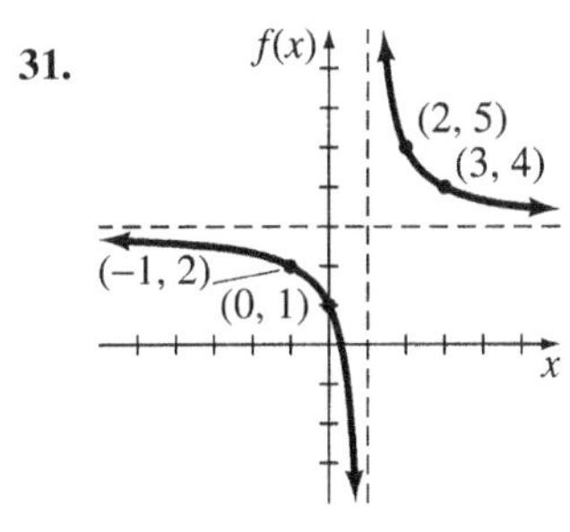


**33.**
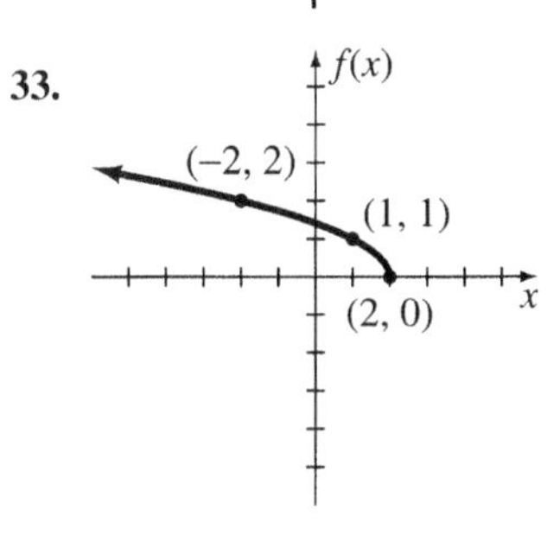


**35.**
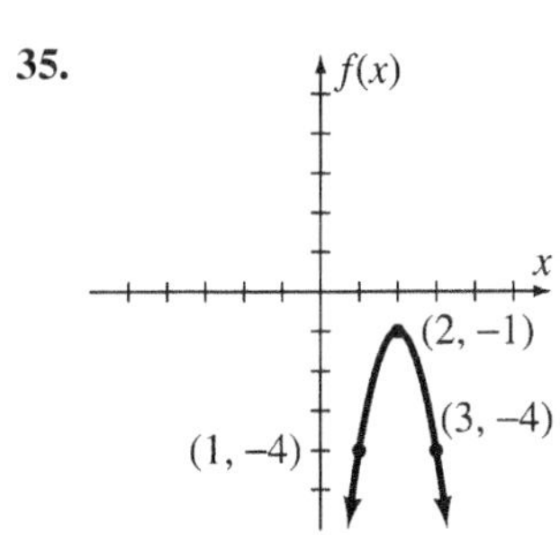


**37.**
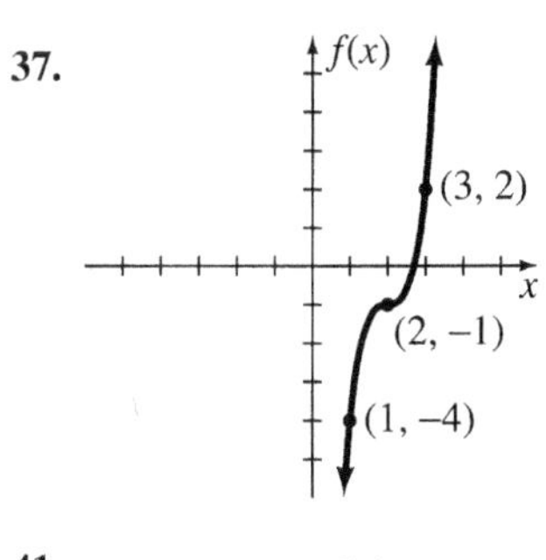


**39.**
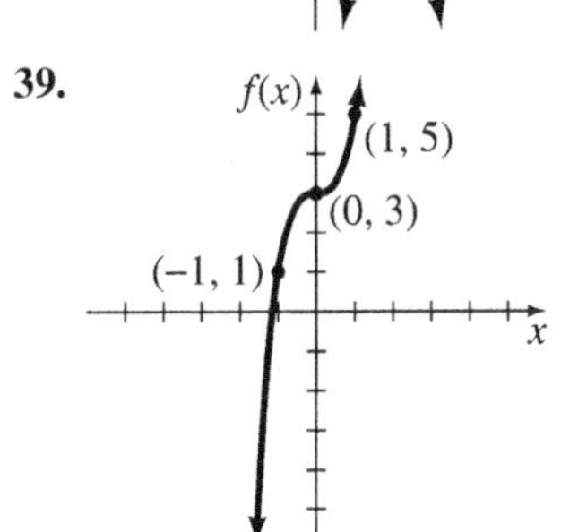


**41.**
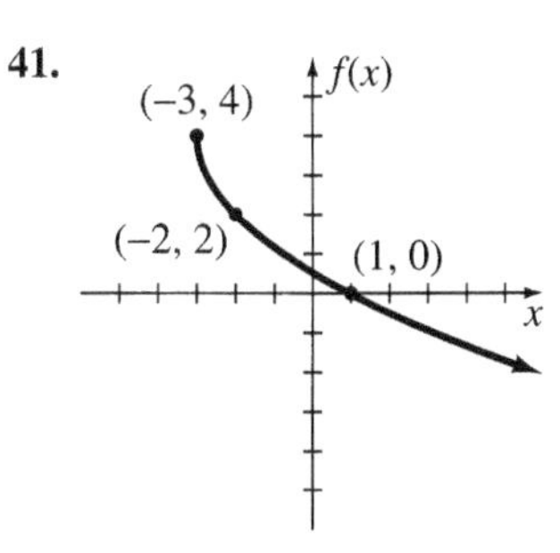


**43.**
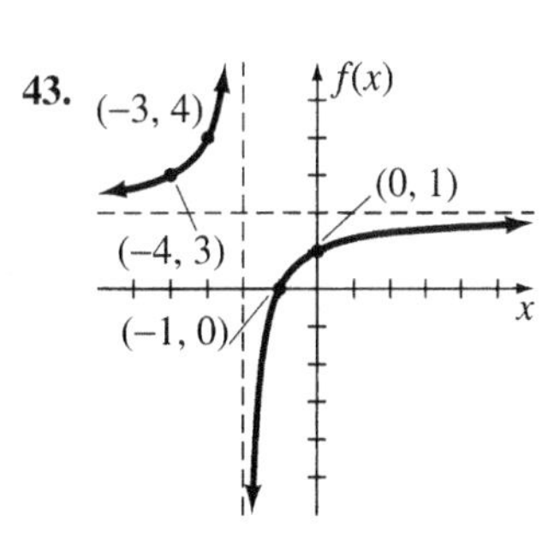


**45.**
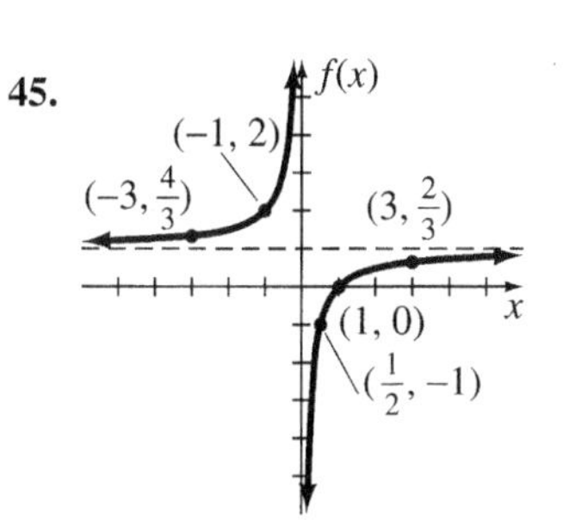


**47.** **49.**
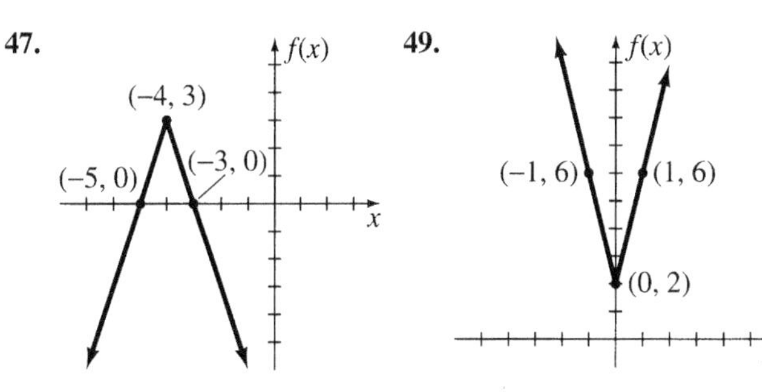


**51.**
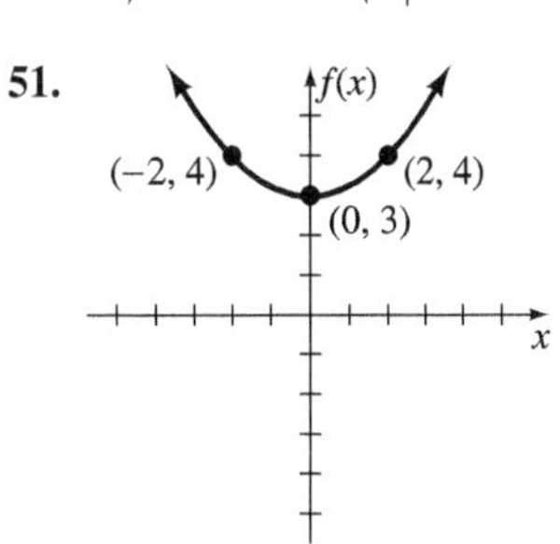


**53.**
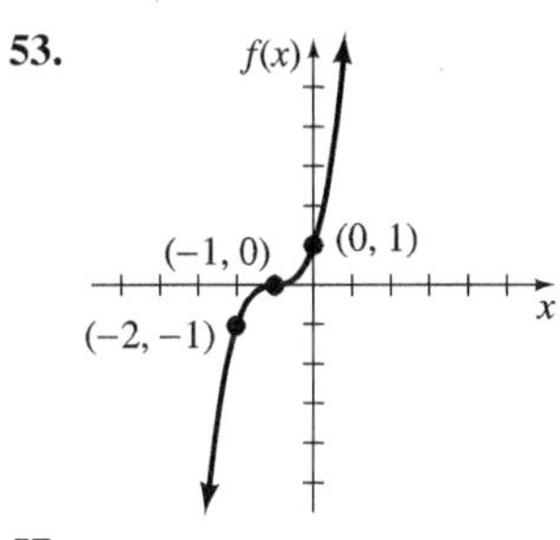


**55.**
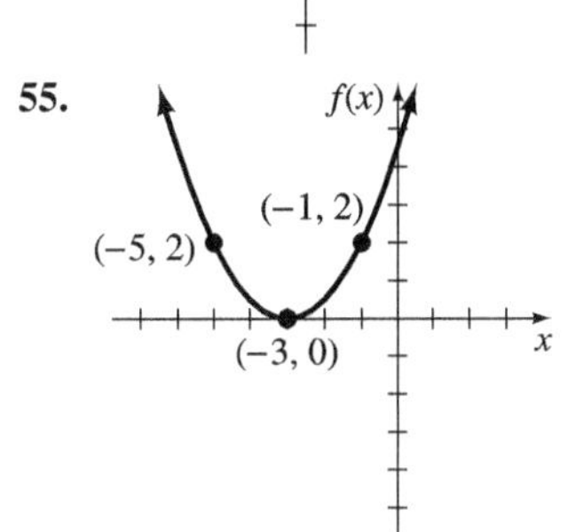


**57.**
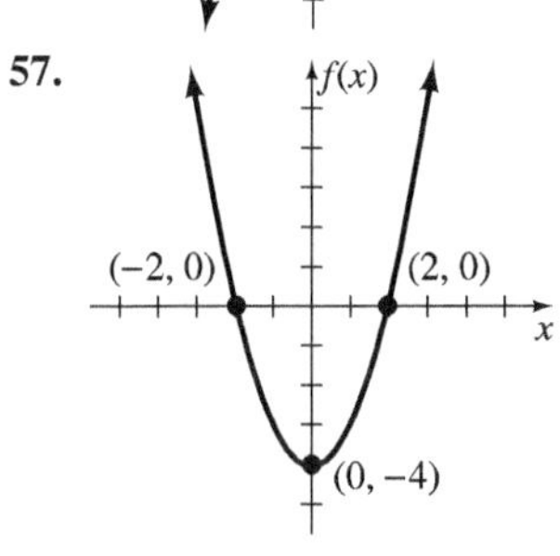


**59.**
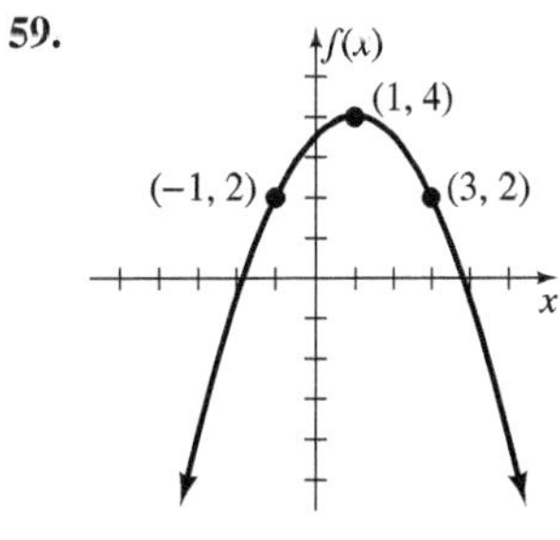


**61. (a)** **61. (c)**
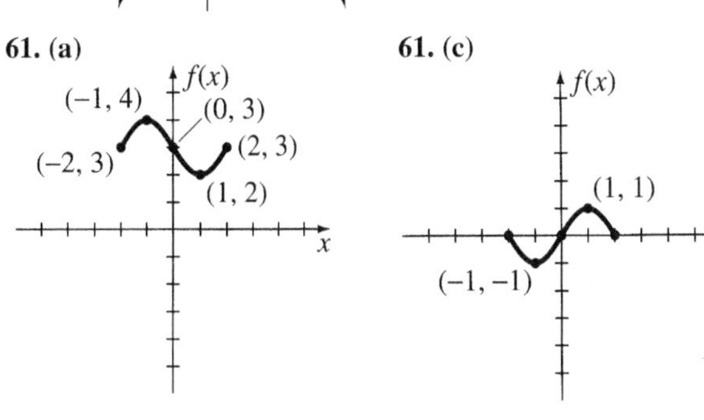

## Problem Set 9.4 (page 473)

**1.** $(f \circ g)(x) = 15x - 3$, $D = \{\text{all reals}\}$ or $D: (-\infty, \infty)$; $(g \circ f)(x) = 15x - 1$, $D = \{\text{all reals}\}$ or $D: (-\infty, \infty)$
**3.** $(f \circ g)(x) = -14x - 7$, $D = \{\text{all reals}\}$ or $D: (-\infty, \infty)$; $(g \circ f)(x) = -14x + 11$, $D = \{\text{all reals}\}$ or $D: (-\infty, \infty)$
**5.** $(f \circ g)(x) = 3x^2 + 11$, $D = \{\text{all reals}\}$ or $D: (-\infty, \infty)$; $(g \circ f)(x) = 9x^2 + 12x + 7$, $D = \{\text{all reals}\}$ or $D: (-\infty, \infty)$
**7.** $(f \circ g)(x) = 2x^2 - 11x + 17$, $D = \{\text{all reals}\}$ or $D: (-\infty, \infty)$; $(g \circ f)(x) = -2x^2 + x + 1$, $D = \{\text{all reals}\}$ or $D: (-\infty, \infty)$
**9.** $(f \circ g)(x) = \dfrac{3}{4x - 9}$, $D = \left\{x \mid x \neq \dfrac{9}{4}\right\}$ or $D: \left(-\infty, \dfrac{9}{4}\right) \cup \left(\dfrac{9}{4}, \infty\right)$; $(g \circ f)(x) = \dfrac{12 - 9x}{x}$, $D = \{x \mid x \neq 0\}$ or $D: (-\infty, 0) \cup (0, \infty)$
**11.** $(f \circ g)(x) = \sqrt{5x + 4}$, $D = \left\{x \mid x \geq -\dfrac{4}{5}\right\}$ or $D: \left[-\dfrac{4}{5}, \infty\right)$; $(g \circ f)(x) = 5\sqrt{x + 1} + 3$, $D = \{x \mid x \geq -1\}$ or $D: [-1, \infty)$
**13.** $(f \circ g)(x) = x - 4$, $D = \{x \mid x \neq 4\}$ or $D: (-\infty, 4) \cup (4, \infty)$; $(g \circ f)(x) = \dfrac{x}{1 - 4x}$, $D = \left\{x \mid x \neq 0 \text{ and } x \neq \dfrac{1}{4}\right\}$ or $D: (-\infty, 0) \cup \left(0, \dfrac{1}{4}\right) \cup \left(\dfrac{1}{4}, \infty\right)$
**15.** $(f \circ g)(x) = \dfrac{2\sqrt{x}}{x}$, $D = \{x \mid x > 0\}$ or $D: (0, \infty)$; $(g \circ f)(x) = \dfrac{4\sqrt{x}}{x}$, $D = \{x \mid x > 0\}$ or $D: (0, \infty)$
**17.** $(f \circ g)(x) = \dfrac{3x + 3}{2}$, $D = \{x \mid x \neq -1\}$ or $D: (-\infty, -1) \cup (-1, \infty)$; $(g \circ f)(x) = \dfrac{2x}{2x + 3}$, $D = \left\{x \mid x \neq 0 \text{ and } x \neq -\dfrac{3}{2}\right\}$ or $D: \left(-\infty, -\dfrac{3}{2}\right) \cup \left(-\dfrac{3}{2}, 0\right) \cup (0, \infty)$
**27.** 124 and $-130$ **29.** 323 and 257 **31.** $\dfrac{1}{2}$ and $-1$
**33.** Undefined and undefined **35.** $\sqrt{7}$ and 0 **37.** 37 and 27

## Problem Set 9.5 (page 479)

**1.** Not a function **3.** Function **5.** Function
**7.** Function **9.** One-to-one function
**11.** Not a one-to-one function
**13.** Not a one-to-one function **15.** One-to-one function
**17.** Domain of $f$: $\{1, 2, 3, 4\}$
Range of $f$: $\{3, 6, 11, 18\}$
$f^{-1}$: $\{(3, 1), (6, 2), (11, 3), (18, 4)\}$
Domain of $f^{-1}$: $\{3, 6, 11, 18\}$
Range of $f^{-1}$: $\{1, 2, 3, 4\}$
**19.** Domain of $f$: $\{-2, -1, 0, 5\}$
Range of $f$: $\{-1, 1, 5, 10\}$
$f^{-1}$: $\{(-1, -2), (1, -1), (5, 0), (10, 5)\}$
Domain of $f^{-1}$: $\{-1, 1, 5, 10\}$
Range of $f^{-1}$: $\{-2, -1, 0, 5\}$
**21.** $f^{-1}(x) = \dfrac{x + 4}{5}$ **23.** $f^{-1}(x) = \dfrac{1 - x}{2}$
**25.** $f^{-1}(x) = \dfrac{5}{4}x$ **27.** $f^{-1}(x) = 2x - 8$
**29.** $f^{-1}(x) = \dfrac{15x + 6}{5}$ **31.** $f^{-1}(x) = \dfrac{x - 4}{9}$
**33.** $f^{-1}(x) = \dfrac{-x - 4}{5}$ **35.** $f^{-1}(x) = \dfrac{-3x + 21}{2}$
**37.** $f^{-1}(x) = \dfrac{3}{4}x + \dfrac{3}{16}$ **39.** $f^{-1}(x) = -\dfrac{7}{3}x - \dfrac{14}{9}$
**41.** $f^{-1}(x) = \dfrac{1}{4}x$

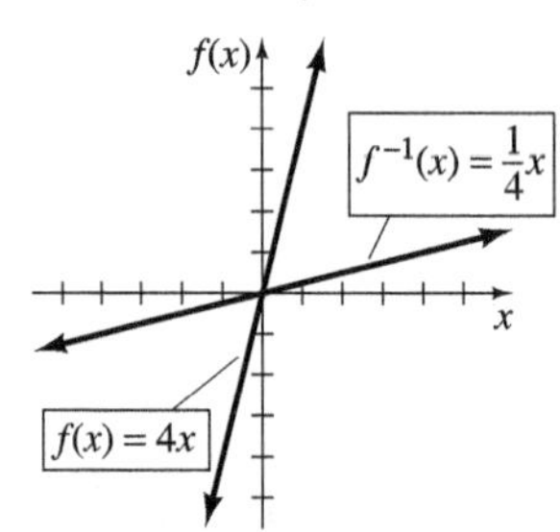


**43.** $f^{-1}(x) = -3x$

f(x) = −1/3 x
f(x)
x
f^{-1}(x) = −3x

**45.** $f^{-1}(x) = \dfrac{x + 3}{3}$

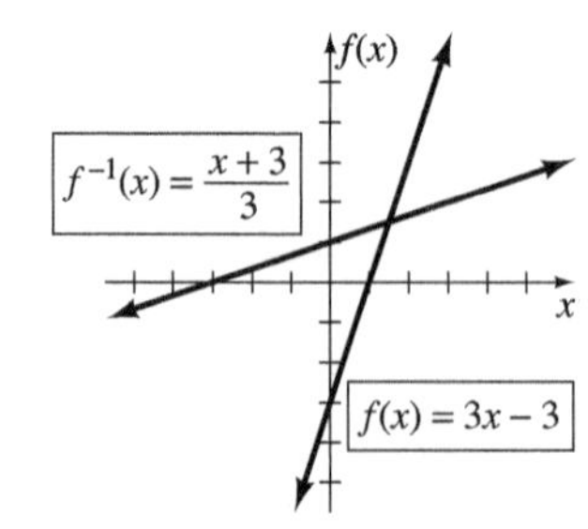


**47.** $f^{-1}(x) = \dfrac{-x - 4}{2}$

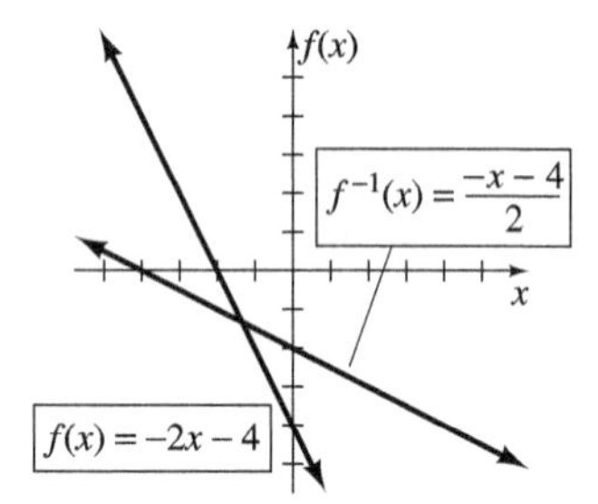


**49.** $f^{-1}(x) = \sqrt{x}, x \geq 0$

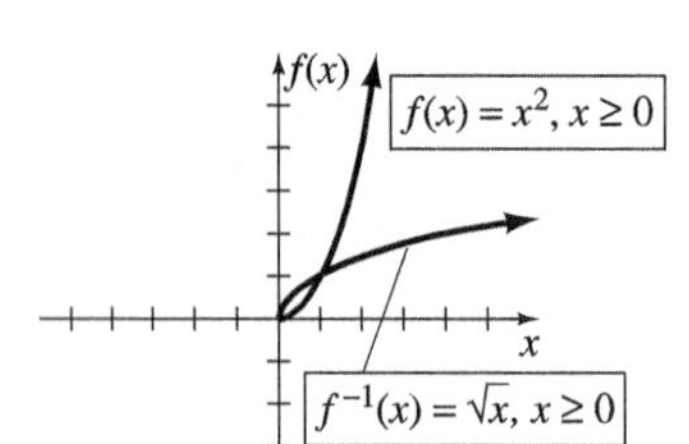


**53. (a)** $f^{-1}(x) = \dfrac{x - 1}{2}$
**(c)** $f^{-1}(x) = \dfrac{-x + 5}{4}$
**(e)** $f^{-1}(x) = \dfrac{1}{2}x$

## Problem Set 9.6 (page 487)

**1.** $y = \frac{k}{x^2}$ **3.** $C = \frac{kg}{t^3}$ **5.** $V = kr^3$ **7.** $S = ke^2$

**9.** $V = khr^2$ **11.** $\frac{2}{3}$ **13.** $-4$ **15.** $\frac{1}{3}$ **17.** $-2$ **19.** 2

**21.** 5 **23.** 9 **25.** 9 **27.** $\frac{1}{6}$ **29.** 112

**31.** 12 cubic centimeters **33.** 28 **35.** 2 seconds
**37.** 12 ohms **39. (a)** \$87.50 **(b)** \$273 **(c)** \$560
**41.** 3560.76 cubic meters **43.** 0.048

## Chapter 9 Sample Problems (page 489)

**1.** $D = \{0, 2, 3\}, R = \{1, 5, 6, 7\}$. It is not a function.

**2.** $f(-2) = 17$ **3.** $\left\{x \neq -\frac{1}{2}\right\}$ or $\left(-\infty, -\frac{1}{2}\right) \cup \left(-\frac{1}{2}, \infty\right)$

**4.** 3

**5.**

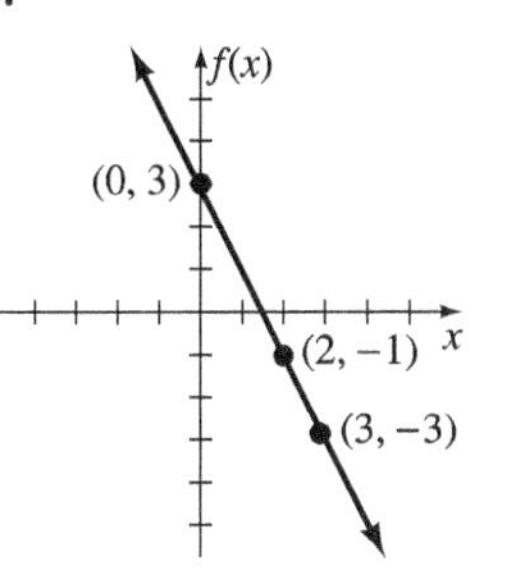


**6.** $C(90) = 38$

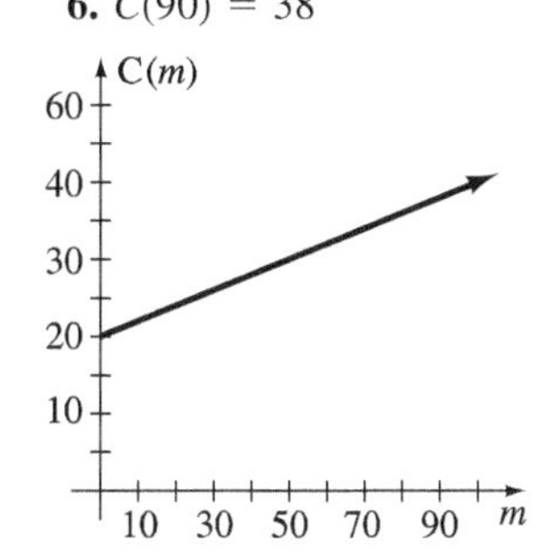

**7.**

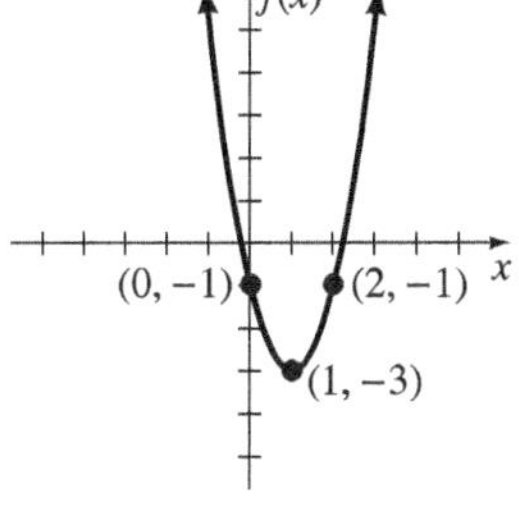
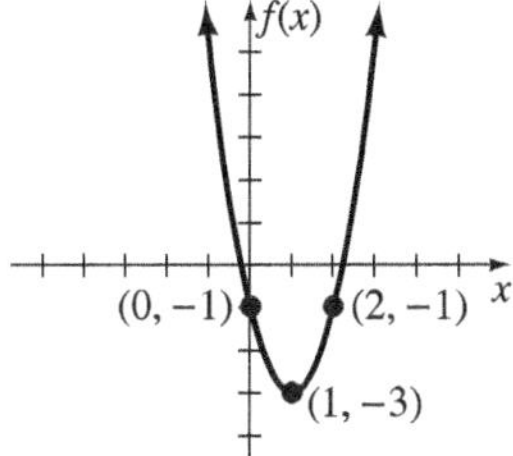


**8.** 35 items

**9.**

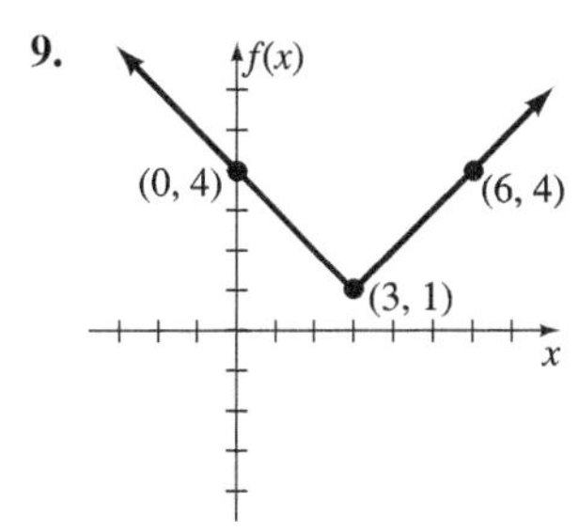


**10.**

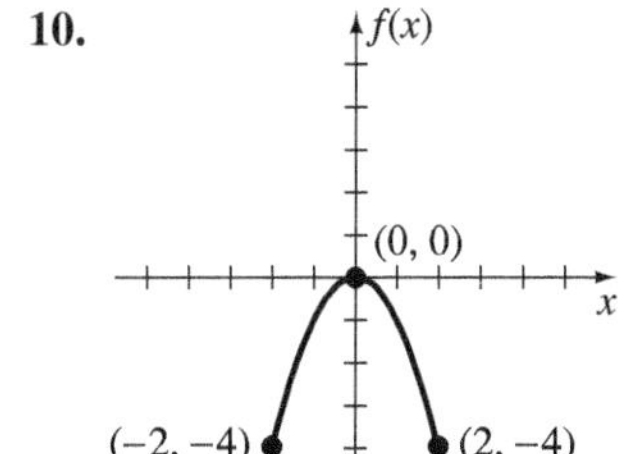


**11.**

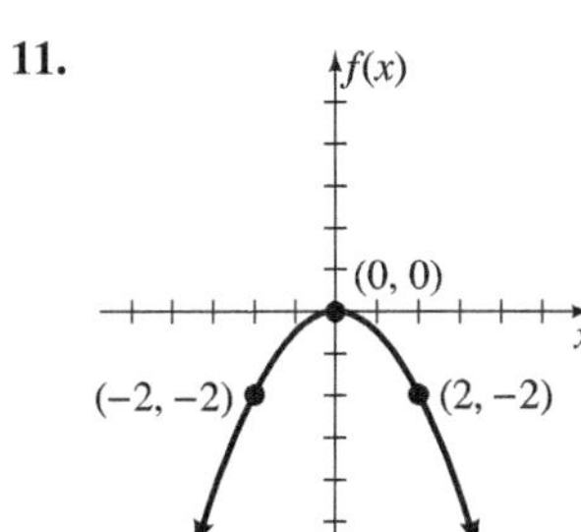


**12.**

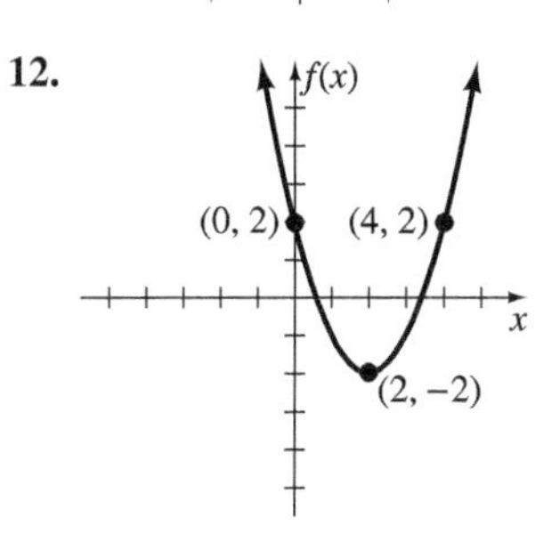


**13.** $x^2 - 2x - 3$ **14.** $(g \circ f)(-2) = 17$
**15.** This is a graph of a function. **16.** Not one-to-one
**17.** $f^{-1} = \{(3, 2), (5, -1), (-2, 4), (7, 1)\}$

**18.** $f^{-1}(x) = \frac{x - 5}{3}$

**19.** Your pay would be \$288.75 for working 35 hours.
**20.** 3.2 hours **21.** $120\pi$ cubic feet

## Chapter 9 Review Problem Set (page 499)

**1.** Yes; $D = \{6, 7, 8, 9\}$, $R = \{2, 3, 4, 5\}$
**2.** No; $D = \{1, 2\}$, $R = \{1, 3, 5, 7\}$
**3.** No; $D = \{0\}$, $R = \{-6, -5, -4, -3\}$
**4.** Yes; $D = \{0, 1, 2, 3\}$, $R = \{8\}$
**5.** $f(2) = -1$; $f(-3) = 14$; $f(a) = a^2 - 2a - 1$

**6.** $f(2) = 0$; $f(-1) = -6$; $f(3a) = \frac{6a - 4}{3a + 2}$

**7.** $D = \{\text{all reals}\}$
**8.** $D = \{x | x \neq 5\}$ or $D: (-\infty, 5) \cup (5, \infty)$
**9.** $D = \{x | x \neq 0, x \neq 4\}$ or
$D: (-\infty, -4) \cup (-4, 0) \cup (0, \infty)$
**10.** $D = \{x | x \leq -5 \text{ or } x \geq 5\}$ or
$D: (-\infty, -5] \cup [5, \infty)$
**11.** 6 **12.** $4a + 1 + 2h$ **13.** \$0.72
**14.** $f(x) = 0.7x$; \$45.50; \$33.60; \$10.85

**15.**

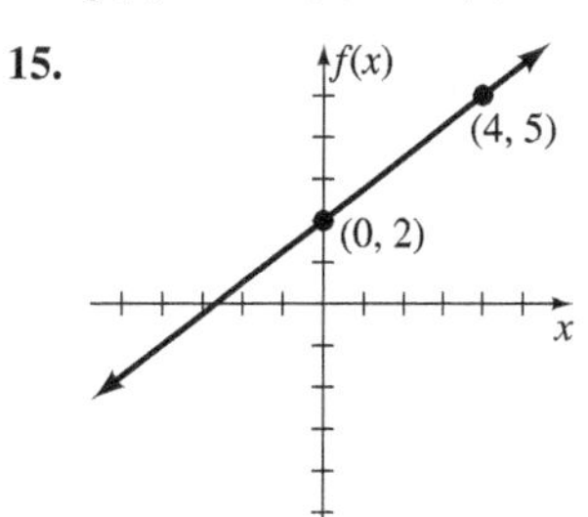


**16.**

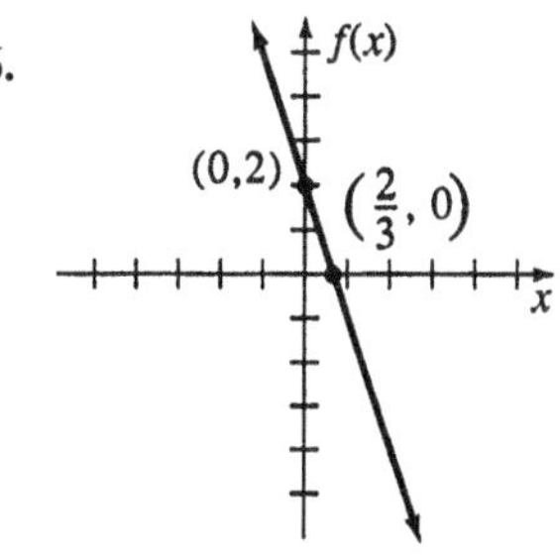


**17.**

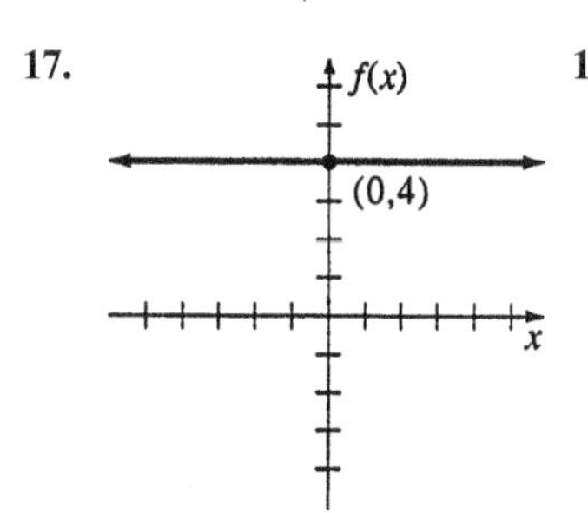


**18.**

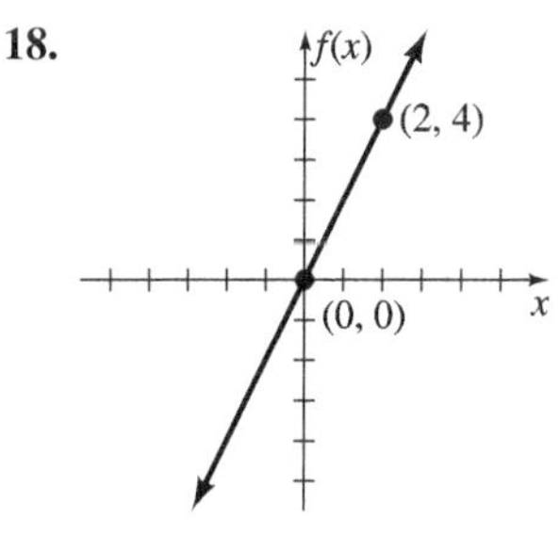


**19.** $f(x) = -2x + 5$

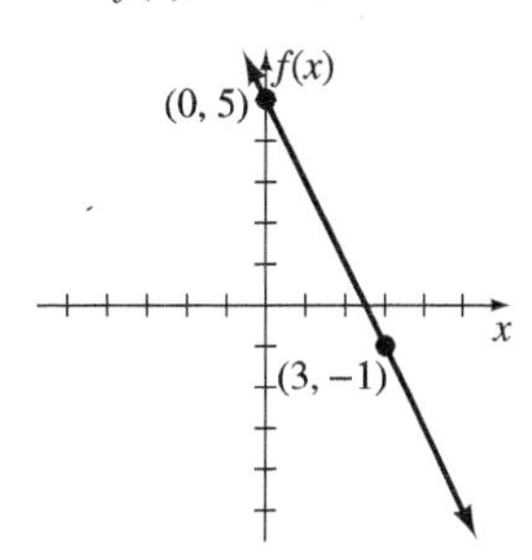


**20.** $f(x) = -3$

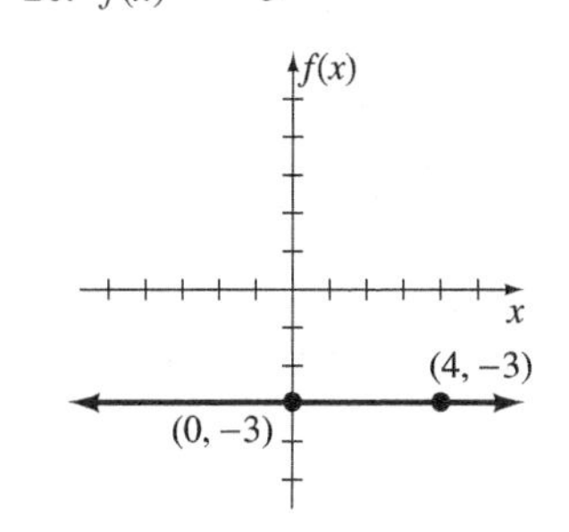

**21.** $f(x) = \frac{3}{2}x$

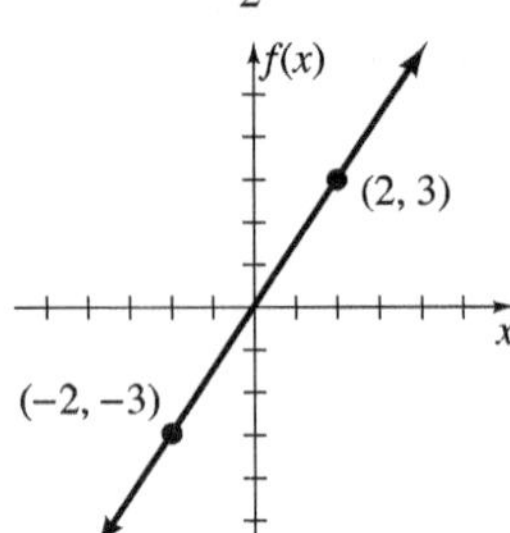


**22.** $f(x) = \frac{4}{3}x - 3$

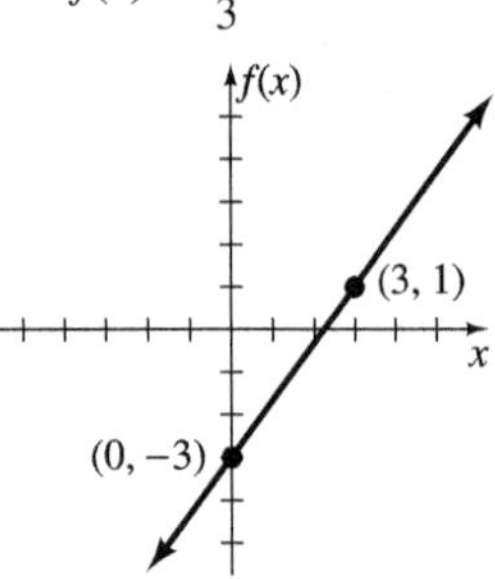


**23.** $f(x) = 6x - 100$; 110

**24.** $f(x) = \frac{10}{3}x + 150$; \$300

**25.**

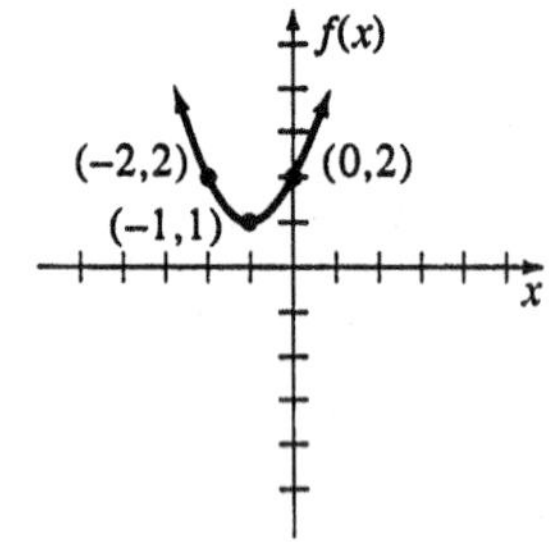


**26.**

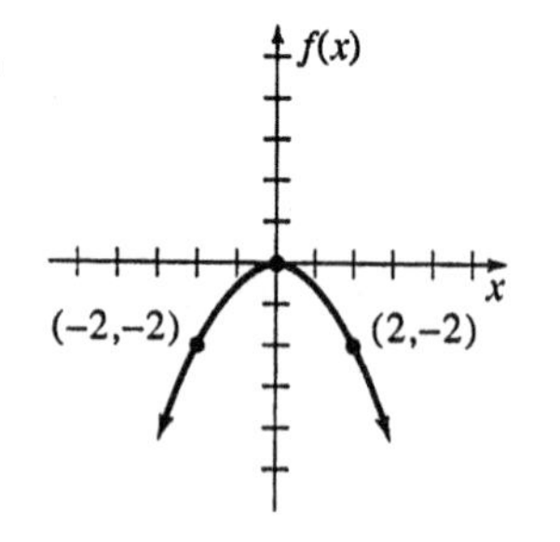


**27.**

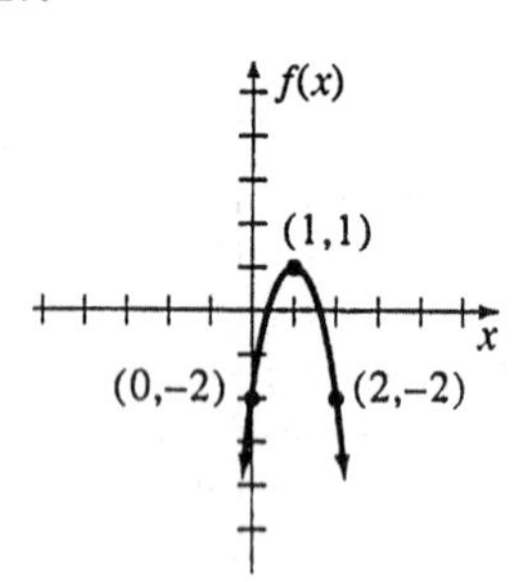


**28.** $f(x) = x^2 + 4x + 5$

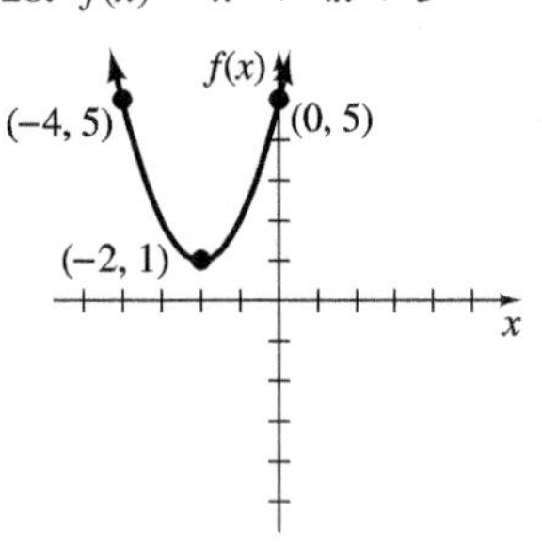


**29.** $f(x) = 2x^2$

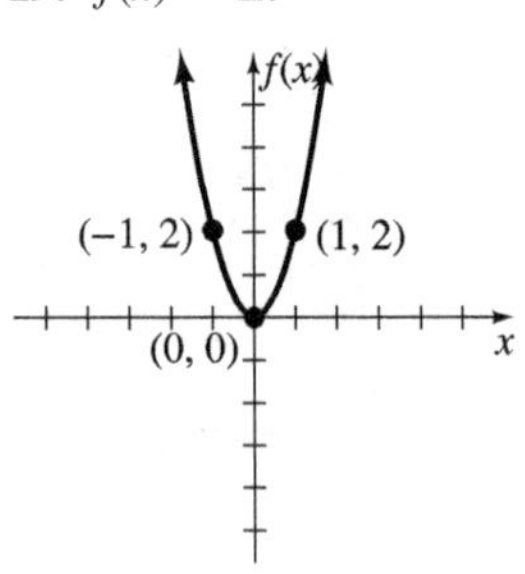


**30.** $f(x) = -2x^2 - 6x + 3$

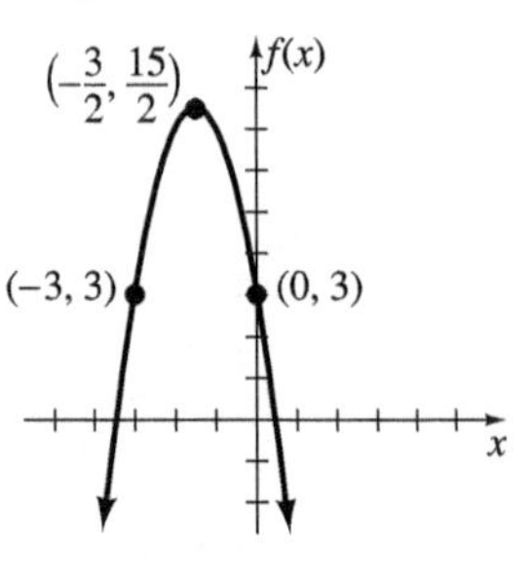


**31.** $f(x) = -x^2 - 2x + 6$

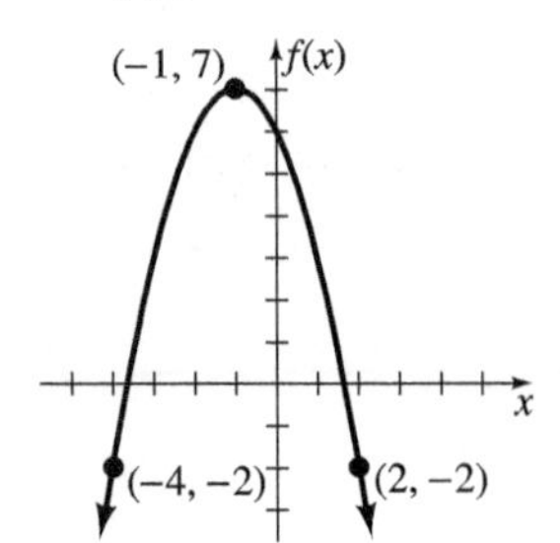


**32.** $f(x) = -x^2 + 3x + 4$

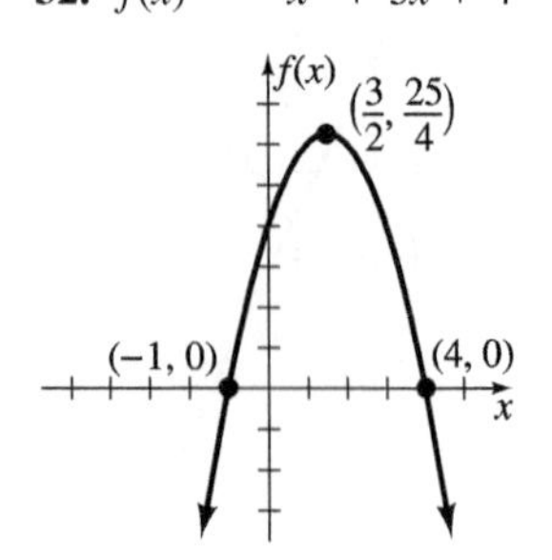


**33.** $f(x) = x^2 + 5x + 3$

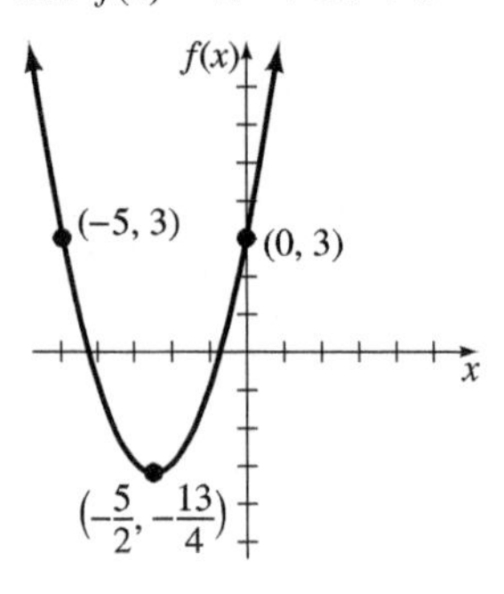


**34.** **(a)** $(-5, -28)$; $x = -5$ **(b)** $\left(-\frac{7}{2}, \frac{67}{2}\right)$; $x = -\frac{7}{2}$

**35.** 20 and 20 **36.** 3, 47 **37.** Width, 15 yd; length, 30 yd
**38.** 25 students

**39.**

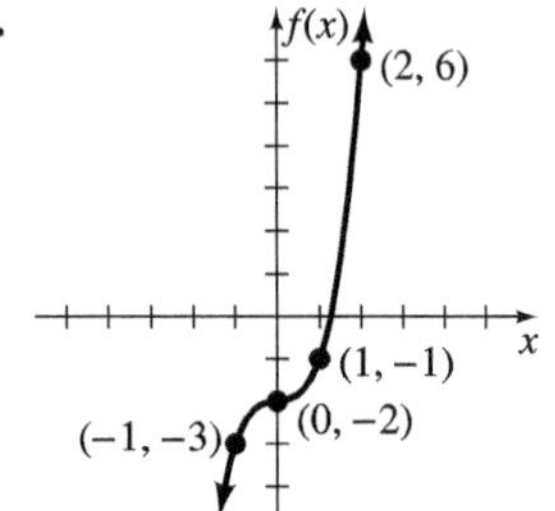


**40.**

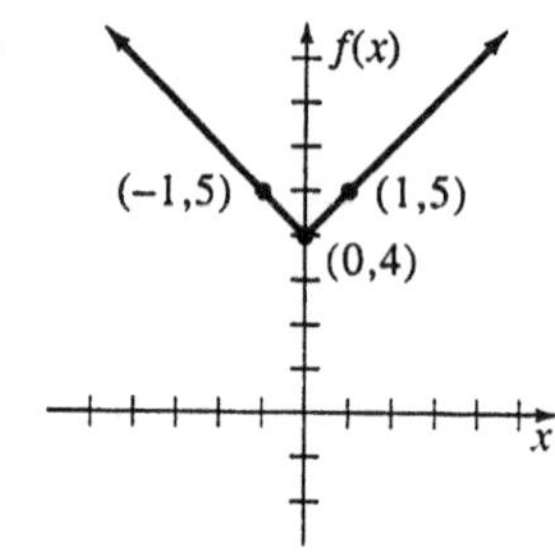


**41.**

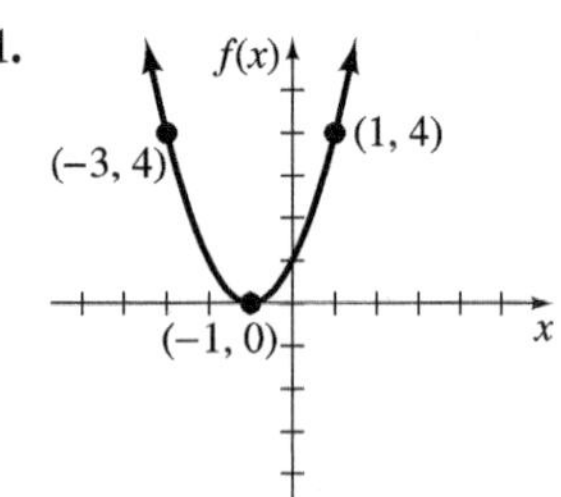


**42.**

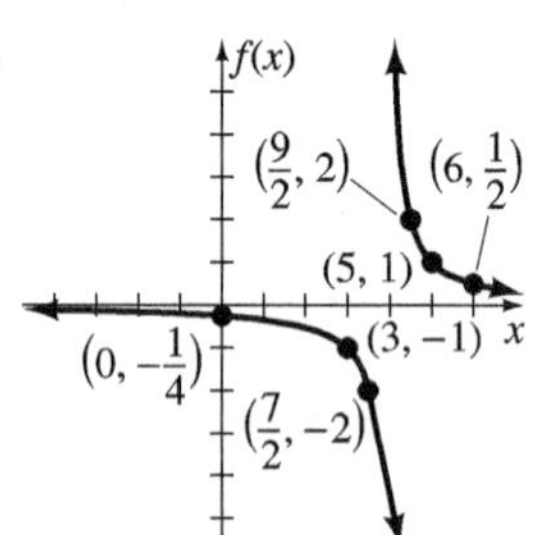


**43.**

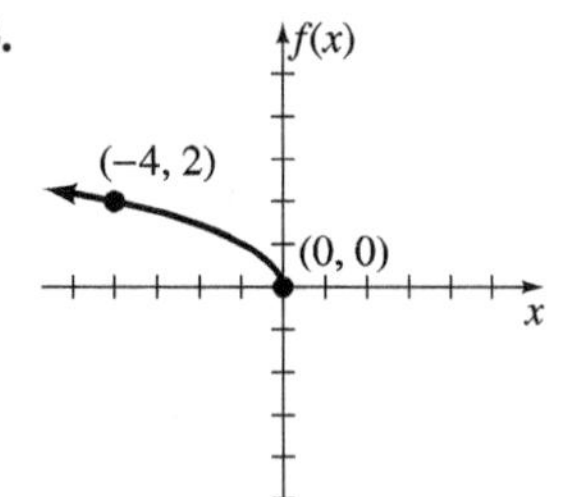


**44.**

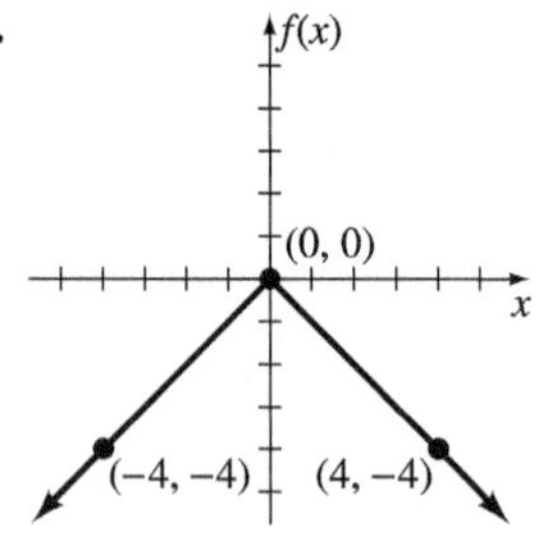


**45.**

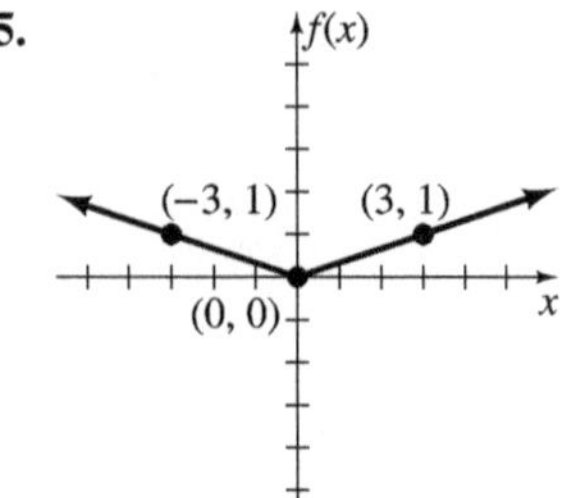


**46.**

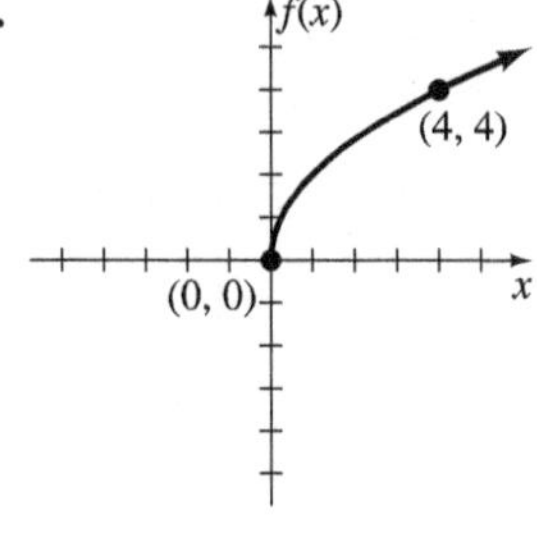

**47.**

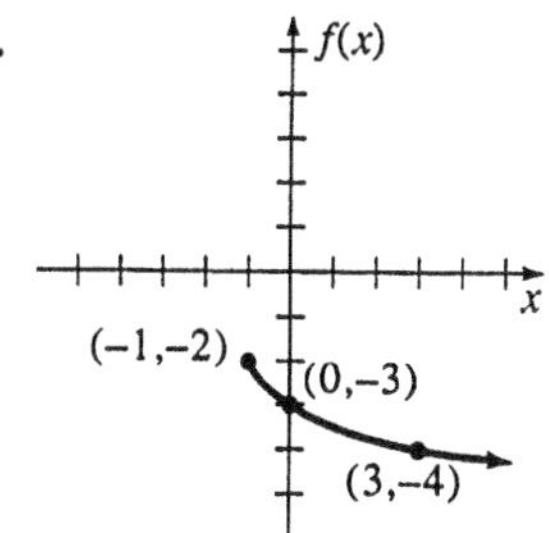


**48.**

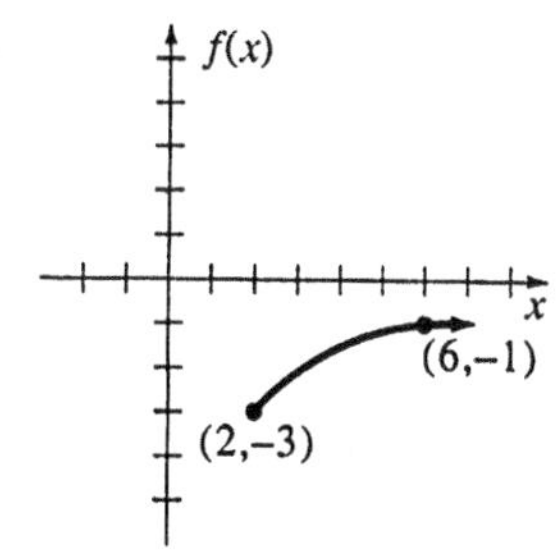


**49.**

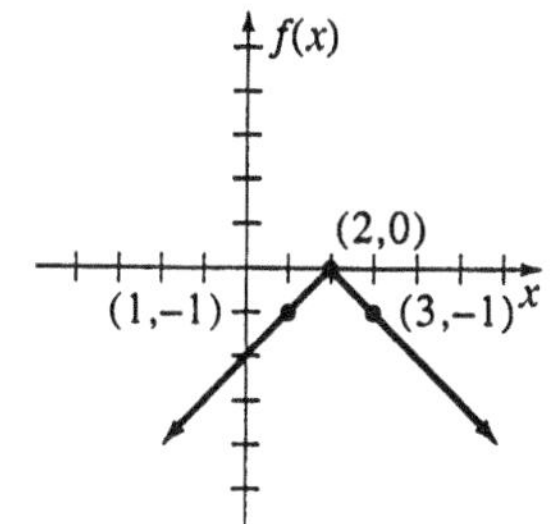


**50.**

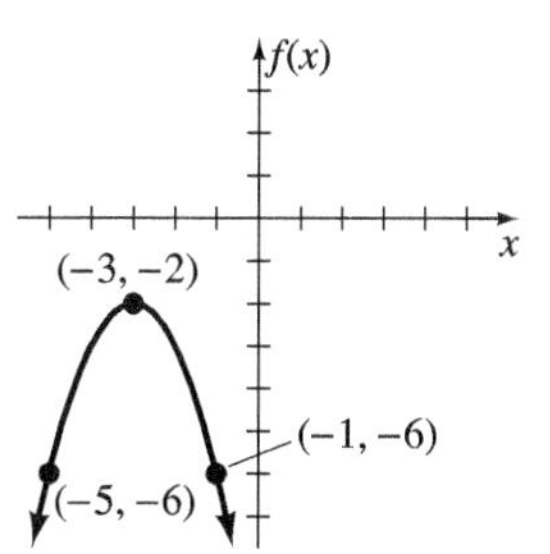


**51.** $f(x) = \frac{1}{x} + 3$

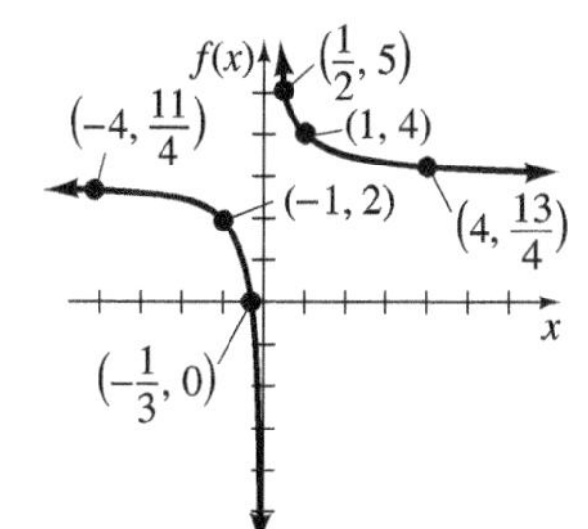


**52.** $f(x) = (x + 2)^3$

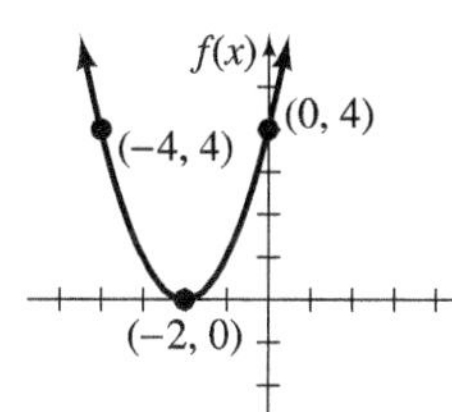


**53.** $f(x) = -(x + 1)^2 + 5$

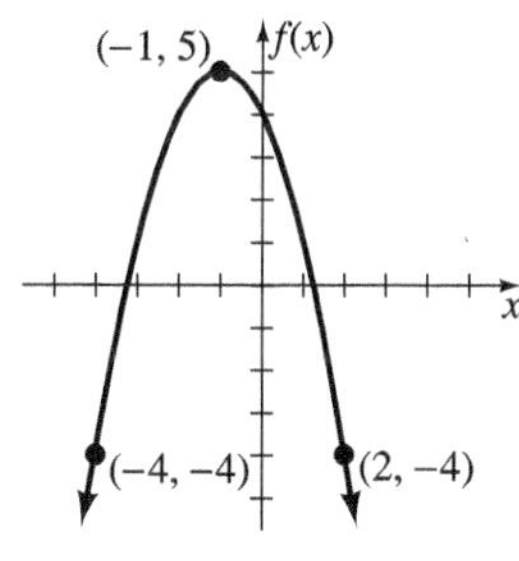


**54.** $f(x) = -x^2 + 4$

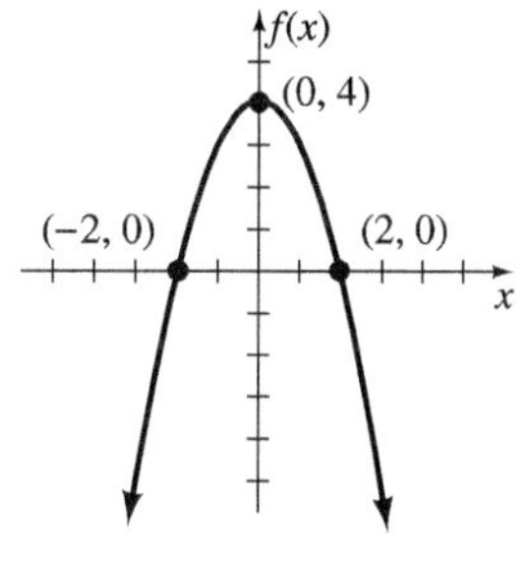


**55.** $f(x) = -\frac{1}{x + 2}$

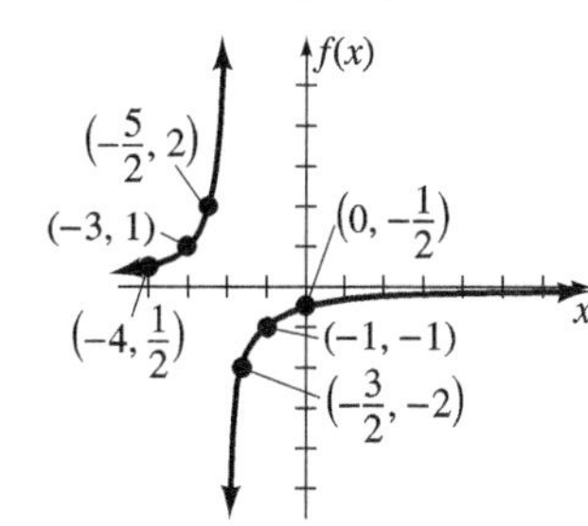


**56.** $f(x) = -(x + 1)^3$

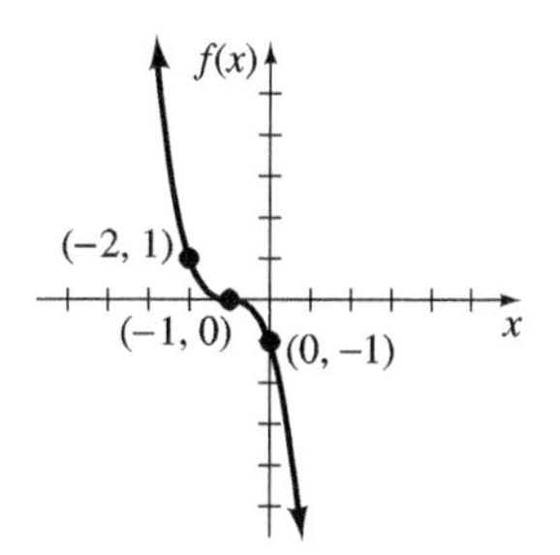


**57.** $(f \circ g)(x) = 6x - 11$; $(g \circ f)(x) = 6x - 13$
**58.** $(f \circ g)(x) = x^2 - 2x - 1$; $(g \circ f)(x) = x^2 - 10x + 27$
**59.** $(f \circ g)(x) = 4x^2 - 20x + 20$; $(g \circ f)(x) = -2x^2 + 5$

**60.** 179 **61.** −6 **62.** Function **63.** Not a function
**64.** Not a function **65.** Not one-to-one **66.** One-to-one
**67.** One-to-one **68.** $f^{-1} = \{(4, -1), (5, 0), (7, 1), (9, 2)\}$
**69.** $f^{-1} = \{(4, 2), (9, 3), (16, 4), (25, 5)\}$
**70.** $f^{-1}(x) = \frac{x + 1}{6}$ **71.** $f^{-1}(x) = \frac{3x - 21}{2}$
**72.** $f^{-1}(x) = \frac{-35x - 10}{21}$ **73.** \$375 **74.** 600 in.$^2$
**75.** 10 min **76.** 128 lb **77.** $k = 9$ **78.** $y = 120$

## Chapter 9 Test (page 501)

**1.** $D = \left\{x \middle| x \neq -4 \text{ and } x \neq \frac{1}{2}\right\}$ or
$D: (-\infty, -4) \cup \left(-4, \frac{1}{2}\right) \cup \left(\frac{1}{2}, \infty\right)$
**2.** $D = \left\{x \middle| x \leq \frac{5}{3}\right\}$ or $D: \left(-\infty, \frac{5}{3}\right]$ **3.** $\frac{11}{6}$ **4.** 11
**5.** $(-6, 3)$ **6.** $6a + 3h + 2$ **7.** $(f \circ g)(x) = -21x - 2$
**8.** $(g \circ f)(x) = 8x^2 + 38x + 48$ **9.** $(f \circ g)(x) = \frac{3x}{2 - 2x}$
**10.** $f^{-1}(x) = \frac{x + 9}{5}$ **11.** $f^{-1}(x) = \frac{-x - 6}{3}$
**12.** $f^{-1}(x) = \frac{15x + 9}{10}$ **13.** −4 **14.** 15
**15.** 6 and 54 **16.** \$96
**17.** The graph of $f(x) = (x - 6)^3 - 4$ is the graph of $f(x) = x^3$ translated 6 units to the right and 4 units downward.
**18.** The graph of $f(x) = -|x| + 8$ is the graph of $f(x) = |x|$ reflected across the $x$ axis and translated 8 units upward.
**19.** The graph of $f(x) = -\sqrt{x + 5} + 7$ is the graph of $f(x) = \sqrt{x}$ reflected across the $x$ axis and translated 5 units to the left and 7 units upward.

**20.**

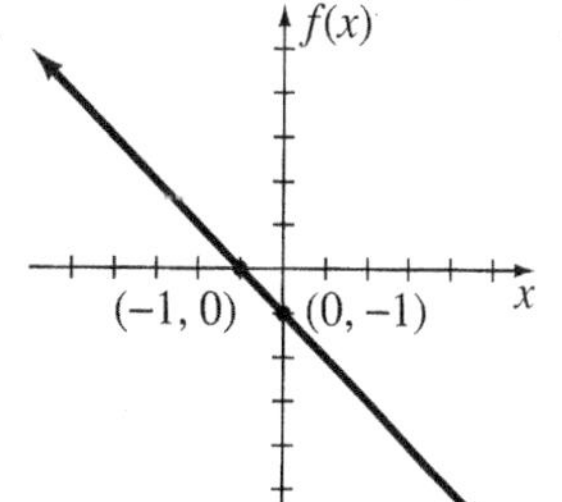


**21.**

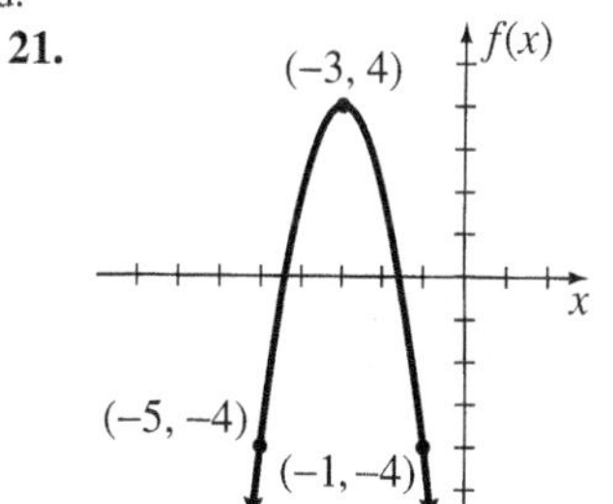


**22.**

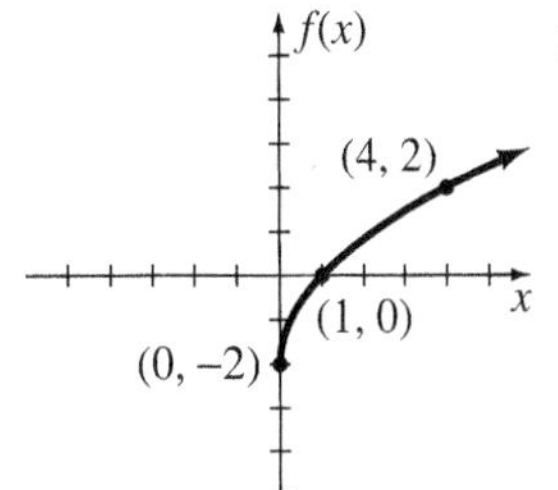


**23.**

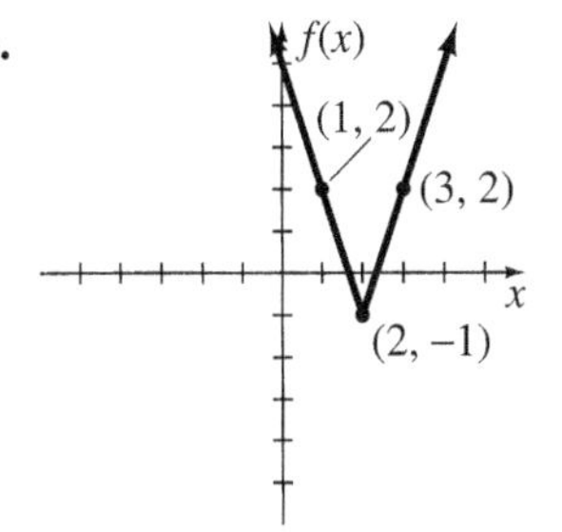

**24.**

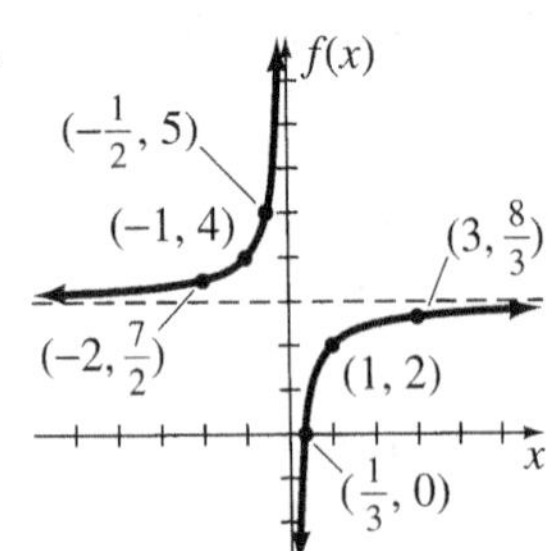


**25.**

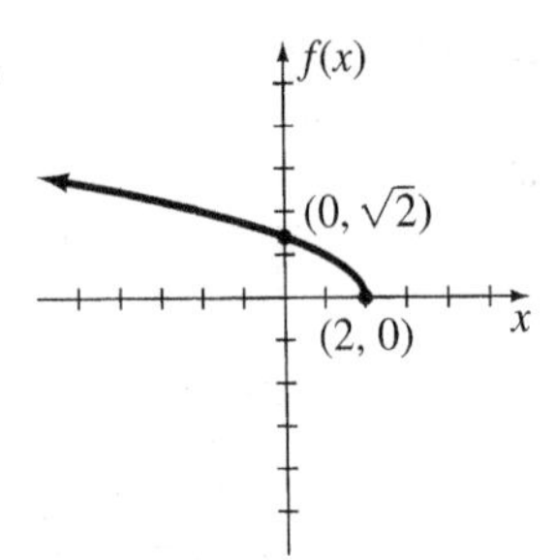


## Chapter 10

### Problem Set 10.1 (page 509)

**1.** $\{(3, 2)\}$ **3.** $\{(-2, 1)\}$ **5.** Dependent **7.** $\{(4, -3)\}$ **9.** Inconsistent **11.** $\{(2, 1)\}$ **13.** $\{(-3, -1)\}$ **15.** $\{(2, 4)\}$

**17.**

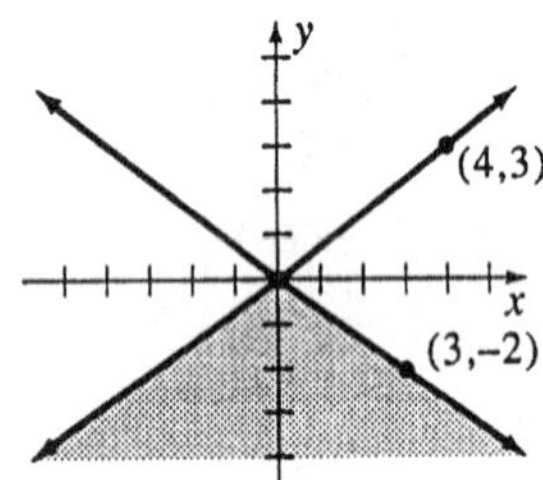


**19.**

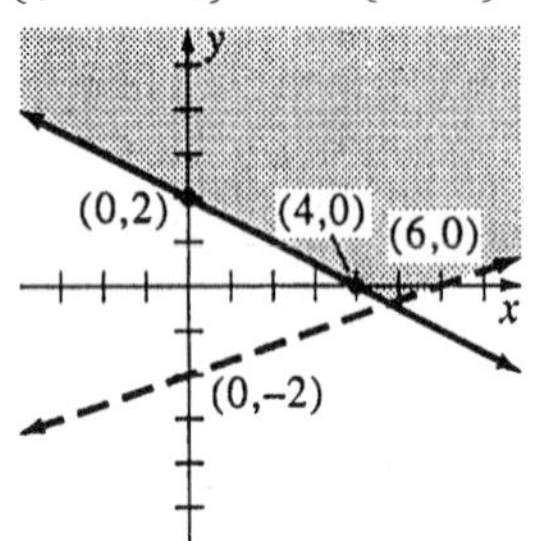


**21.**

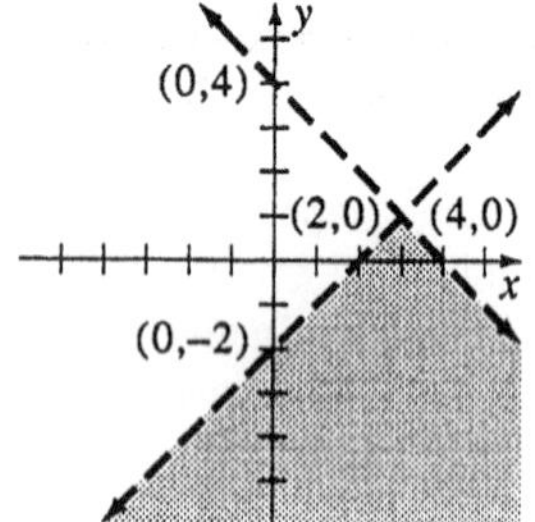


**23.**

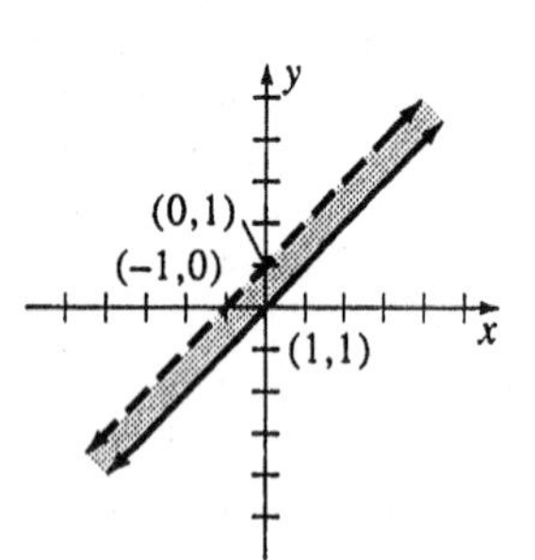


**25.**

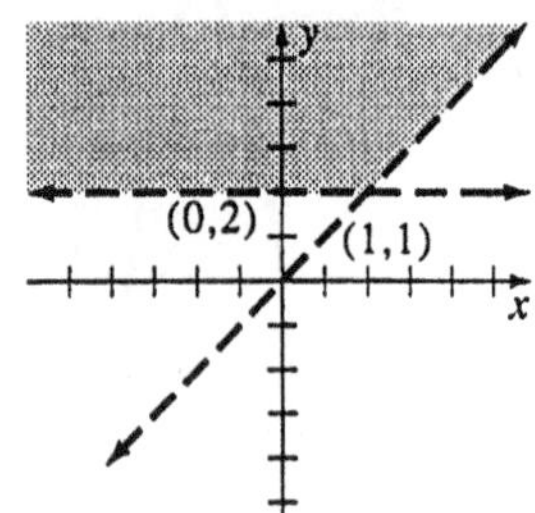


**27.**

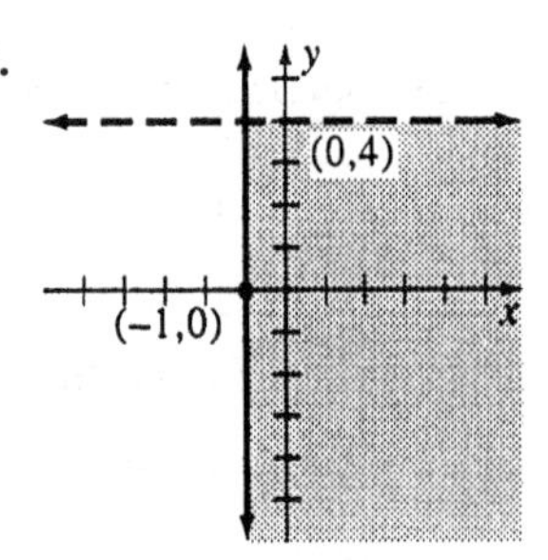


**29.**

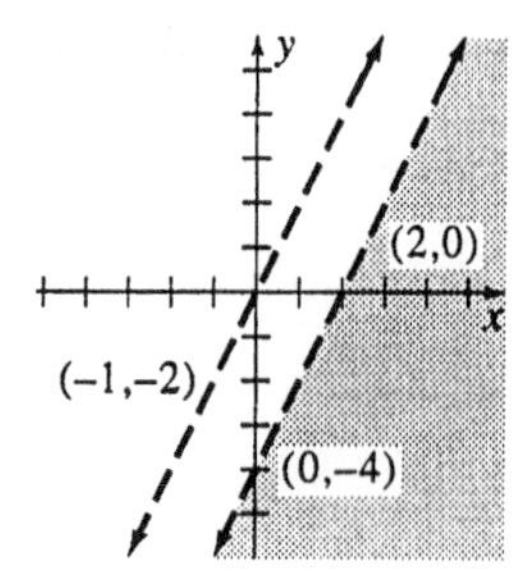


**31.**

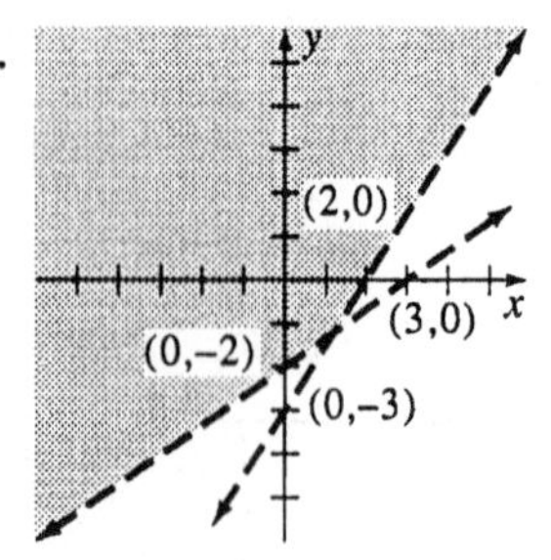


### Problem Set 10.2 (page 514)

**1.** $\{(8, 12)\}$ **3.** $\{(-4, -6)\}$ **5.** $\{(-9, 3)\}$ **7.** $\{(1, 3)\}$ **9.** $\left\{\left(5, \frac{3}{2}\right)\right\}$

**11.** $\left\{\left(\frac{9}{5}, -\frac{7}{25}\right)\right\}$ **13.** $\{(-2, -4)\}$ **15.** $\{(5, 2)\}$

**17.** $\{(-4, -8)\}$ **19.** $\left\{\left(\frac{11}{20}, \frac{7}{20}\right)\right\}$ **21.** $\{(-1, 5)\}$

**23.** $\left\{\left(-\frac{3}{4}, -\frac{6}{5}\right)\right\}$ **25.** $\left\{\left(\frac{5}{27}, -\frac{26}{27}\right)\right\}$

**27.** \$2000 at 7% and \$8000 at 8% **29.** 34 and 97
**31.** 20°, 70° **33.** 42 females
**35.** 20 inches by 27 inches
**37.** 60 five-dollar bills and 40 ten-dollar bills
**39.** 2500 student tickets and 500 nonstudent tickets

### Problem Set 10.3 (page 521)

**1.** $\{(4, -3)\}$ **3.** $\{(-1, -3)\}$ **5.** $\{(-8, 2)\}$ **7.** $\{(-4, 0)\}$

**9.** $\{(1, -1)\}$ **11.** $\varnothing$ **13.** $\{(-2, 3)\}$ **15.** $\left\{\left(\frac{3}{2}, -\frac{1}{3}\right)\right\}$

**17.** $\{(x, y) \mid 3x - y = 4\}$ **19.** $\{(4, -9)\}$ **21.** $\{(7, 0)\}$

**23.** $\{(7, 12)\}$ **25.** $\left\{\left(\frac{7}{11}, \frac{2}{11}\right)\right\}$ **27.** $\varnothing$ **29.** $\{(2, -4)\}$

**31.** $\{(x, y) | y = -2x + 1\}$ **33.** $\{(-2, -4)\}$

**35.** $\left\{\left(-1, -\frac{14}{3}\right)\right\}$ **37.** $\{(-6, 12)\}$ **39.** $\{(2, 8)\}$

**41.** $\{(-1, 3)\}$ **43.** $\{(16, -12)\}$ **45.** $\left\{\left(-\frac{3}{4}, \frac{3}{2}\right)\right\}$

**47.** $\{(5, -5)\}$
**49.** 5 gallons of 10% solution and 15 gallons of 20% solution
**51.** \$1 for a tennis ball and \$2 for a golf ball
**53.** 40 double rooms and 15 single rooms
**55.** 50 cheese pizzas and 35 supreme pizzas
**57.** $\frac{3}{4}$ **59.** 18 centimeters by 24 centimeters

**65.** **(a)** $\{(4, 6)\}$ **(b)** $\{(3, 5)\}$ **(c)** $\{(2, -3)\}$ **(d)** $\left\{\left(\frac{1}{2}, \frac{1}{3}\right)\right\}$

**(e)** $\left\{\left(\frac{1}{4}, -\frac{2}{3}\right)\right\}$ **(f)** $\{(-4, -5)\}$

### Problem Set 10.4 (page 530)

**1.** $\{(-2, 5, 2)\}$ **3.** $\{(4, -1, -2)\}$ **5.** $\{(-1, 3, 5)\}$

**7.** Infinitely many solutions **9.** $\varnothing$ **11.** $\left\{\left(-2, \frac{3}{2}, 1\right)\right\}$

**13.** $\left\{\left(\frac{1}{3}, -\frac{1}{2}, 1\right)\right\}$ **15.** $\left\{\left(\frac{2}{3}, -4, \frac{3}{4}\right)\right\}$ **17.** $\{(-2, 4, 0)\}$

**19.** $\left\{\left(\frac{1}{2}, \frac{1}{3}, \frac{1}{6}\right)\right\}$

**21.** \$1.22 per bottle of catsup, \$1.77 per jar of peanut butter, and \$1.80 per jar of pickles

**23.** $-2$, 6, and 16 **25.** 40°, 60°, and 80°

**27.** \$500 at 4%, \$1000 at 5%, and \$1500 at 6%

## Problem Set 10.5 (page 536)

**1.** $\{(-2, 0), (-4, 4)\}$ **3.** $\{(-1, -1), (2, -4)\}$

**5.** $\{(2, -3), (-2, 3)\}$ **7.** $\{(2, 5), (-2, -5)\}$

**9.** $\{(-6, 7), (-2, -1)\}$ **11.** $\{(1, 1)\}$ **13.** $\{(-5, -3)\}$

**15.** $\{(10, -8)\}$ **17.** $\{(-3, 2)\}$ **19.** $\{(-3, 0), (3, 0)\}$

**21.** $\{(2, 2), (-2, -2)\}$ **23.** $\{(2, -7), (-2, -7)\}$

**25.** $\{(2 + i\sqrt{3}, -2 + i\sqrt{3}), (2 - i\sqrt{3}, -2 - i\sqrt{3})\}$

**27.** $\left\{\left(\frac{12}{5} + \frac{4i}{5}, \frac{6}{5} - \frac{8i}{5}\right), \left(\frac{12}{5} - \frac{4i}{5}, \frac{6}{5} + \frac{8i}{5}\right)\right\}$

## Chapter 10 Sample Problems (page 537)

**1.**

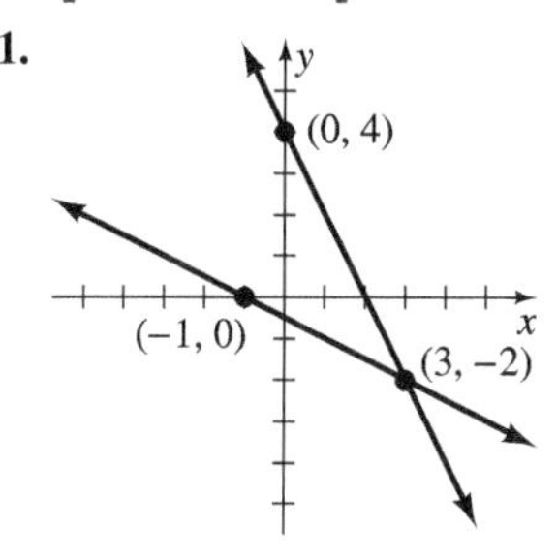


**2.**

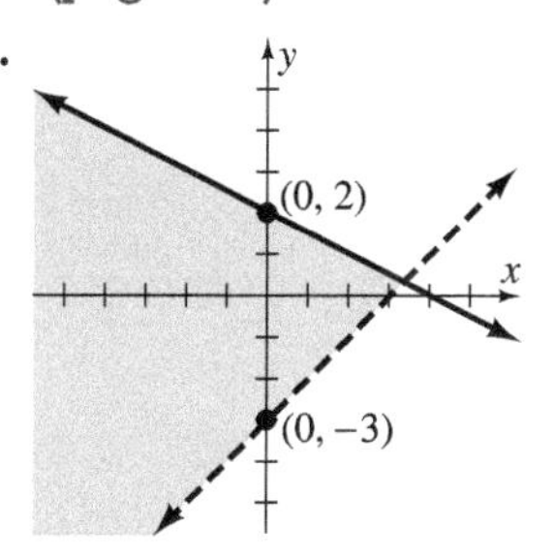


**3.** $\{(1, 6)\}$ **4.** $\{(1, -1)\}$ **5.** $\{(12, -1)\}$

**6.** 160 trucks and 350 cars **7.** $\{(4, 1, 2)\}$

**8.** 10 inches, 18 inches, and 27 inches

**9.** The solution set is $\{(0, 1), (1, 2)\}$

## Chapter 10 Review Problem Set (page 542)

**1.** $\{(2, 6)\}$ **2.** $\{(-3, 7)\}$ **3.** $\{(-9, 8)\}$ **4.** $\left\{\left(\frac{89}{23}, -\frac{12}{23}\right)\right\}$

**5.** $\{(4, -7)\}$ **6.** $\{(-2, 8)\}$ **7.** $\{(0, -3)\}$ **8.** $\{(6, 7)\}$

**9.** $\{(1, -6)\}$ **10.** $\{(-4, 0)\}$ **11.** $\{(-12, 18)\}$ **12.** $\{(24, 8)\}$

**13.** $\left\{\left(\frac{2}{3}, -\frac{3}{4}\right)\right\}$ **14.** $\left\{\left(-\frac{1}{2}, \frac{3}{5}\right)\right\}$ **15.** $\{(2, -4, -6)\}$

**16.** $\{(-1, 5, -7)\}$ **17.** $\{(2, -3, 1)\}$ **18.** $\{(2, -1, -2)\}$

**19.** $\left\{\left(-\frac{1}{3}, -1, 4\right)\right\}$ **20.** $\{(0, -2, -4)\}$

**21.** $\{(1, 1), (-2, 7)\}$

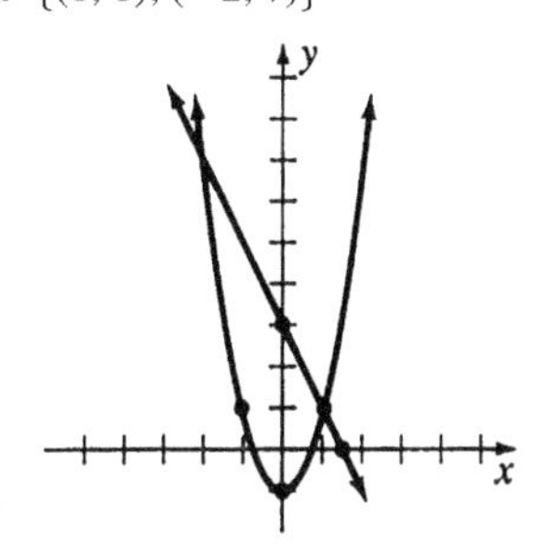


**22.** $\left\{\left(\frac{3 - \sqrt{41}}{2}, \frac{-3 - \sqrt{41}}{2}\right), \left(\frac{3 + \sqrt{41}}{2}, \frac{-3 + \sqrt{41}}{2}\right)\right\}$

**23.** **24.**

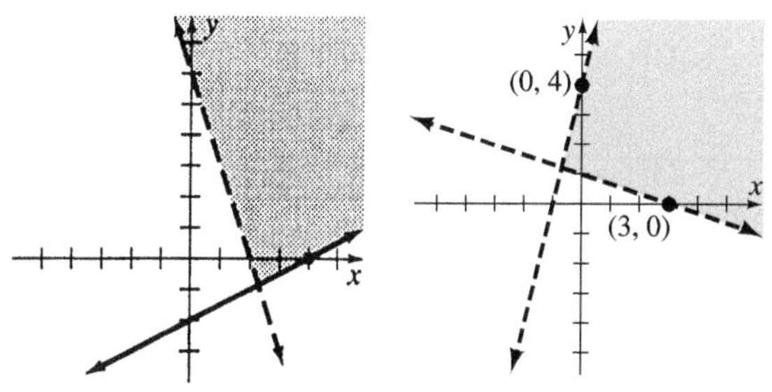


**25.** **26.**

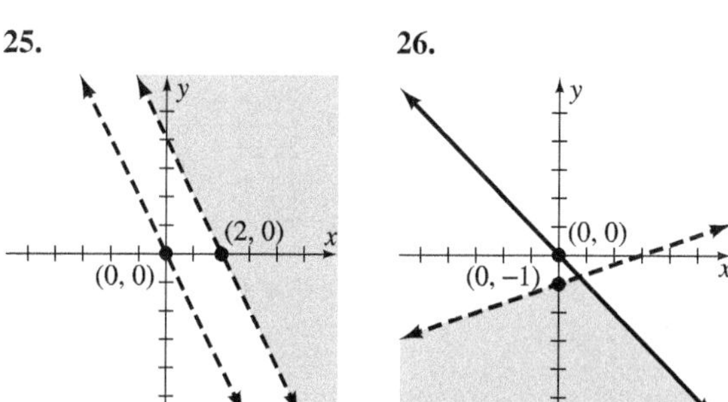


**27.** 2 and 3 or $-3$ and $-2$ **28.** 5 and 3 or 5 and $-3$ or $-5$ and 3 or $-5$ and $-3$

**29.** 2 and 5 or $-3$ and 10 **30.** 6 meters by 9 meters

**31.** \$9 per pound for cashews and \$5 per pound for Spanish peanuts

**32.** \$1.50 for a liter of pop and \$2.25 for a pound of candy

**33.** The fixed fee is \$2, and the additional fee is \$0.10 per pound.

**34.** $6\frac{2}{3}$ qts of 1% and $3\frac{1}{3}$ qts of 4%

**35.** 7 centimeters by 21 centimeters

**36.** \$1,200 at 1% and \$3,000 at 1.5%

**37.** 5 five-dollar bills and 25 one-dollar bills

**38.** 30 review problems and 80 new problems

**39.** 24 five-dollar bills, 30 ten-dollar bills, 10 twenty-dollar bills

**40.** 40°, 60°, 80°

**41.** \$2100 on Bank of U.S., \$1600 on Community Bank, and \$2700 on First National

**42.** 6 inches, 12 inches, 15 inches

## Chapter 10 Test (page 544)

**1.** II **2.** I and IV **3.** III **4.** I **5.** $\{(2, 7)\}$ **6.** $\{(-3, 5)\}$

**7.** $\{(-2, -4)\}$ **8.** $\{(3, -3)\}$ **9.** $\left\{\left(-\frac{1}{2}, 1\right)\right\}$

**10.** $\{(24, 18)\}$ **11.** $\{(1, -2, 3)\}$ **12.** $\{(-1, 2, -3)\}$

**13.** $x = 7$ **14.** $y = -1$ **15.** $z = 0$ **16.** Two **17.** Four

**18.** One **19.** None **20.** $\{(2, 3)\}$

**21.**

**22.** \$1.40 for soda; \$1.00 for water
**23.** 24 quarters **24.** 3 liters
**25.** Width = 15 inches, Length = 26 inches

## Chapters 1–10 Cumulative Review Problem Set (page 545)

**1.** $-6$ **2.** $-8$ **3.** $\frac{13}{24}$ **4.** 24 **5.** $\frac{13}{6}$ **6.** $-90\sqrt{2}$
**7.** $2x + 5\sqrt{x} - 12$ **8.** $-18 + 22\sqrt{3}$
**9.** $2x^3 + 11x^2 - 14x + 4$ **10.** $\frac{x + 4}{x(x + 5)}$ **11.** $\frac{16x^2}{27y}$
**12.** $\frac{16x + 43}{90}$ **13.** $\frac{35a - 44b}{60a^2b}$ **14.** $\frac{2}{x - 4}$
**15.** $2x^2 - x - 4$ **16.** $\frac{5y^2 - 3xy^2}{x^2y + 2x^2}$ **17.** $\frac{2y - 3xy}{3x + 4xy}$
**18.** $\frac{(2n + 3)(n + 3)}{(n + 2)(3n + 13)}$ **19.** $\frac{3a^2 - 2a + 1}{2a - 1}$
**20.** $(5x - 2)(4x + 3)$ **21.** $2(2x + 3)(4x^2 - 6x + 9)$
**22.** $(2x + 3)(2x - 3)(x + 2)(x - 2)$
**23.** $4x(3x + 2)(x - 5)$ **24.** $(y - 6)(x + 3)$
**25.** $(5 - 3x)(2 + 3x)$ **26.** $\frac{81}{16}$ **27.** 4 **28.** $-\frac{3}{4}$ **29.** $-0.3$
**30.** $\frac{1}{81}$ **31.** $\frac{21}{16}$ **32.** $\frac{9}{64}$ **33.** 72 **34.** $-\frac{12}{x^3y}$ **35.** $\frac{8y}{x^5}$
**36.** $-\frac{a^3}{9b}$ **37.** $4\sqrt{15}$ **38.** $-6\sqrt{6}$ **39.** $\frac{5\sqrt{3}}{9}$ **40.** $\frac{2\sqrt{3}}{3}$
**41.** $2\sqrt[3]{7}$ **42.** $\frac{\sqrt[3]{6}}{2}$ **43.** $8xy\sqrt{13x}$ **44.** $\frac{\sqrt{6xy}}{3y}$
**45.** $11\sqrt{6}$ **46.** $-\frac{169\sqrt{2}}{12}$ **47.** $-16\sqrt[3]{3}$
**48.** $\frac{-3\sqrt{2} - 2\sqrt{6}}{2}$ **49.** $\frac{6\sqrt{15} - 3\sqrt{35} - 6 + \sqrt{21}}{5}$
**50.** 0.021 **51.** 300 **52.** 0.0003 **53.** $32 + 22i$
**54.** $-17 + i$ **55.** $0 - \frac{5}{4}i$ **56.** $-\frac{19}{53} + \frac{40}{53}i$
**57.** $-\frac{10}{3}$ **58.** $\frac{4}{7}$ **59.** $2\sqrt{13}$ **60.** $5x - 4y = 19$
**61.** $4x + 3y = -18$ **62.** $(-2, 6)$ and $r = 3$
**63.** $(-5, -4)$ **64.** 8 units
**65.** **66.**

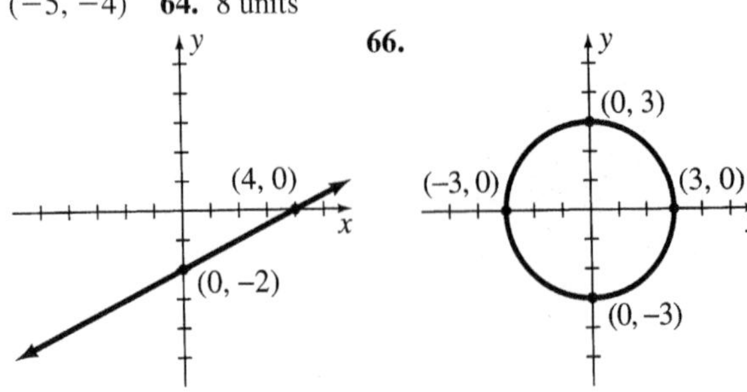


**67.** **68.**

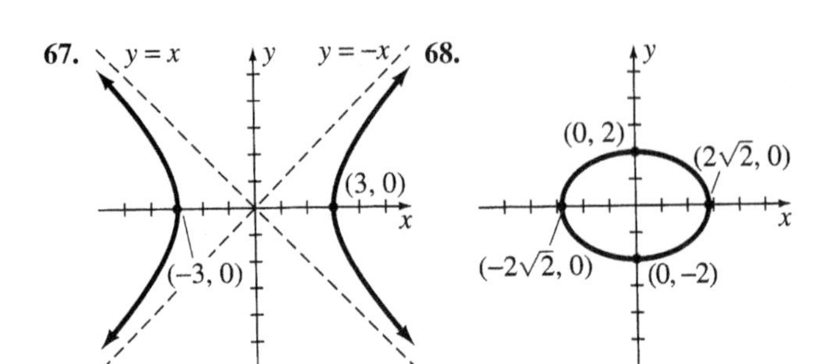


**69.** **70.**

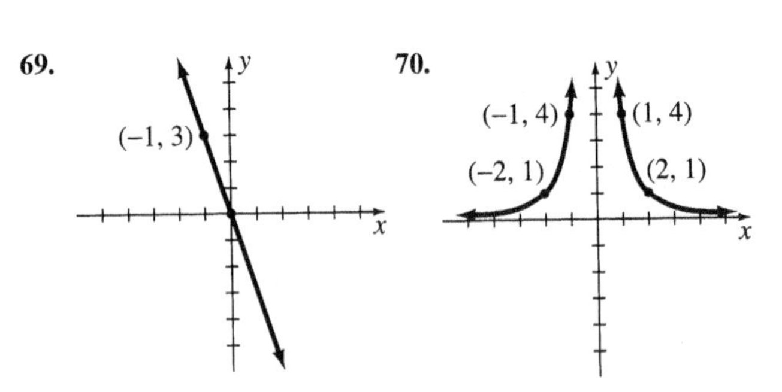


**71.** **72.**

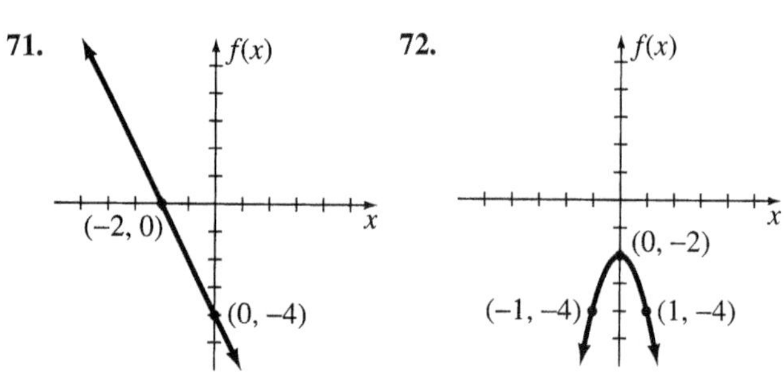


**73.** **74.**

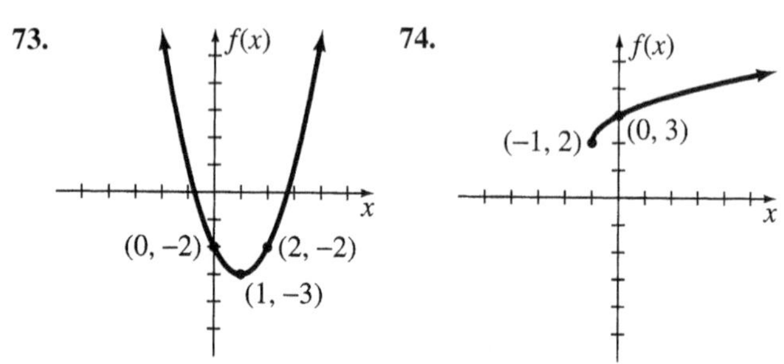


**75.** **76.**

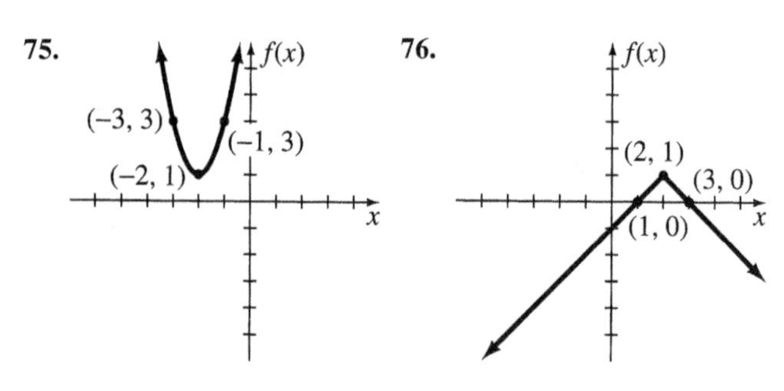


**77.** $(g \circ f)(x) = 2x^2 - 13x + 20$
$(f \circ g)(x) = 2x^2 - x - 4$
**78.** $f^{-1}(x) = \frac{x + 7}{3}$
**79.** $f^{-1}(x) = -2x + \frac{4}{3}$ **80.** $k = -3$ **81.** $y = 1$

**82.** 12 cubic centimeters **83.** $\left\{-\frac{21}{16}\right\}$

**84.** $\left\{\frac{40}{3}\right\}$ **85.** $\{6\}$ **86.** $\left\{-\frac{5}{2}, 3\right\}$ **87.** $\left\{0, \frac{7}{3}\right\}$

**88.** $\{-6, 0, 6\}$ **89.** $\left\{-\frac{5}{6}, \frac{2}{5}\right\}$ **90.** $\left\{-3, 0, \frac{3}{2}\right\}$

**91.** $\{\pm 1, \pm 3i\}$ **92.** $\{-5, 7\}$ **93.** $\{-29, 0\}$ **94.** $\left\{\frac{7}{2}\right\}$

**95.** $\{12\}$ **96.** $\{-3\}$ **97.** $\left\{\frac{1 \pm 3\sqrt{5}}{3}\right\}$

**98.** $\left\{\frac{-5 \pm 4i\sqrt{2}}{2}\right\}$ **99.** $\left\{\frac{3 \pm i\sqrt{23}}{4}\right\}$

**100.** $\left\{\frac{3 \pm \sqrt{3}}{3}\right\}$ **101.** $\{1 \pm \sqrt{34}\}$

**102.** $\left\{\pm\frac{\sqrt{5}}{2}, \pm\frac{\sqrt{3}}{3}\right\}$ **103.** $\left\{\frac{-5 \pm i\sqrt{15}}{4}\right\}$

**104.** $(-\infty, 3)$ **105.** $(-\infty, 50]$ **106.** $\left(-\infty, -\frac{11}{5}\right) \cup (3, \infty)$

**107.** $\left(-\frac{5}{3}, 1\right)$ **108.** $\left[-\frac{9}{11}, \infty\right)$ **109.** $[-4, 2]$

**110.** $\left(-\infty, \frac{1}{3}\right) \cup (4, \infty)$ **111.** $(-8, 3)$

**112.** $(-\infty, 3] \cup (7, \infty)$ **113.** $(-6, -3)$ **114.** $\{(9, 6)\}$

**115.** $\left\{\left(\frac{9}{5}, -\frac{7}{25}\right)\right\}$ **116.** $\left\{\left(\frac{70}{23}, \frac{36}{23}\right)\right\}$ **117.** $\{(-2, 4, 0)\}$

**118.** $\{(-1, 4, -1)\}$ **119.** 17, 19, and 21
**120.** 14 nickels, 20 dimes, and 29 quarters
**121.** 48° and 132° **122.** \$600
**123.** \$1700 at 8% and \$2000 at 9%
**124.** 66 miles per hour and 76 miles per hour
**125.** 69 or less **126.** −3, 0, or 3
**127.** 1 inch **128.** \$1050 and \$1400
**129.** 3 hours **130.** 10°, 60°, and 110°

# Chapter 11

### Problem Set 11.1 (page 555)

**1.** $\{3\}$ **3.** $\{2\}$ **5.** $\{4\}$ **7.** $\{1\}$ **9.** $\{5\}$ **11.** $\{1\}$ **13.** $\left\{\frac{3}{2}\right\}$

**15.** $\left\{\frac{5}{6}\right\}$ **17.** $\{-3\}$ **19.** $\{-3\}$ **21.** $\{0\}$ **23.** $\{1\}$

**25.** $\{-1\}$ **27.** $\left\{-\frac{2}{5}\right\}$ **29.** $\left\{\frac{5}{2}\right\}$ **31.** $\{3\}$ **33.** $\left\{\frac{1}{2}\right\}$

**35.**

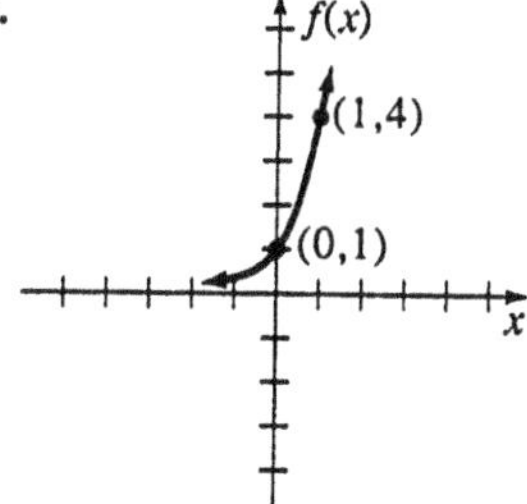


**37.**

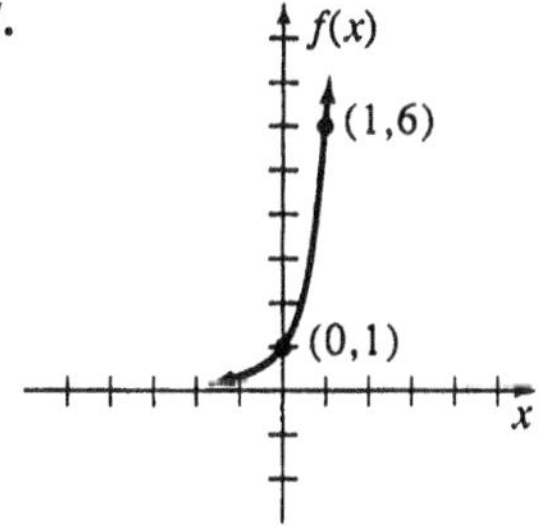


**39.**

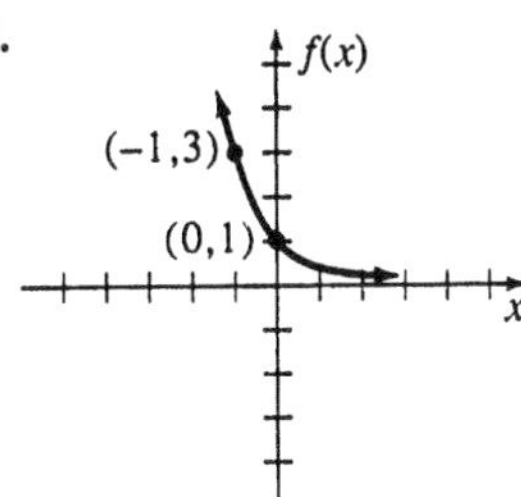


**41.**

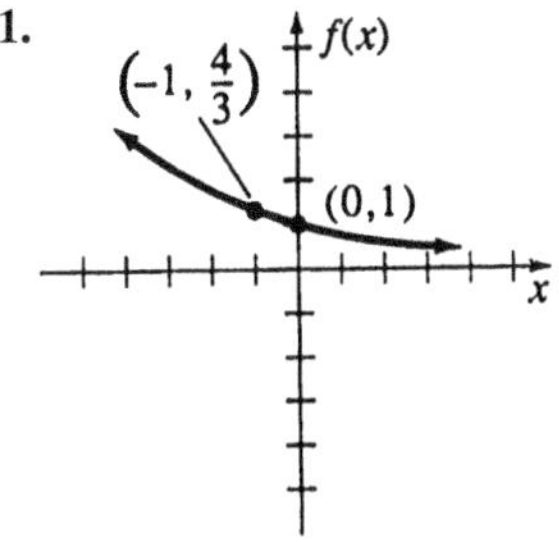


**43.**

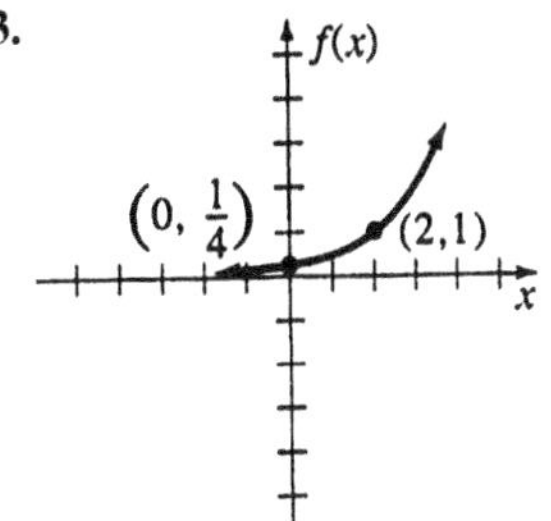


**45.**

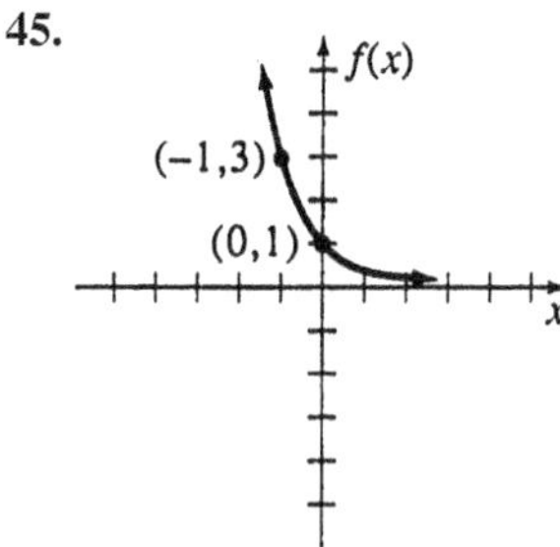


**47.**

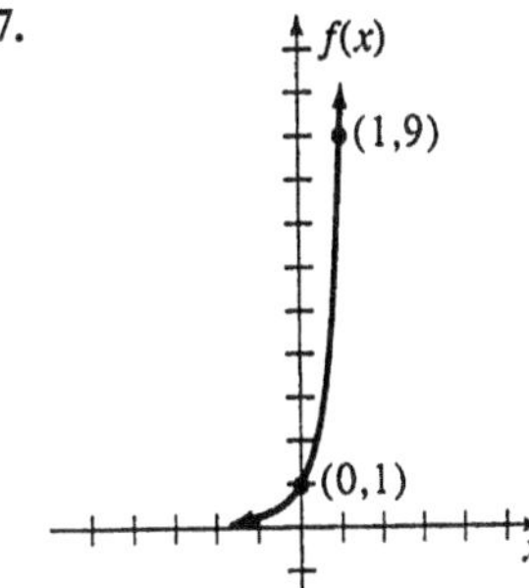


**49.**

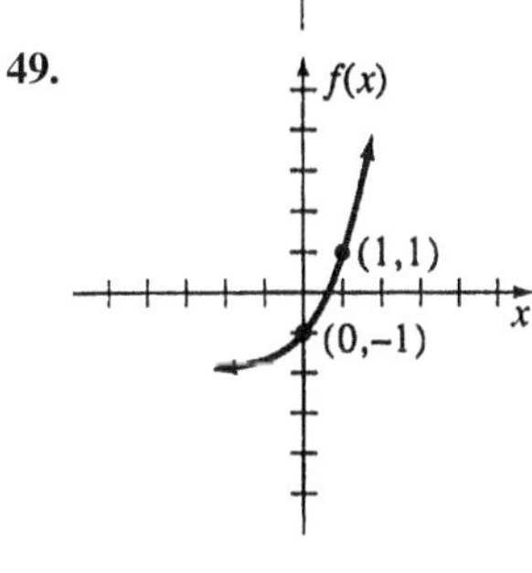


**51.**

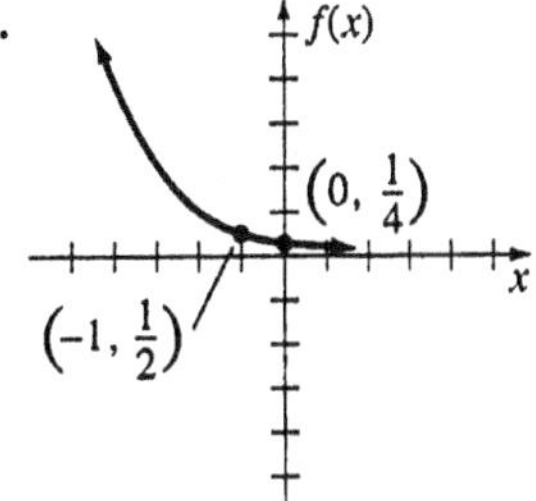


### Problem Set 11.2 (page 562)

**1.** **(a)** \$1.55 **(b)** \$4.17 **(c)** \$2.33 **(d)** \$2.28
**(e)** \$21,900 **(f)** \$246,342 **(g)** \$658
**3.** \$283.70 **5.** \$659.74 **7.** \$1251.16 **9.** \$2234.77

**11.** \$9676.41 **13.** \$13,814.17 **15.** \$567.63
**17.** \$1422.36 **19.** \$5715.30 **21.** \$14,366.56
**23.** \$26,656.96
**25.**

| | 1 yr | 5 yr | 10 yr | 20 yr |
|---|---|---|---|---|
| Compounded annually | \$1060 | 1338 | 1791 | 3207 |
| Compounded semiannually | 1061 | 1344 | 1806 | 3262 |
| Compounded quarterly | 1061 | 1347 | 1814 | 3291 |
| Compounded monthly | 1062 | 1349 | 1819 | 3310 |
| Compounded continuously | 1062 | 1350 | 1822 | 3320 |

**27.** Nora will have \$0.24 more. **29.** 9.54%
**31.** 50 grams; 37 grams

**33.**

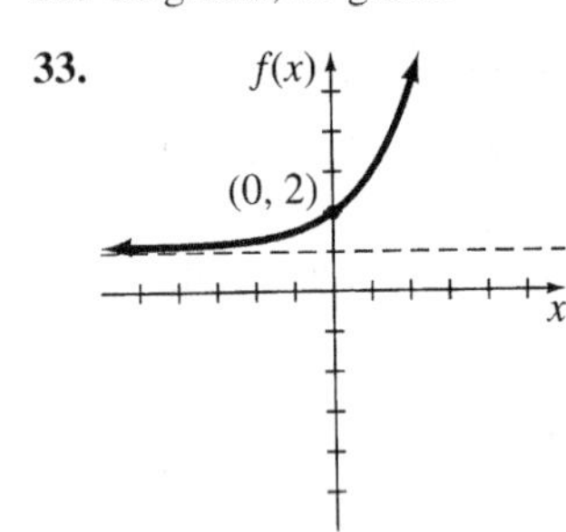


**35.**

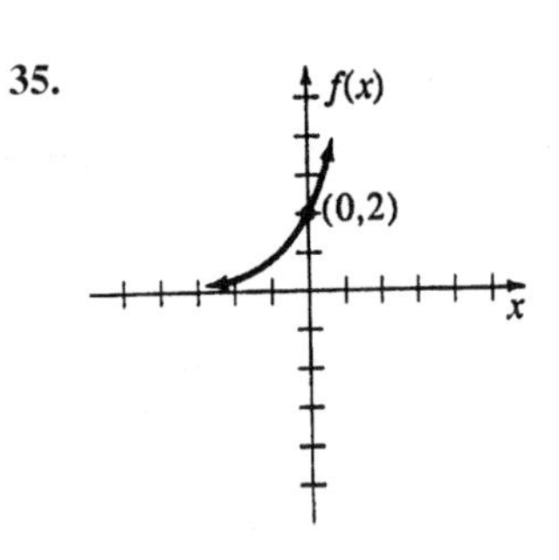


**37.**

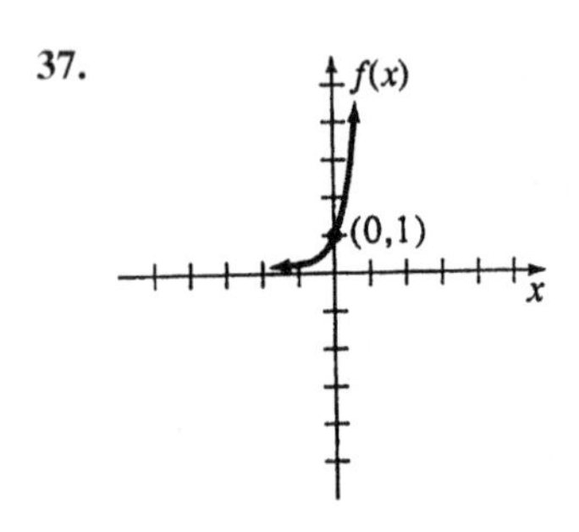


**39.** 2226; 3320; 7389 **41.** 2000
**43.** **(a)** 82,888 **(c)** 96,302
**45.** **(a)** 6.5 pounds per square inch
**(c)** 13.6 pounds per square inch

## Problem Set 11.3 (page 572)

**1.** $\log_2 128 = 7$ **3.** $\log_5 125 = 3$ **5.** $\log_{10} 1000 = 3$

**7.** $\log_2\left(\frac{1}{4}\right) = -2$ **9.** $\log_{10} 0.1 = -1$ **11.** $3^4 = 81$

**13.** $4^3 = 64$ **15.** $10^4 = 10{,}000$ **17.** $2^{-4} = \frac{1}{16}$

**19.** $10^{-3} = 0.001$ **21.** 4 **23.** 4 **25.** 3 **27.** $\frac{1}{2}$ **29.** 0

**31.** −1 **33.** 5 **35.** −5 **37.** 1 **39.** 0 **41.** {49}

**43.** {16} **45.** {27} **47.** $\left\{\frac{1}{8}\right\}$ **49.** {4} **51.** 5.1293

**53.** 6.9657 **55.** 1.4037 **57.** 7.4512 **59.** 6.3219
**61.** −0.3791 **63.** 0.5766 **65.** 2.1531 **67.** 0.3949
**69.** $\log_b x + \log_b y + \log_b z$ **71.** $\log_b y - \log_b z$

**73.** $3\log_b y + 4\log_b z$ **75.** $\frac{1}{2}\log_b x + \frac{1}{3}\log_b y - 4\log_b z$

**77.** $\frac{2}{3}\log_b x + \frac{1}{3}\log_b z$ **79.** $\frac{3}{2}\log_b x - \frac{1}{2}\log_b y$

**81.** $\left\{\frac{9}{4}\right\}$ **83.** {25} **85.** {4} **87.** $\left\{\frac{19}{8}\right\}$ **89.** {9}

**91.** {1} **93.** $\left\{-\frac{22}{5}\right\}$ **95.** ∅ **97.** {−1}

## Problem Set 11.4 (page 578)

**1.** 0.8597 **3.** 1.7179 **5.** 3.5071 **7.** −0.1373 **9.** −3.4685
**11.** 411.43 **13.** 90,095 **15.** 79.543 **17.** 0.048440
**19.** 0.0064150 **21.** 1.6094 **23.** 3.4843 **25.** 6.0638
**27.** −0.7765 **29.** −3.4609 **31.** 1.6034 **33.** 3.1346
**35.** 108.56 **37.** 0.48268 **39.** 0.035994

**41.** **43.**

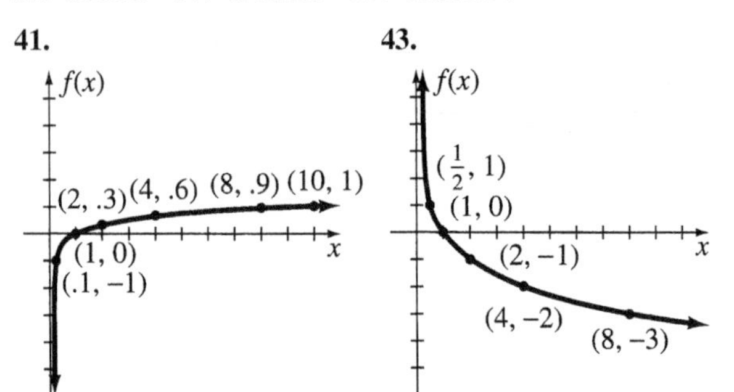


**45.** **47.**

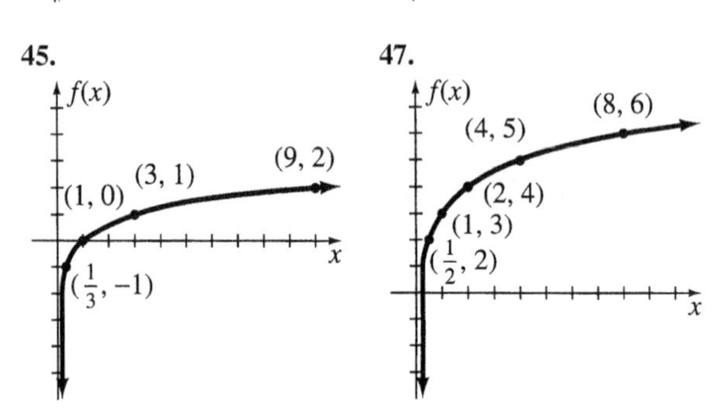


**49.** **51.**

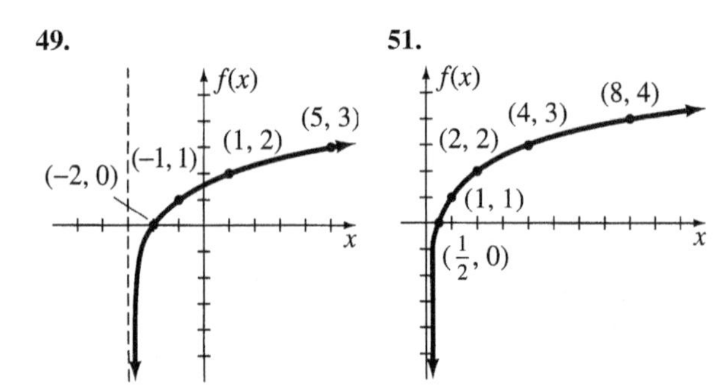


**53.**

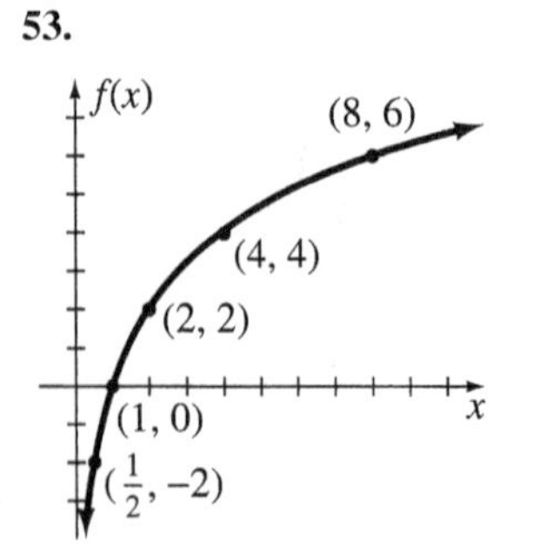


**55.** 0.36 **57.** 0.73 **59.** 23.10 **61.** 7.93

## Problem Set 11.5 (page 586)

**1.** {3.15} **3.** {2.20} **5.** {4.18} **7.** {0.12} **9.** {1.69}

**11.** {4.57} **13.** {2.46} **15.** {4} **17.** $\left\{\frac{19}{47}\right\}$

**19.** {1} **21.** {8} **23.** 4.524 **25.** −0.860 **27.** 3.105
**29.** −2.902 **31.** 5.989 **33.** 7.2 years **35.** 11.6 years

**37.** 6.8 hours **39.** 1.5 hours **41.** 34.7 years **43.** 6.7
**45.** Approximately 8 times

## Chapter 11 Sample Problems (page 588)

**1.**

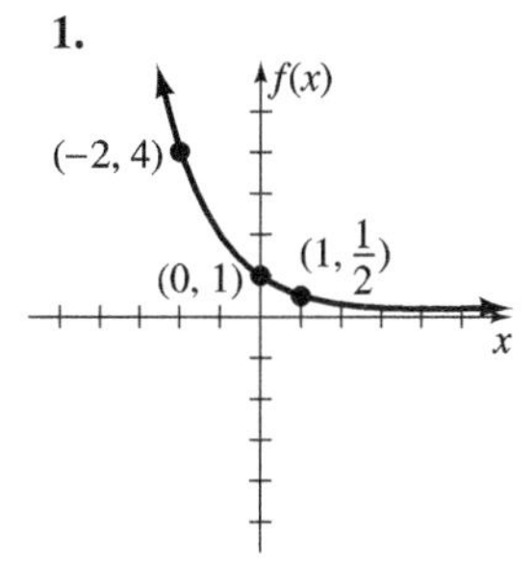


**2.** $\left\{\frac{3}{2}\right\}$ **3.** **(a)** $\log_2 8 = 3$ **(b)** $2^4 = 16$ **4.** $-4$

**5.** 2.1461 **6.** 5.4638 **7.** 3.19

**8.**

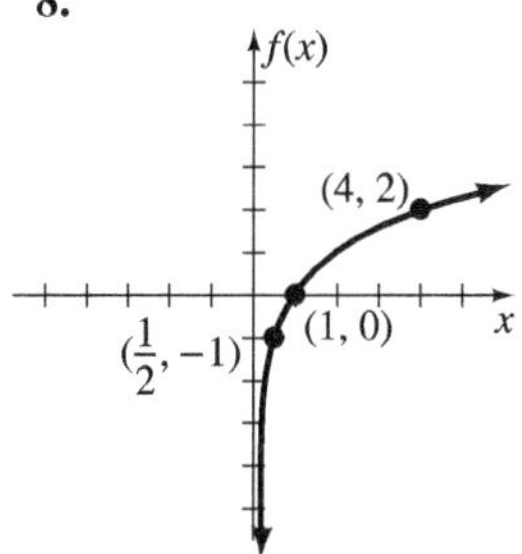


**9.** $3\log_b x - \log_b y - \frac{1}{2}\log_b z$ **10.** {3} **11.** {−1}

**12.** {2.10} **13.** 13,466 **14.** $2391.24 **15.** $2394.43
**16.** 155.4 milligrams **17.** 5.48 hours **18.** 7.5

## Chapter 11 Review Problem Set (page 594)

**1.** **2.**

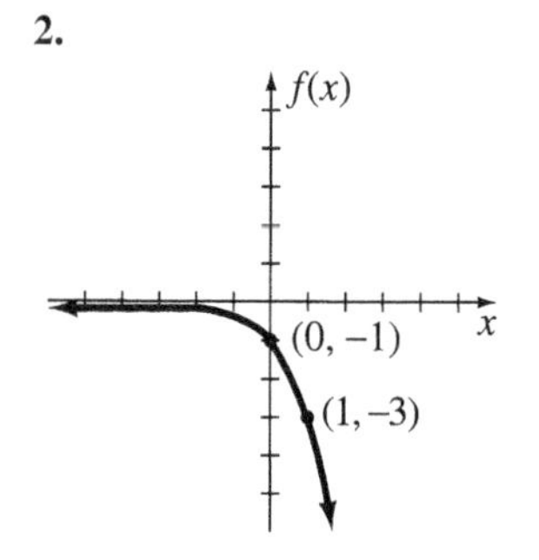


**3.** {−4} **4.** $\left\{-\frac{3}{4}\right\}$ **5.** {3} **6.** $\left\{\frac{1}{2}\right\}$ **7.** $\log_8 64 = 2$

**8.** $\log_a c = b$ **9.** $\log_3\left(\frac{1}{9}\right) = -2$ **10.** $\log_{10}\left(\frac{1}{10}\right) = -1$

**11.** $5^3 = 125$ **12.** $x^z = y$ **13.** $2^{-3} = \frac{1}{8}$ **14.** $10^4 = 10{,}000$

**15.** 7 **16.** 3 **17.** 4 **18.** −3 **19.** 2 **20.** 13 **21.** 0
**22.** −2 **23.** 1.8642 **24.** −2.6289 **25.** −4.2687
**26.** 4.7380 **27.** 4.164 **28.** 2.548 **29.** −3.129 **30.** −8.118

**31.**

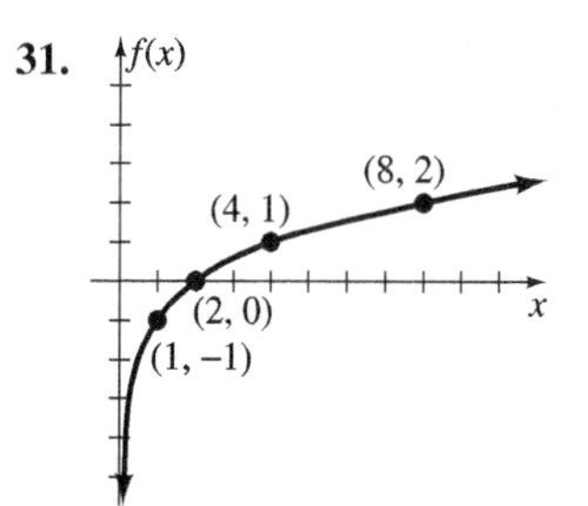


**32.**

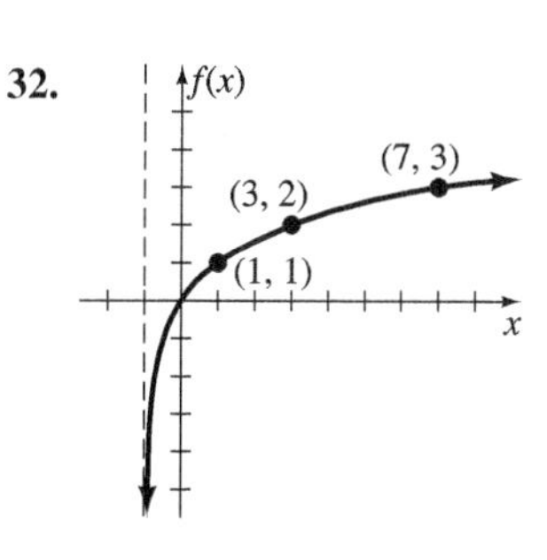


**33.** $\log_b x + \frac{1}{2}\log_b y - 3\log_b z$ **34.** $1 - \log_b c - \log_b d$
**35.** {1.1882} **36.** {2639.4} **37.** {0.0132289}
**38.** {0.077197} **39.** {4} **40.** {9} **41.** $\left\{\frac{1}{2}\right\}$ **42.** {8}
**43.** {5} **44.** $\left\{\frac{45}{8}\right\}$ **45.** {5} **46.** $\left\{\frac{15}{7}\right\}$ **47.** {3.40}
**48.** {1.95} **49.** {5.30} **50.** {5.61}
**51.** **(a)** $933.05 **(b)** $305.96 **(c)** $98.54
**52.** **(a)** $7910 **(b)** $14,766 **(c)** $15,000 **53.** $19,632.75
**54.** $40,305.65 **55.** Approximately 8.8 yr
**56.** Approximately 22.9 yr **57.** $5656.26
**58.** Approximately 19.8 yr **59.** 133 g
**60.** 46.4 hr **61.** 61,070; 67,493; 74,591
**62.** Approximately 4.8 yr **63.** 8.1
**64.** Approximately 5 times as intense

## Chapter 11 Test (page 597)

**1.** $\frac{1}{2}$ **2.** 1 **3.** 1 **4.** −1 **5.** {−3} **6.** $\left\{-\frac{3}{2}\right\}$ **7.** $\left\{\frac{8}{3}\right\}$

**8.** {243} **9.** {2} **10.** $\left\{\frac{2}{5}\right\}$ **11.** 4.1919 **12.** 0.2031

**13.** 0.7325 **14.** 5.4538 **15.** {5.17} **16.** {10.29}

**17.** 4.0069 **18.** $\log_b\left(\frac{x^3y^2}{z}\right)$ **19.** $6342.08

**20.** 13.5 years **21.** 7.8 hours **22.** 4813 grams

**23.**

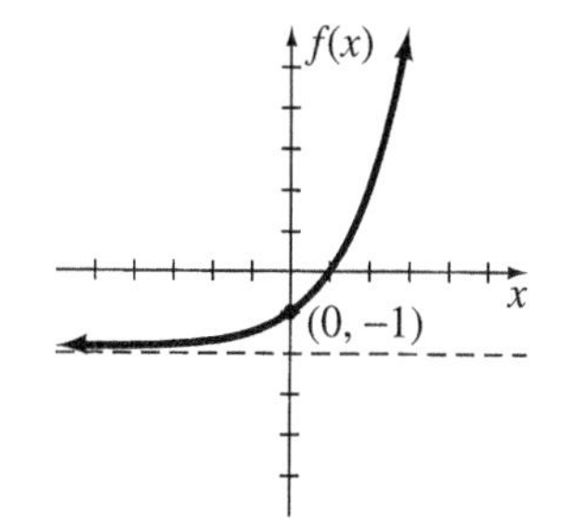


**24.**

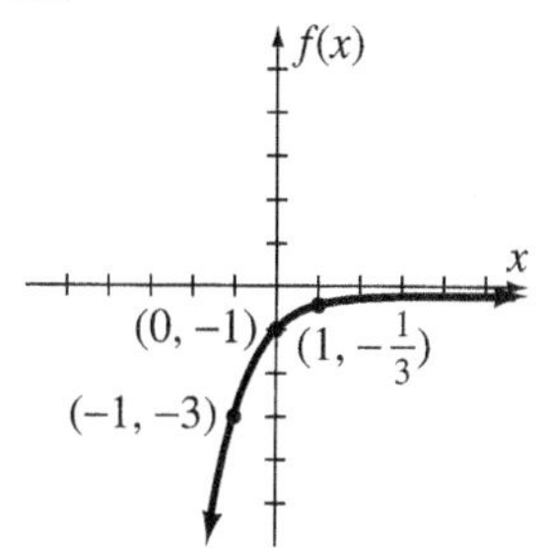


**25.**

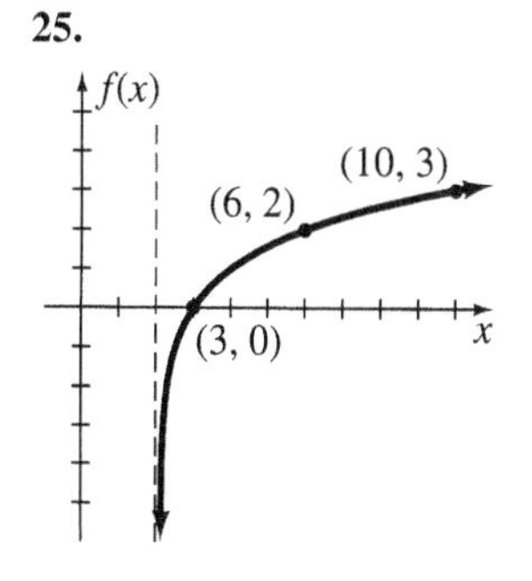

# Appendix A

## Practice Exercises (page 601)

1. $x^8 + 8x^7y + 28x^6y^2 + 56x^5y^3 + 70x^4y^4 + 56x^3y^5 + 28x^2y^6 + 8xy^7 + y^8$
2. $x^7 + 7x^6y + 21x^5y^2 + 35x^4y^3 + 35x^3y^4 + 21x^2y^5 + 7xy^6 + y^7$
3. $81x^4 + 108x^3y + 54x^2y^2 + 12xy^3 + y^4$
4. $x^4 + 8x^3y + 24x^2y^2 + 32xy^3 + 16y^4$
5. $x^5 - 5x^4y + 10x^3y^2 - 10x^2y^3 + 5xy^4 - y^5$
6. $x^4 - 4x^3y + 6x^2y^2 - 4xy^3 + y^4$
7. $x^{10} + 10x^9y + 45x^8y^2 + 120x^7y^3 + 210x^6y^4 + 252x^5y^5 + 210x^4y^6 + 120x^3y^7 + 45x^2y^8 + 10xy^9 + y^{10}$
8. $x^9 + 9x^8y + 36x^7y^2 + 84x^6y^3 + 126x^5y^4 + 126x^4y^5 + 84x^3y^6 + 36x^2y^7 + 9xy^8 + y^9$
9. $64x^6 + 192x^5y + 240x^4y^2 + 160x^3y^3 + 60x^2y^4 + 12xy^5 + y^6$
10. $x^5 + 15x^4y + 90x^3y^2 + 270x^2y^3 + 405xy^4 + 243y^5$
11. $x^5 - 15x^4y + 90x^3y^2 - 270x^2y^3 + 405xy^4 - 243y^5$
12. $64x^6 - 192x^5y + 240x^4y^2 - 160x^3y^3 + 60x^2y^4 - 12xy^5 + y^6$
13. $243a^5 - 810a^4b + 1080a^3b^2 - 720a^2b^3 + 240ab^4 - 32b^5$
14. $16a^4 - 96a^3b + 216a^2b^2 - 216ab^3 + 81b^4$
15. $x^6 + 6x^5y^3 + 15x^4y^6 + 20x^3y^9 + 15x^2y^{12} + 6xy^{15} + y^{18}$
16. $x^{10} + 5x^8y + 10x^6y^2 + 10x^4y^3 + 5x^2y^4 + y^5$
17. $x^7 + 14x^6 + 84x^5 + 280x^4 + 560x^3 + 672x^2 + 448x + 128$
18. $x^6 + 18x^5 + 135x^4 + 540x^3 + 1215x^2 + 1458x + 729$
19. $x^4 - 12x^3 + 54x^2 - 108x + 81$
20. $x^9 - 9x^8 + 36x^7 - 84x^6 + 126x^5 - 126x^4 + 84x^3 - 36x^2 + 9x - 1$
21. $x^{15} + 15x^{14}y + 105x^{13}y^2 + 455x^{12}y^3$
22. $x^{12} + 12x^{11}y + 66x^{10}y^2 + 220x^9y^3$
23. $a^{13} - 26a^{12}b + 312a^{11}b^2 - 2288a^{10}b^3$
24. $x^{20} - 20x^{19}y + 190x^{18}y^2 - 1140x^{17}y^3$
25. $462x^5y^6$ **26.** $56x^5y^3$ **27.** $-160x^3y^3$ **28.** $126x^5y^4$
29. $2000x^3y^2$ **30.** $189a^2b^5$

# Index

# 13

# Right Triangle Trigonometry

## Mathematics at Work

Automotive collision repair technicians repair, repaint, and refinish automotive vehicle bodies; straighten vehicle frames; and replace damaged glass and other automotive parts that cannot be economically repaired. Using modern techniques including diagnostics, electronic equipment, computer support equipment, and other specialized equipment, the technician's primary task is to restore damaged vehicles to their original condition. Training and education for this work are available at many community colleges and trade schools. Various automobile manufacturers and their participating dealers also sponsor programs at postsecondary schools across the United States. Good reading, mathematics, computer, and communications skills are needed.

Voluntary certification is available through the National Institute for Automotive Service Excellence (ASE) and is the recognized standard of achievement for automotive collision repair technicians. For more information, go to the website **www.cengage.com/mathematics/ewen**.

**Automotive Collision Repair Technician**
Automotive collision repair technician repairing an automobile

## OBJECTIVES

- Write the trigonometric ratios for the sine, cosine, and tangent of an angle using the basic terms of a right triangle.
- Find the value of a trigonometric ratio using a scientific calculator.
- Use a trigonometric ratio to find angles.
- Use a trigonometric ratio to find sides.
- Solve a right triangle.
- Solve application problems involving trigonometric ratios and right triangles.

# 13.1 Trigonometric Ratios

Many applications in science and technology require the use of triangles and trigonometry. Early applications of trigonometry, beginning in the second century B.C., were in astronomy, surveying, and navigation. Applications that you may study include electronics, the motion of projectiles, light refraction in optics, and sound.

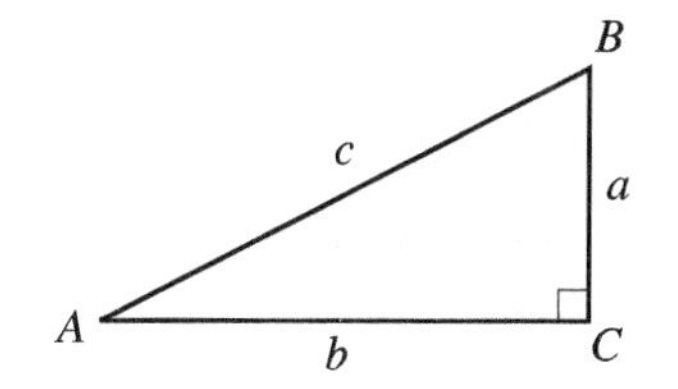


figure **13.1**
Right triangle

In this chapter, we consider only right triangles. A right triangle has one right angle, two acute angles, a hypotenuse, and two legs. The right angle, as shown in Figure 13.1, is usually labeled with the capital letter $C$. The vertices of the two acute angles are usually labeled with the capital letters $A$ and $B$. The hypotenuse is the side opposite the right angle, the longest side of a right triangle, and is usually labeled with the lowercase letter $c$. The legs are the sides opposite the acute angles. The leg (side) opposite angle $A$ is labeled $a$, and the leg opposite angle $B$ is labeled $b$. Note that each side of the triangle is labeled with the lowercase of the letter of the angle opposite that side.

The two legs are also named as the side *opposite* angle $A$ and the side *adjacent to* (or next to) angle $A$ or as the side opposite angle $B$ and the side adjacent to angle $B$. See Figure 13.2.

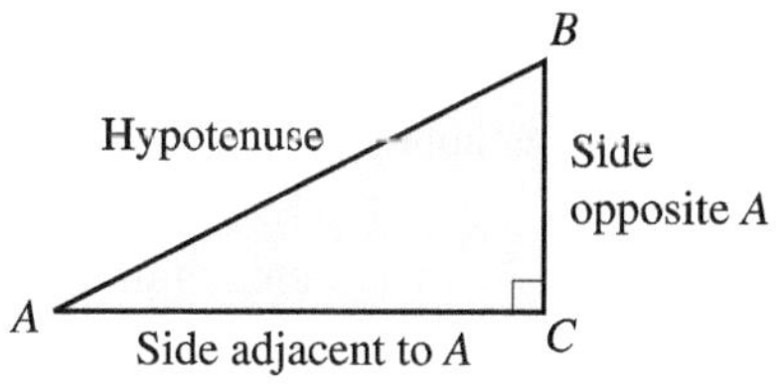


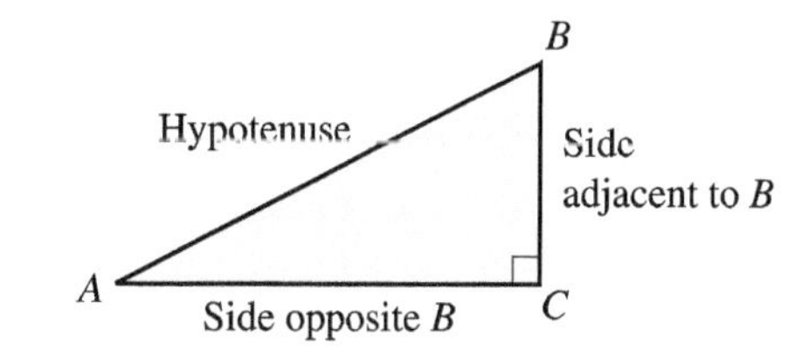


figure **13.2**
Key parts of a right triangle

**Pythagorean Theorem**

In any right triangle,

$$c^2 = a^2 + b^2$$

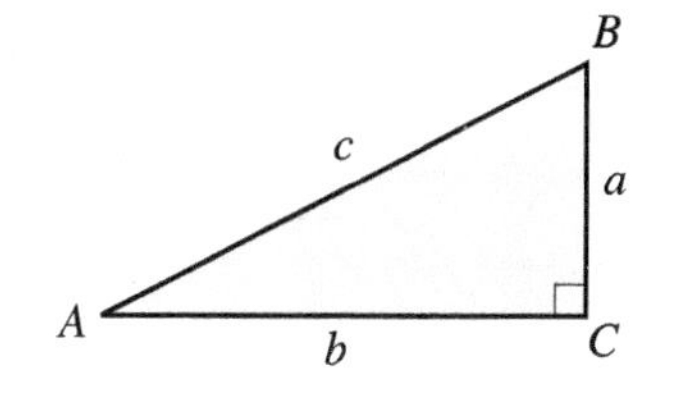

That is, the square of the length of the hypotenuse is equal to the sum of the squares of the lengths of the legs. The following equivalent formulas are often more useful:

$c = \sqrt{a^2 + b^2}$ used to find the length of the hypotenuse

$a = \sqrt{c^2 - b^2}$ used to find the length of leg $a$

$b = \sqrt{c^2 - a^2}$ used to find the length of leg $b$

Recall that the Pythagorean theorem was developed in detail in Section 12.3.

**Example 1** Find the length of side $b$ in Figure 13.3.

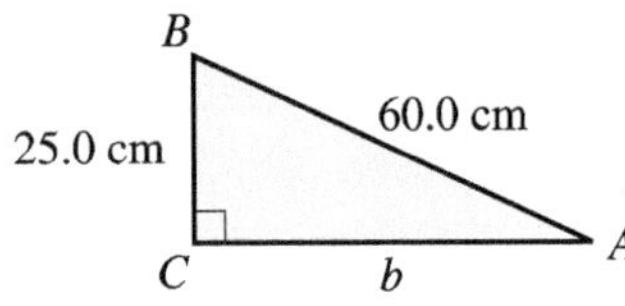


figure 13.3

Using the formula to find the length of leg $b$, we have

$$b = \sqrt{c^2 - a^2}$$
$$b = \sqrt{(60.0\text{ cm})^2 - (25.0\text{ cm})^2}$$
$$= 54.5\text{ cm}$$

**Example 2** Find the length of side $c$ in Figure 13.4.

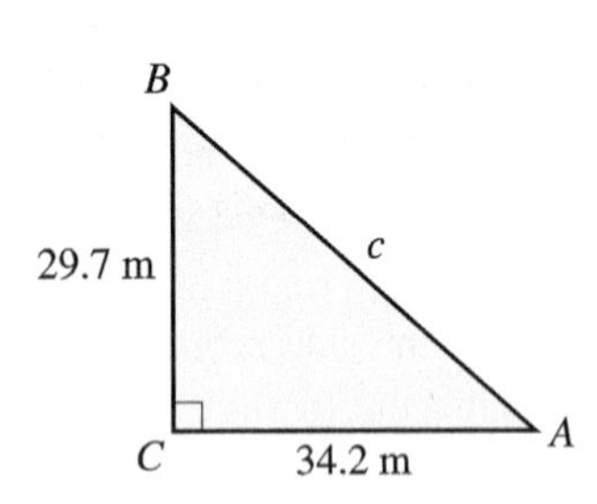


figure 13.4

Using the formula to find the hypotenuse $c$, we have

$$c = \sqrt{a^2 + b^2}$$
$$c = \sqrt{(29.7\text{ m})^2 + (34.2\text{ m})^2}$$
$$= 45.3\text{ m}$$

A **ratio** is the comparison of two quantities by division. The ratios of the sides of a right triangle can be used to find an unknown part—or parts—of that right triangle. Such a ratio is called a **trigonometric ratio** and expresses the relationship between an acute angle and the lengths of two of the sides of a right triangle.

The **sine of angle A**, abbreviated "sin $A$," equals the ratio of the length of the side opposite angle $A$, which is $a$, to the length of the hypotenuse, $c$.

The **cosine of angle A**, abbreviated "cos $A$," equals the ratio of the length of the side adjacent to angle $A$, which is $b$, to the length of the hypotenuse, $c$.

The **tangent of angle A**, abbreviated "tan $A$," equals the ratio of the length of the side opposite angle $A$, which is $a$, to the length of the side adjacent to angle $A$, which is $b$.

That is, in a right triangle (Figure 13.5), we have the following ratios.

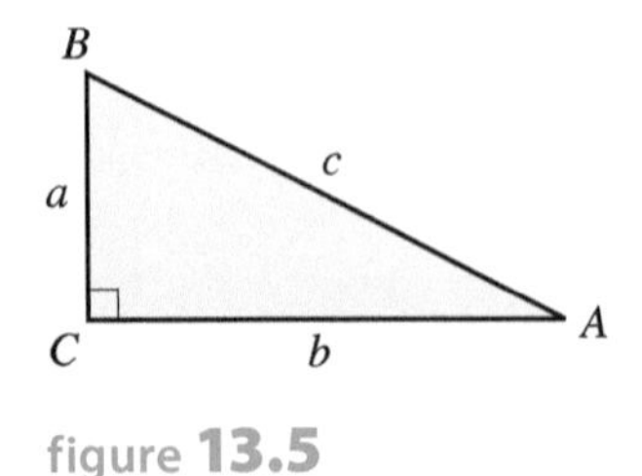


figure 13.5

**Trigonometric Ratios**

$$\sin A = \frac{\text{length of side opposite angle } A}{\text{length of hypotenuse}} = \frac{a}{c}$$

$$\cos A = \frac{\text{length of side adjacent to angle } A}{\text{length of hypotenuse}} = \frac{b}{c}$$

$$\tan A = \frac{\text{length of side opposite angle } A}{\text{length of side adjacent to angle } A} = \frac{a}{b}$$

Similarly, the ratios can be defined for angle $B$.

$$\sin B = \frac{\text{length of side opposite angle } B}{\text{length of hypotenuse}} = \frac{b}{c}$$

$$\cos B = \frac{\text{length of side adjacent to angle } B}{\text{length of hypotenuse}} = \frac{a}{c}$$

$$\tan B = \frac{\text{length of side opposite angle } B}{\text{length of side adjacent to angle } B} = \frac{b}{a}$$

**Example 3** Find the three trigonometric ratios for angle $A$ in the triangle in Figure 13.6.

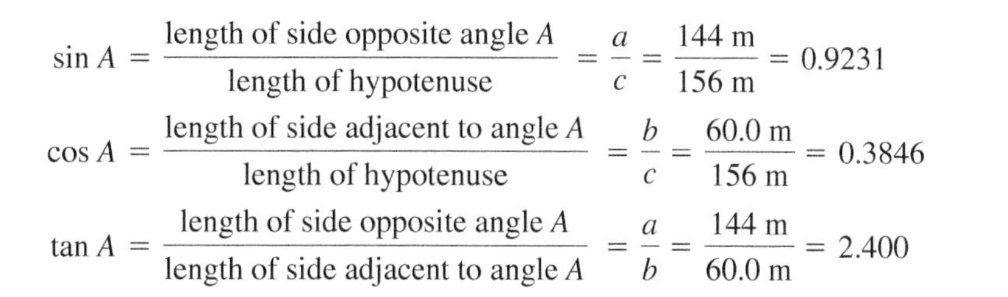

$$\sin A = \frac{\text{length of side opposite angle } A}{\text{length of hypotenuse}} = \frac{a}{c} = \frac{144 \text{ m}}{156 \text{ m}} = 0.9231$$

$$\cos A = \frac{\text{length of side adjacent to angle } A}{\text{length of hypotenuse}} = \frac{b}{c} = \frac{60.0 \text{ m}}{156 \text{ m}} = 0.3846$$

$$\tan A = \frac{\text{length of side opposite angle } A}{\text{length of side adjacent to angle } A} = \frac{a}{b} = \frac{144 \text{ m}}{60.0 \text{ m}} = 2.400$$

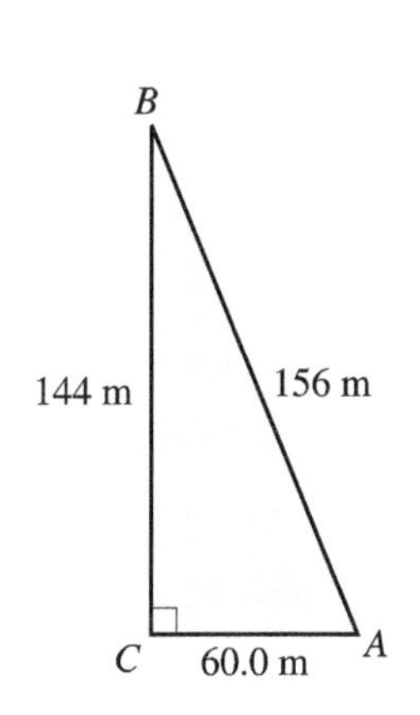


figure **13.6**

The values of the trigonometric ratios of various angles can be found with calculators. You will need a calculator that has sin, cos, and tan keys.

**Very Important Note:** When working with the trigonometric functions on your calculator, make certain that it is set in the degree mode. If your calculator has a DRG key, it is used to change angle measurement modes from degrees to radians to grads. In the degree mode, the circle or one complete revolution is divided into 360°. In the radian mode, the circle or one complete revolution is divided into $2\pi$ rad. In the grad mode, the circle or one complete revolution is divided into $400^g$. We will be working exclusively in the degree mode.

**Example 4** Find sin 37.5° rounded to four significant digits.

SIN 37.5 =

0.608761429

Thus, sin 37.5° = 0.6088 rounded to four significant digits.

**Example 5** Find cos 18.63° rounded to four significant digits.

COS 18.63 =

0.947601273

Thus, cos 18.63° = 0.9476 rounded to four significant digits.

**Example 6** Find tan 81.7° rounded to four significant digits.

TAN 81.7 =

6.854750833

Thus, tan 81.7° = 6.855 rounded to four significant digits.

A calculator may also be used to find the *angle* when the value of the trigonometric ratio is known. The procedure is shown in the following examples.

**NOTE:** The following keys on a calculator have the meanings that follow:

SIN$^{-1}$ "the angle whose sine is"

COS$^{-1}$ "the angle whose cosine is"

TAN$^{-1}$ "the angle whose tangent is"

**Example 7** Find angle $A$ to the nearest tenth of a degree when $\sin A = 0.6372$.

39.583346

**NOTE:** Make certain that your calculator is in the degree mode.

Thus, angle $A = 39.6°$ rounded to the nearest tenth of a degree.

**Example 8** Find angle $B$ to the nearest tenth of a degree when $\tan B = 0.3106$.

17.25479431

Thus, angle $B = 17.3°$ rounded to the nearest tenth of a degree.

**Example 9** Find angle $A$ to the nearest hundredth of a degree when $\cos A = 0.4165$.

COS$^{-1}$ .4165 =

65.3861858

Thus, angle $A = 65.39°$ rounded to the nearest hundredth of a degree.

# EXERCISES 13.1

*Refer to right triangle ABC in Illustration 1 for Exercises 1–10:*

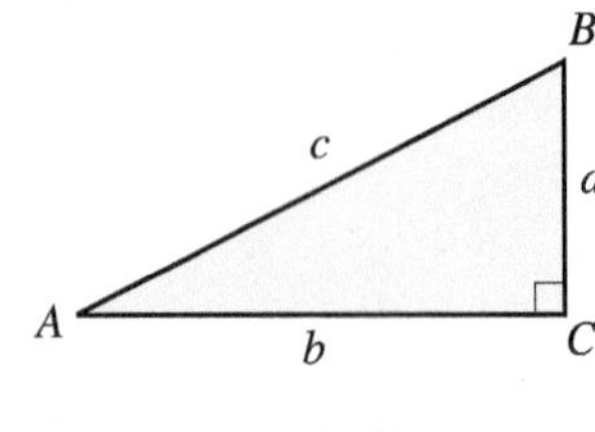


ILLUSTRATION 1

**1.** The side opposite angle $A$ is __?__.

**2.** The side opposite angle $B$ is __?__.

**3.** The hypotenuse is __?__.

**4.** The side adjacent to angle $A$ is __?__.

**5.** The side adjacent to angle $B$ is __?__.

**6.** The angle opposite side $a$ is __?__.

**7.** The angle opposite side $b$ is __?__.

**8.** The angle opposite side $c$ is __?__.

**9.** The angle adjacent to side $a$ is __?__.

**10.** The angle adjacent to side $b$ is __?__.

*Use right triangle ABC in Illustration 1 and the Pythagorean theorem to find each unknown side, rounded to three significant digits:*

**11.** $c = 75.0$ m, $a = 45.0$ m

**12.** $a = 25.0$ cm, $b = 60.0$ cm

**13.** $a = 29.0$ mi, $b = 47.0$ mi

**14.** $a = 12.0$ km, $c = 61.0$ km

**15.** $c = 18.9$ cm, $a = 6.71$ cm

**16.** $a = 20.2$ mi, $b = 19.3$ mi

**17.** $a = 171$ ft, $b = 203$ ft

**18.** $c = 35.3$ m, $b = 25.0$ m

**19.** $a = 202$ m, $c = 404$ m

**20.** $a = 1.91$ km, $b = 3.32$ km

**21.** $b = 1520$ km, $c = 2160$ km

**22.** $a = 203{,}000$ ft, $c = 521{,}000$ ft

**23.** $a = 45{,}800$ m, $b = 38{,}600$ m

**24.** $c = 3960$ m, $b = 3540$ m

*Use the triangle in Illustration 2 for Exercises 25–30:*

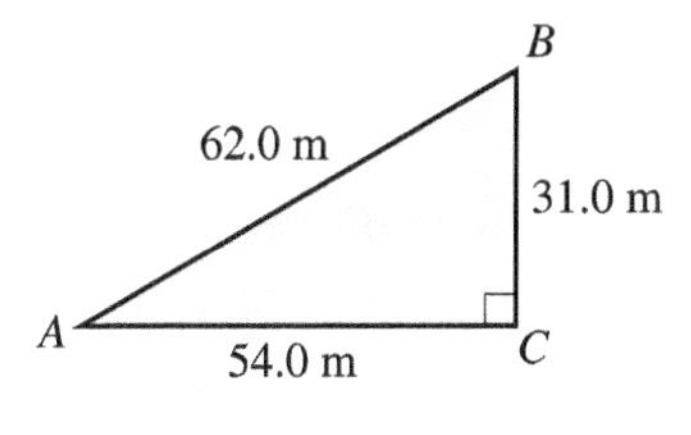


ILLUSTRATION 2

**25.** Find sin $A$. **26.** Find cos $A$. **27.** Find tan $A$.

**28.** Find sin $B$. **29.** Find cos $B$. **30.** Find tan $B$.

*Find the value of each trigonometric ratio rounded to four significant digits:*

**31.** sin 49.6° **32.** cos 55.2° **33.** tan 65.3°

**34.** sin 69.7° **35.** cos 29.7° **36.** tan 14.6°

**37.** sin 31.64° **38.** tan 13.25° **39.** cos 75.31°

**40.** cos 84.83° **41.** tan 3.05° **42.** sin 6.74°

**43.** sin 37.62° **44.** cos 18.94° **45.** tan 21.45°

**46.** sin 11.31° **47.** cos 47.16° **48.** tan 81.85°

*Find each angle rounded to the nearest tenth of a degree:*

**49.** $\sin A = 0.7941$ **50.** $\tan A = 0.2962$

**51.** $\cos B = 0.4602$ **52.** $\cos A = 0.1876$

**53.** $\tan B = 1.386$ **54.** $\sin B = 0.3040$

**55.** $\sin B = 0.1592$ **56.** $\tan B = 2.316$

**57.** $\cos A = 0.8592$ **58.** $\cos B = 0.3666$

**59.** $\tan A = 0.8644$ **60.** $\sin A = 0.5831$

*Find each angle rounded to the nearest hundredth of a degree:*

**61.** $\tan A = 0.1941$ **62.** $\sin B = 0.9324$

**63.** $\cos A = 0.3572$ **64.** $\cos B = 0.2597$

**65.** $\sin A = 0.1506$ **66.** $\tan B = 2.500$

**67.** $\tan B = 3.806$ **68.** $\sin A = 0.4232$

**69.** $\cos B = 0.7311$ **70.** $\cos A = 0.6427$

**71.** $\sin B = 0.3441$ **72.** $\tan A = 0.5536$

**73.** In Exercises 25–30, there are two pairs of ratios that are equal. Name them.

## 13.2 Using Trigonometric Ratios to Find Angles

A trigonometric ratio may be used to find an angle of a right triangle, given the lengths of any two sides.

**Example 1**

In Figure 13.7, find angle $A$ using a calculator, as follows.

We know the sides opposite and adjacent to angle $A$. So we use the tangent ratio:

$$\tan A = \frac{\text{length of side opposite angle } A}{\text{length of side adjacent to angle } A}$$

$$\tan A = \frac{28.5 \text{ m}}{21.3 \text{ m}} = 1.338$$

Next, find angle $A$ to the nearest tenth of a degree when $\tan A = 1.338$. The complete set of operations on a calculator follows.

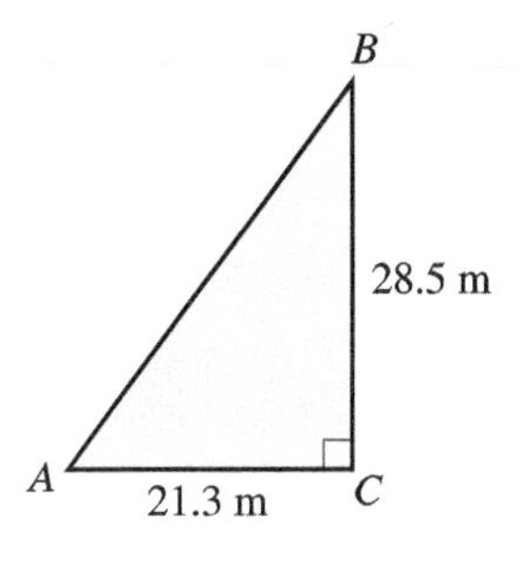


figure **13.7**

53.22672246

Thus, angle $A = 53.2°$ rounded to the nearest tenth of a degree.

When calculations involve a trigonometric ratio, we shall use the following rule for significant digits.

| Angles expressed to the nearest | The length of each side of the triangle contains |
|---|---|
| 1° | Two significant digits |
| 0.1° | Three significant digits |
| 0.01° | Four significant digits |

An example of each case is shown in Figure 13.8.

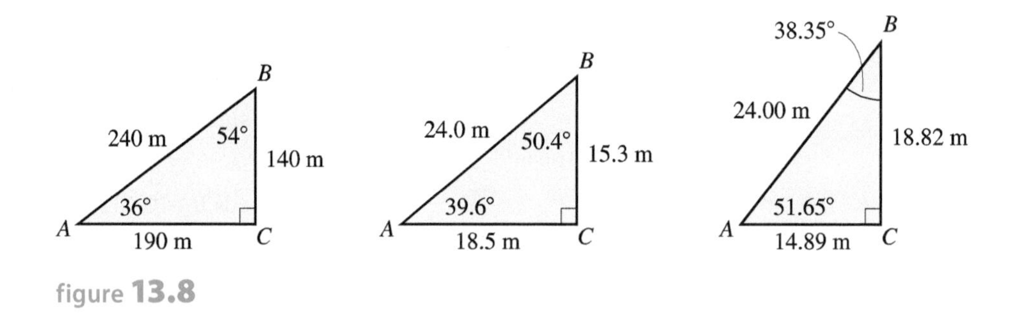


figure **13.8**

**Example 2**

Find angle $B$ in the triangle in Figure 13.9.

We know the hypotenuse and the side adjacent to angle $B$. So let's use the cosine ratio.

$$\cos B = \frac{\text{length of side adjacent to angle } B}{\text{length of hypotenuse}}$$

$$\cos B = \frac{35.20 \text{ cm}}{45.85 \text{ cm}}$$

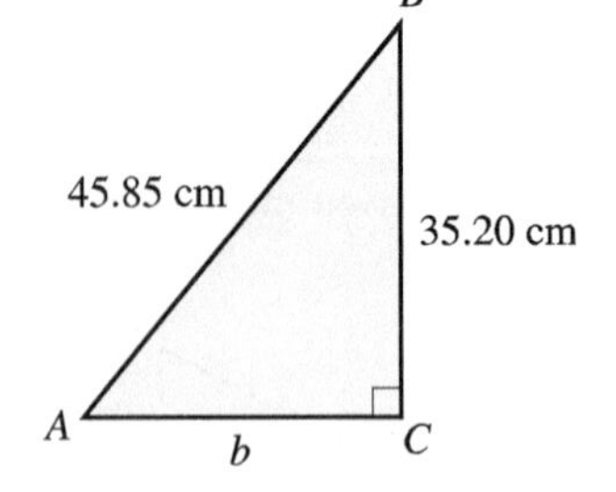


figure **13.9**

Find angle $B$ using a calculator as follows.

35.2 ÷ 45.85 = COS⁻¹ ANS =

39.85033989

Thus, angle $B = 39.85°$ rounded to the nearest hundredth of a degree.

The question is often raised, "Which of the three trig ratios do I use?" First, notice that each trigonometric ratio consists of two sides and one angle, or three quantities in all. To find the solution to such an equation, two of the quantities must be known. We will answer the question in two parts.

### Which Trig Ratio to Use

1. If you are finding an angle, two sides must be known. Label these two known sides as *side opposite* the angle you are finding, *side adjacent* to the angle you are finding, or *hypotenuse*. Then choose the trig ratio that has these two sides.
2. If you are finding a side, one side and one angle must be known. Label the known side and the unknown side as *side opposite* the known angle, *side adjacent* to the known angle, or *hypotenuse*. Then choose the trig ratio that has these two sides.

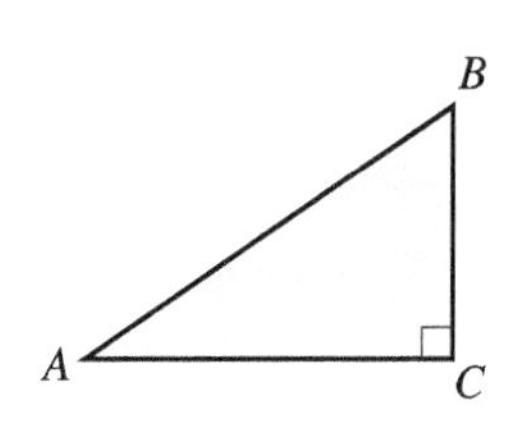

figure **13.10**

A useful and time-saving fact about right triangles (Figure 13.10) is that the sum of the acute angles of any right triangle is 90°.

$$A + B = 90°$$

Why? We know that the sum of the interior angles of any triangle is 180°. A right triangle, by definition, contains a right angle, whose measure is 90°. That leaves 90° to be divided between the two acute angles.

Note, then, that if one acute angle is given or known, the other acute angle may be found by subtracting the known angle from 90°. That is,

$$A = 90° - B$$

$$B = 90° - A$$

**Example 3** Find angle $A$ and angle $B$ in the triangle in Figure 13.11.

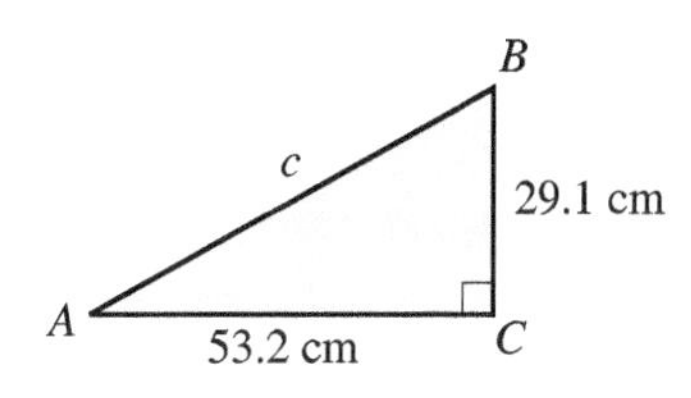

figure **13.11**

$$\tan A = \frac{\text{length of side opposite angle } A}{\text{length of side adjacent to angle } A}$$

$$\tan A = \frac{29.1 \text{ cm}}{53.2 \text{ cm}} = 0.5470$$

$$A = 28.7°$$

Angle $B = 90° - 28.7° = 61.3°$.

## EXERCISES 13.2

*Using Illustration 1, find the measure of each acute angle for each right triangle:*

1. $a = 36.0$ m, $b = 50.9$ m
2. $a = 72.0$ cm, $c = 144$ cm
3. $b = 39.7$ cm, $c = 43.6$ cm
4. $a = 171$ km, $b = 695$ km
5. $b = 13.6$ m, $c = 18.7$ m
6. $b = 409$ km, $c = 612$ km
7. $a = 29.7$ m, $b = 29.7$ m, $c = 42.0$ m
8. $a = 36.2$ mm, $b = 62.7$ mm, $c = 72.4$ mm
9. $a = 2902$ km, $b = 1412$ km
10. $b = 21.34$ m, $c = 47.65$ m
11. $a = 0.6341$ cm, $c = 0.7982$ cm
12. $b = 4.372$ m, $c = 5.806$ m
13. $b = 1455$ ft, $c = 1895$ ft
14. $a = 25.45$ in., $c = 41.25$ in.
15. $a = 243.2$ km, $b = 271.5$ km
16. $a = 351.6$ m, $b = 493.0$ m
17. $a = 16.7$ m, $c = 81.4$ m
18. $a = 847$ m, $b = 105$ m
19. $b = 1185$ ft, $c = 1384$ ft
20. $a = 48.7$ cm, $c = 59.5$ cm

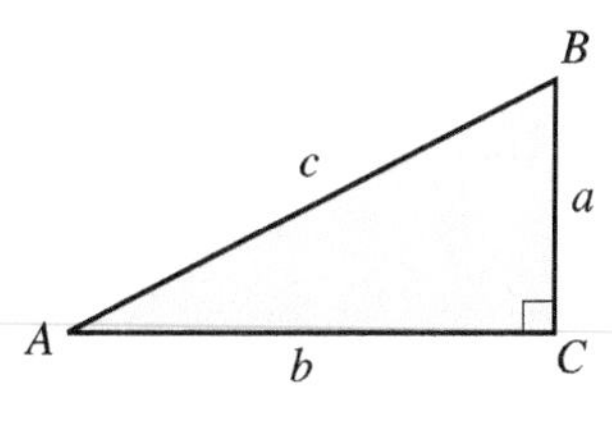

ILLUSTRATION 1

21. $a = 845$ km, $b = 2960$ km
22. $b = 2450$ km, $c = 3570$ km
23. $a = 8897$ m, $c = 9845$ m
24. $a = 58.44$ mi, $b = 98.86$ mi

# 13.3 Using Trigonometric Ratios to Find Sides

We also use a trigonometric ratio to find a side of a right triangle, given one side and the measure of one of the acute angles.

**Example 1** Find side $a$ in the triangle in Figure 13.12.

figure **13.12**

With respect to the known angle $B$, we know the hypotenuse and are finding the adjacent side. So we use the cosine ratio.

$$\cos B = \frac{\text{length of side adjacent to angle } B}{\text{length of hypotenuse}}$$

$$\cos 24.5° = \frac{a}{258 \text{ ft}}$$

$$a = (\cos 24.5°)(258 \text{ ft}) \quad \text{Multiply both sides by 258 ft.}$$

Side $a$ can be found by using a calculator as follows.

COS 24.5 ) * × 258 =

234.7700079

Thus, side $a$ = 235 ft rounded to three significant digits. ■

**Example 2** Find sides $b$ and $c$ in the triangle in Figure 13.13.

If we find side $b$ first, we are looking for the side adjacent to angle $A$, the known angle. We are given the side opposite angle $A$. Thus, we should use the tangent ratio.

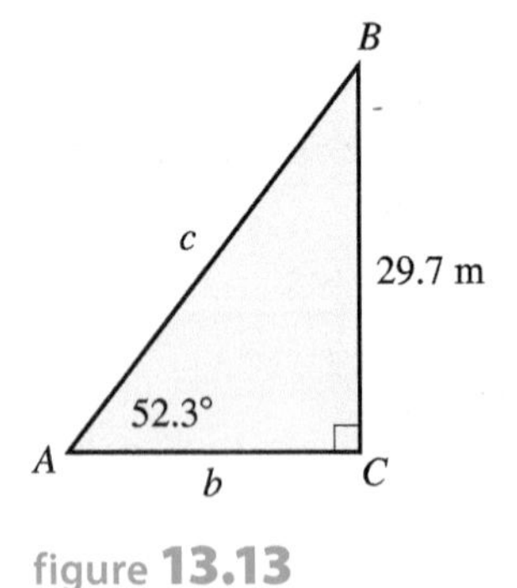

figure **13.13**

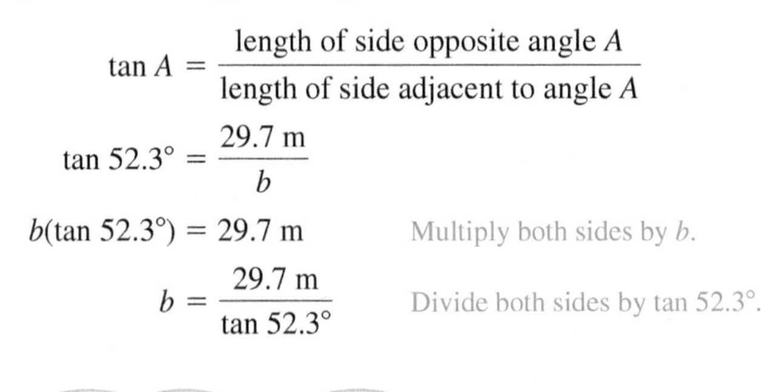

$$\tan A = \frac{\text{length of side opposite angle } A}{\text{length of side adjacent to angle } A}$$

$$\tan 52.3° = \frac{29.7 \text{ m}}{b}$$

$$b(\tan 52.3°) = 29.7 \text{ m} \quad \text{Multiply both sides by } b.$$

$$b = \frac{29.7 \text{ m}}{\tan 52.3°} \quad \text{Divide both sides by tan 52.3°.}$$

29.7 ÷ TAN 52.3 =

22.95476858

Thus, side $b$ = 23.0 m rounded to three significant digits.

*Note: You might need to insert a right parenthesis to clarify the order of operations. The trigonometry keys may also include the left parenthesis.

To find side $c$, we are looking for the hypotenuse, and we have the opposite side given. Thus, we should use the sine ratio.

$$\sin A = \frac{\text{length of side opposite angle } A}{\text{length of hypotenuse}}$$

$$\sin 52.3° = \frac{29.7 \text{ m}}{c}$$

$$c(\sin 52.3°) = 29.7 \text{ m} \qquad \text{Multiply both sides by } c.$$

$$c = \frac{29.7 \text{ m}}{\sin 52.3°} \qquad \text{Divide both sides by } \sin 52.3°.$$

$$= 37.5 \text{ m}$$

The Pythagorean theorem may be used to check your work.

## EXERCISES 13.3

*Find the unknown sides of each right triangle (see Illustration 1):*

1. $a = 36.7$ m, $A = 42.1°$
2. $b = 73.6$ cm, $B = 19.0°$
3. $a = 236$ km, $B = 49.7°$
4. $b = 28.9$ ft, $A = 65.2°$
5. $c = 49.1$ cm, $A = 36.7°$
6. $c = 236$ m, $A = 12.9°$
7. $b = 23.7$ cm, $A = 23.7°$
8. $a = 19.2$ km, $B = 63.2°$
9. $b = 29{,}200$ km, $A = 12.9°$
10. $c = 36.7$ mi, $B = 68.3°$
11. $a = 19.72$ m, $A = 19.75°$
12. $b = 125.3$ cm, $B = 23.34°$
13. $c = 255.6$ mi, $A = 39.25°$
14. $c = 7.363$ km, $B = 14.80°$
15. $b = 12{,}350$ m, $B = 69.72°$
16. $a = 3678$ m, $B = 10.04°$
17. $a = 1980$ m, $A = 18.4°$
18. $a = 9820$ ft, $B = 35.7°$
19. $b = 841.6$ km, $A = 18.91°$
20. $c = 289.5$ cm, $A = 24.63°$

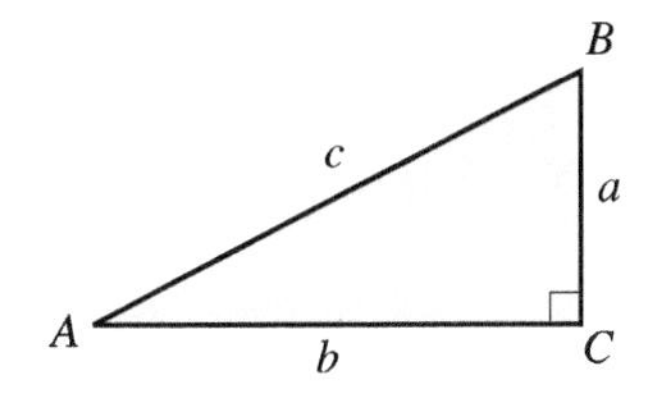


ILLUSTRATION 1

21. $c = 185.6$ m, $B = 61.45°$
22. $b = 21.63$ km, $B = 82.06°$
23. $c = 256$ cm, $A = 25.6°$
24. $a = 18.3$ mi, $A = 71.2°$

## 13.4 Solving Right Triangles

The phrase **solving a right triangle** refers to finding the measures of the various parts of a triangle that are not given. We proceed as we did in the last two sections.

**Example 1** Solve the right triangle in Figure 13.14.

We are given the measure of one acute angle and the length of one leg.

$$A = 90° - B$$

$$A = 90° - 36.7° = 53.3°$$

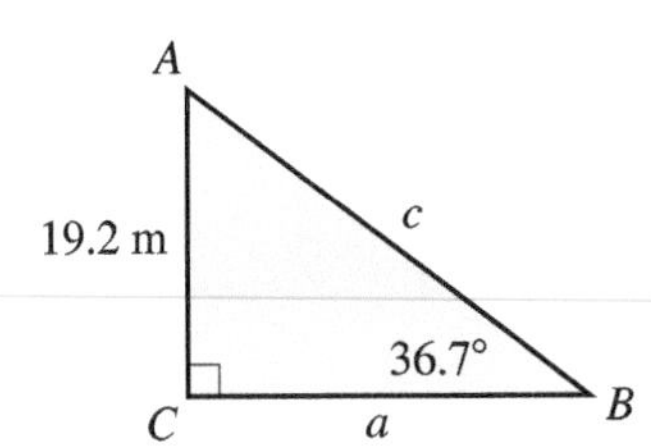


figure 13.14

We then can use either the sine or the cosine ratio to find side $c$.

$$\sin B = \frac{\text{length of side opposite angle } B}{\text{length of hypotenuse}}$$

$$\sin 36.7° = \frac{19.2 \text{ m}}{c}$$

$$c(\sin 36.7°) = 19.2 \text{ m} \qquad \text{Multiply both sides by } c.$$

$$c = \frac{19.2 \text{ m}}{\sin 36.7°} \qquad \text{Divide both sides by } \sin 36.7°.$$

$$= 32.1 \text{ m}$$

Now we may use either a trigonometric ratio or the Pythagorean theorem to find side $a$.

**Solution by a Trigonometric Ratio:**

$$\tan B = \frac{\text{length of side opposite angle } B}{\text{length of side adjacent to angle } B}$$

$$\tan 36.7° = \frac{19.2 \text{ m}}{a}$$

$$a(\tan 36.7°) = 19.2 \text{ m} \qquad \text{Multiply both sides by } a.$$

$$a = \frac{19.2 \text{ m}}{\tan 36.7°} \qquad \text{Divide both sides by } \tan 36.7°.$$

$$= 25.8 \text{ m}$$

**Solution by the Pythagorean Theorem:**

$$a = \sqrt{c^2 - b^2}$$

$$a = \sqrt{(32.1 \text{ m})^2 - (19.2 \text{ m})^2}$$

$$= 25.7 \text{ m}$$

Can you explain the difference in these two results? ■

**Example 2** Solve the right triangle in Figure 13.15.

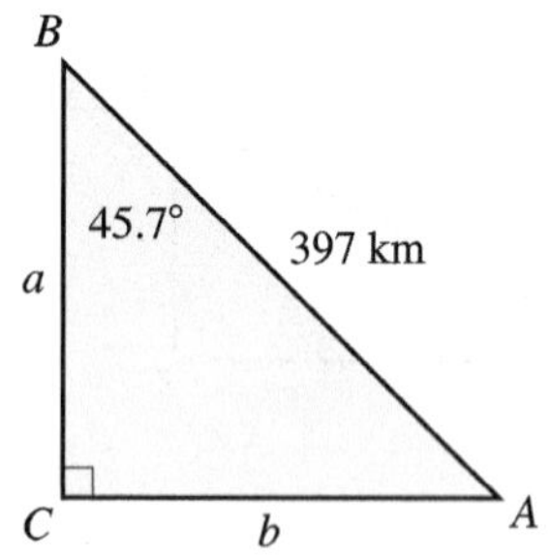

figure 13.15

We are given the measure of one acute angle and the length of the hypotenuse.

$$A = 90° - B$$

$$A = 90° - 45.7° = 44.3°$$

To find side $b$, we must use the sine or the cosine ratio, since the hypotenuse is given.

$$\sin B = \frac{\text{length of side opposite angle } B}{\text{length of hypotenuse}}$$

$$\sin 45.7° = \frac{b}{397 \text{ km}}$$

$$(\sin 45.7°)(397 \text{ km}) = b \qquad \text{Multiply both sides by 397 km.}$$

$$284 \text{ km} = b$$

Again, we can use either a trigonometric ratio or the Pythagorean theorem to find side $a$.

**Solution by a Trigonometric Ratio:**

$$\cos B = \frac{\text{length of side adjacent to angle } B}{\text{length of hypotenuse}}$$

$$\cos 45.7° = \frac{a}{397 \text{ km}}$$

$$(\cos 45.7°)(397 \text{ km}) = a \qquad \text{Multiply both sides by 397 km.}$$

$$277 \text{ km} = a$$

**Solution by the Pythagorean Theorem:**

$$a = \sqrt{c^2 - b^2}$$

$$a = \sqrt{(397 \text{ km})^2 - (284 \text{ km})^2}$$

$$= 277 \text{ km}$$

■

Example **3** Solve the right triangle in Figure 13.16.

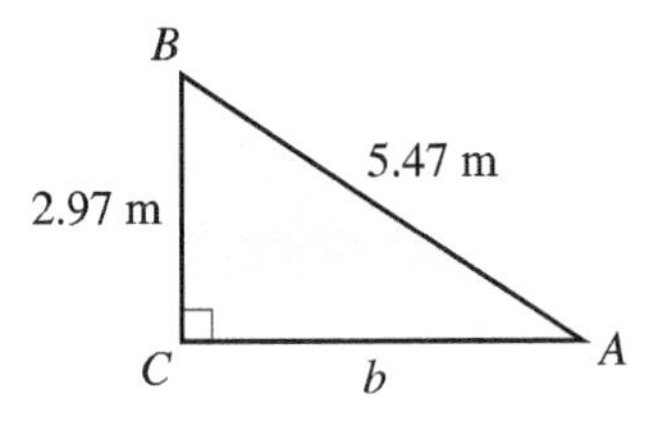


figure **13.16**

We are given two sides of the right triangle.

To find angle $A$ or angle $B$, we could use either the sine or cosine, since the hypotenuse is given.

$$\sin A = \frac{\text{length of side opposite angle } A}{\text{length of hypotenuse}}$$

$$\sin A = \frac{2.97 \text{ m}}{5.47 \text{ m}} = 0.5430$$

$$A = 32.9°$$

Then

$$B = 90° - A$$

$$B = 90° - 32.9° = 57.1°$$

The unknown side $b$ can be found by using the Pythagorean theorem.

$$b = \sqrt{c^2 - a^2}$$

$$b = \sqrt{(5.47 \text{ m})^2 - (2.97 \text{ m})^2}$$

$$= 4.59 \text{ m}$$

## EXERCISES 13.4

*Using Illustration 1, solve each right triangle:*

1. $A = 50.6°$, $c = 49.0$ m
2. $a = 30.0$ cm, $b = 40.0$ cm
3. $B = 41.2°$, $a = 267$ ft
4. $A = 39.7°$, $b = 49.6$ km
5. $b = 72.0$ mi, $c = 78.0$ mi
6. $B = 22.4°$, $c = 46.0$ mi
7. $A = 52.1°$, $a = 72.0$ mm
8. $B = 42.3°$, $b = 637$ m
9. $A = 68.8°$, $c = 39.4$ m
10. $a = 13.6$ cm, $b = 13.6$ cm
11. $a = 12.00$ m, $b = 24.55$ m
12. $B = 38.52°$, $a = 4315$ m
13. $A = 29.19°$, $c = 2975$ ft
14. $B = 29.86°$, $a = 72.62$ m
15. $a = 46.72$ m, $b = 19.26$ m
16. $a = 2436$ ft, $c = 4195$ ft
17. $A = 41.1°$, $c = 485$ m
18. $a = 1250$ km, $b = 1650$ km
19. $B - 9.45°$, $a = 1585$ ft
20. $A = 14.60°$, $b = 135.7$ cm

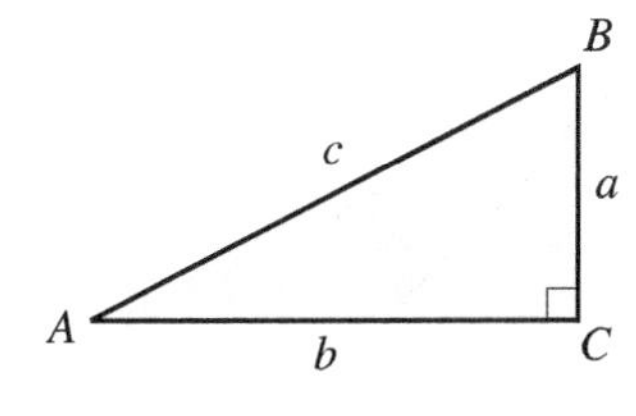


ILLUSTRATION 1

21. $b = 269.5$ m, $c = 380.5$ m
22. $B = 75.65°$, $c = 92.75$ km
23. $B - 81.5°$, $b = 9370$ ft
24. $a = 14.6$ mi, $c = 31.2$ mi

# 13.5 Applications Involving Trigonometric Ratios

Trigonometric ratios can be used to solve many applications similar to those problems we solved in the preceding sections. However, instead of having to find all the parts of a right triangle, we usually need to find only one.

Example **1** The roof in Figure 13.17 has a rise of 7.50 ft and a run of 18.0 ft. Find angle $A$.

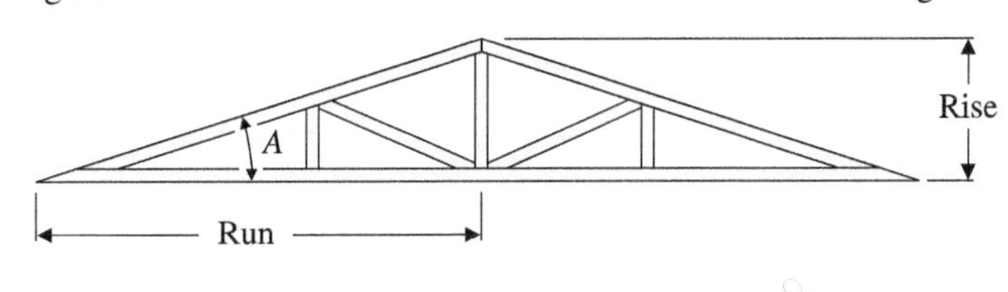


figure **13.17**

We know the length of the side opposite angle $A$ and the length of the side adjacent to angle $A$. So we use the tangent ratio.

$$\tan A = \frac{\text{length of side opposite angle } A}{\text{length of side adjacent to angle } A}$$

$$\tan A = \frac{7.50 \text{ ft}}{18.0 \text{ ft}} = 0.4167$$

$$A = 22.6°$$

The **angle of depression** is the angle between the horizontal and the line of sight to an object that is *below* the horizontal. The **angle of elevation** is the angle between the horizontal and the line of sight to an object that is *above* the horizontal.

In Figure 13.18, angle $A$ is the angle of depression for an observer in the helicopter sighting down to the building on the ground, and angle $B$ is the angle of elevation for an observer in the building sighting up to the helicopter.

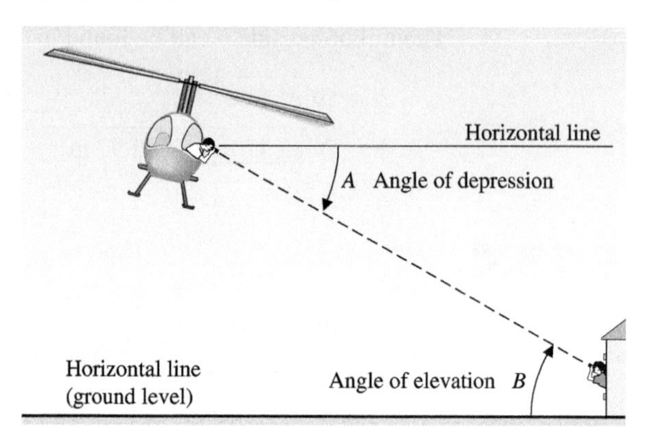


figure **13.18**

**Example 2** A ship's navigator measures the angle of elevation to the beacon of a lighthouse to be 10.1°. He knows that this particular beacon is 225 m above sea level. How far is the ship from the lighthouse?

First, you should sketch the problem, as in Figure 13.19. Since this problem involves finding the length of the side adjacent to an angle when the opposite side is known, we use the tangent ratio.

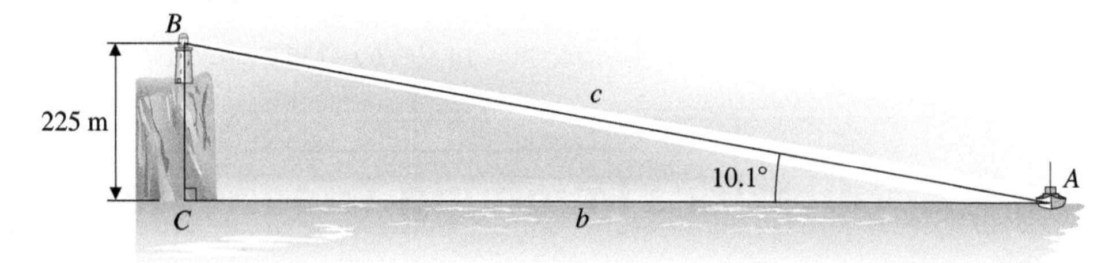


figure **13.19**

$$\tan A = \frac{\text{length of side opposite angle } A}{\text{length of side adjacent to angle } A}$$

$$\tan 10.1° = \frac{225 \text{ m}}{b}$$

$$b(\tan 10.1°) = 225 \text{ m} \quad \text{Multiply both sides by } b.$$

$$b = \frac{225 \text{ m}}{\tan 10.1°} \quad \text{Divide both sides by } \tan 10.1°.$$

$$= 1260 \text{ m}$$

**Example 3** In ac (alternating current) circuits, the relationship between impedance $Z$ (in ohms), the resistance $R$ (in ohms), and the phase angle $\theta$ is shown by the right triangle in Figure 13.20(a). If the resistance is 250 Ω and the phase angle is 41°, find the impedance.

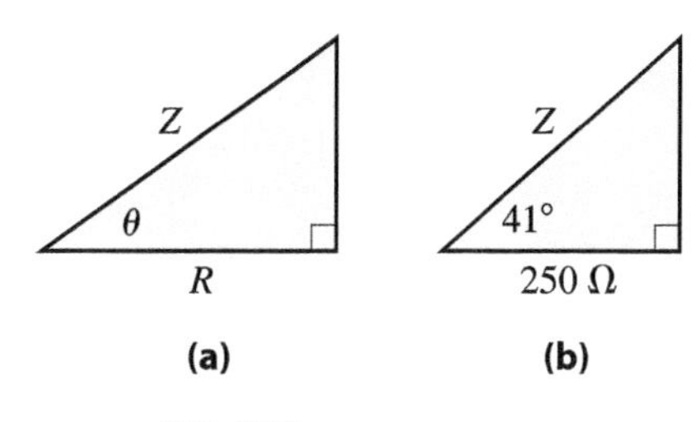


figure **13.20**

Here, we know the adjacent side and the angle and wish to find the hypotenuse (see Figure 13.20(b)). So we use the cosine ratio.

$$\cos\theta = \frac{\text{length of side adjacent to angle } \theta}{\text{length of hypotenuse}}$$

$$\cos 41^\circ = \frac{250\ \Omega}{Z}$$

$$Z(\cos 41^\circ) = 250\ \Omega \qquad \text{Multiply both sides by } Z.$$

$$Z = \frac{250\ \Omega}{\cos 41^\circ} \qquad \text{Divide both sides by } \cos 41^\circ.$$

$$= 330\ \Omega$$

**Example 4** You are machining the part shown in Figure 13.21. Before you begin, you must find angle 1 and length $AB$.

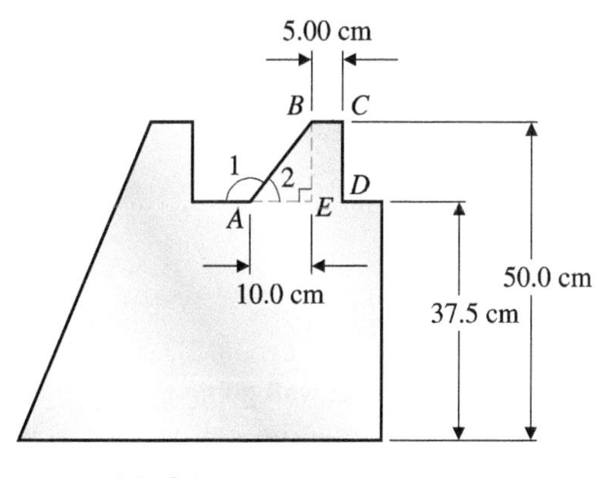


figure **13.21**

First, complete triangle $ABE$ by drawing the dashed lines as shown.

Length $BE$ = length $CD$ = 50.0 cm − 37.5 cm = 12.5 cm

From right triangle $ABE$, we have

$$\tan\angle 2 = \frac{\text{length of side opposite } \angle 2}{\text{length of side adjacent to } \angle 2} = \frac{BE}{AE}$$

$$\tan\angle 2 = \frac{12.5\text{ cm}}{10.0\text{ cm}} = 1.25$$

$$\angle 2 = 51.3^\circ$$

From right triangle $ABE$, we have

$$\sin \angle 2 = \frac{\text{length of side opposite } \angle 2}{\text{length of hypotenuse}}$$

$$\sin 51.3° = \frac{12.5 \text{ cm}}{AB}$$

$$AB(\sin 51.3°) = 12.5 \text{ cm} \quad \text{Multiply both sides by } AB.$$

$$AB = \frac{12.5 \text{ cm}}{\sin 51.3°} \quad \text{Divide both sides by } \sin 51.3°.$$

$$AB = 16.0 \text{ cm}$$

Since $\angle 1$ and $\angle 2$ are supplementary, $\angle 1 = 180° - \angle 2 = 180° - 51.3° = 128.7°$.

## EXERCISES **13.5**

**1.** A conveyor is used to lift paper to a shredder. The most efficient operating angle of elevation for the conveyor is 35.8°. The paper is to be elevated 11.0 m. What length of conveyor is needed?

**2.** Maria is to weld a support for a 23-m conveyor so that it will operate at a $\bar{2}0°$ angle. What is the length of the support? See Illustration 1.

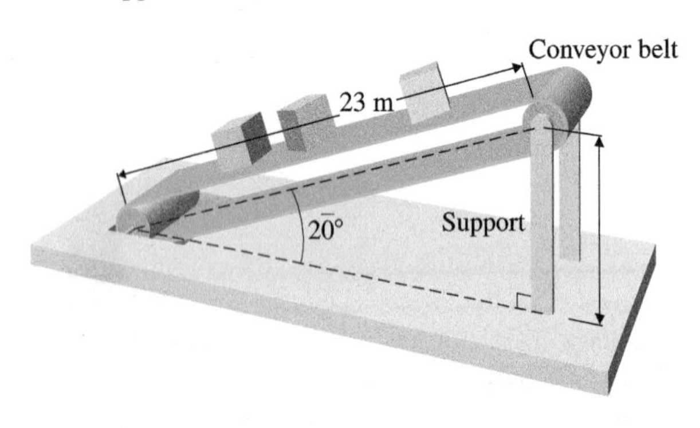


ILLUSTRATION 1

**3.** A bullet is found embedded in the wall of a room 2.3 m above the floor. The bullet entered the wall going upward at an angle of 12°. How far from the wall was the bullet fired if the gun was held 1.2 m above the floor?

**4.** The recommended safety angle of a ladder against a building is 78°. A $1\bar{0}$-m ladder will be used. How high up on the side of the building will the ladder safely reach? (See Illustration 2.)

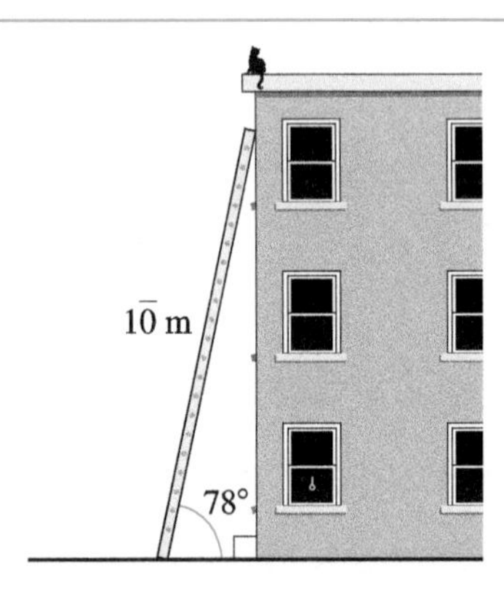


ILLUSTRATION 2

**5.** A piece of conduit 38.0 ft long cuts across the corner of a room, as shown in Illustration 3. Find length $x$ and angle $A$.

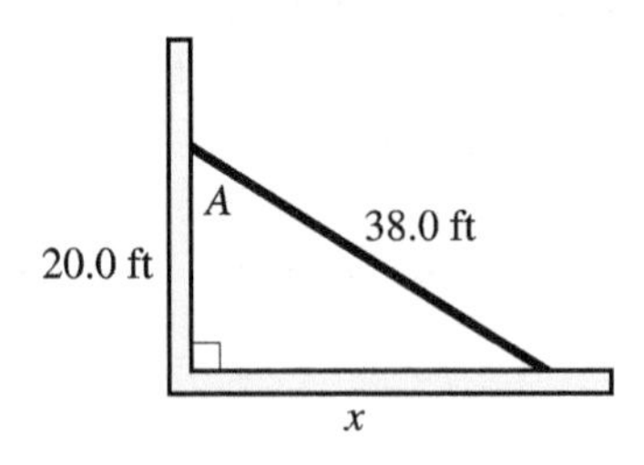


ILLUSTRATION 3

**6.** Find the width of the river in Illustration 4.

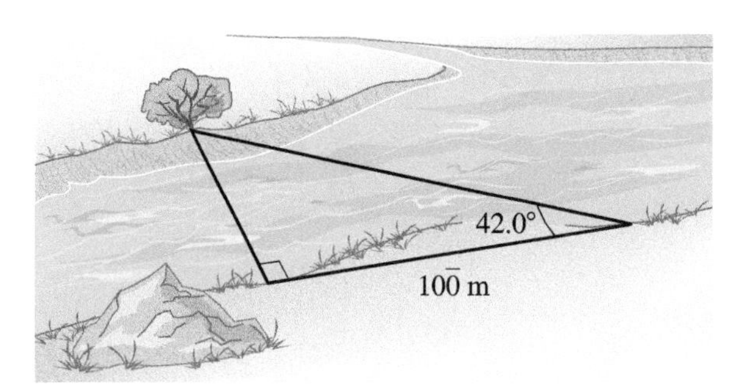


ILLUSTRATION 4

**7.** A roadbed rises 220 ft for each 2300 ft of horizontal. (See Illustration 5.) Find the angle of inclination of the roadbed. (This is usually referred to as *% of grade.*)

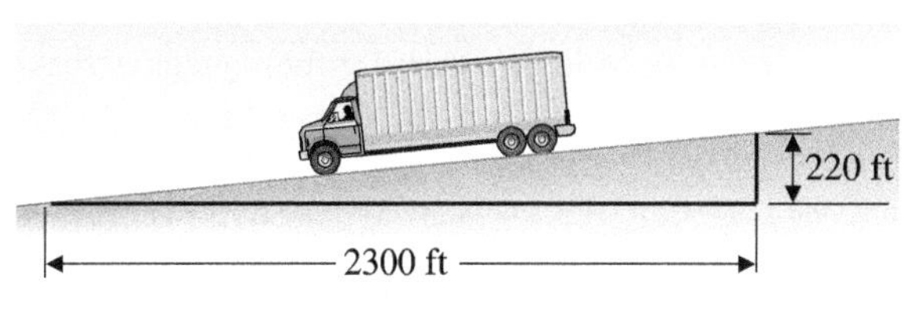


ILLUSTRATION 5

**8.** A smokestack is $18\overline{0}$ ft high. A guy wire must be fastened to the stack 20.0 ft from the top. The guy wire makes an angle of 40.0° with the ground. Find the length of the guy wire.

**9.** A railroad track has an angle of elevation of 1.0°. What is the difference in height (in feet) of two points on the track 1.00 mi apart?

**10.** A machinist needs to drill four holes 1.00 in. apart in a straight line in a metal plate, as shown in Illustration 6. If the first hole is placed at the origin and the line forms an angle of 32.0° with the vertical axis, find the coordinates of the other three holes (*A*, *B*, and *C*).

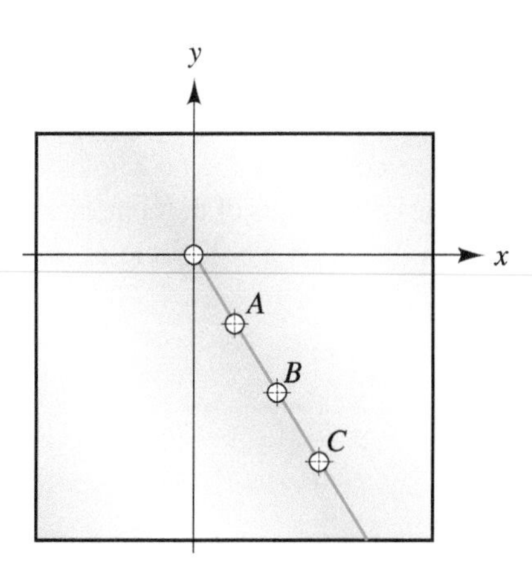


ILLUSTRATION 6

**11.** Enrico has to draft a triangular roof to a house. (See Illustration 7.) The roof is 30.0 ft wide. If the rafters are 17.0 ft long, at what angle will the rafters be laid at the eaves? Assume no overhang.

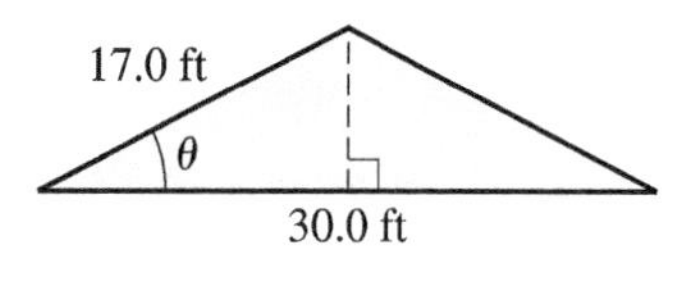


ILLUSTRATION 7

**12.** A small plane takes off from an airport and begins to climb at a 10.0° angle of elevation at $50\overline{0}0$ ft/min. After 3.00 min, how far along the ground will the plane have flown?

**13.** A gauge is used to check the diameter of a crankshaft journal. It is constructed to make measurements on the basis of a right triangle with a 60.0° angle. Distance *AB* in Illustration 8 is 11.4 cm. Find radius *BC* of the journal.

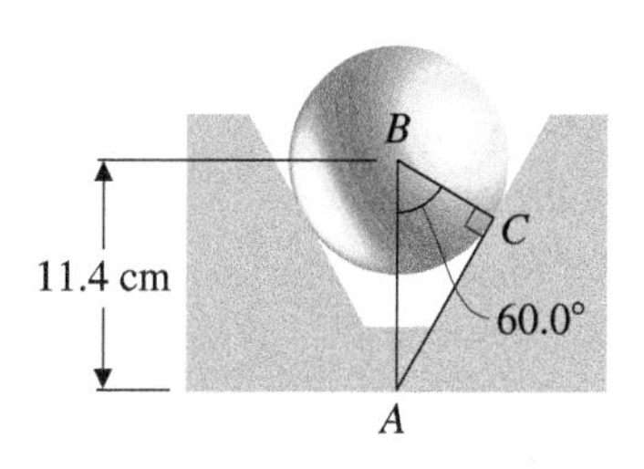


ILLUSTRATION 8

**14.** Round metal duct runs alongside some stairs from the floor to the ceiling. If the ceiling is 9.00 ft high and the angle of elevation between the floor and duct is 37.0°, how long is the duct?

**15.** The cables attached to a TV relay tower are $11\overline{0}$ m long. They meet level ground at an angle of 60.0°, as in Illustration 9. Find the height of the tower.

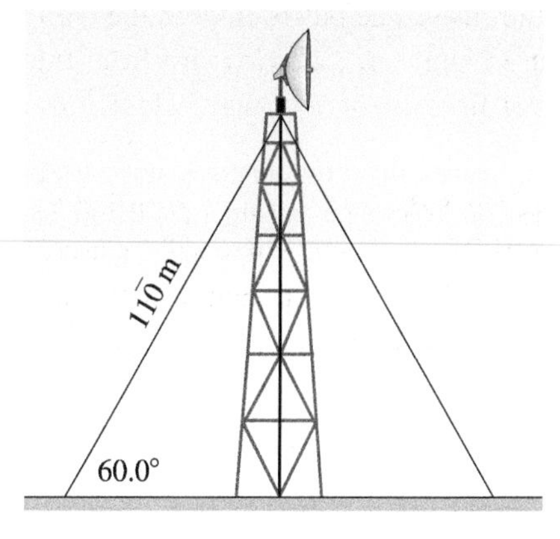


ILLUSTRATION 9

**16.** A lunar module is resting on the moon's surface directly below a spaceship in its orbit, 12.0 km above the moon. (See Illustration 10.) Two lunar explorers find that the angle from their position to that of the spaceship is 82.9°. What distance are they from the lunar module?

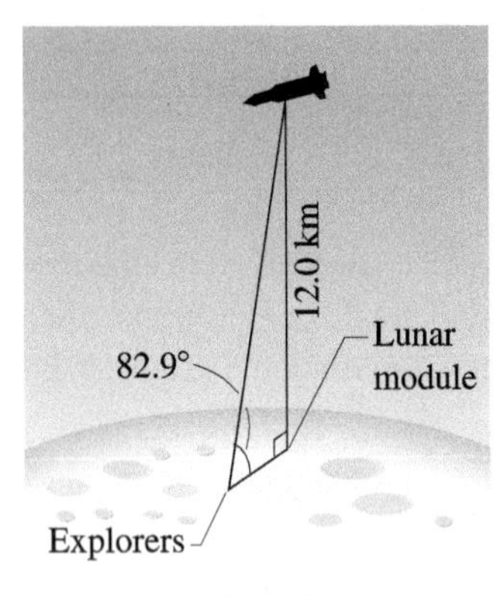


ILLUSTRATION 10

**17.** In ac (alternating current) circuits, the relationship between the impedance $Z$ (in ohms), the resistance $R$ (in ohms), the phase angle $\theta$, and the reactance $X$ (in ohms) is shown by the right triangle in Illustration 11.

**a.** Find the impedance if the resistance is 350 Ω and the phase angle is 35°.

**b.** Suppose the resistance is 550 Ω and the impedance is 700 Ω. What is the phase angle?

**c.** Suppose the reactance is 182 Ω and the resistance is 240 Ω. Find the impedance and the phase angle.

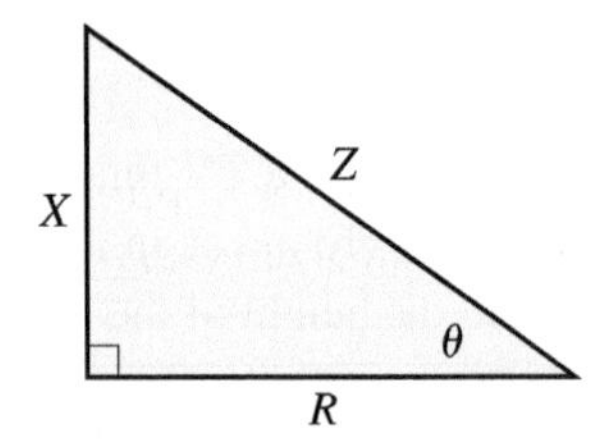


ILLUSTRATION 11

**18.** A right circular conical tank with its point down (Illustration 12) has a height of 4.00 m and a radius of 1.20 m. The tank is filled to a height of 3.70 m with liquid. How many litres of liquid are in the tank? (1000 litres = 1 m$^3$)

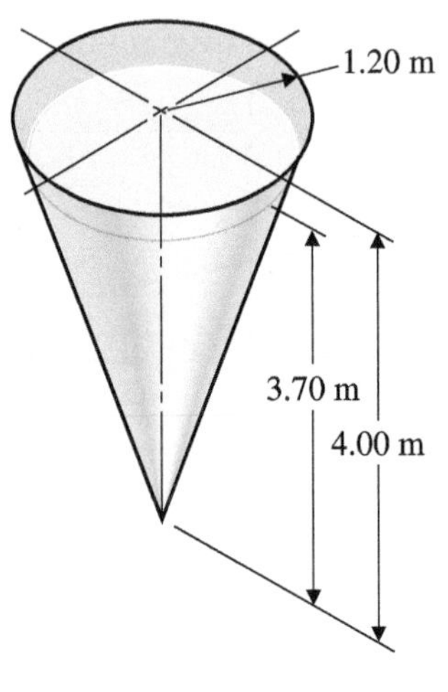


ILLUSTRATION 12

**19.** Use the right triangle in Illustration 13:

**a.** Find the voltage applied if the voltage across the coil is 35.6 V and the voltage across the resistance is 40.2 V.

**b.** Find the voltage across the resistance if the voltage applied is 378 V and the voltage across the coil is 268 V.

**c.** Find the voltage across the coil if the voltage applied is 448 V and the voltage across the resistance is 381 V.

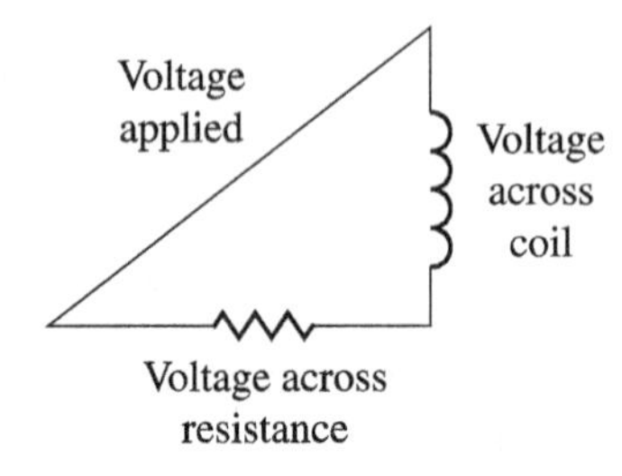


ILLUSTRATION 13

**20.** Machinists often use a coordinate system to drill holes by placing the origin at the most convenient location. A bolt circle is the circle formed by completing an arc through the centers of the bolt holes in a piece of metal. Find the coordinates of the centers of eight equally spaced $\frac{1}{4}$-in. holes on a bolt circle of radius 6.00 in., as shown in Illustration 14.

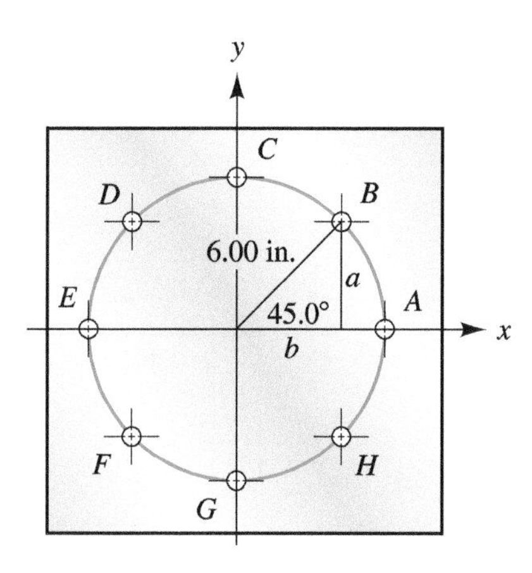


ILLUSTRATION 14

**21.** Twelve equally spaced holes must be drilled on a 14.500-in.-diameter bolt circle. (See Illustration 15.) What is the straight-line center-to-center distance between two adjacent holes?

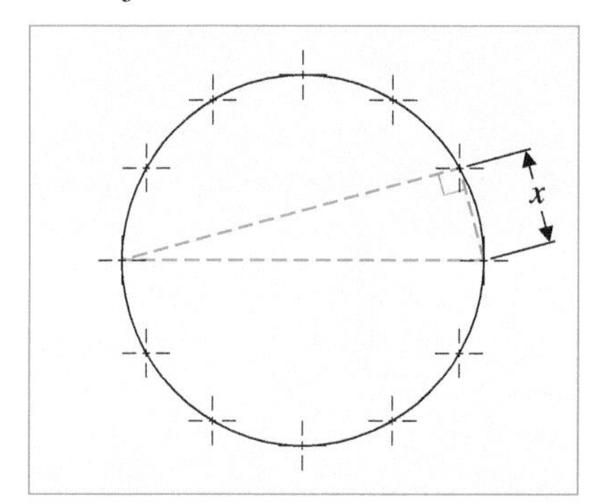


ILLUSTRATION 15

**22.** Dimension $x$ in the dovetail shown in Illustration 16 is a check dimension. Precision steel balls of diameter 0.1875 in. are used in this procedure. What should this check dimension be?

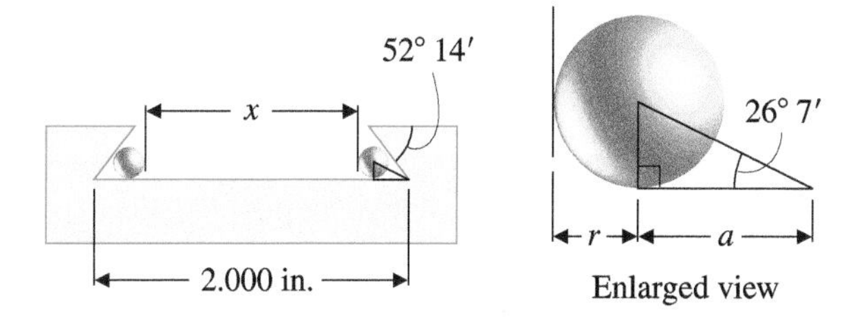


ILLUSTRATION 16

**23.** Find angle $\theta$ of the taper in Illustration 17.

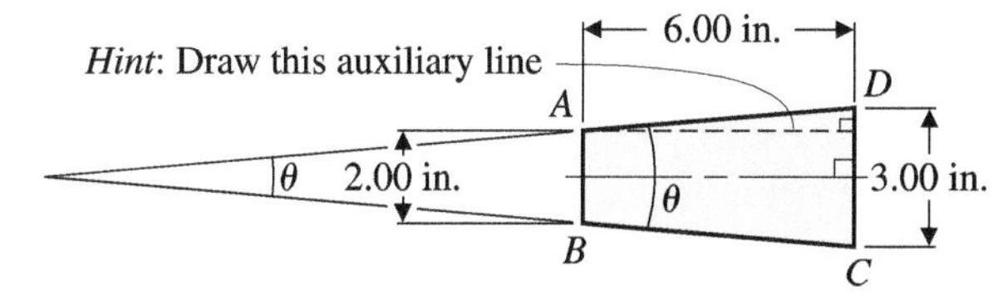


ILLUSTRATION 17

**24.** You need to use a metal screw with a head angle of angle $A$, which is not less than 65° and no larger than 70°. The team leader wants you to find angle $A$ from the sketch shown in Illustration 18 and determine if the head angle will be satisfactory. Find the head angle $A$.

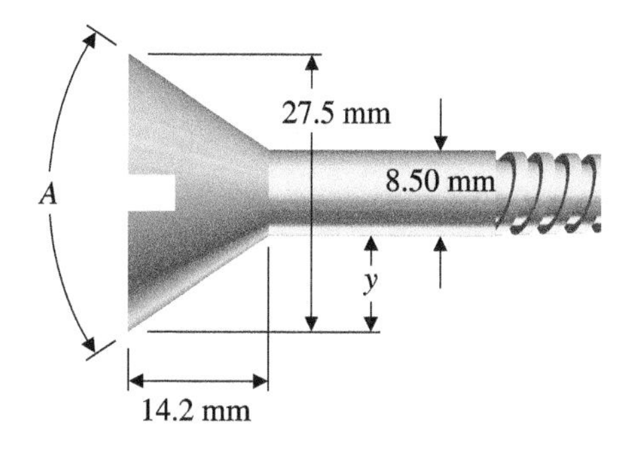


ILLUSTRATION 18

**25.** Find **a.** distance $x$ and **b.** distance $BD$ in Illustration 19. Length $BC = 5.50$ in.

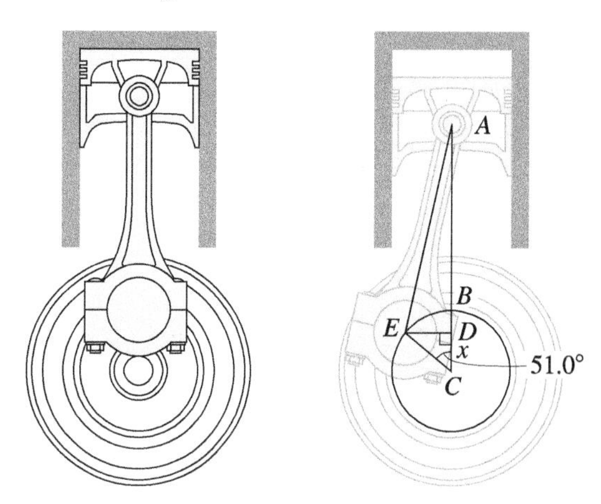


ILLUSTRATION 19

**26.** Find length $AB$ along the roofline of the building in Illustration 20. The slope of the roof is 45.0°.

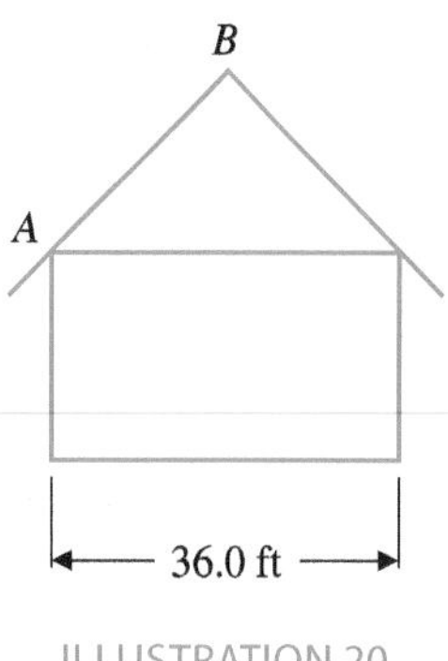


ILLUSTRATION 20

**27.** Find length $x$ and angle $A$ in Illustration 21.

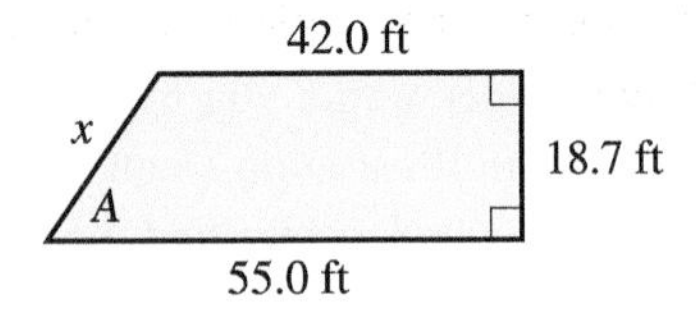


ILLUSTRATION 21

**28.** From the base of a building, measure out a horizontal distance of 215 ft along the ground. The angle of elevation to the top of the building is 63.0°. Find the height of the building.

**29.** A mechanical draftsperson needs to find the distance across the corners of a hex-bolt. See Illustration 22. If the distance across the flats is 2.25 cm, find the distance across the corners.

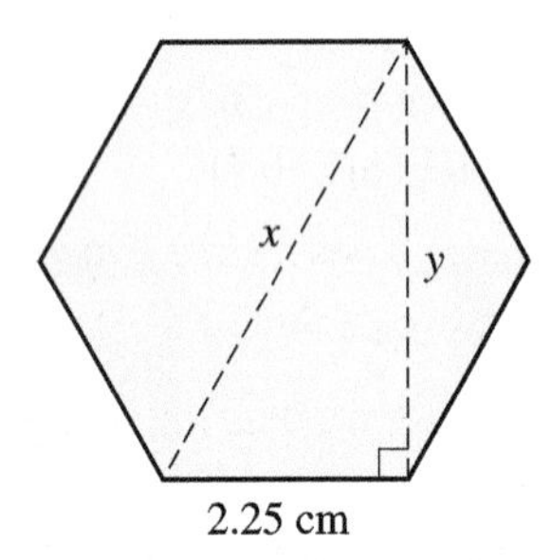


ILLUSTRATION 22

**30.** A hydraulic control valve has two parallel angular passages that must connect to two threaded ports, as shown in Illustration 23 with all lengths in inches. What are the missing dimensions necessary for the location of the two ports?

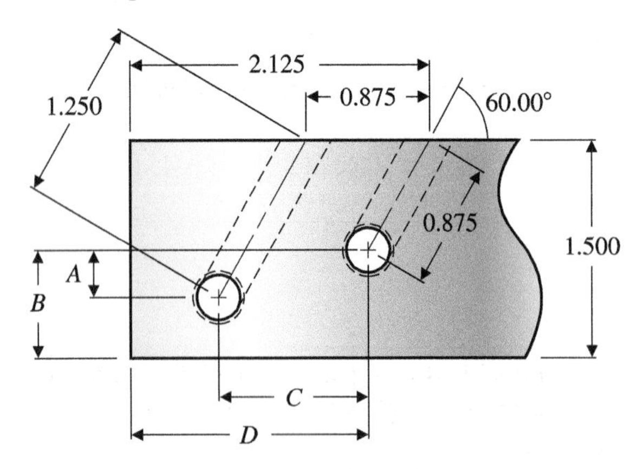


ILLUSTRATION 23

**31.** A benchmark has been covered up with dirt and needs to be found. The CAD drawing in Illustration 24 shows that it is located 33.0 ft from a property stake at an angle of 112.0°. What distance $b$ from the stake parallel to the curb line and distance $a$ in from the curb line is this benchmark?

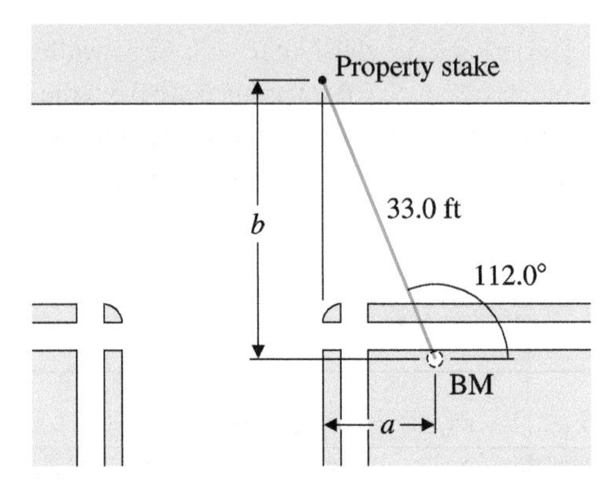


ILLUSTRATION 24

**32.** A mating part is being designed with two pins attached to a flat block to match the item in Illustration 25. The pins must fit into the holes shown. What is the distance from point $C$ (center of small hole) to point $D$ (center of larger hole)? Also find angle $x$.

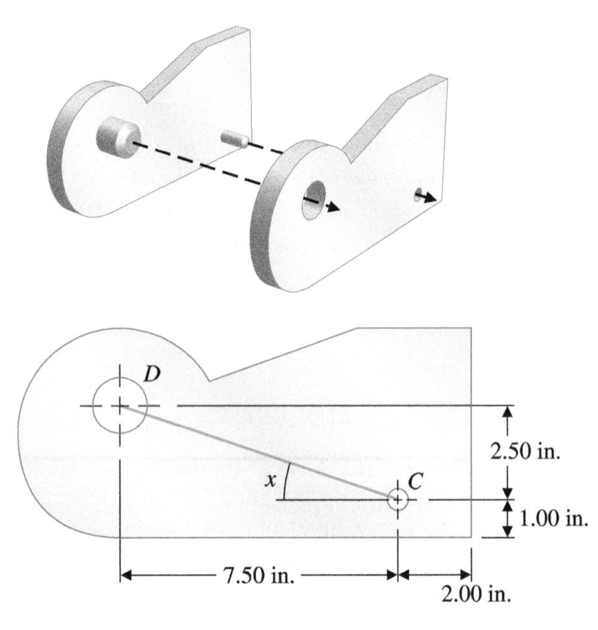


ILLUSTRATION 25

**33.** Solar heating and electric panels should face the sun so that you gain as much sunlight as possible. Fixed panels are mounted at the best winter angle to the horizontal according to the following formula: $\theta = 0.90$ (your latitude in degrees) $+ 29°$. A homeowner in Ithaca, New York (latitude 42.44°N) wishes to mount a solar heating panel that is 4.0 ft wide and 8.0 ft long facing south with its 4.0-ft base mounted on a flat roof at the appropriate winter angle. **a.** Find the height above the roof of the top of the solar panel. **b.** Find the length of the base of the support if the top of the solar panel is supported by vertical posts.

**34.** A lean-to is a simple shelter with three walls, a sloping roof, and an open front facing away from the prevailing winds. The back wall is short compared to the front opening. If the lean-to at a campsite has a front opening that is 6.0 ft tall, a back wall that is 2.0 ft tall, and a floor that is 8.0 ft deep, what angle does the roofline make with the ground?

## CHAPTER 13 Group Activities

**1.** Estimate the height of a building or some other tall object by first marking a point $A$ some distance from the base of the object. Then measure the distance from the base of the object to point $A$. Then measure the angle (as best you can) formed at point $A$ by a horizontal line to the base of the object and a line to the top of the object. Then calculate its height by using right triangle trigonometry. Have different groups repeat the exercise by choosing other points at different distances from the base of the object. Compare the results and explain any differences.

**2.** In Section 13.1, we saw $\sin A$, $\cos A$, $\tan A$, and the Pythagorean theorem $c^2 = a^2 + b^2$. Sometimes it is useful to change equations and formulas involving sine, cosine, and tangent to get other equations and formulas involving sine, cosine, and tangent. Below are some key trigonometric identities involving sine, cosine, and tangent.

$$\tan x = \frac{\sin x}{\cos x} \qquad \sin^2 x + \cos^2 x = 1$$

If we start with an expression that includes sine, cosine, and tangent and change it to another expression, we have what is called an identity.

Verify the following identity by starting with the left-hand side and substituting other basic identities from the chapter until you obtain the expression on the right-hand side.

**Example**

$$(\cos x + \sin x)^2 - 2 = -1(\sin x - \cos x)^2$$

**Verification**

$$\begin{aligned}(\cos x + \sin x)^2 - 2 &= \cos^2 x + 2\cos x \sin x + \sin^2 x - 2\\ &= \sin^2 x + \cos^2 x + 2\cos x \sin x - 2\\ &= -1 + 2\cos x \sin x\\ &= -1(-2\cos x \sin x + 1)\\ &= -1(\sin^2 x - 2\cos x \sin x + \cos^2 x)\\ &= -1(\sin x - \cos x)^2\end{aligned}$$

This was a difficult example. The following will not be as challenging for your group.

Verify the following identity by starting with the left-hand side and substituting basic identities from the chapter until you obtain the right-hand side.

**a.** $\sin x + \tan x = \tan x(\cos x + 1)$

**b.** $\tan x + \cos x = \dfrac{\cos^2 x + \sin x}{\cos x}$

**c.** $\tan x(\cos x + 1) = \tan x + \sin x$

**d.** $\tan x - \sin x = \tan x(1 - \cos x)$

See whether you can develop some other identities.

## CHAPTER 13 Summary

### Glossary of Basic Terms

**Angle of depression.** The angle between the horizontal and the line of sight to an object that is *below* the horizontal. (p. 461)

**Angle of elevation.** The angle between the horizontal and the line of sight to an object that is *above* the horizontal. (p. 461)

**Cosine of angle $A$ ($\cos A$).** Equals the ratio of the length of the side adjacent to angle $A$ to the length of the hypotenuse. (p. 451)

**Ratio.** The comparison of two quantities by division. (p. 451)

**Sine of angle $A$ ($\sin A$).** Equals the ratio of the length of the side opposite angle $A$ to the length of the hypotenuse. (p. 451)

**Solving a right triangle.** Find the measures of the various parts of a triangle that are not given. (p. 458)

**Tangent of angle $A$ ($\tan A$).** Equals the ratio of the length of the side opposite angle $A$ to the length of the side adjacent to angle $A$. (p. 451)

**Trigonometric ratio.** Expresses the relationship between an acute angle and the lengths of two of the sides of a right triangle. (p. 451)

## 13.1 Trigonometric Ratios

1. **Pythagorean theorem**: In any right triangle, $c^2 = a^2 + b^2$. (p. 450)

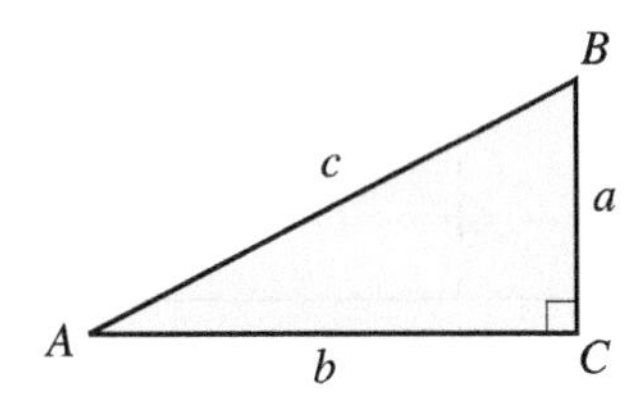


2. **Trigonometric ratios:**

$$\sin A = \frac{\text{length of side opposite angle } A}{\text{length of hypotenuse}}$$

$$\cos A = \frac{\text{length of side adjacent to angle } A}{\text{length of hypotenuse}}$$

$$\tan A = \frac{\text{length of side opposite angle } A}{\text{length of side adjacent to angle } A}$$

(p. 451)

## 13.2 Using Trigonometric Ratios to Find Angles

1. **Trigonometric ratios and significant digits:** When calculations involve a trigonometric ratio, use the following for significant digits. (p. 455)

| Angles expressed to the nearest | The length of each side of the triangle contains |
|---|---|
| 1° | Two significant digits |
| 0.1° | Three significant digits |
| 0.01° | Four significant digits |

2. **Which trig ratio to use:**
   a. If you are finding an angle, two sides must be known. Label these two known sides as *side opposite* the angle you are finding, *side adjacent* to the angle you are finding, or *hypotenuse.* Then choose the trigonometric ratio that has these two sides.
   b. If you are finding a side, one side and one angle must be known. Label the known side and the unknown side as *side opposite* the known angle, *side adjacent* to the known angle, or *hypotenuse.* Then choose the trigonometric ratio that has these two sides. (p. 456)
3. **Right triangle acute angle relationships:** In any right triangle, the sum of the acute angles is 90°; that is, $A + B = 90°$. (p. 456)

## CHAPTER 13 Review

*For Exercises 1–7, see Illustration 1:*

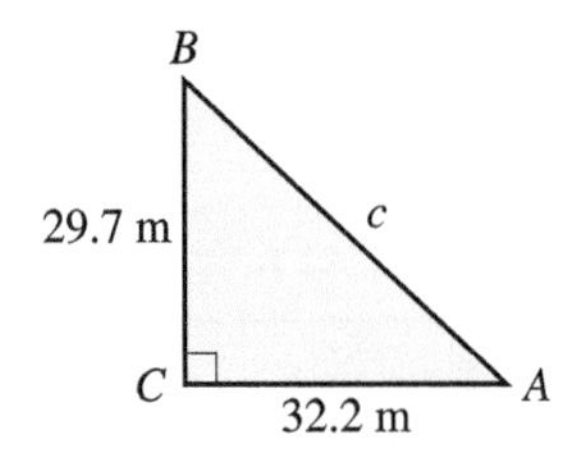


ILLUSTRATION 1

1. What is the length of the side opposite angle $A$ in the right triangle?
2. What is the angle adjacent to the side whose length is 29.7 m?
3. Side $c$ is known as the __?__.
4. What is the length of side $c$?
5. $\frac{\text{length of side opposite angle } A}{\text{length of hypotenuse}}$ is what ratio?
6. $\cos A = \frac{?}{\text{length of hypotenuse}}$
7. $\tan B = \frac{?}{?}$

*Find the value of each trigonometric ratio rounded to four significant digits:*

8. cos 36.2°
9. tan 48.7°
10. sin 23.72°

*Find each angle rounded to the nearest tenth of a degree:*

11. $\sin A = 0.7136$
12. $\tan B = 0.1835$
13. $\cos A = 0.4104$
14. Find angle $A$ in Illustration 2.
15. Find angle $B$ in Illustration 2.

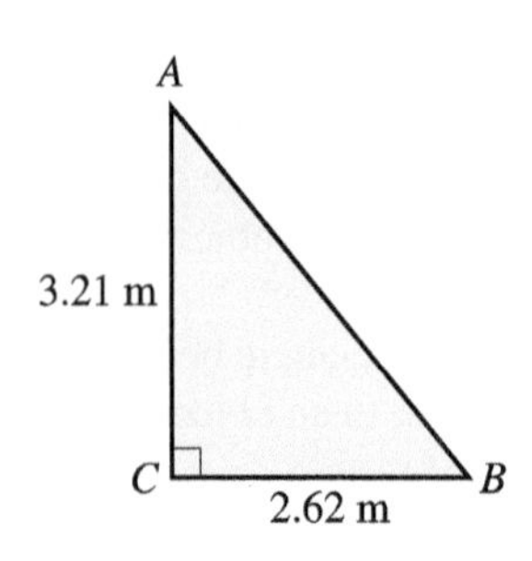


ILLUSTRATION 2

**16.** Find side $b$ in Illustration 3.

**17.** Find side $c$ in Illustration 3.

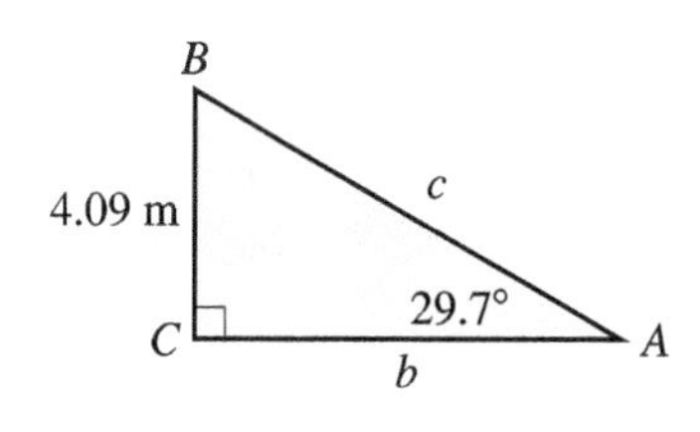


ILLUSTRATION 3

*Solve each right triangle:*

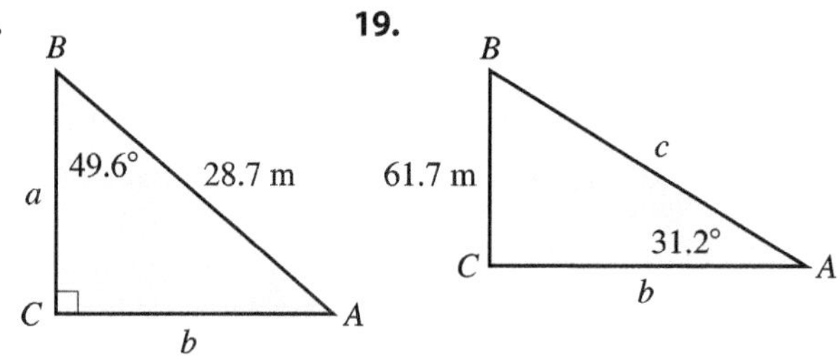


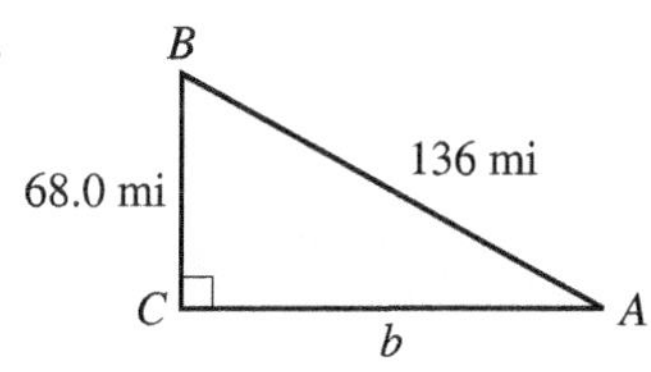


**21.** A satellite is directly overhead one observer station when it is at an angle of 68.0° from another observer station. (See Illustration 4.) The distance between the two stations is $20\bar{0}0$ m. What is the height of the satellite?

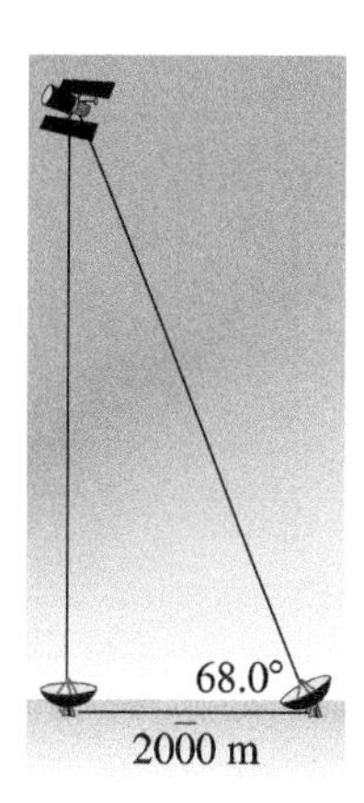


ILLUSTRATION 4

**22.** A ranger at the top of a fire tower observes the angle of depression to a fire on level ground to be 3.0°. If the tower is 275 ft tall, what is the ground distance from the base of the tower to the fire?

**23.** Find the angle of slope of the symmetrical roof in Illustration 5.

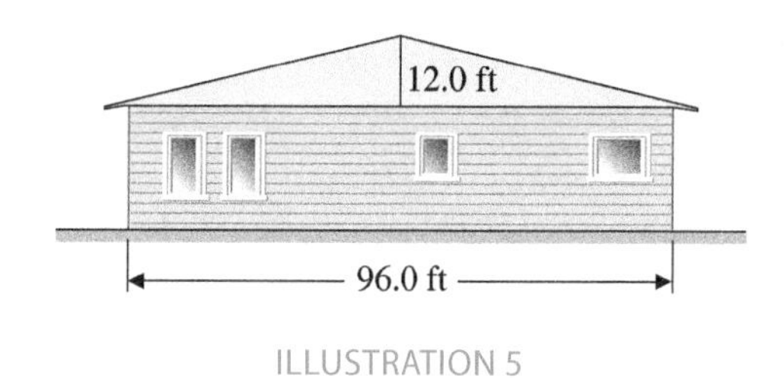


ILLUSTRATION 5

## CHAPTER 13 Test

*Find the value of each trigonometric ratio rounded to four significant digits:*

**1.** sin 35.5° **2.** cos 16.9° **3.** tan 57.1°

*Find each angle rounded to the nearest tenth of a degree:*

**4.** $\cos A = 0.5577$ **5.** $\tan B = 0.8888$

**6.** $\sin A = 0.4166$

**7.** Find angle $B$ in Illustration 1.

**8.** Find side $a$ in Illustration 1.

**9.** Find side $c$ in Illustration 1.

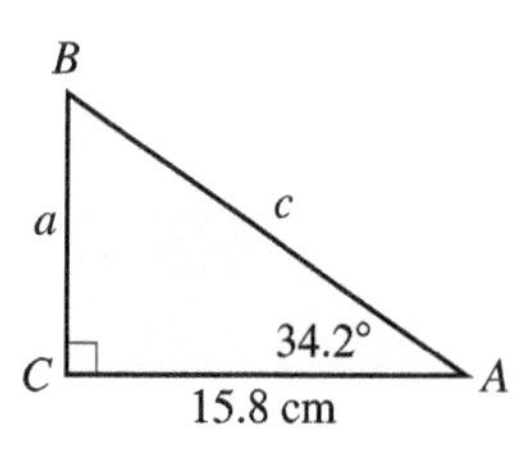


ILLUSTRATION 1

**10.** Find angle $A$ in Illustration 2.

**11.** Find angle $B$ in Illustration 2.

**12.** Find side $b$ in Illustration 2.

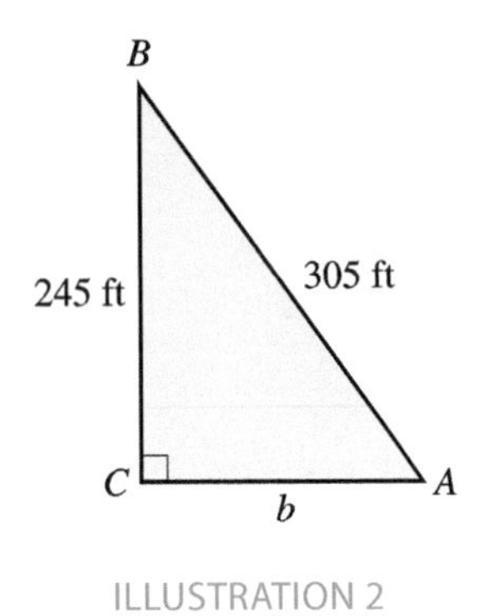


ILLUSTRATION 2

**13.** A tower 50.0 ft high has a guy wire that is attached to its top and anchored in the ground 15.0 ft from its base. Find the length of the guy wire.

**14.** Find length $x$ in the retaining wall in Illustration 3.

**15.** Find angle $A$ in the retaining wall in Illustration 3.

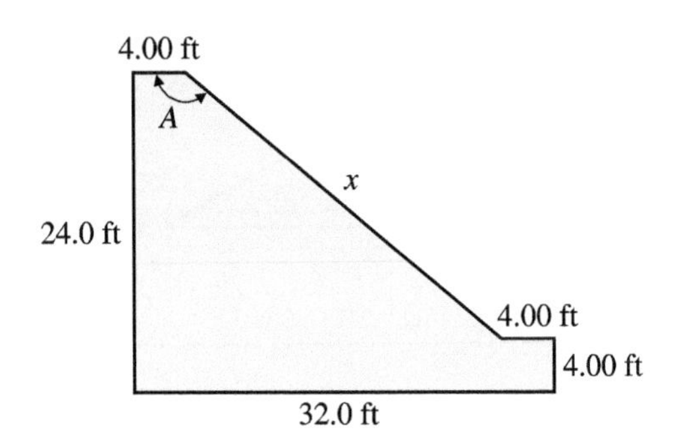


ILLUSTRATION 3

# Chapter 13

## Exercises 13.1 (pages 453–454)

**1.** $a$ **3.** $c$ **5.** $a$ **7.** $B$ **9.** $B$ **11.** 60.0 m **13.** 55.2 mi **15.** 17.7 cm **17.** 265 ft **19.** $35\bar{0}$ m **21.** 1530 km **23.** 59,900 m **25.** 0.5000 **27.** 0.5741 **29.** 0.5000 **31.** 0.7615 **33.** 2.174 **35.** 0.8686 **37.** 0.5246 **39.** 0.2536 **41.** 0.05328 **43.** 0.6104 **45.** 0.3929 **47.** 0.6800 **49.** 52.6° **51.** 62.6° **53.** 54.2° **55.** 9.2° **57.** 30.8° **59.** 40.8° **61.** 10.98° **63.** 69.07° **65.** 8.66° **67.** 75.28° **69.** 43.02° **71.** 20.13° **73.** $\sin A$ and $\cos B$; $\cos A$ and $\sin B$

## Exercises 13.2 (page 456)

**1.** $A = 35.3°$; $B = 54.7°$ **3.** $A = 24.4°$; $B = 65.6°$ **5.** $A = 43.3°$; $B = 46.7°$ **7.** $A = 45.0°$; $B = 45.0°$ **9.** $A = 64.05°$; $B = 25.95°$ **11.** $A = 52.60°$; $B = 37.40°$ **13.** $A = 39.84°$; $B = 50.16°$ **15.** $A = 41.85°$; $B = 48.15°$ **17.** $A = 11.8°$; $B = 78.2°$ **19.** $A = 31.11°$; $B = 58.89°$ **21.** $A = 15.9°$; $B = 74.1°$ **23.** $A = 64.65°$; $B = 25.35°$

## Exercises 13.3 (page 458)

**1.** $b = 40.6$ m; $c = 54.7$ m **3.** $b = 278$ km; $c = 365$ km **5.** $a = 29.3$ cm; $b = 39.4$ cm **7.** $a = 10.4$ cm; $c = 25.9$ cm **9.** $a = 6690$ km; $c = 30{,}\bar{0}00$ km **11.** $b = 54.92$ m; $c = 58.36$ m **13.** $a = 161.7$ mi; $b = 197.9$ mi **15.** $a = 4564$ m; $c = 13{,}170$ m **17.** $b = 5950$ m; $c = 6270$ m **19.** $a = 288.3$ km; $c = 889.6$ km **21.** $a = 88.70$ m; $b = 163.0$ m **23.** $a = 111$ cm; $b = 231$ cm

## Exercises 13.4 (page 460)

**1.** $B = 39.4°$; $a = 37.9$ m; $b = 31.1$ m **3.** $A = 48.8°$; $b = 234$ ft; $c = 355$ ft **5.** $A = 22.6°$; $B = 67.4°$; $a = 30.0$ mi **7.** $B = 37.9°$; $b = 56.1$ mm; $c = 91.2$ mm **9.** $B = 21.2°$; $a = 36.7$ m; $b = 14.2$ m **11.** $A = 26.05°$; $B = 63.95°$; $c = 27.33$ m **13.** $B = 60.81°$; $a = 1451$ ft; $b = 2597$ ft **15.** $A = 67.60°$; $B = 22.40°$; $c = 50.53$ m **17.** $a = 319$ m; $b = 365$ m; $B = 48.9°$ **19.** $b = 263.8$ ft; $c = 1607$ ft; $A = 80.55°$ **21.** $a = 268.6$ m; $A = 44.90°$; $B = 45.10°$ **23.** $a = 14\bar{0}0$ ft; $c = 9470$ ft; $A = 8.5°$

## Exercises 13.5 (pages 463–468)

**1.** 18.8 m **3.** 5.2 m **5.** 58.2°; 32.3 ft **7.** 5° **9.** 92.1 ft **11.** 28.1° **13.** 5.70 cm **15.** 95.3 m **17. a.** 430 Ω **b.** 38° **c.** 37°; $3\bar{0}0$ Ω **19. a.** 53.7 V **b.** 267 V **c.** 236 V **21.** 3.7529 in. **23.** 9.5° **25. a.** 3.46 in. **b.** 2.04 in. **27.** $A = 55.2°$; $x = 22.8$ ft **29.** $x = 4.50$ cm; $y = 3.90$ cm **31.** $a = 12.4$ ft; $b = 30.6$ ft **33. a.** 7.4 ft **b.** 3.1 ft

## Chapter 13 Review (pages 469–470)

**1.** 29.7 m **2.** $B$ **3.** Hypotenuse **4.** 43.8 m **5.** $\sin A$ **6.** Length of side adjacent to $\angle A$ **7.** $\dfrac{\text{length of side opposite } \angle B}{\text{length of side adjacent to } \angle B} = \dfrac{32.2 \text{ m}}{29.7 \text{ m}}$ **8.** 0.8070 **9.** 1.138 **10.** 0.4023 **11.** 45.5° **12.** 10.4° **13.** 65.8° **14.** 39.2° **15.** 50.8° **16.** 7.17 m **17.** 8.25 m **18.** $b = 21.9$ m; $a = 18.6$ m; $A = 40.4°$ **19.** $b = 102$ m; $c = 119$ m; $B = 58.8°$ **20.** $A = 30.0°$; $B = 60.0°$; $b = 118$ mi **21.** 4950 m **22.** 5250 ft **23.** 14.0°